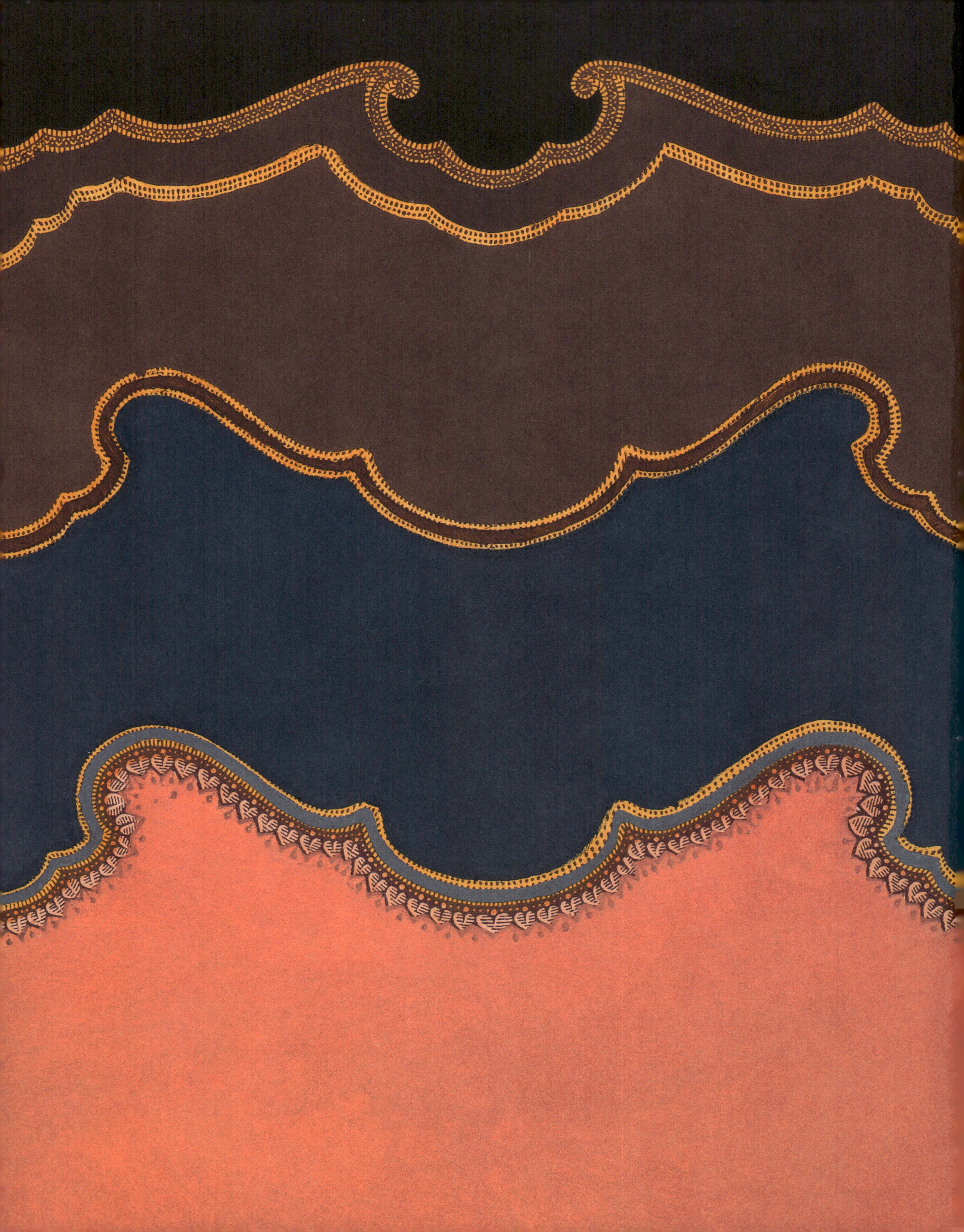

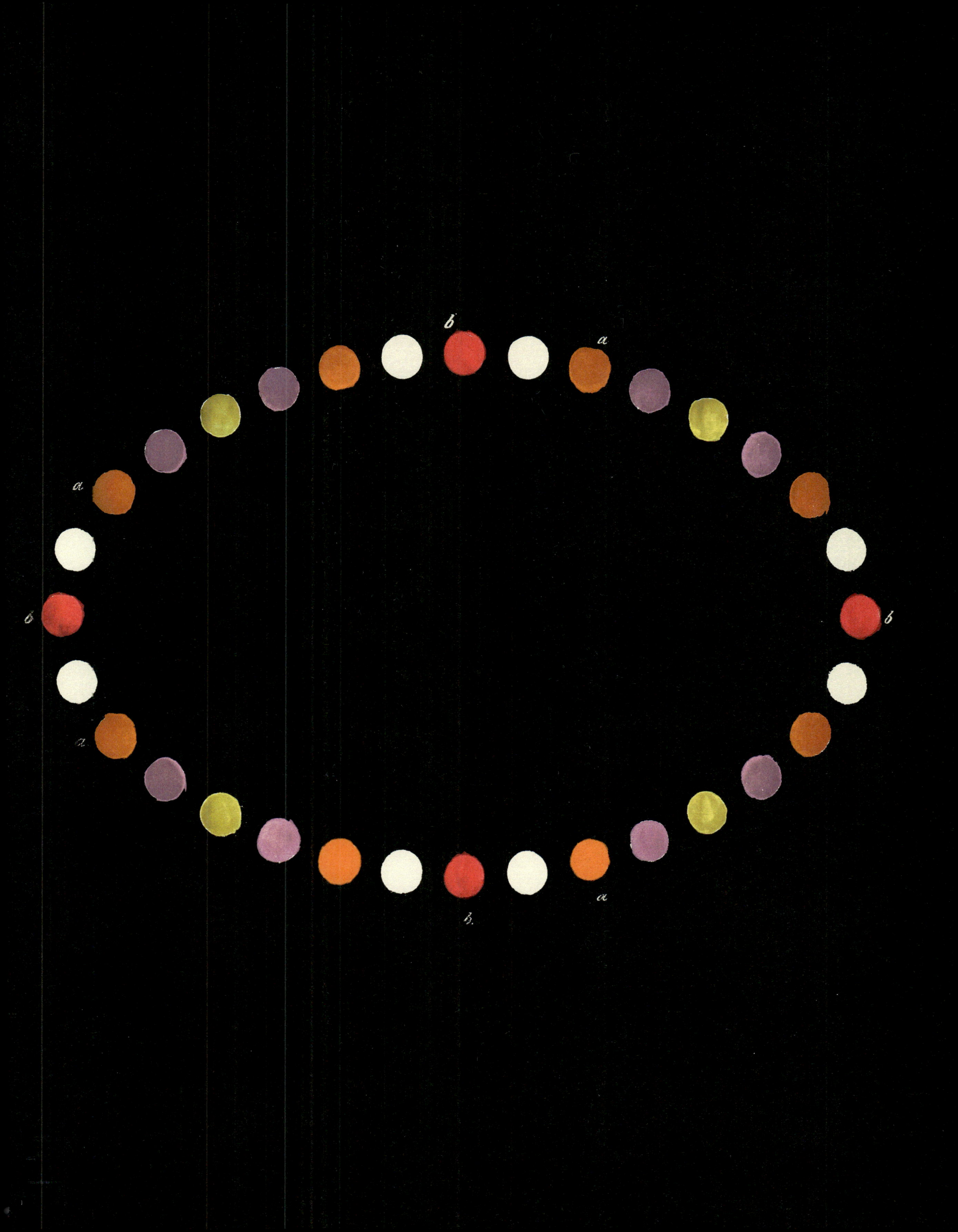
b
a
a
b
b
a
a
b.

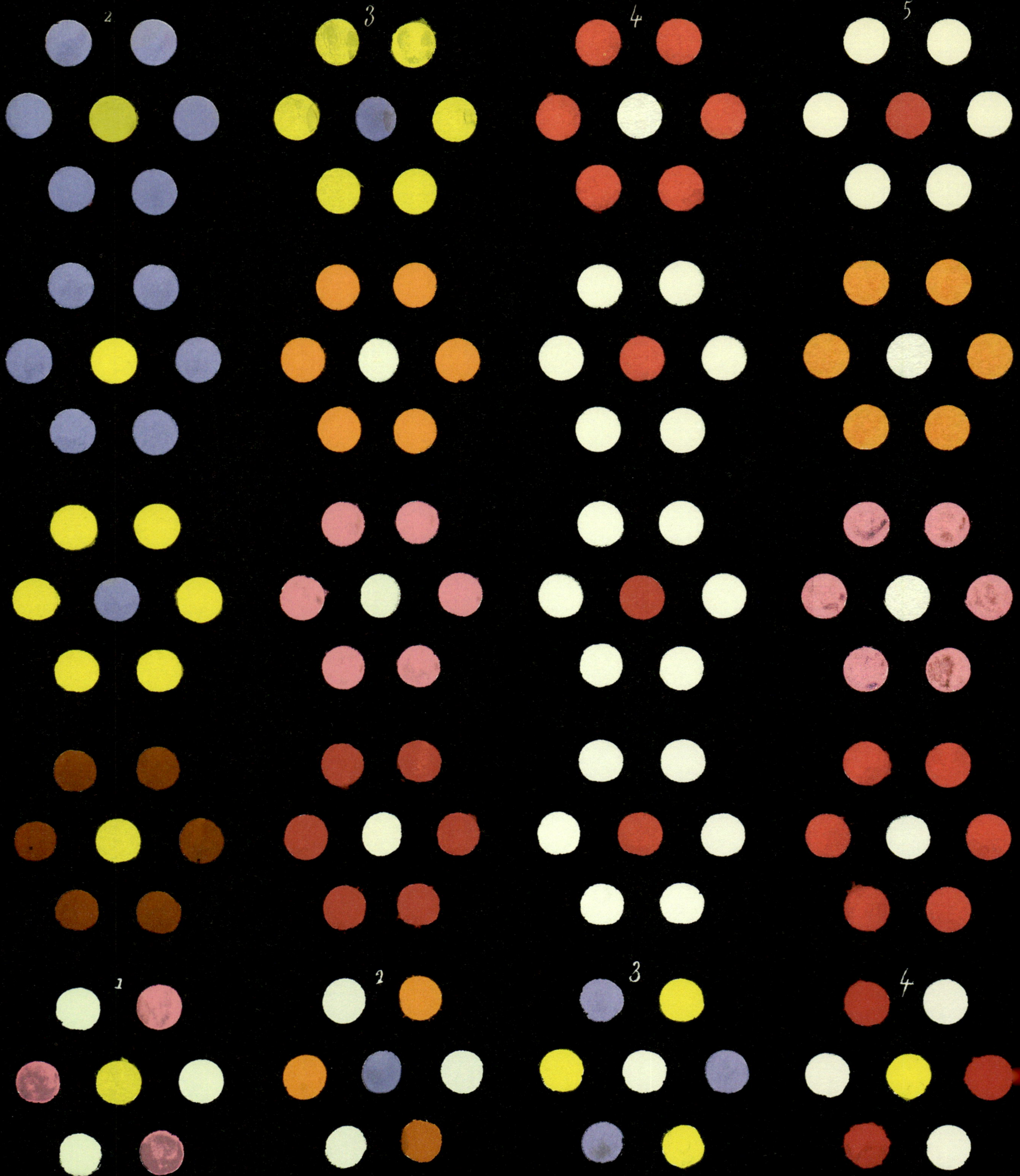
2
3
4
5
1
2
3
4

Alexandra Loske (Ed.)
Sarah Lowengard

THE BOOK OF COLOUR CONCEPTS

1686–1963

Volume 1

TASCHEN

Figure 79

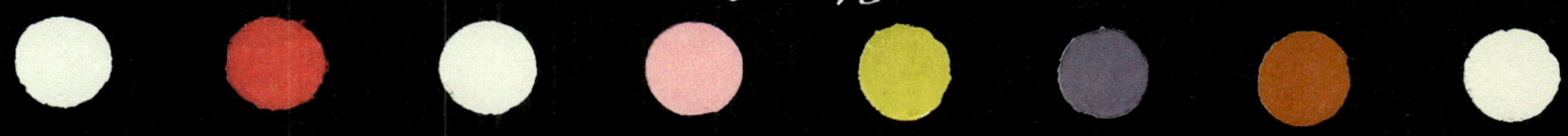

Figure 80

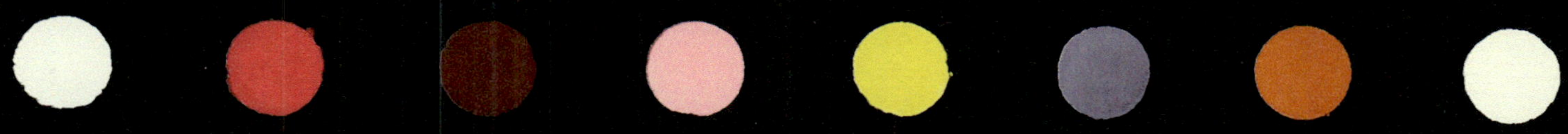

Figure 81

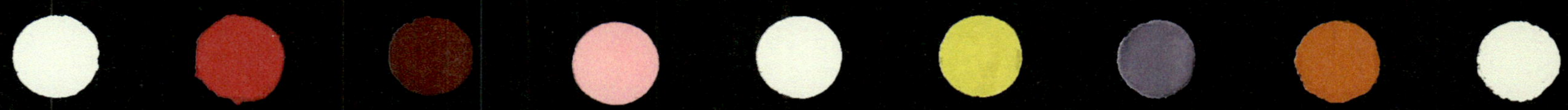

Figure 82

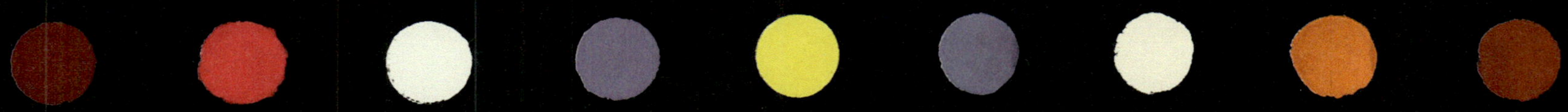

Figure 83

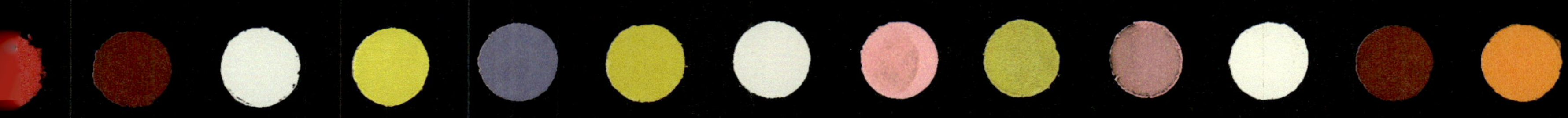

Viridis species

Viridum aeris commune et Hispanicum Chrysocolla lazurij ex reliquijs Malachitarum molitum, succus ex baccalis Merulae ex floribus Caciris succus Rutae &c. Eorum explorata tantum habeo Viridum aeris commune Chrysocollam lazurium succum ex baccalis merulae & ex floribus Caciris.

Caerulei species

Lazurium Smalta. Indicum. Flandrense quod Lagmoss agnominatur. Isatidis Mixtillorum &c. hoc ordine omitti possunt. Isatidis & mixtillorum succi si cui compendium placeret.

Receptissimorum colorum Una exhibitio

Cerussa
Bleiweiß

Creta
Kreide

Nitentium exhibitio

Argentum politum
Bruniret Silber

Aurum politum
Bruniret golde

Argentum madidum
Matt silber

Aurum madidum
Matt gold

Argentum molitum
Gemahlen silber

Aurum molitum
Gemahlen golde

Alexandra Loske

COLOUR CONTROLLED AND SET FREE

The Human History of Capturing Colour in Words and Images

Our world is full of colour. Most of us experience it in colour. The Earth is known as the 'Blue Planet' and optically we disappear into blue: when photographed in 1990 by the space probe *Voyager 1* from 6 billion kilometres away, we appeared, as the astronomer Carl Sagan described it a few years later, as "a pale blue dot", less than a pixel in size and hardly visible, but still of an identifiable colour. We turn into a spot of colour until we disappear from view. Colour, then, is one of the defining and unifying aspects of our lives, a universal and shared human experience, and it is not surprising that it has fascinated us from the beginning of history. The earliest manifestations of human creativity, in the form of carvings and markings on wood, bone and rock, show that we have always tried to make and use colour, chiefly to depict or copy what mattered to and what surrounded us. Although we only have fragmentary and faded visual and physical proof of this earliest human activity, it is possible that even at this early time colour was also used in a symbolic manner. The mineral pigments extracted from the earth by early humans were mixed with organic matter such as fat, grease or bodily fluids, and may well have acquired special or sacred meaning in the process, thus creating a first simple hierarchy of colours. We will never know exactly what our pre-historic ancestors thought about colour, but there is some evidence that certain colours were used in particular contexts.

We can trace an active intellectual engagement with colour throughout human history, from the simple concepts used in early art to the highly developed systems rooted in modern science, philosophy or the arts, categories which often overlap and make cross-references with each other. The aim of this book is to present an overview of human thinking and feeling about colour, the desire to create, arrange and systemise colour, to use and structure it, and to assign it complex meaning. Attempts at classifying and understanding colour have been numerous and have developed in a great variety of manifestations, from dualistic concepts of darkness and light to elaborate geometric patterns and systems that seek to give a specific shape to colour. With the development of language, writing and eventually print culture and sophisticated image reproduction methods, these concepts became more and more refined, complex, visually striking and also better recorded.

This edition begins its visual journey through the history of colour concepts in the 17th century, the time of the earliest surviving illustrated manuscripts and printed sources that focus specifically on colour, and concludes in the mid-20th century, before the advent of the digital age. Just as much as each colour concept included here reflects the times in which it was created, so too does this book through the particular concepts that have been selected. There has been a sharp and growing interest in colour and colour history over the last few decades, and as the result of new research we are discovering new strands together with long-overlooked voices and examples of material culture (writers and educators, books, manuscripts, art and methods) that had either been forgotten or had never been considered as part of the colour canon.

No book on colour can provide an encyclopaedic record with a cultural, historical, geographical and thematic claim for completion. Some titles make deliberate omissions in order to focus on particular cultural or historic themes, others make choices that reflect the personal interests, tastes or expertise of the authors or editors. This book is, metaphorically speaking, a coat of many colours, and is as subjective in its selection as colour perception is deemed to be. The 67 works shown and discussed here are representative of the rich and varied history of colour concepts since the 17th century, and while we have selected many familiar examples of colour concepts that are firmly established in colour history, we hope to have succeeded in putting them in a wider cultural context and to have looked at them with fresh eyes. In addition to these 'classic' examples in their field we have made a particular effort to feature some lesser-known, or more obscure

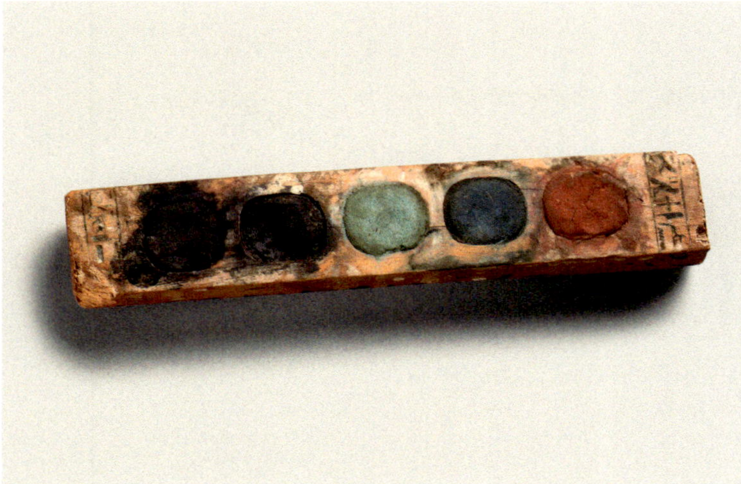

2

Ill. 1 *(Page 6)*
Théodore Turquet de Mayerne
Colour Circles from the De Mayerne Manuscript (detail)
Farbige Kreise aus dem De-Mayerne-Manuskript
Cercles chromatiques du manuscrit de Mayerne
Círculos de colores del Manuscrito de Mayerne
From: *Pictoria, sculptoria et quae subalternarum artium* (London, 1620–1646)
London, British Library

Ill. 2
Paintbox of the Vizier Amenemope, *c.* 1427–1401 B.C.
Malkasten des Wesirs Amenemope
Palette du vizir Amenemopet
Caja de pinturas del visir Amenemope
Boxwood with inscription inlaid in Egyptian blue; 2.2 x 21 x 3.6 cm (⅞ x 8¼ x 1⅜ in.)
Ohio, The Cleveland Museum of Art, Gift of the John Huntington Art and Polytechnic Trust

works that deserve to be considered for their aesthetic and theoretical appeal and importance. One further aim is to give more room to works on colour written, published and illustrated by women in the last two centuries. The historical and geographical range offered here is wide, with examples from Europe, the United States and parts of Asia, but we have also sought to be as inclusive as possible with regard to genre and style. A 20th-century educational colour primer can be as fascinating as Isaac Newton's (1643–1727) colour wheel from 1704 (p. 211), and they both play important roles in colour history.

Why Colour Concepts?

As an expansive and adaptable term, 'colour concept' is well suited to thinking about colour and trying to understand and systemise it. Yet rather than looking to find the perfect term it could perhaps be asked why the need is felt to systemise something that is so intangible, ever-changing, elusive and, to a certain degree, subjective. Until fairly recently, literature on colour didn't distinguish properly between material and immaterial colour, but once the difference was fully understood, it became clear that there are in fact many forms of colour that combine optical colour and the colours of solid materials. Colour easily slips away; it changes with the light, looks different at different times of the day, fades, darkens and can be unpredictable when combined, mixed or juxtaposed. Furthermore, it can be difficult to obtain, as pigment makers and historians can confirm, and it must have been frustrating for early humans to live in a world where the two dominant colours, the blue of the sky and the green of the plant world, were two of the most difficult to obtain. We may live in a world that is abundant with chlorophyll, but few plants make good green pigments. Colour is also infuriatingly fickle, unstable and often toxic, and was not always loved – for as the artist and writer David Batchelor has explained so beautifully in his essay *Chromophobia*, the loathing of colour is a significant part of western cultural history, rooted in "a fear of contamination and corruption by something that is unknown or appears unknowable [... manifesting] itself in the many and varied attempts to purge colour from culture, to devalue colour, to diminish its significance, to deny its complexity". Western attitudes to colour, or, more specifically, to highly saturated colours and polychromy, Batchelor argued, were commonly associated with ideas of the foreign, oriental, primitive, feminine, vulgar, superficial or even pathological, in contrast to the Classical ideals of virginal white or colourless beauty. There is a long history in western culture of colour and darkness signifying dangerous or threatening 'otherness', and such a destabilising force needs either to be avoided or tamed. Creating colour systems or concepts is one way of taming colour, doing so intellectually at first, followed by measured use and application. If we impose an order on colour, even just theoretically, we have at least attempted to take control of something that is mutable and all-pervasive.

Fear, doubt and hatred of colour is of course only one side of the coin though. Our relationship with colour has always been multifaceted, and the last 50 or 60 years in particular – with the invention of colour film and photography, digital colour and the move to an almost fully coloured print culture – have not just made us more at ease with colour, but have resulted in an unprecedented accessibility to it. With hardly any exceptions, colour is now easily available and no longer exclusive, while there are likewise few limits to colour reproduction even if it is still the case that the quality varies.

The Order of Colour: From Paintboxes to Palettes

Where textual sources for colour concepts are lacking, archaeology and material culture can give some insight into colour culture. At

Ill. 3
Unknown Artist
The Wilton Diptych, *c.* 1395–1399
Das Wilton-Diptychon
Diptyque de Wilton
Díptic de Wilton
Tempera on panel, 37 x 53 cm
(14⅝ x 20⅞ in.)
London, National Gallery, Bought with a special grant and contributions from Samuel Courtauld, Viscount Rothermere, C.T. Stoop and The Art Fund, 1929

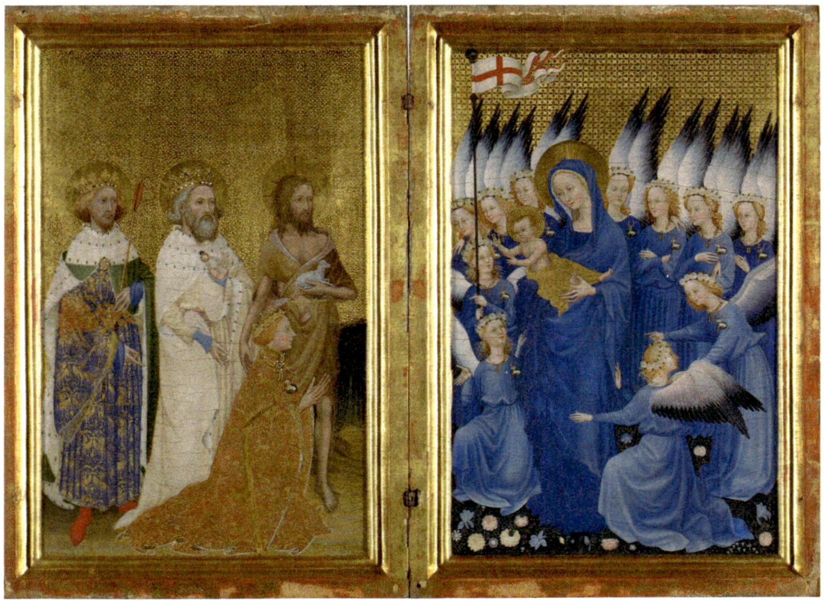

3

the very least, the colours used in art and decoration are an indicator of what pigments were available at certain times in history, and reveal details about techniques, painting materials, colour preferences and – when seen in their wider cultural context – perhaps the symbolic meanings of colours. One of the oldest surviving artists' palettes, a paintbox from the 18th Dynasty of Ancient Egypt that dates from *c.* 1427–1401 B.C. (ill. 2), gives us a vivid and tactile understanding of which pigments were available to its owner, Amenemope, a vizier under Amenhotep II, but perhaps also which colours he preferred, and how he liked to arrange the cakes of red ochre, Egyptian blue, a mixed green and two charcoal blacks in his paintbox. Why did he choose to have two types of black? What did black mean to him, or in Ancient Egyptian culture in general? The arrangement of artists' materials may seem a long way removed from a complex three-dimensional colour model of the 20th century, but the underlying human desire is the same: to give an order to colour, whether for practical reasons, personal preference or even marketing purposes. Each painter's palette, in the case of the actual object with the remains of paint on it (ill. 44) or as depicted in self-portraits, reveals a part of the artist's life and times, their practices and preferences. Even modern paintboxes that can be filled with manufactured watercolour cakes provide information about their users and owners: what size and quality of box could they afford, which colours did they pick, which ones get used the most, which ones are hardly ever used, and how are they arranged? As with the Ancient Egyptian example, a Reeves watercolour box from the early 20th century can also reveal much about what colours were available or were popular at the time. The Reeves paintbox shown here (ill. 35) includes several colours that are named after Wilhelm Ostwald (1853–1932), an important German chemist and colour theorist.

Picturing Colour Before the Modern Era

The main theme of this book concerns the visualisation of colour order and the ideas associated with these systems. While the concepts themselves can be traced back to the Classical world, the first illustrations of colours in the form of diagrams only appeared at the end of the 17th century. Until relatively recently, in material terms Classical culture had appeared to be largely colourless: the main colour association of Ancient Rome and Greece was whiteness, owing to the predominant images of white marble sculpture and buildings that have long lost their coloured surfaces. However, this general impression has been substantially modified by studies in the last few years into the polychrome nature of Greek and Roman sculpture and architecture.

Conceptually and linguistically, several traditions can be distinguished in the history of colour concepts. Empedocles (494–434 B.C.) is considered one of the first to have written on the subject, and from his surviving poetic fragments it is possible to discern a cosmological system based on the four elements of earth, fire, water and air from which he believed colours were derived. In the following century, Aristotle (384–322 B.C.) may have carried out simple optical experiments with coloured glass, and also examined colour vision and contrast. He proposed a polar system that ranged from divine white light, which made colours visible, through to darkness, represented by black. From white there evolved the hierarchical colours of yellow, green, blue and red, associated with earth, water, air and fire respectively. Aristotle's follower Theophrastus (372–287 B.C.) considered how we perceive or sense colours, and did so by referring to Democritus' (460–370 B.C.) view of the four basic colours, red, black, white and a greenish-yellow, which related to certain shapes, surface patterns and other sense-based attributes. For his part, the physician Hippocrates (460–375 B.C.) distinguished four colours that represented the main

Ills. 4–5
Ulrich Pinder
Urine Samples and Uroscopic Colour Wheel
Urinproben und Harnglasscheibe
Échantillons d'urine et Cercle chromatique uroscopique
Muestras de orina y rueda de colores uroscópica
From: *Epiphanie medicorum. Speculum videndi urinas hominum* (Nuremberg, 1506)
Washington, D.C., Folger Shakespeare Library

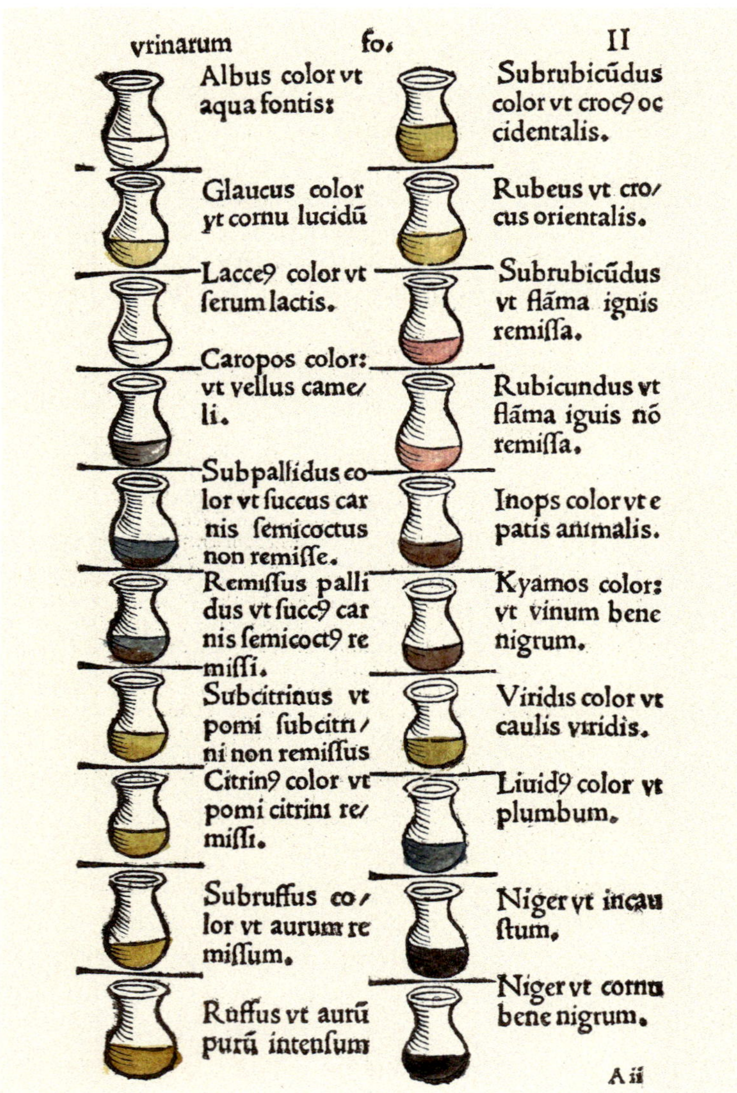

4

fluids or humours of the human body, which in turn were seen as the cause of certain character traits, namely white for phlegmatic, red for sanguine, yellow for choleric and black for melancholic.

Classical colour concepts and theoretical discussions of the subject have typically survived only in fragmented form, and with the additional inconvenience of the linguistic challenges they present. Colour is let down by language generally, since words cannot match the full complexity of colour or colour perception. Most languages only have a dozen or so basic colour terms, whereas the number of colours we can actually see is uncountable. This discrepancy becomes even more problematic when dealing with ancient languages, older texts or translations. There is also a lack of visual representations of colour concepts from the ancient world and earlier cultures, which makes it challenging to understand the earliest of these notions in any comprehensive manner. However, glimpses of the potent symbolic value of colours can be gleaned through certain archaeological finds, such as the theatrical masks found at Pompeii, both in the form of actual objects and as shown in fresco paintings, which were painted in colours expressing the four humours (ill. 7). Despite the scarcity of specific literature on colour, we can nevertheless trace the development of several cultural concepts relating to it by way of the materiality of pigments and dyestuffs. As with the costly ultramarine pigment in medieval and Renaissance art (ill. 3), the material value and rarity of certain colours in the ancient world led to them becoming associated with high-status figures and deities.

Throughout the Middle Ages, the basic Aristotelian framework of colour being situated between darkness and light (black and white) prevailed in the philosophical discourse on the subject, but there was also an increase in the number of more practical texts on colour. In the early 13th century, the Oxford university scholar and bishop of Lincoln Robert Grosseteste (*c.* 1175–1253) translated some of Aristotle's works and wrote his own treatise on colour in response, *De Colore*, in

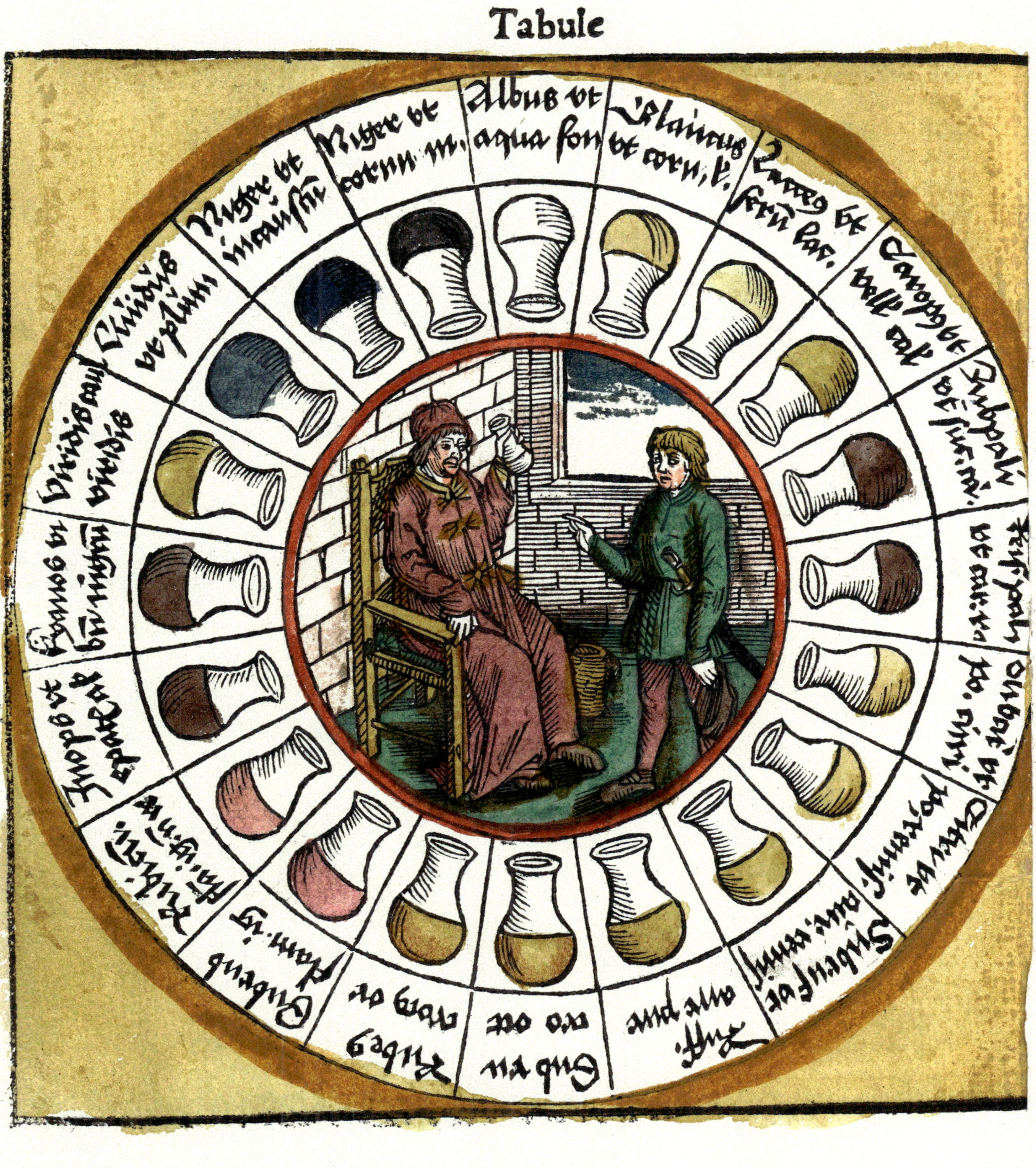

5

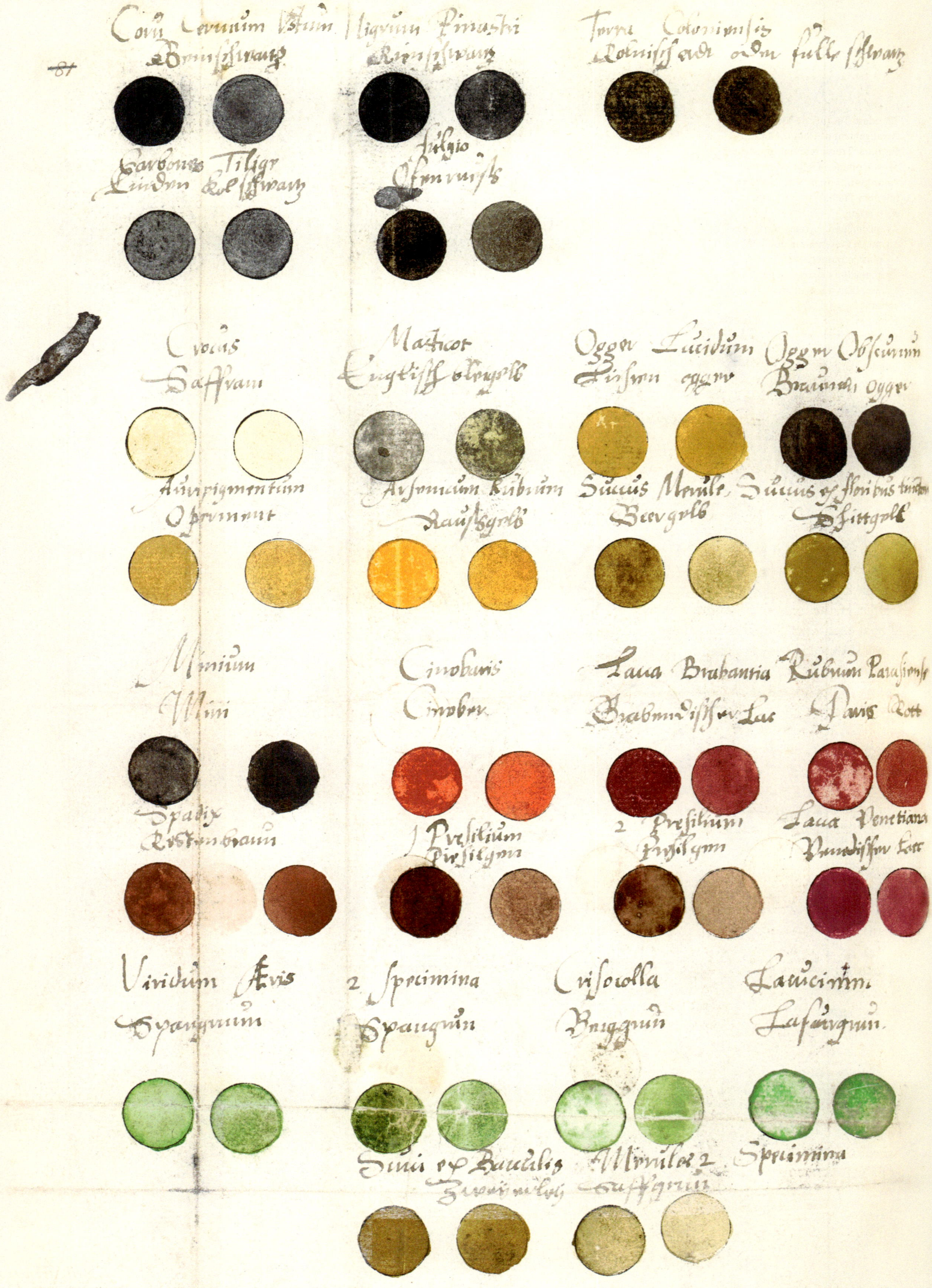
81
Cornu Cervinum Ustum
Beinschwartz
lignum Pinastri
Kienschwartz
Terra Coloniensis
Colnisch erdt
Carbones Tiliae
Linden kolschwartz
fuligo
Ofenrusß
Crocus
Saffran
Masticot
Englisch bleygelb
Ogger Lucidum
Liechten ogger
Ogger Obscurum
Braunen ogger
Auripigmentum
Operment
Arsenicum Rubrum
Rauschgelb
Beergelb
Schittgelb
Minium
Mini
Cinobaris
Cinober
Lacca Brabantia
Brabandischer Lac
Rubeum Parisiense
Paris Rott
Spadix
Kestenbraun
1 Presilium
Presilgen
2 Presilium
Presilgen
Lacca Venetiana
Venedischer Lac
Viridum Æris
Spangrun
2 Specimina
Spangrun
Crisocolla
Berggrun
Lasurgrun
2 Specimina

ILL. 6
Théodore Turquet de Mayerne
Arrangement of Colour Circles, Entitled 'True Presentation of the Most Common Colours'
Anordnung von farbigen Kreisen mit dem Titel „Wahre Aufstellung der Farben"
Agencement de cercles chromatiques, intitulé « Présentation des couleurs les plus courantes »
Presentación de círculos de color, titulada «Presentación real de los colores más comunes»
From: *Pictoria, sculptoria et quae subalternarum artium*
(London, 1620–1646)
London, British Library

ILL. 7
Two Faces
Zwei Gesichter
Deux visages
Dos rostros
Fresco, 1st century A.D.
Pompeii, House of the Golden Bracelet

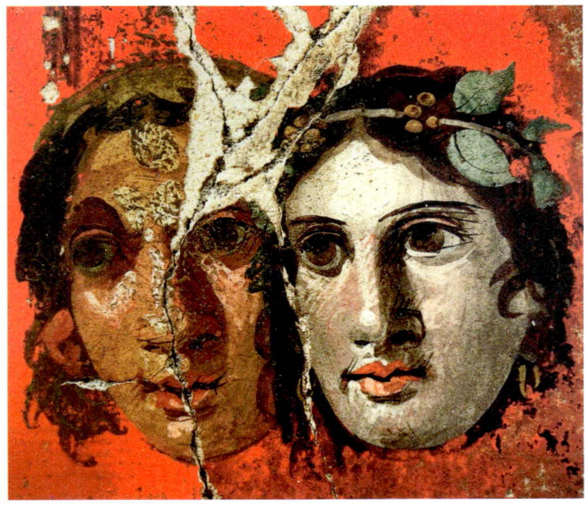

7

which he developed Aristotle's theories of colour being derived from light, while speculating on the amount of light that might be contained within individual colours. From around the time of the 15th century, colour gradually became more commercialised with the increase in global exchange and trade. Knowledge was now being captured in a more systematic way, as can be seen with the appearance of the first treatises on artistic techniques, painters' materials and artists' lives. Many of these texts are very technical and would not have reached a wide readership, especially in the pre-print age, but even so a number of written records have survived and these open a window on to the medieval world of artisanal colour, such as the somewhat earlier *De diversis artibus* (*On Diverse Arts*) by Theophilus Presbyter (fl. *c.* 1070–1125), which is essentially a book of colour recipes dating from the early 12th century. Many more artisanal writings followed in the 15th and 16th centuries, and while they are focused on technique and materiality, and do not provide philosophical discussions of colour concepts, they are nevertheless invaluable for the study of colour terminology and pigment history. Occasionally, we encounter an illustrated text that includes images of raw ingredients used to make pigments and paints, for example, the 15th- or 16th-century *Livre des simples médecines* (*The Book of Simple Medicines*, ills. 19, 20), which shows various organic and mineral substances such as squid ink, lapis lazuli and a mummy (for mummy brown) that were also believed to have therapeutic properties. The illustrations are straightforward and pictorial, but the objects are carefully arranged in the style of a collector's tray or a cabinet of curiosities.

Hippocrates' concept of the four humours that pervade the human body and spirit, and which could be represented by four colours, also continued to inform basic colour concepts throughout the Middle Ages and into the Renaissance. While images and artwork that deal specifically with colour theory and concepts are still unknown in this period, the first examples of colour being used to illustrate and enable order and classification in other contexts do begin to appear. For example, Ulrich Pinder's (d. 1519) uroscopic colour wheel from 1506 (ill. 5) depicts 20 different shades of urine, which were used to diagnose medical conditions. The pictorial urine flasks were hand-coloured, and this is an early example of a colour wheel based on nature's colours, not dissimilar to those included in Moses Harris's (1730–*c.* 1788) books on insects nearly three centuries later. In a similar manner, a wealth of encyclopaedic texts such as bestiaries, herbals, lapidaries and concordances may provide further direct or indirect information about the corresponding period's knowledge and use of colour.

The publication in 1435 of Leon Battista Alberti's (1404–1472) *De pictura* (*On Painting*) coincided with the rise of oil paint as a medium. On the subject of colour, which he considered an important and valuable element of painting, he proposed that all colours could be created by mixing together just four: red, green, blue and grey. He was a painter's writer, less concerned with coloured light than with physical colour, so the colour grey to which he gave such prominent status must be seen in the context of shading and creating three-dimensional effects in painting. Many texts written on the wider debate about the roles of drawing (*disegno*) and colour (*colore*) followed on from Alberti and his contemporaries, and there are also thoughts about colour from one of the greatest artists, Leonardo da Vinci (1452–1519), who in his various notebooks and collated writings mused on colour vision and perception, on the appearance of colour and the practical application of colour theory in painting. He proposed a colour concept, greatly influenced by Aristotle and Alberti, based on six "simple colours": red, yellow, green and blue, flanked by white and black. None of these medieval and early Renaissance texts include colour diagrams, so it is not possible to tell whether the proposed systems were envisaged as being linear or circular, two- or three-dimensional, but they are nevertheless fascinating precursors of modern colour theory.

ILL. 8
Louis Desplaces
An Allegorical Monument to Isaac Newton and His Theories on Prisms
Allegorisches Denkmal für Isaac Newton und seine Prismentheorien
Monument allégorique en l'honneur de Isaac Newton et de ses théories sur les prismes
Monumento alegórico a Isaac Newton y sus teorías sobre los prismas
From: *Tombeaux des princes, grands capitaines et autres hommes illustres* (Paris, 1741)
London, Wellcome Collection

ILL. 9
Angelica Kauffman
Colour, 1778–1780
Die Farbe / La Couleur / Color
Oil on canvas, 126 x 148.5 cm (49 ½ x 58 ½ in.)
London, Royal Academy of Arts

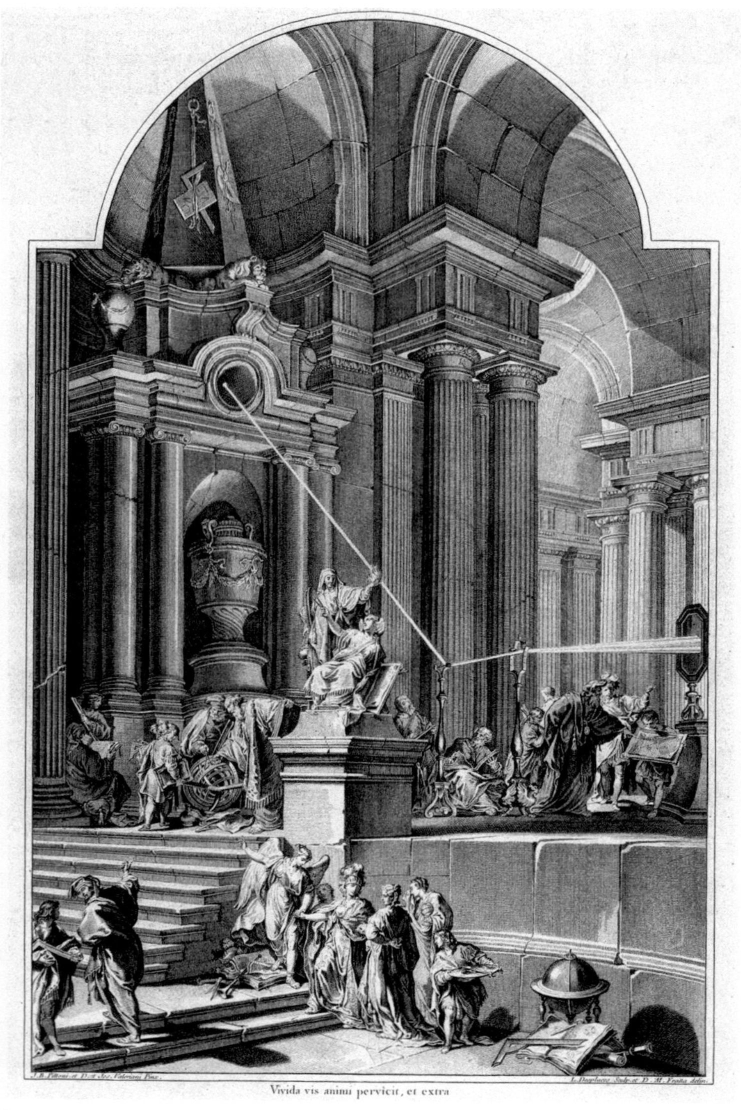

8

A New Chromatic Age: Colour from the 17th Century Onward

Although this book begins its journey through colour concepts in the later 17th century, when print culture had become well established, for various reasons many of the early examples of colour literature are still in manuscript form. The nascent spirit of the Enlightenment helped promote the exchange of intellectual ideas across geographical and cultural boundaries, but while colour symbolism and meta-systems linked to religious beliefs, alchemy and folklore continued to exist, this was the age when colour became scientific and intellectualised. The general desire at this time to understand and classify the natural world included colour – as an optical phenomenon, a tool for depicting nature, and an expressive medium in the fine and decorative arts.

In a roughly (but not strictly) chronological order, each chapter of this book focuses on a specific theme or strand in the story of colour concepts. To be included, the featured books and manuscripts had to be concerned predominantly with colour in theory or practice, or else use colour in a coherent and inventive way, for instance as a tool to illustrate related concepts and ideas.

As noted, many works by women are featured, in order to address the historical gender imbalance in the field of colour literature, including educational books on colour and painting from the early 20th century, as, for example, the U.S. educator Bonnie Snow's (*c.* 1862–1925, ills. 11, 12, 18) *The Theory and Practice of Color* (co-authored with Hugo B. Froehlich, 1918, pp. 552–559). One of the earliest examples of such a neglected source is Mary Gartside (*c.* 1755–1819, pp. 218–229), a British flower painter (ill. 21) and teacher of watercolour painting, and the first woman known to have published a book on colour in English: *An Essay on Light and Shade, on Colours, and on Composition in General* (1805). For the second edition (1808) she changed the title to *An Essay on a New Theory of Colours*, thereby stressing the

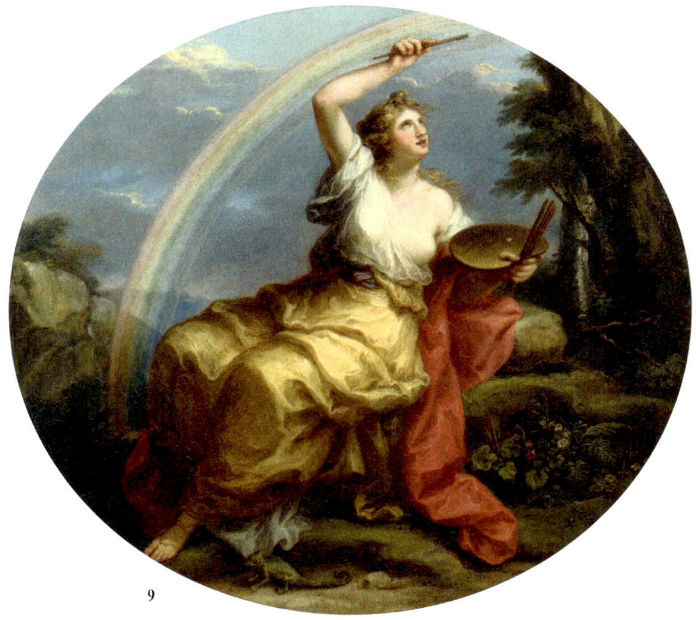

9

seriousness of her publication. Other works include the genre-bending publications by women from the wider circles of theosophists and spiritualists, such as *The New Science of Color* (1915) by the multinational and multi-cultural poet and actor Beatrice Irwin (1877–1956), who had a great interest in the potential of combining artificial coloured light with music (p. 701). Now little known, her ideas on synaesthesia influenced several famous 20th-century artists, including Georgia O'Keeffe (1887–1986), who may well have indirectly made references to Irwin in some of her work (ills. 50, 51). There are certain women in colour history who deserve to be included in the canon without having written a stand-alone theory of colour. Hilma af Klint (1862–1944), discussed in detail later (pp. 704–715), is one such, as is the British painter Winifred Nicholson (1893–1981), whose artistic output, in combination with her notes, letters and two short, published essays on colour, make her a great 20th-century colour theorist, despite never gathering her ideas and colour concepts into a concise format issued under one cover. Similar to Gartside, she appears to have developed an interest in colour theory through flower painting, and her earlier compositions were often set against illuminated landscapes or framed by windows, thus already demonstrating an interest in the interaction of light and colour (ills. 22, 55). This obvious interest in prismatic colour found a new lease of creativity when, in the mid-1970s when she was in her 80s, Nicholson was given two glass prisms. In the last few years of her life she created a fascinating series of so-called "prismatic paintings" in which she explored her experiments with prisms and light in visual form, often choosing a completely abstract style (ill. 56).

The first chapter of the present book, "Pyramids of Colour: Early Charts and Tables", offers a closer look at the earliest illustrated books in western culture. Our very first example is Richard Waller's (1655–1715) colour chart from 1686 with its hand-painted dots (pp. 80–85), which was probably derived from similar charts produced previously by the alchemist and physician at the English court Théodore Turquet de Mayerne (1573–1655) that were included in the so-called "De Mayerne manuscript" (*Pictoria, sculptoria et quae subalternarum artium*), from 1620–1646 (ills. 1, 6). Both of these charts are relatively simple in their presentation of colours in the form of tables and neat circles. As such, they are not representations of intellectual colour concepts, but should instead be seen in the wider context of painting handbooks, given that they provide much information on pigments and other artists' materials that were available at the time, and how they were mixed and classified. The chapter also shows how more complex charts and tables were created in the next century and subsequently found their way into print culture. The intellectualisation of colour concepts at this period is evident, as lists of colours become more elaborate and extensive and the first attempts at three-dimensional concepts are made. All such early charts, tables and diagrams were hand-coloured, despite the advances in colour printing, which means that each copy is an individual work of art. The practice of hand-colouring continued until well into the 19th century, and in some cases it wasn't replaced by mechanical colour reproduction until the 20th century.

In Chapter 2, "The Shape of Colour: Circles, Wheels and Globes", we examine how the simple visual concepts of colour order were developed further throughout the 18th and 19th centuries, which resulted in the emergence of a fascinating variety of geometric shapes and forms. Circles appear to have been the most prominent choice, leading to many variations of what we may now consider the classic colour wheel. Other shapes were used too though, while often a colour diagram will feature pictorial additions and embellishments, or a colour wheel will be accompanied by explanatory charts and tables, as, for example, in *The Natural System of Colours* (*c.* 1769–1776, pp. 138–143) by Moses Harris and the aforementioned works by Mary Gartside.

Aesthetically, 18th- and 19th-century colour diagrams are among the most impressive images in the history of colour literature, but despite their visual power they rarely exist without the written or printed word.

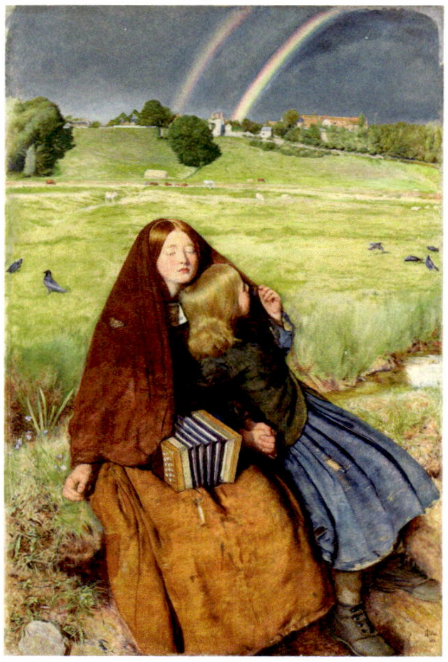

10

ILL. 10
John Everett Millais
The Blind Girl, 1856
Das blinde Mädchen
La Jeune Aveugle
La chica ciega
Oil on canvas, 81.2 x 52.3 cm
(32 x 20⅝ in.)
Birmingham Museums, Presented by the Rt. Hon. William Kendrick, 1892

Language, although no match for the visible colour spectrum, forms an important element in most of these diagrams, sometimes with intriguing and poetic results, as, for example, in the *Temperamentenrose* (1798/99, p. 155) by Johann Wolfgang von Goethe (1749–1832) and Friedrich Schiller (1759–1805). One challenge commonly faced by these diagrams was how to convey the three-dimensional aspect of colour order in a two-dimensional medium, and it is at this point that the first colour globes or balls make their appearance (see, for example, Philipp Otto Runge's (1777–1810) *Farben-Kugel* from 1810 (p. 163, ill. 13). Unfortunately, no three-dimensional models of colour globes survive from this time, but in 2016 the French-German artist Eva Bodinet constructed a model based on Runge's *Farben-Kugel* (ills. 14, 15), and in the past few years several commercial manufacturers have created ready-made 3D colour globes. Runge is also a poignant example of the increased sensibility and spirituality that was perceived in relation to colour during the Romantic era. Many artists, philosophers and writers of the later 18th and early 19th centuries adapted colour concepts to suit their individual belief systems and assigned them meanings that were highly symbolic. Between 1802 and 1810, Runge created a series of images showing the times of day, including *Der kleine Morgen* (*Little Morning*), which were representation of colour theory adapted to express his Christian beliefs (ill. 17).

In Chapter 2 a widening range of colour studies also becomes apparent as the 19th century progresses, alongside the ways in which colour literature builds on and makes reference to earlier texts. Better documentation from this period reveals how some of these early works on colour influenced specific artists and teachers. In the 1820s, the artist J. M. W. Turner (1775–1851) taught the basics of colour theory as part of his role as Professor of Perspective at the Royal Academy of Arts in London. For this he created two lecture slides showing colour circles (ills. 25, 26), with the aim of explaining the difference between material and immaterial colours. These circles with their overlapping triangles of colour at the centre were very clearly influenced by Moses Harris's colour circles, which Turner had studied. Throughout his life, Turner had a keen interest in new pigments and colour theory, while he also owned and read many publications on colour and made the interaction between colour and light a key theme in his art (ill. 27). Later in life, and just a few years after the first part of Goethe's main work on colour (*Zur Farbenlehre*, 1810–1812, pp. 236–249) had been published in English translation (*Goethe's Theory of Colours*, 1840), Turner painted two canvases in direct reference to the lessons in this text, and with a further possible nod to Harris (ill. 43).

The following Chapter 3, "New Elucidations: The Rise of Colour Theory", elaborates further on some of the most important works on colour from the 19th century, and provides reasons for the sharp rise in the number of new theories and lengthy philosophical, scientific and practical treatises on colour. This chromatic revolution is intrinsically linked to the industrialisation of colour, specifically to new manufacturing, production and printing methods, but it has its origins in the early 18th century. An undisputed milestone in the history of colour concepts was written by Isaac Newton, and when *Opticks* was published in 1704 is was the fruit of many decades of research (pp. 210–217). The book included the first widely disseminated colour diagram, in the form of a wheel, which was based on Newton's experiments into splitting pure white light into spectral colours, using a prism. Newton's impact on scientific thinking throughout the 18th and into the 19th century was immeasurable, and his theories were adapted, admired, developed and criticised (ill. 8). The fascination with the rainbow spectrum and with other findings of Newton's were, and still are frequently noted in the visual arts, as, for example, in Angelica Kauffman's (1741–1807) allegorical painting *Colour* (ill. 9), commissioned by the Royal Academy of Arts. Although an allegorical image, and thus not representing a colour concept *per se*, this painting illustrates an important point in colour history in that collectively the set that was commissioned from Kauffman

ILLS. 11–12
Bonnie Snow
Instructions and Diagrams on the Painting of Windows
Anleitungen und Diagramme zum Malen von Fenstern
Instructions et graphiques pour la peinture des fenêtres
Instrucciones y diagramas sobre la pintura de ventanas
From: *Text Books of Art Education*, vols. 1 (left) and 2 (right), New York, 1904
Sussex, collection of the author

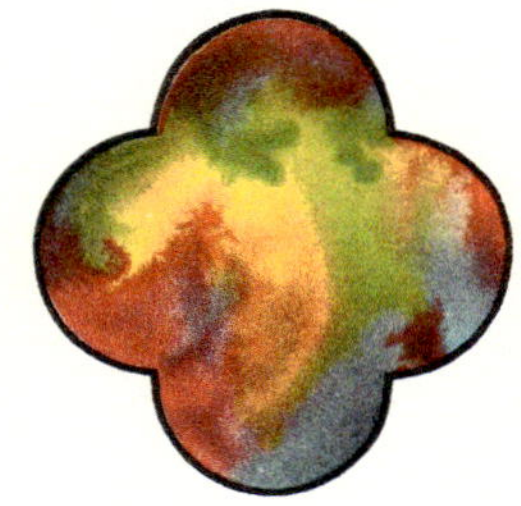

ART EDUCATION—BOOK TWO 56

What pleasure to paint a window like this, beautiful in shape, beautiful in color!

Draw a two-inch square. On this lay tablets — the inch square in the center, with semicircles about it. Draw around it. Draw around the semicircles.

Paint the shape with water. Drop in the fresh colors and let the water blend them.

Finish with the dark edge.

12

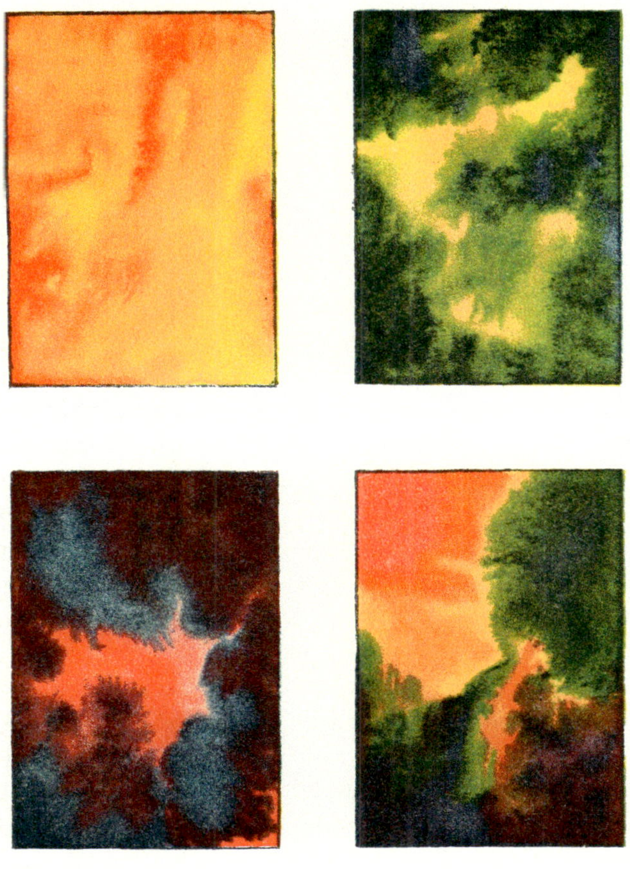

ART EDUCATION—BOOK ONE 55

11

represented the significant "Elements of Art", namely Invention, Composition, Design and Colour. The female figure representing Colour dips her paintbrush into Newton's rainbow, pulling prismatic colour on to a physical palette, and in the process adapting optical colour theory to material colour while also relating it to the teaching practices at the newly founded Academy.

As the 19th century continued, Romantic concepts and attitudes, such as Runge's spiritual colour symbolism, gradually gave way to more technical and scientific studies, many of them linked to the manufacture of new pigments and dyestuffs. The quality of these new colours became a subject of study for the great British 'colourman' George Field (*c.* 1777–1854, pp. 258–273), who not only developed a colour theory of his own but also tested, produced and patented pigments. His colours were in demand with many 19th-century artists, including members of the Pre-Raphaelite Brotherhood and their followers, and their availability thus coincided with a renewed interest in medieval painting techniques and materials. The jewel-like brilliance and daring polychromy of paintings such as *Mariana* (1851, ill. 38) by John Everett Millais (1829–1896) reflect this interest in good-quality pigments and a strong awareness of current colour concepts. It has even been suggested that Dante Gabriel Rossetti's (1828–1882) painting *Ecce Ancilla Domini! (The Annunciation*, 1849/50, ill. 29) could have been inspired by Field's proposed colour symbolism. The intellectual attention given to colour literature and theory is widely evident among 19th-century western artists, and both Impressionism and Pointillism were directly influenced by Michel-Eugène Chevreul's (1786–1889) ground-breaking theory of simultaneous contrast in colours (from 1839 onward), which he developed while working for the Gobelins tapestry factory in Paris (pp. 282–307). The Post-Impressionist Vincent van Gogh (1853–1890) used the principles of colour contrast to astounding effect in his art, and it is significant that among his personal possessions was a red lacquer box containing several balls of coloured

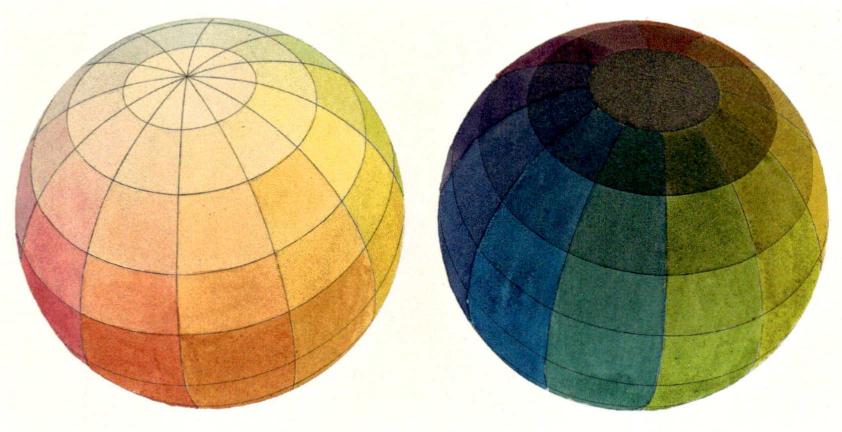

13

lengths of wool, twisted and combined in different variations just as if he had been creating colour harmonies in this way according to Chevreul's laws (ill. 37). There is also a great awareness of colour contrast in the work of James McNeill Whistler (1834–1903), and while his direct engagement with colour literature is not recorded his interest in tonal harmony and colour symbolism is clear in the titles he gave to his paintings, such as "harmonies", "nocturnes" and "symphonies" of specific colours (ills. 40, 41).

Chapters 4 and 5 deal with more technical and practical colour concepts. "Colours as Guiding Lights: Nomenclatures and Standards" is concerned with systematic lists of colours, often in relation to the natural world and taxonomical classification, which present fascinating attempts to create reliable nomenclatures and standards for use in art, manufacturing and the sciences. Some of these systems also aimed to provide a chronicle of historical colour, such as Elizabeth Burris-Meyer's (1899–1969) *Historical Color Guide* from 1938, which she later adapted for use in interior decoration, fashion and make-up (pp. 388–409). "Theory and Practice: The Teaching of Colour" focuses on the history of practical manuals of colour for painters and printers, from the early 19th to the early 20th century. The earlier colour manuals are still much indebted to the development of painting in watercolour, and many of them were intended for amateurs as painting materials became more affordable at the start of the 19th century. Painting was an acceptable pastime for women in the 18th and 19th centuries, and for that reason a good proportion of these manuals were aimed specifically at a female readership, although they were predominantly written by men. From the beginning of the 20th century, however, more women became active in writing educational and instructive colour manuals or primers, grasping their opportunities for writing, publishing and teaching in the field. Until now this has been a little-researched area of women's creative engagement, and because teaching manuals, especially those for younger children, were not considered serious texts they have typically not been well recorded and not many copies survive.

The early 20th century was marked as well by another surge in books about colour, many of them building on the works of the great colour theorists of the previous century but developing their concepts further, as is demonstrated in Chapter 6, "Colour Breaking Free: The Early 20th Century". Colour in this period cannot be properly evaluated without taking into account a range of scientific and artistic achievements, including the various advances in the understanding of colour vision and specifically the anatomy of the human eye, the electrification of the world and the resulting possibilities provided by artificial lighting, and the radical new use of colour as an expressive tool in the arts, for painting in particular. European avant-garde groups such as Der Blaue Reiter (The Blue Rider), the Fauvists and many of the Expressionists and early abstract artists actively engaged in colour theory. They embraced the symbolic and visual power of colour and depicted the world in subjective, non-realistic tones. For some, concepts of colour became the main subject of certain works, as seen, for example, in Franz Marc's (1880–1916) abstract painting *Kämpfende Formen* (*Fighting Forms*, ill. 47) from 1914.

The highly coloured and increasingly mechanised first decades of the 20th century, when material colour could be bought ready-made and neatly packaged in almost every bright shade imaginable, were crying out for new attempts to give colour a definitive order. Some of the more influential results are discussed in Chapter 6, such as the complex work of Albert Henry Munsell (1858–1918) in America (pp. 584–599) and Wilhelm Ostwald in Europe (pp. 600–609). Both men created highly educational and adaptable systems that were based on a three-dimensional concept of colour, and indeed some of these three-dimensional models that were used in teaching have survived (see, for example, Ostwald's *Colour Solid* in the archives of Winsor & Newton, p. 430). The early 20th century was also a period of improved research

ILL. 13
Philipp Otto Runge
Colour Globe / Farben-Kugel
Sphère chromatique / Esfera de color
(detail)
From: *Farben-Kugel, oder, Construction des Verhältnisses aller Mischungen der Farben zu einander, und ihrer vollständigen Affinität* (Hamburg, 1810)
Hamburger Kunsthalle

ILLS. 14–15
Eva Bodinet
Colour Globe, 2016
Farbglobus
Sphère chromatique
Esfera de color
Sussex, collection of the author

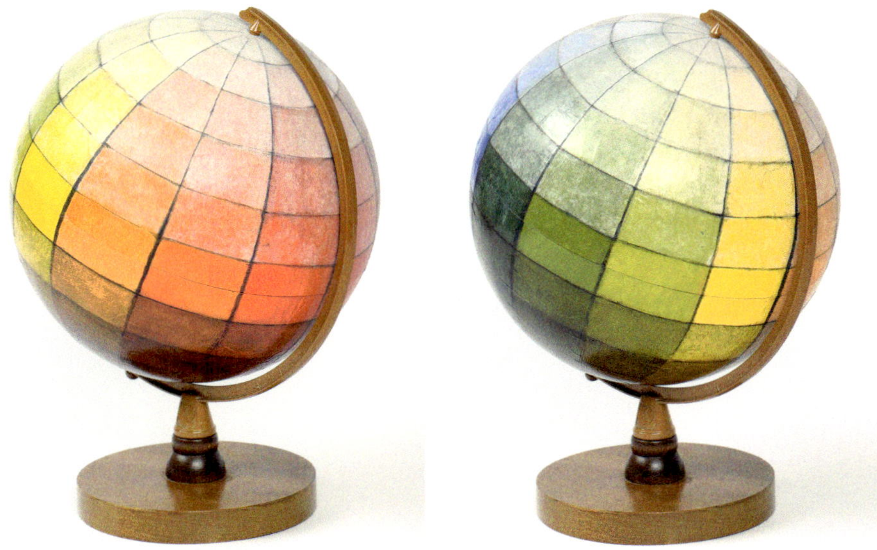

14–15

and publishing opportunities for women, which resulted in several notable examples of scientific studies by women on colour vision and psychology, colour in literature and poetry, historical colour theory and other aspects of colour history. Some of these books published between 1900 and the 1930s are among the most inventive and beautiful in colour history, as, for example, Emily Noyes Vanderpoel's (1842–1939) *Color Problems* (1902, pp. 562–583).

Chapter 7, "The Sound of Colour: Spiritualism, Occultism and Music", examines the synaesthetic associations between colour and music, and, by extension, movement, dance and spirituality. To some degree, the highly individual and often pseudo-scientific colour concepts published by the Theosophists, Anthroposophists and members of similar groups and movements around the start of the last century were a response to an increasingly mechanised world, but it is useful to review each of their colour concepts carefully and judge them not by their scientific accuracy but by their creative inventiveness and the social and historical circumstances that underpin them. You don't have to believe in colour auras, as illustrated and described by Annie Besant (1847–1933) and Charles Webster Leadbeater (1854–1934), but their attempt to give form and colour to thoughts and emotions is a fascinating comment on human nature at the time (pp. 648–687).

Analogies between colour and music have been a prominent theme in the history of colour concepts, and most of the texts mentioned in this chapter of the book discussed colour in relation to sound, music, poetry and dance, while many of them also combined mysticism with science, individual religious beliefs with colour theory, nature with spiritualism and historic texts on colour with current painting practice. This may well sound a bewildering and idiosyncratic mixture, and it often makes for difficult reading, but the visual representations of these concepts are captivating. The Australian artist Roy de Maistre (1894–1968) created many paintings on the subject, including a five-metre-long scroll that represented a "colour translation" of Haydn's *Trio in B Flat* (p. 60, ills. 57, 67). The work of the Swedish mystic and anthroposophist Hilma af Klint was deeply rooted in occultism and colour symbolism, and in recent years she has been rediscovered and rightly praised for her visionary style. Although she never formulated a distinct colour theory, many of her large-scale paintings incorporate coloured circles, pyramids and images of optical colour, thereby creating a remarkably individual visual chromatic vocabulary in which realism was combined with abstraction (ills. 39, 48, 49).

This book cannot make any claim to being a global overview of historic colour concepts. It is, perhaps inevitably, predominantly western-centric in its selection of works, but Chapter 8 aims to redress this imbalance by looking east, beyond our more familiar western view, to provide a glimpse into Japanese and Russian colour concepts of the 19th and 20th centuries. It is not the author's intention to claim expertise in these regions' colour concepts, but rather to acknowledge their aesthetic appeal and to highlight how colour concepts from geographically distant parts of the world may relate to each other. Comparing something as abstract as colour is a potent example of what connects humans across the globe. One intriguing example of cross-cultural colour exchange can be noted in the afterlife of one of George Field's colour diagrams from the 1830s which recurs in a Japanese textbook for schoolchildren, *Gakko hitsuyo irozu mondo*, that was first published in 1876 (pp. 740–743).

The final chapter, "Interaction and Abstraction: Bauhaus and Beyond", considers colour in its most pure, reduced and elemental form as found in art, design, didactic literature and teaching in the early to mid-20th century. The road to abstraction was long and gradual, with contributions from various pioneers including Turner, Gartside, Whistler, af Klint and several artists from various European avant-garde groups, but the Bauhaus and its wider circle of influence became one of the leading representatives of modernism, progressiveness and, in stylistic terms, abstraction. Founded by the architect

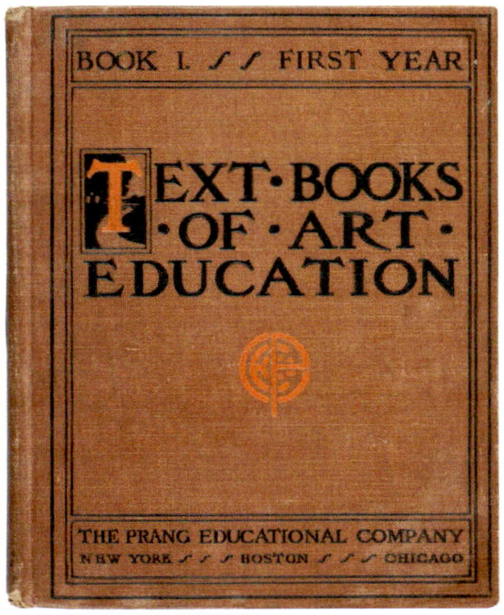

16

ILL. 16
Bonnie Snow
Cover / Buchdeckel / Couverture / Cubierta
From: *Text Books of Art Education*,
vol. 1 (New York, 1904)
Sussex, collection of the author

ILL. 17
Philipp Otto Runge
Little Morning, 1809/10
Der kleine Morgen
Le Petit Matin
La mañana
Oil on canvas, 109 x 85.5 cm
(43 x 33½ in.)
Hamburger Kunsthalle

ILL. 18
Bonnie Snow
Floral Colour Wheel and Colour Chart
Aus Blüten zusammengesetzter Farbkreis und Farbdiagramm
Cercle chromatique floral et Cercle chromatique
Rueda de colores floral y diagrama cromático
From: *Text Books of Art Education*,
vol. 2 (New York, 1904)
Sussex, collection of the author

Walter Gropius (1883–1969) in Weimar in 1919, the short-lived school of art and design had its roots in German avant-garde and spiritualist groups, as well as the English Arts and Crafts movement. It was concerned with simplicity and functionality of design, based on giving artists a thorough training and on quality of craftsmanship. Members, teachers and students in the Bauhaus included Ludwig Mies van der Rohe, Paul Klee, Johannes Itten and Josef Albers, as well as several women, whose roles in the Bauhaus story have only recently been fully acknowledged, such as Anni Albers (ill. 64), Marianne Brandt, Gertrud Arndt (ill. 59), Gertrud Grunow (ills. 61, 62) and Gunta Stölzl (ills. 65, 66). In 1925 the Bauhaus moved to Dessau, where Gropius designed the iconic Bauhaus building (see ill. 63), but in 1933, after a year's residency in Berlin, it was forced to close after having come under increasing pressure from the Nazis. Some of the members emigrated to England or the United States, where they were able to maintain ties with each other and continue their work. Although colour was only one element of the Bauhaus courses, it led to many related publications on colour, some of which are featured in this book's closing chapter, including two classic works on colour that were published nearly 30 years after the school's closure: Josef Albers's (1888–1976) *Interaction of Color* (1963, pp. 824–837) and Johannes Itten's (1888–1967) *Kunst der Farbe* (*The Art of Color*, 1961, pp. 804–823). These and other, lesser-known works on colour by associated authors and artists are examples of a near-perfect synthesis of thinking about colour in terms of living and working with it, teaching it and applying it to art and design, while at the same time reflecting the modernist ideals of simplicity, geometry and abstraction. In art, the colour concepts and aesthetics of the Bauhaus are mirrored by the wider move towards abstraction in the 20th century, and the similarities in the use of colour with the extremely geometric work of modern masters such as Piet Mondrian (ill. 68), Henri Matisse and Kazimir Malevich are obvious.

The Timelessness of Colour Concepts

The abstract beauty of colour diagrams, of colour concepts made visual, gives them a timeless appeal. The simple, geometric shapes and forms created to form the framework of early colour charts and tables feel immediately accessible, because the same or similar forms still exist today in art, advertising and design, while colour circles, wheels, triangles, cubes and even more complex diagrams or organic forms are relatable across centuries. Colour theory and colour concepts continue to be a motif and source of inspiration for contemporary artists. Since 2009, the Icelandic-Danish artist Olafur Eliasson (b. 1967) has created a series of large colour wheels in which the various colours bled seamlessly into one another (ill. 70), in similar fashion to one of Chevreul's colour wheels in his *Atlas* from 1861. A set of them, entitled *Turner colour experiments*, was displayed at Tate Britain in proximity to some of Turner's paintings. Just as Turner referenced Goethe and Harris in some of his work, so Eliasson referred back to colour theorists and artists before him, aiming to create his own new colour theory in art.

Human beings have always tried to tame and control colour, this ungraspable yet all-pervading element of our vision. Our attempts at capturing colour, by drawing circles, squares and lines around it, by squeezing it into lists or moulding it into three-dimensional shapes, are ultimately expressions of human thinking and creativity. The perfect colour concept may not yet have been formulated, but turning existing ideas about colour into words, images, diagrams and three-dimensional objects has resulted in a rich history of colour as visual art. Individual expertise, personal choices, new research discoveries and fresh thinking make every new book on colour history exciting, and the authors hope that the selection of colour concepts presented in these two volumes will provide some unexpected and inspiring views and stories that illuminate the history of colour further.

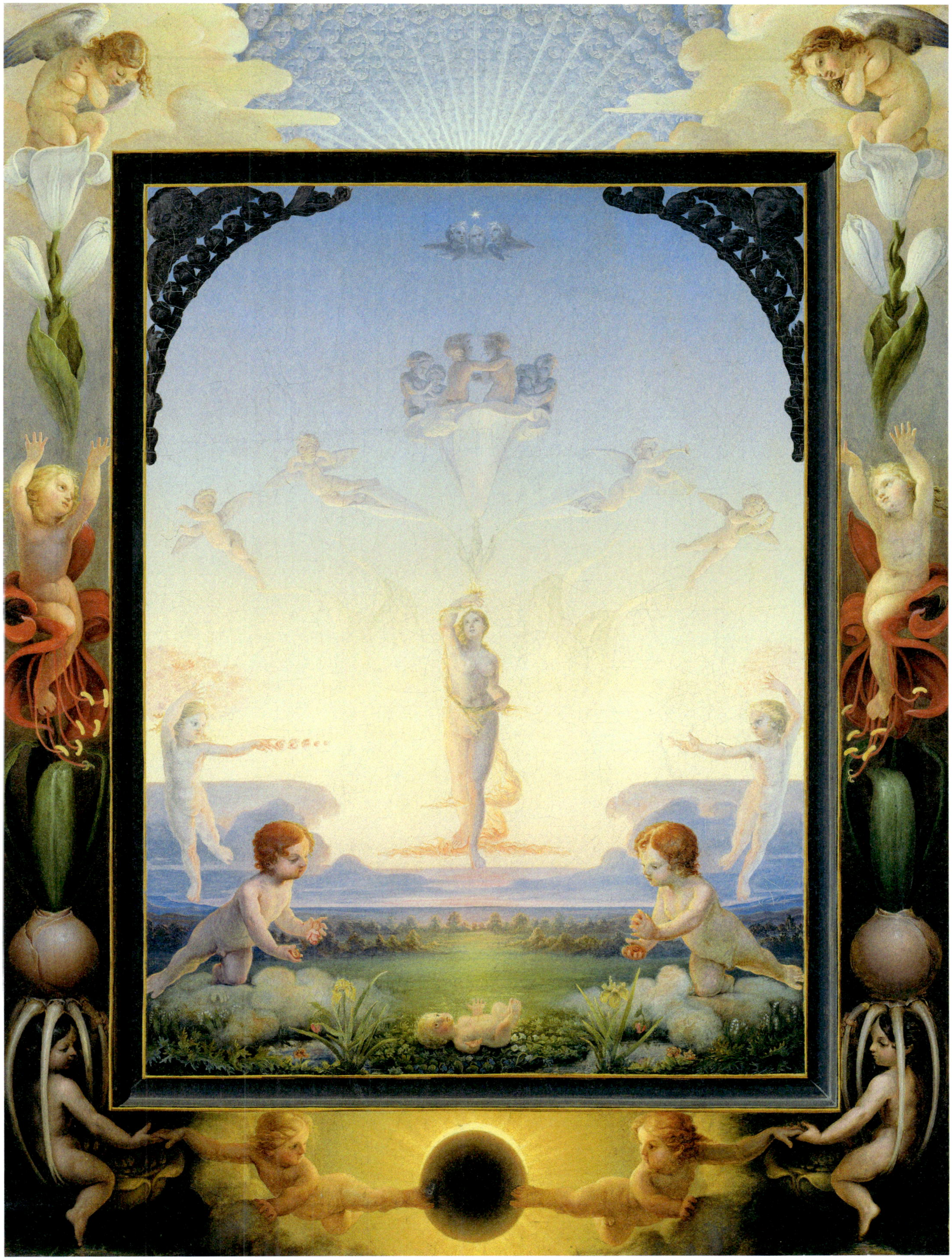

17

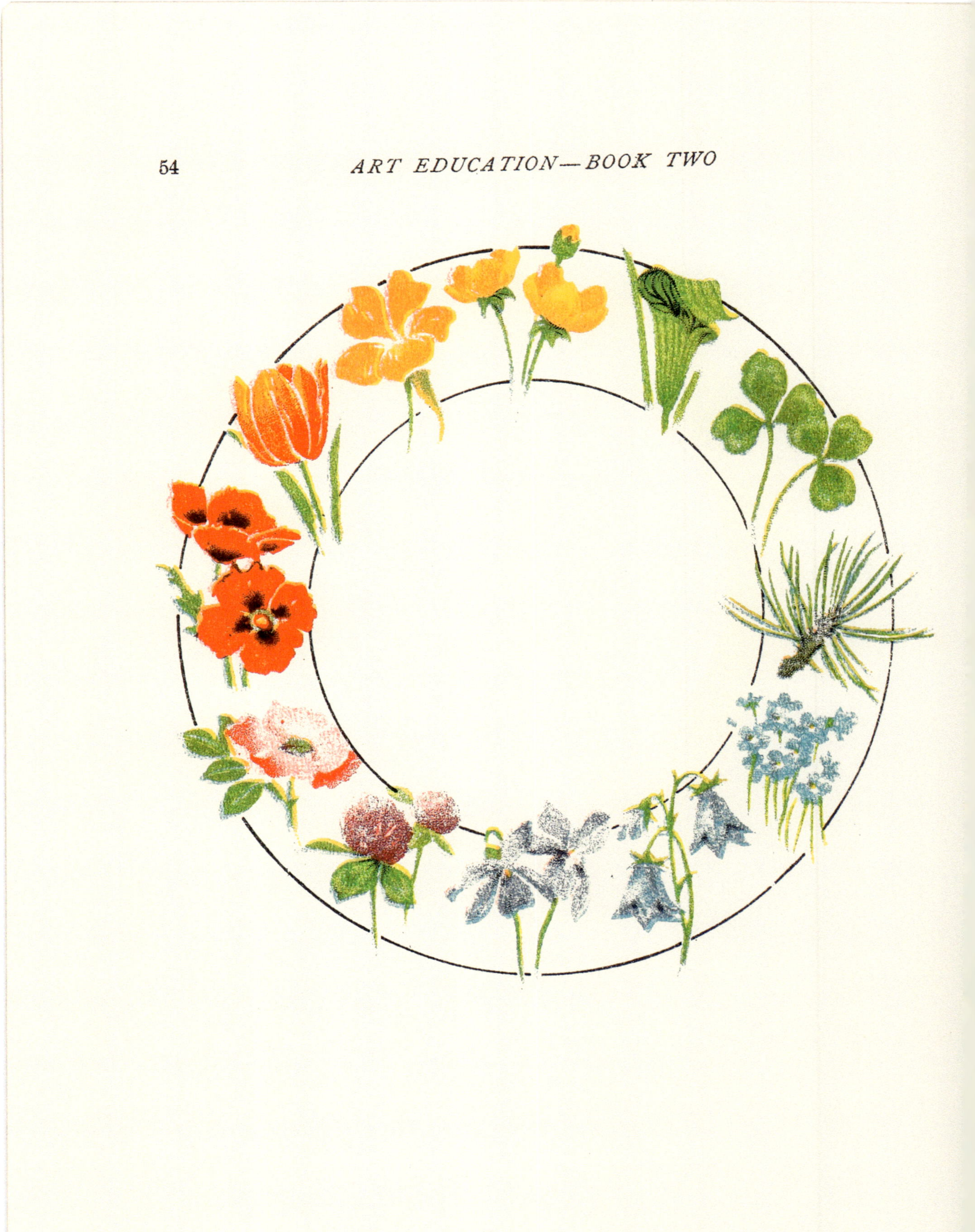

18

TEXT BOOKS OF ART EDUCATION

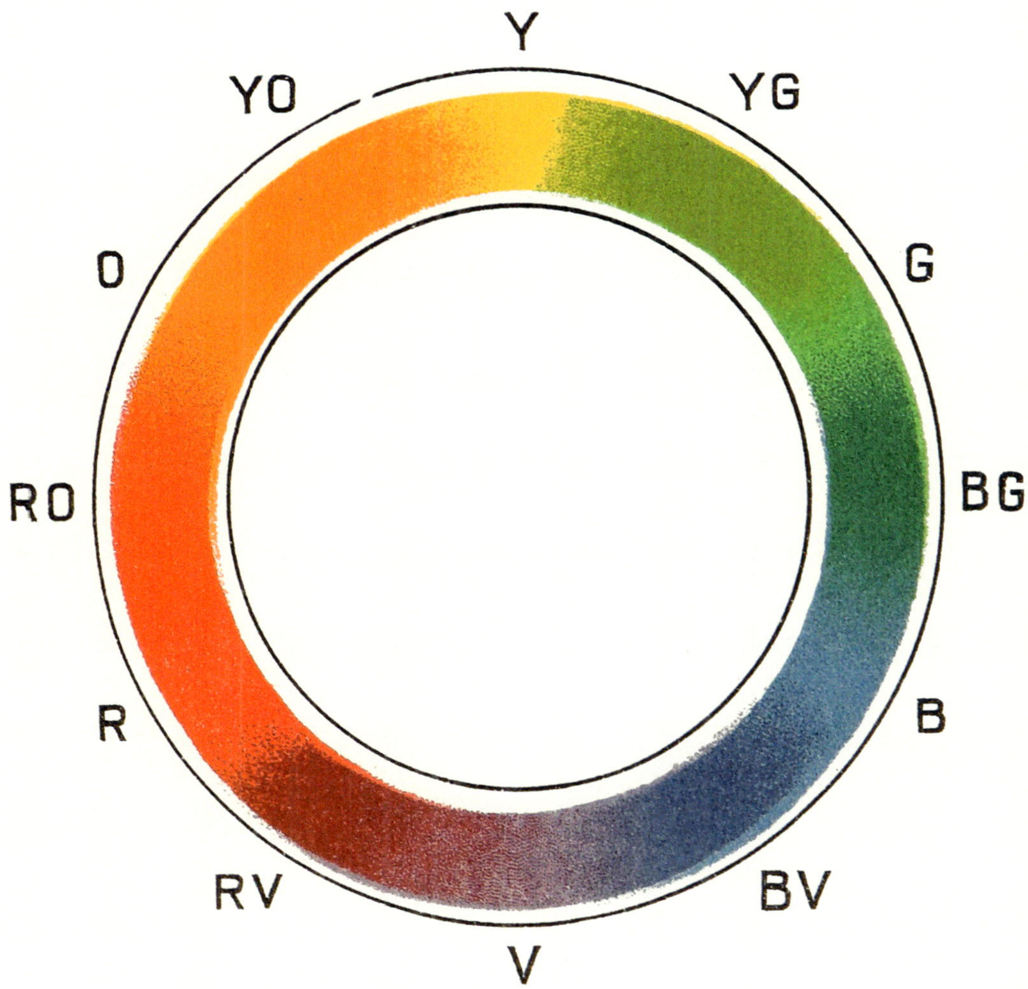

COLOR CHART

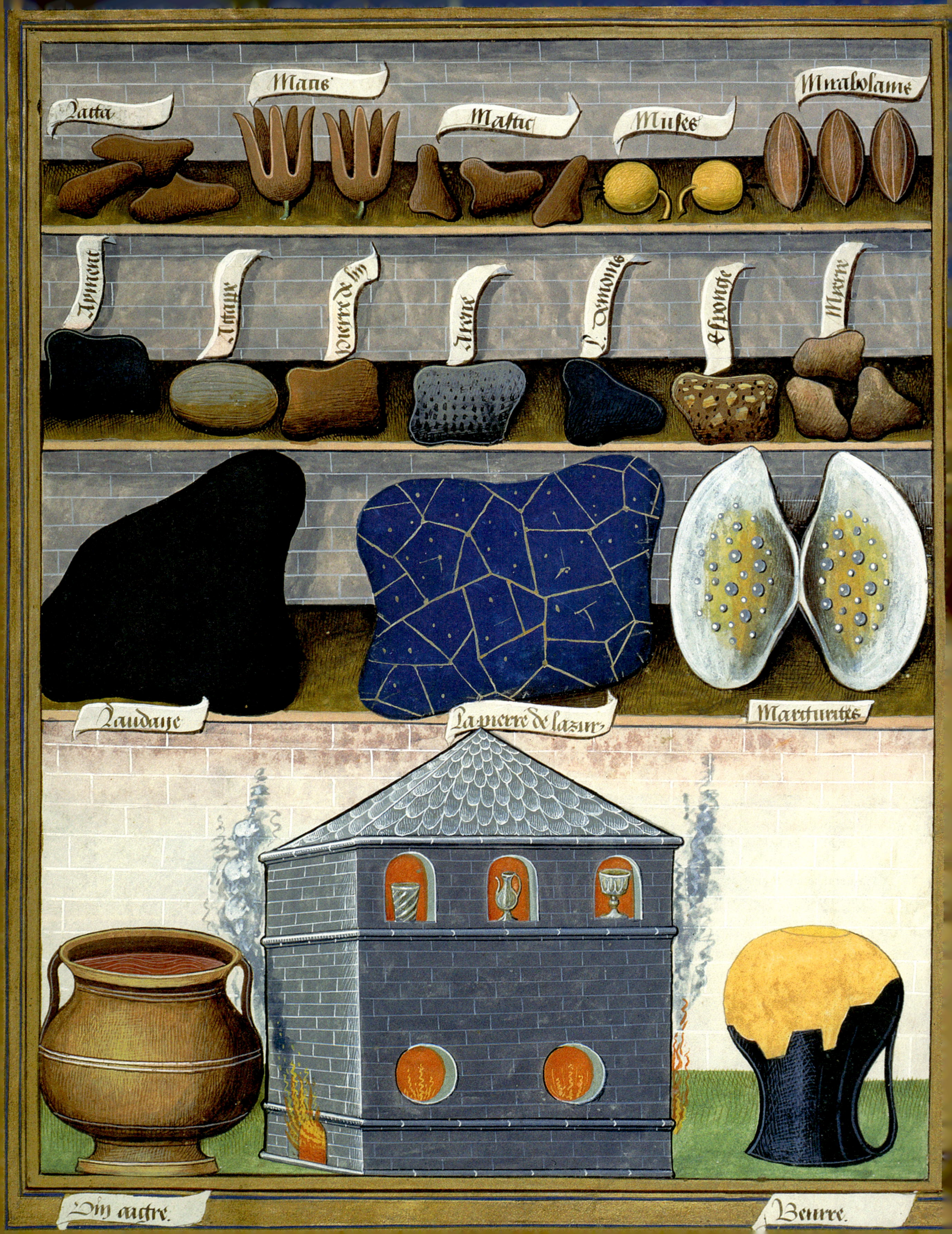

Datta
Mans
Mastic
Muses
Mirabolame
Aymant
Attaur
Pierre de ly
Arene
S. Amome
Esponge
Myrre
Laudane
La pierre de lazur
Margurites
Vin aigre.
Beurre.

Alexandra Loske

DIE KONTROLLE UND ENTFESSELUNG DER FARBEN

Eine Geschichte des menschlichen Bestrebens, Farbe in Worten und Bildern zu erfassen

Unsere Welt ist voller Farbe. Die meisten erleben sie in Farbe. Die Erde ist als „Blauer Planet" bekannt, und optisch lösen wir uns in Blau auf: Als wir 1990 aus etwa sechs Milliarden Kilometern Entfernung von der Raumsonde Voyager 1 fotografiert wurden, erschienen wir, wie der Astronom Carl Sagan es einige Jahre später formulierte, als „blassblauer Punkt": weniger als ein Pixel groß und kaum sichtbar, doch noch immer mit erkennbarer Färbung. Wir werden zum Farbfleck, bevor wir ganz aus dem Blickfeld verschwinden. Demnach ist Farbe einer der prägenden und verbindenden Aspekte unseres Lebens, eine wahrhaft universelle und von allen Menschen geteilte Erfahrung, und es überrascht nicht, dass sie uns schon immer fasziniert. Die frühesten Manifestationen menschlicher Kreativität in Form von Schnitzereien und Markierungen in und auf Holz, Knochen und Felsen zeigen, dass wir von jeher versucht haben, Farben herzustellen und zu verwenden, um vor allen Dingen abzubilden oder wiederzugeben, was uns beschäftigt und was uns umgibt. Wenngleich die Bilder und physischen Beweise dieser frühesten menschlichen Betätigung bruchstückhaft und verblasst sind, ist nicht auszuschließen, dass Farbe sogar in dieser Zeit auch als Symbol genutzt wurde. Die mineralischen Pigmente, die der frühe Mensch aus dem Boden gewann, wurden mit organischer Materie wie Fett, Schmalz oder Körperflüssigkeiten vermischt. Dabei könnten sie gut spezifische oder sakrale Bedeutungen angenommen haben, aus denen sich eine erste einfache Farbhierarchie entwickelte. Wir werden nie genau wissen, was unsere prähistorischen Vorfahren über Farbe dachten, doch gibt es Hinweise, dass bestimmte Farben in speziellen Zusammenhängen verwendet wurden.

Eine theoretische Auseinandersetzung mit Farbe lässt sich durch die gesamte Menschheitsgeschichte zurückverfolgen, von den einfachen Ausprägungen der frühen Kunst bis zu den hoch entwickelten Systemen, die in der modernen Wissenschaft, Philosophie oder Kunst verwurzelt sind – Kategorien, die sich häufig überschneiden und aufeinander beziehen. Seit jeher kreisen menschliche Gefühle und Gedanken um Farben und widmen sich dem Verlangen, Farben herzustellen, sie zu ordnen und zu systematisieren, sie zu nutzen und zu strukturieren und ihnen vielschichtige Bedeutungen zuzuweisen. Die zahlreichen Bemühungen, sie zu klassifizieren und zu begreifen, schlugen sich in vielgestaltigen Erscheinungsformen nieder, von dualistischen Konzepten aus Dunkelheit und Licht bis zu ausgefeilten geometrischen Mustern und Systemen, die versuchen, Farbe mittels spezifischer Formen zu kategorisieren. Die Entwicklung der Sprache, der Schrift und schließlich des Druckwesens mit seinen ausgeklügelten Methoden der Bildreproduktion hat dazu geführt, dass diese Ideen immer anspruchsvoller, komplizierter, visuell verblüffender und zu guter Letzt auch besser dokumentiert wurden.

In diesem Buch beginnt die visuelle Reise durch die Geschichte der Farbtheorie zur Zeit der frühesten illustrierten Manuskripte und gedruckten Quellen aus dem 17. Jahrhundert, die sich speziell auf Farbe konzentrieren, und endet Mitte des 20. Jahrhunderts vor Anbruch des Digitalzeitalters. So wie jedes hier beschriebene Farbsystem ist auch das vorliegende Buch selbst mit seiner Auswahl an bestimmten Theorien ein Spiegel der Zeit, in der es verfasst wurde. Im Verlauf der letzten Jahrzehnte hat das Interesse an Farbe und Farbgeschichte deutlich zugenommen. Infolge aktueller Forschungen entdecken wir neue Stränge und lange Zeit übersehene Stimmen und Beispiele einer materiellen Kultur (Schriftsteller und Pädagogen, Bücher, Manuskripte, Kunst und Methoden), die entweder in Vergessenheit geraten waren oder bisher nie als Teil des Farbkanons galten.

Kein Buch über Farbe kann den Anspruch erheben, eine ebenso umfassende wie kulturell, historisch, geografisch und thematisch vollständige Dokumentation vorzulegen. Manche Titel nehmen bewusst Auslassungen vor, um den Fokus auf bestimmte kulturelle oder historische Themen zu lenken, andere treffen ihre Auswahl

ILLS. 19–20 *(Pages 24, 26)*
Robinet Testard
Selected Minerals and Organic Matter
Anordnung von Mineralen und organischen Stoffen
Agencement de matières minérales et organiques
Presentación de minerales y materia orgánica
From: Matthaeus Platearius, *Le Livre des simples médecines* (*c.* 1500)
St. Petersburg, Russian National Library

ILL. 21
Mary Gartside
Mixed Flowers on a Slab, Including Roses, and Convolvulus, *c.* 1785
Gemischte Blumen auf einer Platte, darunter Rosen und Winden
Mélange de fleurs sur un bloc de pierre, avec roses et volubilis
Conjunto de flores sobre una losa, incluidas rosas y campanillas
Watercolour and bodycolour on card, 36.1 x 27.7 cm (14 ¼ x 10 ⅞ in.)
Cambridge, Fitzwilliam Museum

ILL. 22
Winifred Nicholson
Easter Monday, *c.* 1950
Ostermontag
Lundi de Pâques
Lunes de Pascua
Oil on panel, 174 x 174 cm (68 ½ x 68 ½ in.)
Carlisle, Tullie House Museum and Art Gallery

anhand der persönlichen Interessen, Vorlieben oder Fachkenntnisse der Autoren und Herausgeber. Dieses Buch ist, bildlich gesprochen, ein bunter Strauß und in seiner Auswahl so subjektiv wie die Farbwahrnehmung selbst. Die 67 hier gezeigten und besprochenen

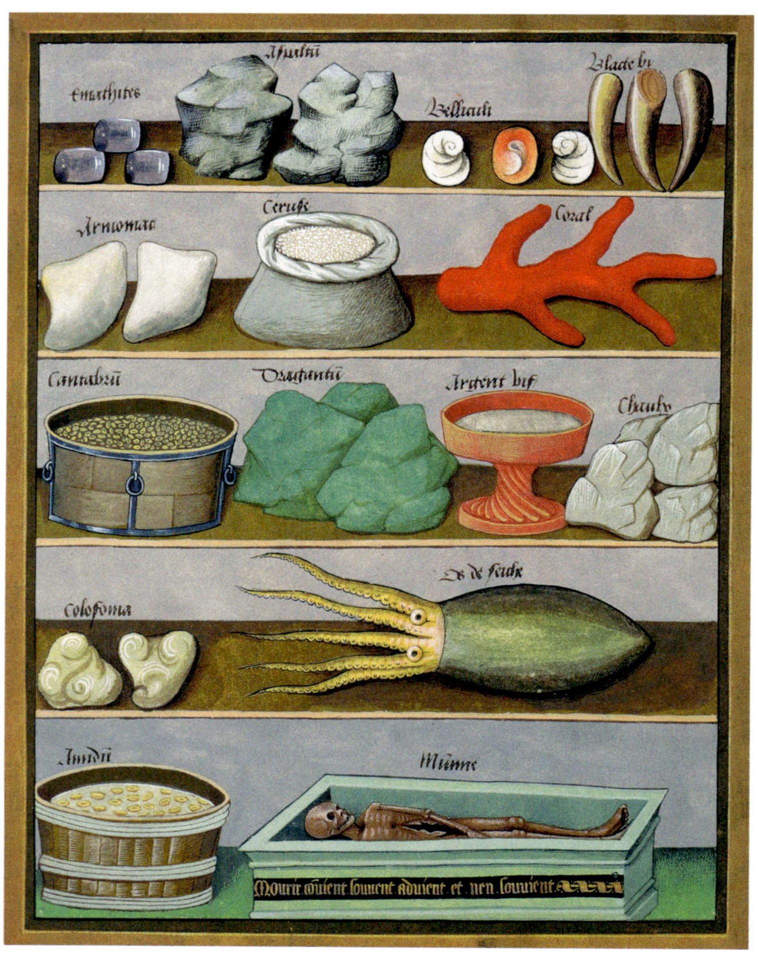

20

Werke stehen stellvertretend für die bewegte und abwechslungsreiche Geschichte der Farbtheorien seit dem 17. Jahrhundert. Während wir viele vertraute Beispiele für Farbsysteme ausgewählt haben, die in der Farbgeschichte fest verankert sind, konnten wir diese hoffentlich erfolgreich in einen größeren kulturellen Zusammenhang stellen und mit neuen Augen betrachten. Neben diesen „Klassikern" ihres Fachs haben wir uns besonders darum bemüht, einige weniger bekannte oder ungewöhnlichere Werke zu präsentieren, die aufgrund ihrer ästhetischen und theoretischen Wirkung und Bedeutung Aufmerksamkeit verdienen. Darüber hinaus wollten wir Farbtraktaten mehr Raum geben, die in den letzten zweihundert Jahren von Frauen verfasst, publiziert und bebildert wurden. Die große historische und geografische Bandbreite dieses Buches reicht von Europa über die USA bis in asiatische Regionen, doch auch verschiedene Gattungen und Stile sollten so umfassend wie möglich abgedeckt werden. Eine lehrreiche Farbfibel aus dem 20. Jahrhundert kann genauso faszinierend sein wie Isaac Newtons (1643–1727) Farbkreis von 1704 (S. 211), und beide spielen in der Geschichte der Farben eine wichtige Rolle.

Warum Farbtheorien?

Der weit gefasste und dehnbare Begriff „Farbtheorie" lässt sich sehr gut auf das Nachdenken über Farben und den Versuch anwenden, sie zu verstehen und zu systematisieren. Doch anstatt nach dem perfekten Begriff zu suchen, könnte man vielleicht besser fragen, warum wir überhaupt das Bedürfnis verspüren, etwas so Ungreifbares, sich ständig Veränderndes, schwer Definierbares und in gewissem Maße Subjektives systematisch zu erfassen. Bis vor nicht allzu langer Zeit fand in der Farbliteratur keine saubere Trennung zwischen stofflicher und immaterieller Farbe statt. Doch nachdem man den Unterschied vollständig durchdrungen hatte, wurde klar, dass tatsächlich viele Formen

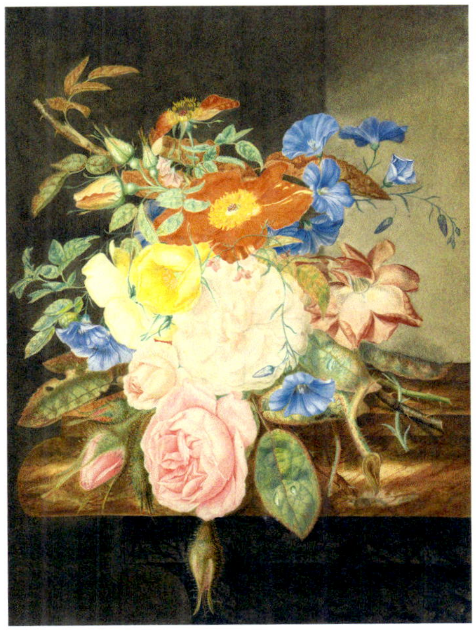

21

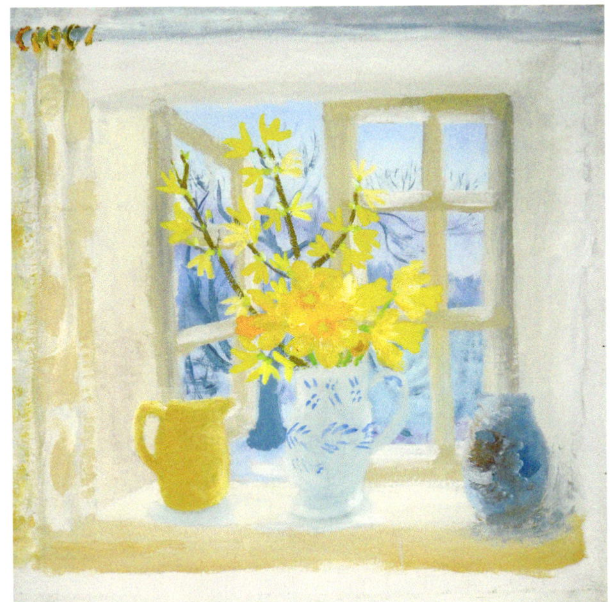

22

von Farbe existieren, in denen sich immaterielle Farbe und die Farben von Feststoffen miteinander verbinden. Farben entgleiten uns leicht, sie verändern sich mit dem Licht, sehen je nach Tageszeit anders aus, verblassen, werden dunkler und können unkalkulierbar sein, wenn sie kombiniert, gemischt oder nebeneinandergesetzt werden. Darüber hinaus sind sie schwer nachzuahmen, wie Pigmenthersteller und Historiker bestätigen können. Für den frühen Menschen muss es frustrierend gewesen sein, in einer Welt zu leben, deren vorherrschende Farben – das Blau des Himmels und das Grün der Pflanzenwelt – zugleich die beiden am schwierigsten herzustellenden waren. Wir mögen in einer Welt leben, die reich an Chlorophyll ist, doch nur wenige Pflanzen ergeben brauchbare grüne Pigmente. Farbe ist außerdem aufreizend launisch, instabil, oft giftig und wurde nicht immer geliebt – denn wie der Künstler und Schriftsteller David Batchelor in seinem Essay *Chromophobie* so wunderbar erklärt hat, ist die Abscheu vor der Farbe ein bedeutender Bestandteil der westlichen Kulturgeschichte und tief verwurzelt in einer „Angst vor Verunreinigung und Verderben durch etwas, das unbekannt ist und anscheinend auch unbekannt bleiben wird." Diese Angst oder Aversion „äußert sich in den zahlreichen unterschiedlichen Versuchen, Farbe aus der Kultur auszutreiben, zu entwerten, ihre Bedeutung zu schmälern, ihre Komplexität zu leugnen." Westliche Einstellungen zur Farbe oder, genauer gesagt, zu stark gesättigten Farben und zur Polychromie, so Batchelor, gingen im Gegensatz zu den klassischen Idealen von jungfräulichem Weiß oder farbloser Schönheit oft mit Vorstellungen einher, die das Fremde, Orientalische, Primitive, Weibliche, Vulgäre, Oberflächliche oder gar Pathologische einschließen. Die westliche Kultur blickte insofern auf eine lange Geschichte von Farbe und Dunkelheit als Sinnbilder für ein gefährliches oder bedrohliches „Anderes" zurück, und eine derart destabilisierende Kraft galt es entweder zu meiden oder zu bändigen. Farbsysteme oder Farbtheorien zu entwickeln, ist dabei eine Möglichkeit, Farbe zunächst gedanklich zu zähmen, um sie anschließend wohlüberlegt einzusetzen und aufzutragen. Wenn wir Farben eine Ordnung auferlegen, und sei es nur theoretisch, dann haben wir zumindest versucht, die Kontrolle über ein veränderliches und allgegenwärtiges Phänomen zu übernehmen.

Doch natürlich sind Furcht vor, Zweifel an und Hass auf Farbe nur eine Seite der Medaille. Unser Verhältnis zu Farbe war stets facettenreich. Vor allem die letzten fünfzig bis sechzig Jahre – in denen Erfindungen wie die Farbfotografie, der Farbfilm und das Farbfernsehen massentauglich wurden, digitale Farben aufkamen und das Druckwesen nahezu vollständig auf Farbe umstellte – haben nicht nur dazu geführt, dass wir uns mit Farbe wohlfühlen, sondern haben uns diese auch so zugänglich gemacht wie niemals zuvor. Fast ausnahmslos ist Farbe heute ein bequem verfügbares Gut, das niemanden mehr ausschließt, und auch der Farbreproduktion sind trotz weiterhin auftretender Qualitätsschwankungen kaum noch Grenzen gesetzt.

Die Ordnung der Farben: Vom Malkasten bis zur Palette

Wo Textquellen zur Farbtheorie fehlen, ermöglichen Archäologie und materielle Kultur Einblicke in die Farbkultur. Zumindest die in Kunst und Dekoration verwendeten Farben deuten darauf hin, welche Pigmente zu bestimmten Zeiten der Geschichte verfügbar waren, und geben detailliert Auskunft über Techniken, Malmittel, Farbvorlieben und – wenn man sie im größeren kulturellen Zusammenhang betrachtet – vielleicht auch über symbolische Farbbedeutungen. Eine der ältesten erhaltenen Künstlerpaletten, ein auf etwa 1427 bis 1401 vor unserer Zeitrechnung datierter Malkasten aus der 18. altägyptischen Dynastie (Abb. 2), vermittelt anschaulich und ganz unmittelbar, welche Pigmente ihrem Besitzer Amenemope, einem Wesir unter Amenhotep II., zur Verfügung standen, aber womöglich auch, welche Farben er bevorzugte und wie er die in Mulden gepressten Pigmente

Ills. 23–24
Johann Wolfgang von Goethe
Chromatic Card Game;
Two Picture Cards
Farbkartenspiel; zwei Bildkarten
Jeu de cartes chromatiques;
deux cartes illustrées
Juego de cartas cromáticas;
dos cartas ilustradas
From: *Beyträge zur Optik* (Weimar, 1791)
Hand-coloured woodcuts, 5.8 x 10 cm
(2¼ x 4 in.)
Klassik Stiftung Weimar

Roter Ocker, Ägyptisch Blau, Grün und zwei verschiedene Sorten Kohlenschwarz in seinem Kasten gern arrangierte. Wofür brauchte er zweimal Schwarz? Was bedeutete Schwarz für ihn oder die altägyptische Kultur insgesamt? Diese Anordnung von Künstlermaterialien scheint Lichtjahre entfernt von einem dreidimensionalen Farbmodell des 20. Jahrhunderts, doch das darunterliegende menschliche Bedürfnis ist dasselbe: den Farben eine Ordnung zu geben, ob aus praktischen Gründen, persönlicher Vorliebe oder gar zu Werbezwecken. Jede Malerpalette – als tatsächliches Objekt mit Farbresten darauf (Abb. 44) oder als Teil eines Selbstporträts – verrät etwas über Leben und Zeit des Künstlers, seine Malpraxis und seine Präferenzen. Selbst moderne Malkästen, die mit industriell gefertigten

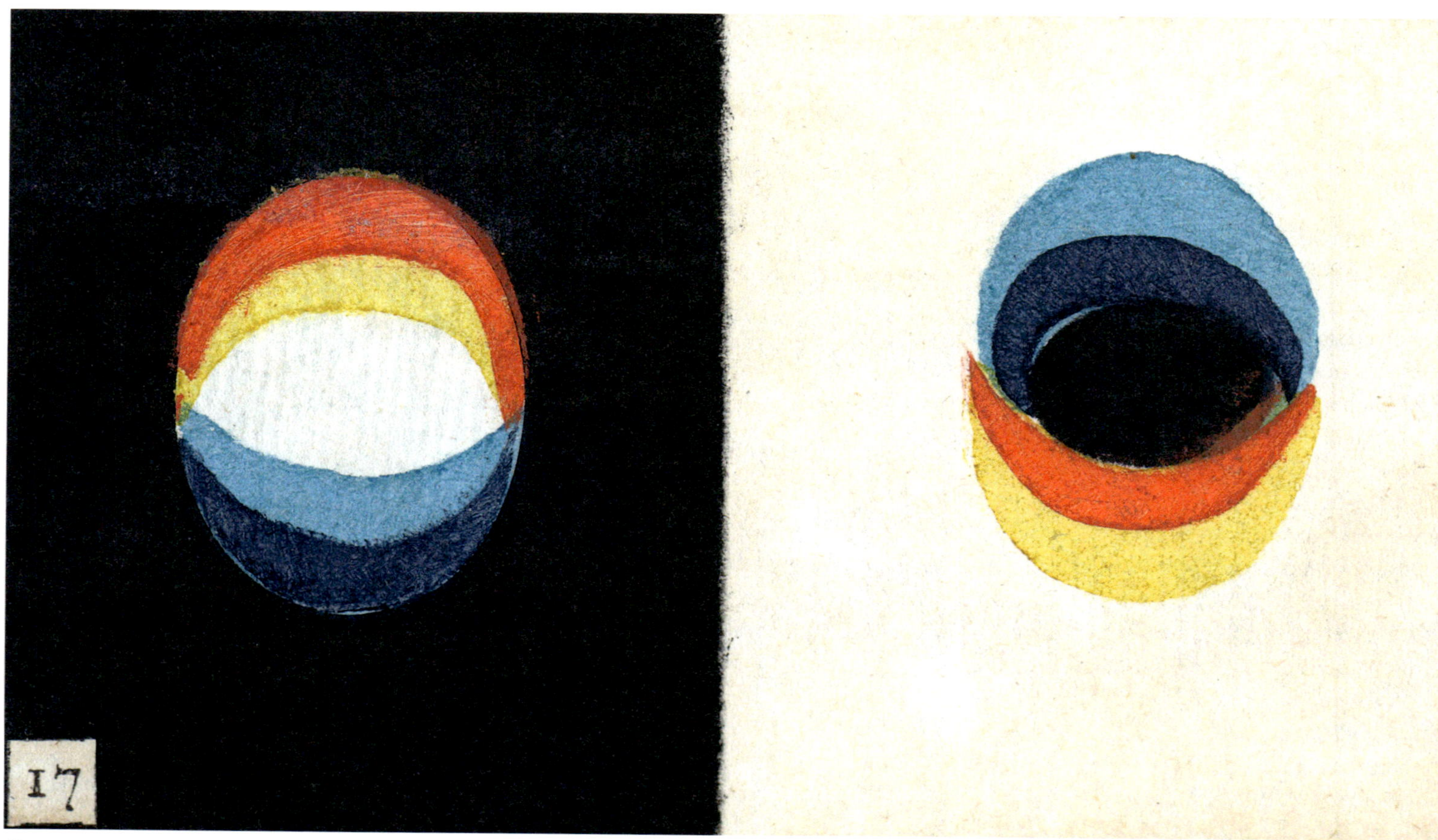

23

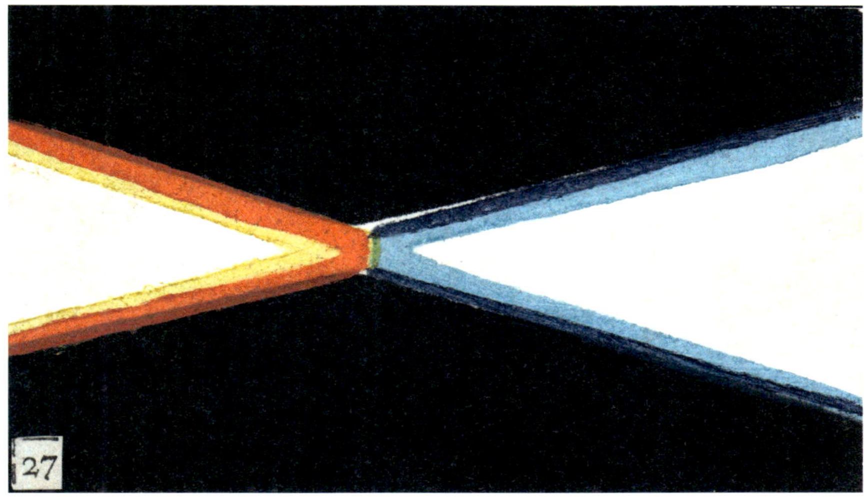

24

Wasserfarbtabletten bestückt werden können, geben Aufschluss über ihre Anwender und Besitzer: Welche Kastengröße und -qualität konnten sie sich leisten, welche Farben haben sie gewählt, welche wurden am meisten benutzt, welche so gut wie nie, und wie wurden die Farben angeordnet? Ähnlich wie das Beispiel aus dem alten Ägypten kann auch ein von Reeves in London hergestellter Aquarellmalkasten (Abb. 36) aus dem frühen 20. Jahrhundert viel darüber aussagen, welche Farben zur damaligen Zeit verfügbar oder populär waren. Das hier abgebildete Exemplar enthält mehrere Farben, deren Namensgeber der bedeutende deutsche Chemiker und Farbtheoretiker Wilhelm Ostwald (1853–1932) war.

Abhandlungen über Farbe bis zur Frühen Neuzeit

Dieses Werk ist in erster Linie der Visualisierung von Farbordnungen und den damit verbundenen Ideen gewidmet. Während die Theorien selbst bis in die Antike zurückreichen, tauchten die frühesten Farbillustrationen in Form von Diagrammen erst Ende des 17. Jahrhunderts auf. Noch bis vor relativ kurzer Zeit erschien die antike Kultur unter Materialgesichtspunkten als überwiegend farblos. Aufgrund der vorherrschenden Bilder von weißen Skulpturen und Bauwerken, die ihre Farbe seit Langem eingebüßt hatten, wurden das alte Rom und die griechische Antike hauptsächlich mit der Farbe Weiß assoziiert. Dieser allgemeine Eindruck wurde jedoch von Studien aus den letzten Jahren widerlegt und durch ein polychromes Bild der griechischen und römischen Bildhauerei und Baukunst ersetzt.

In konzeptueller und sprachlicher Hinsicht lassen sich in der Geschichte der Farbtheorien mehrere Traditionen unterscheiden. Der griechische Philosoph Empedokles (494–434 v. Chr.) soll als einer der Ersten über das Thema geschrieben haben. Aus den erhaltenen Fragmenten seiner Dichtung ist ein kosmologisches System herauszulesen, das auf den vier Elementen Erde, Feuer, Wasser und Luft basiert, von denen er die Farben ableiten zu können glaubte. Im folgenden Jahrhundert könnte Aristoteles (384–322 v. Chr.) einfache optische Experimente mit farbigem Glas durchgeführt haben. Er untersuchte neben dem Farbensehen auch die Farbkontraste und schlug ein polares System vor – vom göttlichen weißen Licht, das Farbe sichtbar machte, bis zur Dunkelheit, die sich in Schwarz niederschlug. Aus Weiß entstand die Hierarchie der Farben Gelb, Grün, Blau und Rot, die in dieser Reihenfolge mit Erde, Wasser, Luft und Feuer assoziiert wurden. Der Aristoteles-Schüler Theophrast (372–287 v. Chr.) dachte darüber nach, wie wir Farben erkennen oder empfinden, und bezog sich dabei auf Demokrit (460–370 v. Chr.) und dessen Lehre von den vier Grundfarben Rot, Schwarz, Weiß und Grüngelb, die auf bestimmten Formen, Oberflächenmustern und anderen sinnesbezogenen Eigenschaften beruhten. Der griechische Arzt Hippokrates (460–375 v. Chr.) unterschied vier Farben, die die vier Grundsäfte oder Flüssigkeiten des menschlichen Körpers repräsentierten, von denen man ihrerseits annahm, dass sie für bestimmte Charakterzüge verantwortlich waren. So stand Weiß für phlegmatisch, Rot für sanguinisch, Gelb für cholerisch und Schwarz für melancholisch.

Klassische Farbsysteme und theoretische Erörterungen des Themas sind in der Regel nur bruchstückhaft erhalten und zudem mit großen sprachlichen Herausforderungen befrachtet. Sprache wird Farben grundsätzlich nicht gerecht, weil Worte außerstande sind, der vollen Komplexität von Farbe oder Farbwahrnehmung zu entsprechen. Die meisten Sprachen verfügen über etwa ein Dutzend grundlegende Farbbezeichnungen, während die Vielfalt der Farben, die wir tatsächlich sehen, unendlich ist. Diese Diskrepanz wird noch problematischer, wenn es um alte Sprachen, ältere Texte oder Übersetzungen geht. Der zusätzliche Mangel an visuellen Darstellungen von Farbsystemen aus der Antike und vorangegangenen Kulturen erschwert ein auch nur annähernd umfassendes

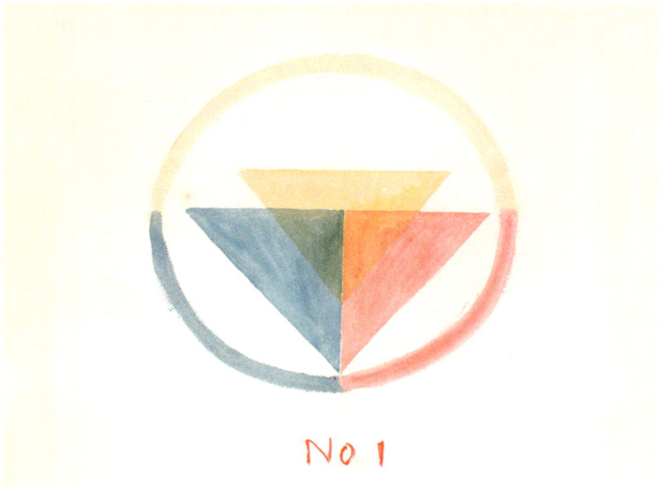

25

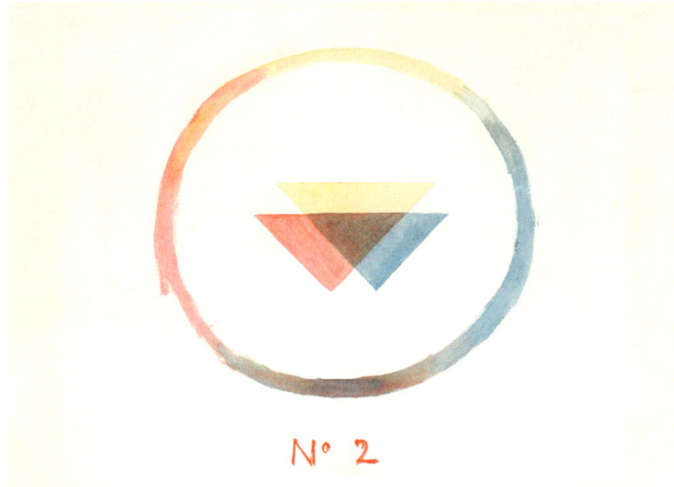

26

Verständnis von den ältesten dieser Vorstellungen. Gleichwohl vermitteln bestimmte archäologische Funde flüchtige Eindrücke vom Symbolwert der Farben. Die in Pompeji ausgegrabenen Theatermasken etwa, sowohl in Objektform als auch auf Fresken, waren in Farben bemalt, die stellvertretend für die vier Temperamente standen (Abb. 7). Obwohl nur vereinzelt spezielle Farbliteratur überliefert ist, können wir anhand der materiellen Natur von Pigmenten und Farbstoffen die Entwicklung von Farbtheorien in verschiedenen Kulturen nachverfolgen. Ähnlich wie beim teuren Ultramarinpigment in der Kunst des Mittelalters und der Renaissance (Abb. 3) führten der Materialwert und die Seltenheit bestimmter Farben in der Antike dazu, dass sie mit hochrangigen Figuren und Gottheiten assoziiert wurden.

Während im philosophischen Farbdiskurs des Mittelalters das aristotelische Grundschema einer Farbenreihe zwischen Dunkelheit und Licht (Schwarz und Weiß) vorherrschte, nahm auch die Zahl der eher praktisch orientierten Texte zu. Im frühen 13. Jahrhundert übersetzte Robert Grosseteste (um 1175–1253), Gelehrter an der Universität Oxford und Bischof von Lincoln, einige aristotelische Schriften und verfasste als Antwort darauf seine eigene Abhandlung über Farbe. In *De Colore* entwickelte er Aristoteles' Theorie der Farbe als Ableitung des Lichts weiter und stellte Mutmaßungen über die Lichtmenge an, die in einzelnen Farben enthalten sein könnte. Etwa ab dem 15. Jahrhundert, getrieben vom wachsenden weltweiten Austausch und Handel, wurde Farbe zunehmend zum Wirtschaftsgut. Wie sich mit Erscheinen der ersten Abhandlungen zu künstlerischen Praktiken, Malmaterialien und Künstlerbiografien zeigte, wurde Wissen jetzt systematischer erfasst. Viele dieser Texte waren sehr technisch und erreichten keine große Leserschaft, erst recht nicht vor Anbruch des Druckzeitalters. Und doch haben sich allerlei schriftliche Aufzeichnungen erhalten, die ein Fenster zur mittelalterlichen Welt der handwerklich hergestellten Farben öffnen, etwa das bereits deutlich früher erschienene *De diversis artibus* („Über verschiedene Künste") von Theophilus Presbyter (tätig um 1070–1125), bei dem es sich im Kern um eine Sammlung von Farbrezepten aus dem frühen 12. Jahrhundert handelt (eine deutsche Gesamtausgabe erschien 1999 unter dem Titel *Theophilus Presbyter und das mittelalterliche Kunsthandwerk*). Viele weitere kunstpraktisch ausgerichtete Schriften folgten im 15. und 16. Jahrhundert, und obwohl sie sich auf Techniken und Materialien konzentrierten und keine philosophischen Auseinandersetzungen mit Farbtheorien liefern, sind sie für Studien zur Farbterminologie und Pigmentgeschichte von unschätzbarem Wert. Gelegentlich begegnen wir einem illustrierten Text mit Abbildungen von Rohzutaten für Pigmente und Malfarben. Das im 15. oder 16. Jahrhundert erschienene *Livre des simples médecines* beispielsweise (Abb. 19, 20) zeigt verschiedene organische und mineralische Substanzen wie Sepiatinte, Lapislazuli und eine Mumie (für Mumienbraun), denen man auch heilende Wirkungen zuschrieb. Die Illustrationen sind bildhaft und schlicht, die einzelnen Objekte jedoch sorgfältig angeordnet wie in einem Setzkasten oder Kuriositätenkabinett.

Die hippokratische Lehre von den vier Säften, die den menschlichen Körper und Geist durchdringen und von vier Farben symbolisiert werden, beeinflusste ebenfalls bis ins Mittelalter und die Frührenaissance die Entwicklung einfacher Farbsysteme. Während Bilder und Kunstwerke, die sich speziell mit Farbtheorien und -systemen beschäftigen, in dieser Periode noch unbekannt sind, beginnen erste Beispiele für die Verwendung von Farbe aufzutauchen, um mit ihrer Hilfe Ordnung und Systematik zu ermöglichen und abzubilden. Die 1506 von Ulrich Pinder (gest. 1519) vorgelegte Harnglasscheibe (Abb. 5) zum Beispiel zeigt zwanzig verschiedene Urinfarben, die herangezogen wurden, um Erkrankungen zu diagnostizieren. Mit ihren handkolorierten Abbildungen der einzelnen Harngläser stellt diese Urinschautafel ein frühes Beispiel für einen auf den Farben der Natur

ILLS. 25–26
J. M. W. Turner
Lecture Diagrams: Colour Circles No. 1 and 2, *c.* 1824–1828
Vorlesungsdiagramme: Farbkreise Nr. 1 und 2
Graphiques pédagogiques: cercles chromatiques n°s 1 et 2
Diagramas de clase: círculos de colores n.ºs 1 y 2
Graphite and watercolour on paper;
No. 1: 55.5 x 76.2 cm (21⅞ x 30 in.),
No. 2: 55 x 75.8 cm (21⅝ x 29⅞ in.)
London, Tate Britain

ILL. 27
J. M. W. Turner
The Scarlet Sunset, *c.* 1830–1840
Der scharlachrote Sonnenuntergang
Coucher de soleil écarlate
Puesta de sol escarlata
Watercolour and gouache on paper,
13.4 x 18.9 cm (5¼ x 7½ in.)
London, Tate Britain

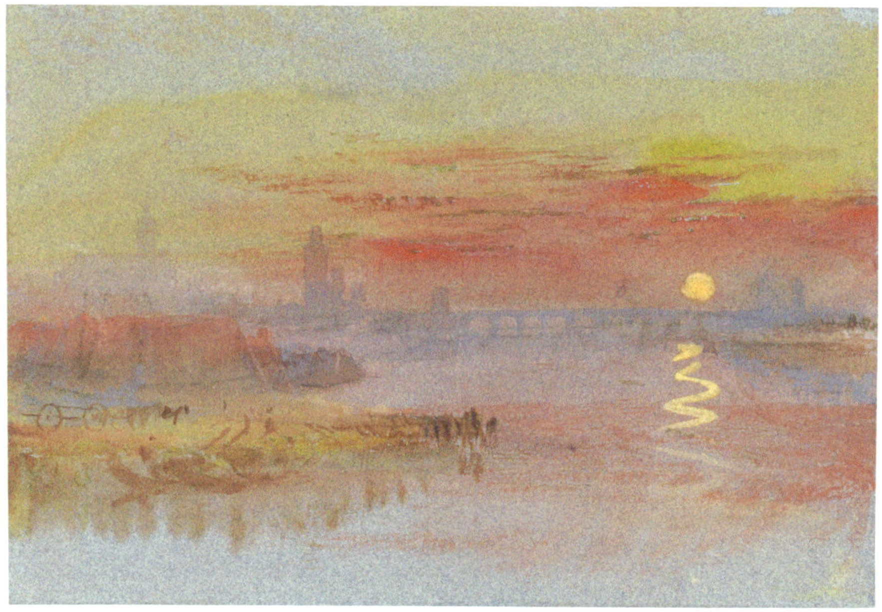

27

basierenden Farbkreis dar, der den fast dreihundert Jahre später veröffentlichten Farbkreisen in den Insektenbüchern eines Moses Harris (1730–um 1788) nicht unähnlich war. Auf vergleichbare Weise liefern zahlreiche enzyklopädische Textsammlungen wie Bestiarien, Kräuterbücher, Lapidarien und Konkordanzen direkt oder indirekt weitere Aufschlüsse zum Wissensstand und Farbeinsatz ihrer jeweiligen Zeit.

Die Veröffentlichung der von Leon Battista Alberti (1404–1472) verfassten bedeutenden kunsttheoretischen Abhandlung *Über die Malkunst*, die 1435 unter dem Titel *De pictura* erschien, fiel zusammen mit dem Aufstieg der Ölfarbe als Malmittel. Was die Farbe betraf, die er als wichtiges und wertvolles Element der Malerei ansah, ging Alberti davon aus, dass sich alle Farben herstellen ließen, indem man ganze vier von ihnen mischte: Rot, Grün, Blau und Grau. Als Kunsttheoretiker war er weniger mit farbigem Licht als mit physischer Farbe beschäftigt, sodass man die von ihm so herausragend positionierte Farbe Grau im Kontext von Schattierungen und dreidimensionalen Effekten in der Malerei betrachten muss. Viele Texte, die sich mit der ausführlichen Debatte über die Rollen von Zeichnung (*disegno*) und Farbe (*colore*) befassten, ergaben sich aus den Schriften Albertis und seiner Zeitgenossen. Auch Leonardo da Vinci (1452–1519), einer der größten Künstler der Geschichte, dachte über Farbensehen und Farbwahrnehmung, die Erscheinung von Farben und die praktische Anwendung von Farbtheorien in der Malerei nach, wie den Aufzeichnungen in seinen diversen Notizbüchern und gesammelten Schriften zu entnehmen ist. Seine stark von Aristoteles und Alberti beeinflusste Farbenordnung basierte auf sechs „Grundfarben": Rot, Gelb, Grün und Blau, flankiert von Weiß und Schwarz. Da keiner dieser Texte aus dem Mittelalter und der Frührenaissance Farbdiagramme enthält, lässt sich nicht bestimmen, ob die vorgebrachten Systeme linear oder kreisförmig, zwei- oder dreidimensional gedacht waren, doch auch ohne dieses Wissen stellen sie faszinierende Vorläufer der modernen Farbenlehre dar.

Das Zeitalter der Polychromie: Farbe ab dem 17. Jahrhundert

Obwohl die Reise dieses Werkes durch die Farbsysteme erst beginnt, als das Druckwesen im späteren 17. Jahrhundert bereits fest etabliert war, liegen zahlreiche frühe Beispiele der Farbliteratur aus unterschiedlichen Gründen nur in Manuskriptform vor. Der aufkeimende Geist der Aufklärung förderte den Austausch geistigen Gedankenguts über geografische und kulturelle Grenzen hinweg. Doch während Farbsymboliken und Metasysteme, die an religiöse Überzeugungen, Alchemie und Folklore geknüpft waren, in dieser Zeit fortbestanden, wurde die Farbe zum Gegenstand von Theorie und Wissenschaft. Das allgemeine Verlangen, die Natur zu verstehen und einzuteilen, schloss die Farbe mit ein – als optisches Phänomen, als Instrument der Naturdarstellung und als Ausdrucksmedium in Kunst und Kunsthandwerk.

In annähernd (aber nicht streng) chronologischer Reihenfolge konzentriert sich jedes Kapitel dieses Werkes auf einen bestimmten Gegenstand oder Strang in der Geschichte der Farbsysteme. Die hier vorgestellten Bücher oder Manuskripte beschäftigen sich vornehmlich mit Farbe in Theorie und Praxis oder befassen sich auf schlüssige und originelle Weise mit Farbe, etwa um auf sie bezogene Vorstellungen und Ideen zu veranschaulichen.

Um die historische Ungleichheit der Geschlechter in den Darstellungen über Farbliteratur zu korrigieren, werden in vorliegender Arbeit viele Werke aus Frauenhand präsentiert, darunter Farb- und Mallehrbücher aus dem frühen 20. Jahrhundert wie das 1918 von der US-amerikanischen Pädagogin Bonnie Snow (um 1862–1925, Abb. 11, 12, 18) gemeinsam mit Hugo B. Froehlich publizierte *The Theory and Practice of Color*, S. 552–559). Als eines der frühesten Beispiele für eine kaum beachtete Autorin ist Mary Gartside (um 1755–1819, S. 218–229) zu nennen. Die britische Blumenmalerin (Abb. 21) und Lehrerin für Aquarellmalerei veröffentlichte 1805 mit *An Essay on Light and Shade, on*

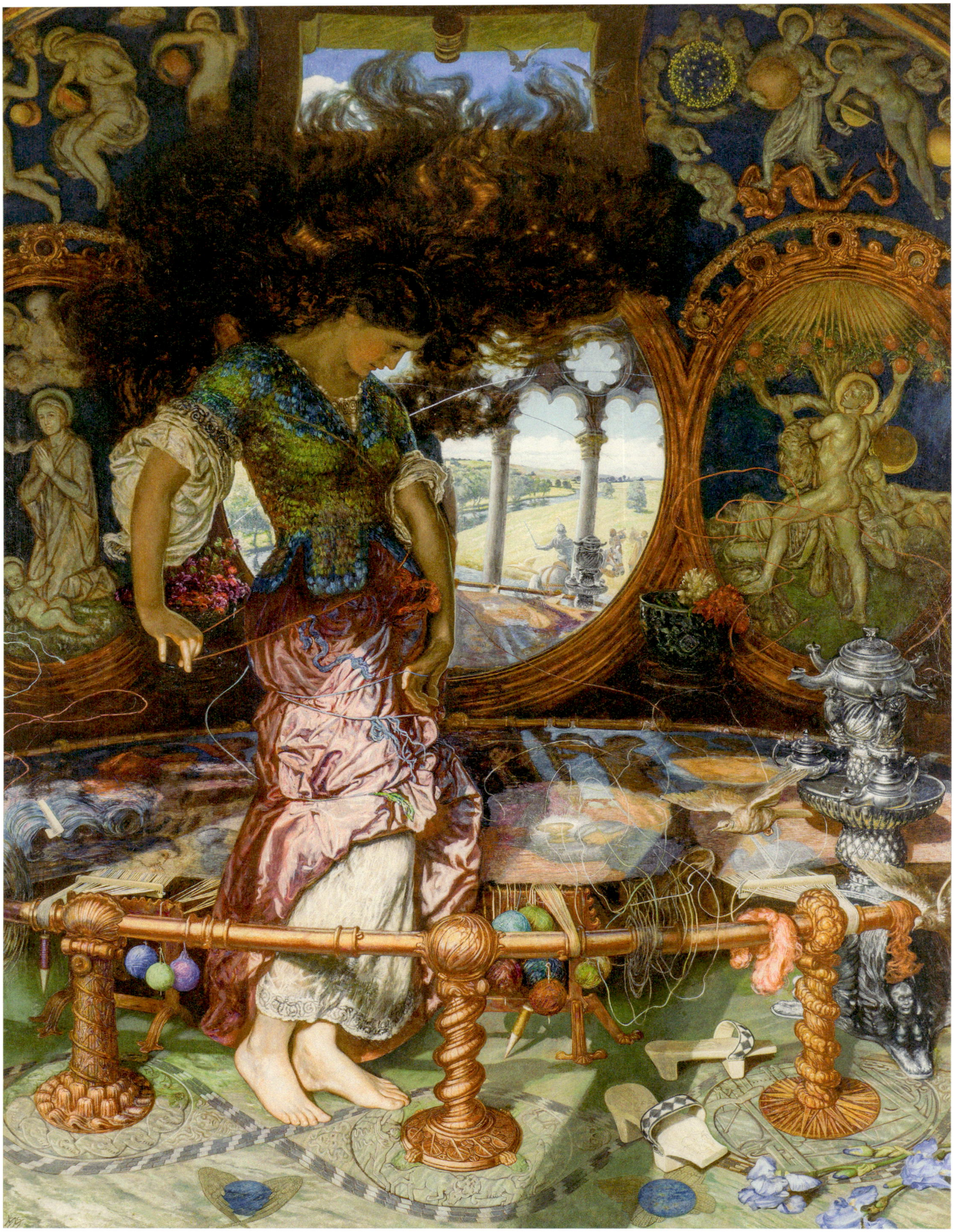

28

Ill. 28
William Holman Hunt
The Lady of Shalott, *c.* 1888–1905
Die Dame von Shalott
La Dame de Shalott
La dama de Shalott
Oil on canvas, 188.3 x 146.4 cm
(74 ⅛ x 57 ⅝ in.)
Hartford, Connecticut, Wadsworth Atheneum Museum of Art, The Ella Gallup Sumner and Mary Catlin Sumner Collection Fund

Ill. 29
Dante Gabriel Rossetti
Ecce Ancilla Domini!
The Annunciation, 1849/50
Die Verkündigung
L'Annonciation
La Anunciación
Oil on canvas, 72.4 x 41.9 cm
(28 ½ x 16 ½ in.)
London, Tate Britain

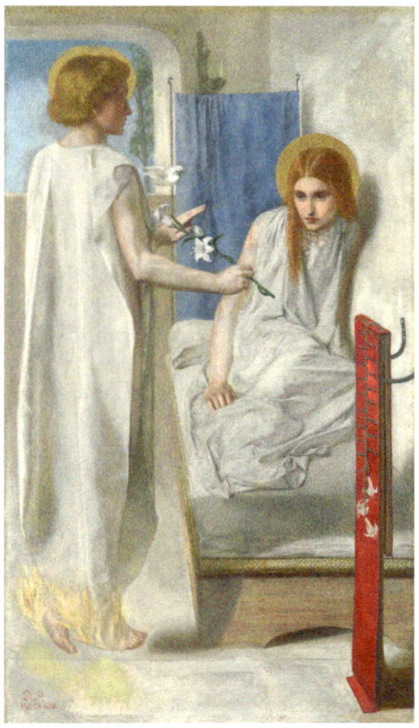

29

Colours, and on Composition in General („Ein Aufsatz über Licht und Schatten, über Farben und über Komposition im Allgemeinen") als erste bisher bekannte Frau ein englischsprachiges Buch über Farbe. Für die zweite Fassung von 1808 änderte sie den Titel in *An Essay on a New Theory of Colours* („Ein Aufsatz über eine neue Farbenlehre") und hob damit die Ernsthaftigkeit ihrer Publikation hervor. Vorgestellt werden zudem gattungssprengende Publikationen von Frauen aus dem erweiterten Kreis der Theosophen und Spiritisten, etwa *The New Science of Color* („Die neue Farbwissenschaft", 1915) von Beatrice Irwin (1877–1956). Die global und kulturübergreifend tätige Lyrikerin und Schauspielerin interessierte sich sehr für das Potenzial der Verknüpfung von farbigem Kunstlicht und Musik (S. 701). Mit ihren heute kaum noch bekannten synästhetischen Vorstellungen beeinflusste sie verschiedene Kunstschaffende des 20. Jahrhunderts, darunter auch Georgia O'Keeffe (1887–1986), in deren Arbeiten zuweilen ein indirekter Bezug auf Irwin anklingt (Abb. 50, 51). Bestimmte Frauen der Farbgeschichte verdienen in den Kanon aufgenommen zu werden, ohne eine eigenständige Farbenlehre formuliert oder verfasst zu haben. Hilma af Klint (1862–1944), auf die wir später ausführlich zurückkommen (S. 704–715), gehört ebenso dazu wie die britische Malerin Winifred Nicholson (1893–1981). Nicholsons künstlerische Leistungen, ergänzt um ihre Notizen, Briefe und zwei kurze Aufsätze über Farbe, machen sie zu einer wichtigen Instanz der Farbtheorie im 20. Jahrhundert, obwohl sie ihre Ideen und Farbvorstellungen nie präzise zwischen zwei Buchdeckeln zusammengetragen hat. Ähnlich wie Mary Gartside scheint sie ihr Interesse an der Farbtheorie über die Blumenmalerei entwickelt zu haben. Schon in ihren frühen, oft vor beleuchteten Landschaften angesiedelten oder von Fenstern umrahmten Blumenkompositionen bewies sie ein Interesse am Zusammenspiel von Licht und Farbe (Abb. 22, 55). Dieses offensichtliche Interesse an prismatischen Farben erfuhr neuen kreativen Auftrieb, als die damals schon über achtzigjährige Künstlerin Mitte der 1970er-Jahre zwei Glasprismen geschenkt bekam. In ihren letzten Lebensjahren schuf Nicholson eine faszinierende Serie von „prismatischen Gemälden", in denen sie ihre Experimente mit Prismen und Licht visuell ausforschte und dafür häufig einen ganz und gar abstrakten Malstil wählte (Abb. 56).

Das erste Kapitel des vorliegenden Buches trägt den Titel „Farbpyramiden: Frühe Tafeln und Tabellen" und nimmt die ältesten illustrierten Bücher der westlichen Kultur genauer in den Blick. Unser allererstes Beispiel ist Richard Wallers (1655–1715) Tafel der Pigmente von 1686 mit handgemalten Punkten (S. 80–85). Sie ging vermutlich auf ähnliche Tafeln zurück, die zuvor der Alchemist und Arzt am englischen Hof, Théodore Turquet de Mayerne (1573–1655), erstellt und in das sogenannte De-Mayerne-Manuskript (*Pictoria sculptoria et quae subalternarum artium*, 1620–1646) aufgenommen hatte (Abb. 1, 6). In beiden Fällen erfolgte die vergleichsweise schlichte Präsentation der Farben über Tabellen und sorgfältig gesetzte Kreise. Insofern stellen die Tafeln keine elaborierten Farbsysteme dar, sondern sind vielmehr im größeren Zusammenhang von Malhandbüchern zu betrachten – auch vor dem Hintergrund, dass sie zahlreiche Informationen über damals verfügbare Pigmente und andere Künstlermaterialien enthalten und Auskunft darüber geben, wie diese gemischt und klassifiziert wurden. Das Kapitel führt außerdem vor Augen, wie im Jahrhundert darauf komplexere Tafeln und Tabellen erstellt wurden, die anschließend ihren Weg in die Druckerpresse fanden. In dieser Periode wurden die Farbsysteme zunehmend ausgearbeitet und differenziert. Farblisten werden aufwendiger und umfangreicher, Versuche einer Farbordnung erstmals dreidimensional. Da diese frühen Tafeln, Tabellen und Diagramme trotz der Fortschritte im Farbdruck ausnahmslos von Hand koloriert wurden, ist jede Ausfertigung ein individuelles Kunstwerk. Die Praxis der Handkolorierung setzte sich bis weit ins 19. Jahrhundert fort und wurde bisweilen erst im 20. Jahrhundert durch mechanische Farbreproduktionsverfahren ersetzt.

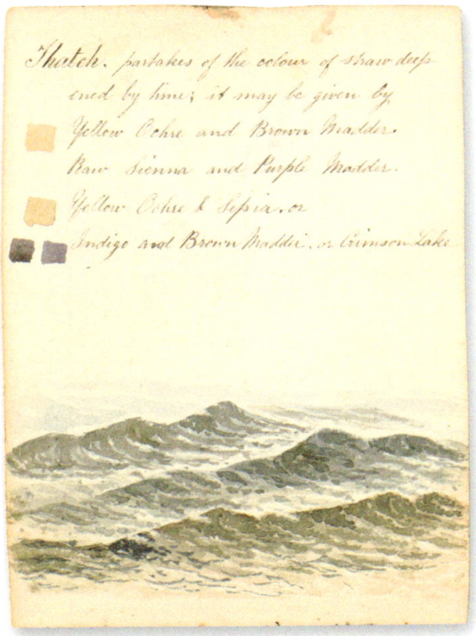

30

In Kapitel 2, „Farbanordnung: Kreise, Räder und Kugeln“, untersuchen wir, wie die einfachen visuellen Konzepte der Farbordnung im 18. und 19. Jahrhundert zu einer faszinierenden Vielfalt geometrischer Formen und Strukturen weiterentwickelt wurden. Wenngleich auch andere Formen zum Einsatz kamen, führten Kreise als die wohl meistgewählte Form zu etlichen Variationen des aus heutiger Sicht klassischen Farbkreises. Oft weisen Farbdiagramme ergänzende Illustrationen und Ausschmückungen auf, oder ein Farbkreis wird von erläuternden Tafeln und Tabellen begleitet, wie zum Beispiel in *The Natural System of Colours* („Das natürliche System der Farben“, um 1769–1776, S. 138–143) von Moses Harris und in den oben erwähnten Werken von Mary Gartside.

Ästhetisch zählen die Farbdiagramme des 18. und 19. Jahrhunderts zu den eindrucksvollsten Darstellungen in der Geschichte der Farbliteratur. Doch trotz ihrer visuellen Kraft kommen sie selten ohne das geschriebene oder gedruckte Wort aus. Obwohl die Sprache dem sichtbaren Farbspektrum hoffnungslos unterlegen ist, bildet sie in den meisten dieser Diagramme einen wichtigen Baustein mit zuweilen verblüffenden und poetischen Resultaten, so etwa in der von Johann Wolfgang von Goethe (1749–1832) und Friedrich Schiller (1759–1805) entwickelten *Temperamentenrose* (1798/99, S. 155). Eine Hürde, die alle Diagramme gemeinhin zu bewältigen hatten, war die Wiedergabe des dreidimensionalen Aspekts der Farbordnung in einem zweidimensionalen Medium. An diesem Punkt treten die ersten Farbgloben oder Farbkugeln in Erscheinung, etwa die 1810 von Philipp Otto Runge (1777–1810) erdachte *Farben-Kugel* (S. 163, Abb. 13). Bedauerlicherweise haben sich keine räumlichen Modelle der damaligen Farbgloben erhalten. 2016 jedoch konstruierte die französisch-deutsche Künstlerin Eva Bodinet ein auf Runges *Farbenkugel* basierendes Modell (Abb. 14, 15), und in den letzten Jahren brachten mehrere kommerzielle Hersteller fertige Farbgloben auf den Markt. Runges Beispiel steht auch stellvertretend für die gesteigerte Empfindsamkeit und Spiritualität im Verhältnis zur Farbe, wie sie in der Romantik anzutreffen ist. Zahlreiche Künstler, Philosophen und Schriftsteller des späten 18. und frühen 19. Jahrhunderts passten Farbtheorien an ihre persönlichen Glaubenssätze an und wiesen ihnen hochgradig symbolische Bedeutungen zu. Zwischen 1802 und 1810 schuf Runge eine Serie von Tageszeitenbildern wie *Der kleine Morgen*, die als anschaulich gemachte Farbenlehre seine christlichen Überzeugungen ausdrücken sollten (Abb. 17).

Kapitel 2 zeigt auch, wie im Verlauf des 19. Jahrhunderts die Bandbreite der Farbstudien immer größer wird, während die Farbliteratur auf früheren Texten aufbaut und sich auf sie bezieht. In den 1820er-Jahren zum Beispiel lehrte der Künstler Joseph Mallord William Turner (1775–1851) in seiner Funktion als Professor für Perspektive an der Royal Academy of Arts in London die Grundlagen der Farbtheorie. Um den Unterschied zwischen stofflichen und immateriellen Farben zu erläutern, fertigte er zwei große Vorlesungsdiagramme mit Farbkreisen an (Abb. 25, 26). Mit ihren überlappenden Farbdreiecken im Zentrum waren sie eindeutig von den Farbkreisen des Entomologen Moses Harris beeinflusst, die Turner studiert hatte. Zeitlebens interessierte sich Turner intensiv für neue Pigmente und Farbtheorien, las und besaß zahlreiche Publikationen über Farbe und machte das Zusammenspiel von Farbe und Licht zum zentralen Thema seiner Kunst (Abb. 27). In späteren Jahren – und nur kurz nachdem der erste Teil von *Zur Farbenlehre* (1810–1812, S. 236–249), Goethes Hauptwerk über Farbe, in einer englischen Übersetzung (*Goethe's Theory of Colours*, 1840) erschienen war – schuf Turner zwei Gemälde, die sich unmittelbar auf die darin formulierten Aussagen bezogen und möglicherweise erneut auch auf Harris anspielten (Abb. 43).

Das folgende Kapitel 3, „Neue Erklärungen: Der Aufstieg der Farbtheorie“, geht ausführlich auf einige der wichtigsten Farbtraktate des 19. Jahrhunderts ein und erläutert die Gründe für die drastische Zunahme an neuen Theorien und umfangreichen philosophischen, wissenschaftlichen und praktischen Abhandlungen über Farbe. Diese

ILLS. 30–34 (*Pages 34–36*)
Attributed to George Barnard
Colour Samples and Notes from a Sketchbook, mid-19th century
Farbmuster und Anmerkungen aus einem Skizzenbuch
Échantillons de couleurs et Extraits d'un carnet de croquis
Muestras de color y anotaciones de un cuaderno de bocetos
Sussex, collection of the author

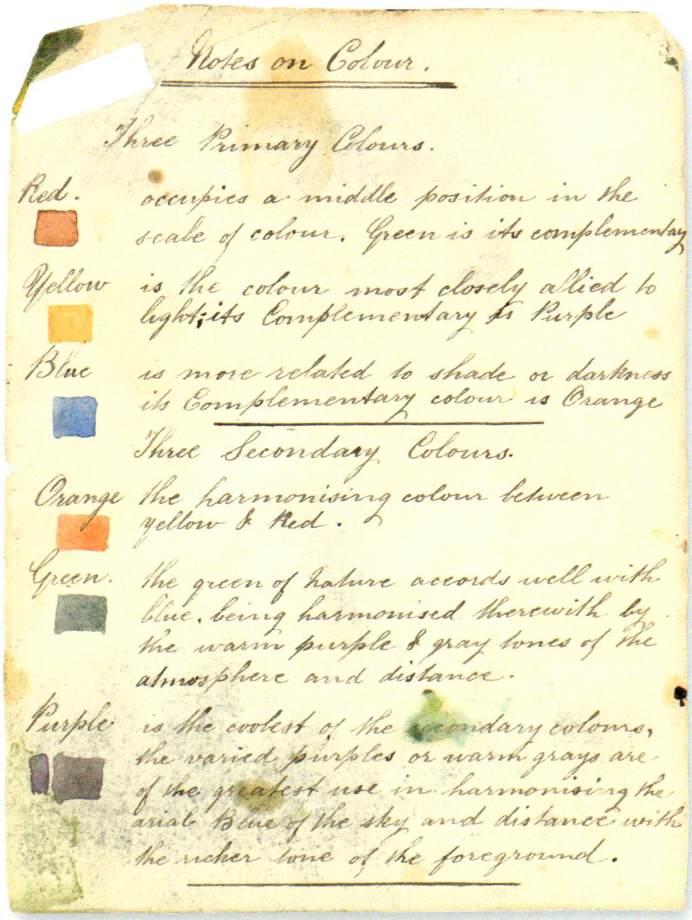

Notes on Colour.

Three Primary Colours.

Red. occupies a middle position in the scale of colour. Green is its complementary

Yellow is the colour most closely allied to light; its Complementary is Purple

Blue is more related to shade or darkness its Complementary colour is Orange

Three Secondary Colours.

Orange the harmonising colour between yellow & Red.

Green. the green of nature accords well with blue, being harmonised therewith by the warm purple & gray tones of the atmosphere and distance.

Purple is the coolest of the secondary colours, the varied purples or warm grays are of the greatest use in harmonising the aerial Blue of the sky and distance with the richer tone of the foreground.

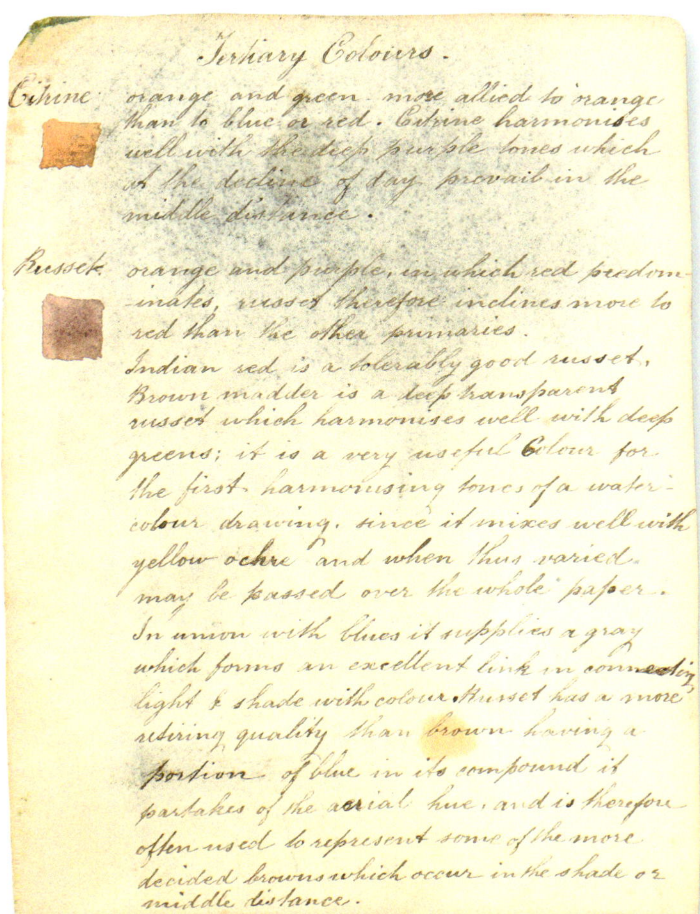

Tertiary Colours.

Citrine: orange and green more allied to orange than to blue or red. Citrine harmonises well with the deep purple tones which at the decline of day prevail in the middle distance.

Russet. orange and purple, in which red predominates, russet therefore inclines more to red than the other primaries. Indian red is a tolerably good russet, Brown madder is a deep transparent russet which harmonises well with deep greens; it is a very useful Colour for the first harmonising tones of a water-colour drawing, since it mixes well with yellow ochre and when thus varied may be passed over the whole paper. In union with blues it supplies a gray which forms an excellent link in connecting light & shade with colour. Russet has a more retiring quality than brown having a portion of blue in its compound it partakes of the aerial hue, and is therefore often used to represent some of the more decided browns which occur in the shade or middle distance.

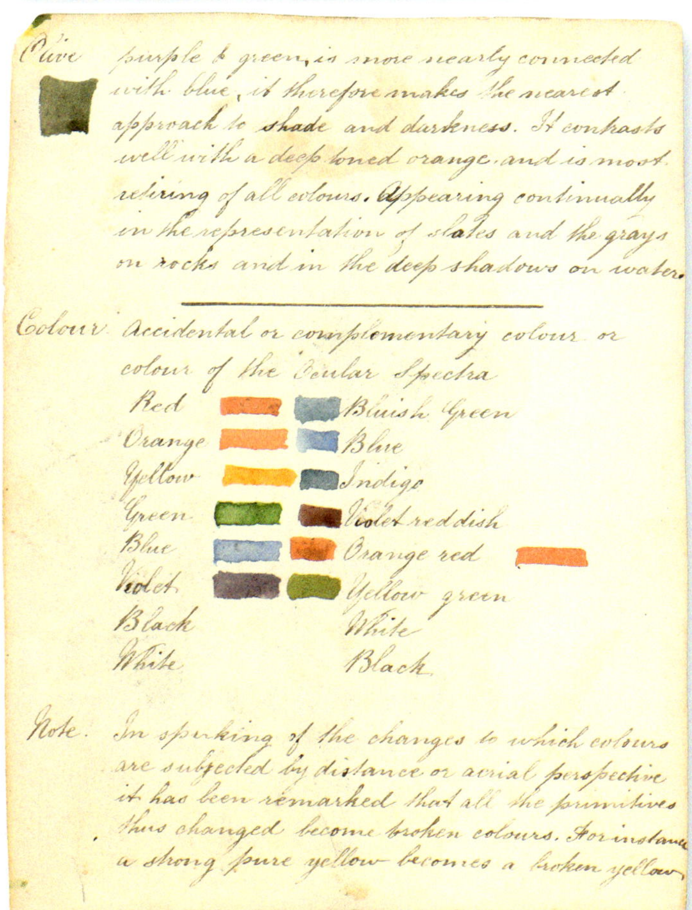

Olive purple & green, is more nearly connected with blue, it therefore makes the nearest approach to shade and darkness. It contrasts well with a deep toned orange and is most retiring of all colours. Appearing continually in the representation of slates and the grays on rocks and in the deep shadows on water.

Colour Accidental or complementary colour or colour of the Ocular Spectra

Red	Bluish Green
Orange	Blue
Yellow	Indigo
Green	Violet reddish
Blue	Orange red
Violet	Yellow green
Black	White
White	Black

Note. In speaking of the changes to which colours are subjected by distance or aerial perspective it has been remarked that all the primitives thus changed become broken colours. For instance a strong pure yellow becomes a broken yellow

31–33

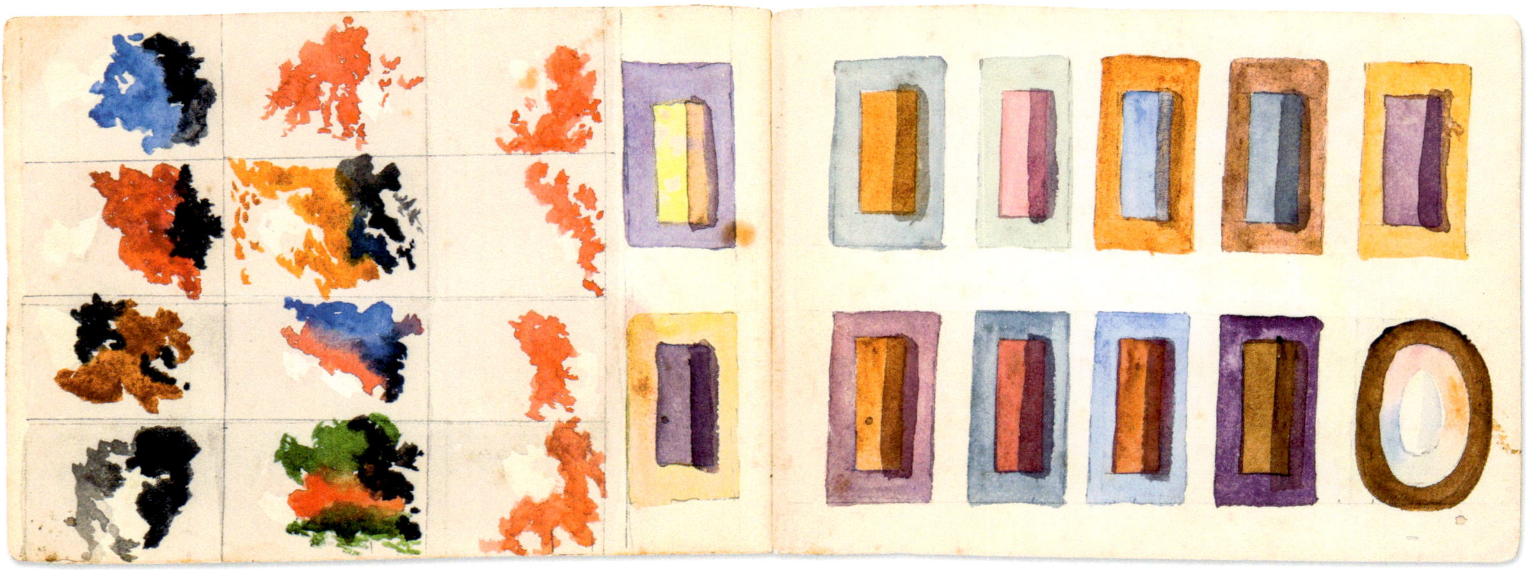

34

Farbrevolution ist untrennbar mit der Industrialisierung der Farbe verbunden, insbesondere mit neuen Fertigungs-, Produktions- und Druckverfahren, hat ihren Ursprung jedoch im frühen 18. Jahrhundert. Ein unumstrittener Meilenstein in der Geschichte der Farbsysteme wurde von Isaac Newton verfasst, der 1704 mit den *Opticks* die Früchte seiner jahrzehntelangen Farbforschungen veröffentlichte (S. 210–217). Das Buch enthielt – in Form eines Farbkreises – das erste an eine breite Leserschaft gerichtete Farbdiagramm. Es beruhte auf Newtons Versuchen, das weiße Licht mithilfe eines Prismas in Spektralfarben zu zerlegen. Wie wir heute wissen, war Newtons Einfluss auf die Wissenschaft des 18. und 19. Jahrhunderts unermesslich. Seine Theorien wurden adaptiert, bewundert, weiterentwickelt und kritisiert (Abb. 8). Die Faszination, die von den Regenbogenfarben und anderen Entdeckungen Newtons ausging, zeigt sich in der bildenden Kunst, etwa in dem allegorischen Gemälde *Colour* (Abb. 9), das Angelika Kauffmann (1741–1807) schuf. Obwohl es als Allegorie angelegt ist und damit an sich kein Farbsystem darstellt, veranschaulicht das Werk einen bedeutenden Aspekt der Farbgeschichte, denn zusammen repräsentieren diese vier Gemälde die maßgeblichen „Elemente der Kunst“: Erfindung, Komposition, Zeichnung und Farbe. Letztere wurde damit endlich auf eine Ebene mit den drei übrigen Kategorien gehoben. In *Colour* taucht die Farbe in Gestalt einer weiblichen Figur ihren Pinsel in Newtons Regenbogen und zieht die prismatischen Farben auf eine physische Palette herunter. Auf diese Weise überträgt sie die Spektralfarbentheorie auf die stofflichen Farben, bringt sie jedoch gleichzeitig mit den Unterrichtspraktiken an der neu gegründeten Akademie in Verbindung.

Im fortschreitenden 19. Jahrhundert machten romantische Ideen und Einstellungen wie Runges Farbsymbolik allmählich Studien mit einer eher technischen und wissenschaftlichen Ausrichtung Platz, von denen viele in Zusammenhang mit der Herstellung neuer Pigmente und Farbstoffe standen. Die Qualität der neuen Farben wurde zum Forschungsfeld des bedeutenden britischen „Farbenmannes“ George Field (um 1777–1854, S. 258–273), der nicht nur eine eigene Farbenlehre entwickelte, sondern auch Pigmente erprobte, herstellte und patentieren ließ. Seine Farben waren bei zahlreichen Künstlern des 19. Jahrhunderts gefragt, darunter auch bei Mitgliedern der Präraffaeliten und ihren Anhängern, und ihre Verfügbarkeit fiel zusammen mit einem wiedererwachten Interesse an mittelalterlichen Malmethoden und materialien. Die juwelengleiche Leuchtkraft und gewagte Vielfarbigkeit von Gemälden wie *Mariana* (1851, Abb. 38) von John Everett Millais (1829–1896) spiegelten dieses Interesse an hochwertigen Pigmenten wider und verrieten ein ausgeprägtes Bewusstsein für die damals aktuellen Farbsysteme. Man vermutete sogar, dass das Gemälde *Ecce Ancilla Domini! (The Annunciation*, 1849/50, Abb. 29) von Dante Gabriel Rossetti (1828–1882) unmittelbar von der Symbolik der Farben nach George Field inspiriert war. Die Aufmerksamkeit, die man der Farbliteratur und -theorie entgegenbrachte, lässt sich in zahlreichen Werken westlicher Künstler des 19. Jahrhunderts erkennen. Sowohl der Impressionismus als auch der Pointillismus wurden (ab 1839) direkt von der bahnbrechenden Theorie der simultanen Farbkontraste beeinflusst, die Michel-Eugène Chevreul (1786–1889) während seiner Tätigkeit für die staatliche Gobelinmanufaktur in Paris entwickelt hatte (S. 282–307). Der Post-Impressionist Vincent van Gogh (1853–1890) machte sich die Prinzipien des Farbkontrasts in seiner Malerei mit erstaunlicher Wirkung zunutze. Es ist bezeichnend, dass sich in seinem persönlichen Nachlass ein rotes Lackkästchen mit mehreren Knäueln aus bunten Wollfäden befand, die auf unterschiedliche Weisen aufgewickelt und kombiniert waren, so als hätte van Gogh auf diese Art harmonische Farbkombinationen nach den Gesetzen Chevreuls geschaffen (Abb. 37). Auch das Werk von James Abbott McNeill Whistler (1834–1903) strahlt ein großes Bewusstsein für Farbkontraste aus. Wenngleich seine unmittelbare Beschäftigung mit Farbliteratur nicht dokumentiert ist, äußert sich Whistlers

Interesse an Farbharmonien und symbolischer Farbverwendung in den Titelbegriffen seiner Gemälde, die er – anstelle beschreibender Bezeichnungen – als „Harmonien", „Nachtstücke" oder „Symphonien" in bestimmten Farben bezeichnete (Abb. 40, 41).

In den Kapiteln 4 und 5 geht es um eher technisch und praktisch ausgerichtete Farbsysteme. „Farben als Orientierung: Nomenklaturen und Standards" beschäftigt sich mit systematischen Farblisten, die oft in Bezug zur Natur und zur taxonomischen Klassifikation stehen und faszinierende Versuche darstellen, verlässliche Terminologien und Standards zur Anwendung in Kunst, Industrie und Wissenschaft zu etablieren. Einige dieser Systeme zielten auch darauf ab, eine Chronik historischer Farben mitzuliefern, so etwa der 1938 erschienene *Historical Color Guide* on Elizabeth Burris-Meyer (1899–1969), den die Autorin später zum Leitfaden für Innenausstattung, Mode und Schönheitspflege umarbeitete (S. 388–409). Kapitel 5, „Theorie und Praxis: Unterweisungen in Farbe" konzentriert sich auf die Geschichte der praktischen Farbhandbücher für Maler und Drucker von Anfang des 19. bis Anfang des 20. Jahrhunderts. Die älteren Handbücher sind noch stark der Entwicklung der Aquarellmalerei verpflichtet und waren angesichts der zu Beginn des 19. Jahrhunderts erschwinglicher werdenden Malmaterialien überwiegend für Amateure bestimmt. Im 18. und 19. Jahrhundert war das Malen ein anerkannter Zeitvertreib für Frauen. Darum richtete sich ein Großteil der Handbücher, obwohl mehrheitlich von Männern abgefasst, speziell an eine weibliche Leserschaft. Seit Beginn des 20. Jahrhunderts ergriffen jedoch immer mehr Frauen die Gelegenheit, über dieses Thema zu schreiben, zu veröffentlichen und zu unterrichten, und verfassten ihrerseits Farblehrbücher und Farbfibeln. Bis heute ist dieses Feld der kreativen Beschäftigung von Frauen kaum erforscht, und weil Lehrbücher, insbesondere für jüngere Kinder, als Textgattung vernachlässigt wurden, sind sie in der Regel unzureichend dokumentiert und nur in wenigen Fällen erhalten.

Neben den Handbüchern kam im frühen 20. Jahrhundert eine weitere Flut von Büchern über Farbe auf den Markt. Wie in Kapitel 6, „Farbe sprengt ihre Ketten: Das frühe 20. Jahrhundert", dargelegt wird, knüpften viele von ihnen an die Werke der großen Farbtheoretiker des vorangegangenen Jahrhunderts an, entwickelten deren Konzepte jedoch weiter. Farbe lässt sich in dieser Periode nicht adäquat einschätzen, ohne eine Reihe von wissenschaftlichen und künstlerischen Errungenschaften zu berücksichtigen. Dazu zählen die Fortschritte in der Erforschung des Farbensehens und speziell der Anatomie des menschlichen Auges, die Elektrifizierung der Welt, die daraus resultierenden Möglichkeiten der künstlichen Beleuchtung und der radikal neue Einsatz von Farbe als Mittel des Ausdrucks in der Kunst, besonders in der Malerei. Vertreter avantgardistischer Gruppen und Strömungen in Europa – wie Der Blaue Reiter, die Fauvisten, viele Expressionisten und frühe abstrakte Künstler – befassten sich mit Farbtheorien. Sie machten sich die symbolische und visuelle Kraft der Farbe zu eigen und bildeten die Welt in subjektiven, nicht realistischen Tönen ab. Für manche unter ihnen wurden Farbauffassungen zum Hauptmotiv bestimmter Werke, zu sehen etwa in dem 1914 von Franz Marc (1880–1916) geschaffenen abstrakten Gemälde *Kämpfende Formen* (Abb. 47).

Die überaus bunten, von einer zunehmenden Mechanisierung geprägten ersten Dekaden des 20. Jahrhunderts, in denen man stoffliche Farben gebrauchsfertig und sauber verpackt in fast allen erdenklichen und leuchtenden Schattierungen kaufen konnte, schrien förmlich nach neuen Versuchen, den Farben eine ultimative Ordnung zu geben. Einige der einflussreicheren Arbeiten, etwa von Albert Henry Munsell (1858–1918) in den USA (S. 584–599) und Wilhelm Ostwald in Europa (S. 600–609), werden in Kapitel 6 behandelt. Beide Männer entwickelten höchst lehrreiche und anpassungsfähige Systeme, die auf einer dreidimensionalen Farbtheorie beruhten; tatsächlich haben einige der räumlichen Modelle, die sie für ihre

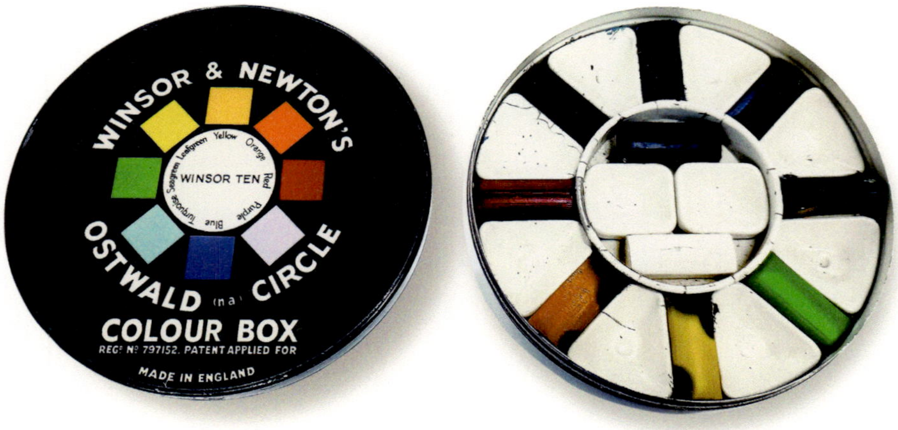

35

Unterrichtszwecke nutzten, die Zeiten überdauert (siehe zum Beispiel Ostwalds *Farbkörper* in den Archiven von Winsor & Newton, S. 430). Dass sich im frühen 20. Jahrhundert auch die Forschungs- und Veröffentlichungsmöglichkeiten für Frauen verbesserten, führte zu bemerkenswerten wissenschaftlichen Studien von Frauen zu Themen wie Farbensehen und Psychologie, Farbe in Literatur und Dichtung, historische Farbtheorie und zu weiteren Aspekten der Farbgeschichte. Einige dieser zwischen 1900 und den 1930er-Jahren erschienenen Bücher, wie die *Color Problems* („Farbproblematiken", 1902, S. 562–583) von Emily Noyes Vanderpoel (1842–1939), zählen zu den ideenreichsten und schönsten Publikationen in der Geschichte der Farben.

Kapitel 7, „Der Klang der Farbe: Spiritismus, Okkultismus und Musik", untersucht die synästhetischen Verbindungen zwischen Farbe und Musik und bezieht dabei Bewegung, Tanz und Spiritualität mit ein. In gewissem Maße waren die von Theosophen, Anthroposophen und Mitgliedern ähnlicher Gruppen und Bewegungen um die Wende zum letzten Jahrhundert publizierten, hochgradig individuellen und oft pseudowissenschaftlichen Farbtheorien eine Reaktion auf eine zunehmend mechanisierte Welt. Dennoch ist es ratsam und hilfreich, jede dieser Theorien gewissenhaft zu begutachten und sie nicht nach ihrer wissenschaftlichen Exaktheit, sondern nach ihrer kreativen Innovationskraft und den ihr zugrunde liegenden gesellschaftlichen und historischen Umständen zu beurteilen. Man muss nicht an die farbigen Auren glauben, die von Annie Besant (1847–1933) und Charles Webster Leadbeater (1854–1934) illustriert und beschrieben wurden, doch ihr Versuch, Gedanken und Gefühlen Form und Farbe zu geben, kommentiert auf faszinierende Weise kulturelle Einflüsse der Zeit (S. 648–687).

Analogien zwischen Farbe und Musik bilden in der Geschichte der Farbsysteme ein herausragendes Thema. Die Mehrzahl der in diesem Kapitel vorgestellten Texte behandelt Farbe im Zusammenhang mit Klang, Musik, Poesie und Tanz, viele von ihnen kombinieren zudem Mystik mit Wissenschaft, individuelle religiöse Überzeugungen mit Farbtheorie, Natur mit Spiritualität und historische Texte über Farbe mit modernen Malpraktiken. Was sich nach einer verwirrenden und eigentümlichen Mischung anhören mag und oft schwierig zu lesen ist, besticht in der bildlichen Wiedergabe der einzelnen Konzepte aufs Äußerste. Der australische Künstler Roy de Maistre (1894–1968) schuf zahlreiche Gemälde zum Thema, unter anderem eine fünf Meter lange Bildrolle mit einer „Farbübersetzung" des Trios in B-Dur von Joseph Haydn (S. 60, Abb. 57, 67). Das Werk der schwedischen Mystikerin und Anthroposophin Hilma af Klint, die erst in den letzten Jahren wiederentdeckt und für ihren visionären Stil zu Recht gerühmt wurde, war tief im Okkultismus und in der Farbsymbolik verwurzelt. Obwohl sie nie eine eigene Farbenlehre formulierte, enthalten viele ihrer großformatigen Gemälde farbige Kreise, Pyramiden und Darstellungen immaterieller Farbe. In dem von ihr geschaffenen, bemerkenswert individuellen Farbvokabular verbanden sich Realismus und Abstraktion (Abb. 39, 48, 49).

Das vorliegende Buch kann keinerlei Anspruch erheben, einen umfassenden Überblick über die Geschichte der Farbtheorie zu liefern. In seiner Werkauswahl ist es vorwiegend – und vielleicht unausweichlich – auf die westliche Hemisphäre fixiert. Diesem Ungleichgewicht soll jedoch entgegengewirkt werden, indem Kapitel 8 nach Osten blickt, über unsere vertraute westliche Sichtweise hinaus, und einen Eindruck von den japanischen und russischen Farbsystemen des 19. und 20. Jahrhunderts vermittelt. Dabei geht es weniger um eine wissenschaftliche Untersuchung von Farbkonzepten dieser Regionen, als vielmehr um die Würdigung von deren ästhetischen Reiz. Zudem soll nachvollziehbar werden, dass zwischen Farbtheorien aus geografisch weit auseinanderliegenden Teilen der Erde Zusammenhänge bestehen können. Etwas so Abstraktes wie Farbe zu vergleichen, führt uns überzeugend vor Augen, was Menschen weltweit verbindet. Ein faszinierendes Beispiel für den interkulturellen Farbaustausch bildet

Ill. 35
Colour Box Depicting Wilhelm Ostwald's Colour Circle
Farbkasten mit Wilhelm Ostwalds Farbkreis
Boîte de couleurs reprenant le cercle chromatique d'Ostwald
Caja de colores con el círculo de color de Wilhelm Ostwald
London, Slade School of Fine Art, Winsor and Newton Archive

Ill. 36
Reeves Watercolour Paintbox, *c.* 1930
Aquarellmalkasten von Reeves
Boîte d'aquarelle Reeves
Caja de acuarelas de Reeves
Sussex, collection of the author

das Nachleben eines Farbdiagramms von George Field aus den 1830er-Jahren, das in dem erstmals 1876 erschienenen japanischen Schulbuch *Gakko hitsuyo irozu mondo* (S. 740–743) wieder auftaucht.

Das letzte Kapitel, „Interaktion und Abstraktion: Vom Bauhaus in die Gegenwart", behandelt Farbe in ihrer reinsten, reduziertesten und elementarsten Form, wie man sie von Anfang bis Mitte des 20. Jahrhunderts in Kunst, Design, Lehrbüchern und Unterricht fand. Den langen, allmählichen Weg zur Abstraktion bereiteten neben Pionieren wie Turner, Gartside, Whistler oder af Klint auch Avantgardisten aus verschiedenen europäischen Künstlergruppen. Doch vor allem das Bauhaus mit seinem erweiterten Einflussbereich entwickelte sich zum führenden Vertreter von Modernismus, Fortschrittlichkeit und, stilistisch gesprochen, Abstraktion. 1919 von dem Architekten Walter Gropius (1883–1969) in Weimar gegründet, hatte die kurzlebige Schule für Kunst, Architektur und Design ihre Wurzeln in avantgardistischen und spirituellen Gruppen in Deutschland und in der englischen Arts-and-Crafts-Bewegung. Einfaches und funktionales Design, beruhend auf einer gründlichen künstlerischen Ausbildung und solidem Handwerk, bildete den Kern des Lehrbetriebs. Zu den Mitgliedern, Lehrern und Schülern am Bauhaus gehörten Ludwig Mies van der Rohe, Paul Klee, Johannes Itten und Josef Albers, aber auch mehrere Frauen, deren Rollen in der Geschichte des Bauhauses erst vor Kurzem umfassend gewürdigt wurden, darunter Anni Albers (Abb. 64), Marianne Brandt, Gertrud Arndt (Abb. 59), Gertrud Grunow (Abb. 61, 62) und Gunta Stölzl (Abb. 65, 66). 1925 zog das Bauhaus nach Dessau um, wo Gropius das zur Ikone gewordene Bauhausgebäude entwarf (vgl. Abb. 63), doch schon 1933, nach der Vertreibung aus Dessau und lediglich einem Jahr mit Sitz in Berlin, musste das Bauhaus unter dem wachsenden Druck der Nationalsozialisten endgültig schließen. Einige seiner Mitglieder emigrierten nach England oder in die USA, wo sie Verbindung zueinander halten und ihre Arbeit fortsetzen konnten. Obwohl Farbe nur ein Bestandteil der Bauhausklassen war, entstanden zahlreiche einschlägige Publikationen über Farbe, von denen einige im Schlusskapitel dieses Buches vorgestellt werden. Darunter sind zwei Klassiker der Farbliteratur, die knapp dreißig Jahre nach der Schließung der Schule veröffentlicht wurden: *Interaction of Color* (1963, S. 824–837) von Josef Albers (1888–1976), das

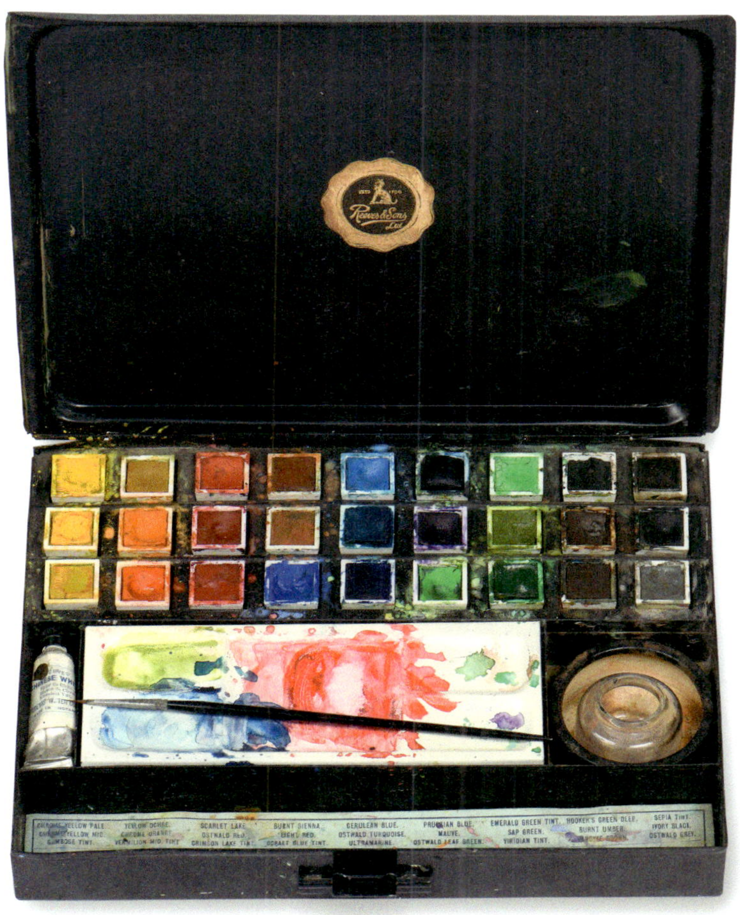

36

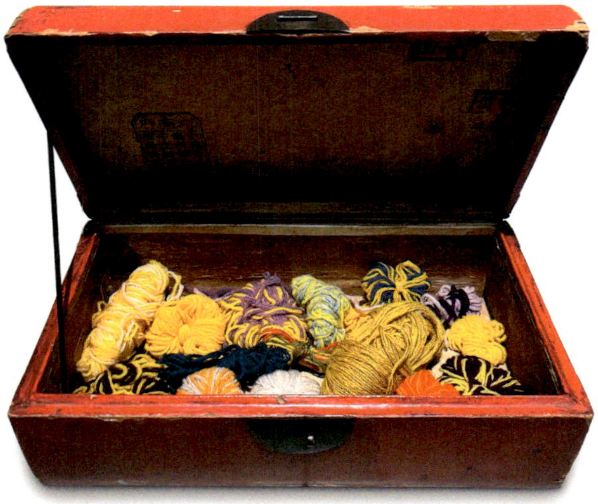

37

ILL. 37
Red Lacquer Box with Lengths of Coloured Wool Belonging to Vincent van Gogh
Rotes Lackkästchen mit farbigen Wollknäueln aus dem Besitz von Vincent van Gogh
Boîte de laque rouge de pelotes de laine ayant appartenu à Vincent van Gogh
Caja lacada roja con ovillos de lana de colores perteneciente a Vincent van Gogh
Amsterdam, Van Gogh Museum

ILL. 38
John Everett Millais
Mariana, 1851
Oil on mahogany,
59.7 x 49.5 cm (23½ x 19½ in.)
London, Tate Britain

1970 unter gleichem Titel in einer deutschen Übersetzung erschien, und *Kunst der Farbe* (1961, S. 804–823) von Johannes Itten (1888–1967). Diese und andere, weniger bekannte Farbabhandlungen von gleichgesinnten Autoren, sind Beispiele für eine nahezu vollkommene Synthese des Nachdenkens über Farbe – über das Leben und Arbeiten mit ihr, über den Farbunterricht und über die Anwendung von Farbe in Kunst und Design –, während sie gleichzeitig die modernen Ideale von Einfachheit, Geometrie und Abstraktion reflektieren. In der Kunst des 20. Jahrhunderts spiegeln sich die Farbtheorie und -ästhetik des Bauhauses nicht nur in der verbreiteten Annäherung an die Abstraktion wider; auch die Ähnlichkeiten im Farbgebrauch mit den extrem minimalistischen und geometrischen Werken von Meistern der Moderne wie Piet Mondrian (Abb. 68), Henri Matisse und Kasimir Malewitsch sind unübersehbar.

Die Zeitlosigkeit von Farbtheorien

Die abstrakte Schönheit von Farbdiagrammen, von sichtbar gemachten Farbtheorien, verleiht ihnen einen zeitlosen Reiz. Die einfachen geometrischen Formen und Figuren, die geschaffen wurden, um frühen Farbtafeln und -tabellen ein Gerüst zu geben, erscheinen unmittelbar zugänglich, weil die gleichen oder ähnliche Formen in Kunst, Werbung und Design bis heute existieren. Farbkreise, Farbräder, Farbdreiecke, Farbwürfel und komplexere Farbdiagramme oder organische Formen bleiben über Jahrhunderte hinweg nachvollziehbar. Für zeitgenössische Künstler sind Farbtheorien und Farbsysteme Motiv und Inspirationsquelle zugleich. Seit 2009 hat der isländisch-dänische Künstler Ólafur Elíasson (geb. 1967) eine Reihe von großen Farbkreisen erschaffen, in denen die Farben, ähnlich wie in einem der Farbkreise aus dem 1861 erschienenen *Atlas* von Michel-Eugène Chevreul, nahtlos ineinander übergehen (Abb. 70). Eine Auswahl daraus mit dem Titel *Turner colour experiments* wurde in der Tate Britain in Nachbarschaft zu einigen Turner-Gemälden gezeigt. So wie sich Turner in manchen Werken auf Goethe und Harris bezog, so rekurrierte Elíasson in dem Bestreben, eine eigene Farbenlehre zu entwickeln, auf Farbtheoretiker und Künstler vor ihm.

Schon immer hat der Mensch versucht, Farbe, dieses ungreifbare, aber alles durchdringende Element unseres Sehens, zu bändigen und zu beherrschen. Unsere Versuche aber, Farbe einzufangen, indem wir Kreise, Quadrate und Linien um sie ziehen, indem wir sie in Listen zwängen oder in räumliche Formen pressen, sind letztlich Ausdruck des menschlichen Denkens und der menschlichen Kreativität. Die perfekte Farbtheorie mag noch nicht formuliert sein, doch vorhandene Ideen über Farbe in Worte, Bilder, Diagramme und dreidimensionale Objekte zu verwandeln, hat zu einer langen Geschichte der Farbe als Mittel und Inhalt der bildenden Kunst geführt. Individuelle Sachkenntnis, persönliche Vorlieben, aktuelle Forschungsergebnisse und neue Denkansätze machen jedes neue Buch über Farbgeschichte zum Abenteuer. Die Autorinnen hoffen, dass die in diesen beiden Bänden präsentierte Auswahl von Farbsystemen unerwartete und inspirierende Einblicke und Episoden bereithält, die ein neues Licht auf die Farbgeschichte werfen.

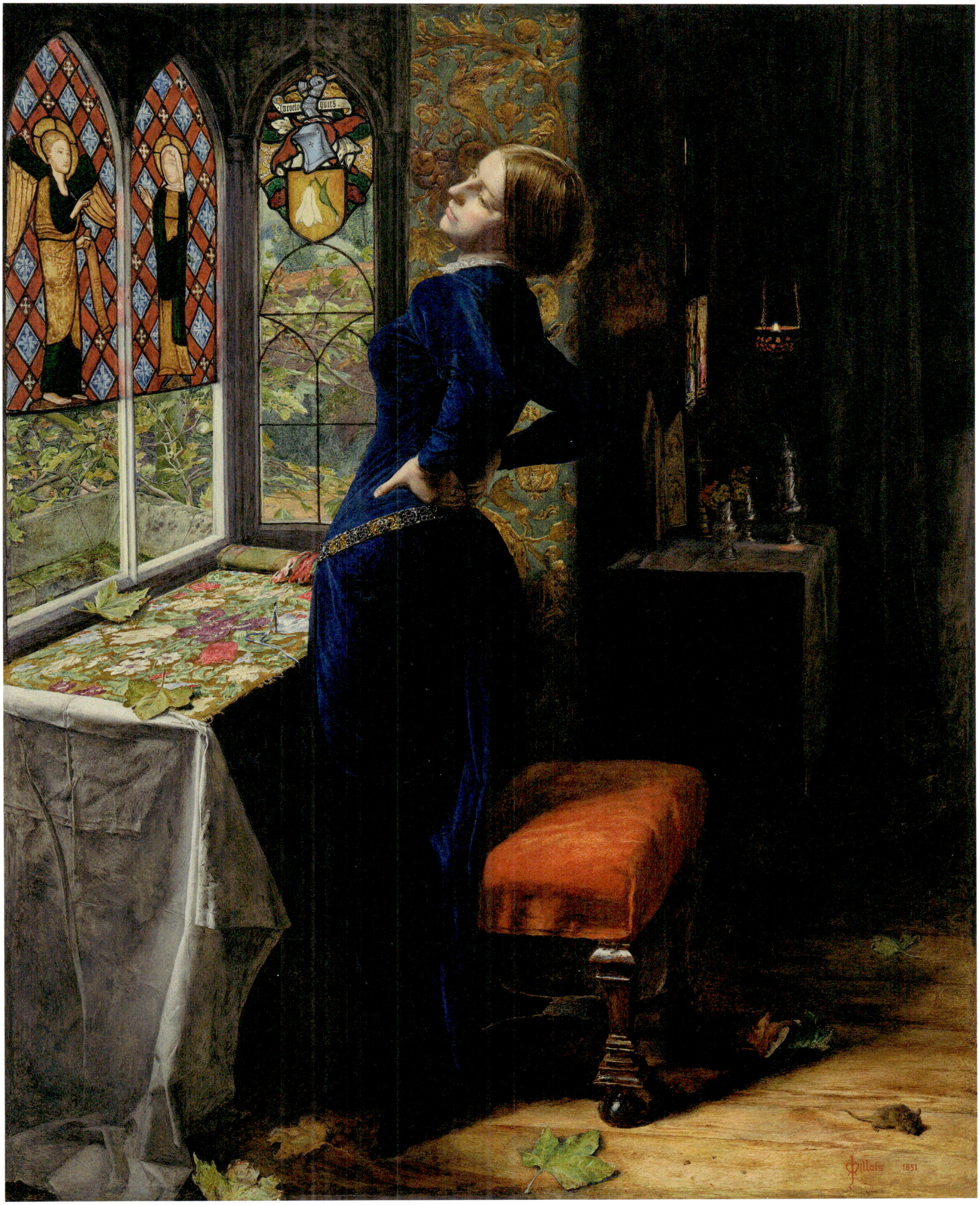

38

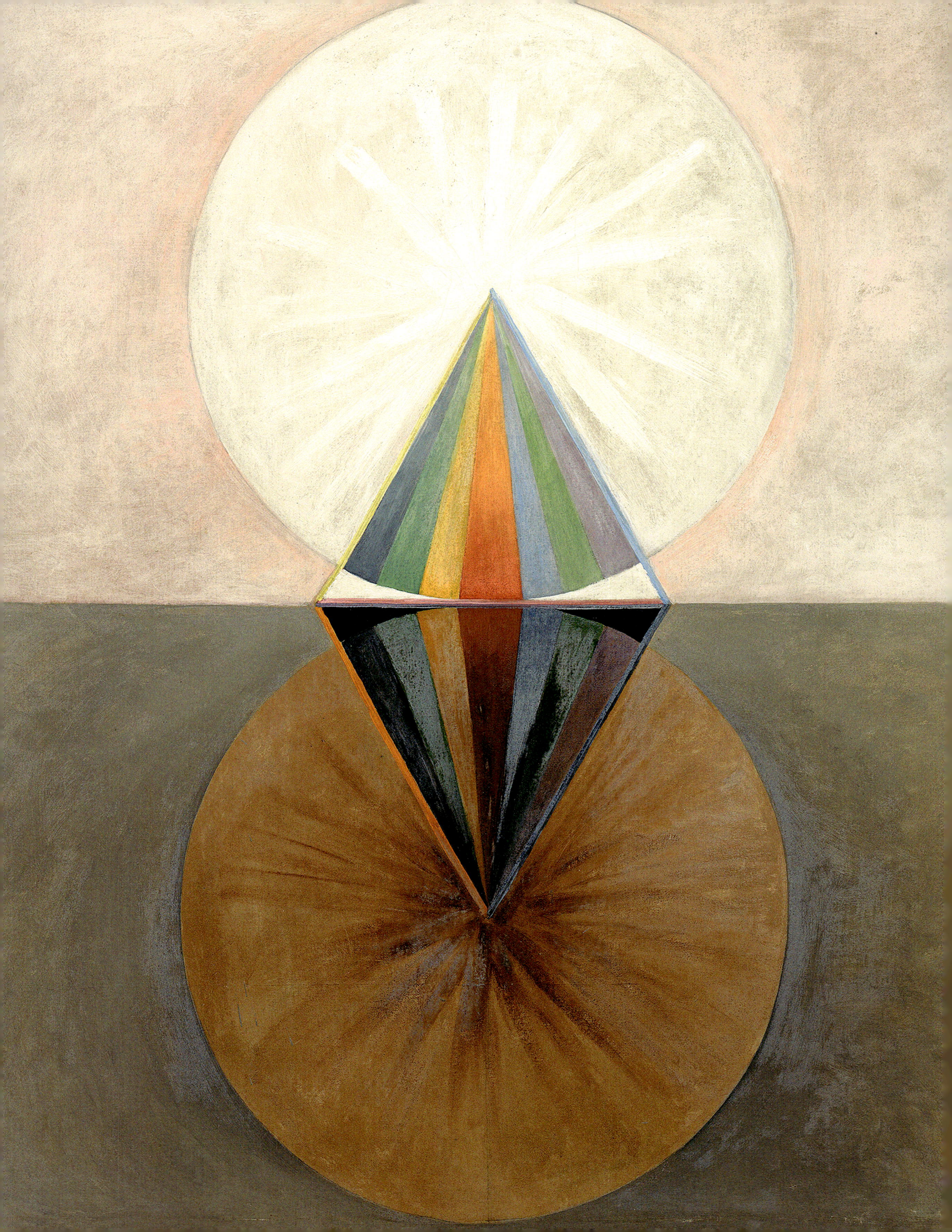

Alexandra Loske

LA COULEUR MAÎTRISÉE ET LIBÉRÉE

Comment l'être humain s'empare de la couleur par le mot et l'image

Notre monde regorge de couleurs. La plupart d'entre nous le vivent en couleurs. Nous disons de notre Terre qu'elle est la « planète bleue » et, d'un point de vue optique, nous nous noyons dans le bleu. Lorsqu'en 1990, la sonde spatiale *Voyager 1* nous a photographiés depuis une distance de six milliards de kilomètres, nous avions l'apparence, comme l'a décrit l'astronome Carl Sagan quelques années plus tard, d'un « point bleu pâle », mesurant moins d'un pixel et à peine visible, mais d'une teinte identifiable néanmoins. Nous sommes cette tache de couleur avant de disparaître. Ainsi la couleur est-elle l'un des aspects qui définissent et unifient notre existence et une caractéristique véritablement universelle que partagent tous les êtres humains. Il n'est donc pas surprenant qu'elle nous fascine depuis la nuit des temps. Les premières manifestations de la créativité humaine, sous la forme de gravures et de dessins sur bois, os ou roche, montrent que nous avons toujours cherché à fabriquer la couleur et à l'employer, avant tout pour représenter ou imiter ce qui a de l'importance à nos yeux et ce qui nous entoure. Bien que les traces physiques et visuelles de ces premières créations humaines soient fragmentaires et altérées, il est possible que, même aux temps primitifs, la couleur ait aussi été utilisée de manière symbolique. Les pigments minéraux extraits de la terre par les premiers humains étaient mélangés à des substances organiques, comme les graisses végétale et animale ou les humeurs du corps, et, ce faisant, pourraient bien avoir acquis une signification particulière ou sacrée, d'où serait née une hiérarchie chromatique simple. Si nous ne saurons jamais avec précision ce que nos ancêtres préhistoriques pensaient de la couleur, nous avons des preuves que certaines teintes étaient employées dans des contextes particuliers.

Il est possible de retracer, au fil de l'histoire humaine, l'intérêt intellectuel que nous nourrissons concrètement pour la couleur, depuis les conceptions rudimentaires des premières formes d'art jusqu'aux systèmes sophistiqués fondés sur la science moderne, la philosophie ou les arts, autant de catégories qui se recoupent et se répondent fréquemment. Cet ouvrage a pour objet de proposer un panorama de la réflexion et du ressenti de l'être humain à l'égard de la couleur, de son désir de créer, d'organiser et de systématiser la couleur, de l'employer et de la structurer, de lui attribuer un sens complexe. Les innombrables tentatives de classification et de compréhension de la couleur ont suscité toutes sortes de manifestations, depuis les conceptions dualistes de l'obscurité et de la lumière jusqu'aux motifs et systèmes géométriques élaborés qui visent à donner une forme spécifique à la couleur. Avec le développement du langage, de l'écriture, puis de l'imprimerie et des méthodes sophistiquées de reproduction des images, ces conceptions ont gagné en raffinement et en complexité, pris des formes visuelles saisissantes, et sont aussi mieux conservées.

Notre voyage visuel à travers l'histoire des conceptions de la couleur commence au XVIIe siècle, époque dont datent les documents les plus anciens consacrés spécifiquement à la couleur – manuscrits illustrés et sources imprimées –, et s'achève au milieu du XXe siècle, avant l'avènement de l'ère numérique. Chacune des conceptions se fait l'écho du temps de sa création tout comme le présent ouvrage, à travers celles que nous avons retenues. Depuis une vingtaine d'années, la couleur et son histoire suscitent un intérêt marqué et grandissant. Grâce à de nouvelles recherches, nous découvrons des tendances, des voix et des exemples jusque-là négligés de la culture matérielle (écrivains, pédagogues, ouvrages, manuscrits, œuvres d'art, méthodes), oubliés ou jamais incorporés aux canons de la couleur.

Aucun ouvrage consacré à la couleur ne saurait offrir une vue d'ensemble encyclopédique ni prétendre à l'exhaustivité sur les plans culturel, historique, géographique et thématique. Certaines publications laissent délibérément des aspects de côté afin de mettre en lumière tel ou tel thème culturel ou historique, d'autres reflètent les intérêts, goûts ou savoirs de leur auteur ou directeur d'ouvrage. Métaphoriquement bigarré, le présent ouvrage présente, par ses choix, la

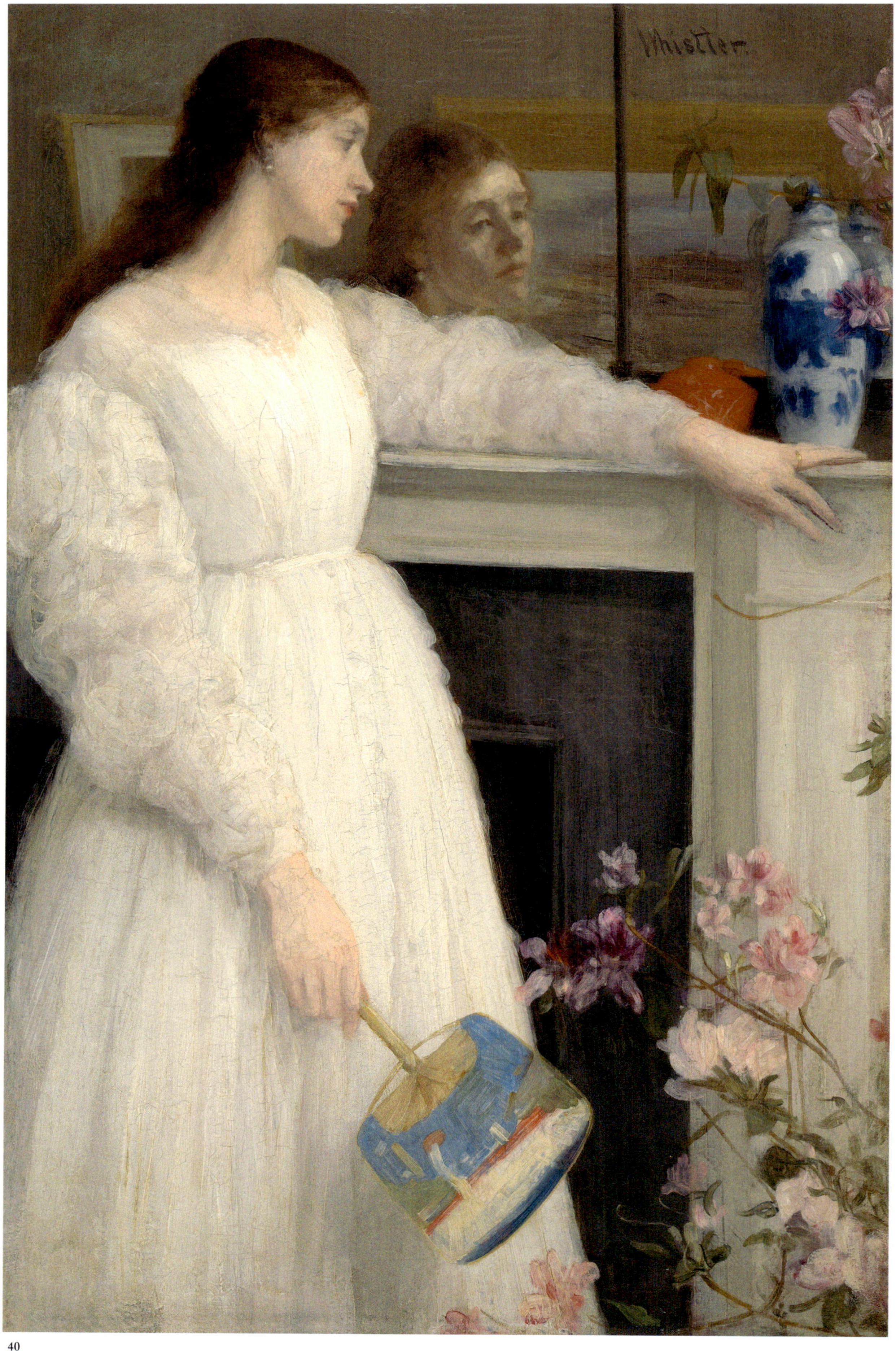

40

ILL. 39 *(Page 42)*
Hilma af Klint
The Swan (The SUW Series, Group IX: Part I, No. 12), 1914/15
Der Schwan / Le Cygne / El cisne
Oil on canvas, 151.5 x 151 cm
(59 ⅝ x 59 ½ in.)
Stockholm, The Hilma af Klint Foundation

ILL. 40
James McNeill Whistler
Symphony in White, No. 2: The Little White Girl, 1864
Symphonie in Weiß, Nr. 2: Das kleine weiße Mädchen
Symphonie en blanc n° 2 : la jeune fille en blanc
Sinfonía en blanco, n.° 2: La niña blanca
Oil on canvas, 76.5 x 51.1 cm
(30 ⅛ x 20 ⅛ in.)
London, Tate Britain

ILL. 41
James McNeill Whistler
Harmony in Blue and Silver: Trouville, 1865
Harmonie in Blau und Silber: Trouville
Harmonie en bleu et argent: Trouville
Armonía en azul y plata: Trouville
Oil on canvas, 51 x 76.7 cm
(20 ⅛ x 30 ¼ in.)
Boston, Isabella Stewart Gardner Museum

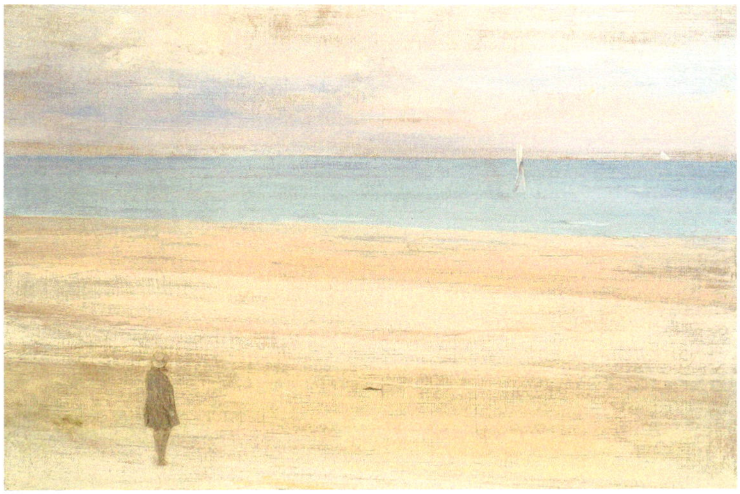

41

même subjectivité que celle qu'on attribue à la perception des couleurs. Les 67 œuvres présentées et analysées ici sont représentatives de l'histoire riche et variée des conceptions de la couleur depuis le XVIIe siècle. Si nous avons choisi de nombreux exemples connus et bien établis dans l'histoire de la couleur, nous espérons être parvenues à les situer dans un contexte culturel plus vaste, en les abordant avec un œil neuf. Outre ces exemples considérés comme « caractéristiques » de leur domaine, nous nous sommes efforcées de répertorier des œuvres méconnues ou ignorées, qui méritent d'être prises en compte pour leur intérêt et leur importance esthétiques et théoriques. Nous avons également cherché à mettre en lumière des œuvres écrites, publiées et illustrées par des femmes au cours des deux siècles passés. Notre spectre historique et géographique est vaste et nos exemples proviennent d'Europe, des États-Unis et de certaines régions d'Asie. Mais nous avons aussi voulu inclure autant de genres et de styles que possible. Un manuel élémentaire du XXe siècle peut être aussi passionnant que le cercle chromatique élaboré par Isaac Newton (1643–1727) en 1704 (p. 211), chacun d'eux jouant un rôle essentiel dans l'histoire de la couleur.

Des conceptions de la couleur, pour quoi faire ?

L'expression « conception de la couleur », malléable et appliquée à beaucoup d'aspects, convient bien à la réflexion sur ce sujet et aux tentatives de compréhension et de systématisation des couleurs. Pourtant, au lieu de chercher la meilleure appellation, on pourrait se demander pourquoi l'on éprouve le besoin de systématiser quelque chose d'aussi intangible, changeant, insaisissable et, dans une certaine mesure, subjectif. Jusqu'à une période relativement récente, les écrits consacrés à la couleur ne distinguaient pas clairement la couleur matérielle de la couleur immatérielle. Et dès lors que cette différence a été bien comprise, il est apparu qu'il existe en réalité de nombreuses formes chromatiques associant la couleur immatérielle aux teintes des matériaux solides. La couleur nous échappe facilement. Elle change en fonction de la lumière, prend une teinte différente selon le moment de la journée, pâlit, s'assombrit et peut se révéler imprévisible lorsqu'elle est combinée, mélangée ou juxtaposée. Qui plus est, elle peut être difficile à obtenir, comme peuvent le confirmer les fabricants de pigments et les historiens. Les premiers humains ont même dû être contrariés par le fait d'habiter un environnement où les deux couleurs dominantes, le bleu du ciel et le vert de la végétation, sont parmi les plus difficiles à fabriquer. Nous avons beau vivre dans un monde où la chlorophylle abonde, peu de végétaux permettent de réaliser de bons pigments verts. Sans compter que l'inconstante, l'instabilité et la toxicité fréquente de la couleur peuvent être exaspérantes. Et l'on n'a pas toujours aimé la couleur. Comme l'analyse brillamment l'artiste et écrivain David Batchelor dans *Chromophobia*, l'histoire de la culture occidentale se caractérise particulièrement par sa haine de la couleur, ancrée dans « la peur de la contamination et de la corruption par quelque chose qui est inconnu et semble inconnaissable [... et qui] se manifeste par les tentatives nombreuses et variées de débarrasser la culture de la couleur, de dévaluer celle-ci, de dénigrer son importance, de nier sa complexité ». Les attitudes de l'Occident à l'égard de la couleur ou, plus particulièrement, des couleurs très saturées et de la polychromie s'accompagnent généralement, selon Batchelor, de connotations en rapport avec l'étranger, l'oriental, le primitif, le féminin, le vulgaire, le superficiel, voire le pathologique, à l'opposé des notions classiques de blanc virginal ou de beauté incolore. La culture occidentale voit depuis fort longtemps la couleur et la noirceur comme les signifiants d'une altérité dangereuse ou menaçante, force déstabilisante qu'il s'agit d'éviter ou de dompter. La création de systèmes ou de conceptions chromatiques est une manière d'apprivoiser la couleur, intellectuellement d'abord, puis grâce à une utilisation et à une application mesurées. En imposant un ordre à la couleur, fût-ce de façon théorique, au moins tentons-nous de maîtriser une chose impermanente, mais omniprésente.

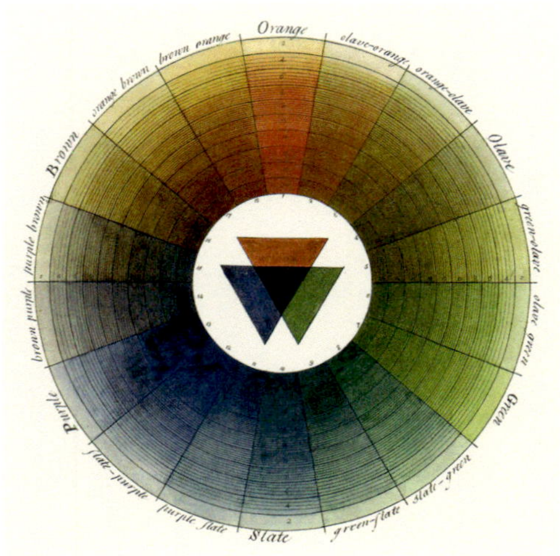

42

Ill. 42
Moses Harris
Colour Circle, Prismatic
Farbkreis, prismatisch
Cercle chromatique, couleurs prismatiques
Círculo de color, prismático
From: *The Natural System of Colours* (London, *c.* 1769–1776)
Winterthur, Stadtbibliothek, Farbsammlung Werner Spillmann

Ill. 43
J. M. W. Turner
Light and Colour (Goethe's Theory) – The Morning after the Deluge – Moses Writing the Book of Genesis, *c.* 1843
Licht und Farbe (Goethes Lehre) – der Morgen nach der Sintflut – Moses beim Schreiben der Genesis
Lumière et couleur (la théorie de Goethe) – le lendemain du déluge – Moïse écrivant le livre de la Genèse
Luz y color (teoría de Goethe). La mañana después del diluvio. Moisés escribiendo el libro del Génesis
Oil on canvas, 78.7 x 78.7 cm (31 x 31 in.)
London, Tate Britain

La peur, la défiance et la haine de la couleur ne sont, bien sûr, qu'un aspect des choses. Notre rapport à la couleur a toujours été multiforme. Les cinquante ou soixante dernières années, en particulier – dans le sillage de l'invention de la photographie et du cinéma en couleurs, de la couleur numérique, et la généralisation de l'impression en couleurs – nous ont accoutumés à la couleur comme jamais auparavant. À de très rares exceptions près, la couleur est désormais facilement accessible et n'est plus l'apanage de quelques-uns. La reproduction en couleurs connaît peu de limites, même si la qualité varie encore.

L'ordre chromatique : de la boîte de couleurs à la palette

Là où les sources textuelles peuvent manquer, l'archéologie et la culture matérielle offrent une manière de comprendre les attitudes culturelles vis-à-vis de la couleur. À tout le moins, les teintes employées en peinture et en ornementation sont des indices qui nous renseignent sur les pigments disponibles à certaines époques et révèlent des détails relatifs aux techniques, aux matériaux picturaux, aux préférences chromatiques et éventuellement, d'un point de vue culturel plus large, aux significations symboliques assignées aux couleurs. L'une des plus anciennes palettes d'artiste qui soient parvenues jusqu'à nous – une palette de la 18e dynastie égyptienne, datant de vers 1427–1401 avant J.-C. (ill. 2) – nous donne une idée tangible et tactile des pigments dont disposait son propriétaire, Amenemopet, vizir d'Amenhotep II. Elle nous renseigne aussi peut-être sur ses goûts en matière de couleur et sur sa manière de disposer dans sa palette les godets d'ocre rouge, de bleu égyptien et de vert mélangé, ainsi que deux sortes de noir. Pourquoi avait-il deux noirs différents ? Quelle était la signification du noir pour lui ou dans la culture de l'Égypte ancienne ? L'ordonnancement des matériaux de peinture peut sembler bien éloigné de la complexité d'un modèle chromatique tridimensionnel du XXe siècle. Pourtant, le désir sous-jacent est le même : attribuer à la couleur un ordre, que ce soit pour des raisons pratiques ou des goûts individuels, voire à des fins commerciales. La palette d'un peintre, qu'il s'agisse de l'objet proprement dit et des restes de peinture qu'il porte (ill. 44) ou tel qu'il est représenté dans les autoportraits, dévoile une partie de l'existence de l'artiste, de ses pratiques et de ses préférences. Les boîtes de couleurs modernes, que l'on peut remplir de tablettes d'aquarelle industrielle, renseignent également aussi sur leurs utilisateurs et possesseurs : quelle taille et quelle qualité de boîte l'artiste avait-il ou elle les moyens de se procurer, quelles teintes a-t-il ou elle employées, quelles sont celles qui ont été le plus ou le moins utilisées, comment étaient-elles disposées ? De même que l'exemple de l'Égypte ancienne, une boîte d'aquarelle Reeves (ill. 36) du début du XXe siècle peut en dire beaucoup sur la disponibilité des couleurs et leur popularité à cette époque. La boîte Reeves illustrée ici comprend plusieurs couleurs baptisées du nom de Wilhelm Ostwald (1853–1932), grand chimiste et théoricien de la couleur allemand.

La représentation de la couleur avant l'époque moderne

Cet ouvrage a pour objet principal les systèmes de représentation visuelle de l'ordre chromatique et les idées qui les sous-tendent. Si les conceptions proprement dites remontent à l'époque classique, les premières illustrations de couleurs sous la forme de graphiques n'apparaissent qu'à la fin du XVIIe siècle. Jusqu'à une période relativement récente, la culture classique nous a semblé essentiellement incolore sur le plan matériel : le blanc est la teinte primordiale associée à la Rome et à la Grèce antiques, en raison des sculptures et des édifices de marbre blanc qui ont depuis longtemps perdu leurs surfaces colorées. Cependant, cette impression générale est bouleversée depuis

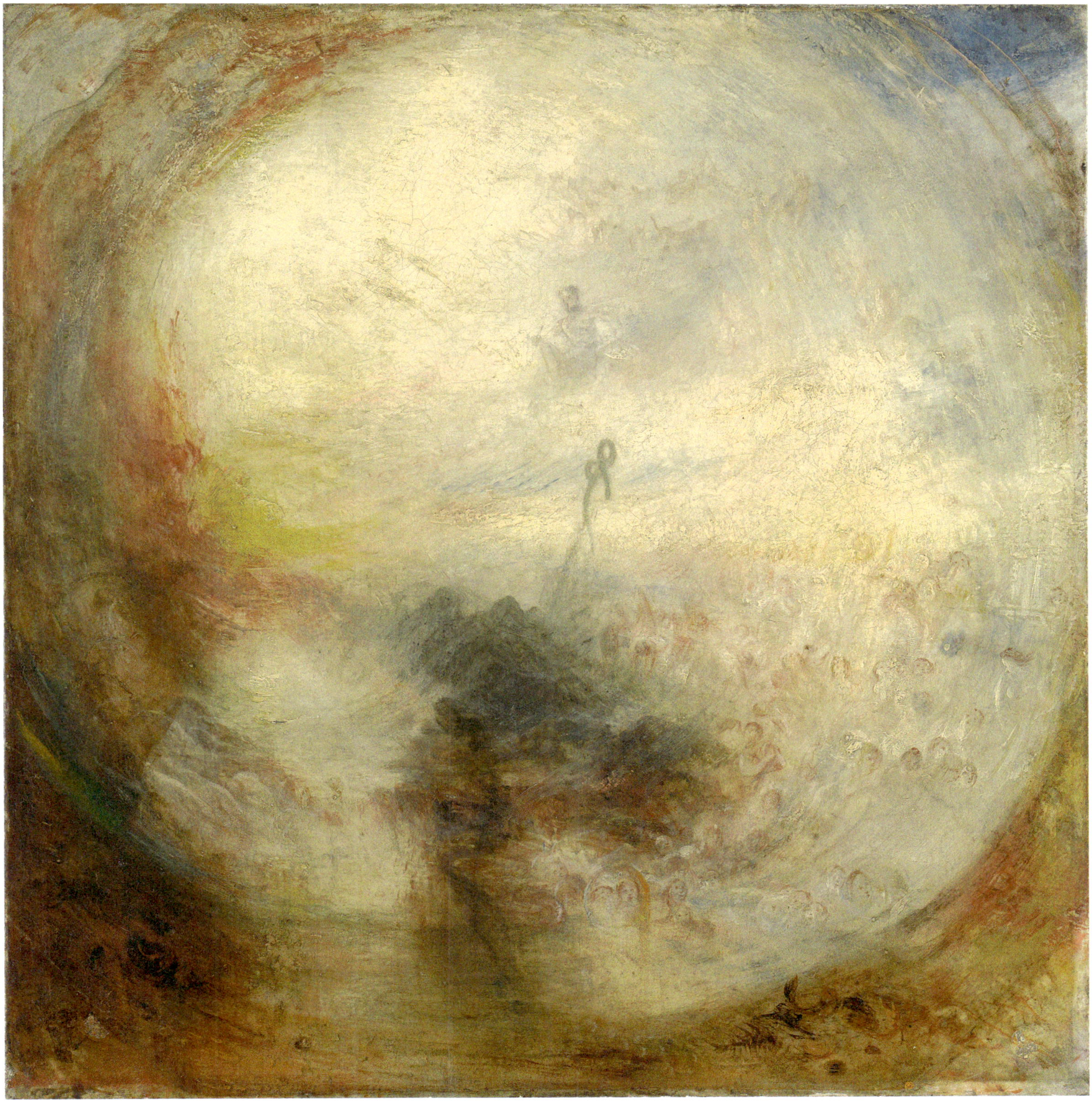

43

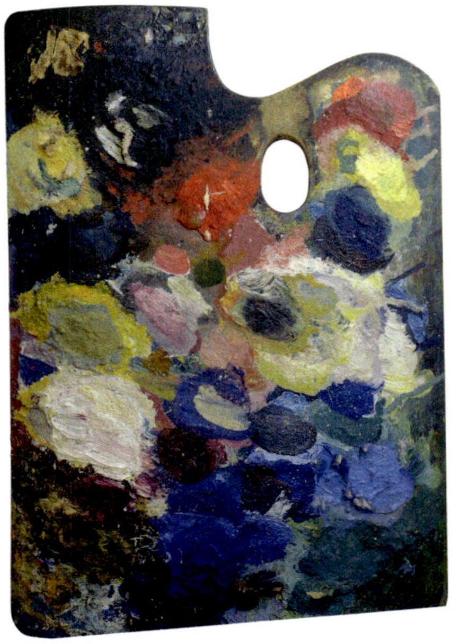

44

ILL. 44
Matthias Schaller
Wassily Kandinsky, 2007
Palette from the series
Das Meisterstück,
191 x 141.6 cm (75 ¼ x 55 ¾ in.)
Private collection

ILL. 45
Wassily Kandinsky
Color Study. Squares with Concentric Circles, 1913
Farbstudie. Quadrate mit konzentrischen Ringen
Étude de couleurs. Carrés et cercles concentriques
Estudio de color. Cuadrados con círculos concéntricos
Watercolour, gouache and crayon on paper, 23.8 x 31.4 cm (9 ⅜ x 12 ⅜ in.)
Munich, Lenbachhaus

quelques années par des études consacrées à la nature polychrome de la sculpture et des architectures grecque et romaine.

Sur les plans conceptuel et linguistique, on distingue plusieurs traditions dans l'histoire des conceptions de la couleur. Empédocle (494–434 av. J.-C.) serait l'un des premiers à avoir écrit sur le sujet. Les fragments de ses poèmes qui sont parvenus jusqu'à nous permettent de discerner une cosmologie fondée sur les quatre éléments – terre, feu, eau et air –, dont Empédocle pensait que les couleurs étaient dérivées. Au siècle suivant, Aristote (384–322 av. J.-C.) aurait réalisé des expériences optiques simples avec du verre coloré afin d'étudier la vision en couleurs et le contraste. Il propose un système polaire dont le spectre va de la lumière blanche divine, qui rend les couleurs visibles, à l'obscurité, figurée par le noir. À partir du blanc naît la hiérarchie des teintes jaune, verte, bleue et rouge, associées respectivement à la terre, à l'eau, à l'air et au feu. Élève d'Aristote, Théophraste (372–287 av. J.-C.) s'intéresse à la manière dont nous percevons ou ressentons les couleurs et se réfère à la conception des quatre couleurs – rouge, noir, blanc et jaune verdâtre – de Démocrite (460–370 av. J.-C.), associées à certaines formes et surfaces, ainsi qu'à d'autres aspects liés à nos sens. De son côté, le médecin Hippocrate (460–375 av. J.-C.) distingue quatre couleurs représentant les principaux fluides ou humeurs du corps humain, vues comme la cause de certains traits de caractère : blanc pour les flegmatiques, rouge pour les sanguins, jaune pour les colériques et noir pour les mélancoliques.

Les conceptions de la couleur et les discussions théoriques de l'époque classique n'existent plus que sous des formes fragmentées et comportent, en outre, l'inconvénient de présenter des difficultés linguistiques. La couleur est généralement mal servie par le langage car les mots sont impuissants à rendre toute la complexité des couleurs et de leur perception. La plupart des langues ne possèdent qu'une douzaine de termes pour qualifier les couleurs, alors que celles que nous sommes capables de voir sont innombrables. Cet écart est encore plus problématique en ce qui concerne les langues mortes, les textes anciens et leurs traductions. Nous manquons aussi de représentations visuelles des conceptions chromatiques du monde antique et des cultures premières, ce qui ne facilite pas la compréhension exhaustive des notions les plus anciennes. Néanmoins, il est possible d'avoir un aperçu de la puissante valeur symbolique des couleurs grâce à certaines découvertes archéologiques, comme les masques de théâtre trouvés à Pompéi, sous la forme d'objets ou d'images visibles sur les fresques, peints dans des teintes exprimant les quatre humeurs (ill. 7). En dépit de la rareté des sources écrites consacrées à la couleur, nous pouvons retracer le développement de plusieurs conceptions culturelles en la matière grâce à la matérialité des pigments et des colorants. Comme dans le cas du pigment bleu outremer de la peinture médiévale et de la Renaissance (ill. 3), la valeur matérielle et la rareté de certaines teintes ont conduit le monde antique à les associer à des figures de premier plan et à des divinités.

Si, tout au long du Moyen Âge, le système aristotélicien qui encadre la couleur entre l'obscurité et la lumière (le noir et le blanc) prévaut dans le discours philosophique sur le sujet, les manuels pratiques consacrés à la couleur se font aussi de plus en plus nombreux. Au début du XIIIe siècle, Robert Grosseteste (vers 1175–1253), érudit de l'université d'Oxford et évêque de Lincoln, en Angleterre, traduit certains ouvrages d'Aristote et, pour y répondre, rédige son propre traité des couleurs, *De colore*. Il y expose les théories chromatiques du philosophe grec et s'interroge sur la quantité de lumière que chaque couleur est susceptible de contenir. Vers le XVe siècle, la couleur devient peu à peu une marchandise, avec l'essor des échanges commerciaux. Les connaissances sont organisées de manière de plus en plus systématique et des premiers traités portant sur les techniques picturales, les matériaux des peintres et la vie des artistes voient le jour. Souvent très techniques, ils ont touché peu de lecteurs, en particulier avant le développement de l'imprimerie. Les manuscrits parvenus jusqu'à nous

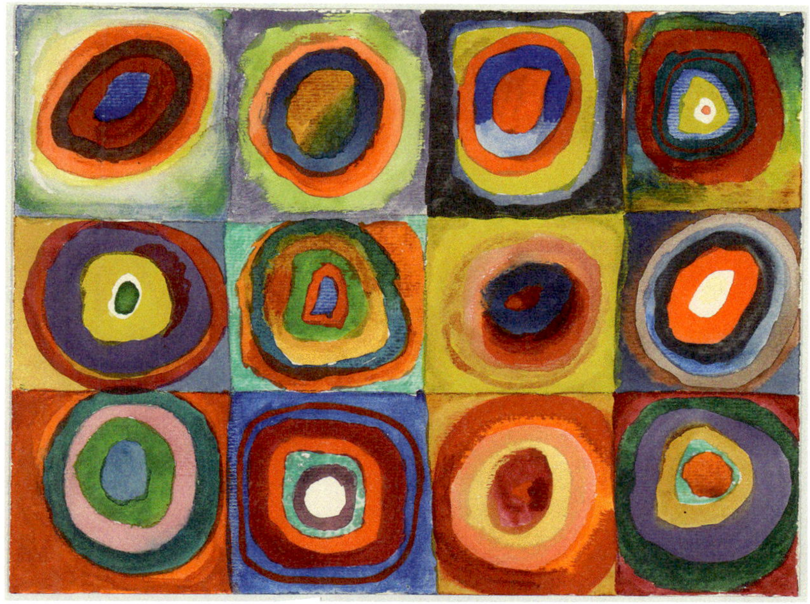

45

nous éclairent sur la fabrication artisanale des couleurs à l'époque médiévale. C'est notamment le cas d'un ouvrage du début du XIIe siècle, *De diversis artibus* (*Traité des divers arts*), dans lequel Théophile le moine (actif vers 1070–1125) propose principalement des recettes de couleurs. Bien d'autres ouvrages pratiques suivront au cours des XVe et XVIe siècles. Mettant l'accent sur les techniques et les matériaux, ils n'offrent aucune discussion philosophique des conceptions de la couleur, mais sont des sources précieuses pour l'étude de la terminologie chromatique et de l'histoire des pigments. On tombe parfois sur un texte comprenant des illustrations de matières premières employées pour fabriquer pigments et peintures. C'est, par exemple, *Le Livre des simples médecines* (ills. 19, 20) – encre de calmar, lapis-lazuli et momie (pour le brun de momie) – dont on estimait aussi qu'elles possédaient des propriétés thérapeutiques. Il s'agit d'illustrations simples, mais les objets sont soigneusement disposés comme sur un plateau de collectionneur ou dans un cabinet de curiosités.

Les quatre humeurs qui, selon Hippocrate, sont omniprésentes dans le corps et l'esprit humain et peuvent être figurées par quatre couleurs, continuent de nourrir les conceptions chromatiques rudimentaires du Moyen Âge à la Renaissance. Si, au cours de cette époque, l'on ne connaît aucune image ou œuvre d'art traitant spécifiquement des théories et conceptions de la couleur, apparaissent néanmoins les premiers exemples d'utilisation des couleurs pour en illustrer et faciliter l'ordonnancement et la classification. Ainsi, le cercle chromatique uroscopique réalisé en 1506 (ill. 5) par Ulrich Pinder (mort en 1519) représente vingt nuances d'urine, utilisées pour les diagnostics médicaux. Coloriées à la main, ces images de flacons d'urine offrent un des premiers exemples de cercle chromatique fondé sur les couleurs naturelles, assez semblable à ceux que l'on trouvera près de trois siècles plus tard dans les livres consacrés aux insectes de l'entomologiste anglais Moses Harris (1730–vers 1788). Dans la même veine, une foule d'écrits encyclopédiques tels que bestiaires, herbiers, lapidaires et index, comportent chacun des renseignements directs ou indirects sur les connaissances et l'usage des couleurs de leur époque.

En 1435, la publication de *De pictura* (*De la peinture*) par Leon Battista Alberti (1404–1472), coïncide avec l'avènement de la peinture à l'huile. Estimant que la couleur est un élément important et précieux de la peinture, Alberti avance que le mélange de quatre couleurs seulement – rouge, vert, bleu et gris – peut donner naissance à toutes les teintes. S'intéressant moins à la lumière colorée qu'à la couleur physique, il écrit à l'attention des peintres, de sorte que le gris, auquel il accorde une place de choix, doit être considéré sous l'angle des ombres et de l'effet de volume en peinture. De nombreux ouvrages portant sur le débat plus vaste des rôles respectifs du dessin (*disegno*) et de la couleur (*colore*) suivent les conceptions d'Alberti et de ses contemporains, L'un des plus grands artistes, Léonard de Vinci (1452–1519), nous a livré ses pensées sur le sujet dans divers carnets et recueils consacrés à la vision et à la perception des couleurs, à leur apparence et à l'application pratique de la théorie chromatique à la peinture. Il propose une conception simple, fortement influencée par Aristote et Alberti et fondée sur six « couleurs simples »: rouge, jaune, vert et bleu qu'encadrent le blanc et le noir. Aucun de ces textes datant du Moyen Âge et du début de la Renaissance n'étant illustré, il est impossible de savoir si les systèmes proposés étaient envisagés sous une forme linéaire ou circulaire, ou encore en deux ou trois dimensions. Néanmoins, ce sont des précurseurs fascinants de la théorie moderne des couleurs.

Une nouvelle ère chromatique : la couleur depuis le XVIIe siècle

À la fin du XVIIe siècle, alors que l'imprimerie est déjà bien établie, pour diverses raisons, une grande partie des premiers écrits de l'époque sur la couleur sont des manuscrits. L'esprit naissant des Lumières favorise alors l'échange d'idées par-delà les frontières géographiques et

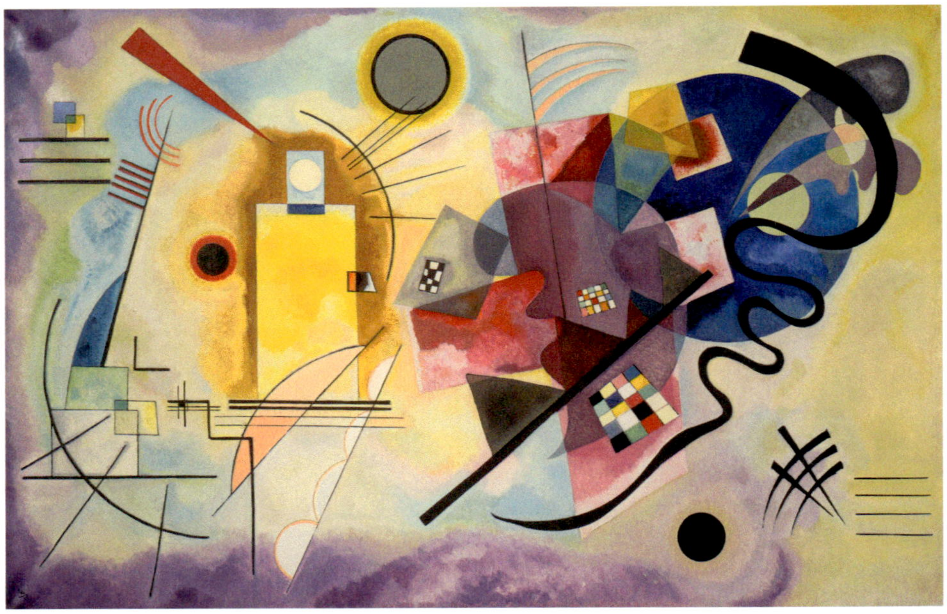

46

culturelles. Si les symbolismes et métasystèmes chromatiques associés aux croyances religieuses, à l'alchimie et aux traditions populaires sont encore tenaces, la couleur commence à prendre une dimension scientifique et intellectualisée. On souhaite comprendre et catégoriser le monde naturel, y compris la couleur, considérée comme un phénomène optique, un instrument de description de la nature et un moyen d'expression dans les beaux-arts et les arts décoratifs.

Respectant un ordre relativement chronologique, chaque chapitre de cet ouvrage s'intéresse à un thème ou un aspect spécifique de l'histoire des concepts chromatiques. Les livres et manuscrits mentionnés ont été sélectionnés selon les critères suivants : avoir pour sujet principal la couleur sur le plan théorique ou pratique, ou employer la couleur de manière cohérente et inventive comme, par exemple, pour l'illustration de concepts et d'idées relatifs à la couleur.

Nous présentons de nombreuses œuvres dues à des femmes, ceci pour remédier au déséquilibre historique entre les hommes et les femmes ayant produit des écrits sur la couleur. Il s'agit notamment de livres éducatifs consacrés à la couleur en peinture au début du XXe siècle, comme, par exemple, *The Theory and Practice of Color* (« La Théorie et pratique de la couleur », 1918, p. 552–559), de la pédagogue américaine Bonnie Snow (vers 1862–1925 ; coécrit avec Hugo B. Froehlich, ills. 11, 12, 18). Citons aussi Mary Gartside (vers 1755–1819, p. 218–229), peintre de fleurs britannique qui enseigna l'aquarelle et dont l'œuvre est négligée (ill. 21). Elle est la première femme à avoir fait paraître un ouvrage consacré à la couleur en anglais : *An Essay on Light and Shade, on Colours, and on Composition in General* (« Essai sur la lumière et l'ombre, sur les couleurs, et sur la composition en général », 1805). Pour l'édition suivante (1808), elle donna un nouveau titre à son ouvrage, *An Essay on a New Theory of Colours* (« Essai sur une nouvelle théorie des couleurs »), afin de souligner le sérieux de sa publication. Des femmes ont également signé des ouvrages sortant des sentiers battus, de par leur appartenance à des cercles théosophes et spirites, comme *The New Science of Color* (« La nouvelle science de la couleur », 1915) de Beatrice Irwin (1877–1956), poétesse et actrice appartenant à plusieurs cultures, qui s'est beaucoup intéressée au potentiel que recèle l'association de la lumière artificielle colorée et de la musique (p. 701). Aujourd'hui méconnues, ses réflexions sur la synesthésie ont influencé plusieurs artistes célèbres du XXe siècle, comme Georgia O'Keeffe (1887–1986), dont certaines des œuvres pourraient bien faire allusion aux idées d'Irwin (ills. 50, 51). Certaines femmes sont dignes de figurer au panthéon de l'histoire de la couleur sans pour autant avoir écrit une théorie à part entière sur le sujet. Hilma af Klint (1862–1944) en fait partie (p. 704–715), de même que Winifred Nicholson (1893–1981). Par ses œuvres, ses notes, ses lettres et deux courts textes publiés à propos de la couleur, Nicholson est peut-être l'une des plus grandes théoriciennes du XXIe siècle en ce domaine, bien qu'elle n'ait jamais réuni ses idées et conceptions en un seul volume concis. À l'instar de Gartside, c'est en peignant des fleurs qu'elle semble s'être intéressée à la théorie chromatique. Ses premières compositions ont souvent pour décor des paysages très éclairés ou encadrés par des fenêtres, indice précoce de son goût pour l'interaction de la lumière et de la couleur (ills. 22, 55). Cet intérêt manifeste pour les couleurs prismatiques trouve un nouveau débouché artistique lorsque, au milieu des années 1970, Nicholson, alors octogénaire, reçoit en cadeau deux prismes de verre. Au cours des dernières années de sa vie, elle réalise une série fascinante de « peintures prismatiques », où elle donne une forme visuelle aux expériences qu'elle a effectuées avec les prismes et la lumière, en recourant souvent à un style totalement abstrait (ill. 56).

Le premier chapitre, « Pyramides de la couleur : premiers graphiques et tableaux », s'intéresse aux tout premiers livres illustrés occidentaux. Notre exemple inaugural est le tableau des couleurs créé en 1686 par Richard Waller (1655–1715), avec ses points peints à la main (p. 80–85), et probablement inspiré de tableaux similaires réalisés auparavant par Théodore Turquet de Mayerne (1573–1655), alchimiste et médecin à la

ILL. 46
Wassily Kandinsky
Yellow-Red-Blue, 1925
Gelb-Rot-Blau
Jaune-rouge-bleu
Amarillo, rojo y azul
Oil on canvas, 127 x 200 cm (50 x 78¾ in.)
Paris, Centre Georges Pompidou

ILL. 47
Franz Marc
Fighting Forms, 1914
Kämpfende Formen
Formes combattantes
Formas que luchan
Oil on canvas, 91 x 131.5 cm
(35⅞ x 51¾ in.)
Munich, Pinakothek der Moderne

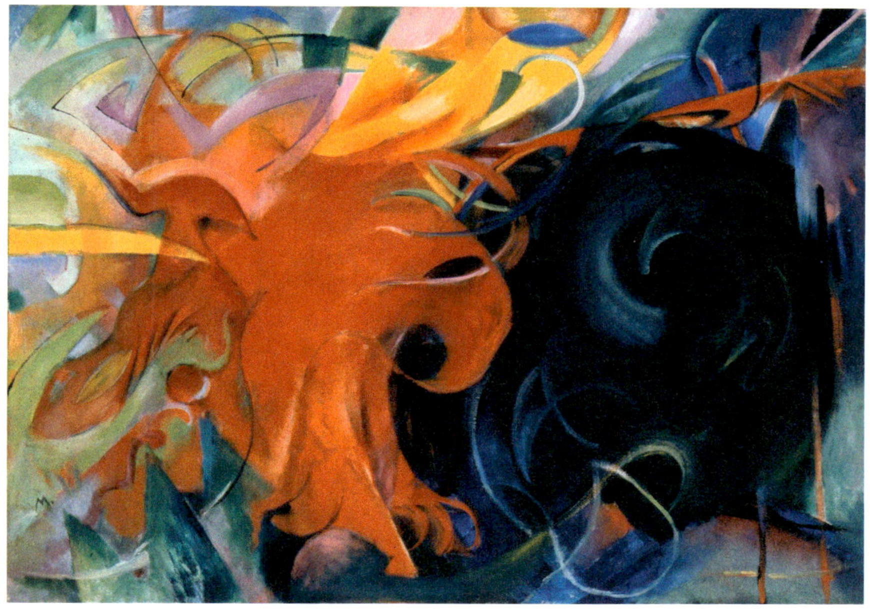

47

cour d'Angleterre, lesquels figurent dans le « manuscrit de Mayerne » (*Pictoria, sculptoria et quae subalternarum artium*, 1620–1646, ills. 1, 6). Dans les deux cas, les couleurs sont présentées de façon relativement simple, sous forme de tableaux et de cercles bien tracés. S'il ne s'agit pas de représentations de conceptions chromatiques nées de l'intellect, il faut les situer dans le contexte plus large des manuels de peinture car ils offrent une mine de renseignements sur les pigments et d'autres matériaux dont les peintres disposaient à l'époque et sur la manière dont ceux-ci les mélangeaient et les organisaient. Ce chapitre montre également comment des graphiques et tableaux plus complexes sont créés au siècle suivant et diffusés via des ouvrages et autres documents imprimés. À cette époque, l'intellectualisation des concepts chromatiques est flagrante : les répertoires de couleurs sont de plus en plus élaborés et détaillés et l'on réalise les premières tentatives de conceptions tridimensionnelles. En dépit des progrès de l'imprimerie en couleurs, tous ces graphiques et tableaux sont coloriés à la main. Par conséquent, chaque exemplaire constitue une œuvre d'art à part entière. L'application des couleurs à la main est une pratique qui se poursuit au XIXe siècle et, dans certains cas, elle ne sera remplacée par la reproduction polychrome mécanique qu'au XXe siècle.

Le chapitre 2, « La forme de la couleur : cercles, disques et sphères », étudie comment les conceptions visuelles simples de la couleur sont approfondies au cours des XVIIIe et XIXe siècles et donnent naissance à une passionnante diversité de formes géométriques. Le cercle est, semble-t-il, privilégié et constitue l'origine de nombreuses variantes de ce qui est devenu le cercle chromatique classique. On fait aussi appel à d'autres formes ; souvent, un graphique de couleurs comporte des ajouts et des enrichissements picturaux, ou bien un cercle chromatique s'accompagne de diagrammes et tableaux explicatifs comme, par exemple, dans *The Natural System of Colours* (« Le système naturel des couleurs », vers 1769–1776, p. 138–143) de Moses Harris, ainsi que dans les œuvres de Mary Gartside.

Sur le plan esthétique, les graphiques chromatiques des XVIIIe et XIXe siècles font partie des plus belles images de la littérature consacrée à la couleur. Cependant, malgré leur puissance visuelle, elles s'accompagnent généralement de textes manuscrits ou imprimés. Bien qu'elle peine à exprimer tout le spectre des couleurs visibles, la langue joue un rôle important dans la plupart de ces graphiques, parfois avec des résultats enchanteurs et poétiques, notamment dans *Die Temperamentrose* (« La rose des tempéraments » 1798/99), p. 155) de Johann Wolfgang von Goethe (1749–1832) et Friedrich Schiller (1759–1805), par exemple. L'une des difficultés courantes consiste à figurer en deux dimensions l'aspect tridimensionnel de la couleur. C'est à ce stade qu'apparaissent les premières sphères chromatiques comme, par exemple, la *Farben-Kugel* (1810, p. 163, ill. 13) de Philipp Otto Runge (1777–1810). Malheureusement, aucun modèle de sphère chromatique de l'époque n'est parvenu jusqu'à nous. Toutefois, en 2016, l'artiste franco-allemande Eva Bodinet a créé une maquette inspirée de la *Farben-Kugel* (ills. 14, 15) et, ces dernières années, plusieurs fabricants ont réalisé pour le commerce des sphères chromatiques en 3D prêtes à l'emploi. Runge offre un exemple touchant de la sensibilité et de la spiritualité accrues des romantiques dans leur perception de la couleur. Bien des artistes, des philosophes et des écrivains de la fin du XVIIIe siècle et du début du XIXe ont adapté certaines conceptions chromatiques à leurs croyances personnelles en leur attribuant des significations hautement symboliques. De 1802 à 1810, Runge crée une suite d'images qui figurent les moments de la journée, notamment *Der kleine Morgen* (« *Le Petit Matin* », ill. 17), et représentent une adaptation de la théorie des couleurs à l'expression de ses convictions chrétiennes.

Le chapitre 2 évoque l'expansion des études chromatiques au cours du XIXe siècle, parallèlement aux écrits sur la couleur s'appuyant sur les textes antérieurs. Nous nous intéressons notamment au peintre Joseph Mallord William Turner (1775–1851) qui, dans les années 1820,

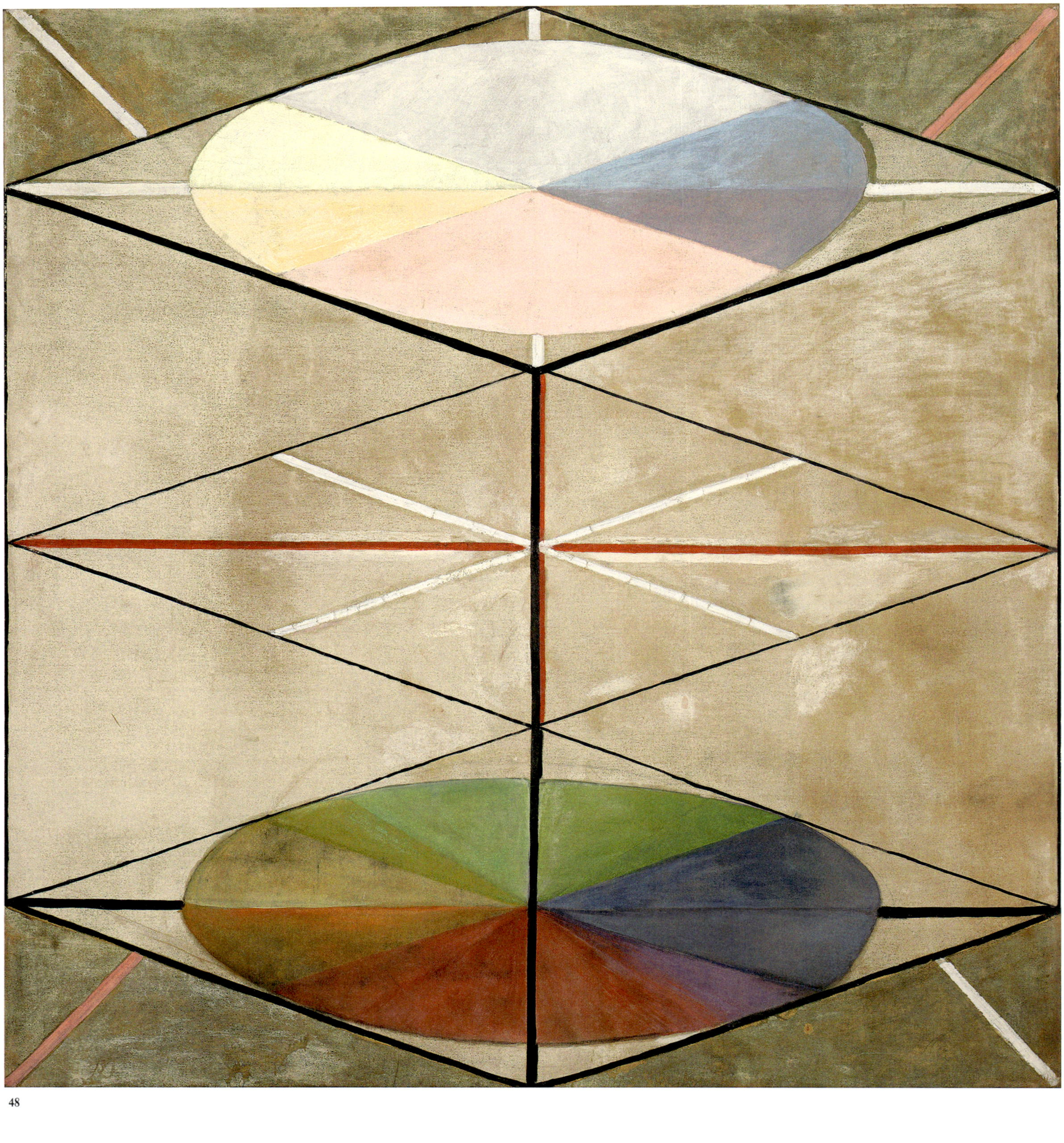

48

ILL. 48
Hilma af Klint
The Swan (The SUW Series, Group IX: Part I, No. 23), 1914/15
Der Schwan / Le Cygne / El cisne
Oil on canvas, 152.5 x 150 cm
(60 x 59 in.)
Stockholm, The Hilma af Klint Foundation

ILL. 49
Hilma af Klint
The Swan (The SUW Series, Group IX: Part I, No. 17), 1915
Der Schwan / Le Cygne / El cisne
Oil on canvas, 150.5 x 151 cm
(59¼ x 59½ in.)
Stockholm, The Hilma af Klint Foundation

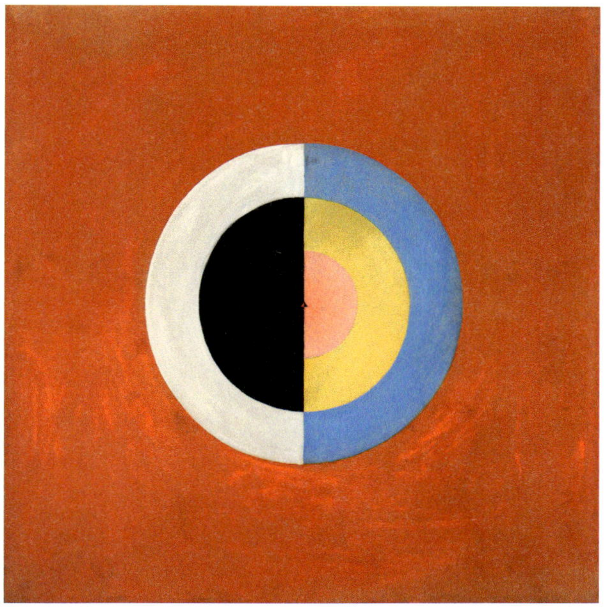

49

enseigne les fondamentaux de la théorie des couleurs, dans ses fonctions de professeur de perspective à la Royal Academy of Arts de Londres. Pour ce faire, il réalise deux grands dessins représentant des cercles chromatiques (ills. 25, 26), afin d'expliquer ce qui distingue les couleurs matérielles des couleurs immatérielles. Ces cercles, qui entourent des triangles de couleur superposés, doivent beaucoup à ceux de Moses Harris, que Turner a étudiés. Tout au long de sa vie, le peintre anglais ne cessera de réfléchir à la création de nouveaux pigments et à la théorie chromatique. Il possède et lit de nombreux ouvrages sur le sujet et fait de l'interaction de la couleur et de la lumière l'un des thèmes majeurs de son œuvre (ill. 27). Par la suite et quelques années après la parution de la première partie du principal ouvrage de Goethe sur le sujet *Zur Farbenlehre* (*Traité des couleurs*, 1810–1812, p. 236–249) dans une traduction en anglais (*Goethe's Theory of Colours*, 1840), Turner peint deux toiles qui font directement référence aux leçons de Goethe et, peut-être aussi, allusion à Harris (ill. 43).

Le chapitre 3, « Nouveaux éclaircissements : les progrès de la théorie des couleurs », aborde de manière plus approfondie certaines des plus grandes œuvres du XIXe siècle portant sur la couleur et offre des explications quant à la brusque augmentation des théories nouvelles et des longs traités philosophiques, scientifiques et pratiques sur le sujet. Si cette révolution chromatique est intimement liée à l'industrialisation de la couleur, en particulier aux nouvelles méthodes de fabrication, de réalisation et d'impression, ses racines se situent au début du XVIIIe siècle. L'un des ouvrages ayant incontestablement fait date dans l'histoire des conceptions chromatiques est dû à Isaac Newton. En 1704, celui-ci fait paraître *Opticks*, fruit de décennies de recherches sur la couleur (p. 210–217). On y trouve, sous la forme d'un disque, le premier graphique des couleurs ayant été diffusé à grande échelle : il s'inspire des expériences du scientifique anglais sur la fragmentation de la lumière blanche en spectres de couleurs à l'aide d'un prisme. Comme on le sait aujourd'hui, l'influence de Newton sur la pensée scientifique des XVIIIe et XIXe siècles est inestimable et ses théories ont été adaptées, admirées, prolongées et critiquées (ill. 8). La fascination exercée par le spectre de l'arc-en-ciel et par d'autres découvertes de Newton a été, et est encore, souvent présente dans les arts visuels. En témoigne, par exemple, *Colour* (ill. 9), peinture allégorique d'Angelica Kauffmann (1741–1807) commandée par la Royal Academy of Arts. Allégorie et non figuration d'une conception chromatique en tant que telle, cette œuvre illustre un moment important de l'histoire de la peinture car l'ensemble commandé à Kauffmann représente les grands « Éléments de l'art » – l'invention, la composition, la conception et la couleur – et place la couleur sur un pied d'égalité avec les trois autres catégories. La femme qui incarne la couleur trempe son pinceau dans l'arc-en-ciel de Newton et en extrait la couleur prismatique pour la poser sur une palette bien concrète. Ce faisant, elle adapte la théorie des couleurs immatérielles à la couleur matérielle, tout en l'associant à l'enseignement pratiqué à la Royal Academy of Arts qui vient d'être fondée.

Au fil du XIXe siècle, les conceptions et attitudes romantiques, comme le symbolisme chromatique spirituel de Runge, cèdent peu à peu le pas à des études plus techniques et scientifiques, dont beaucoup sont liées à la fabrication de pigments et de colorants nouveaux. La qualité de ces nouvelles teintes est l'objet d'une étude de George Field (vers 1777–1854, p. 258–273), grand coloriste britannique qui a élaboré sa propre théorie des couleurs, mais aussi expérimenté, fabriqué et fait breveter des pigments. Ses couleurs sont très demandées par de nombreux peintres de ce siècle, notamment par les membres de la Confrérie préraphaélite et leurs disciples. La disponibilité de ces teintes coïncide avec un regain d'intérêt pour les techniques et les matériaux de la peinture médiévale. Des peintures comme *Mariana* (1851, ill. 38) de John Everett Millais (1829–1896), remarquable par un éclat rappelant celui des bijoux et une polychromie audacieuse,

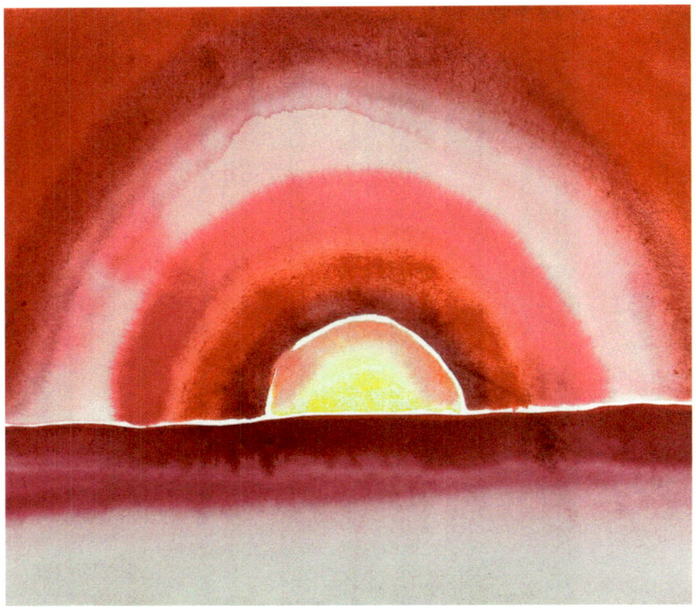

50

ILL. 50
Georgia O'Keeffe
Sunrise, 1916
Sonnenaufgang
Lever de soleil
Amanecer
Watercolour on paper,
22.5 x 30.5 cm (8⅞ x 12 in.)
Private collection

ILL. 51
Georgia O'Keeffe
Music, Pink and Blue No. 2, 1918
Musik, Rosa und Blau Nr. 2
Musique, rose et bleu nº 2
Música, rosa y azul n.º 2
Oil on canvas, 88.9 x 76 cm (35 x 29⅞ in.)
New York, Whitney Museum of American Art; Gift of Emily Fisher Landau in honor of Tom Armstrong

illustrent cet intérêt pour les pigments de bonne qualité et une conscience aiguë des théories sur la couleur alors en vogue. D'aucuns avancent même qu'une œuvre de Dante Gabriel Rossetti (1828–1882), *Ecce Ancilla Domini! (The Annunciation*, 1849/50, ill. 29), aurait été directement inspirée par le symbolisme des couleurs proposé par George Field. L'intérêt intellectuel accordé aux écrits et aux théories consacrés à la couleur est flagrant chez les peintres occidentaux du XIXᵉ siècle. L'impressionnisme et le pointillisme sont sous l'influence de la théorie pionnière du contraste simultané des couleurs qu'a élaborée Michel-Eugène Chevreul (1786–1889) à partir de 1839, lorsqu'il travaillait à la Manufacture des Gobelins, à Paris (p. 282–307). Le postimpressionniste Vincent van Gogh (1853–1890) applique les principes de Chevreul dans ses œuvres d'une manière stupéfiante. Il est révélateur que, parmi les biens personnels de l'artiste, se trouve une boîte de laque rouge contenant plusieurs pelotes de laine de couleur, emmêlées et associées selon différentes variations, comme s'il avait ainsi créé des harmonies chromatiques suivant les lois de Chevreul (ill. 37). L'œuvre de James McNeill Whistler (1834–1903) exprime également une grande conscience du contraste des couleurs. Si nous ne disposons d'aucune trace de son intérêt pour les ouvrages consacrés à la couleur, son goût de l'harmonie tonale et du symbolisme chromatique est évident dans les titres qu'il a donnés à ses peintures, où figurent des termes comme « harmonies », « nocturnes » et « symphonies » de couleurs particulières (ills. 40, 41), au détriment de toute description narrative.

Les chapitres 4 et 5 s'intéressent plus en détail aux conceptions techniques et pratiques de la couleur. « La couleur éclairante : nomenclatures et normes » traite des répertoires de couleurs, souvent en rapport avec le monde naturel et les classifications taxinomiques, qui présentent des tentatives séduisantes de création de nomenclatures et de normes fiables, utilisables dans les arts, la fabrication et les sciences. Certains répertoires visent aussi à faire l'histoire de la couleur, comme l'*Historical Color Guide* (« Guide historique de la couleur », 1938) d'Elizabeth Burris-Meyer (1899–1969), que celle-ci a ensuite adapté pour être utilisé dans la décoration intérieure, la mode et le maquillage (p. 388–409). Le chapitre 5 retrace l'histoire des manuels pratiques destinés aux peintres et aux imprimeurs, du début du XIXᵉ siècle au début du XXᵉ. Les premiers manuels décrivant les couleurs doivent encore beaucoup à l'évolution de la peinture à l'aquarelle et nombre d'entre eux s'adressent aux amateurs car les fournitures artistiques sont de plus en plus abordables au début du XIXᵉ siècle. La peinture étant devenue un passe-temps féminin convenable aux XVIIIᵉ et XIXᵉ siècles, une grande partie de ces manuels est destinée tout particulièrement aux femmes, bien que leurs auteurs soient majoritairement des hommes. Mais à partir du début du XXᵉ siècle, de plus en plus de femmes se consacrent à la rédaction de manuels pédagogiques et s'emparent des possibilités qui s'offrent à elles d'écrire, de publier et d'enseigner sur ce thème. Jusqu'à maintenant, ce domaine de la création féminine a été très peu étudié. Peu pris au sérieux, ces ouvrages pédagogiques, notamment ceux qui s'adressent aux jeunes enfants, ont été mal répertoriés et il n'en subsiste que peu d'exemplaires.

Le début du XXᵉ siècle est marqué par une nouvelle vague de livres consacrés à la couleur, dont beaucoup s'appuient sur les ouvrages des grands théoriciens du siècle précédent et en développent les conceptions. Tel est le sujet du chapitre 6, « La libération de la couleur au début du XXᵉ siècle ». On ne saurait analyser les concepts chromatiques de cette époque sans tenir compte de toutes sortes de réalisations scientifiques et artistiques, qui comprennent notamment les progrès effectués dans la compréhension de la vision en couleurs et, tout particulièrement, l'anatomie de l'œil humain, l'électrification et les possibilités qu'offrent l'éclairage électrique, ainsi que l'emploi radicalement nouveau de la couleur comme moyen d'expression artistique, notamment en peinture. Des collectifs européens d'avant-garde comme Der Blaue Reiter (Le Cavalier bleu) et le fauvisme

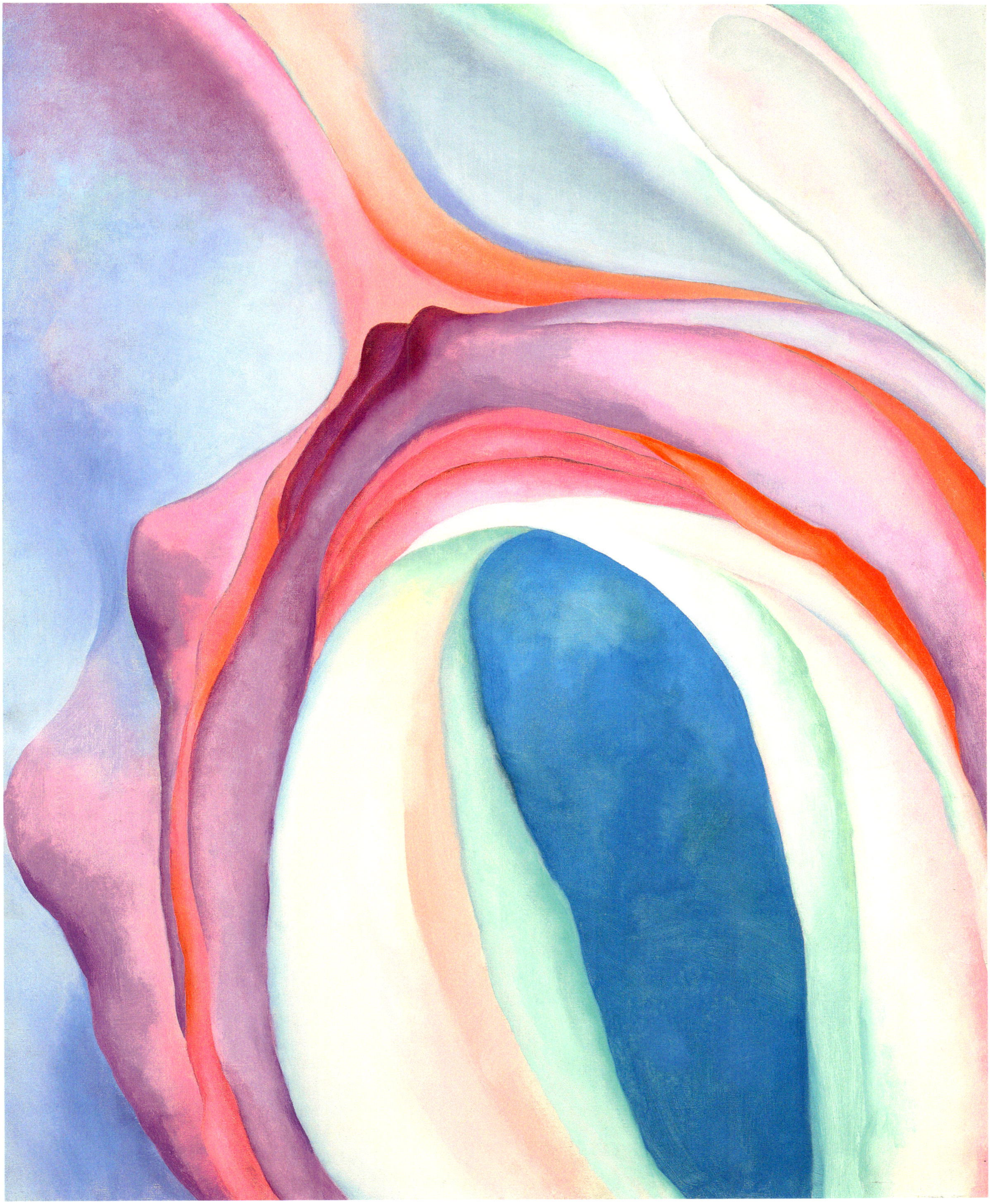

51

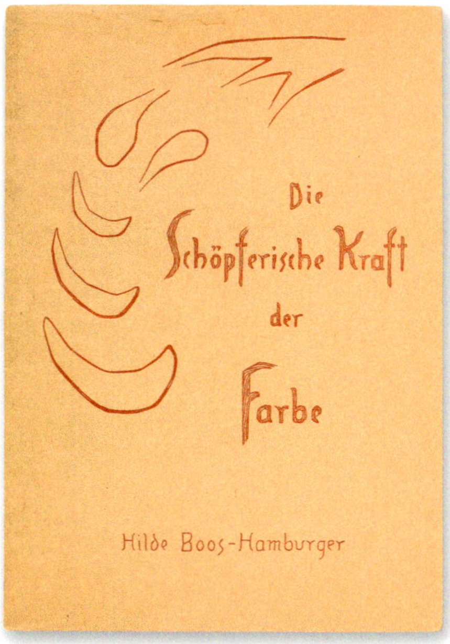

52

ILLS. 52–54
Hilde Boos-Hamburger
Cover and Colour Samples
Buchumschlag und Farbmuster
Couverture et Échantillons de couleurs
Cubierta y muestras de color
From: *Die Schöpferische Kraft der Farbe* (Basel, 1942)
Sussex, collection of the author

s'intéressent activement à la théorie de la couleur, de même que de nombreux expressionnistes et les premiers peintres abstraits. Ils s'approprient le pouvoir symbolique et visuel de la couleur et représentent le monde qui nous entoure dans des tonalités subjectives et dépourvues de réalisme. Certains artistes font même des conceptions chromatiques le thème de leurs œuvres, comme, par exemple, Franz Marc (1880–1916) avec une peinture abstraite intitulée *Kämpfende Formen* (*Combat de formes*, ill. 47) datant de 1914.

Au cours des premières décennies du XXe siècle, très colorées et de plus en plus mécanisées, la couleur matérielle s'achète désormais sous une forme prête à l'emploi et dans de belles boîtes qui contiennent presque toutes les teintes vives qu'on puisse imaginer. Cette période incite fortement à tenter derechef d'établir un ordre chromatique une fois pour toutes. Les productions les plus marquantes dans ce domaine sont évoquées au chapitre 6 comme, par exemple, les œuvres d'Albert Henry Munsell (1858–1918) aux États-Unis (p. 584–599) et de Wilhelm Ostwald (p. 600–609) en Europe. Ces deux hommes sont les auteurs de systèmes très pédagogiques et adaptables, fondés sur une conception tridimensionnelle de la couleur. Certaines de leurs maquettes en volume, employées dans leurs enseignements, sont même parvenues jusqu'à nous (voir, par exemple, le « Corps coloré » d'Ostwald, conservé dans les archives de Winsor & Newton, p. 430). Au début du XXe siècle, les femmes bénéficient de nouvelles possibilités de recherche et de publication donnant lieu à plusieurs études scientifiques remarquables, que celles-ci réalisent sur la vision et la psychologie des couleurs, la couleur en littérature et en poésie, l'histoire des théories des couleurs, ainsi que d'autres aspects historiques. Publiés entre 1900 et 1930, certains de ces ouvrages figurent parmi les plus beaux et les plus inventifs de l'histoire de la couleur, comme, par exemple, *Color Problems* (« Problèmes de couleurs », 1902, p. 562–583) d'Emily Noyes Vanderpoel (1842–1939).

Le chapitre 7, « Le son de la couleur : spiritisme, occultisme et musique », étudie les associations synesthésiques de la couleur et de la musique et, par extension, avec le mouvement, la danse et la spiritualité. Dans une certaine mesure, les conceptions très personnelles et pseudoscientifiques des théosophes, anthroposophes et membres de groupes et mouvements de nature semblable au début du XXe siècle sont suscitées par une réaction à un monde de plus en plus mécanisé. Toutefois, il faut rendre compte avec soin de chacune de ces conceptions chromatiques et ne pas les évaluer à l'aune d'une précision scientifique, mais selon leur inventivité artistique et les circonstances sociales et historiques qui les sous-tendent. Il n'est pas nécessaire de croire en l'existence d'auras colorées, telles que les illustrent et décrivent Annie Besant (1847–1933) et Charles Webster Leadbeater (1854–1934). Pourtant, les tentatives de ces auteurs pour donner forme et couleur aux pensées et aux émotions constituent des points de vue fascinants sur la nature humaine (p. 648–687).

L'histoire des concepts chromatiques est dominée par les analogies entre couleur et musique. La plupart des écrits cités dans ce chapitre étudient les rapports des couleurs avec les sons, la musique, la poésie et la danse. Un grand nombre associe également mysticisme et science, convictions religieuses personnelles et théorie chromatique, nature et spiritisme, textes historiques sur la couleur et pratiques picturales de leur époque. On peut voir en cette énumération un mélange singulier et déconcertant. Ces ouvrages sont souvent difficiles à lire, mais les représentations visuelles de ces conceptions sont attrayantes. Le peintre australien Roy de Maistre (1894–1968) est l'auteur de nombreuses peintures sur ce thème, notamment un rouleau de cinq mètres de long représentant une « traduction en couleurs » du *Trio en si bémol majeur* de Haydn (p. 60 et ills. 57, 67). L'œuvre d'Hilma af Klint, mystique et anthroposophe suédoise, s'enracine dans l'occultisme et le symbolisme chromatique. On assiste depuis ces dernières années à une redécouverte de son œuvre, dont le style

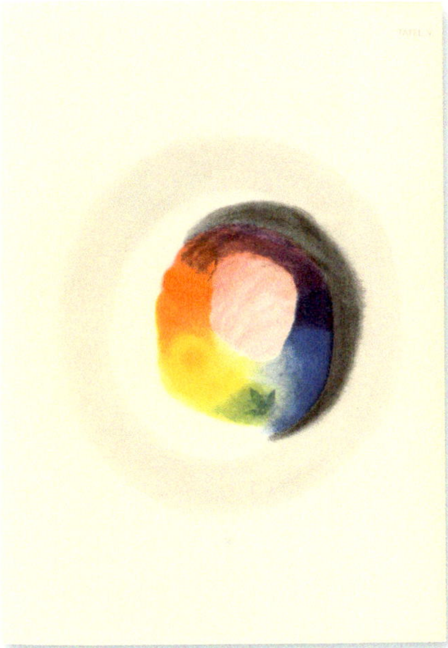

53

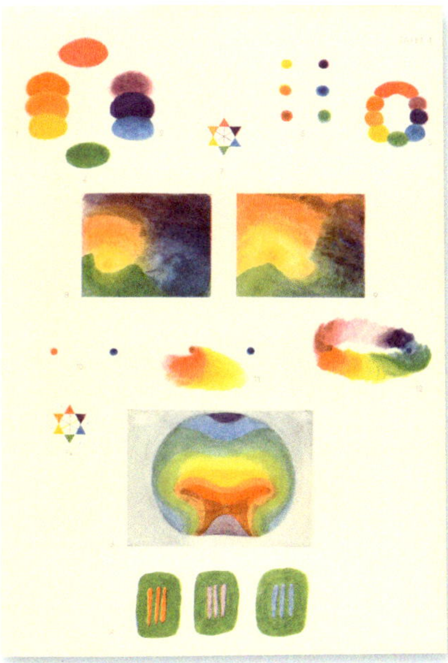

54

visionnaire est salué, à juste titre. Bien qu'elle n'ait jamais rédigé de théorie chromatique de son cru, nombre de ses peintures de grandes dimensions se composent de cercles et pyramides chromatiques et d'images de la couleur immatérielle. En résulte un langage visuel et chromatique remarquablement personnel, où cohabitent réalisme et abstraction (ills. 39, 48, 49).

Le présent ouvrage ne prétend pas exhaustif. Si les sources retenues sont principalement occidentales, chose quasi inévitable, le chapitre 8 vise à rétablir l'équilibre en dépassant les points de vue usuels en Occident et en s'intéressant aux conceptions orientales. Il donne un aperçu des approches japonaises et russes de la couleur en vigueur durant les XIXe et XXe siècles. Loin de nous ériger en spécialistes des conceptions de la couleur de ces régions du monde, nous en reconnaissons néanmoins l'attrait esthétique et souhaitons montrer comment des concepts chromatiques géographiquement éloignés peuvent entrer en dialogue. Comparer quelque chose d'aussi abstrait que la couleur révèle avec beaucoup de force ce qui relie les êtres humains de par le monde. Un exemple captivant des échanges interculturels sur la couleur nous est donné par la pérennité des graphiques chromatiques créés par George Field dans les années 1830, qui réapparaissent dans un manuel scolaire japonais, *Gakko hitsuyo irozu mondo*, paru pour la première fois en 1876 (p. 740–743).

Le dernier chapitre, « Interaction et abstraction : le Bauhaus et après », évoque la couleur sous sa forme la plus pure, la plus réduite et la plus élémentaire dans la peinture, le design, les ouvrages didactiques et l'enseignement de la première moitié du XXe siècle. Le cheminement vers l'abstraction est long et jalonné d'étapes nourries des contributions de divers pionniers, comme Turner, Gartside, Whistler, af Klint et d'autres artistes membres de différentes avant-gardes européennes. Mais c'est le Bauhaus qui, avec ses cercles d'influence, devient l'un des phares du modernisme, du progressisme et, d'un point de vue stylistique, de l'abstraction. Fondée en 1919 à Weimar par l'architecte Walter Gropius (1883–1969), cette éphémère école d'art et de design trouve ses origines dans l'avant-garde allemande et les cercles spiritistes, ainsi que dans le mouvement anglais Arts and Crafts. Elle met l'accent sur la simplicité et la fonctionnalité et s'attache à offrir aux artistes une formation complète en soulignant la qualité de la facture. Parmi les membres, enseignants et étudiants du Bauhaus, on compte notamment Ludwig Mies van der Rohe, Paul Klee, Johannes Itten et Josef Albers, ainsi que plusieurs femmes, comme Anni Albers (ill. 64), Marianne Brandt, Gertrud Arndt (ill. 59), Gertrud Grunow (ills. 61, 62) et Gunta Stölzl (ills. 65, 66), dont le rôle que celles-ci ont joué dans l'histoire de l'école n'est véritablement reconnu que depuis peu. En 1925, le Bauhaus est transféré à Dessau, où Gropius a conçu le fameux bâtiment qui l'accueille (voir ill. 63). En 1933, au terme d'une année à Berlin, où l'école a de nouveau déménagé, celle-ci est contrainte à la fermeture sous la pression croissante des nazis. Certains de ses membres émigrent en Angleterre ou aux États-Unis, où ils parviennent à maintenir leurs relations et à poursuivre leur travail. Si la couleur n'est qu'un élément parmi d'autres des cours dispensés au Bauhaus, elle suscite de nombreuses publications, dont deux ouvrages qui font référence sur le sujet, parus près de trente ans après la fermeture de l'école : *Interaction of Color* (*L'Interaction des couleurs*, 1963, p. 824–837) de Josef Albers (1888–1976) et *Kunst der Farbe* (*L'Art de la couleur*, 1961, p. 804–823) de Johannes Itten (1888–1967). Ces œuvres, ainsi que des publications moins célèbres dues à des écrivains et artistes associés au Bauhaus, témoignent d'une synthèse quasi parfaite de la pensée sur la couleur, prenant en compte la vie et le travail parmi les couleurs, l'enseignement de la couleur et son application à la peinture et au design, sans négliger la réflexion portant sur les idéaux modernistes de simplicité, de géométrie et d'abstraction. En peinture, les conceptions chromatiques et l'esthétique du Bauhaus se retrouvent dans l'évolution plus large vers l'abstraction qui a lieu au XXe siècle. Les œuvres

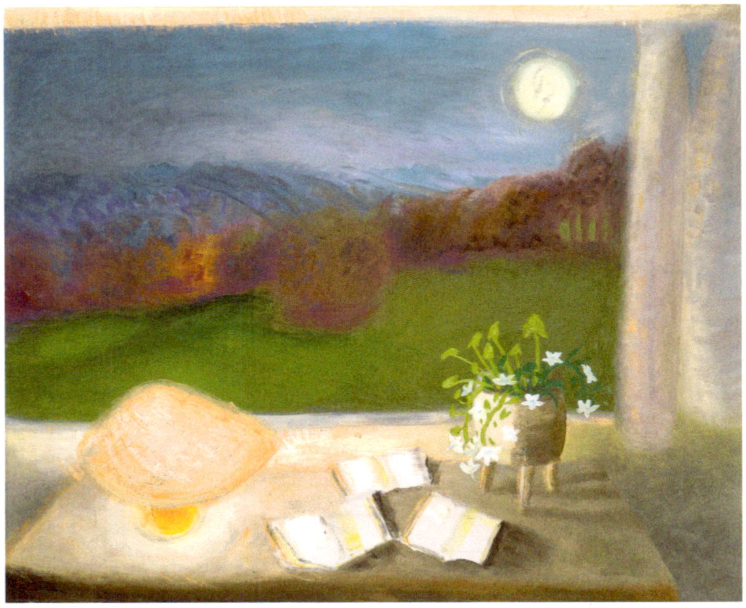

55

Ill. 55
Winifred Nicholson
The Hunter's Moon, 1955
Der Jägermond
La Lune du chasseur
La luna del cazador
Oil on canvas, 111.8 x 137.2 cm
(44 x 54 in.)
London, Tate Modern

Ill. 56
Winifred Nicholson
Prismatic No. 5, 1979
Prismatisch Nr. 5
Peinture prismatique n° 5
Pintura prismática n.° 5
Oil on canvas, 61 x 61 cm
(24 x 24 in.)
Private collection

extrêmement minimalistes et géométriques de maîtres de la peinture moderne comme Piet Mondrian (ill. 68), Henri Matisse et Kazimir Malévitch présentent des similitudes manifestes dans leur usage de la couleur.

Intemporalité des conceptions chromatiques

Par leur beauté abstraite et la manière visuelle dont les conceptions chromatiques sont rendues, les représentations de la couleur exercent une fascination intemporelle. Les formes simples et géométriques créées pour façonner l'aspect des premiers graphiques et tableaux chromatiques nous parlent immédiatement, car des formes identiques ou similaires peuplent la peinture, la publicité et le design d'aujourd'hui. De même, les cercles, disques, triangles et cubes, ainsi que les graphiques ou formes organiques encore plus complexes, entretiennent un dialogue à travers les siècles. Les théories et conceptions des couleurs demeurent un thème et une source d'inspiration chez les artistes contemporains. Depuis 2009, l'artiste islando-danois Olafur Eliasson (né en 1967) a réalisé une série de grands disques chromatiques, dans lesquels les couleurs se fondent imperceptiblement les unes dans les autres (ill. 70), à la manière des disques de Chevreul dans son *Atlas* de 1861. Une sélection des disques d'Eliasson a été exposée à la Tate Britain, à Londres, sous le titre *Turner colour experiments* (« Expériences chromatiques de Turner »), en regard de peintures de Turner. De même que, dans certaines de ses œuvres, Turner fait allusion à Goethe et à Harris, Eliasson évoque les théoriciens de la couleur et les peintres qui l'ont précédé, mais cherche également à proposer une théorie des couleurs dans l'art de son cru.

De tout temps, l'être humain a tenté d'apprivoiser et de maîtriser la couleur, cet aspect insaisissable mais omniprésent de notre vision, même si nos efforts pour circonscrire la couleur à des cercles, des carrés et des traits, pour la faire entrer de force dans des listes ou lui faire prendre des formes en trois dimensions, ne sont en fin de compte que des expressions de la pensée et de la créativité humaines. Peut-être reste-t-il à formuler la conception chromatique parfaite. Néanmoins, la transformation d'idées sur la couleur en mots, images, diagrammes et volumes a produit une riche histoire de la couleur, véritable art visuel à part entière. Les savoirs individuels, les choix personnels, les découvertes et le renouvellement de la pensée font que chaque nouvel ouvrage sur le sujet est une réalisation stimulante. Nous espérons que notre sélection de conceptions et théories de la couleur offrira une variété inattendue d'opinions et de récits, ainsi que des sources d'inspiration, et qu'elle éclairera l'histoire de la couleur d'une lumière nouvelle.

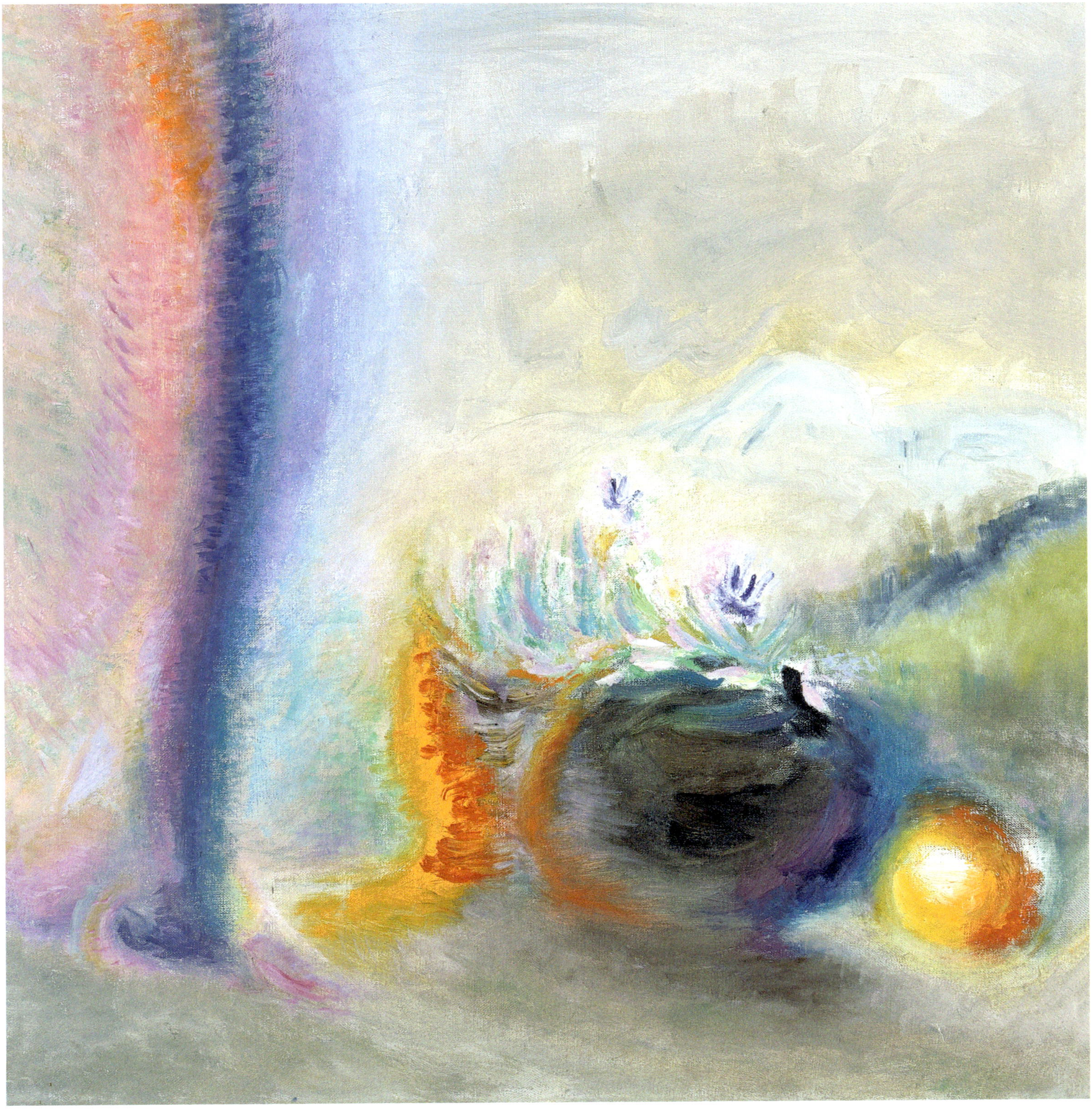

56

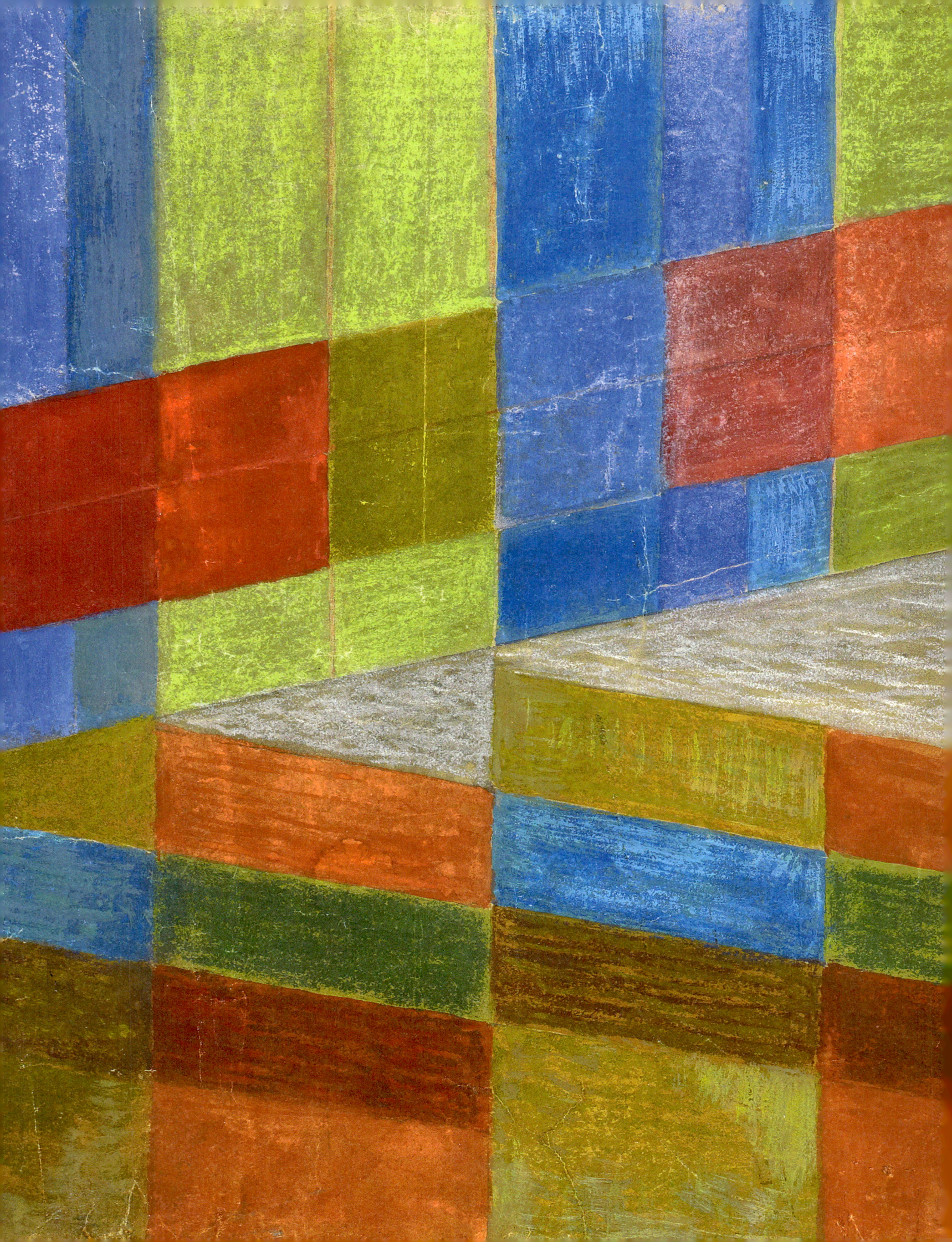

Alexandra Loske

COLOR LIBRE Y CONTROLADO

La historia humana de la captura del color en palabras e imágenes

Nuestro mundo está lleno de color. Lo experimentamos en color. Conocemos la Tierra como el «planeta azul», y ópticamente desaparecemos en el azul: cuando la sonda espacial Voyager 1 nos fotografió en 1990 desde 6000 millones de kilómetros de distancia, aparecimos, en palabras del astrónomo Carl Sagan unos años después, como «un pálido punto azul» apenas visible de un tamaño inferior a un píxel, pero aun así de un color identificable. Nos convertimos en una mancha de color hasta desaparecer de la vista. El color es, por lo tanto, uno de los aspectos definitorios y unificadores de nuestras vidas, una verdadera experiencia humana universal y compartida, y no es de extrañar que nos haya fascinado desde el principio de los tiempos. Las primeras manifestaciones de la creatividad humana, en forma de tallas e incisiones en madera, hueso y roca, muestran que siempre hemos intentado hacer y usar colores, principalmente para representar o copiar lo que nos importaba y nos rodeaba. Aunque solo tenemos pruebas visuales y físicas débiles y fragmentarias de esa actividad humana más antigua, es posible que incluso en esa época inicial el color también se usara de manera simbólica. Los pigmentos minerales extraídos de la tierra por los primeros humanos se mezclaban con materias orgánicas como grasa, sebo o fluidos corporales, y muy bien pudieron adquirir un significado especial o sagrado en el proceso, lo que dio paso a una primera jerarquía de colores. Nunca sabremos con exactitud qué pensaban nuestros ancestros prehistóricos sobre el color, pero hay pruebas de que ciertos colores se usaban en contextos particulares.

Podemos ver señales de un compromiso intelectual activo con el color a lo largo de la historia humana, desde los conceptos simples utilizados en el arte primitivo hasta los sistemas altamente desarrollados arraigados en la ciencia, la filosofía o las artes modernas, ámbitos que a menudo se superponen y establecen referencias cruzadas entre sí. El objetivo de este libro es presentar una visión general de los pensamientos y sentimientos de los humanos sobre el color, sobre su deseo de crear, organizar y sistematizar el color, de usarlo y estructurarlo, y de asignarle un significado complejo. Ha habido numerosos intentos de clasificarlo y comprenderlo que han dado pie a una gran variedad de manifestaciones, desde conceptos dualistas de oscuridad y luz hasta patrones y sistemas geométricos elaborados que buscaron dar una forma específica al color. Con el desarrollo del lenguaje, la escritura y, finalmente, la cultura de la impresión y los sofisticados métodos de reproducción de imágenes, estos conceptos se volvieron cada vez más refinados, complejos y visualmente llamativos, y quedaron mejor registrados.

Este libro comienza su viaje visual a través de la historia de los conceptos de color en el siglo XVII, la época de los primeros manuscritos ilustrados y fuentes impresas que han llegado hasta nuestros días y que se centran específicamente en el color, y concluye a mediados del siglo XX, antes de la llegada de la era digital. Al igual que cada concepto de color aquí incluido refleja la época en que apareció, también lo hace este libro por los conceptos que selecciona. Durante las últimas dos décadas ha habido un interés agudo y creciente en el color y la historia del color y, como resultado de nuevas investigaciones, estamos descubriendo nuevos aspectos y voces y ejemplos de cultura material que se habían pasado por alto durante mucho tiempo (escritores, educadores, libros, manuscritos, obras de arte y métodos), que se habían olvidado o que nunca se habían considerado parte del canon del color.

Ningún libro puede proporcionar un registro enciclopédico de las dimensiones cultural, histórica, geográfica y temática del color. Algunos títulos hacen omisiones deliberadas para enfocarse en temas culturales o históricos particulares; otros seleccionan la información de tal manera que refleja los intereses personales, gustos o conocimientos del autor o de los editores. Este libro es, metafóricamente hablando, una túnica de muchos colores, y pretende ser tan subjetivo en su selección como se considera que es la percepción del color. Las 67 obras que aquí se muestran y analizan son representativas de la rica y variada historia de los conceptos de color desde el siglo XVII, y aunque hemos

Page 60
Roy de Maistre
Colour Music
Detail of ill. 57

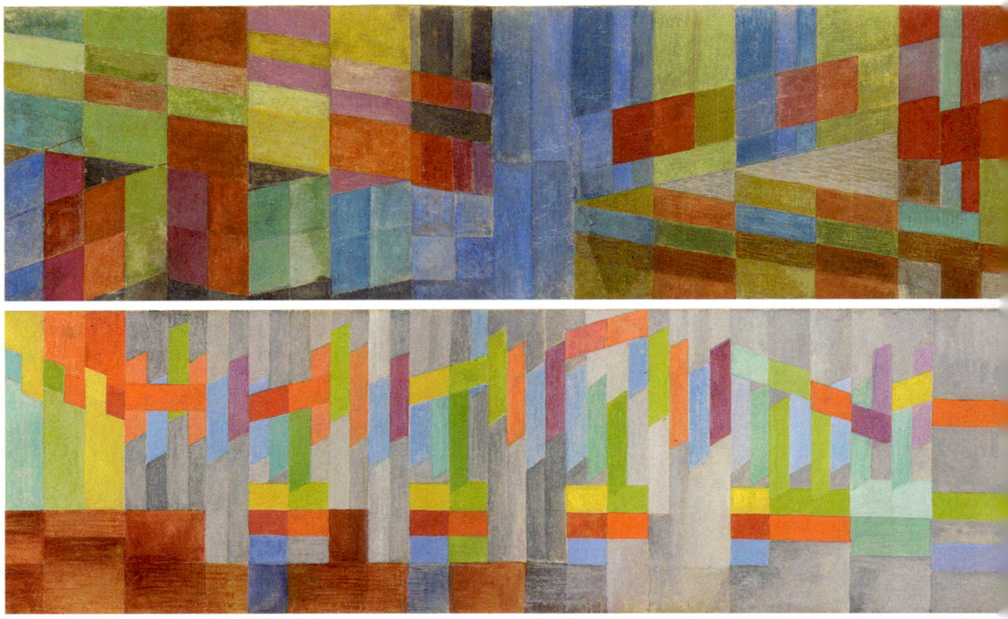

57

Ill. 57
Roy de Maistre
Colour Music, *c.* 1934
Farbenmusik
Musique chromatique
Música del color
Gouache on piano roll,
31 x 1,606 cm (12 ¼ x 632 ¼ in.)
New South Wales Art Gallery,
Gift of Sir John Rothenstein in memory
of the artist, 1969

seleccionado muchos ejemplos familiares que están firmemente establecidos en la historia del color, esperamos haber logrado ponerlos en un contexto cultural más amplio y haberlos contemplado con nuevos ojos. Además de estos ejemplos «clásicos» en su campo, hemos hecho un esfuerzo particular para presentar algunos trabajos menos conocidos o más oscuros que merecen consideración por su atractivo e importancia estética y teórica. Otro objetivo es dar más espacio a las obras sobre el color escritas, publicadas e ilustradas por mujeres en los dos últimos siglos. El alcance histórico y geográfico que aquí se ofrece es amplio, con ejemplos de Europa, Estados Unidos y partes de Asia, pero también hemos buscado ser lo más inclusivos posible en cuanto a género y estilo. Una sencilla tabla de colores educativa del siglo XX puede ser tan fascinante como la rueda de colores de Isaac Newton (1643–1727) de 1704 (pág. 211), y ambas desempeñan un papel importante en la historia del color.

¿Por qué «conceptos de color»?

Expansivo y adaptable, «concepto de color» es un término muy adecuado para pensar en el color y tratar de comprenderlo y sistematizarlo. Sin embargo, en lugar de buscar el término perfecto, tal vez hay que preguntarse por qué se siente la necesidad de sistematizar algo tan intangible, cambiante, esquivo y, en cierto modo, subjetivo. Hasta hace relativamente poco, los textos sobre el color no distinguían de forma adecuada entre el color material y el inmaterial, pero una vez que se entendió por completo la diferencia, quedó claro que, de hecho, hay muchas formas de color que combinan el color inmaterial y los colores de los materiales sólidos. El color es escurridizo: cambia con la luz, se ve diferente en función del momento del día, se desvanece, oscurece y puede ser impredecible cuando se combina, mezcla o yuxtapone. Además, puede ser difícil de obtener, como saben muy bien los fabricantes de pigmentos y los historiadores; para los primeros humanos, tuvo que ser frustrante vivir en un mundo donde los dos colores dominantes, el azul del cielo y el verde del mundo vegetal, eran dos de los más difíciles de obtener. Vivimos en un mundo donde abunda la clorofila, pero pocas plantas producen buenos pigmentos verdes. El color también es exasperantemente voluble, inestable y, a menudo, tóxico, y no siempre se apreció, ya que, como el artista y escritor David Batchelor ha explicado tan bellamente en su ensayo *Chromophobia*, el odio al color es una parte importante de la historia cultural occidental, arraigado en «un miedo a la contaminación y corrupción por algo que es desconocido o parece incognoscible [que se manifiesta] en los muchos y variados intentos de purgar el color de la cultura, devaluarlo, disminuir su significado y negar su complejidad». Las actitudes occidentales hacia el color o, de manera más específica, hacia los colores altamente saturados y la policromía, argumenta Batchelor, solían estar asociadas con ideas sobre lo extranjero, oriental, primitivo, femenino, vulgar, superficial o incluso patológico, en contraste con el ideal clásico de la belleza virginal blanca o incolora. Hay una larga historia en la cultura occidental del color y de la oscuridad que lo vincula con una «otredad» peligrosa o amenazante, una fuerza desestabilizadora a evitar o domesticar. La creación de sistemas o conceptos de color es una forma de lograr lo último, intelectualmente al principio, y con un uso y una aplicación comedidos a continuación. Si imponemos un orden en el color, aunque solo sea de forma teórica, por lo menos hemos intentado tomar el control de algo que es mutable y omnipresente.

Por supuesto, el miedo, la duda y el odio al color solo son una cara de la moneda. Nuestra relación con él siempre ha sido polifacética, y los últimos 50 o 60 años en particular, como consecuencia de la invención de la fotografía y la película en color, el color digital y el paso a una cultura de la impresión casi totalmente en color, han hecho que nos sintamos más cómodos con él y nos sea mucho más accesible. Sin apenas excepciones, ese acceso es hoy fácil y no exclusivo, con pocos límites para la reproducción del color, aunque su calidad todavía varíe.

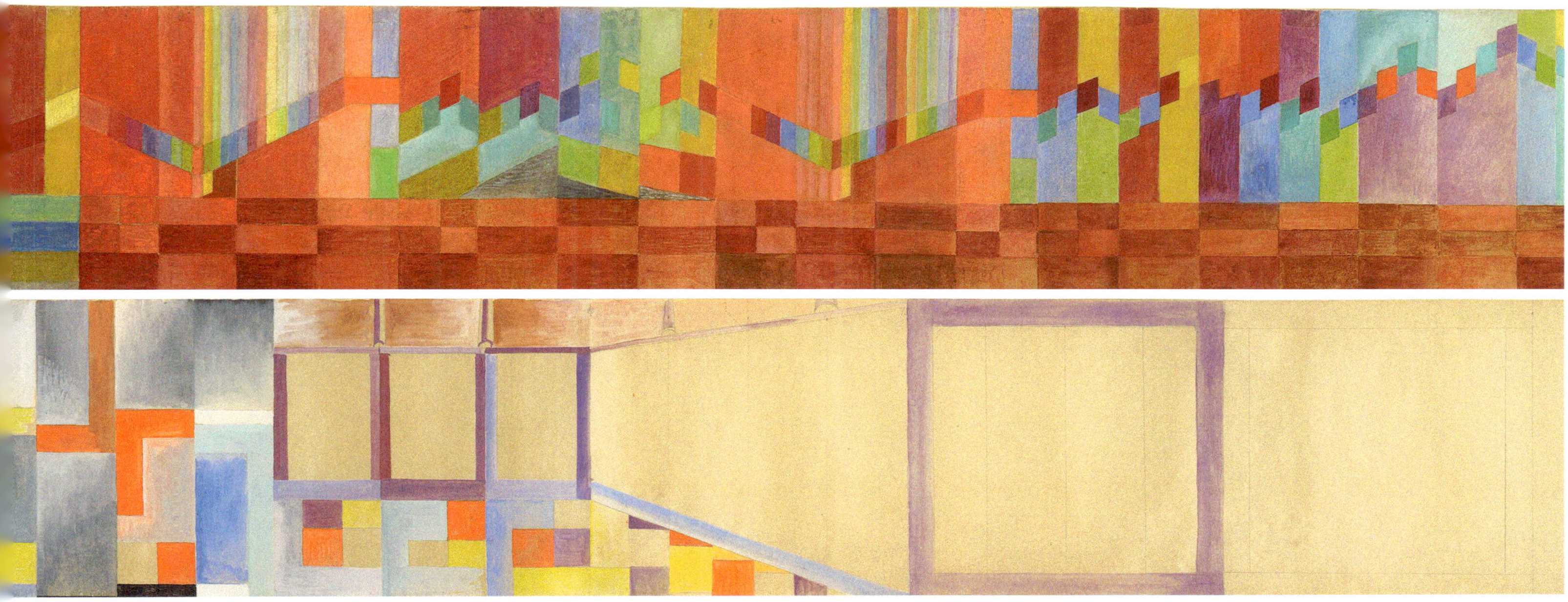

El orden del color: de las cajas de pinturas a las paletas

Allí donde no hay fuentes escritas para los conceptos de color, la arqueología y la cultura material pueden darnos una idea de la cultura del color. Los colores utilizados en el arte y la decoración nos indican, como mínimo, qué pigmentos estaban disponibles en ciertos momentos de la historia, además de revelarnos detalles sobre técnicas y materiales pictóricos, preferencias de color y, cuando se ven en su contexto cultural más amplio, quizás su significado simbólico. Una de las paletas de artistas más antiguas que se conservan, una caja de pinturas de la dinastía XVIII del antiguo Egipto que data de c. 1427–1401 a. C. (il. 2), nos permite saber de una forma vívida y táctil qué pigmentos tenía a su disposición su dueño, Amenemope, visir durante el reinado de Amenhotep II, pero quizás también qué colores prefería y cómo le gustaba ordenar las pastillas de ocre rojo, azul egipcio, una mezcla de verde y dos carboncillos distintos en su caja de pinturas. ¿Por qué eligió tener dos tipos de negro? ¿Qué significaba el negro para él, o en la cultura del antiguo Egipto en general? La disposición de los materiales de los artistas puede parecer muy alejada de un modelo de color tridimensional complejo del siglo XX, pero el deseo humano subyacente es el mismo: dar un orden al color, ya sea por razones prácticas, preferencia personal o incluso fines comerciales. La paleta de cada pintor, tanto en el caso del objeto real con restos de pintura (il. 44) como en la empleada en los autorretratos, revela una parte de la vida y época del artista, de sus prácticas y preferencias. Incluso las modernas cajas de acuarelas industriales brindan información sobre sus usuarios y propietarios: de qué tamaño y calidad eran las cajas que podían permitirse, qué colores elegían, cuáles usaban más, cuáles no usaban casi nunca y cómo los disponían. Al igual que con el ejemplo del antiguo Egipto, una caja de acuarelas de Reeves (il. 36) de principios del siglo XX también puede revelar mucho sobre qué colores estaban disponibles o eran populares en ese momento. La caja de pinturas de Reeves que aquí se muestra incluye varios colores que llevan el nombre de Wilhelm Ostwald (1853–1932), un importante químico alemán y teórico del color.

La representación del color antes de la era moderna

El asunto principal de este libro es la visualización del orden de los colores y las ideas ligadas a los sistemas desarrollados con tal fin. Si bien los conceptos se remontan al mundo clásico, las primeras ilustraciones de diagramas cromáticos aparecieron a finales del siglo XVII. Hasta hace relativamente poco, la cultura clásica parecía, en términos materiales, en gran parte incolora: las antiguas Grecia y Roma se asociaban con la blancura debido a las imágenes de esculturas de mármol y edificios predominantemente blancos que perdieron hace mucho los colores que los cubrían. Sin embargo, los estudios de los últimos años sobre la policromía de la escultura y la arquitectura griegas y romanas han modificado esa impresión de forma sustancial.

En lo conceptual y lingüístico, se pueden distinguir varias tradiciones en la historia de los conceptos de color. Se considera que Empédocles (494–434 a. C.) fue uno de los primeros autores en escribir sobre el tema. Los fragmentos que se conservan de su poesía permiten discernir un sistema cosmológico basado en los cuatro elementos (tierra, fuego, agua y aire), de los que creía que derivaban los colores. En el siglo siguiente, Aristóteles (384–322 a. C.) pudo haber realizado experimentos ópticos sencillos con vidrios de colores y estudiado la visión coloreada y el contraste de colores. Propuso un sistema polar que iba desde la luz blanca divina, que hacía visibles los colores, hasta la oscuridad, representada por el negro. Del blanco evolucionaron los colores jerárquicos: amarillo, verde, azul y rojo, asociados a la tierra, al agua, al aire y al fuego, respectivamente. Un seguidor de Aristóteles, Teofrasto (372–287 a. C.), consideró cómo percibimos o sentimos los colores, y lo hizo en referencia a la visión de Demócrito (460–370 a. C.) de los cuatro colores básicos (rojo, negro, blanco y un

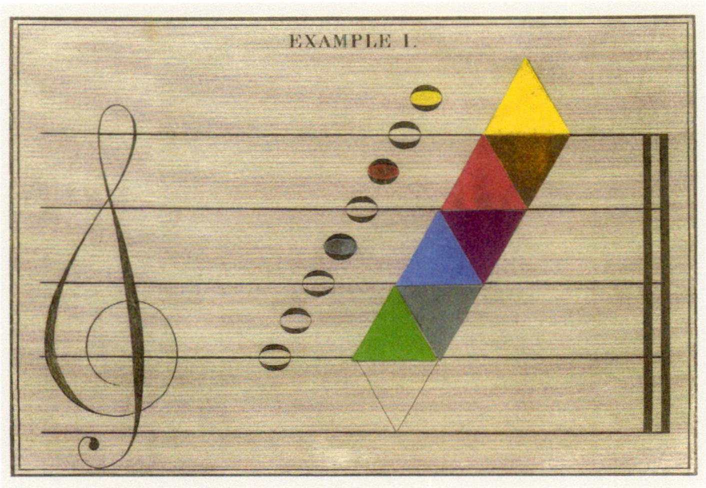

58

ILL. 58
David Ramsay Hay
A Comparison of Colour and Sound
Ein Vergleich von Farbe und Ton
Comparaison de la couleur et du son
Una comparación de color y sonido
From: *The Laws of Harmonious Colouring, Adapted to Interior Decorations, Manufacturers, and Other Useful Purposes* (London, 1838)
Los Angeles, Getty Research Institute

ILL. 59
Gertrud Arndt
Four Study Sheets from the Teaching of Paul Klee, 1923/24
Vier Studienblätter aus dem Unterricht von Paul Klee
Quatre feuilles d'études des cours de Paul Klee
Cuatro bocetos de los apuntes de Paul Klee
Collage, watercolour and ink over pencil on cardboard; 33 x 22 cm (13 x 8⅝ in.)
Berlin, Bauhaus-Archiv / Museum für Gestaltung

amarillo verdoso), que relacionó con ciertas formas, patrones superficiales y otros atributos basados en los sentidos. Por su parte, el médico Hipócrates (460–375 a. C.) distinguió cuatro colores que representaban los principales fluidos o humores del cuerpo humano, que a su vez eran vistos como causantes de ciertos rasgos de carácter, a saber: el blanco en los flemáticos, el rojo en los sanguíneos, el amarillo en los coléricos y el negro en los melancólicos.

Por lo general, solo nos han llegado fragmentos de los conceptos de color y de las discusiones teóricas sobre el tema de época clásica, con el inconveniente añadido de los desafíos lingüísticos que presentan. El lenguaje suele ser una herramienta imperfecta para transmitir la esencia de un color, ya que las palabras no pueden igualar su complejidad o percepción. La mayoría de los idiomas solo cuentan con una docena de términos para los colores básicos, mientras que el número de colores que podemos ver en realidad es incontable. Esta discrepancia se vuelve aún más problemática cuando se trata de lenguas extintas, textos antiguos o traducciones. También hay una carencia de representaciones visuales de los conceptos de color del mundo antiguo y de las primeras culturas, lo que dificulta la comprensión integral de las primeras ideas sobre este asunto. Sin embargo, se puede vislumbrar el potente valor simbólico de los colores a través de ciertos hallazgos arqueológicos, como las máscaras teatrales encontradas en Pompeya, tanto en forma de objetos reales como en pinturas al fresco, pintadas con colores que expresan los cuatro humores (il. 7). A pesar de la escasez de trabajos centrados en el color, podemos rastrear el desarrollo de varios conceptos culturales relacionados con él mediante la materialidad de los pigmentos y colorantes. Al igual que sucede con el costoso pigmento azul de ultramar en el arte medieval y renacentista (il. 3), el valor material y la rareza de ciertos colores en el mundo antiguo hizo que se los asociara con deidades y figuras de alto estatus.

A lo largo de la Edad Media, el marco aristotélico básico que sitúa el color entre la oscuridad y la luz (el negro y el blanco) prevaleció en el discurso filosófico, pero también hubo un aumento en el número de textos sobre el color de carácter más práctico. A principios del siglo XIII, el estudioso de la Universidad de Oxford y obispo de Lincoln Robert Grosseteste (c. 1175–1253) tradujo algunas obras de Aristóteles y escribió en respuesta su propio tratado sobre el color, *De Colore*, donde desarrolló la teoría de Aristóteles de que el color deriva de la luz, al tiempo que especuló sobre la cantidad de luz que cada color podía contener. Hacia el siglo XV, el aumento de los intercambios comerciales también supuso un aumento gradual de la comercialización de los colores. El conocimiento se recogía ahora de una forma más sistemática, como se puede comprobar con la aparición de los primeros tratados sobre las técnicas artísticas, los materiales empleados por los pintores y las vidas de los artistas. Muchos de esos textos son muy técnicos y sin duda no llegaron a un gran número de lectores, en especial antes de la aparición de la imprenta, pero aun así han sobrevivido varios registros escritos que abren una ventana al mundo medieval del color artesanal, como el algo anterior *De diversis artibus* de Teófilo Presbítero (*fl.* c. 1070–1125), que es esencialmente un libro de recetas de colores que data de principios del siglo XII. En los siglos XV y XVI aparecieron muchos textos sobre artesanía centrados en la técnica y los materiales que no participaban en discusiones filosóficas sobre los conceptos de color, pero que tienen un valor incalculable para el estudio de la terminología del color y la historia de los pigmentos. De vez en cuando, encontramos un texto ilustrado que incluye imágenes de las materias primas empleadas para hacer pigmentos y pinturas, por ejemplo, el *Livre des simples médecines*, del siglo XV o XVI (ils. 19, 20), que muestra varias sustancias orgánicas y minerales, como tinta de calamar, lapislázuli y una momia (para el marrón momia), de los que se creía que también tenían propiedades terapéuticas. Las ilustraciones son sencillas y descriptivas, pero los objetos están cuidadosamente dispuestos al estilo de una bandeja expositora de coleccionista o un gabinete de curiosidades.

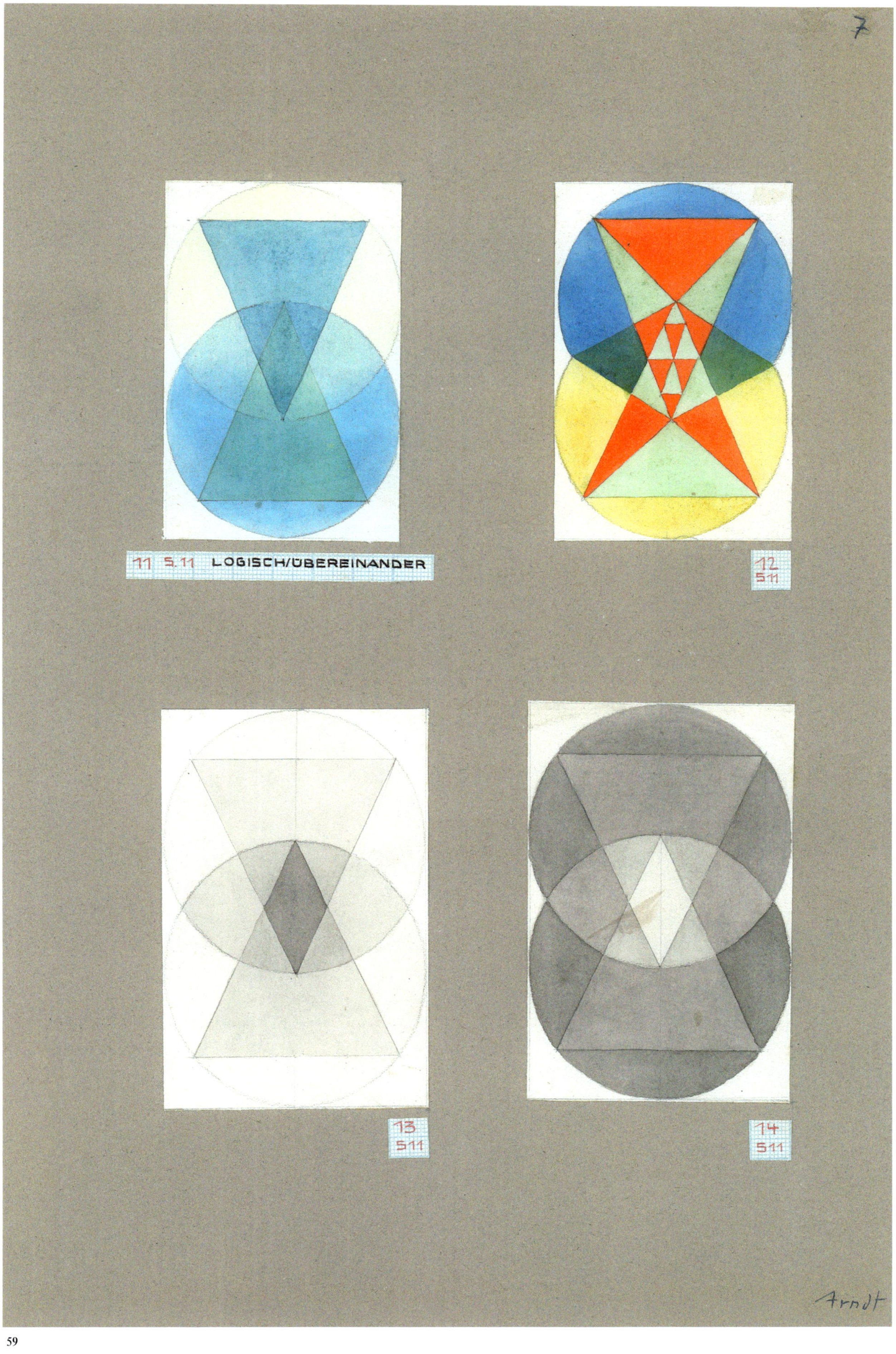

59

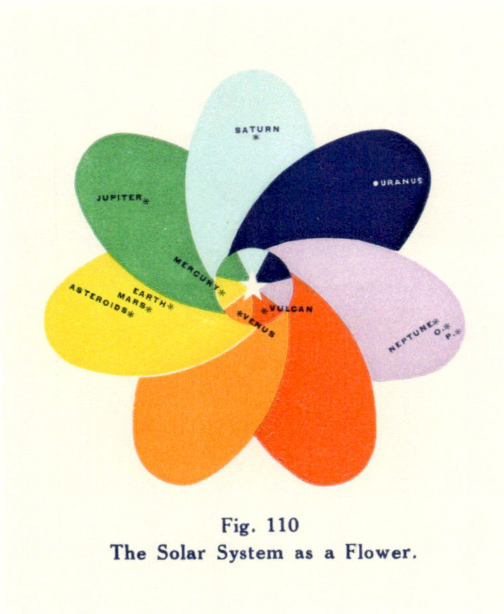

60

Ill. 60
C. Jinarajadasa
The Solar System as a Flower
Das Sonnensystem als Blume
Le système solaire vu comme une fleur
El sistema solar como una flor
From: *First Principles of Theosophy* (Chennai, 1922)
Sussex, collection of the author

Ill. 61
Alfred Arndt
Colour Circle from the Lessons of Gertrud Grunow, *c.* 1921
Farbkreis aus dem Unterricht von Gertrud Grunow
Cercle chromatique tiré des cours de Gertrud Grunow
Círculo de colores de las lecciones de Gertrud Grunow
Watercolour, ink, pencil and silver paper collage; 46.7 x 46 cm (18⅜ x 18⅛ in.)
Berlin, Bauhaus-Archiv / Museum für Gestaltung

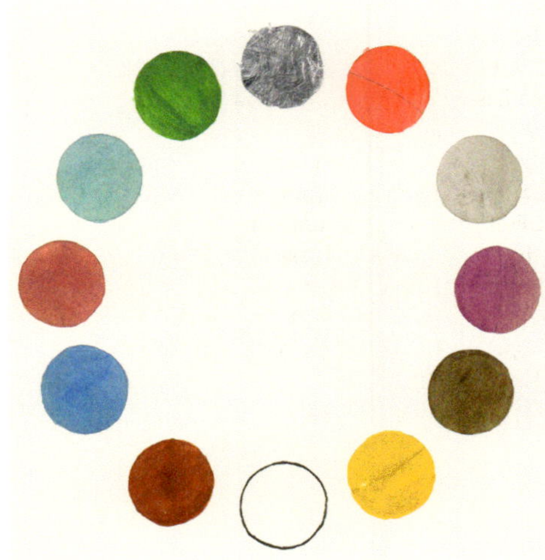

61

La idea de Hipócrates de los cuatro humores que impregnan el cuerpo y el espíritu humanos, que podrían representarse mediante cuatro colores, siguió dando forma a los conceptos básicos de color durante la Edad Media y el Renacimiento. Las imágenes y las obras de arte que lidian específicamente con la teoría y los conceptos de color aún se desconocen en este período, pero en otros contextos aparecen los primeros ejemplos del uso del color para ilustrar y facilitar el orden y la clasificación. Por ejemplo, la rueda uroscópica de Ulrich Pinder (m. 1519) de 1506 (il. 5) representa 20 tonos diferentes de orina, que servían para diagnosticar enfermedades. Los frascos de orina se colorearon a mano, y este es en sí mismo un ejemplo de primera época de una rueda de colores basada en los colores de la naturaleza, no muy diferente a los incluidos en los libros de Moses Harris (1730–c. 1788) sobre insectos casi tres siglos después. De manera similar, una gran cantidad de textos enciclopédicos, como bestiarios, herbarios, lapidarios y concordancias, pueden proporcionar más información, directa o indirectamente, sobre el conocimiento y el uso del color en el período correspondiente.

La publicación en 1435 de *De pictura* (*De la pintura*) de Leon Battista Alberti (1404–1472) coincidió con el auge de la pintura al óleo como medio. Sobre el tema del color, que consideraba un elemento importante y valioso de la pintura, proponía que todos los colores podían crearse mezclando solo cuatro: rojo, verde, azul y gris. Alberti era un autor para pintores, menos preocupado por la luz coloreada que por el color físico, por lo que el gris, al que otorgó un estatus tan prominente, debe verse en el contexto del sombreado y de la creación de efectos tridimensionales en la pintura. Muchos textos sobre el asunto más amplio de los roles del dibujo (*disegno*) y del color (*colore*) son una continuación de las ideas de Alberti y sus contemporáneos. Uno de los artistas más grandes, Leonardo da Vinci (1452–1519), también nos dejó sus ideas sobre el color. En sus diversos cuadernos y recopilaciones de escritos reflexionó sobre la visión y percepción del color, la apariencia del color y la aplicación práctica de la teoría del color en la pintura. Propuso un concepto de color sencillo, muy influenciado por Aristóteles y Alberti, basado en seis «colores simples»: rojo, amarillo, verde y azul, flanqueados por el blanco y el negro. Ninguno de estos textos medievales y del protorrenacimiento incluye diagramas cromáticos, por lo que no es posible saber si los sistemas propuestos se concibieron como lineales o circulares, bidimensionales o tridimensionales; sin embargo, son precursores fascinantes de la teoría moderna del color.

Una nueva era cromática: el color del siglo XVII en adelante

Este libro comienza su viaje por los conceptos de color a finales del siglo XVII, cuando la cultura de la impresión estaba bien establecida, aunque por varias razones muchos de los primeros textos sobre el color aparecieron en forma de manuscrito. El espíritu naciente de la Ilustración ayudó a promover el intercambio de ideas a través de las fronteras geográficas y culturales, pero mientras que el simbolismo del color y los metasistemas vinculados a las creencias religiosas, la alquimia y el folclore continuaron existiendo, el color se volvió científico e intelectualizado. El deseo general existente de comprender y clasificar el mundo natural incluía el color como fenómeno óptico, herramienta para representar la naturaleza y medio expresivo en las bellas artes y las artes decorativas.

En un orden cronológico aproximado (pero no estricto), cada capítulo de este libro se centra en un tema o aspecto específico de la historia de los conceptos de color. Para ser incluidos, los libros y manuscritos presentados debían estar relacionados predominantemente con el color en la teoría o la práctica, o utilizar el color de manera coherente e inventiva, por ejemplo, como herramienta para ilustrar conceptos e ideas relacionados.

Ill. 62
Ilse Bettenheim-Hoernecke
Colour Circle and Colour Forms from the Lessons of Gertrud Grunow, 1924
Farbkreis mit Farbformen aus dem Unterricht von Gertrud Grunow
Cercle chromatique et formes colorées tirés des cours de Gertrud Grunow
Círculo de colores con formas de colores de las lecciones de Gertrud Grunow
Watercolour and pen over graphite preparatory drawing; 29.9 x 41.2 cm (11 x 16¼ in.)
Klassik Stiftung Weimar

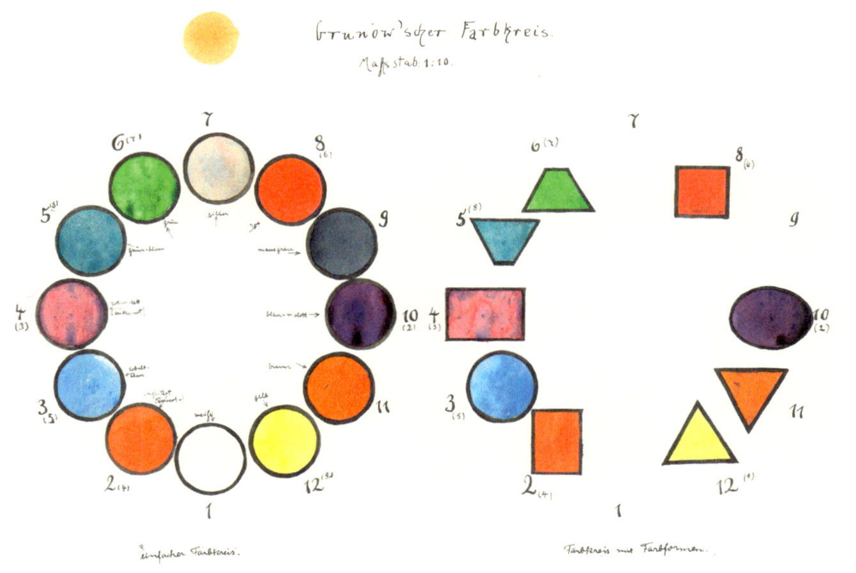

62

Este libro incorpora muchas obras de mujeres con el fin de abordar el desequilibrio histórico de género en el campo de las publicaciones sobre el color, incluidos los libros educativos y la pintura de principios del siglo XX, como, por ejemplo, el libro de la educadora estadounidense Bonnie Snow (c. 1862–1925, ils. 11, 12, 18) *The Theory and Practice of Color* (1918, págs. 552–559). Uno de los primeros ejemplos de una fuente tan poco considerada es Mary Gartside (c. 1755–1819, págs. 218–229), una pintora de flores británica (il. 21) y profesora de pintura con acuarela, y la primera mujer de la que se sabe que publicó un libro sobre el color en inglés: *An Essay on Light and Shade, on Colours, and on Composition in General* (1805). Con motivo de la segunda edición (1808), cambió el título por *An Essay on a New Theory of Colours*, que subrayaba la seriedad de su trabajo. Otras obras incluyen las innovadoras publicaciones de mujeres de los círculos más amplios de teósofos y espiritistas, como *The New Science of Color* (1915) de la poeta y actriz multinacional y multicultural Beatrice Irwin (1877–1956), quien tenía un gran interés en el potencial de la combinación de la música con la luz coloreada artificial (pág. 701). Sus ideas sobre la sinestesia, hoy poco conocidas, influyeron en varios artistas famosos del siglo XX, incluida Georgia O'Keeffe (1887–1986), quien bien pudo haber hecho referencias indirectas a Irwin en algunas de sus obras (ils. 50, 51). Hay ciertas mujeres en la historia del color que merecen ser incluidas en el canon sin haber escrito una teoría del color propiamente dicha. Hilma af Klint (1862–1944), de quien se hablará en detalle más adelante (págs. 704–715), es una de ellas, al igual que la pintora británica Winifred Nicholson (1893–1981). La producción artística de Nicholson, en combinación con sus notas, cartas y dos ensayos breves publicados sobre el color, la convierten quizás en uno de los teóricos del color más importantes, a pesar de que nunca reunió sus ideas y conceptos de color en un mismo tomo bajo un mismo título. Al igual que Gartside, parece que desarrolló un interés en la teoría del color a través de la pintura de flores, y sus primeras composiciones a menudo se ubican ante paisajes iluminados o aparecen enmarcadas por ventanas, lo que demuestra un interés en la interacción de la luz y el color (ils. 22, 55). Este interés obvio en los colores prismáticos encontró una nueva oportunidad para la creatividad cuando, a mediados de la década de 1970, con 80 años, Nicholson recibió dos prismas de vidrio. Al final de su vida creó una fascinante serie de «pinturas prismáticas» en las que exploró de forma visual sus experimentos con prismas y la luz, a menudo en un estilo completamente abstracto (il. 56).

El capítulo con el que arranca el presente libro, «Pirámides de color: primeros gráficos y tablas», se detiene en los primeros libros ilustrados de la cultura occidental. Nuestro ejemplo inicial es la tabla de pigmentos, con sus puntos pintados a mano, de Richard Waller (1655–1715) datada de 1686 (págs. 80–85), quien probablemente se basó en tablas similares producidas por el alquimista y médico de la corte inglesa Théodore Turquet de Mayerne (1573–1655), incluidas en el llamado «Manuscrito de Mayerne» (*Pictoria, sculptoria et quae subalternarum artium*) de 1620–1646 (ils. 1, 6). Ambos presentan los colores por medio de tablas y círculos nítidos relativamente simples. No se trata de representaciones de conceptos de color intelectuales, sino que deben verse en el contexto más amplio de los manuales de pintura, dado que brindan mucha información sobre pigmentos y otros materiales empleados por los artistas disponibles en ese momento, y sobre cómo se mezclaban y clasificaban. El capítulo también muestra cómo se crearon gráficos y tablas más complejos en el siglo siguiente, y cómo más adelante estos se abrieron paso en la cultura de la impresión. La intelectualización de los conceptos de color en este período es evidente, ya que las listas de colores se vuelven más elaboradas y extensas, y se hacen los primeros intentos de conceptos tridimensionales. Todos estos primeros gráficos, tablas y diagramas se colorearon a mano, a pesar de los avances de la impresión en color, y cada copia es una obra de arte en sí misma. La práctica de colorear a mano continuó hasta bien entrado el siglo XIX y, en algunos casos, no se reemplazó por la reproducción mecánica del color hasta el siglo XX.

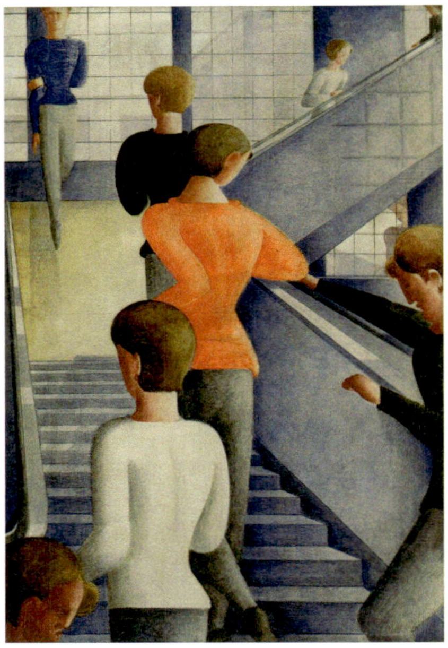

63

ILL. 63
Oskar Schlemmer
Bauhaus Stairway, 1932
Bauhaustreppe
L'Escalier du Bauhaus
Escalera de la Bauhaus
Oil on canvas, 162.3 x 114.3 cm
(63⅞ x 45 in.)
New York, Museum of Modern Art,
Gift of Philip Johnson

ILL. 64
Anni Albers
Design for a 1926 unexecuted wallhanging, date unknown
Entwurf für einen Wandbehang aus dem Jahr 1926
Étude pour une tenture murale non réalisée de 1926
Diseño para un tapiz mural no realizado de 1926
Gouache with pencil on reprographic paper; 38.1 x 24.8 cm (15 x 9¾ in.)
Bethany, Connecticut, The Josef and Anni Albers Foundation

En el capítulo 2, «La forma del color: círculos, ruedas y esferas», examinamos cómo los conceptos visuales simples del orden de los colores se desarrollaron todavía más a lo largo de los siglos XVIII y XIX; como resultado, aparecieron una fascinante variedad de formas y figuras geométricas. Los círculos parecen haber sido la opción favorita, lo que dio lugar a muchas variaciones de lo que ahora podemos considerar la rueda de colores clásica. Sin embargo, también se usaron otras formas, mientras que a menudo un diagrama cromático presentaba adiciones pictóricas y adornos, o una rueda de colores iba acompañada de gráficos y tablas explicativos, como, por ejemplo, en el caso de *The Natural System of Colours* (c. 1769–1776, págs. 138–143) de Moses Harris y las obras antes mencionadas de Mary Gartside.

Desde el punto de vista estético, los diagramas cromáticos de los siglos XVIII y XIX se encuentran entre las imágenes más impresionantes de la historia de los libros sobre el color, pero, a pesar de su poder visual, rara vez existen sin la compañía de la palabra escrita o impresa. El lenguaje, si bien no puede capturar el espectro de color visible, es un elemento importante en la mayoría de estos diagramas, a veces con resultados intrigantes y poéticos, como, por ejemplo, la *Temperamentenrose* (1798/1799, pág. 155) de Johann Wolfgang von Goethe (1749–1832) y Friedrich Schiller (1759–1805). Un desafío al que solían enfrentarse estos diagramas era cómo transmitir el aspecto tridimensional del orden de los colores en un medio bidimensional, y es aquí donde aparecen las primeras esferas o bolas de colores (véase, por ejemplo, el *Farben-Kugel* de Philipp Otto Runge (1777–1810) de 1810, pág. 163, il. 13). Por desgracia, no nos han llegado modelos tridimensionales de esferas de colores de esa época, pero en 2016 la artista francoalemana Eva Bodinet construyó un modelo basado en el *Farben-Kugel* de Runge (ils. 14, 15), y en los últimos años varios fabricantes han comercializado esferas de color en 3D. Runge es también un ejemplo conmovedor de la mayor sensibilidad y espiritualidad que se percibía en el color durante la era romántica. Muchos artistas, filósofos y escritores de finales del siglo XVIII y principios del XIX adaptaron los conceptos de color a sus sistemas de creencias individuales y les asignaron significados altamente simbólicos. Entre 1802 y 1810, Runge creó una serie de imágenes que mostraban las horas del día, incluida *Der kleine Morgen* (*La mañana*), que eran representaciones de la teoría del color adaptadas para expresar sus creencias cristianas (il. 17).

El capítulo 2 también recoge la variedad cada vez mayor de estudios sobre el color que aparecen a medida que avanza el siglo XIX, así como las formas en que estos se basan en textos anteriores y hacen referencia a ellos. Por ejemplo, en la década de 1820, el artista Joseph Mallord William Turner (1775–1851) enseñó los conceptos básicos de la teoría del color como profesor de Perspectiva en la Royal Academy of Arts de Londres. Para ello, preparó dos grandes dibujos que mostraban círculos de color (ils. 25, 26) con el objetivo de explicar la diferencia entre colores materiales e inmateriales. Estos círculos con sus triángulos de color superpuestos en el centro estaban muy claramente influidos por los círculos de color de Moses Harris, que Turner había estudiado. A lo largo de su vida, Turner mostró un gran interés por los nuevos pigmentos y la teoría del color, poseyó y leyó muchas publicaciones sobre la temática e hizo de la interacción entre el color y la luz un tema clave en su obra (il. 27). Solo unos años después de que se publicara la primera parte de la principal obra de Goethe sobre el color (*Zur Farbenlehre*, 1810–1812, págs. 236–249) traducida al inglés (*Goethe's Theory of Colours*, 1840), Turner pintó dos lienzos en referencia directa a las lecciones de ese texto, con un posible guiño adicional a Harris (il. 43).

El capítulo 3, «Nuevas elucidaciones: el surgimiento de la teoría del color», profundiza en algunos de los trabajos más importantes sobre el color del siglo XIX y ofrece las razones del fuerte aumento en el número de nuevas teorías, estudios filosóficos extensos y tratados científicos y prácticos sobre el color. Esta revolución cromática está intrínsecamente ligada a la industrialización del color, en concreto

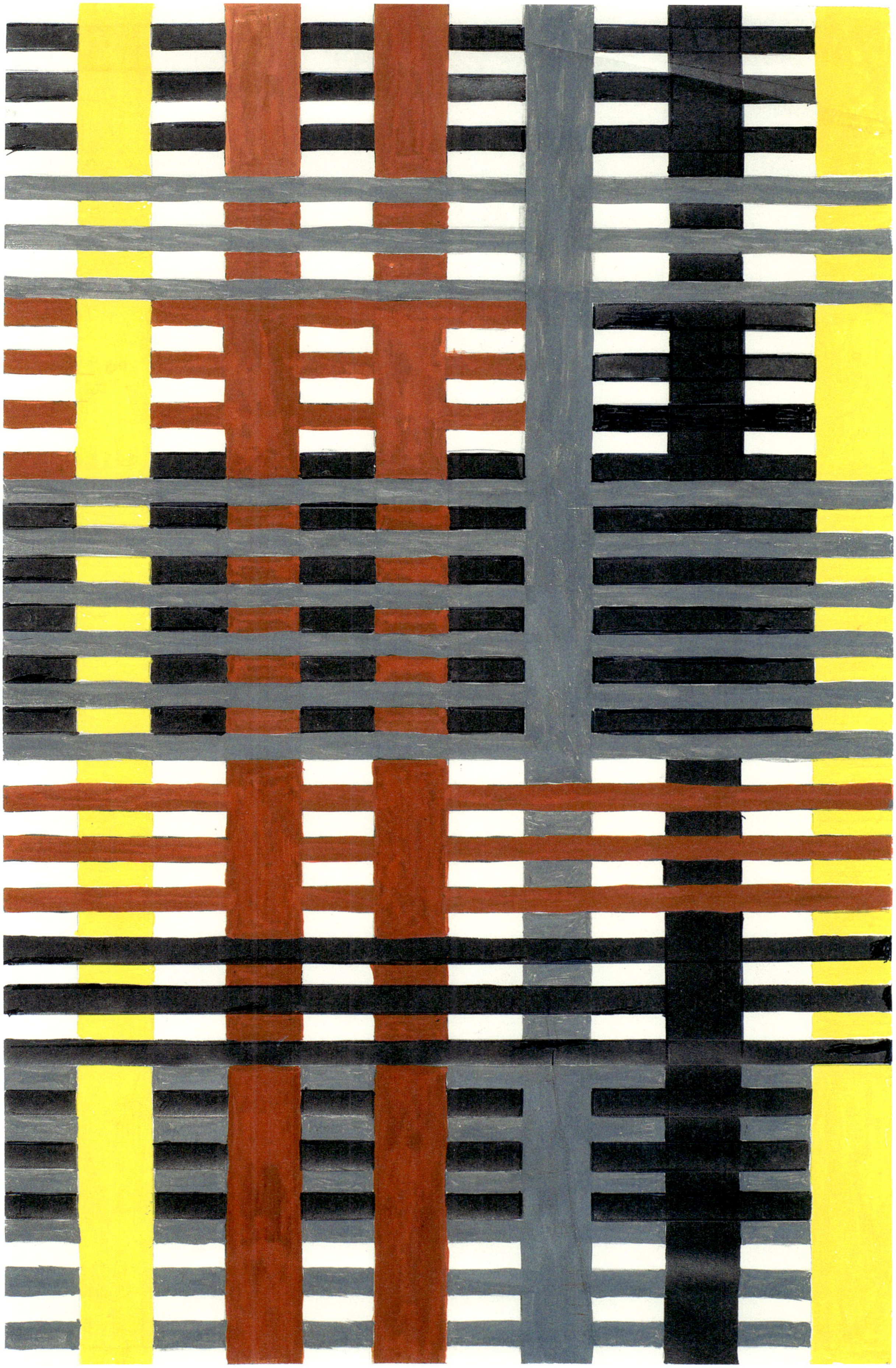

64

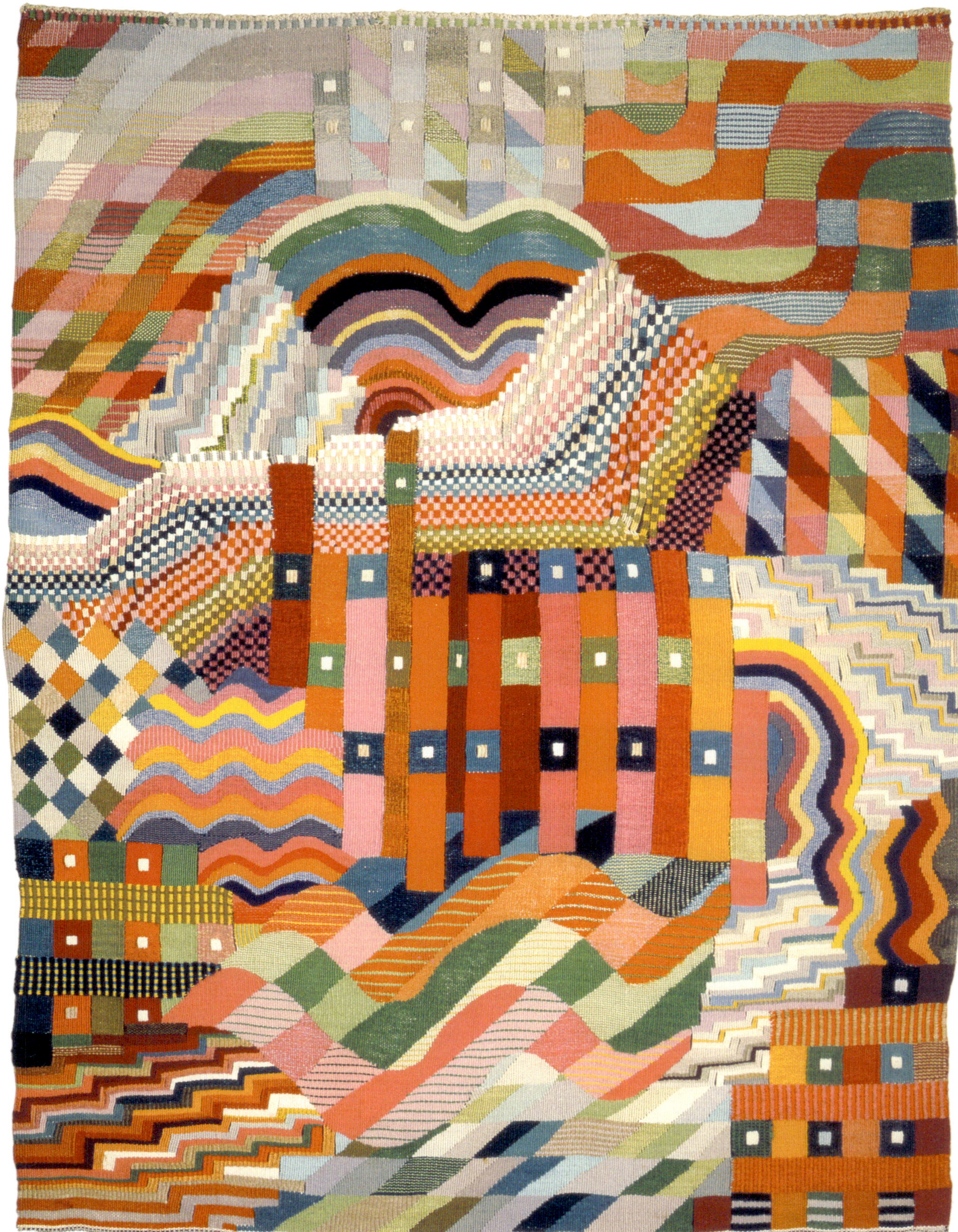

65

Ill. 65
Gunta Stölzl
Red-green Slit Tapestry, 1927/28
Schlitzgobelin Rot-Grün
Tapisserie tissée rouge-vert
Tapiz slit rojo-verde
Cotton, wool, silk and linen;
150 x 110 cm (59 x 43¼ in.)
Berlin, Bauhaus-Archiv /
Museum für Gestaltung

Ill. 66
Gunta Stölzl
Design for a Jacquard-woven Wall Hanging, *c.* 1927
Entwurf für einen jacquardgewebten Wandbehang
Motif pour une tenture murale tissée en jacquard
Diseño para un tapiz de tejido Jacquard
Watercolour, pencil, ink and gouache;
12 x 12 cm (4¾ x 4¾ in.)
London, Victoria & Albert Museum

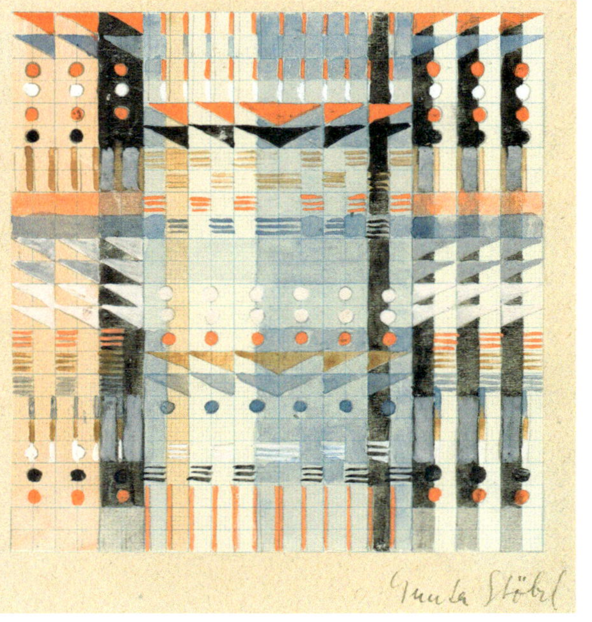

66

a los nuevos métodos de fabricación, producción e impresión, pero tiene su origen a principios del siglo XVIII, cuando Newton escribió un hito indiscutible en la historia de los conceptos de color: *Opticks* (1704), el fruto de muchas décadas de investigación (págs. 210–217). El libro incluía el primer diagrama cromático ampliamente difundido, en forma de rueda, que se basaba en sus experimentos para dividir la luz blanca pura en colores espectrales con el uso de un prisma. Como es sabido, el impacto de Newton en el pensamiento científico a lo largo de los siglos XVIII y XIX fue inconmensurable, y sus teorías fueron adaptadas, admiradas, desarrolladas y criticadas (il. 8). La fascinación por el espectro del arcoíris y otros descubrimientos de Newton tuvo su reflejo, que todavía se encuentra con frecuencia, en las artes visuales, como, por ejemplo, en la pintura alegórica *Colour* (il. 9) de Angelica Kauffman (1741–1807). Aunque es una imagen alegórica y, por lo tanto, no representa un concepto de color *per se*, ilustra un punto importante en la historia del color, ya que el conjunto que se encargó a Kauffman debía representar los «elementos del arte» significativos, a saber: la invención, la composición, el dibujo y el color, este último situado por fin al mismo nivel que los otros tres. La figura femenina que representa el color sumerge su pincel en el arcoíris de Newton y coloca el color prismático en una paleta física; en el proceso, adapta la teoría del color inmaterial al color material, al tiempo que lo relaciona con las prácticas docentes en la Academia recién fundada.

A medida que el siglo XIX avanzaba, los conceptos y actitudes románticos, como el simbolismo del color espiritual de Runge, dieron paso gradualmente a estudios más técnicos y científicos, muchos de ellos vinculados a la fabricación de nuevos pigmentos y tintes. La calidad de estos nuevos colores se convirtió en un tema de análisis para el gran «colorista» británico George Field (c. 1777–1854, págs. 258–273), quien no solo desarrolló una teoría del color propia sino que también probó, produjo y patentó pigmentos. Muchos artistas del siglo XIX, incluidos miembros de la Hermandad prerrafaelita y sus seguidores, demandaban sus colores, y su disponibilidad coincidió con un interés renovado en las técnicas y los materiales de la pintura medieval. Pinturas como *Mariana* (1851, il. 38) de John Everett Millais (1829–1896) reflejan, con su brillo propio de una joya y su atrevida policromía, ese interés por los pigmentos de buena calidad y una gran conciencia de los conceptos de color del momento. Incluso se ha sugerido que la pintura de Dante Gabriel Rossetti (1828–1882) *Ecce Ancilla Domini!* (*The Annunciation*, 1849/50, il. 29) podría haberse inspirado directamente en el simbolismo de los colores propuesto por George Field. La atención intelectual prestada a la teoría y los estudios sobre el color es muy evidente entre los artistas occidentales del siglo XIX, y tanto el impresionismo como el puntillismo se vieron directamente influenciados por la innovadora teoría de Michel-Eugène Chevreul (1786–1889) del contraste simultáneo de los colores (a partir de 1839), que desarrolló mientras trabajaba para la fábrica de tapices Gobelins en París (págs. 282–307). El posimpresionista Vincent van Gogh (1853–1890) utilizó los principios del contraste de colores con efectos asombrosos en su arte, y es significativo que entre sus posesiones personales hubiera una caja lacada roja que contenía varios ovillos de lana de colores, retorcidos y combinados en diferentes variaciones como si hubiera estado creando armonías de color de acuerdo con las leyes de Chevreul (il. 37). También hay una gran conciencia del contraste de los colores en el trabajo de James McNeill Whistler (1834–1903), y aunque no hay constancia de que este estuviera especialmente interesado en los estudios sobre el color, su preocupación por la armonía tonal y el simbolismo del color se refleja con claridad en los títulos que dio a sus pinturas, donde aparecen palabras como «armonías», «nocturnos» y «sinfonías» de colores específicos (ils. 40, 41) a expensas de las descripciones narrativas.

Los capítulos 4 y 5 tratan conceptos de color más técnicos y prácticos. «Colores como luces guía» se ocupa de listas sistemáticas de colores, a menudo en relación con el mundo natural y la clasificación

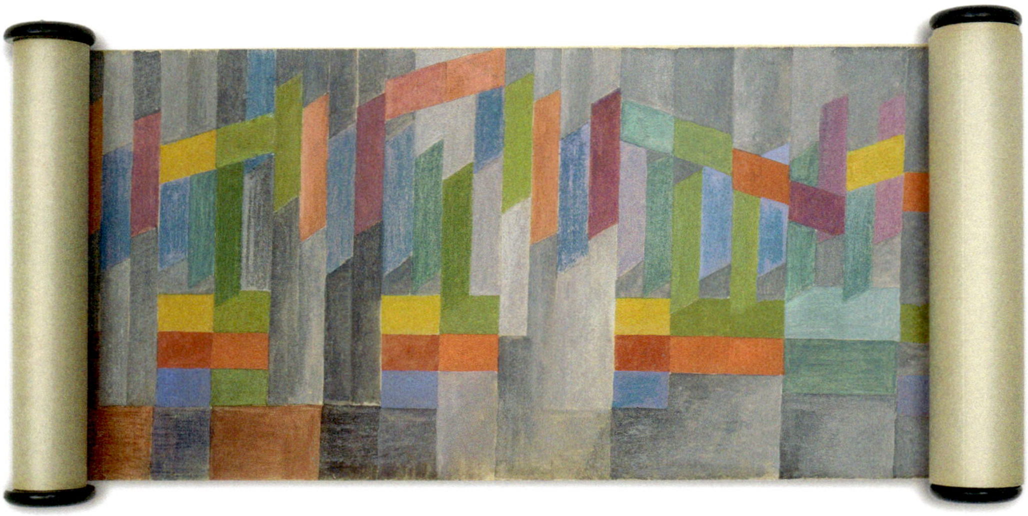

67

taxonómica, que ofrecen intentos fascinantes de crear nomenclaturas y estándares fiables para su uso en el arte, la industria y las ciencias. Algunos de estos sistemas también tenían como objetivo proporcionar una crónica del color a lo largo de la historia, como *Historical Color Guide* (1938) de Elizabeth Burris-Meyer (1899–1969), que luego adaptó para su uso en decoración de interiores, moda y maquillaje (págs. 388–409). El capítulo 5 se centra en la historia de los manuales prácticos de color para pintores e impresores, desde principios del siglo XIX hasta principios del XX. Los manuales de color anteriores todavía estaban muy en deuda con el desarrollo de la pintura en acuarela; a medida que los materiales para la pintura se volvieron más asequibles a principios del siglo XIX, muchos se destinaron a aficionados. La pintura era un pasatiempo aceptable para las mujeres en los siglos XVIII y XIX, por lo que buena parte de esos manuales estaban dirigidos al público femenino, aunque la gran mayoría estaban escritos por hombres. Sin embargo, desde principios del siglo XX, cada vez más mujeres se implicaron activamente en la escritura de manuales o compendios sobre el color educativos e instructivos, y aprovecharon las oportunidades que se les presentaron para escribir, publicar y enseñar. Hasta ahora, esta ha sido un área poco investigada del compromiso creativo de las mujeres, y dado que los manuales de enseñanza, en especial aquellos dirigidos a los niños más pequeños, no se consideraban textos serios, generalmente no se han documentado bien y no sobreviven muchas copias.

Los inicios del siglo XX también estuvieron marcados por otra oleada de libros sobre el color, muchos de ellos basados en los trabajos de los grandes teóricos del siglo anterior, cuyos conceptos desarrollaron, como se demuestra en el capítulo 6, «La liberación del color: principios del siglo XX». El color en ese período no puede evaluarse de forma adecuada sin tener en cuenta una serie de logros científicos y artísticos, incluidos los diversos avances en la comprensión de la visión del color y en concreto de la anatomía del ojo humano, la electrificación del mundo y las posibilidades resultantes proporcionadas por la iluminación artificial y el uso radicalmente nuevo del color como herramienta expresiva en las artes, en particular en la pintura. Grupos de vanguardia europeos como Der Blaue Reiter (El jinete azul), los fauvistas y muchos de los expresionistas y primeros artistas abstractos se involucraron de forma activa en la teoría del color. Abrazaron el poder simbólico y visual del color, y representaron el mundo en tonos subjetivos y no realistas. Para algunos, los conceptos de color se convirtieron en el tema principal de ciertas obras, como se ve, por ejemplo, en la pintura abstracta de Franz Marc (1880–1916) *Kämpfende Formen* (*Formas que luchan*, il. 47) de 1914.

Las primeras décadas del siglo XX, altamente coloreadas y cada vez más mecanizadas, en las que pigmentos de casi todos los tonos imaginables se podían comprar listos para usar perfectamente empaquetados, pedían a gritos nuevos intentos de darle un orden definitivo al color. Algunos de los resultados más influyentes se analizan en el capítulo 6, como el trabajo de Albert Henry Munsell (1858–1918) en Estados Unidos (págs. 584–599) y Wilhelm Ostwald en Europa (págs. 600–609). Ambos crearon sistemas altamente educativos y adaptables que se basaban en un concepto tridimensional del color y, de hecho, algunos de los modelos tridimensionales que utilizaron en su enseñanza han sobrevivido hasta el día de hoy (véase, por ejemplo, *Colour Solid* de Ostwald en los archivos de Winsor & Newton, pág. 430). Para las mujeres, el comienzo del siglo XX fue también una época de mejores oportunidades para la investigación y la publicación, lo que dio como resultado varios ejemplos destacados de estudios científicos sobre la visión del color y la psicología, el color en la literatura y la poesía, la teoría histórica del color y otros aspectos de la historia del color. Algunos de esos libros publicados entre 1900 y la década de 1930 se encuentran entre los más ingeniosos y hermosos de la historia del color, como *Color Problems* (1902, págs. 562–583) de Emily Noyes Vanderpoel (1842–1939).

ILL. 67
Roy de Maistre
Colour Music (enclosed roll), *c.* 1934
Farbenmusik (beidseitig aufgerollt)
Musique chromatique (rouleau)
Música del color (rollo cerrado)
Gouache on piano roll, 31 x 1,606 cm
(12¼ x 632¼ in.)
New South Wales Art Gallery,
Gift of Sir John Rothenstein in memory
of the artist, 1969

ILL. 68
Piet Mondrian
Composition with Red,
Yellow and Blue, *c.* 1937–1942
Komposition mit Rot, Gelb und Blau
Composition en rouge, jaune et bleu
Composición en rojo, amarillo y azul
Oil on canvas, 72.5 x 69 cm
(25⅝ x 27¼ in.)
London, Tate Modern

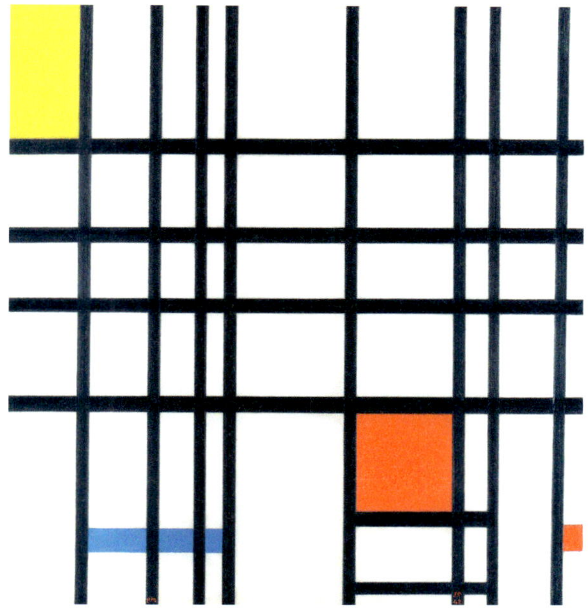

68

El capítulo 7, «El sonido del color: espiritualismo, ocultismo y música», examina las asociaciones sinestésicas entre el color y la música y, por extensión, el movimiento, la danza y la espiritualidad. Hasta cierto punto, los conceptos de color altamente individuales y a menudo pseudocientíficos publicados por teósofos, antropósofos y miembros de grupos y movimientos similares a principios del siglo pasado fueron una respuesta a un mundo cada vez más mecanizado, pero es útil revisarlos con atención y juzgarlos no por su precisión científica sino por su inventiva creativa y las circunstancias sociales e históricas que los sustentan. No hay que creer en las auras de color, como las ilustran y describen Annie Besant (1847–1933) y Charles Webster Leadbeater (1854–1934), pero sus intentos de dar forma y color a los pensamientos y las emociones son un comentario fascinante sobre la naturaleza humana de ese momento (págs. 648–687).

Las analogías entre el color y la música son un tema destacado en la historia de los conceptos de color. La mayoría de los textos mencionados en este capítulo abordan el color en relación con el sonido, la música, la poesía y la danza, mientras que muchos de ellos también combinan el misticismo con la ciencia, las creencias religiosas individuales con la teoría del color, la naturaleza con el espiritualismo y los textos históricos sobre el color con la práctica pictórica del momento. Esto bien puede parecer una mezcla desconcertante e idiosincrásica, y a menudo dificulta la lectura, pero las representaciones visuales de estos conceptos son extremadamente cautivadoras. El artista australiano Roy de Maistre (1894–1968) creó muchas pinturas sobre el tema, entre ellas un rollo de cinco metros de largo que representa un «traducción al color» del *Trío en si bemol* de Haydn (pág. 60 e ils. 57, 67). El trabajo de la mística y antropósofa sueca Hilma af Klint, que en los últimos años ha sido redescubierta y elogiada con razón por su estilo visionario, está profundamente arraigado en el ocultismo y el simbolismo del color. Aunque nunca formuló su propia teoría del color, muchas de sus pinturas de gran formato incorporan círculos de colores, pirámides e imágenes de colores inmateriales, lo que da como resultado un vocabulario cromático visual marcadamente propio que combina realismo y abstracción (ils. 39, 48, 49).

Este libro no pretende ser una descripción global de los conceptos históricos de color. En la selección de obras predominan, quizás de forma inevitable, las pertenecientes al arte occidental, pero el capítulo 8 tiene como objetivo corregir ese desequilibrio y llevar la mirada más allá de lo familiar, a los conceptos de color japoneses y rusos de los siglos XIX y XX. No es intención de las autoras atribuirse un conocimiento experto en ese ámbito, sino más bien reconocer su atractivo estético y resaltar cómo los conceptos de color de partes del mundo geográficamente distantes pueden relacionarse entre sí. La comparación de algo tan abstracto ofrece poderosos ejemplos de lo que conecta a los humanos, al tiempo que permite observar un ejemplo intrigante de intercambio transcultural en la repercusión de uno de los diagramas cromáticos de George Field de la década de 1830, que reaparece en un libro de texto japonés para colegiales, *Gakko hitsuyo irozu mondo*, publicado por primera vez en 1876 (págs. 740–743).

El capítulo final, «Interacción y abstracción: la Bauhaus y más allá», considera el color en su forma más pura, reducida y elemental, tal como se encuentra en el arte, el diseño, los textos didácticos y la enseñanza a principios y mediados del siglo XX. El camino hacia la abstracción fue largo y gradual, con contribuciones de varios pioneros, incluidos Turner, Gartside, Whistler, Af Klint y otros artistas de diferentes grupos de vanguardia europeos, pero la Bauhaus y su círculo más amplio de influencia se convirtió en uno de los principales representantes del modernismo, la innovación y, en términos estilísticos, la abstracción. Fundada por el arquitecto Walter Gropius (1883–1969) en Weimar en 1919, esta efímera escuela de arte y diseño tenía sus raíces en los grupos vanguardistas y espiritistas alemanes, así como en el movimiento inglés Arts and Crafts. Se preocupaba por la sencillez y la funcionalidad del diseño, y se basó en la formación completa de los

69

ILL. 69
Yves Klein
Blue Sponge Relief: RE 19, 1958
Blaues Schwammrelief
Relief éponge bleu
Relieve de esponja azul
Pigment in synthetic resin on sponge, pebbles and board, 200 x 165 cm (78¾ x 65 in.)
Cologne, Museum Ludwig

ILL. 70
Olafur Eliasson
Colour experiment no. 57, 58, 59, 60, 62, 63, 2014
Farbexperiment
Expérience chromatique
Experimento de color
Oil on canvas, ϕ 190 cm (74¾ in.)

Pages 76/77
Shopfront of Myland's Paints, Stockwell Road, London, 1906
Schaufenster von Mylands, Stockwell Road, London
Devanture du magasin Mylands, Stockwell Road, Londres
Fachada de la tienda de Mylands, Stockwell Road, Londres
Photograph

artistas y la calidad de la artesanía. Entre sus miembros, profesores y estudiantes se encontraron Ludwig Mies van der Rohe, Paul Klee, Johannes Itten y Josef Albers, así como varias mujeres, cuyo papel en la historia de la Bauhaus se ha reconocido recientemente, como Anni Albers (il. 64), Marianne Brandt, Gertrud Arndt (il. 59), Gertrud Grunow (il. 61, 62) y Gunta Stölzl (ils. 65, 66). En 1925, la escuela se trasladó a Dessau, donde Gropius diseñó su icónica sede (véase il. 63), pero en 1933, tras un año en Berlín, se vio obligada a cerrar debido a la creciente presión de los nazis. Algunos de sus miembros emigraron a Inglaterra o Estados Unidos, donde pudieron mantener lazos entre ellos y continuar con su trabajo. Aunque el color era solo un elemento de los cursos de la Bauhaus, dio lugar a muchas publicaciones relacionadas, algunas de las cuales se presentan en el capítulo final de este libro. Estas incluyen dos obras clásicas sobre el color que se publicaron casi 30 años después del cierre de la escuela: *Interaction of Color* (1963, págs. 824–837) de Josef Albers (1888–1976) y *Kunst der Farbe* (*El arte del color*, 1961, págs. 804–823) de Johannes Itten (1888–1967). Estos y otros trabajos menos conocidos sobre el color de estudiosos y artistas asociados son ejemplos de una síntesis casi perfecta de la reflexión sobre el color fruto de vivir y trabajar con él, de enseñarlo y aplicarlo en el arte y el diseño, al tiempo que evidencian los ideales modernistas de simplicidad, geometría y abstracción. En el arte, los conceptos de color y la estética de la Bauhaus se reflejan en el movimiento más amplio hacia la abstracción, y las similitudes en el uso del color con el trabajo extremadamente minimalista y geométrico de maestros modernos como Piet Mondrian (il. 68), Henri Matisse y Kazimir Malevich son evidentes.

La atemporalidad de los conceptos de color

La belleza abstracta de los diagramas cromáticos, de los conceptos de color en imágenes, tiene un atractivo atemporal. Las formas geométricas simples y los marcos de los primeros gráficos y tablas de colores se perciben como accesibles al instante porque formas iguales o similares existen en el arte, la publicidad y el diseño de hoy, mientras que los círculos de colores, las ruedas, los triángulos, los cubos y los diagramas o formas orgánicas aún más complejas encuentran precedentes a lo largo de los siglos. La teoría del color y los conceptos de color siguen siendo motivo y fuente de inspiración para los artistas contemporáneos. Desde 2009, el artista islandés-danés Olafur Eliasson (n. 1967) creó una serie de grandes ruedas cromáticas en las que los distintos colores se mezclan a la perfección (il. 70), de manera similar a una de las ruedas cromáticas de Chevreul en su Atlas de 1861. Un conjunto de estas, titulado *Turner colour experiments*, se exhibió en la Tate Britain junto con algunas de las pinturas de Turner. Así como Turner hizo referencia a Goethe y Harris en algunas de sus obras, Eliasson se refirió a los teóricos del color y a los artistas anteriores a él con el objetivo de crear su propia teoría del color.

Los seres humanos siempre han tratado de domesticar y controlar el color, ese elemento inasible pero omnipresente de nuestra visión, aunque nuestros intentos por capturarlo dibujando círculos, cuadrados y líneas a su alrededor, comprimiéndolo en listas o moldeándolo en formas tridimensionales, son en última instancia expresiones del pensamiento y la creatividad humanos. Es posible que aún no se haya formulado el concepto de color perfecto, pero convertir las ideas existentes en palabras, imágenes, diagramas y objetos tridimensionales ha dado como resultado una rica historia del color como arte visual. La experiencia individual, las elecciones personales, los descubrimientos de nuevas investigaciones y el pensamiento innovador hacen que cada nuevo libro sobre la historia del color sea emocionante, y las autoras esperan que la selección de conceptos de color que se presentan en este volumen brinde algunas historias y puntos de vista inesperados e inspiradores que arrojen luz sobre la historia del color.

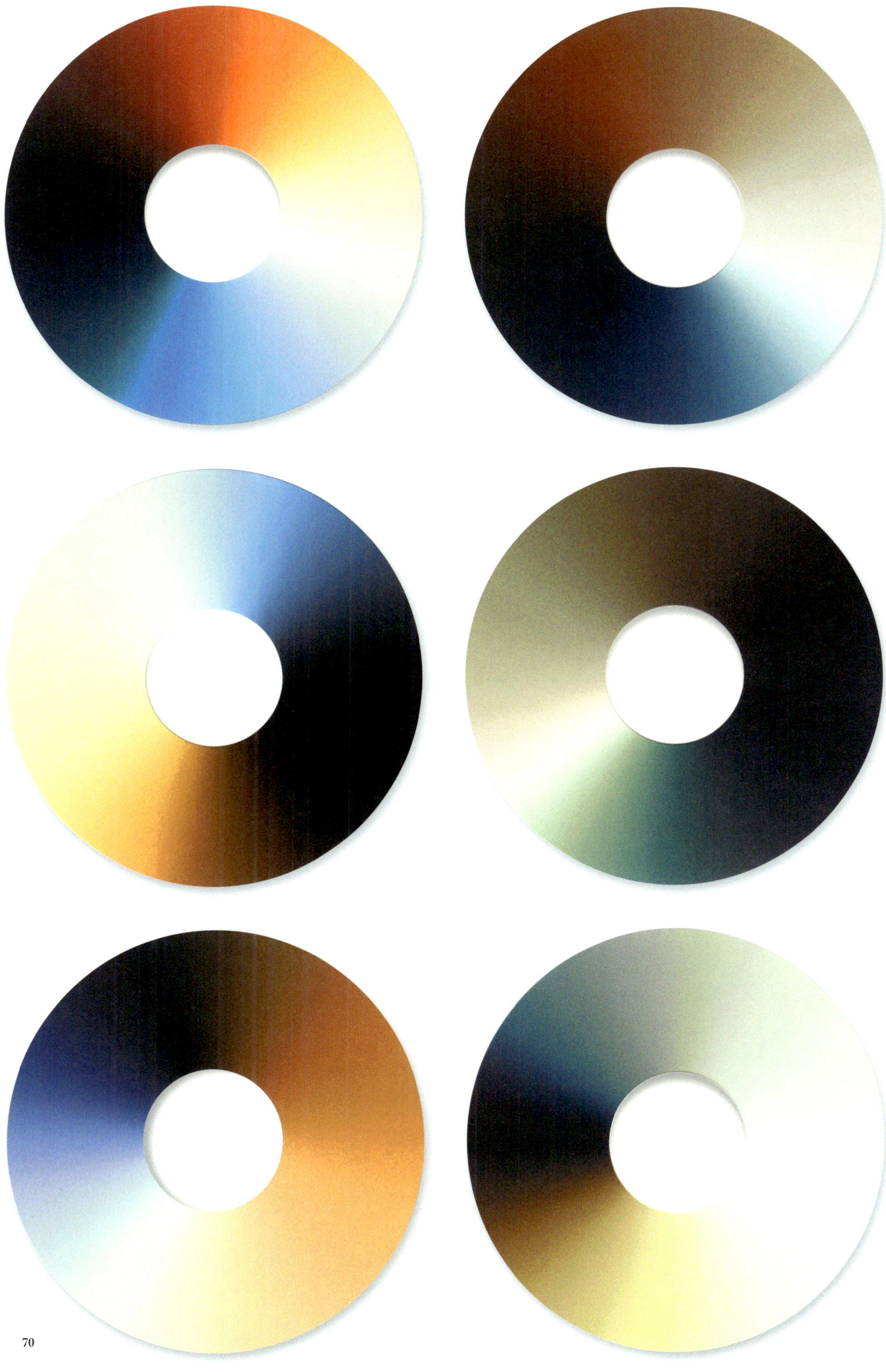

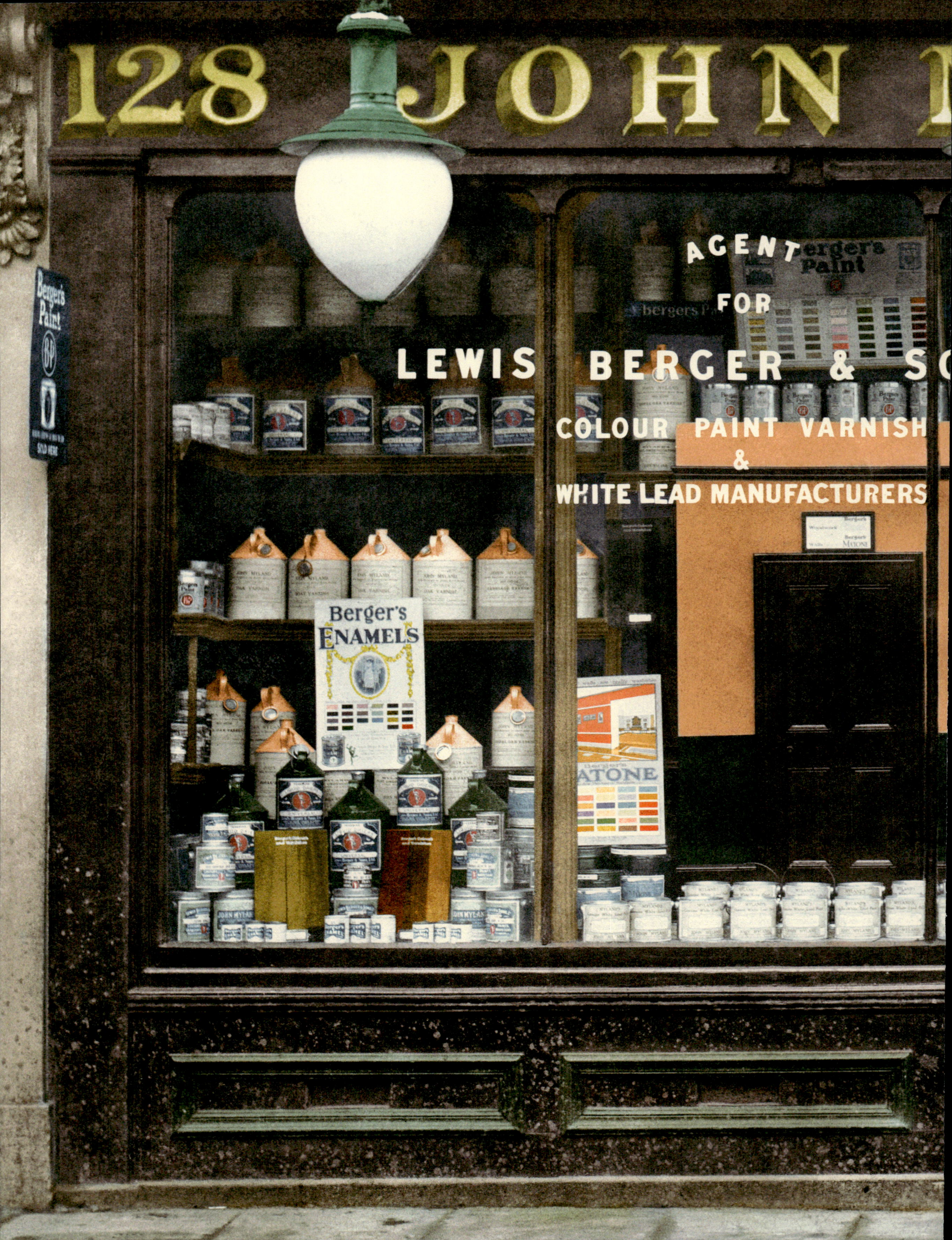
128 JOHN
AGENT
FOR
LEWIS BERGER &
COLOUR PAINT VARNISH
&
WHITE LEAD MANUFACTURERS
Berger's
ENAMELS
Berger's Paint

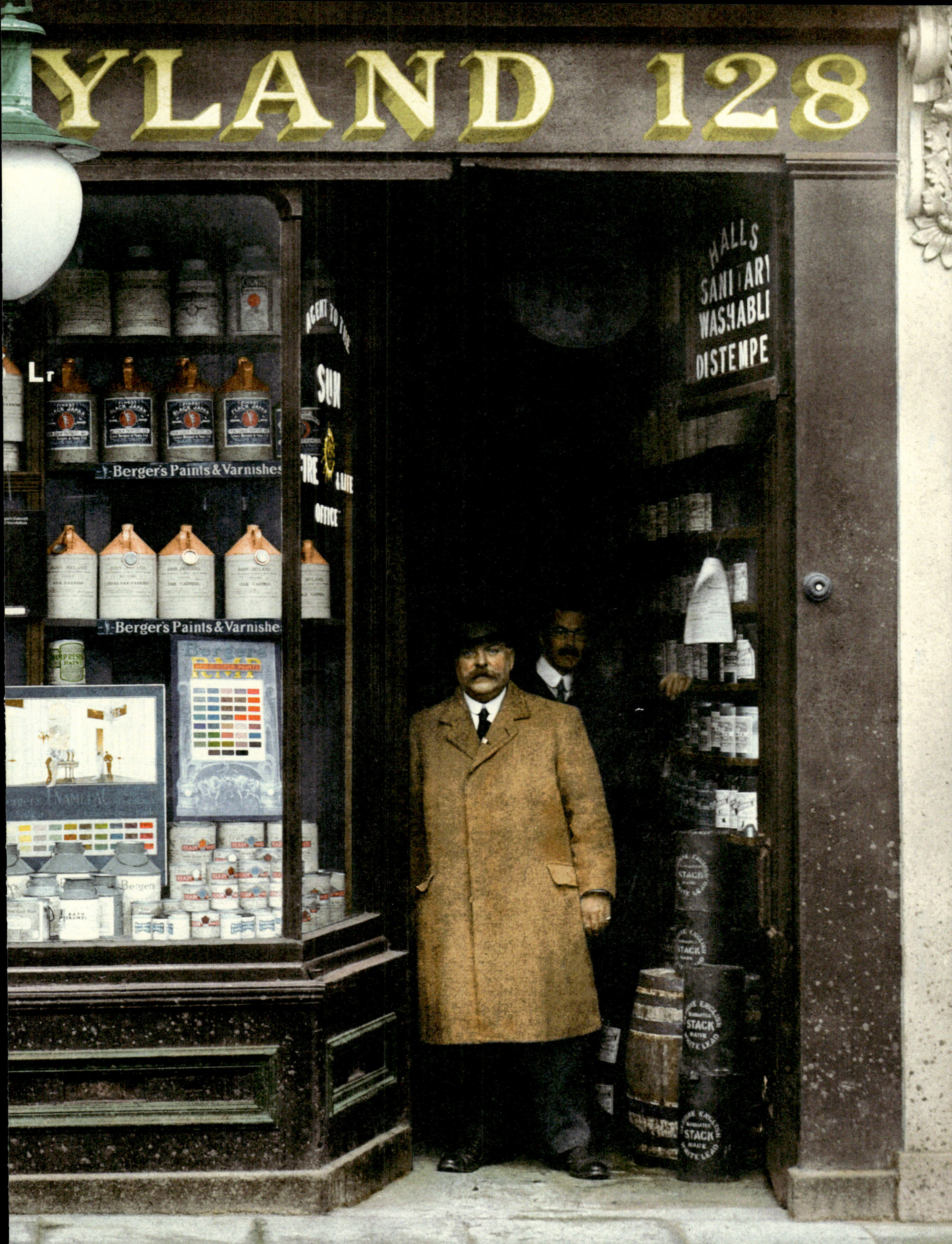
YLAND 128
HALLS
SANITARY
WASHABLE
DISTEMPE
AGENT TO THE
SUN
FIRE & LIFE
OFFICE
Berger's Paints & Varnishes
Berger's Paints & Varnishes

Regiſter.

N1: Bij fol: 25:

N1: Bij fol: 26:

N1: Bij fol: 27:

N1: Bij fol: 28:

N1: Bij fol: 29

N1: Bij fol: 30:

N1: Bij fol: 31:

N1: Bij fol: 32:

N1: Bij fol: 33:

N1: Bij fol: 34:

N1: Bij fol: 35:

N1: Bij fol: 36:

A:3:X:

Chapter 1

Pyramids of Colour: EARLY CHARTS AND TABLES

1686–1799

Richard Waller (1655–1715)

A CATALOGUE OF SIMPLE AND MIXT COLOURS

with a Specimen of Each Colour Prefixt to Its Proper Name

1 chart, from the journal Philosophical Transactions of the Royal Society,
20.5 x 33.5 cm | 8 ⅛ x 13 ¼ in., London, 1686
London, The Royal Society

Richard Waller's "catalogue" of colours is a splendid example of the Enlightenment desire to understand and classify the world, and to put newly acquired and systemised knowledge into a visual format. Waller was an English naturalist, translator and draughtsman, and perhaps best known for editing the posthumous work of Robert Hooke (1635–1703). A member of the Royal Society of London, Waller became its Vice-president when Isaac Newton served as President; its journal, *Philosophical Transactions of the Royal Society*, was launched in 1665 and was the first such scientific periodical in the world. Being himself a multi-lingual polymath, Waller was well-placed within the Enlightenment network in Europe, and also kept himself well informed about the latest discussions and research on colour.

The original colour chart Waller presented to members of the Society in 1686 was painted on glass and is now lost. He had been inspired by *Nomenclatura et species colorum* (Nomenclature and species of colours), a colour chart in several languages by the Swedish draughtsman Elias Brenner (1647–1717) that was published in 1680, and possibly also by Théodore Turquet de Mayerne's colour lists in the "De Mayerne manuscript" of 1620–1646 (pp. 6, 12), and as a result Waller sought to create a reliable and much-needed standard of colours in four languages: Latin, Greek, French and English (although he acknowledged his limitations and was not able to provide full names for all colours). Soon after his presentation the chart was published in the *Philosophical Transactions* with the title 'Tabula colorum physiologica tam mixtorum quam simplicium' (Table of physiological colours both mixed and simple), as a neat, engraved grid with 119 blots of colour, laid out in systematic order with yellows and reds in the column on the left and white and blues along the top. The rest of the grid is filled with one-to-one mixtures of these "simple colours". The chart was accompanied by several pages of descriptions in Latin of various colourants, and is very clearly intended to have been a practical chart of pigments and their mixtures for application in painting and illustration, rather than a philosophical system, even if Waller's general arrangement suggests he had an awareness of Aristotelian ideas of colour and colours being situated between darkness and light. Although Waller's chart is sometimes, and rightly, referred to as one of the first printed visual colour systems to be represented in colour, it is important to point out that the blots were in fact hand-applied with a small taper pipe, designed by Waller, and thus the different shades vary greatly in each of the few surviving copies. While illustration in print media in the form of engravings was entering a new, exciting phase at this time, printing in colour was still very much in its infancy.

Minium		Igneus. Πυῤῥοειδὴς C. du feu. Fire Col.	
Ochra usta		Ruffus. Carret C.	
Cinnabaris		Miniatus. Μιλτώδης du Vermillon.	
Carmin		Coccinus. Κόκκινος C. d'Escarlate. Crimson.	
Lacca		Rubinus. Ruby-red.	
Sanguis Draconis		Sanguineus. Αἱματώδης C. du Sang. Blood-red.	
Rubrica		Ferrugineus. Σιδηροειδὴς Rust Co.	
tramentū Fuliginosū		Piceus. Πίσσινος C. du poix. Pitchy.	

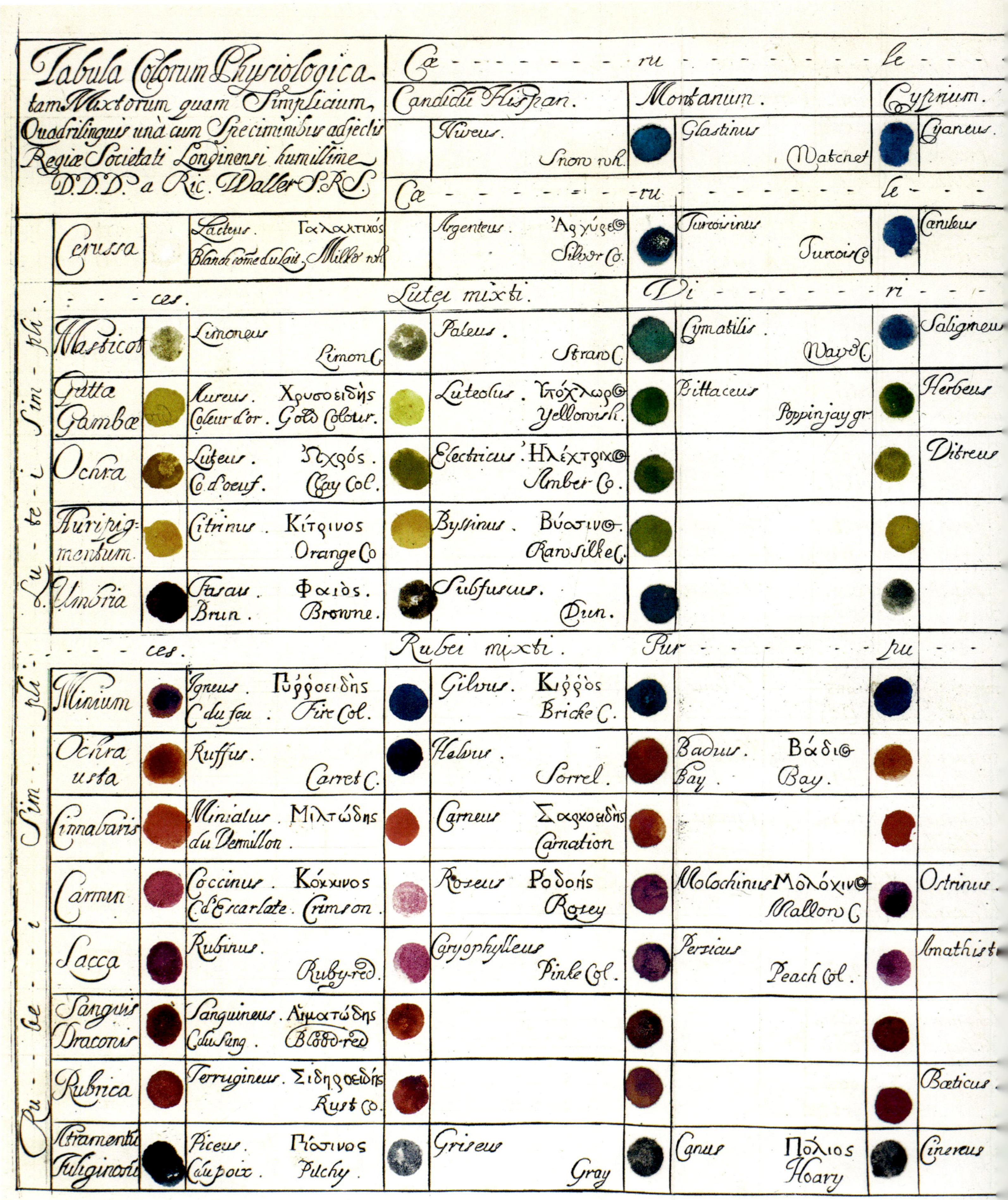
Tabula Colorum Physiologica tam Mixtorum quam Simplicium, Quadrilinguis unà cum Speciminibus adjectis Regiæ Societati Londinensi humillime D.D.D. a Ric. Waller S.R.S.
Cæ - ru - le
Candidi Hispan.
Montanum.
Cyprium.
Niveus. Snow wh.
Glastinus Watchet
Cyaneus.
Cæ - ru - le
Cerussa
Lacteus. Γαλακτικός Blanch come du Lait. Milke wh.
Argenteus. Ἀργύρεος Silver Co.
Turcoisinus Turcois Co.
Cæruleus
- ces. Lutei mixti. Vi - ri -
Masticot
Limoneus Limon C.
Paleus. Straw C.
Cymatilis. Wave C.
Salignew
Gutta Gambæ
Aureus. Χρυσοειδής Coleur d'or. Gold Colour.
Luteolus. Ὑπόχλωρος Yellowish.
Psittaceus Poppinjay gr
Herbeus
Ochra
Luteus. Ὠχρός. Co. d'oeuf. Clay Col.
Electricus. Ἠλέκτρινος Amber Co.
Vitreus
Auripigmentum
Citrinus. Κίτρινος Orange Co
Byssinus. Βύσσινος. Raw silke C.
Umbria
Fuscus. Φαιός. Brun. Browne.
Subfuscus. Dun.
- ces. Rubei mixti. Pur - pu -
Minium
Igneus. Πυῤῥοειδής C. du feu. Fire Col.
Gilvus. Κιῤῥός Bricke C.
Ochra usta
Ruffus. Carret C.
Helvus. Sorrel.
Badius. Βάδιος Bay. Bay.
Cinnabaris
Miniatus. Μιλτώδης du Vermillon.
Carneus Σαρκοειδής Carnation
Carmin
Coccinus. Κόκκινος C. d'Escarlate. Crimson.
Roseus Ῥόδοεις Rosey
Molochinus Μολόχινος Mallow C.
Ostrinus.
Lacca
Rubinus. Ruby-red.
Caryophylleus Pinke Col.
Persicus Peach Col.
Amathist
Sanguis Draconis
Sanguineus. Αἱματώδης C. du Sang. Blood-red
Rubrica
Ferrugineus. Σιδηροειδής Rust Co.
Bæticus.
Atramentum Fuliginosum
Piceus. Πίσσινος Co. du poix. Pitchy.
Griseus Gray
Canus Πόλιος Hoary
Cinereus
Sim - pli - ces Lu - te - i . Sim - pli - ces Ru - be - i

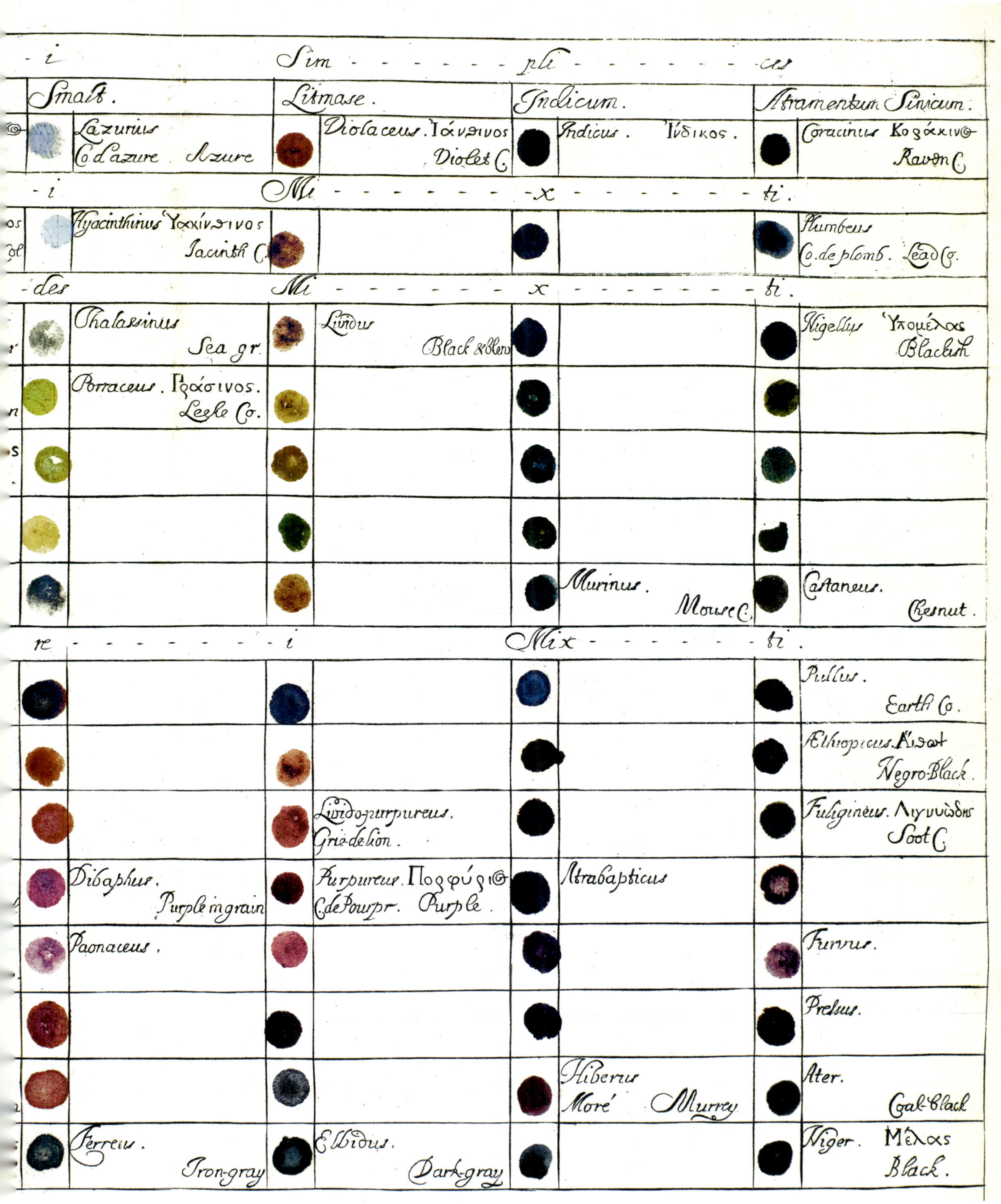

Smalt.	Litmase.	Indicum.	Atramentum Sinicum.
-i Sim - - - - - - -	pli - - - - - - -		ces
Lazurius Co. d'azure. Azure	Violaceus. Ἰάνθινος Violet C.	Indicus. Ἰνδικός.	Coracinus Κοράκινος Raven C.
-i Mi - - - - - - -	-x - - - - - - -		ti.
Hyacinthinus Ὑακίνθινος Jacinth C.			Plumbeus Co. de plomb. Lead Co.
-des Mi - - - - - - -	x - - - - - - -		ti.
Thalassinus Sea gr.	Lividus Black & blew		Nigellus Ὑπομέλας Blackish
Porraceus. Πράσινος. Leeke Co.			
		Murinus. Mouse C.	Castaneus. Chesnut.
re - - - - - - - i		Mix - - - - - - -	ti.
			Pullus. Earth Co.
			Æthiopicus. Αἴθοψ Negro Black.
	Livido-purpureus. Gris-de-lin.		Fuligineus. Λιγνυώδης Soot C.
Dibaphus. Purple in grain	Purpureus. Πορφύρεος C. de Pourpr. Purple.	Atrabapticus	
Paonaceus.			Furvus.
			Pressus.
		Hiberus Moré Murrey	Ater. Gal-Black
Ferreus. Iron-gray	Ellidus. Dark-gray		Niger. Μέλας Black.

[24]

Cum denique ex obſervatione noſtra manifeſtum ſit, revera tinctam eſſe gemmam Chalcedoniam, quamvis fortuito acciderit, neque repetito proceſſu ſimile quid evenerit, merebitur tamen meditationem, an ex aſtrorum fluxu, aliave abſcondita potius vi venerit, & tentamen, an ex mixtura ſalium & ſuccorum acrium poſſit imitando produci ejuſmodi Tinctura, & quidem ſine Igne, ut ſplendor & Pelluciditas gemmæ non deſtruatur, durities autem maneat, adeoque ipſa gemmæ pretioſitas non tantum ſervetur, ſed & per tincturam novam creſcat.

Hujus Tincturæ figuram & modum, ſi forte quem delectet, addere placuit, qualis noctu oppoſita Candelæ viſa fuit, aliquomodo nec ſatis accurate delineatum. Tab. 1. Fig. 6.

A CATALOGUE *of Simple* and *Mixt Colours*, with a *Specimen* of each *Colour* prefixt *to* its proper *Name:* By R. *WALLER*, Fellow of the *Royal Society.*

HAving ſometime ſince ſeen a TABLE of the *Simple Colo rs* made uſe of in *Limning* and *Painting*, Printed in the Year 1680, at *St ckh lm*; I have here endeavoured to give a more *Philoſophical*, and uſeful one by the addition of ſome *mixt Colours:* Not that I pretend to give the *Shades* of all the *mi t Colours*, which were indeed infinite as the Compoſitions and Proportions of them may be unlimited; but I have mixt each of the *Simple Yellows* and *Reds* with each of the *ſimple Blews*, and theſe *Mixtures* give moſt of the *mean* Colours, *viz. Greens, Purples*, &c. To know what each of theſe *mixt Colours* is compounded of, you need but look to the Top of the *Table* directly over the *Colour* enquired after, where you may find the One *Ingredient*, and at the Side in the

Katalog der einfachen und gemischten Farben, mit einem Muster jeder Farbe ihrem Eigennamen vorangestellt

Richard Wallers „Katalog" der Farben ist ein ausgezeichnetes Beispiel für den in der Aufklärung vorherrschenden Wunsch, sowohl die Welt zu verstehen und zu klassifizieren als auch neu erworbenes und systematisiertes Wissen in ein bildhaftes Format zu übertragen. Waller war ein englischer Naturforscher, Übersetzer und Zeichner und vielleicht am besten bekannt für die postume Herausgabe des Werks von Robert Hooke (1635–1703). Er war Mitglied der Royal Society of London und während der Präsidentschaft von Isaac Newton ihr Vizepräsident. Das Journal der Vereinigung, die *Philosophical Transactions of the Royal Society*, wurde 1665 gegründet und war weltweit eine der ersten wissenschaftlichen Fachzeitschriften ihrer Art. Als mehrsprachiger Universalgelehrter hatte Waller einen festen Platz im Netzwerk der europäischen Aufklärung und hielt sich regelmäßig über die neuesten Debatten und Ergebnisse in der Farbforschung auf dem Laufenden.

Die ursprüngliche Farbtabelle, die Waller den Mitgliedern der Royal Society 1686 präsentierte, war auf Glas gemalt und ist heute verschollen. Inspiriert von der *Nomenclatura et species colorum* (Nomenklatur und Gattungen der Farben), einer 1680 von dem schwedischen Zeichner Elias Brenner (1647–1717) veröffentlichten mehrsprachigen Farbtabelle, und vielleicht auch von der Farbenliste im „De-Mayerne-Manuskript" von Théodore Turquet de Mayerne aus den Jahren 1620 bis 1646 (S. 6, 12), wollte Waller einen zuverlässigen und als dringend benötigt empfundenen Farbstandard in vier Sprachen erstellen: Lateinisch, Griechisch, Französisch und Englisch (wenngleich er seine Grenzen eingestand und nicht für alle Farben vollständige Namen beibringen konnte). Kurz nach ihrer Präsentation erschien diese Tabelle unter dem Titel *Tabula colorum physiologica tam mixtorum quam simplicium* (Tafel der physiologischen Farben, sowohl gemischt als auch einfach) in den *Philosophical Transactions*. In dem akkurat gestochenen Raster mit 119 systematisch angeordneten Farbtupfen befanden sich die Gelb- und Rottöne in der linken Spalte und die Weiß- und Blautöne in der oberen Reihe. Der Rest des Rasters war mit Mischungen dieser „einfachen Farben" im Verhältnis eins zu eins aufgefüllt. Begleitet wurde die Tabelle von einer mehrseitigen Beschreibung verschiedener Farbmittel in lateinischer Sprache, aus der sehr deutlich hervorgeht, dass sie als praktische Übersicht über Pigmente und ihre Mischungen für Malerei und Illustration gedacht war und nicht als philosophisches System, selbst wenn Wallers prinzipielle Farbanordnung nahelegt, dass er sich der aristotelischen Vorstellungen von Farbe und der Positionierung der Farben zwischen Dunkelheit und Licht bewusst war. Obwohl Wallers Farbtabelle bisweilen, und zu Recht, als eines der ersten in Farbe wiedergegebenen gedruckten Farbsysteme bezeichnet wird, ist zu betonen, dass die Tupfen von Hand gemalt waren und die unterschiedlichen Schattierungen in den wenigen noch erhaltenen Exemplaren daher stark voneinander abweichen. Während für die Bebilderung von Druckmedien mithilfe von Stichen in dieser Zeit eine neue, aufregende Phase begann, steckte der Farbdruck noch sehr in den Kinderschuhen.

Catalogue des couleurs simples et mélangées, avec un spécimen de chaque couleur associé à son nom

Le « Catalogue des couleurs simples et mélangées, avec un spécimen de chaque couleur associé à son nom » de Richard Waller illustre magnifiquement le souhait des Lumières de comprendre et de catégoriser le monde qui nous entoure et de donner une forme visuelle aux nouvelles connaissances acquises et systématisées. Naturaliste, traducteur et dessinateur anglais, Waller est peut-être plus connu pour l'édition posthume de l'œuvre du scientifique Robert Hooke (1635–1703). Membre de la Royal Society de Londres, Waller en devient vice-président sous la présidence d'Isaac Newton. Sa revue, fondée en 1665 – *Philosophical Transactions of the Royal Society* – est l'une des premières revues scientifiques au monde. Érudit polyglotte et intellectuel de premier plan dans l'Europe des Lumières, Waller se tient informé des recherches et débats consacrés à la couleur.

Aujourd'hui perdu, le tableau des couleurs que présente Waller aux membres de la Royal Society en 1686 est peint sur une plaque de verre. Son auteur s'est inspiré de *Nomenclatura et species colorum* (« Nomenclature et espèces des couleurs »), tableau multilingue publié en 1680 par le dessinateur suédois Elias Brenner (1647–1717), et peut-être aussi des listes de couleurs du « manuscrit de Mayerne » de 1620–1646 (p. 6, 12), dû à Théodore Turquet de Mayerne. Waller cherche ainsi à réaliser une norme chromatique fiable, devenue nécessaire, en quatre langues : latin, grec, français et anglais (il reconnaît cependant ses limites et ne peut indiquer les appellations complètes de toutes les couleurs). Peu après sa présentation, Waller fait paraître ce tableau dans les *Philosophical Transactions*, sous le titre « Tabula colorum physiologica tam mixtorum quam simplicium » (« Tableau de couleurs physiologiques tant mélangées que simples »). Celui-ci prend la forme d'une gravure montrant une grille clairement lisible, qui réunit 119 taches de couleur et organise de façon systématique les jaunes et les rouges dans la colonne de gauche, les blancs et les bleus sur la première ligne. Le reste de la grille comprend la combinaison par paires de ces « couleurs simples ». Ce tableau s'accompagne de plusieurs pages de description en latin des diverses matières colorantes. Il est manifestement destiné à être utilisé comme tableau pratique des pigments et de leurs mélanges par les peintres et illustrateurs et non comme système philosophique, même si la disposition générale adoptée par Waller témoigne de sa connaissance des idées d'Aristote sur la notion de couleur et la répartition des teintes entre obscurité et lumière. Bien que ce tableau soit parfois, et à juste titre, considéré comme l'un des tout premiers systèmes chromatiques imprimés en couleurs, il faut souligner que les taches ont été peintes à la main et que les différentes teintes varient énormément d'un exemplaire à l'autre parmi ceux qui nous sont parvenus. À cette époque, l'illustration imprimée sous forme de gravure connaît une nouvelle étape prometteuse, mais l'impression en couleurs n'en est encore qu'à ses balbutiements.

Catálogo de colores simples y mixtos con un ejemplo de cada color asociado a su nombre

El «catálogo» de colores de Richard Waller es un ejemplo espléndido del deseo de la Ilustración de comprender y clasificar el mundo, y de presentar visualmente el conocimiento recién adquirido y sistematizado. Waller fue un naturalista, traductor y dibujante inglés, aunque es quizás más conocido por editar la obra póstuma de Robert Hooke (1635–1703). Miembro de la Royal Society de Londres, se convirtió en su vicepresidente cuando Isaac Newton ejerció el cargo de presidente; la revista de la institución, *Philosophical Transactions of the Royal Society*, se lanzó en 1665 y fue una de las primeras publicaciones científicas de ese tipo en el mundo. Waller, él mismo un erudito multilingüe, estaba bien situado en la red ilustrada europea, y se mantuvo bien informado sobre los últimos debates e investigaciones sobre el color.

La tabla de colores original que Waller presentó a los miembros de la Royal Society en 1686 estaba pintada sobre vidrio y se ha perdido. Waller se había inspirado en *Nomenclatura et species colorum* (Nomenclatura y tipos de colores), una tabla de colores multilingüe publicada en 1680 por el dibujante sueco Elias Brenner (1647–1717), y posiblemente también en las listas de colores del «manuscrito de Mayerne» de 1620–1646 (págs. 6, 12) de Théodore Turquet de Mayerne; como resultado, Waller buscó crear un estándar de colores fiable y muy necesario en cuatro idiomas: latín, griego, francés e inglés (aunque reconoció sus limitaciones y no pudo proporcionar los nombres completos de todos los colores). Poco después de su presentación, la tabla se publicó en *Philosophical Transactions* con el título «Tabula colorum physiologica tam mixtorum quam simplicium» (Tabla de colores fisiológicos, tanto mixtos como simples) como una ordenada cuadrícula grabada que incluye 119 muestras de colores dispuestas en orden sistemático, con amarillos y rojos en la columna de la izquierda y blanco y azules en la parte superior. El resto de la cuadrícula la ocupan mezclas a partes iguales de esos «colores simples». La tabla, que va acompañada de varias páginas de descripciones en latín de distintos colorantes, está claramente pensada como índice práctico de pigmentos y de mezclas para su aplicación en pintura e ilustración, más que como sistema filosófico, aunque su disposición general sugiera que Waller estaba familiarizado con las ideas aristotélicas sobre la posición de los colores entre la oscuridad y la luz. Aunque el gráfico de Waller se menciona a veces, y con razón, como uno de los primeros sistemas de color visuales impresos editados en color, es importante señalar que las muestras se pintaron a mano y, por lo tanto, sus tonos varían mucho de uno a otro de los pocos ejemplares que se conservan. Si bien el uso del grabado como técnica de ilustración entraba en ese momento en una fase nueva y emocionante, la impresión en color seguía en pañales.

A. Boogert (forename and dates unknown)

KLAER LIGHTENDE SPIEGEL DER VERFKONST

A Clearly Lit Mirror of Painting

19 pages, 16 x 9.5 cm / 6 ¼ x 3 ¾ in., Delft, 1692
Aix-en-Provence, Bibliothèque Méjanes

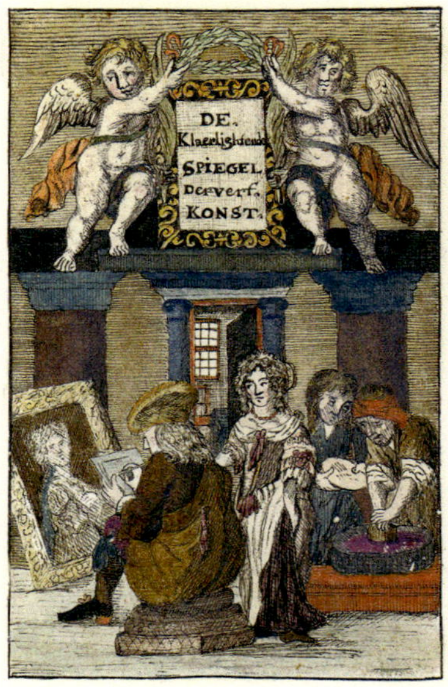

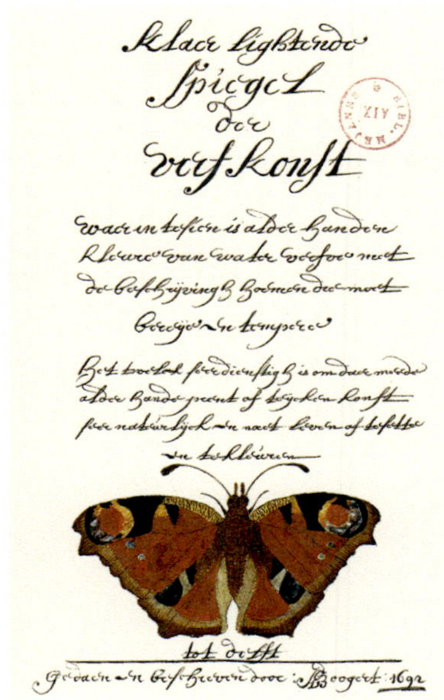

This is one of the most beautiful early books about colour in existence, yet it is shrouded in mystery, and was, as far as is known, never actually published. It survives as a manuscript originally written in Delft in 1692, comprising nearly 900 pages of elegant longhand text in Dutch, illustrated on almost every page with colour samples in watercolour and gouache. The otherwise unidentified author, A. Boogert, added that name to the title-page and to the end of the introduction. The relatively small bound manuscript has been in the collection of the Bibliothèque Méjanes for many years but was brought to wider attention in 2014 by the book historian Erik Kwakkel via social media. Since coming to light, the manuscript has been a source of widespread fascination and the subject of much academic research, although very little has been published about it so far. Comparisons have been drawn with colour guides produced by paint-makers from around the same time, and it is certainly tempting to think of it as a 17th-century colour catalogue for designers, or a list of paints available in Delft in the 1690s, but the real story behind the manuscript and its purpose might well be more complex. It is a striking object, largely because of its brilliant and well-preserved colour samples, which were painted directly on to the paper, but also because of the notably modern-looking, abstract simplicity of their design. The samples are all presented in clearly defined squares or rectangles that collectively occupy most of the illustrated pages, beginning with a section of 42 pages showing one colour on each page in three stages of dilution, followed by 323 plates showing colours mixed in five ratios, resulting in a total of 1,615 samples. A detailed "register" of colours completes the manuscript.

Boogert presented this work throughout in the manner of a finished printed book, repeating the title *Klaer Lightende Spiegel der Verfkonst* (literally, 'A clearly lit / illuminated mirror of painting', but it could be read as 'An enlightened reflection on painting') at the top of each illustrated double-page, which suggests the author had every intention for it to be published. It is hard to imagine how the plates could have been reproduced in print format at this early date, and that was perhaps part of the reason why it was never properly issued. It is nevertheless clear that Boogert intended it to be a comprehensive and practical painting compendium that would be helpful in colouring engravings and drawing from nature, while the text also included advice on how to prepare paints. This may explain the prominent image of a butterfly depicted on the title-page, a symbol of the complexities of colour in the natural world, and also the iconography of the frontispiece, which shows an artist (or engraver) copying a large portrait painting, behind whom an assistant is preparing pigment by grinding it with a muller while another person is waiting to transfer the colour to a palette. Boogert's book was perhaps too complex in its design for late 17th-century print culture, but it certainly strikes a chord with the 21st-century world of digital and social media.

1

2

3

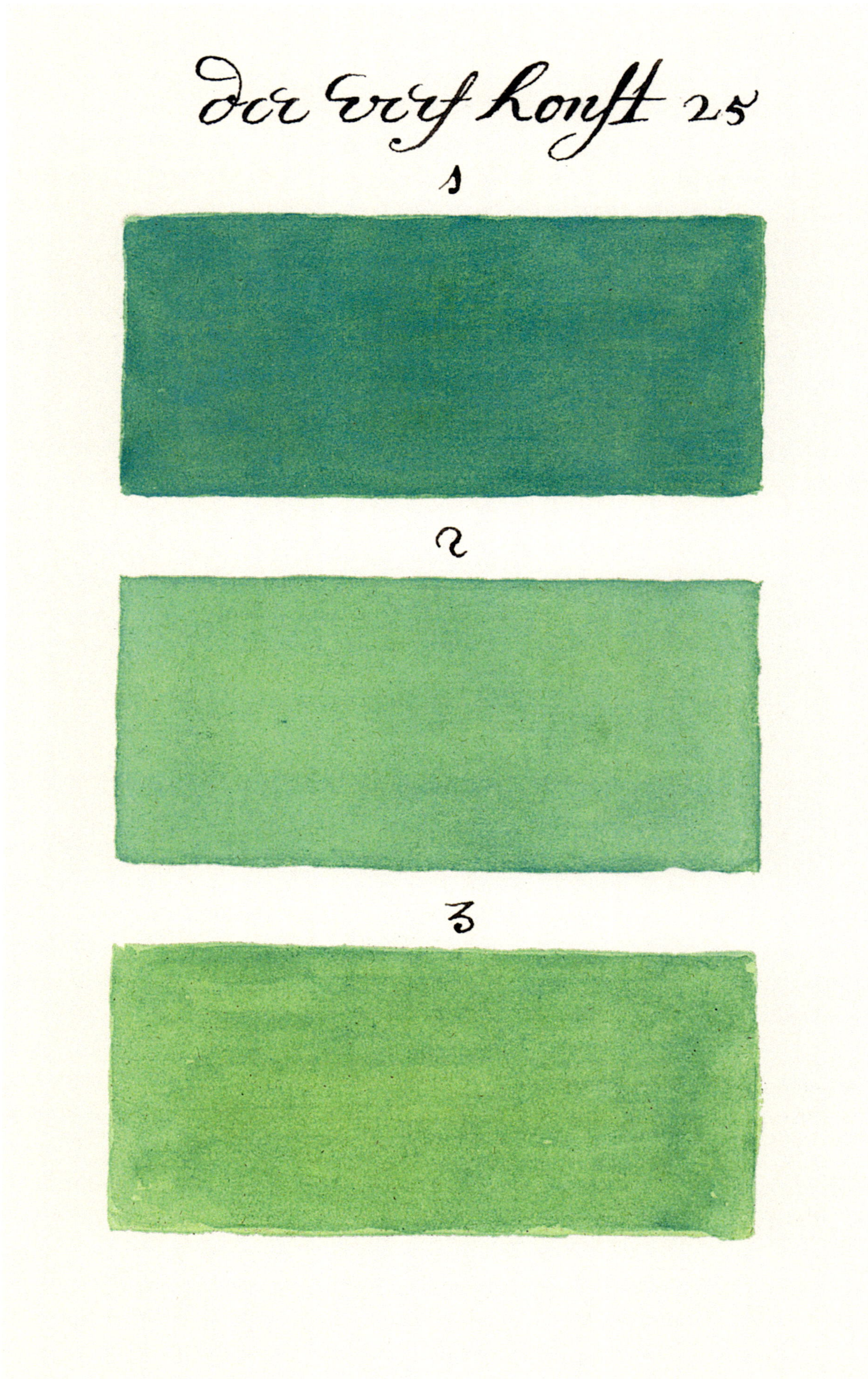
der verf konst 25
1
2
3

Klar leuchtender Spiegel der Malkunst

Dieses Buch zählt zu den schönsten noch existierenden Frühwerken über Farbe. Doch es ist von Geheimnissen umwittert und wurde, soweit man weiß, niemals wirklich veröffentlicht. Erhalten ist das Originalmanuskript, das 1692 in Delft verfasst wurde, fast neunhundert Seiten in eleganter niederländischer Schreibschrift umfasst und auf fast jeder Seite mit Farbmustern in Wasserfarbe und Gouache bebildert ist. Der ansonsten unbekannte Autor setzte den Namen A. Boogert auf die Titelseite und ans Ende der Einleitung. Das vergleichsweise kleine, gebundene Manuskript befand sich über viele Jahr in der Bibliothèque Méjanes, bevor der Buchhistoriker Erik Kwakkel es 2014 über die sozialen Medien einer breiten Öffentlichkeit zugänglich machte. Seit es auf diese Weise bekannt wurde, fasziniert das Manuskript Wissenschaftler, Buchliebhaber, Designer und Farbhistoriker gleichermaßen und wurde zum Gegenstand zahlreicher wissenschaftlicher Untersuchungen, wenngleich bislang nur sehr wenig darüber publiziert wurde. Man hat es mit Handbüchern von Farbenherstellern aus etwa derselben Zeit verglichen, doch so verlockend es auch sein mag, dieses Buch als Farbenkatalog für Gestalter aus dem 17. Jahrhundert oder als Liste der in den 1690er-Jahren in Delft verfügbaren Farben zu betrachten, könnte die wahre Geschichte hinter dem Manuskript und seiner Zielsetzung weitaus vielschichtiger sein. Vor allem wegen seiner leuchtenden und gut erhaltenen Farbmuster, die direkt auf das Papier gemalt wurden, aber auch wegen der gestalterisch bemerkenswert modern wirkenden, abstrakten Einfachheit dieser Farbflächen ist es ein faszinierendes Objekt. Die Muster werden ausnahmslos in gleichmäßig angeordneten Quadraten oder Rechtecken präsentiert, die zusammen den größten Teil

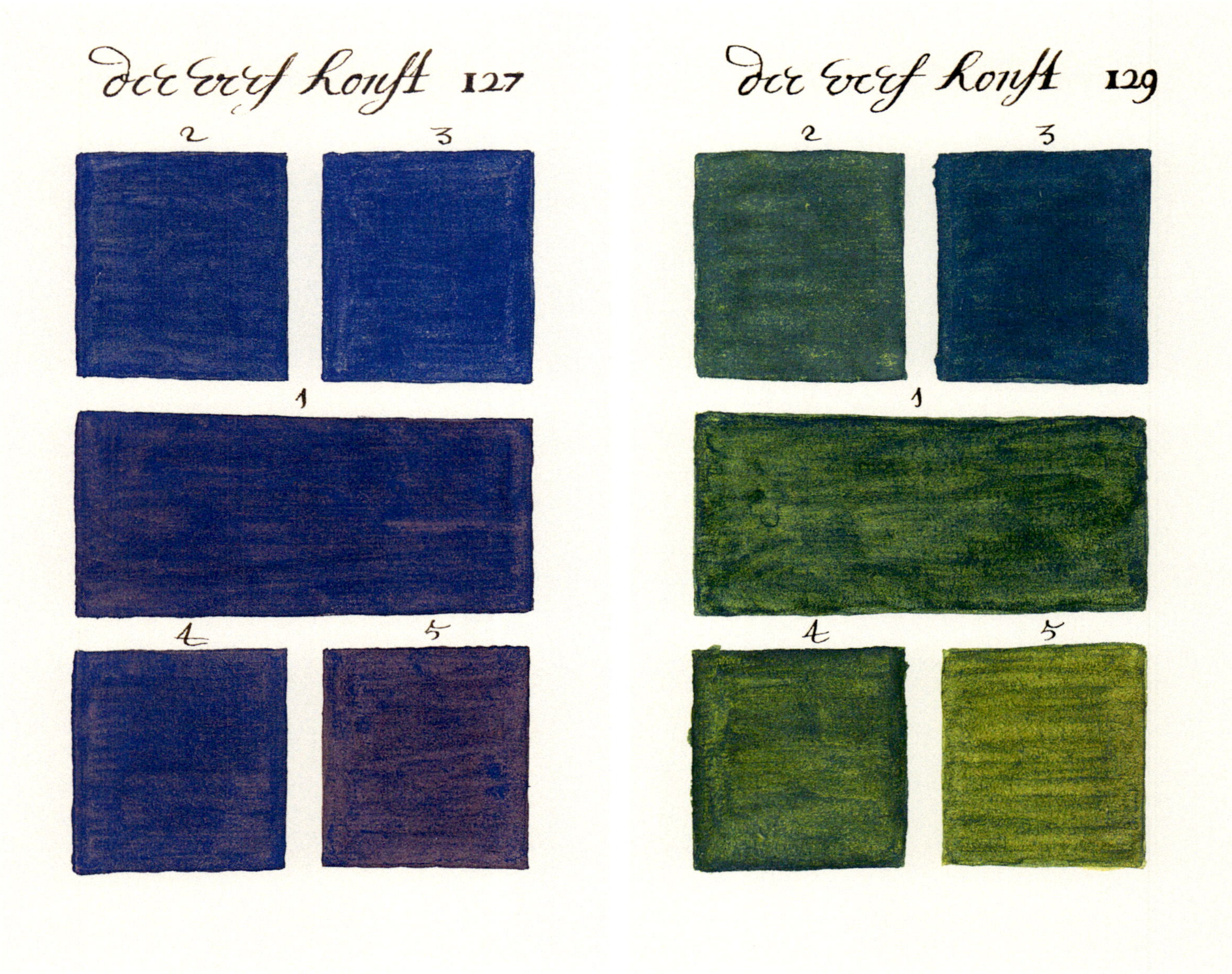

der illustrierten Seiten einnehmen. Sie beginnen mit einem 42-seitigen Abschnitt, der auf jeder Seite eine Farbe in drei Verdünnungszuständen zeigt. Auf ihn folgen 323 Tafeln mit Farben in fünf Mischungsverhältnissen und insgesamt 1.615 Mustern. Ein ausführliches „Register“ der Farben rundet das Manuskript ab.

Boogert präsentierte das Werk durchgehend in der Art eines fertig gedruckten Buches und wiederholte den Titel *Klaer Lightende Spiegel der Verfkonst* (wörtlich „Klar leuchtender/illuminierter Spiegel der Malkunst“, aber auch zu lesen als „Eine aufgeklärte Betrachtung über das Malen“) am Kopf jeder illustrierten Doppelseite, was auf seine feste Veröffentlichungsabsicht hindeutet. Es ist schwer vorstellbar, wie die Tafeln zu diesem frühen Zeitpunkt im Druckformat hätten reproduziert werden sollen, und womöglich war das einer der Gründe, warum das Buch nie als solches erschien. Dennoch steht außer Frage, dass es von Boogert als umfassendes und praktisches Malhandbuch konzipiert war, das bei der Kolorierung von Stichen und beim Zeichnen nach der Natur hilfreich sein sollte, während der Text auch Ratschläge für die Herstellung von Farben enthält. Das mag die prominente Abbildung eines Schmetterlings auf der Titelseite erklären, eines Symbols für die Vielfalt der Farben in der Natur, und auch die Ikonografie des Frontispizes, das einen Künstler (oder Kupferstecher) beim Kopieren eines großen Porträtgemäldes zeigt, hinter dem ein Assistent mit einem Mahlstein Pigmente zubereitet, während eine weitere Person bereitsteht, um die Farbe auf eine Palette zu übertragen. Boogerts Buch war in seiner Gestaltung vielleicht zu komplex für das Druckwesen des späten 17. Jahrhunderts, doch es trifft mit Sicherheit den Nerv der digitalen und sozialen Medienwelt des 21. Jahrhunderts.

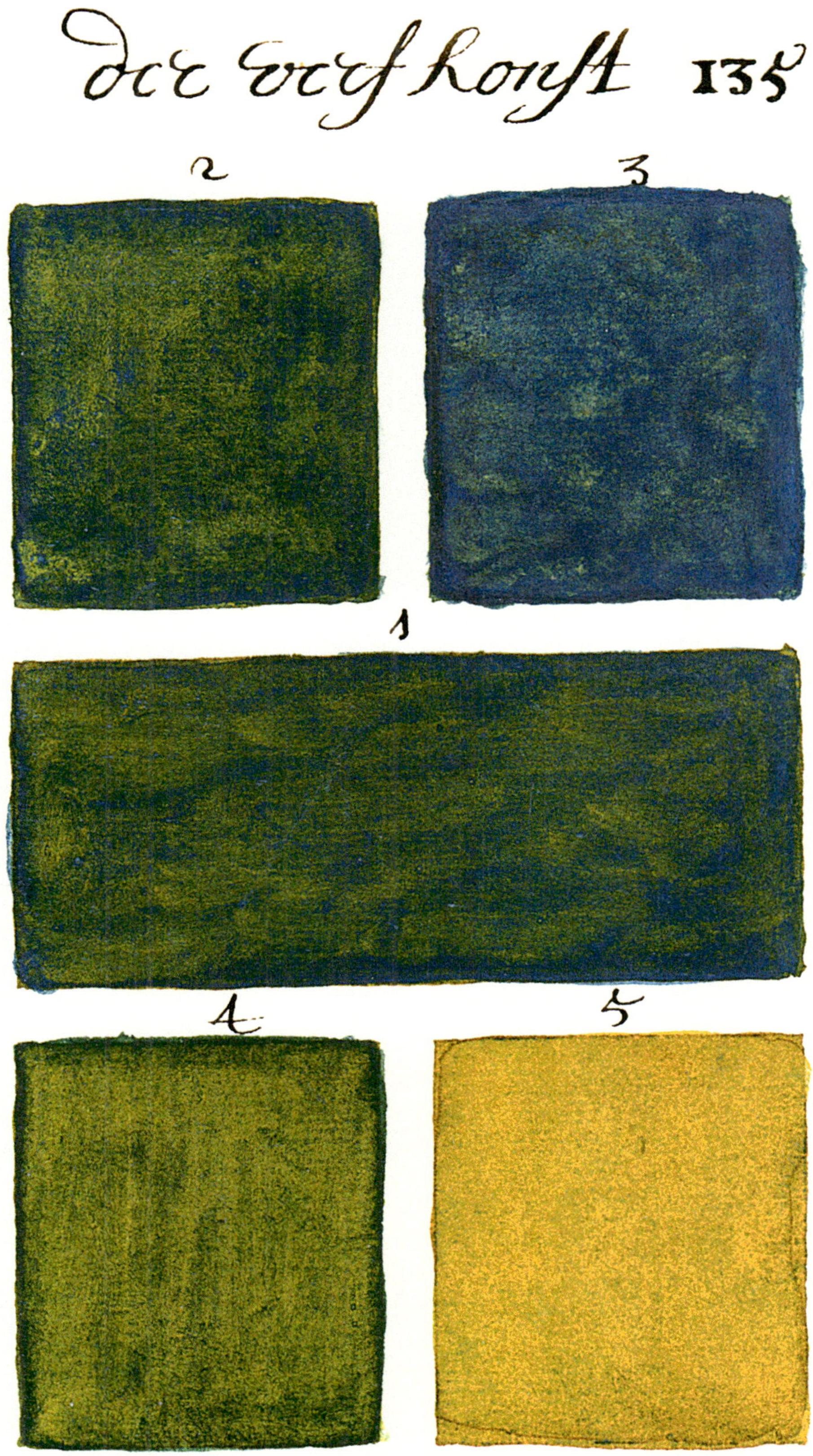
der verfkonst 135
2
3
1
4
5

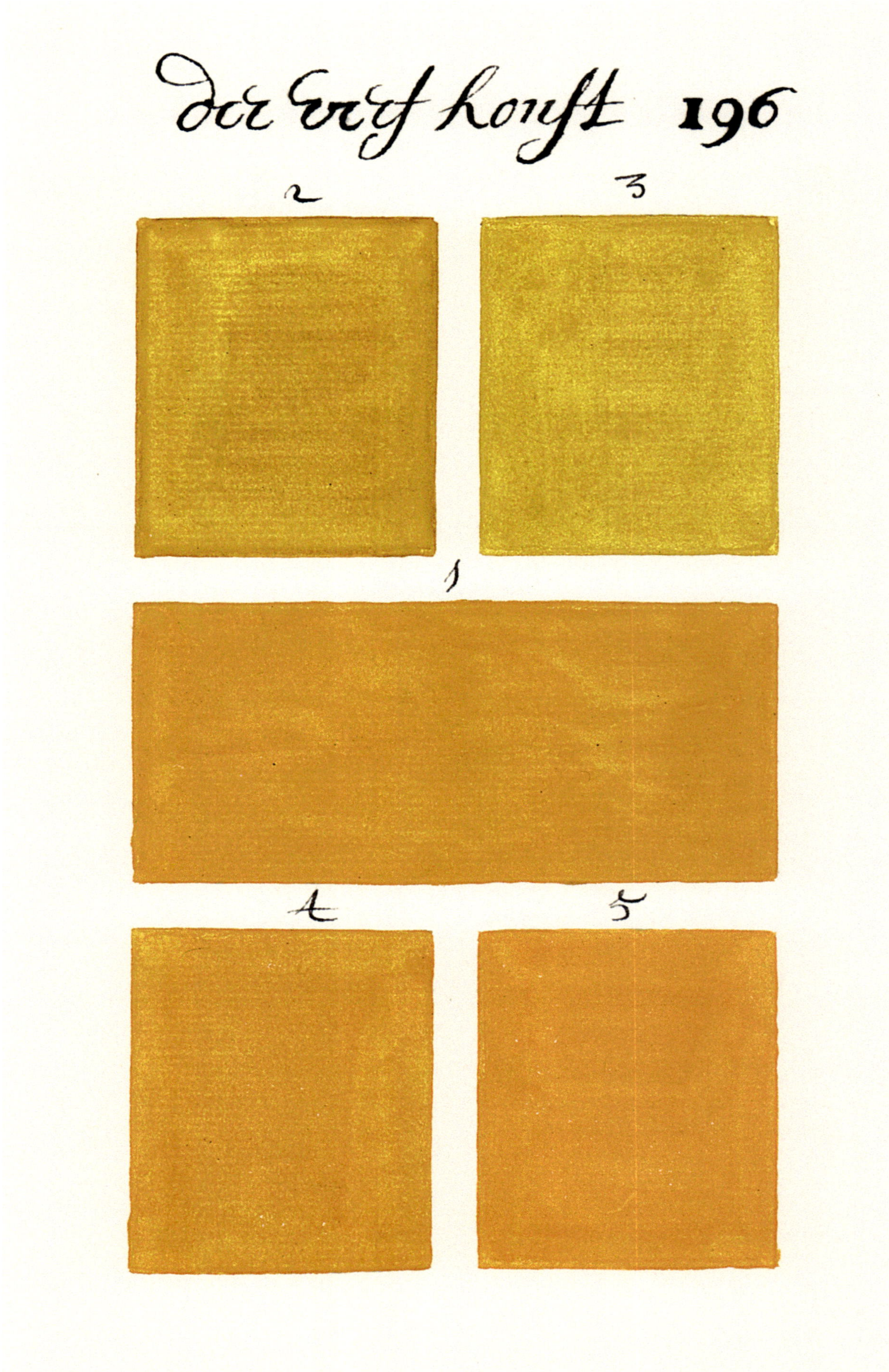
der verf konst 196
2
3
1
4
5

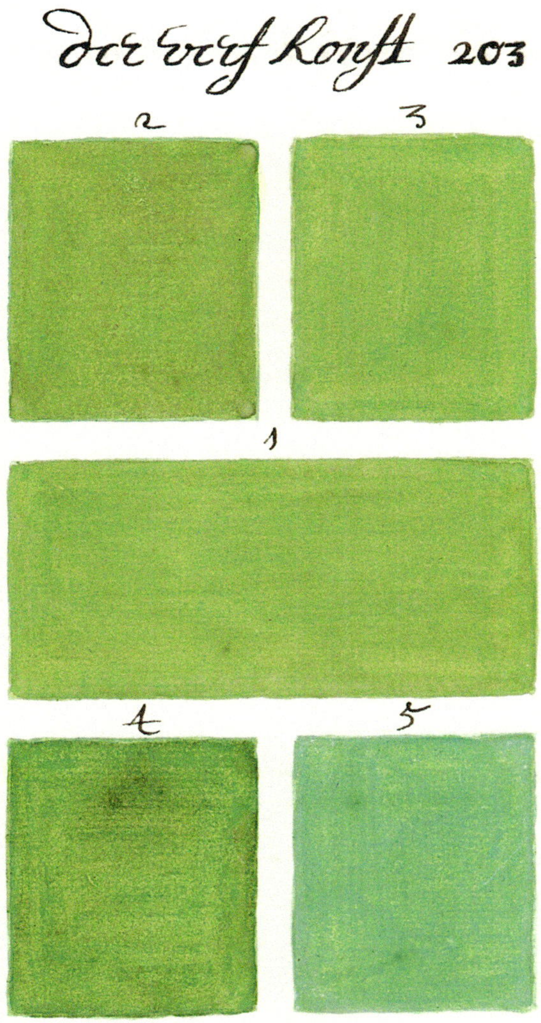

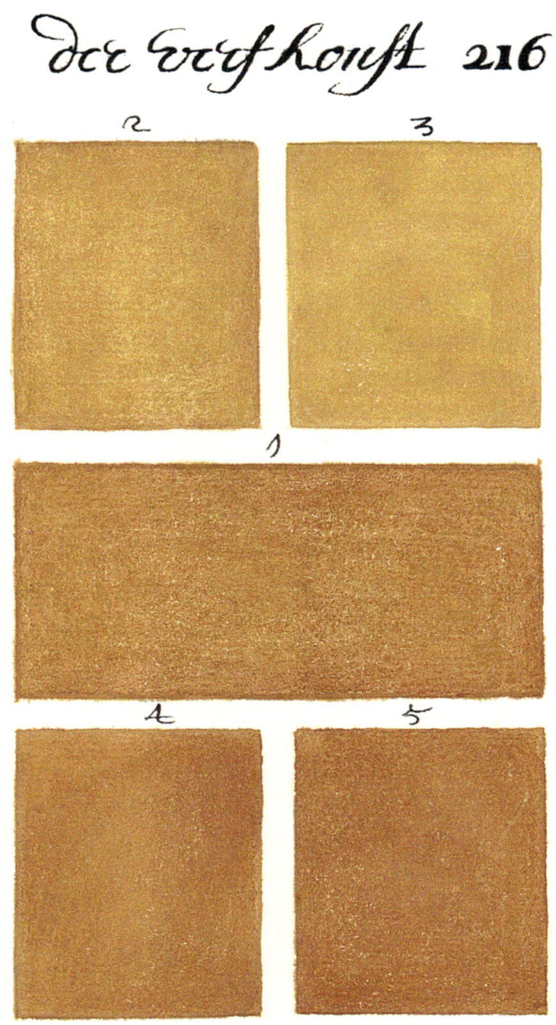

Miroir bien éclairé de la peinture

Il s'agit là d'un des plus beaux ouvrages primitifs sur la couleur ayant survécu. Pourtant, il constitue une énigme car, en l'état actuel des connaissances, il n'a jamais été publié. Il n'en existe qu'un manuscrit rédigé à Delft en 1692, comprenant près de 900 pages avec un texte en néerlandais à l'écriture élégante, presque chaque page étant illustrée d'échantillons de couleurs à l'aquarelle et à la gouache. Le nom de l'auteur, A. Boogert, dont on ne sait rien d'autre, figure sur la page de titre et à la fin de l'introduction. Ce manuscrit relié aux dimensions relativement réduites est conservé à la bibliothèque Méjanes d'Aix-en-Provence depuis de nombreuses années. Mais ce n'est qu'en 2014 que l'historien du livre Erik Kwakkel l'a mis en lumière sur les réseaux sociaux. Depuis lors, l'ouvrage fascine les spécialistes, les bibliophiles, les graphistes et les historiens de la couleur et fait l'objet de nombreuses recherches universitaires, bien que fort peu d'études aient été publiées jusqu'ici à son sujet. Ce manuscrit a été comparé à des guides chromatiques réalisés par des fabricants de peintures de la même époque. Il est certes tentant d'y voir un catalogue de couleurs destiné aux illustrateurs du XVIIe siècle, ou une liste des peintures disponibles à Delft dans les années 1690. Pourtant, son histoire et sa destination véritables pourraient bien être plus complexes. C'est un objet hors du commun, principalement en raison de ces échantillons de couleurs vives fort bien conservés, peints à même le papier, mais aussi pour la simplicité abstraite et remarquablement moderne de sa mise en page. Ces échantillons sont tous présentés sous forme de carrés ou de rectangles bien délimités qui remplissent la plupart des pages illustrées : la première section de 42 pages, à raison d'une teinte par page selon

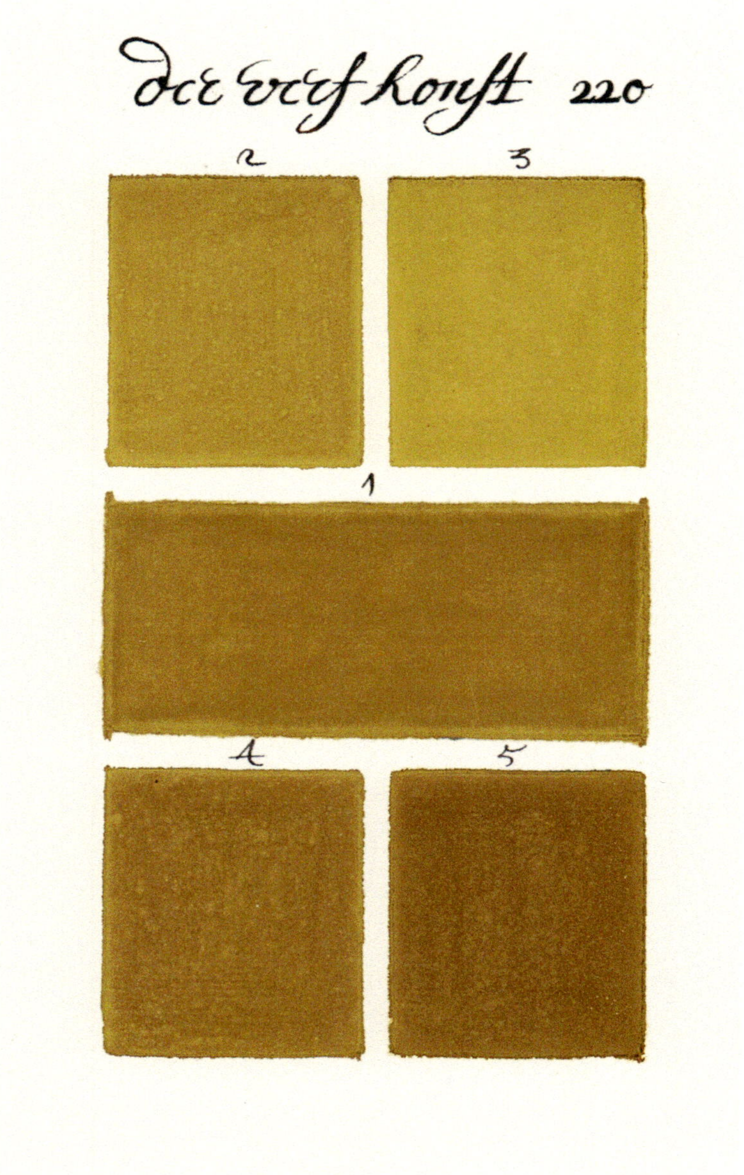

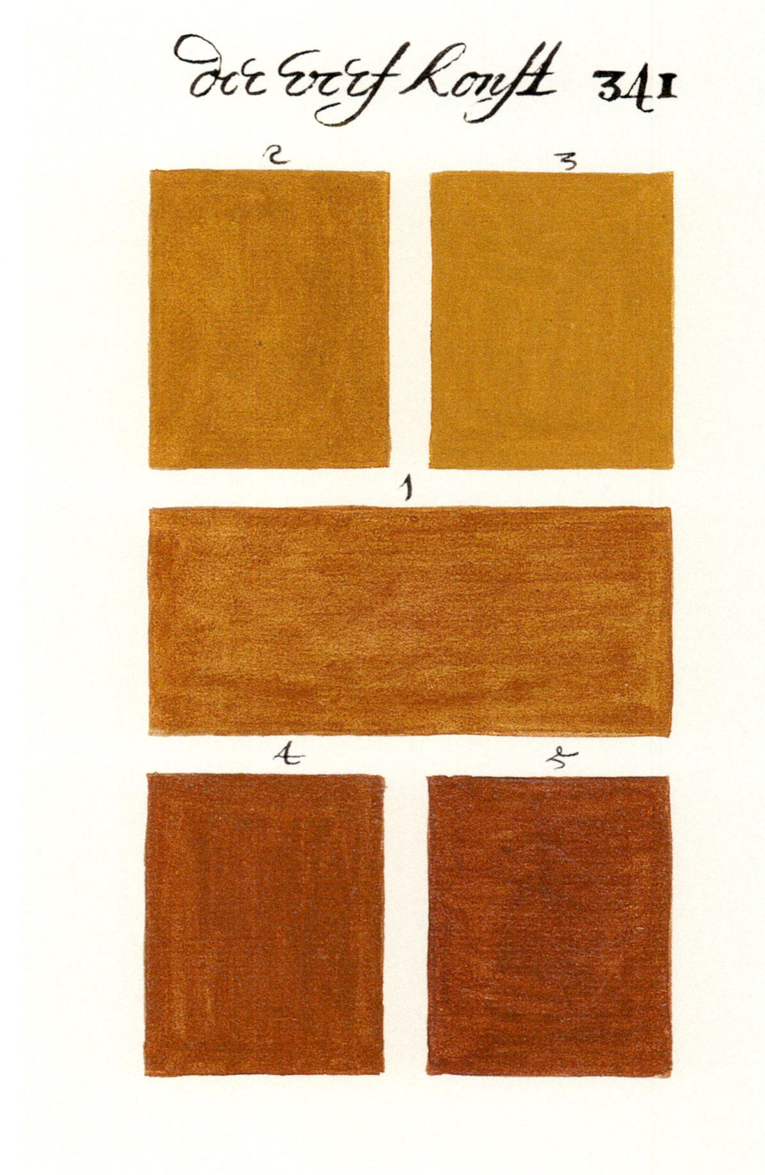

trois états de dilution, est suivie de 323 planches de couleurs avec cinq nuances, pour un total de 1 615 échantillons. Un « registre » de couleurs complète le manuscrit.

Boogert présente cette œuvre à la manière d'un livre achevé et imprimé, dont il répète le titre *Klaer Lightende Spiegel der Verfkonst* (littéralement, « Miroir bien éclairé de la peinture », ou encore « Réflexion éclairée sur la peinture ») en haut de chaque double page illustrée, signe que son auteur avait bien l'intention de le voir publié. Il est difficile d'imaginer comment les planches auraient pu être reproduites par impression à cette époque, raison possible pour laquelle l'ouvrage n'a jamais paru. Néanmoins, il est évident que Boogert souhaitait en faire un précis complet et pratique sur la peinture, qui soit utile pour colorier des gravures et des dessins effectués d'après nature et qui s'accompagne d'un texte comportant des conseils sur la manière de préparer les peintures. Ceci expliquerait la présence sur la page de titre d'une grande image de papillon, symbole des complexités chromatiques du monde naturel, ainsi que l'iconographie du frontispice : on y voit un peintre (ou un graveur) occupé à copier un grand portrait, derrière lequel un assistant prépare un pigment en le concassant avec un pilon, tandis qu'un autre personnage attend de transférer la couleur sur une palette. Peut-être la conception de l'ouvrage de Boogert était-elle trop complexe pour les techniques d'impression du XVIIe siècle. Elle semble plus en phase avec le monde numérique et les réseaux sociaux du XXIe siècle.

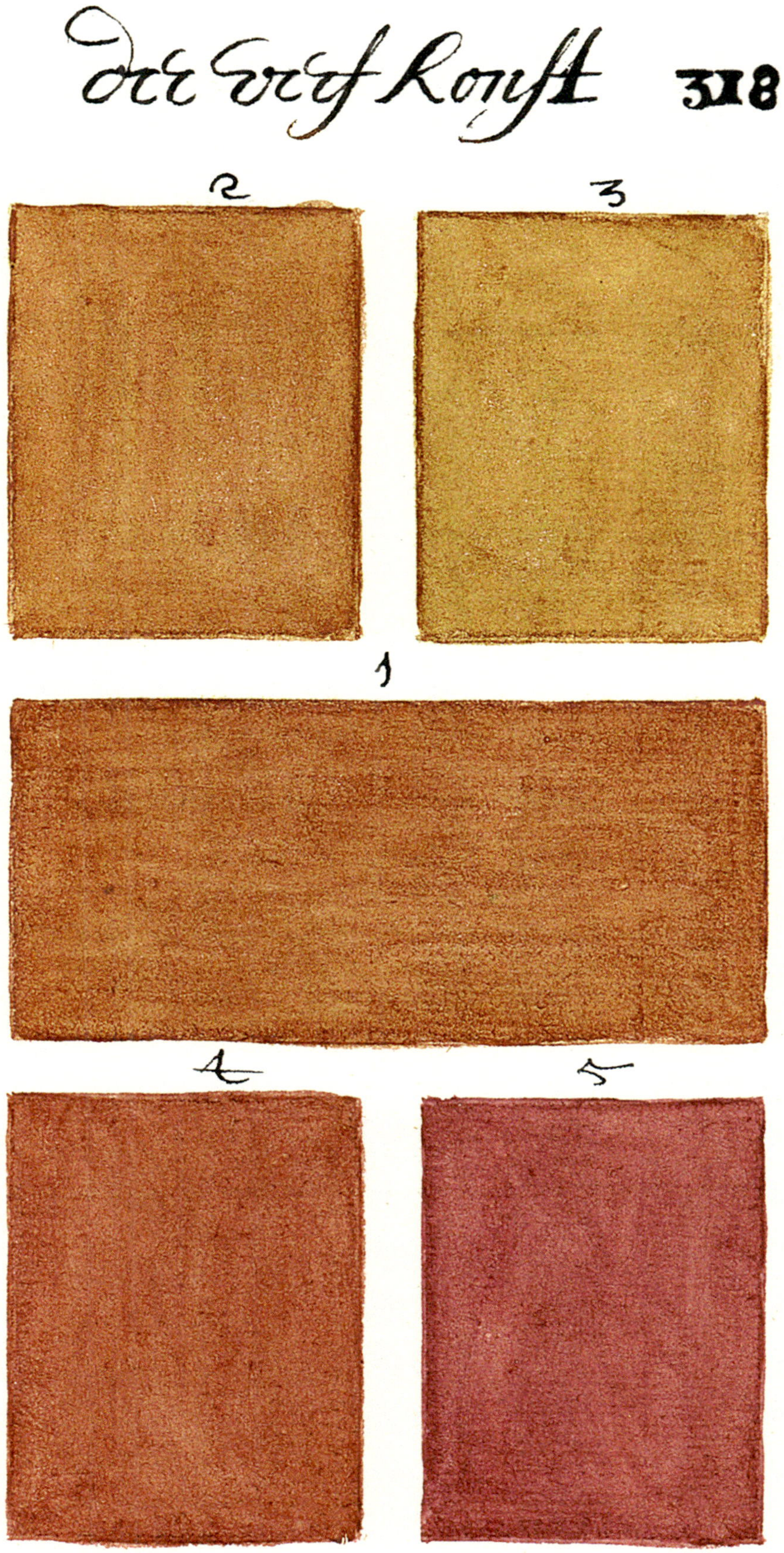
der verf konst
318
2
3
1
4
5

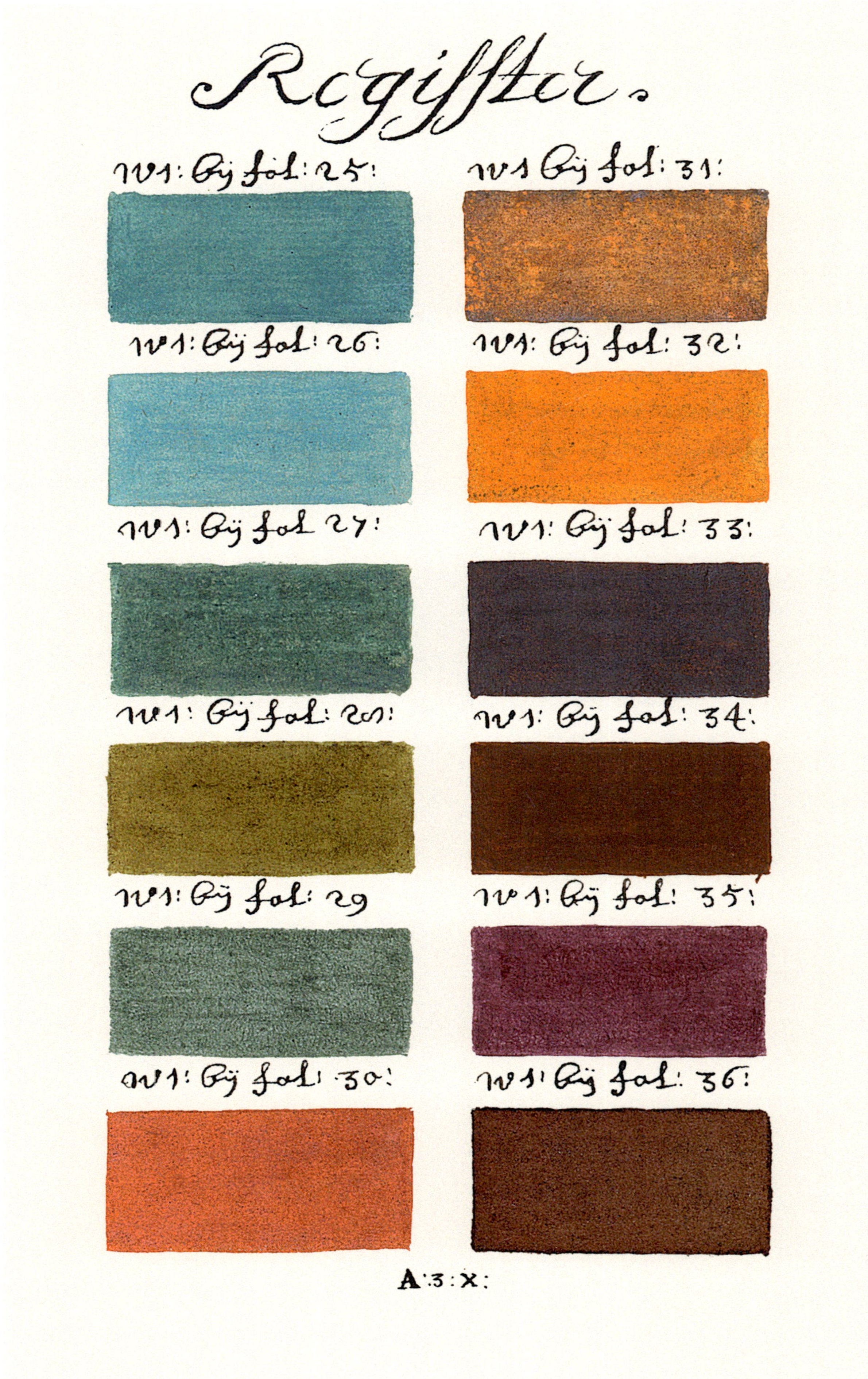
Regiſter.
N°1: Bij fol: 25:
N°1: Bij fol: 31:
N°1: Bij fol: 26:
N°1: Bij fol: 32:
N°1: Bij fol: 27:
N°1: Bij fol: 33:
N°1: Bij fol: 28:
N°1: Bij fol: 34:
N°1: Bij fol: 29
N°1: Bij fol: 35:
N°1: Bij fol: 30:
N°1: Bij fol: 36:
A:3:X:

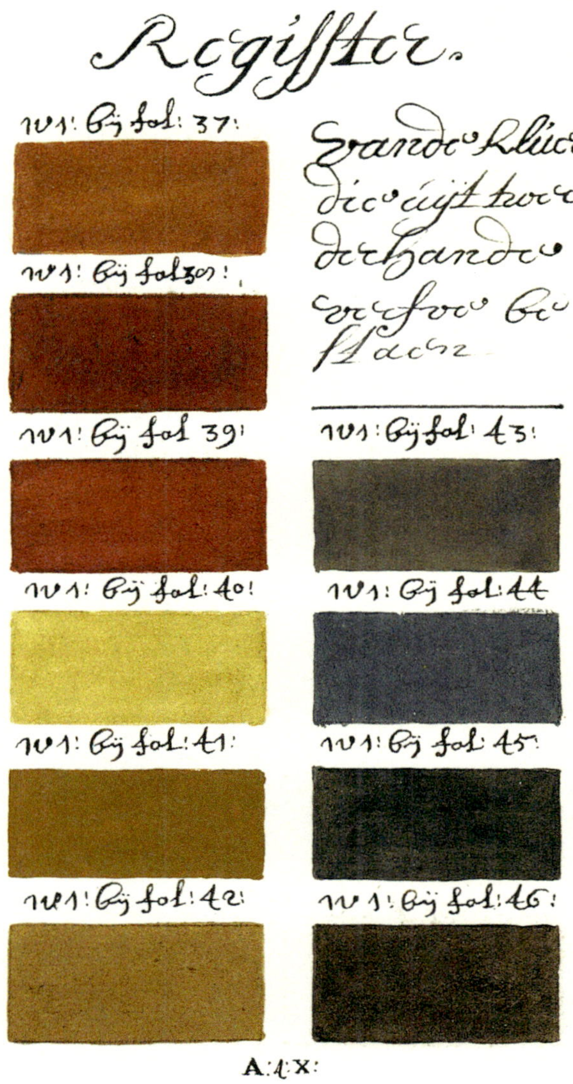
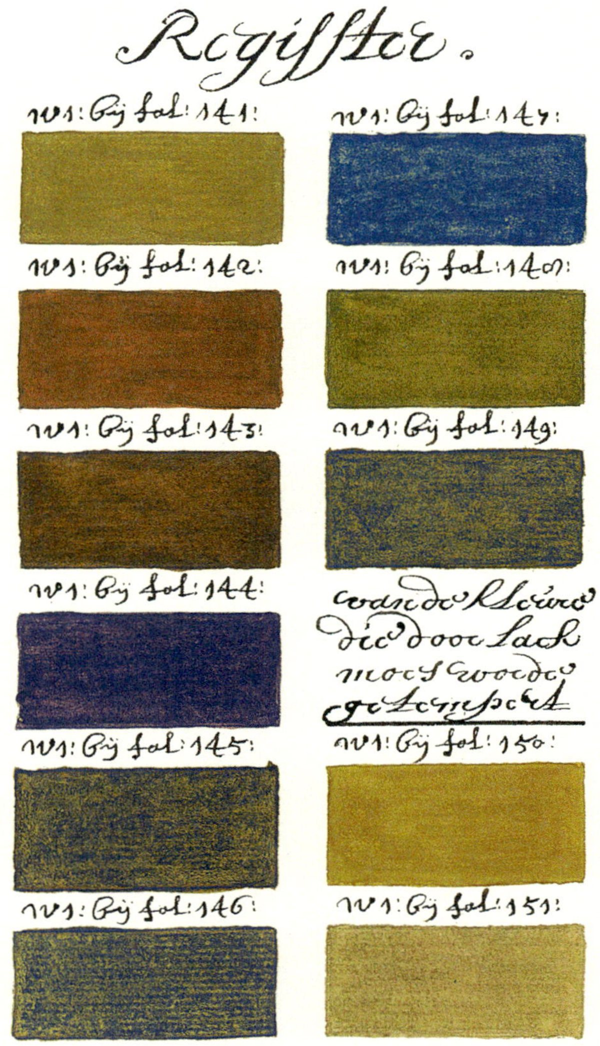

Un espejo de la pintura claramente iluminado

Este es uno de los libros de primera época más bellos que existen sobre el color. Está envuelto en el misterio y, hasta donde se sabe, nunca se publicó. Sobrevive en forma de un manuscrito escrito originalmente en Delft en 1692. Consta de casi 900 páginas de elegante texto en holandés, casi todas ellas ilustradas con muestras de color en acuarela y guache. Su desconocido autor, A. Boogert, agregó ese nombre a la portada y al final de la introducción. Este manuscrito encuadernado relativamente pequeño ha formado parte durante muchos años de la colección de la Bibliothèque Méjanes en Aix-en-Provence, pero el historiador de libros Erik Kwakkel llamó la atención sobre él en 2014 a través de las redes sociales. Desde que salió así a la luz, ha fascinado a académicos, bibliófilos, diseñadores e historiadores del color por igual, y se ha convertido en objeto de mucha investigación académica, aunque hasta ahora se haya publicado muy poco. Se han realizado comparaciones con guías de colores producidas por pintores de la misma época, y ciertamente es tentador pensar en él como un catálogo de colores del siglo XVII para diseñadores o como una lista de pinturas disponibles en Delft en la década de 1690, pero su verdadera historia y propósito podrían ser más complejos. Se trata de un objeto llamativo, en gran parte por sus muestras de color brillantes y bien conservadas, pintadas directamente sobre el papel, pero también por la simplicidad abstracta de su diseño, de aspecto notablemente moderno. Todas las muestras se presentan en cuadrados o rectángulos claramente definidos que en conjunto ocupan la mayor parte de las páginas ilustradas, que comienzan con una sección de 42 páginas que muestra un color en cada página en tres etapas de dilución a la que siguen 323 páginas con colores

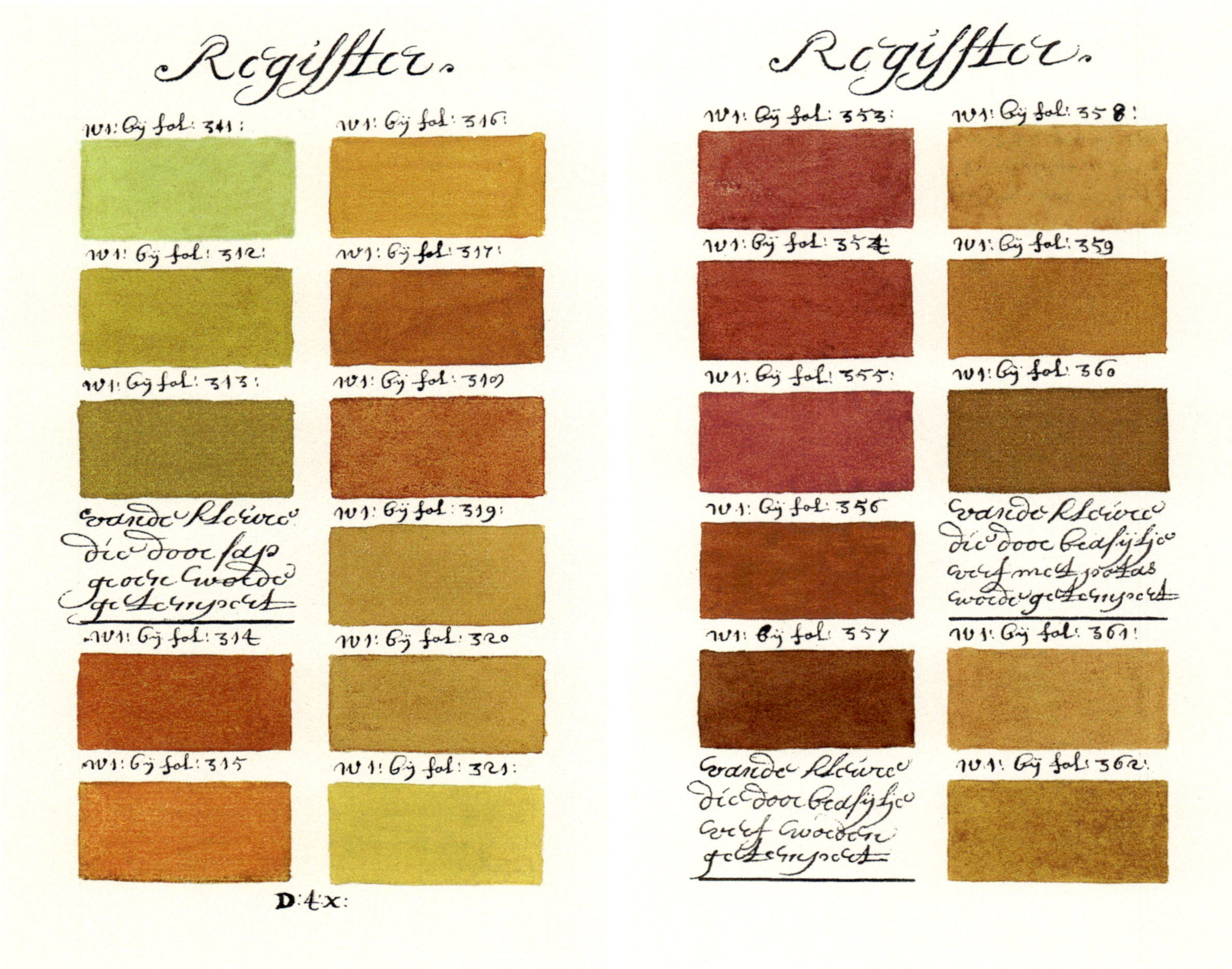

mezclados en cinco proporciones, lo que da como resultado un total de 1615 muestras. Un detallado «registro» de colores completa el manuscrito.

Boogert presentó este trabajo como un libro impreso terminado que repite el título, *Klaer Lightende Spiegel der Verfkonst* (literalmente, «Un espejo de la pintura claramente iluminado», pero que puede interpretarse como «Una reflexión ilustrada sobre la pintura») en el encabezado de cada doble página ilustrada, lo que sugiere que el autor tenía toda la intención de que se publicara. Es difícil imaginar cómo las láminas podrían haberse reproducido en formato impreso en una fecha tan temprana, lo que quizá explique en parte por qué eso nunca sucedió; sin embargo, está claro que Boogert pretendía que fuera un compendio de pintura completo y útil para colorear grabados y dibujar del natural, mientras que el texto también incluye consejos sobre cómo preparar pinturas. Eso puede explicar la imagen prominente de una mariposa en la portada, un símbolo de las complejidades del color en el mundo natural, y también la iconografía del frontispicio, que muestra a un artista (o grabador) copiando un gran retrato; detrás de él, un ayudante se encarga de moler el pigmento mientras otra persona espera para transferir el color a una paleta. El libro de Boogert fue quizás demasiado complejo en su diseño para la cultura impresa de finales del siglo XVII, pero está claro que despierta emociones en el mundo de las redes sociales y digitales del siglo XXI.

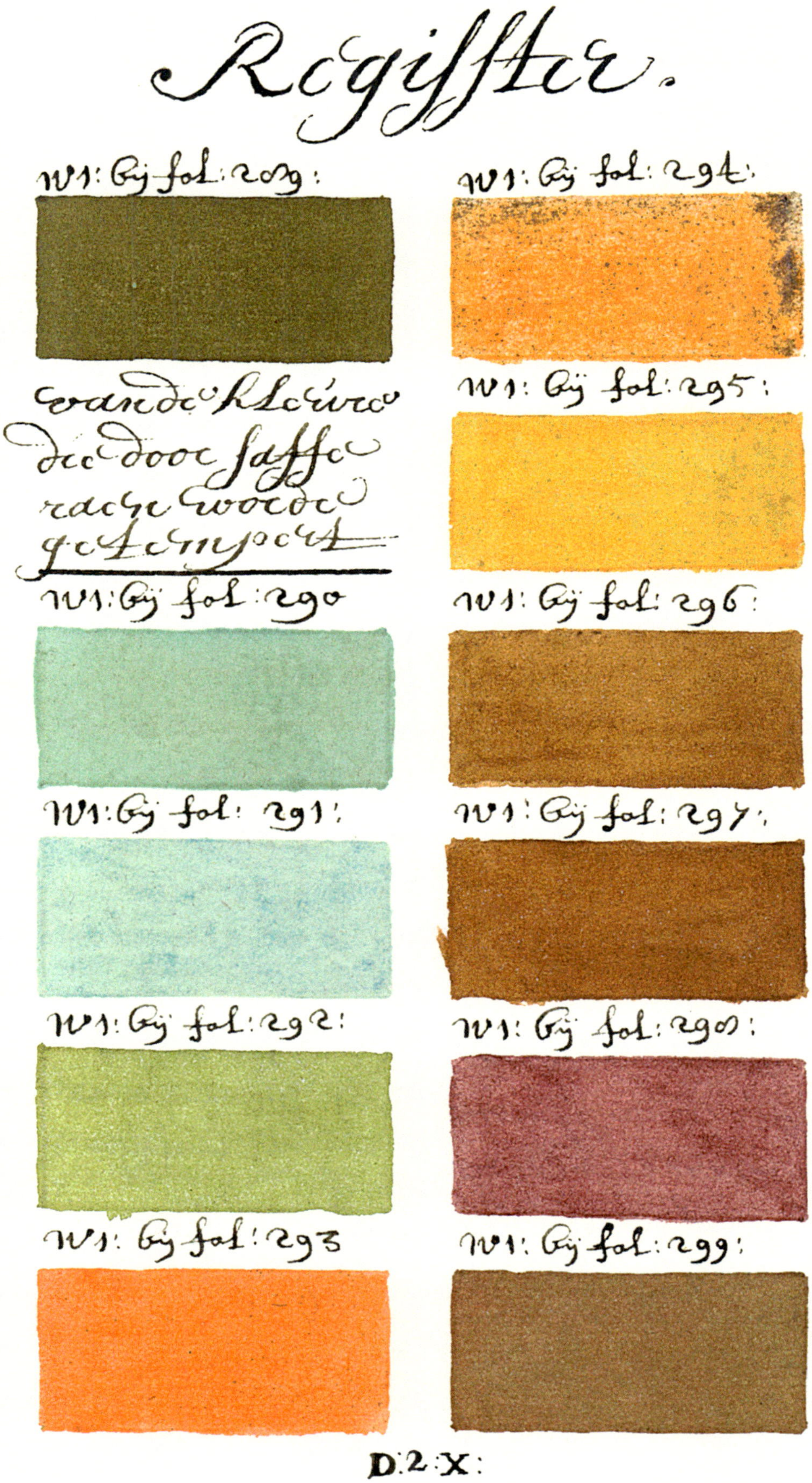
Regiſster.

N1: Bij fol: 289:

van de kleuren
die door saffe
rache worde
getempert

N1: Bij fol: 290

N1: Bij fol: 291:

N1: Bij fol: 292:

N1: Bij fol: 293

N1: Bij fol: 294:

N1: Bij fol: 295:

N1: Bij fol: 296:

N1: Bij fol: 297:

N1: Bij fol: 298:

N1: Bij fol: 299:

D:2:X:

Antoine Gautier de Montdorge (1701 or 1707–1768), Jacob Christoph Le Blon (1667–1741)

L'ART D'IMPRIMER LES TABLEAUX

The Art of Printing Pictures
1 plate, 20 x 23.5 cm / 7 ⅞ x 9 ¼ in., Paris, 1756
Washington, D.C., Smithsonian Libraries

Artists' palettes feature prominently in representations of the visual arts, sometimes as symbols of creativity and style, sometimes as signifiers of specific techniques or of the colours used by or available to artists and theorists. William Hogarth (1697–1764) placed one of them in a central position in an important illustration in his philosophical treatise *The Analysis of Beauty* (1753), while they often appear of course in self-portraits of painters, where they indicate the artists' working methods and colour preferences. The palette in *L'art d'imprimer les tableaux* has a different meaning, however, but one that is still linked to technique and materiality.

Jacob Christoph Le Blon was a German engraver and printer who had studied and worked in Rome, Switzerland, Amsterdam and Paris, during which time he gained significant international knowledge and experience. He is widely credited with having invented the basic process of three- and four-colour printing, and first applied the technique, which he had begun to develop in London from 1715, when he published a bilingual English and French work in 1725 entitled *Coloritto, or The Harmony of Colouring in Painting / L'Harmonie du coloris dans la peinture*. The title-page proudly announced that the short text contained "some colour'd figures", and the book itself, in addition to being a demonstration of the new printing method, is also in part a technical essay on copying and reproducing works of art. Le Blon, who appears to have had the ability to break down an existing image mentally into three layers each of a different colour, summarised his strictly trichromatic system in plain language: "Painting can represent all *visible* Objects with three Colours, *Yellow*, *Red*, and *Blue*; for all other Colours can be compos'd of these *Three*, which I call *Primitive*". This now extremely rare work was reprinted in France in 1756, 15 years after Le Blon's death, edited and expanded by his friend and supporter Antoine Gautier de Montdorge who issued it with the title *L'art d'imprimer les tableaux; traité d'après les écrits, les opérations, & les instructions verbales de J. C. Le Blon*. Both works include a fold-out plate that realistically depicts a painter's palette with 15 paint colours, each marked with a letter to identify them when they are discussed in one of the book's chapters. The visual appearance of the colours on the palette is flawed, but the image still represents both Le Blon's advice on colour mixing and composition together with his pioneering printing technique. While his printing methods were not wholly successful, they nevertheless paved the way for chromolithography and modern three- and four-tone colour printing.

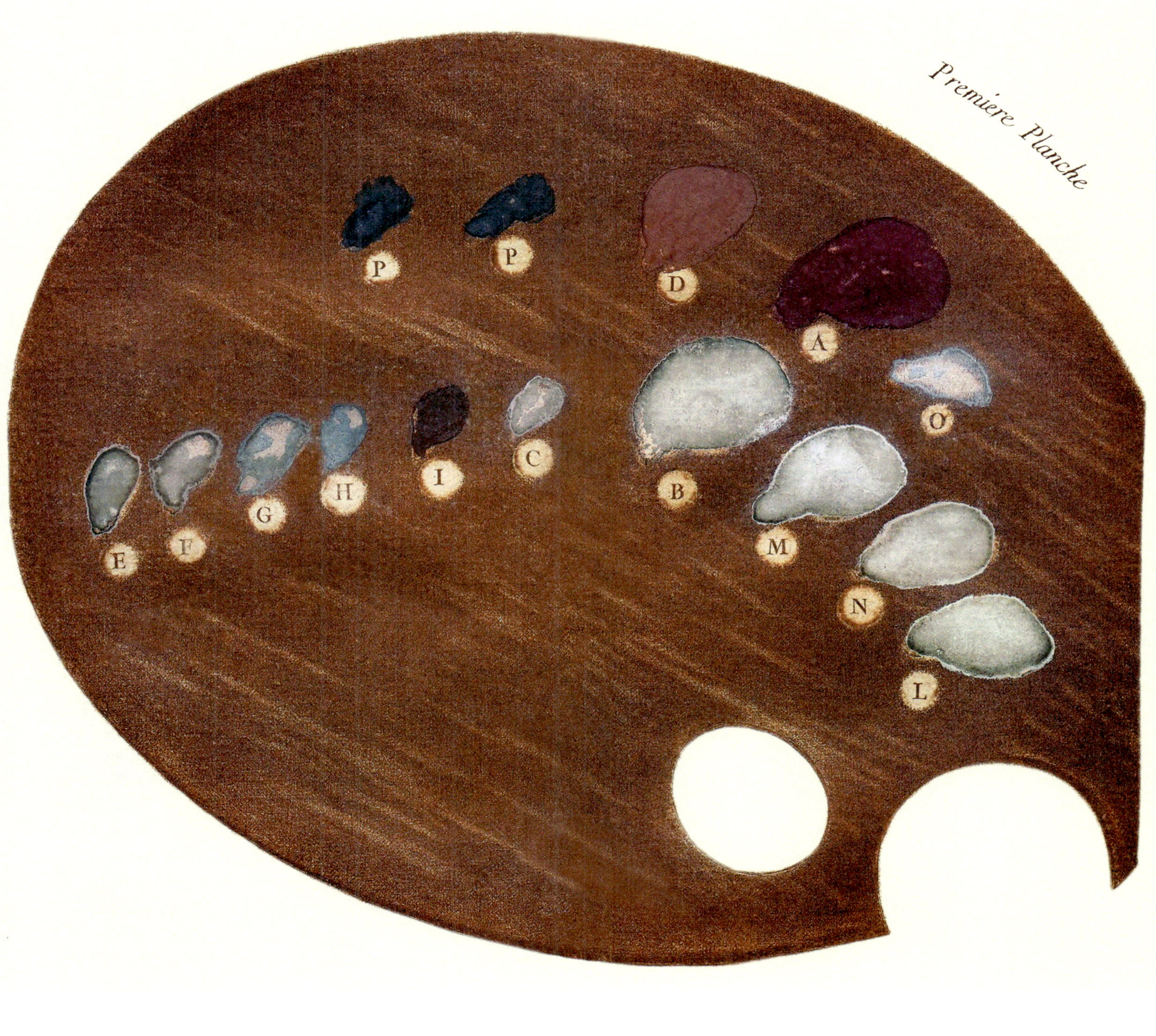
Première Planche
P
P
D
A
O
C
I
H
G
F
E
B
M
N
L

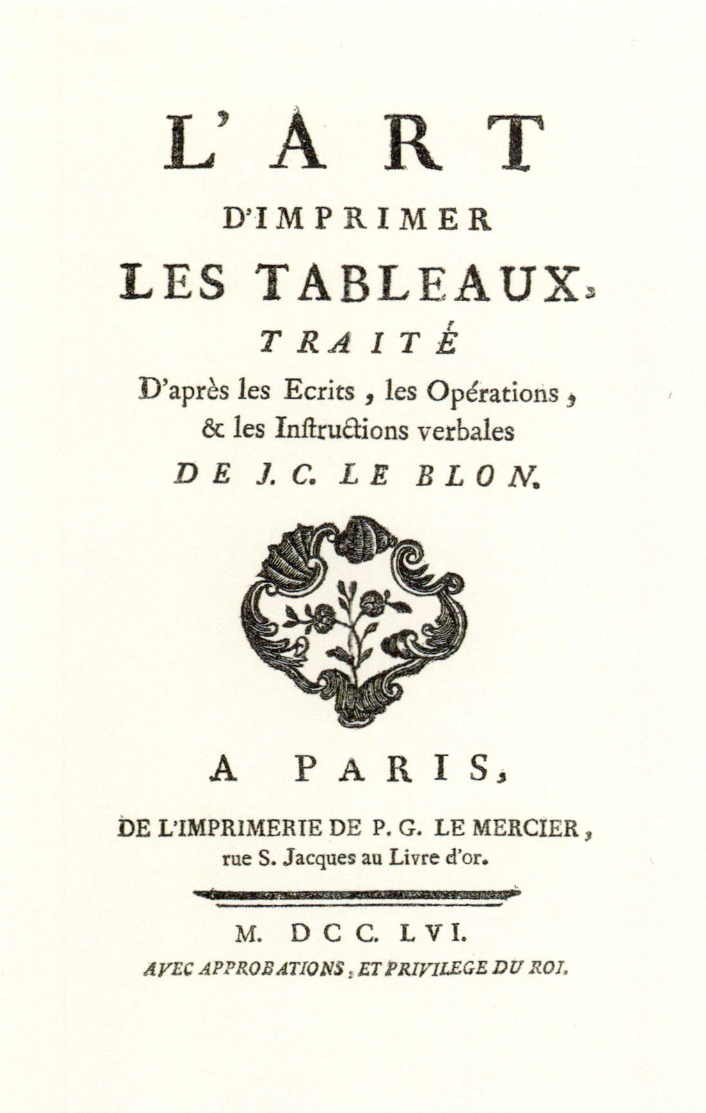
L'ART
D'IMPRIMER
LES TABLEAUX,
TRAITÉ
D'après les Ecrits, les Opérations,
& les Inſtructions verbales
DE J. C. LE BLON.

A PARIS,
DE L'IMPRIMERIE DE P. G. LE MERCIER,
rue S. Jacques au Livre d'or.

M. DCC. LVI.
AVEC APPROBATIONS, ET PRIVILEGE DU ROI.

Die Kunst, Bilder zu drucken

Künstlerpaletten spielen in Darstellungen der bildenden Künste eine bedeutende Rolle. Mal stehen sie stellvertretend für Kreativität und Stil, mal verweisen sie auf bestimmte Techniken oder die Farben, die ein Künstler oder Theoretiker einsetzte oder zur Verfügung hatte. William Hogarth (1697–1764) platzierte eine Palette in zentraler Position auf dem Rahmen einer wichtigen Illustration für sein philosophisches Traktat *The Analysis of Beauty* (1753). Häufig tauchen Paletten auch in Selbstporträts von Malerinnen und Malern auf, wo sie auf deren Arbeitsweisen und Farbvorlieben hindeuten. Die Palette in *L'art d'imprimer les tableaux* besitzt hingegen eine andere Bedeutung, die aber dennoch mit Technik und Materialität zu tun hat.

Der deutsche Kupferstecher und Drucker Jacob Christoph Le Blon hatte von seinen Studien- und Arbeitsaufenthalten in Rom, der Schweiz, Amsterdam und Paris einen bedeutenden internationalen Wissens- und Erfahrungsschatz mitgebracht. Ihm wird im Allgemeinen angerechnet, das grundlegende Verfahren für den Drei- und Vierfarbendruck erfunden zu haben. Die Technik, die er 1715 in London zu entwickeln begonnen hatte, wandte er erstmals 1725 an, als er ein zweisprachiges, auf Englisch und Französisch verfasstes Werk mit dem Titel *Coloritto, or The Harmony of Colouring in Painting / L'Harmonie du coloris dans la peinture* veröffentlichte. Auf der Titelseite wurde stolz verkündet, dass der kurze Text „einige farbige Abbildungen" enthalte. Das Buch selbst war nicht nur eine Demonstration des neuen Druckverfahrens, sondern in Teilen auch eine technische Abhandlung über das Kopieren von Kunst und ihre Reproduktion. Le Blon, der anscheinend die Fähigkeit besaß, ein vorhandenes Bild gedanklich in drei Schichten aus jeweils einer Farbe zu zerlegen, fasste sein streng trichromatisches System in einfachen Worten zusammen: „Die Malerei kann alle *sichtbaren* Gegenstände mit drei Farben, *Gelb*, *Rot* und *Blau*, darstellen; denn alle anderen Farben können aus diesen *dreien* zusammengesetzt werden, die ich als *Urfarben* bezeichne." Dieses heute extrem seltene Werk wurde 1765 und damit 15 Jahre nach Le Blons Tod in Frankreich nachgedruckt – überarbeitet und erweitert von seinem Freund und Förderer Antoine Gautier de Montdorge, der es unter dem Titel *L'art d'imprimer les tableaux; traité d'après les écrits, les opérations, & les instructions verbales de J. C. Le Blon* herausgab. Beide Werke enthalten eine Ausklapptafel, die eine wirklichkeitsgetreue Abbildung einer Malpalette mit 15 Malfarben zeigt. Darauf ist jede Farbe mit einem Buchstaben versehen, um sie der jeweiligen Besprechung in einem der Buchkapitel zuordnen zu können. Die Farben auf der Palette sind inzwischen etwas verblasst, doch das Bild repräsentiert noch immer sowohl Le Blons Hinweise zur Farbmischung und Komposition als auch sein neuartiges Druckverfahren. Wenngleich Letzteres nur bedingt erfolgreich war, ebnete es den Weg für den Farbdruck und die modernen Verfahren des Drei- und Vierfarbendrucks.

L'art d'imprimer les tableaux

La palette du peintre figure souvent dans les représentations des arts visuels, tantôt comme symbole de la créativité et du style, tantôt comme signifiant de techniques spécifiques ou des couleurs qu'emploient l'artiste et le théoricien, ou qui leur sont disponibles. William Hogarth (1697–1764) lui accorde une place centrale dans une illustration importante de son traité philosophique, *The Analysis of Beauty* (*Analyse de la beauté*, 1753). Bien entendu, cet outil est aussi fréquemment visible dans les autoportraits de peintres et, ce faisant, renseigne sur leurs méthodes de travail et leurs choix de couleurs. La palette de *L'art d'imprimer les tableaux* revêt un sens différent, qui entretient néanmoins un rapport avec la technique et la matérialité.

Graveur et imprimeur allemand qui s'est formé et a travaillé à Rome, en Suisse, à Amsterdam et à Paris, Jacob Christoph Le Blon acquiert en tous ces lieux savoir et expérience. On lui attribue généralement l'invention du procédé d'impression en trichromie et en quadrichromie, qu'il a élaboré à Londres à partir de 1715 et appliqué à la réalisation d'un ouvrage bilingue anglais-français, paru en 1725 et intitulé *Coloritto, or The Harmony of Colouring in Painting / L'Harmonie du coloris dans la peinture*. La page de titre annonce fièrement que ce court texte contient « quelques figures en couleurs ». Outre sa présentation de la nouvelle méthode d'impression, l'ouvrage est aussi en partie un essai technique consacré aux copies d'œuvres d'art et à leur reproduction. Capable, semble-t-il, de décomposer mentalement une image en trois couches de couleurs différentes, Le Blon résume son système strictement trichromatique dans un langage simple : « La Peinture peut représenter tous les Objets *visibles* avec trois Couleurs, savoir le *Jaune*, le *Rouge* & le *Bleu* ; car toutes les autres Couleurs se peuvent composer de ces *trois*, que je nomme Couleurs *primitives*. » Aujourd'hui extrêmement rare, cet ouvrage fut réimprimé en France en 1756, quinze ans après la mort de Le Blon, édité et enrichi par son ami et bienfaiteur Antoine Gautier de Montdorge, qui le publia sous le titre *L'art d'imprimer les tableaux. Traité d'après les écrits, les opérations, & les instructions verbales de J. C. Le Blon*. Les deux ouvrages contiennent une planche dépliable représentant avec réalisme une palette de peintre composée de 15 teintes, chacune marquée d'une lettre qui l'identifie lorsqu'il en est question dans l'un des chapitres. Si l'aspect visuel des couleurs de la palette est imparfait, l'image n'en communique pas moins les conseils de Le Blon en ce qui concerne le mélange et la composition des couleurs, ainsi que sa technique d'impression novatrice. Ses méthodes d'impression n'ont pas toujours été couronnées de succès, mais elles ont ouvert la voie à la chromolithographie, puis à la trichromie et à la quadrichromie modernes.

El arte de imprimir cuadros

Las paletas de los artistas ocupan un lugar destacado en las representaciones de las artes visuales, a veces como símbolos de creatividad y estilo, otras como indicadores de técnicas específicas o de los colores que los artistas y teóricos usaron o tuvieron a su disposición. William Hogarth (1697–1764) situó una en una posición central en una importante ilustración de su tratado filosófico *The Analysis of Beauty* (*El análisis de la belleza*, 1753); por supuesto, también suelen aparecer en autorretratos de pintores como indicadoras de sus métodos de trabajo y de los colores que preferían. La paleta en *L'art d'imprimer les tableaux* tiene, sin embargo, un significado diferente ligado a la técnica y la materialidad.

Jacob Christoph Le Blon fue un grabador e impresor alemán que estudió y trabajó en Roma, Suiza, Ámsterdam y París, lo que le permitió adquirir unos conocimientos y una experiencia internacional relevantes. Se le atribuye la invención del proceso básico de la impresión en tres y cuatro colores, así como la primera aplicación de la técnica, que comenzó a desarrollar en Londres a partir de 1715, cuando publicó una obra bilingüe en inglés y francés en 1725 titulada *Coloritto, or The Harmony of Colouring in Painting / L'Harmonie du coloris dans la peinture*. La portada anunciaba con orgullo que el breve texto contenía «algunas figuras coloreadas», y el libro en sí, además de ser una demostración del nuevo método de impresión, es también en parte un ensayo técnico sobre la copia y reproducción de obras de arte. Le Blon, quien al parecer tuvo la capacidad de descomponer mentalmente una imagen en tres capas de colores diferentes, resumió su sistema estrictamente tricromático en un lenguaje sencillo: «La pintura puede representar todos los objetos *visibles* con tres colores: *amarillo*, *rojo* y *azul*; porque todos los demás colores pueden componerse con estos *tres*, a los que llamo *primitivos*». Esta obra, hoy extremadamente rara, se reimprimió en Francia en 1756, 15 años después de la muerte de Le Blon, editada y ampliada por su amigo y seguidor Antoine Gautier de Montdorge, quien la publicó con el título *L'art d'imprimer les tableaux; traité d'après les écrits, les opérations, & les instructions verbales de J. C. Le Blon*. Ambas obras incluyen una lámina desplegable que representa de manera realista una paleta de pintor con 15 colores distintos, cada uno marcado con una letra que permite su identificación cuando se analiza en los capítulos del libro. La apariencia visual de los colores en la paleta es hasta cierto punto defectuosa, pero la imagen sigue representando los consejos de Le Blon sobre la mezcla y composición de los colores y sobre su técnica de impresión pionera. Si bien sus métodos de impresión no fueron del todo exitosos, allanaron el camino para la cromolitografía y la impresión moderna en color en tres y cuatro tonos.

Jacob Christian Schäffer (1718–1790)

ENTWURF EINER ALLGEMEINEN FARBENVEREIN

A Proposal for a General Correlation of Colours
2 plates, 23.5 x 19.5 cm | 9 ¼ x 7 ⅝ in., Regensburg, 1769
Fife, University of St Andrews Library, Department of Special Collections

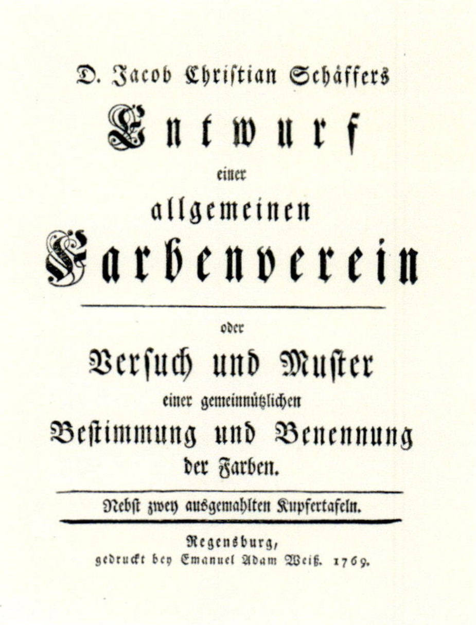
D. Jacob Christian Schäffers
Entwurf
einer
allgemeinen
Farbenverein
oder
Versuch und Muster
einer gemeinnützlichen
Bestimmung und Benennung
der Farben.
Nebst zwey ausgemahlten Kupfertafeln.
Regensburg,
gedruckt bey Emanuel Adam Weiß. 1769.

This slim volume of only 22 pages of text and two hand-coloured copperplate engravings was privately published in 1769 by the German naturalist and illustrator Jacob Christian Schäffer. In the same way as several other naturalists in the 18th century, for example, his English near-contemporary the entomologist Moses Harris (pp. 138–147), Schäffer had a heightened understanding of the importance of colour in the accurate and detailed representation of the natural world. He proposed a system of extensive colour charts and tables that would depict and describe nuances of the seven "Hauptfarben" (principal colours), which he identified as red, yellow, blue, green, brown, white and black. The inclusion of brown, white and black as distinct colours is indicative of the way he understood the colours of the natural world, and clearly identifies this work as dealing with physical rather than optical colour.

Schäffer presented his seven main colours in the first of his two plates, with each one depicted in bands within an engraved picture frame in the manner of an emblem or a coat of arms. Stylistically, the frames betray the historic period in which Schäffer's work was published, and while their ornate, Rococo forms detract from the abstract appearance that is so typical of many colour diagrams they also provide a visual contrast to the plain bands of colour contained within. More importantly, Schäffer introduced a hierarchical system by giving the frames different sizes and placing the larger ones, namely red, yellow and blue, as a central column in the composition. Of these three, red, at the top, is represented by eight bands that correspond to the red pigments then available (including cinnabar, brazilwood and cochineal), yellow, in the middle, by seven, and blue at the bottom by four. The pairs of green and brown, and white and black, are placed to the left and right of the central column, in frames half the size and differentiated by only two bands, apart from green which has three. The different shades depicted are listed in the accompanying text, and in a few instances the colour names are descriptive rather than denoting the name of a pigment, such as "Haarfarbe" (hair colour) as a type of brown. This is an arrangement very much in keeping with Enlightenment attempts at classification, in this case essentially a simple exercise in grouping colours into a straightforward taxonomic system.

The second plate (p. 105) presents an example of Schäffer's proposed larger system of reference, based on the colours shown in the first plate. Taking red, the most prominent colour in the first plate, a grid of 150 numbered boxes is set out in which all the natural tints can be listed, as mixed from the available pigments depicted previously. Only a fifth or so of the boxes are coloured in, however, leaving readers of the book to fill in the empty ones and create their own charts (see also Johann Heinrich Meynier, pp. 124–129).

Roth.

1.		26.		51.		76.		101.		126.	
2.		27.		52.		77.		102.		127	
3.		28.		53.		78.		103.		128.	
4.		29.		54.		79.		104.		129	
5.		30.		55.		80.		105.		130.	
6.		31.		56.		81.		106.		131.	
7.		32.		57.		82.		107.		132.	
8.		33.		58.		83.		108.		133.	
9.		34.		59.		84.		109		134.	
10.		35.		60.		85		110.		135.	
11.		36.		61.		86.		111.		136.	
12.		37.		62.		87.		112.		137.	
13.		38.		63.		88.		113		138.	
14		39.		64.		89.		114.		139.	
15.		40.		65.		90.		115.		140.	
16.		41.		66.		91.		116.		141.	
17.		42.		67.		92.		117.		142.	
18.		43.		68.		93.		118.		143.	
19.		44.		69.		94.		119.		144.	
20.		45.		70.		95.		120.		145.	
21.		46.		71.		96.		121.		146.	
22.		47.		72.		97.		122.		147.	
23		48.		73.		98.		123.		148.	
24		49.		74.		99.		124.		149.	
25.		50.		75.		100.		125.		150.	

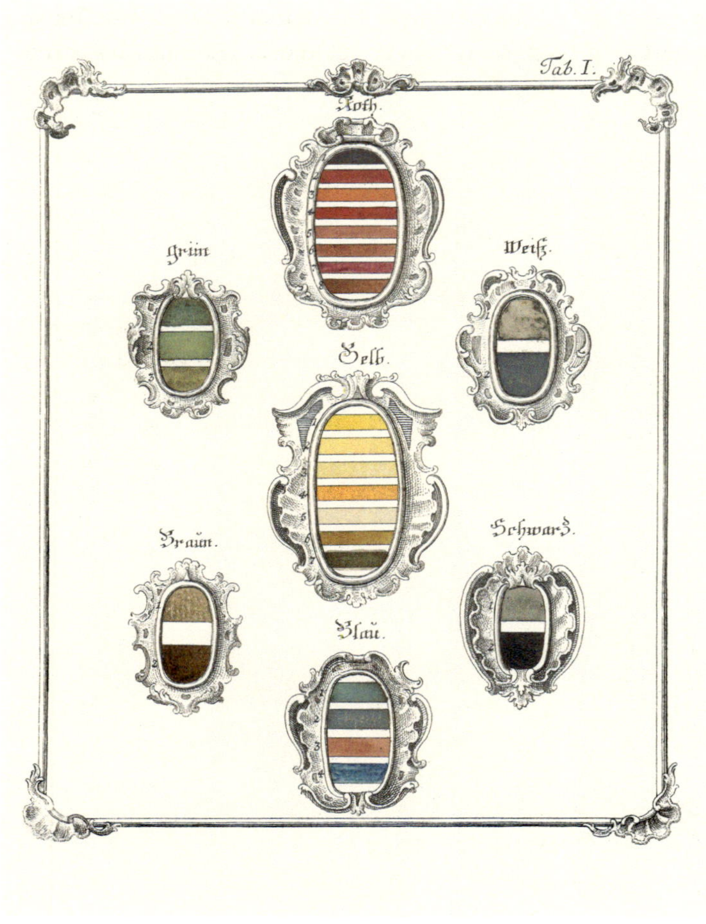

Entwurf einer allgemeinen Farbenverein

Dieses schmale, nur 22 Textseiten und zwei handkolorierte Kupfertafeln umfassende Bändchen wurde 1769 von dem deutschen Naturforscher und Illustrator Jacob Christian Schäffer privat herausgegeben. Wie verschiedene andere Naturkundler des 18. Jahrhunderts, etwa der fast gleichaltrige englische Entomologe Moses Harris (S. 138–147), wusste Schäffer besonders um die Bedeutung von Farben für die akkurate und detaillierte Wiedergabe der Natur. Er schlug ein System ausführlicher Farbtafeln und -tabellen mit Abbildungen und Beschreibungen von Abstufungen der sieben „Hauptfarben" Rot, Gelb, Blau, Grün, Braun, Weiß und Schwarz vor. Die Aufnahme von Braun, Weiß und Schwarz als eigene Farben lässt Schäffers Verständnis von den Naturfarben erkennen und zeigt ganz klar, dass sein Werk nicht den immateriellen, sondern den stofflichen Farben gewidmet ist.

Schäffer präsentierte seine sieben Hauptfarben auf der ersten seiner zwei Tafeln. Jede Farbe ist mit mehreren Streifen in einem gestochenen Rahmen dargestellt, der Ähnlichkeit mit einem Schild oder Wappen hat. Stilistisch sind die Rahmen ein Abbild der historischen Epoche, in der Schäffers Büchlein erschien. Ihre verschnörkelten Rokokoformen lenken von dem abstrakten Erscheinungsbild ab, das so typisch für zahlreiche Farbdiagramme ist, und bilden zugleich einen visuellen Kontrast zu den schlicht gestreiften Farbmustern. Noch wichtiger aber ist, dass Schäffer ein hierarchisches Farbsystem entwickelte, indem er unterschiedlich große Rahmen entwarf und die größeren, mit den Farben Rot, Gelb und Blau, übereinander in der Blattmitte anordnete. Ganz oben steht die Farbe Rot mit acht Streifen, die den damals verfügbaren roten Pigmenten entsprachen (einschließlich Zinnober, Rotholz und Koschenille), in der Mitte folgt Gelb mit sieben und ganz unten Blau mit vier Streifen. Grün und Braun sind links, Weiß und Schwarz rechts der Mittelachse in halb so großen Rahmen platziert und bis auf das dreistreifige Grün nur in jeweils zwei Streifen unterteilt. Der Begleittext listet die Namen der unterschiedlichen Farbtöne auf und greift mitunter anstelle von Pigmentbezeichnungen zu deskriptiven Farbnamen, etwa wenn ein Brauntyp als „Haarfarbe" tituliert wird. Diese Ordnung steht im Einklang mit Versuchen der Aufklärung, die Welt zu klassifizieren, was sich hier im Wesentlichen als simple Übung äußert, Farben in ein einfaches taxonomisches System einzugruppieren.

Die zweite Tafel (S. 105) erschließt das von Schäffer vorgeschlagene größere Bezugsystems, das auf den Farben der ersten Tafel basiert. Am Beispiel von Rot, der prominentesten Farbe aus Tafel eins, wird ein aus 150 nummerierten Feldern bestehendes Raster aufgespannt, in das sich alle natürlichen, aus den zuvor abgebildeten Pigmenten gemischten Farbtöne eintragen lassen. Allerdings ist nur etwa ein Fünftel der Felder ausgemalt, was den Lesern die Gelegenheit gab, die Leerstellen zu füllen und ihre eigenen Tabellen zu erstellen (siehe auch Johann Heinrich Meynier, S. 124–129).

Ébauche d'association des couleurs

En 1769, Jacob Christian Schäffer, naturaliste et illustrateur allemand, publie à compte d'auteur cette plaquette de 22 pages comprenant deux gravures au burin coloriées à la main. À l'instar de plusieurs naturalistes du XVIIIe siècle comme, par exemple, l'entomologiste anglais Moses Harris, son quasi contemporain (p. 138–147), Schäffer a une conscience aiguë de l'importance de la couleur dans la représentation précise et détaillée du monde naturel. Il propose un système de grands graphiques et tableaux chromatiques rendant compte des nuances des sept « Hauptfarben » (couleurs principales) qui, pour lui, sont le rouge, le jaune, le bleu, le vert, le brun, le blanc et le noir. La présence du brun, du blanc et du noir témoigne de sa perception des couleurs du monde naturel et du fait que son œuvre traite de la couleur physique et non de la couleur optique.

Schäffer présente ses sept couleurs dans la première planche, chacune figurée par des bandes et entourée d'un cadre gravé à la façon d'un emblème ou d'un blason. Du point de vue du style, les cadres rococos trahissent l'époque de la publication. Ces ornements empêchent le regard de se concentrer sur l'aspect abstrait si caractéristique de tant de graphiques de couleurs mais offrent néanmoins un contraste visuel avec les bandes de couleurs unies qu'ils enchâssent. Surtout, Schäffer instaure un système hiérarchique en donnant aux cadres des dimensions différentes et en disposant les plus grands – rouge, jaune et bleu – dans la colonne centrale de la composition. En haut, le rouge est figuré par huit bandes correspondant aux pigments rouges disponibles à l'époque (cinabre, pernambouc et cochenille); le jaune, au centre, par sept bandes; et le bleu, en bas, par quatre. Les paires vert-brun et noir-blanc sont placées de part et d'autre de la colonne centrale, dans des cadres moitié moins grands, et ne comportent que deux bandes, à l'exception du vert qui en a trois. Les différentes teintes décrites sont répertoriées dans le texte d'accompagnement. Dans certains cas, les appellations de couleur sont plus descriptives qu'indicatrices d'un nom de pigment, comme « Haarfarbe » (couleur cheveu) pour désigner une teinte de brun. Cette disposition, tout à fait dans l'esprit des tentatives de classification des Lumières, n'est au fond qu'un simple exercice d'organisation des couleurs selon un système taxinomique sommaire.

La seconde planche (p. 105) offre un exemple du système de référence élargi que propose Schäffer à partir des couleurs répertoriées dans la première planche. Pour le rouge, couleur dominante de la première planche, une grille de 150 cases numérotées permet de faire la liste de toutes les nuances naturelles, élaborées avec les pigments décrits précédemment. Toutefois, environ une case sur cinq seulement est coloriée, afin de permettre au lecteur de remplir les cases vides et de créer son propre tableau (voir aussi Johann Heinrich Meynier, p. 124–129).

Propuesta de una correlación general de los colores

En 1769, el naturalista e ilustrador alemán Jacob Christian Schäffer publicó de forma privada este delgado volumen compuesto por 22 páginas de texto y dos grabados realizados con planchas de cobre y coloreados a mano. Al igual que otros naturalistas del siglo XVIII, como el entomólogo inglés Moses Harris (págs. 138–147), quien casi fue su contemporáneo, Schäffer tenía una mayor comprensión de la importancia del color en la representación precisa y detallada del mundo natural. Propuso un sistema compuesto por extensos gráficos y tablas de colores para la representación y descripción de los matices de los siete *Hauptfarben* (colores principales), que identificó como el rojo, el amarillo, el azul, el verde, el marrón, el blanco y el negro. La inclusión del marrón, del blanco y del negro como colores distintos es indicativa del modo en que entendía los colores del mundo natural, y evidencia que este trabajo trata del color físico en lugar del óptico.

En la primera lámina, Schäffer presentó sus siete colores principales en bandas dentro de un marco grabado a modo de emblema o escudo de armas. En lo estilístico, los marcos no se corresponden con el período histórico en el que se publicó el trabajo de Schäffer, y aunque el ornamento de sus formas rococó empaña la apariencia abstracta tan típica de muchos diagramas de color, también ofrece un contraste visual con las bandas de color simples que contienen. Más importante aún es el sistema jerárquico que Schäffer introdujo al dar a los marcos diferentes tamaños y colocar los más grandes (a saber, el rojo, el amarillo y el azul) en el centro de la composición. De estos tres, el rojo, en la parte superior, está representado por ocho bandas, que se corresponden con los pigmentos rojos entonces disponibles (incluidos el cinabrio, el palo brasil y la cochinilla); el amarillo, en el medio, está representado por siete bandas; y el azul, en la parte inferior, por cuatro. Las parejas de verde y marrón y de blanco y negro aparecen a izquierda y derecha de la columna central en marcos de la mitad de tamaño con solo dos bandas, con la excepción del verde, que tiene tres. Los diferentes tonos representados se enumeran en el texto adjunto; en algunos casos, los nombres de los colores son descriptivos en lugar de indicar el nombre de un pigmento, como *Haarfarbe* (color de cabello) en referencia a un tipo de marrón. Se trata de una disposición muy acorde con los intentos de clasificación de la Ilustración: una simple agrupación de colores en un sistema taxonómico directo.

La segunda lámina (pág. 105) presenta un ejemplo del sistema de referencia más amplio propuesto por Schäffer, que se basa en los colores que se muestran en la primera. Toma el rojo, el color más destacado en la primera lámina, y establece una cuadrícula de 150 casillas numeradas donde se pueden incluir todos los tintes naturales fruto de la mezcla de los pigmentos disponibles representados con anterioridad. Sin embargo, solo aproximadamente una quinta parte de los recuadros están coloreados, y deja en manos de los lectores llenar los vacíos y crear sus propios gráficos (véase también Johann Heinrich Meynier, págs. 124–129).

Johann Heinrich Lambert (1728–1777)

BESCHREIBUNG EINER MIT DEM CALAUSCHEN WACHSE AUSGEMALTEN FARBENPYRAMIDE

A Description of a Colour Pyramid Painted Using Calau's Wax

2 plates, 22 x 17.5 cm | 8 ⅝ x 6 ⅞, Berlin, 1772

Fife, University of St Andrews Library, Department of Special Collections

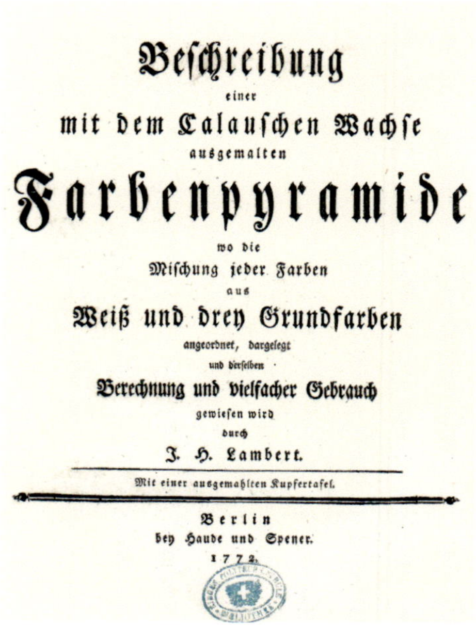
Beschreibung
einer
mit dem Calauschen Wachse
ausgemalten
Farbenpyramide
wo die
Mischung jeder Farben
aus
Weiß und drey Grundfarben
angeordnet, dargelegt
und derselben
Berechnung und vielfacher Gebrauch
gewiesen wird
durch
J. H. Lambert.
Mit einer ausgemahlten Kupfertafel.
Berlin
bey Haude und Spener.
1772.

On 18 November 1758 the German astronomer and mathematician Johann Heinrich Lambert attended a lecture at the Göttingen Academy of Sciences given by his colleague Tobias Mayer (see pp. 112–115), whose proposal was to identify all the colours that could be distinguished by the human eye and compile them into a calculable numbering system. Mayer visualised his concept in the form of a double tetrahedron, a pair of pyramids based on equilateral triangles, with the primary colours yellow, blue and red at the corners. These colours extend, by way of 12 gradations, towards white at the top of the upper pyramid and black at the tip of the bottom pyramid, resulting in a total of 819 identifiable colour mixtures. As physical representatives for the principal colours Mayer selected cinnabar for red, king's yellow and mountain blue (azurite). His system was impressive and mathematically sound, but flawed in practical terms. By the time of his death in 1762, at the age of just 39, he had never had the opportunity to test his colour triangles and pyramids fully using real pigments, but his followers Lambert and Georg Christoph Lichtenberg saw to it that his work was made available and developed it further.

A review of Mayer's lecture appeared in print soon after he had delivered it, and Lambert reprinted it in full in his own *Beschreibung einer mit dem Calauschen Wachse ausgemalten Farbenpyramide* in which he proceeded to interpret and adapt Mayer's system. As acknowledged in the book's title, Lambert put Mayer's theoretical outlines into practice by collaborating with the Berlin artist Benjamin Calau (1724–1785), and together they coloured in Mayer's system, using carmine (red), Prussian blue and gamboge (yellow) for the main colourants, in conjunction with Calau's wax medium. For his *Farbenpyramide* (colour pyramid), Lambert removed Mayer's lower pyramid (with the shading towards black) and amended the diagram to a single eight-tiered pyramid, complete with realistically illustrated architectural detail, comprising 107 colour samples, plus white. The 45 darkest colours, of the seven gradations between the primaries, appear in the bottom triangle, with diminishing numbers of identifiable mixtures on the tiers moving up towards the tip of the structure, where a single square on the seventh triangle represents whiteness or light. Lambert's design places the pyramid on a base of 12 pigments, named in the text as "Naples yellow, king's yellow, orpiment, azurite, smalt, indigo, lamp black, juice green, mountain green, verdigris, cinnabar and Florentine lake". He also provided a key for all the 107 colours shown in his diagram.

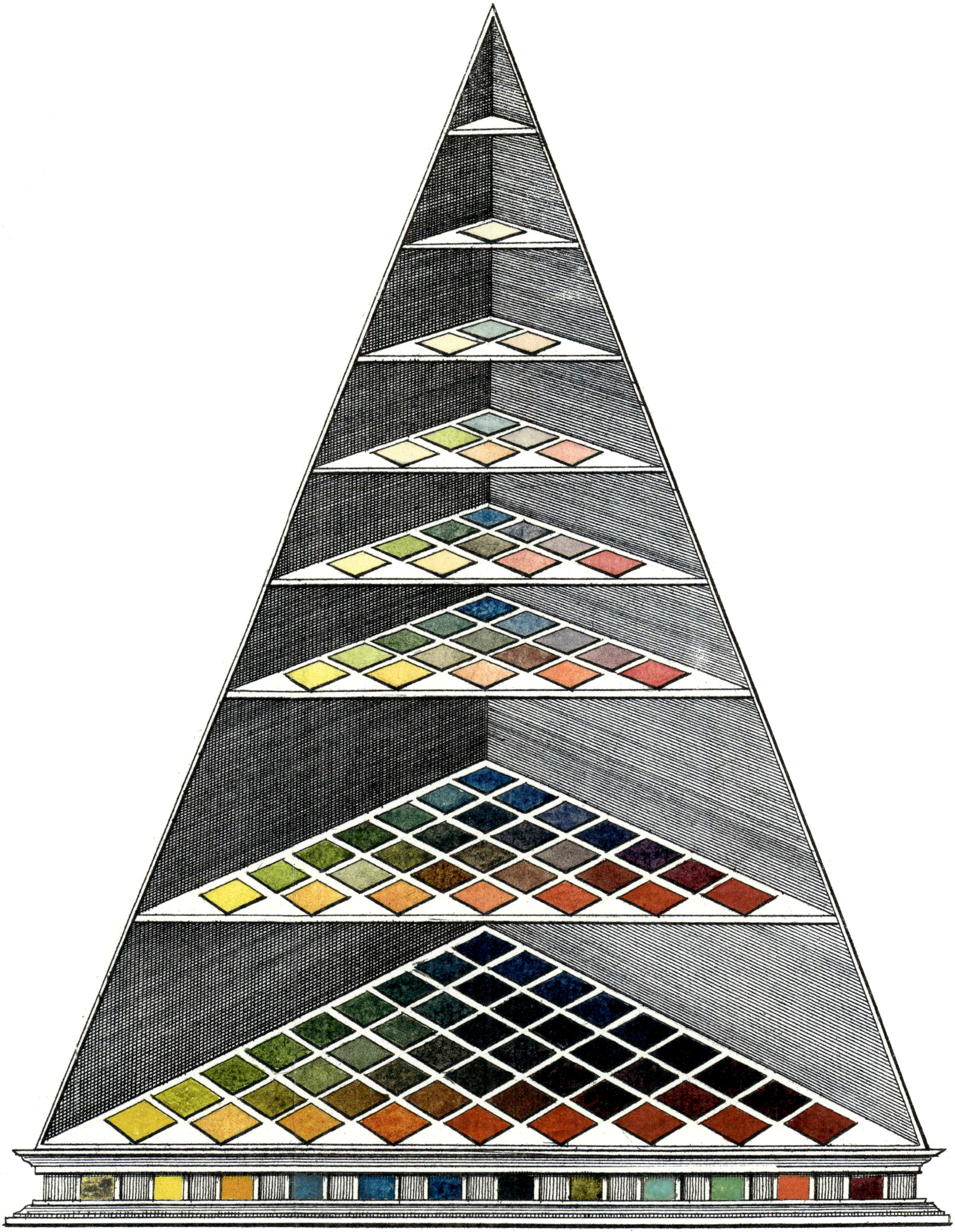

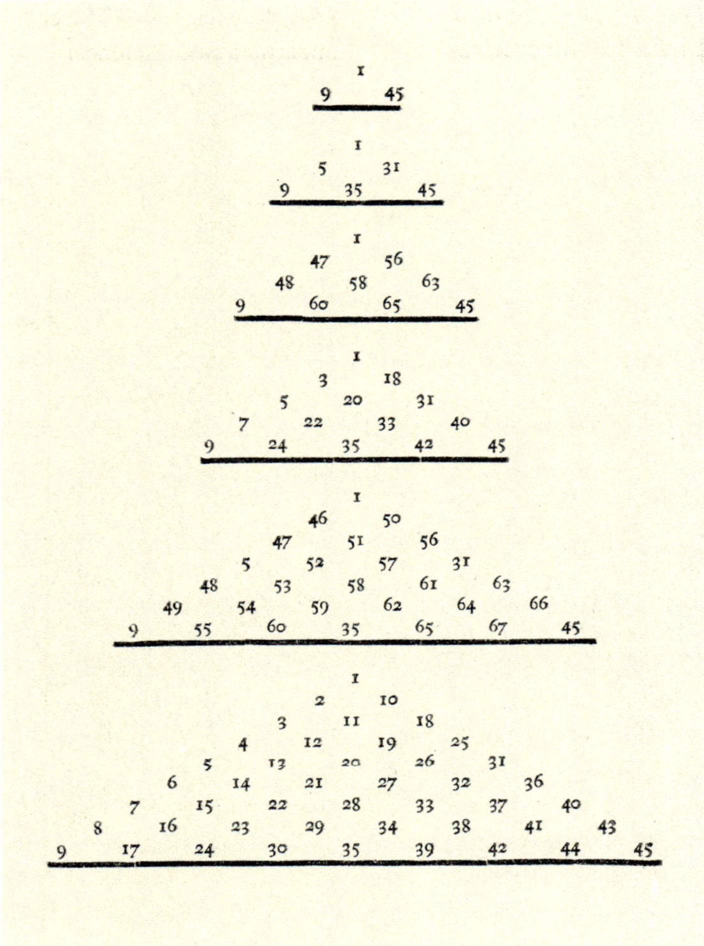

Beschreibung einer mit dem Calauschen Wachse ausgemalten Farbenpyramide

Am 18. November 1758 besuchte der deutsche Astronom und Mathematiker Johann Heinrich Lambert einen Vortrag seines Kollegen Tobias Mayer (siehe S. 112–115) in der Gesellschaft der Wissenschaften zu Göttingen. Mayer schlug vor, alle Farben, die das menschliche Auge unterscheiden konnte, zu bestimmen und in einem berechenbaren Nummernsystem zusammenzufassen. Er veranschaulichte sein Konzept mithilfe eines doppelten Tetraeders aus zwei Pyramiden mit einem gleichseitigen Dreieck als gemeinsamer Grundfläche. Den Ecken dieses Dreiecks wies er die Grundfarben Gelb, Blau und Rot zu. Diese breiten sich in zwölf Abstufungen bis in die weiße Spitze der oberen und die schwarze Spitze der unteren Pyramide aus und bilden damit insgesamt 819 bestimmbare Farbmischungen. Als Pigmente für die Hauptfarben wählte Mayer Zinnober für Rot, Königsgelb und Bergblau (Azurit). Sein System war beeindruckend und mathematisch fehlerfrei, wies jedoch praktische Schwachstellen auf. Bis zu seinem frühen Tod 1762 im Alter von nur 39 Jahren blieb ihm die Gelegenheit verwehrt, seine Farbdreiecke und -pyramiden mit echten Pigmenten auf die Probe zu stellen. Erst seine Schüler, Lambert und Georg Christoph Lichtenberg, machten sein Werk zugänglich und entwickelten es weiter.

Kurz nachdem Mayer seinen Vortrag gehalten hatte, wurde dieser in einer überarbeiteten Fassung veröffentlicht, die Lambert in seiner *Beschreibung einer mit dem Calauschen Wachse ausgemalten Farbenpyramide*, einer Interpretation und Weiterführung des Mayer'schen Farbsystems, vollständig nachdruckte. Wie im Buchtitel angegeben, setzte Lambert die theoretischen Skizzen Mayers in Kooperation mit dem Berliner Künstler Benjamin Calau (1724–1785) in die Praxis um. Gemeinsam malten sie Mayers System aus, wobei sie Karmin (Rot), Preußischblau und Gummigutt (Gelb) in Verbindung mit Calaus Wachsmedium (dem „Calauschen Wachse" des Buchtitels) als Farbmittel für die Primärfarben verwendeten. Für seine Farbenpyramide entfernte Lambert das untere Tetraeder Mayers (mit den ins Schwarz laufenden Schattierungen) und wandelte das Diagramm in eine Pyramide mit acht Etagen, realistisch wiedergegebenen architektonischen Details und insgesamt 107 quadratischen Farbmustern plus Weiß um. Die 45 dunkelsten Farbquadrate, in jeweils sieben Abstufungen aus den drei Primärfarben gemischt, finden sich im unteren Dreieck. Mit jedem darüberliegenden Dreieck nimmt die Anzahl der bestimmbaren Farbmischungen kontinuierlich ab, bis ein einziges Quadrat auf dem siebten Dreieck das Weiße oder das Licht darstellt. In Lamberts Zeichnung steht die Pyramide auf einem Sockel mit zwölf Pigmentmustern, die im Text (entsprechend ihrer Anordnung von links nach rechts) als „Neapolitanisch Gelb", „Königsgelb oder Aurum", „Rauschgelb", „Bergblau", „Schmalte", „Indigo", „Lampenschwarz", „Saftgrün", „Berggrün", „Grünspan", „Zinnober" und „Florentinerlack" bezeichnet sind. Der Text lieferte zudem einen Benennungsschlüssel und die Mischungsverhältnisse für alle 107 Farben des Diagramms mit.

Description d'une pyramide des couleurs peinte avec la cire de Calau

Le 18 novembre 1758, l'astronome et mathématicien Johann Heinrich Lambert assiste à une conférence donnée à l'Académie des sciences de Göttingen par son collègue Tobias Mayer (voir p. 112–115), qui propose d'identifier toutes les couleurs perçues par l'œil humain et de les compiler dans un système numéroté calculable. Mayer visualise sa conception sous la forme d'un double tétraèdre, deux pyramides créées à partir de triangles équilatéraux, aux angles desquels sont placées les couleurs primaires jaune, bleu et rouge. Selon une suite de 12 gradations, les couleurs tendent vers le blanc au sommet de la pyramide supérieure et vers le noir à la pointe de la pyramide inférieure, produisant un total de 819 mélanges chromatiques identifiables. Afin de représenter physiquement les couleurs principales, Mayer a choisi le cinabre pour le rouge, l'orpiment vif pour le jaune, l'azurite pour le bleu. Impressionnant et solide sur le plan mathématique, son système n'en est pas moins imparfait du point de vue pratique. Lorsqu'il meurt en 1762, à l'âge précoce de 39 ans, il n'a jamais pu mettre en pratique ses triangles et pyramides de couleurs en utilisant de véritables pigments. Mais ses disciples, Lambert et Georg Christoph Lichtenberg, feront en sorte de diffuser son œuvre et de la perfectionner.

Peu après la conférence de Mayer, un compte rendu est publié, que Lambert reproduit intégralement dans cette « Beschreibung einer mit dem Calauschen Wachse ausgemalten Farbenpyramide », où il interprète et adapte le système de Mayer. Comme l'évoque son titre, Lambert met en pratique les prémisses théoriques de ce dernier en collaborant avec le peintre berlinois Benjamin Calau (1724–1785). Ensemble, ils colorient le système à l'aide de carmin (rouge), de bleu de Prusse et de gomme-gutte (jaune) pour les principaux colorants, alliés à la cire de Calau (« Calausche Wachse »). Pour sa *Farbenpyramide* (pyramide de couleurs), Lambert supprime la pyramide inférieure de Mayer (avec ses gradations vers le noir) et modifie le graphique pour en faire une pyramide unique à huit niveaux, dotée de détails architecturaux illustrés avec réalisme, qui réunit 107 échantillons de couleurs, plus le blanc. Les 45 couleurs les plus sombres, déclinées suivant les sept gradations entre les couleurs primaires, figurent dans le triangle du bas. Le nombre de mélanges identifiables se réduit d'un niveau supérieur à l'autre en allant vers le sommet de la structure : dans le septième triangle, un carré unique représente le blanc ou la lumière. La pyramide de Lambert repose sur un socle de 12 pigments, désignés dans le texte par les appellations « jaune de Naples, orpiment vif, orpiment, azurite, smalt, indigo, noir de fumée, vert sève, vert malachite, chrysocolle, cinabre et laque de Florence ». Lambert donne aussi les légendes des 107 couleurs rassemblées dans son graphique.

Descripción de una pirámide de colores pintada con cera de Calau

El 18 de noviembre de 1758, el astrónomo y matemático alemán Johann Heinrich Lambert asistió a una conferencia en la Academia de Ciencias de Göttingen impartida por su colega Tobias Mayer (véanse págs. 112–115), quien proponía identificar todos los colores que el ojo humano podía distinguir y compilarlos en un sistema numérico que permitiera realizar cálculos. Mayer imaginó un concepto de color con forma de doble tetraedro, un par de pirámides basadas en triángulos equiláteros con los colores primarios amarillo, azul y rojo en los vértices de sus bases. A través de 12 gradaciones, esos colores tienden hacia el blanco en la cúspide de la pirámide superior y el negro en la cúspide de la inferior, lo que da como resultado un total de 819 mezclas de colores identificables. Como representantes físicos de los colores principales, Mayer seleccionó el cinabrio para el rojo, el amarillo del rey y el azul montaña (azurita). Su sistema era impresionante y matemáticamente sólido, pero defectuoso en términos prácticos. En el momento de su muerte en 1762, con apenas 39 años, no había tenido la oportunidad de poner a prueba por completo sus triángulos y pirámides de color con pigmentos reales, pero sus seguidores Lambert y Georg Christoph Lichtenberg se encargaron de que su trabajo estuviera disponible y lo desarrollaron todavía más.

Poco después de que Mayer impartiera su conferencia se publicó una reseña, que Lambert reimprimió por completo en su *Beschreibung einer mit dem Calauschen Wachse ausgemalten Farbenpyramide*, donde procedió a interpretar y adaptar el sistema de Mayer. Como se reconoce en el título del libro, Lambert puso en práctica el planteamiento teórico de Mayer en colaboración con el artista berlinés Benjamin Calau (1724–1785), y ambos colorearon el sistema de Mayer con carmín (rojo), azul de Prusia y gutagamba (amarillo) como colorantes principales, junto con las ceras de Calau (las «Calauschen Wachse» del título). Para su *Farbenpyramide* (pirámide de color), Lambert eliminó la pirámide inferior de Mayer (que tiende al negro) y modificó el diagrama hasta convertirlo en una sola pirámide de ocho niveles (con detalles arquitectónicos ilustrados de manera realista) que comprende 107 muestras de color, además del blanco. Los 45 colores más oscuros, de las siete gradaciones existentes entre los primarios, aparecen en el triángulo inferior, y a medida que los niveles se elevan hacia la cúspide de la estructura, donde un solo cuadrado en el séptimo triángulo representa la blancura o luz, decrece el número de mezclas identificables. El diseño de Lambert coloca la pirámide sobre una base de 12 pigmentos, nombrados en el texto como amarillo de Nápoles, amarillo del rey, oropimente, azurita, esmalte, índigo, negro de humo, verde jugo, verde montaña, crisocola, cinabrio y laca florentina. También proporcionó una clave para los 107 colores que se muestran en su diagrama.

Tobias Mayer (1723–1762),
Georg Christoph Lichtenberg (1742–1799)

OPERA INEDITA

Unpublished Works

2 plates, 30 x 23.4 cm | 11 ¾ x 9 ¼ in., Göttingen, 1775
Zürich, ETH-Bibliothek

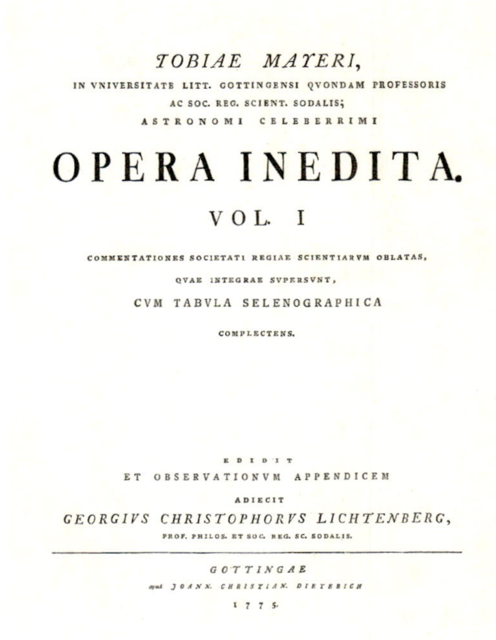
TOBIAE MAYERI,
IN VNIVERSITATE LITT. GOTTINGENSI QVONDAM PROFESSORIS
AC SOC. REG. SCIENT. SODALIS;
ASTRONOMI CELEBERRIMI
OPERA INEDITA.
VOL. I
COMMENTATIONES SOCIETATI REGIAE SCIENTIARVM OBLATAS,
QVAE INTEGRAE SVPERSVNT,
CVM TABVLA SELENOGRAPHICA
COMPLECTENS.
EDIDIT
ET OBSERVATIONVM APPENDICEM
ADIECIT
GEORGIVS CHRISTOPHORVS LICHTENBERG,
PROF. PHILOS. ET SOC. REG. SC. SODALIS.
GOTTINGAE
apud JOANN. CHRISTIAN. DIETERICH
1775.

The astronomer and mathematician Tobias Mayer was also a cartographer, and therefore habitually familiar with representing three-dimensional objects in two-dimensional form. Among his greatest achievements are the lunar tables he compiled, which detailed stages of the moon's movement, and a map of the moon, while his work also included one of the first three-dimensional representations of the range of colours visible to the human eye. As described previously, Mayer proposed a mathematical concept of identifiable colour mixtures, visualised in the shape of a double tetrahedron. The base triangle, at the intersection of the two tetrahedra, contained 91 colour mixtures. Each of these was identified using a combination of the initial letters of the German words for the principal colours (G for yellow, B for blue, R for red), and a series of numbers to indicate the ratios of the principal colours when combined (with the unmixed, or unbroken main colourants, situated at the three vertices, assigned a value of 12). From this system, formulas could be derived for mixing, calculating or identifying colour combinations with correct proportions.

Mayer was never able to test his colour concept fully or see it published, but Lambert (see pp. 108–111) and the physicist and author Georg Christoph Lichtenberg were fascinated with his work and occupied themselves with it for many years after Mayer's death. Lichtenberg collated and edited Mayer's various papers, including an essay on colour, *De affinitate colorum commentatio* (translated in 1971 by Eric Forbes as *Treatise on the Relationship of Colours*), which he published in 1775 in the collection *Opera Inedita* together with several other essays by Mayer also in Latin. The plates in this book relate to the author's colour theory, one of which is an image based on Mayer's own design showing the base triangle of his tetrahedron system with its 91 labelled colours. The other plate shows a hand-coloured triangle drawn by Lichtenberg that makes an intelligent simplification of Mayer's concept. This version followed Mayer's geometrical scheme by using hexagonal shapes to denote each of the separate colourants, but reduced the intermediate grades between the primary colours from eleven to five. Lichtenberg and Lambert were not the last to engage so closely with Mayer's triangle, and in 1781 the German colour-maker August Ludewig Pfannenschmid (dates unknown) included a version of his own in his small treatise on colour mixing, *Versuch einer Anleitung zum Mischen aller Farben aus blau, gelb und roth, nach beiliegendem Triangel.*

Tab. III

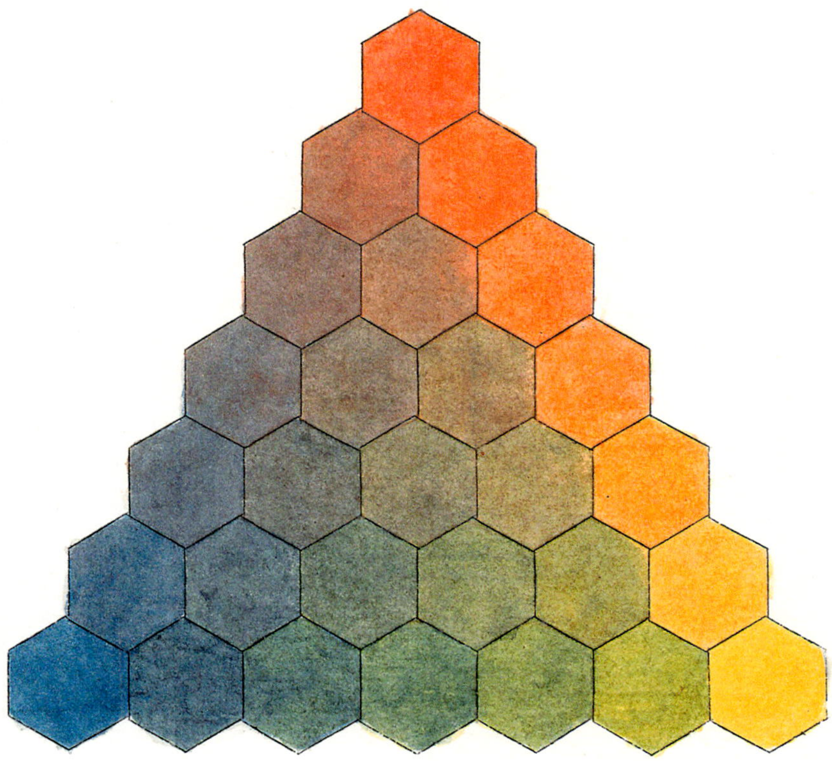

Tab. II.

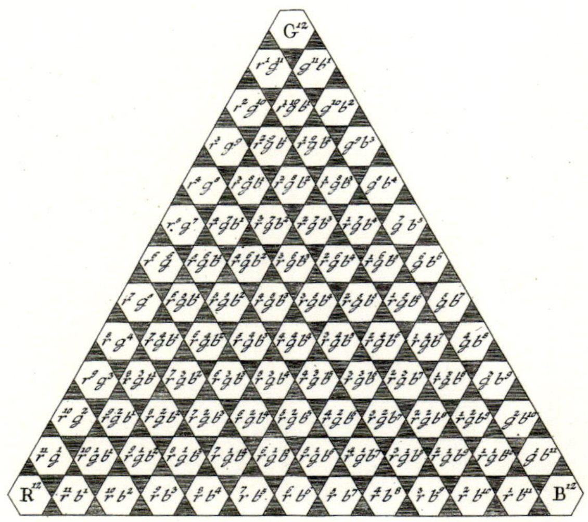

Unveröffentlichte Werke

Der Astronom und Mathematiker Tobias Mayer war auch Kartograf und als solcher mit der Darstellung dreidimensionaler Objekte in zweidimensionaler Form vertraut. Zu seinen größten Errungenschaften zählen neben einer Mondkarte die von ihm erstellten Mondtafeln mit detaillierten Angaben zur Umlaufbahn des Mondes. Darüber hinaus gehört zu seinen Werken eine der ersten dreidimensionalen Darstellungen des für das menschliche Auge sichtbaren Farbspektrums. Wie bereits beschrieben, legte Mayer ein mathematisches System bestimmbarer Farbmischungen vor, das er anhand eines doppelten Tetraeders verständlich machte. Das Basisdreieck an der Schnittstelle der beiden Tetraeder enthielt 91 Mischungen. Jede von ihnen wurde mithilfe einer Kombination aus den Anfangsbuchstaben der Bezeichnungen für die Hauptfarben (G für Gelb, B für Blau, R für Rot) und mehreren Zahlen für das Mischungsverhältnis der jeweils beteiligten Hauptfarben identifiziert (wobei die reinen oder ungebrochenen Ausgangsfarben in den drei Ecken den Wert 12 erhielten). Aus diesem Schema ließen sich Formeln ableiten, um die Farbkombinationen in den richtigen Anteilen zu mischen, zu berechnen oder zu bestimmen.

Mayer konnte seine Farbtheorie nie in Gänze überprüfen und erlebte auch ihre Veröffentlichung nicht, doch Lambert (siehe S. 108–111) und der Physiker und Schriftsteller Georg Christoph Lichtenberg waren fasziniert von seinem Werk, mit dem sie sich noch Jahre nach Mayers Tod beschäftigten. Lichtenberg trug die verschiedenen Schriften Mayers zusammen und bereitete sie auf, darunter auch einen Aufsatz über Farbe mit dem Titel *De affinitate colorum commentatio* (*Treatise on the Relationship of Colours*), den er 1775 zusammen mit anderen, ebenfalls in lateinischer Sprache verfassten Aufsätzen Mayers in der Textsammlung *Opera inedita* veröffentlichte. Die Bildtafeln, die das Buch enthält, beziehen sich auf die Farbtheorie des Autors. Eine der Abbildungen beruht auf Mayers eigenem Entwurf und zeigt das mit 91 Farbbezeichnungen ausgefüllte Basisdreieck seines Tetraedersystems. Die andere Tafel, ein von Lichtenberg gezeichnetes und handkoloriertes Dreieck, stellt eine kluge Vereinfachung des Mayer'schen Konzepts dar. Die Version folgt Mayers geometrischem Schema, indem sie mithilfe sechseckiger Formen jedes einzelne Farbmittel kennzeichnet, die Anzahl der Zwischenstufen zwischen den Grundfarben jedoch von elf auf fünf verringert. Lichtenberg und Lambert waren nicht die Letzten, die sich so eingehend mit Mayers Farbdreieck befassten. 1781 nahm der deutsche Farbenhersteller August Ludewig Pfannenschmid (Lebensdaten unbekannt) eine eigene Version des Dreiecks in seine kurze Abhandlung zur Farbenmischung auf, den *Versuch einer Anleitung zum Mischen aller Farben aus blau, gelb und roth, nach beiliegendem Triangel.*

Œuvres inédites

Astronome et mathématicien, Tobias Mayer est aussi cartographe et, à ce titre, habitué des représentations en deux dimensions d'objets tridimensionnels. Ses plus belles réussites comprennent des tables lunaires, indiquant le détail des phases de la Lune, et une carte de notre satellite, ainsi que l'une des premières figurations en trois dimensions de la gamme des couleurs visibles par l'œil humain. Comme l'explique la notice précédente, Mayer est l'auteur d'un concept mathématique consacré aux mélanges de couleurs identifiables et représenté sous la forme d'un double tétraèdre. À l'intersection des deux tétraèdres, le triangle de base contient 91 mélanges de couleurs. Chaque mélange est identifié par l'initiale du mot allemand des couleurs principales (G, jaune ; B, bleu ; R, rouge). Une série de chiffres indique les proportions des couleurs principales au sein de chaque mélange (situés aux trois sommets, les colorants non mélangés ou homogènes ont une valeur de 12). Grâce à ce système, il est possible de créer des formules pour mélanger, calculer ou identifier des combinaisons chromatiques aux proportions correctes.

Mayer ne pourra jamais mettre véritablement à l'épreuve sa conception ni la voir publiée. Mais Lambert (voir p. 108–111) et Georg Christoph Lichtenberg, physicien et écrivain, tous deux fascinés par l'œuvre de Mayer, s'y sont intéressés pendant de nombreuses années après la mort de ce dernier. Lichtenberg réunit et établit l'édition de divers textes de Mayer, notamment un essai sur la couleur, *De affinitate colorum commentatio* (traduit en anglais en 1971 par Eric Forbes sous le titre *Treatise on the Relationship of Colours*), qu'il publie en 1775 dans un recueil intitulé *Opera Inedita*, avec d'autres écrits de Mayer en latin. Les planches reproduites ici sont associées à sa théorie des couleurs : l'une d'elles s'inspire d'un dessin de Mayer montrant le triangle de base de son système tétraédrique à 91 couleurs identifiées. L'autre planche représente un triangle dessiné de la main de Lichtenberg, simplification habile du concept de Mayer. Grâce à des formes hexagonales désignant chacune des colorants individuels, cette version suit le modèle géométrique de Mayer, mais réduit de onze à cinq les gradations intermédiaires entre les couleurs primaires. Lichtenberg et Lambert ne sont pas les derniers à s'être autant intéressés au triangle de Mayer. En 1781, le coloriste allemand August Ludewig Pfannenschmid (dates inconnues) en donne sa version dans *Treatise on the Relationship of Colours* un petit traité portant sur le mélange des couleurs, *Versuch einer Anleitung zum Mischen aller Farben aus blau, gelb und roth, nach beiliegendem Triangel.*

Opera inedita

El astrónomo y matemático Tobias Mayer también fue cartógrafo; como tal, estaba familiarizado con la representación bidimensional de objetos tridimensionales. Entre sus mayores logros, se encuentran las tablas lunares que compiló, que detallan las etapas del movimiento de la luna e incluyen un mapa de la luna. Su trabajo también incluyó una de las primeras representaciones tridimensionales de la gama de colores visibles para el ojo humano. Como ya se ha contado, Mayer propuso un concepto matemático de mezclas de colores identificables presentado en forma de doble tetraedro. El triángulo base, en el punto de unión de ambos tetraedros, contiene 91 mezclas de colores. Para identificarlas, empleó una combinación de letras y números: las iniciales de las palabras alemanas para los colores principales (G para amarillo, B para azul y R para rojo) y unas cifras que indican las proporciones de los colores principales en cada combinación (con los colorantes principales sin mezclar en los tres vértices, con un valor asignado de 12). Con este sistema se debían poder derivar fórmulas para mezclar, calcular o identificar combinaciones de colores en las proporciones correctas.

Mayer nunca pudo probar completamente su concepto de color ni verlo publicado, pero Lambert (véase pág. 108–111) y el físico y escritor Georg Christoph Lichtenberg quedaron fascinados con su trabajo y se ocuparon de él muchos años después de su muerte. Lichtenberg recopiló y editó varios artículos de Mayer, incluido un ensayo sobre el color, *De affinitate colorum commentatio* (traducido al inglés en 1971 por Eric Forbes como *Treatise on the Relationship of Colours*), que publicó en 1775 en la colección *Opera Inedita* junto con varios otros ensayos de Mayer, también en latín. Las láminas de este libro se relacionan con la teoría del color del autor. Una de ellas es una imagen basada en el diseño de Mayer que muestra el triángulo base de su sistema tetraédrico con sus 91 colores etiquetados. Otra muestra un triángulo coloreado a mano dibujado por Lichtenberg, que simplifica inteligentemente el concepto de Mayer. Esta versión siguió el esquema geométrico de Mayer usando formas hexagonales para indicar cada uno de los colorantes por separado, pero redujo los grados intermedios entre los colores primarios de once a cinco. Lichtenberg y Lambert no fueron los últimos en implicarse a fondo con el triángulo de Mayer, pues en 1781 el fabricante de colores alemán August Ludewig Pfannenschmid (fechas desconocidas) incluyó una versión propia en su pequeño tratado sobre la mezcla de colores, *Versuch einer Anleitung zum Mischen aller Farben aus blau, gelb und roth, nach beiliegendem Triangel.*

Johann Ferdinand von Schönfeld (1750–1821)

WIENER FARBENKABINET

oder vollständiges Musterbuch aller Natur-, Grund- und Zusammensetzungsfarben

A Viennese Cabinet of Colours
10 plates, 24.1 x 21 cm / 9 ½ x 8 ¼ in., Vienna, Prague, 1794
Washington, D.C., Smithsonian Libraries

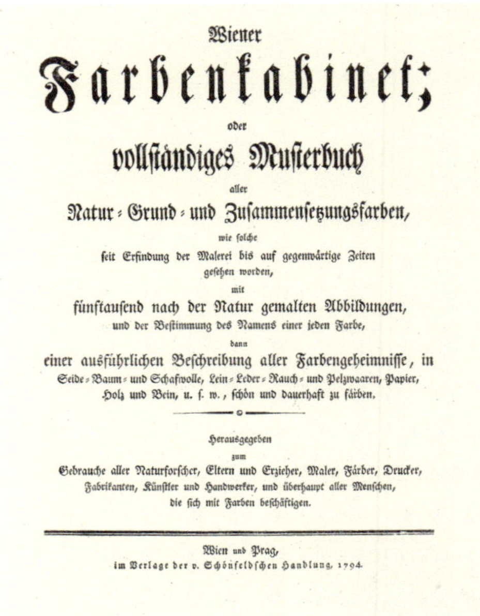
Wiener
Farbenkabinet;
oder
vollständiges Musterbuch
aller
Natur- Grund- und Zusammensetzungsfarben,
wie solche
seit Erfindung der Malerei bis auf gegenwärtige Zeiten
gesehen worden,
mit
fünftausend nach der Natur gemalten Abbildungen,
und der Bestimmung des Namens einer jeden Farbe,
dann
einer ausführlichen Beschreibung aller Farbengeheimnisse, in
Seide- Baum- und Schafwolle, Lein- Leder- Rauch- und Pelzwaaren, Papier,
Holz und Bein, u. s. w., schön und dauerhaft zu färben.

Herausgegeben
zum
Gebrauche aller Naturforscher, Eltern und Erzieher, Maler, Färber, Drucker,
Fabrikanten, Künstler und Handwerker, und überhaupt aller Menschen,
die sich mit Farben beschäftigen.

Wien und Prag,
im Verlage der v. Schönfeldschen Handlung. 1794.

To appreciate the scope of this extraordinary book about colour it is instructive to look further at its long and detailed full title: "a complete book of samples of all natural, basic and combined colours, how these have been perceived since painting was invented through until the present day, containing 5,000 painted samples after nature, and with a name assigned to each (...) followed by a detailed description of all colour secrets, for dyeing silk (etc. ...) for use by all naturalists, parents and teachers, painters, dyers, printers, manufacturers, artists and craftspeople, and any other persons who deal with colour". This ambitious range of purpose and content of what is essentially a colour manual was perhaps only matched by Michel-Eugène Chevreul's publications nearly half a century later (pp. 282–307).

This is a large-format, two-volume work with more than 430 pages of text, and yet the identity of the author remains unknown, and von Schönfeld was in fact the publisher. From the introduction it seems possible that the text was put together as a collaborative effort by members of the Viennese Academy of Fine Arts, and perhaps their associates. The text itself is divided into 14 sections, beginning with essays on the colours black, blue, yellow, red, green, brown and white, relating to the colours themselves and as used by painters, printers, dyers and various other people working with crafts. There are further sections on watercolours and inks, miniature painting, painting materials and techniques, as well as recipes for lacquers and other finishes, followed by an extensive index ("Real-Register").

A simple table for mixing colours appears on page 264 of volume 1, similar to the later *Farb-Tabelle* by Johann Heinrich Meynier (pp. 124–129), but the tables that make the *Farbenkabinet* so very special are attached at the end of volume 2, where there are dozens of plates each with etched grids of 48 boxes, on to which narrow hand-painted colour strips have been pasted, following the colour range discussed previously. The exact number of plates that were intended to be included cannot be stated for certain as only a handful of copies of the *Farbenkabinet* survive in public libraries worldwide, and none seem complete. Based on the figure of 5,000 colour samples given on the book's title-pages, this would correspond to 105 plates; the Smithsonian Libraries copy, a selection of plates from which are shown here, contains 54 plates, making a total of 2,592 colour samples. Leafing through these pages with their thousands of strips of colour painted and pasted in more than 200 years ago is a bewildering optical experience, not dissimilar in its way to looking at modern abstract paintings by Bridget Riley (b. 1931) or Frank Stella (b. 1936). The *Farbenkabinet* is a book of superlatives, and solid proof that by the end of the 18th century colour had become properly respected as an important element of the arts, design, manufacture and education.

Sechste Stammtafel. Blauschwärzliche und blaue Farben.

Nr. 241. Tiefstahlblau.	Nr. 257. Schöndunkelblau.	Nr. 273. Himmelblau.
Nr. 242. Finsterstahlblau.	Nr. 258. Starklasurblau.	Nr. 274. Lebhaftblau.
Nr. 243. Dunkelstahlblau.	Nr. 259. Kräftiglasurblau.	Nr. 275. Himmelschön.
Nr. 244. Starkstahlblau.	Nr. 260. Gemeinlasurblau.	Nr. 276. Saphirblau.
Nr. 245. Kräftigstahlblau.	Nr. 261. Ganzlasurblau.	Nr. 277. Lieblichblau.
Nr. 246. Vollstahlblau.	Nr. 262. Mittellasurblau.	Nr. 278. Sanftblau.
Nr. 247. Ganzstahlblau.	Nr. 263. Kornblumenblau.	Nr. 279. Feinblau.
Nr. 248. Gemeinstahlblau.	Nr. 264. Helllasurblau.	Nr. 280. Lichtblau.
Nr. 249. Mittelstahlblau.	Nr. 265. Lichtlasurblau.	Nr. 281. Hellblau.
Nr. 250. Stahlblau.	Nr. 266. Mattlasurblau.	Nr. 282. Blaßblau.
Nr. 251. Hellstahlblau.	Nr. 267. Schwachlasurblau.	Nr. 283. Zartblau.
Nr. 252. Lichtstahlblau.	Nr. 268. Blaßlasurblau.	Nr. 284. Bleichblau.
Nr. 253. Mattstahlblau.	Nr. 269. Bleichlasurblau.	Nr. 285. Wasserfarbe.
Nr. 254. Blaßstahlblau.	Nr. 270. Hochlasurblau.	Nr. 286. Diamantfarbe.
Nr. 255. Bleichstahlblau.	Nr. 271. Weißlichlasurblau.	Nr. 287. Blaulichweiß.
Nr. 256. Weißlichstahlblau.	Nr. 272. Höchsteslasurblau.	Nr. 288. Weißblau.

Zehnte Stammtafel. Blaurothe Farben.

Nr. 433. Hyacinthenblau.	Nr. 449. Dunkellila.	Nr. 465. Dunkelpfirsichblüthroth.
Nr. 434. Gemeinhyacinthenblau.	Nr. 450. Starklila.	Nr. 466. Starkpfirsichblüthroth.
Nr. 435. Mittelhyacinthenblau.	Nr. 451. Kräftiglila.	Nr. 467. Kräftigpfirsichblüthroth.
Nr. 436. Ganzhyacinthenblau.	Nr. 452. Vollila.	Nr. 468. Vollpfirsichblüthroth.
Nr. 437. Hellhyacinthenblau.	Nr. 453. Ganzlila.	Nr. 469. Ganzpfirsichblüthroth.
Nr. 438. Lichthyacinthenblau.	Nr. 454. Gemeinlila.	Nr. 470. Gemeinpfirsichblüthroth.
Nr. 439. Schwachhyacinthenblau.	Nr. 455. Mittellila.	Nr. 471. Mittelpfirsichblüthroth.
Nr. 440. Cichorienblau.	Nr. 456. Lila.	Nr. 472. Pfirsichblüthroth.
Nr. 441. Hellcichorienblau.	Nr. 457. Hellila.	Nr. 473. Hellpfirsichblüthroth.
Nr. 442. Lichtcichorienblau.	Nr. 458. Lichtlila.	Nr. 474. Lichtpfirsichblüthroth.
Nr. 443. Blaßcichorienblau.	Nr. 459. Blaßlila.	Nr. 475. Blaßpfirsichblüthroth.
Nr. 444. Bleichcichorienblau.	Nr. 460. Bleichlila.	Nr. 476. Bleichpfirsichblüthroth.
Nr. 445. Mattcichorienblau.	Nr. 461. Schwachlila.	Nr. 477. Schwachpfirsichblüthroth.
Nr. 446. Schwachcichorienblau.	Nr. 462. Mattlila.	Nr. 478. Mattpfirsichblüthroth.
Nr. 447. Hochcichorienblau.	Nr. 463. Hochlila.	Nr. 479. Hochpfirsichblüthroth.
Nr. 448. Höchstescichorienblau.	Nr. 464. Höchsteslila.	Nr. 480. Höchstespfirsichblüthroth.

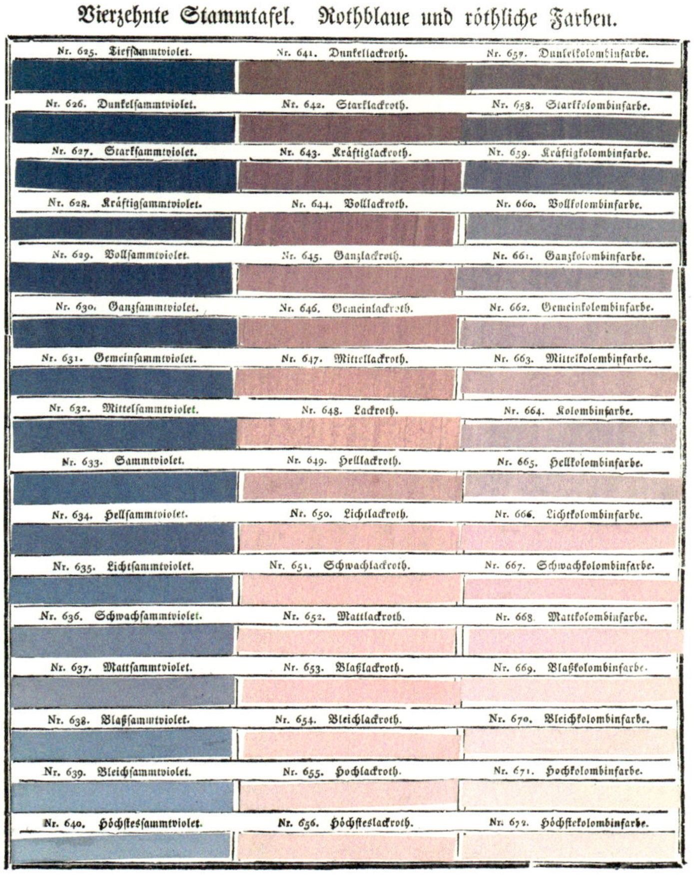

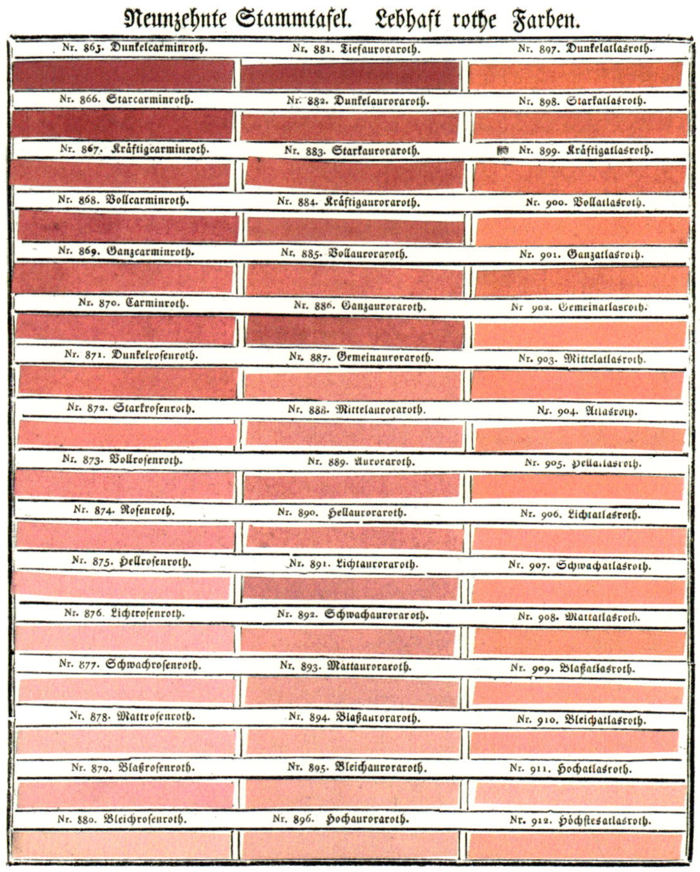

Wiener Farbenkabinet

Um die Reichweite dieses außergewöhnlichen Buches über Farbe würdigen zu können, hilft ein Blick auf seinen detailreichen Titel: „Musterbuch aller Natur-, Grund- und Zusammensetzungsfarben, wie solche seit Erfindung der Malerei bis auf gegenwärtige Zeiten gesehen worden, mit fünftausend nach der Natur gemalten Abbildungen, und der Bestimmung des Namens einer jeden Farbe, dann einer ausführlichen Beschreibung aller Farbengeheimnisse, in Seide [...] u. s. w. [...] zu färben. Herausgegeben zum Gebrauche aller Naturforscher, Eltern und Erzieher, Maler, Färber, Drucker, Fabrikanten, Künstler und Handwerker, und überhaupt aller Menschen, die sich mit Farben beschäftigen.“ An diese ambitionierte Fülle von Absichten und Inhalten eines eigentlichen Farbhandbuchs kamen vermutlich nur Michel-Eugène Chevreuls Publikationen knapp fünfzig Jahre später heran (S. 282–307).

Der Verfasser dieses großformatigen zweibändigen Werks mit mehr als 430 Textseiten ist unbekannt. Schönfeld war de facto der Herausgeber. Der Einleitung nach scheint es möglich, dass der Text von Migliedern der Wiener vereinigten Akademie der bildenden Künste und deren Mitarbeitern zusammengestellt wurde. Er ist in 14 Abschnitte unterteilt und beginnt mit Abhandlungen über die Farben Schwarz, Blau, Gelb, Rot, Grün, Braun und Weiß, in denen die Farben an sich und ihre Verwendung durch Maler, Drucker, Färber und andere Künstler und Handwerker erörtert werden. Weitere Abschnitte sind den Wasserfarben und Tuschen, der Miniaturmalerei, Malmaterialien und -techniken sowie Rezepten für Lacke und andere Firnisse gewidmet, gefolgt von einem ausführlichen „Real-Register“.

Eine Tafel zur Farbenmischung, die der späteren *Farb-Tabelle* von Johann Heinrich Meynier (S. 124–129) ähnelt, erscheint in Band 1 auf Seite 264. Die Tafeln aber, die das *Farbenkabinet* so besonders machen, sind am Ende des zweiten Bandes angehängt, wo schmale, handbemalte Farbstreifen, die dem Farbspektrum folgen, auf Dutzende gestochene Raster mit jeweils 48 Kästchen aufgeklebt wurden. Die genaue Anzahl der ursprünglich geplanten Farbtafeln ist nicht sicher zu bestimmen, da die nur wenigen, in öffentlichen Bibliotheken erhaltenen Exemplare des *Farbenkabinets* alle unvollständig sind. Legt man die fünftausend im Titel genannten Farbmuster zugrunde, dann müsste es 105 Tafeln geben. Das Exemplar der Smithsonian Libraries, aus dem hier einige Tafeln gezeigt werden, enthält 54 Tafeln mit insgesamt 2.592 Mustern. Diese Seiten mit ihren abertausend vor mehr als zweihundert Jahren bemalten und eingeklebten Farbstreifen durchzublättern, ist ein verblüffendes optisches Erlebnis und der Betrachtung moderner abstrakter Gemälde von Bridget Riley (geb. 1931) oder Frank Stella (geb. 1936) nicht unähnlich. Das *Farbenkabinet* ist ein Beweis dafür, dass Farbe im ausgehenden 18. Jahrhundert als wichtiges Element in Kunst, Gestaltung, Industrie und Bildung hoch geschätzt wurde.

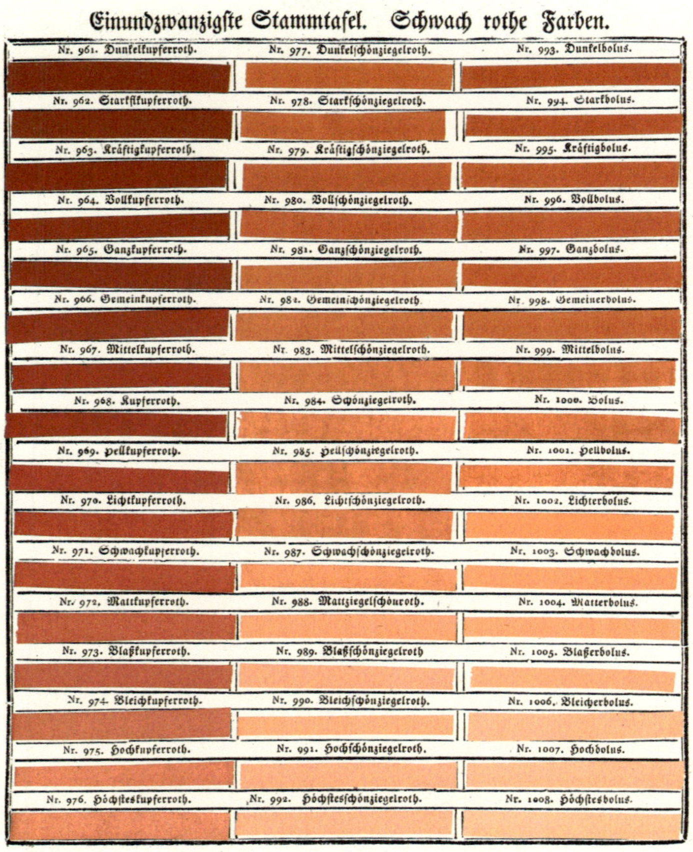

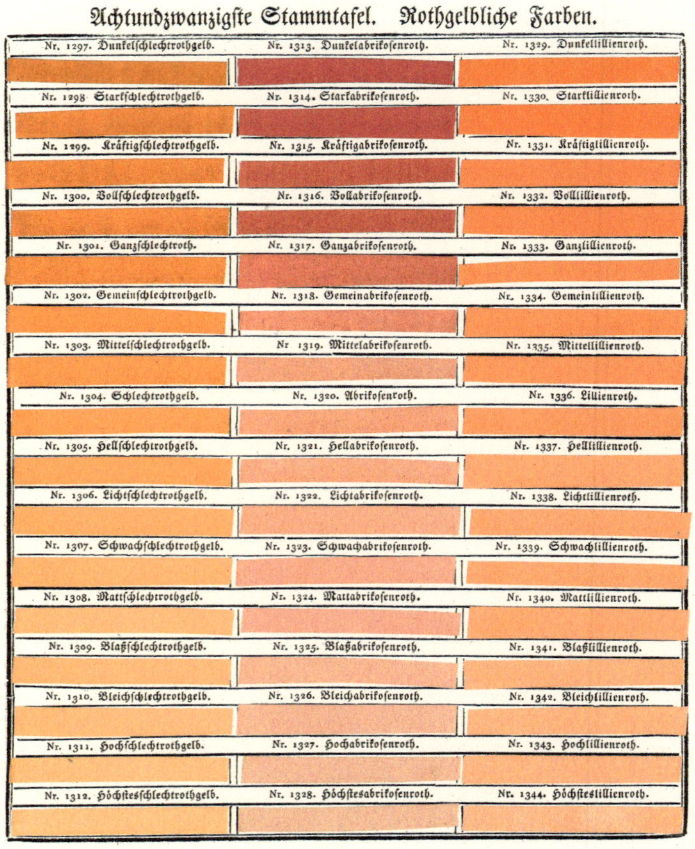

Cabinet des couleurs viennois

Pour bien saisir l'envergure de ce livre extraordinaire, il est utile de prêter attention à son titre : « Cabinet de couleurs viennois, ou Ouvrage complet d'échantillons de toutes les couleurs naturelles, fondamentales et combinées, comment celles-ci sont perçues depuis l'invention de la peinture jusqu'à nos jours, avec 5 000 échantillons peints d'après nature, chacun portant une désignation [...] suivis d'une description détaillée de tous les secrets de la couleur, pour la teinture de la soie [etc. ...] à l'usage de tous les naturalistes, parents et professeurs, peintres, teinturiers, imprimeurs, fabricants, peintres et artisans et toute personne faisant usage de la couleur. » L'ambition des objectifs et du contenu de ce qui est essentiellement un manuel des couleurs n'a peut-être trouvé d'équivalent qu'avec les publications de Chevreul, près d'un demi-siècle plus tard (p. 282–307).

Il s'agit d'un ouvrage de grand format, en deux volumes, avec plus de 430 pages de texte, mais dont l'auteur demeure inconnu. L'introduction laisse penser que le texte est dû à un collectif de membres de l'École des beaux-arts de Vienne et, peut-être, de confrères. Divisé en 14 sections, le texte s'ouvre par des essais dédiés aux couleurs noir, bleu, jaune, rouge, vert, brun et blanc, portant sur la nature de ces couleurs et leur emploi par les peintres, les imprimeurs, les teinturiers et autres artisans. Suivent des sections traitant des aquarelles et des encres, de la miniature, des techniques et matériaux picturaux, des recettes de laques et autres matériaux de finition, ainsi qu'un index exhaustif.

La page 264 du volume 1 figure un simple tableau pour le mélange des couleurs, semblable à la *Farb-Tabelle* que concevra plus tard Johann Heinrich Meynier (p. 124–129). Mais ce sont les tableaux en fin du volume 2 qui font du *Farbenkabinet* un ouvrage si singulier. Il s'agit de dizaines de planches illustrées à l'eau-forte, comprenant cha-cune une grille de 48 cases, sur lesquelles de fines bandes de couleurs peintes à la main ont été collées, selon la gamme chromatique évoquée plus haut. Il est impossible de mentionner le nombre exact de planches qui devaient composer l'ouvrage car il ne reste que très peu d'exemplaires du *Farbenkabinet*, conservés dans des bibliothèques publiques du monde entier, mais dont aucun ne semble complet. Le chiffre de 5 000 échantillons figurant sur les pages de titre laisse penser qu'il devait y avoir 105 planches. L'exemplaire des Bibliothèques de la Smithsonian Institution, dont proviennent les planches reproduites ici, en comporte 54, pour un total de 2 592 échantillons. Contempler ces pages et leurs milliers de bandes de papier est une expérience optique stupéfiante, qui n'est pas sans rappeler l'effet produit par les peintures abstraites de Bridget Riley (née en 1931) ou de Frank Stella (né en 1936). Le *Farbenkabinet* est la preuve tangible qu'à la fin du XVIIIe siècle, la couleur s'est désormais acquis le respect qui lui est dû comme élément essentiel des arts, du dessin, de la fabrication et de l'éducation.

Zweyunddreißigste Stammtafel. Bräunlichgelbe Farben.

Nr. 1489. Dunkelwachsgelb.	Nr. 1505. Dunkelisabellgelb.	Nr. 1521. Dunkelbrandgelb.
Nr. 1490 Starkwachsgelb.	Nr. 1506. Starkisabellgelb.	Nr. 1522. Starkbrandgelb.
Nr. 1491. Kräftigwachsgelb.	Nr. 1507. Kräftigisabellgelb.	Nr. 1523. Kräftigbrandgelb.
Nr. 1492. Vollwachsgelb.	Nr. 1508. Vollisabellgelb.	Nr. 1524. Vollbrandgelb.
Nr. 1493. Ganzwachsgelb.	Nr. 1509. Ganzisabellgelb.	Nr. 1525. Ganzbrandgelb.
Nr. 1494. Gemeinwachsgelb.	Nr. 1510. Gemeinisabellgelb.	Nr. 1526. Gemeinbrandgelb.
Nr. 1495. Mittelwachsgelb.	Nr. 1511. Mittelisabellgelb.	Nr. 1527. Mittelbrandgelb.
Nr. 1496. Wachsgelb.	Nr. 1512. Isabellgelb.	Nr. 1528. Brandgelb.
Nr. 1497. Hellwachsgelb.	Nr. 1513. Hellisabellgelb.	Nr. 1529. Hellbrandgelb.
Nr. 1498. Lichtwachsgelb.	Nr. 1514. Lichtisabellgelb.	Nr. 1530. Lichtbrandgelb.
Nr. 1499. Schwachwachsgelb.	Nr. 1515. Schwachisabellgelb.	Nr. 1531. Schwachbrandgelb.
Nr. 1500. Mattwachsgelb.	Nr. 1516. Mattisabellgelb.	Nr. 1532. Mattbrandgelb.
Nr. 1501. Blaßwachsgelb.	Nr. 1517. Blaßisabellgelb.	Nr. 1533 Blaßbrandgelb.
Nr. 1502. Bleichwachsgelb.	Nr. 1518. Bleichisabellgelb.	Nr. 1534. Bleichbrandgelb.
Nr. 1503. Hochwachsgelb.	Nr. 1519. Hochisabellgelb.	Nr. 1535. Hochbrandgelb.
Nr. 1504. Höchsteswachsgelb.	Nr. 1520. Höchstesisabellgelb.	Nr. 1536. Höchstesbrandgelb.

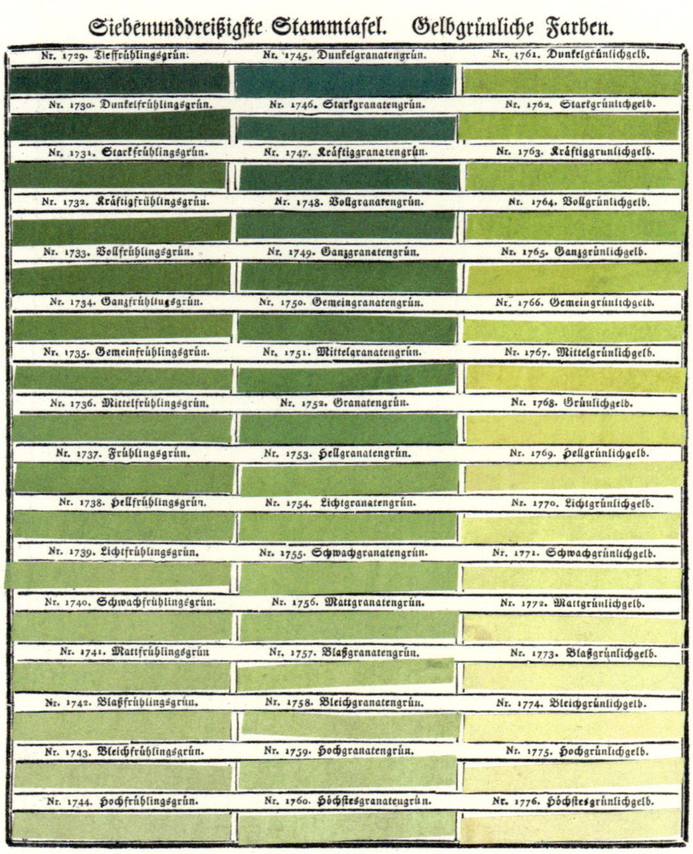

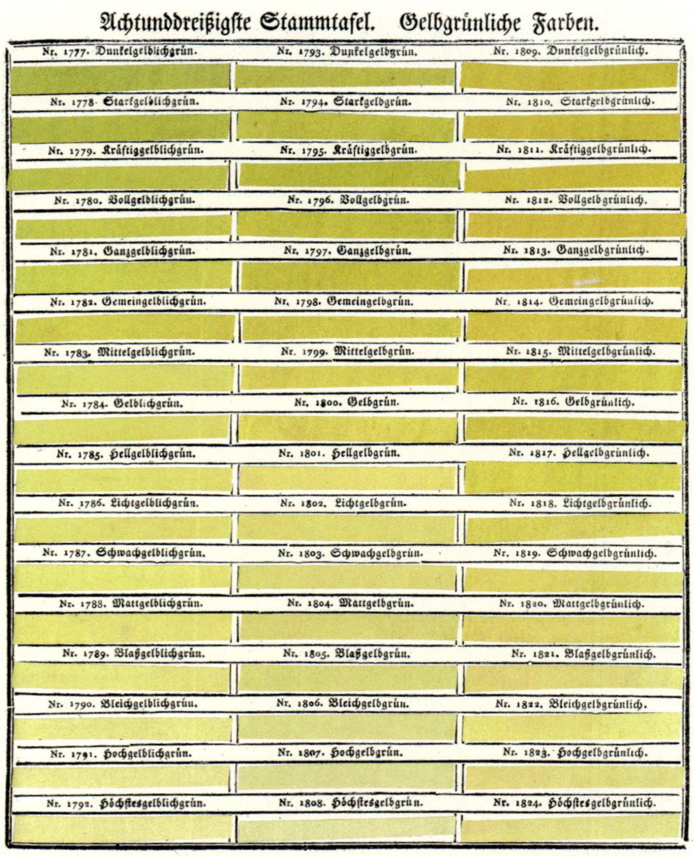

Gabinete de colores vienés

El largo y detallado título de este extraordinario libro ayuda a entender su alcance: «Un completo libro con muestras de todos los colores naturales, básicos y combinados, sobre cómo se han percibido desde que se inventó la pintura hasta la actualidad, que contiene 5000 muestras pintadas copiadas del natural, con un nombre asignado a cada una [...] seguido de una descripción detallada de todos los secretos del color, para teñir seda [etc.] para uso de todos los naturalistas, padres y maestros, pintores, tintoreros, impresores, fabricantes, artistas y artesanos, y cualquier otra persona que se ocupe del color». La ambición del propósito y contenido de lo que es en esencia un manual sobre el color quizás solo pueda compararse con las publicaciones de Michel-Eugène Chevreul casi medio siglo después (págs. 282–307).

Se trata de una obra de gran formato en dos volúmenes con más de 430 páginas de texto. Sin embargo, se desconoce la identidad del autor, de hecho, Von Schönfeld fue el editor. Según la introducción, parece posible que el texto fuera fruto de un esfuerzo colaborativo de miembros de la Academia de Bellas Artes de Viena, y quizás de sus asociados. El libro se divide en 14 secciones. Empieza con ensayos sobre los colores negro, azul, amarillo, rojo, verde, marrón y blanco que abordan los colores en sí y su uso por parte de pintores, impresores, tintoreros y otros artesanos. Hay más secciones sobre acuarelas y tintas, pintura de miniaturas, materiales y técnicas de pintura, así como recetas de lacas y otros acabados, seguidas de un extenso índice («Real-Register»).

En la página 264 del volumen 1 aparece una tabla sencilla sobre la mezcla de colores, similar a la posterior *Farb-Tabelle* de Johann Heinrich Meynier (págs. 124–129), pero las tablas que hacen que *Farbenkabinet* sea tan especial se adjuntan al final del volumen 2, donde hay docenas de láminas, cada una de ellas con tablas grabadas de 48 casillas en las que se pegaron tiras de color pintadas a mano que siguen la gama cromática comentada anteriormente. No se puede afirmar con certeza el número exacto de láminas que se pretendía incluir, ya que solo se conservan unas pocas copias del *Farbenkabinet* en las bibliotecas públicas de todo el mundo, y ninguna parece estar completa. Según la cifra de 5000 muestras de color que figura en la portada del libro, eso correspondería a 105 láminas; la copia de las Bibliotecas Smithsonian, de la que aquí se incluye una selección de láminas, contiene 54 láminas, lo que hace un total de 2592 muestras de color. Hojear esas páginas, con sus miles de tiras de color pintadas y pegadas hace más de 200 años, es una experiencia óptica desconcertante, no muy diferente a mirar las pinturas abstractas modernas de Bridget Riley (n. 1931) o Frank Stella (n. 1936). El *Farbenkabinet* es un libro superlativo y una prueba sólida de que, a finales del siglo XVIII, el color se había ganado el respeto como elemento importante de las artes, el diseño, la manufactura y la educación.

Sechsundvierzigste Stammtafel. Grüne Farben.

Nr. 2161. Dunkelgrünbläulich.	Nr. 2177. Dunkelberggrün.	Nr. 2193. Dunkelspahngrün.
Nr. 2162 Starkgrünbläulich.	Nr. 2178. Starkberggrün.	Nr. 2194. Starkspahngrün.
Nr. 2163. Kräftiggrünbläulich.	Nr. 2179. Kräftigberggrün.	Nr. 2195. Kräftigspahngrün.
Nr. 2164. Vollgrünbläulich.	Nr. 2180. Vollberggrün.	Nr. 2196. Vollspahngrün.
Nr. 2165. Ganzgrünbläulich.	Nr. 2181. Ganzberggrün.	Nr. 2197. Ganzspahngrün.
Nr. 2166. Gemeingrünbläulich.	Nr. 2182. Gemeinberggrün.	Nr 2198. Gemeinspahngrün.
Nr. 2167. Mittelgrünbläulich.	Nr 2183. Mittelberggrün.	Nr. 2199. Mittelspahngrün.
Nr. 2168. Grünbläulich.	Nr. 2184. Berggrün.	Nr. 2200. Spahngrün.
Nr. 2169. Hellgrünbläulich.	Nr. 2185. Hellberggrün.	Nr. 2201. Hellspahngrün.
Nr. 2170. Lichtgrünbläulich.	Nr. 2186. Lichtberggrün.	Nr. 2202. Lichtspahngrün.
Nr. 2171. Schwachgrünbläulich.	Nr. 2187. Schwachberggrün.	Nr. 2203. Schwachspahngrün.
Nr. 2172. Mattgrünbläulich.	Nr. 2188. Mattberggrün.	Nr. 2204. Mattspahngrün.
Nr. 2173. Blaßgrünbläulich.	Nr. 2189. Blaßberggrün.	Nr. 2205. Blaßspahngrün.
Nr. 2174. Bleichgrünbläulich.	Nr. 2190. Bleichberggrün.	Nr. 2206. Bleichspahngrün.
Nr. 2175. Hochgrünbläulich	Nr. 2191. Hochberggrün.	Nr. 2207. Hochspahngrün.
Nr. 2176. Höchstesgrünbläulich.	Nr. 2192. Höchstesberggrün.	Nr. 2208 Höchstesspahngrün.

Johann Heinrich Meynier (1764–1825)

FARB-TABELLE. FÜR ILLUMINISTEN UND LIEBHABER DER WASSERMALEREY

A Colour Table. For Illustrators and Enthusiasts of Watercolour
1 chart, from Die Kunst zu Tuschen und mit Wasserfarben sowohl in Miniatur, als in Gouache und in Aberlischer- oder Aquarell-Manier, Landschaften, Porträte, und andere Gegenstände zu mahlen, *36.1 x 50.8 cm / 14 ¼ x 20 in., Leipzig, 1799*
New Jersey, Princeton University Library, Graphic Arts Collection

The final colour chart in this chapter comes from a book that has seldom been discussed in the context of colour history, possibly because it has never been considered a serious work on colour but it is nevertheless typical of its time and shines a useful light on the sort of people who were painting in the later 18th century, as well as what they were painting and what materials were available. Johann Heinrich Meynier was a German writer and educator amongst other professions, but not apparently a specialist in colour or painting. He was of Huguenot background and highly educated, and is best known for his popular fiction for young readers and general books on nature and history. His publishing output was huge and cannot be fully assessed since he used several different spellings of his name, and more than a dozen pseudonyms.

Meynier's *Die Kunst zu Tuschen und mit Wasserfarben (...) zu mahlen* was aimed at the growing readership of amateur painters with their newly affordable art materials. Painting in watercolour had become increasingly popular in many countries across Europe since the 1760s, when the paints had become available in ready-made moistened cakes that could be easily carried around in boxes and tins. The full title of Meynier's book also refers to an "Aberlian manner" of painting, as promoted by the Swiss painter Johann Ludwig Aberli (1723–1786) and which amounted to a quick method of colouring etchings in with washes, to make them appear like original paintings. Meynier wrote his lengthy manual on painting for leisure with a general audience in mind, something he was criticised for by an early reviewer who bemoaned the fact that many people at the time seemed to think that all it took to understand and produce art was a basic knowledge of materials and technique.

The book's large fold-out chart, entitled *Farb-Tabelle. Für Illuministen und Liebhaber der Wassermalerey*, was an incentive to amateur artists to practise mixing their own colours and to create individualised reference tables of these combinations. As such, the chart published in the book was uncoloured, and a few examples of blank charts still survive. Meynier's system proposed a total of 784 colour mixtures, without giving precise ratios of mixture, ranging diagonally from the lightest colour (king's yellow) in the top left corner, to black ink in the bottom right. The accompanying text lists each mixture again, together with notes about unsuitable "dirty" combinations, such as dark ochre and dragon's blood, or indigo and bistre.

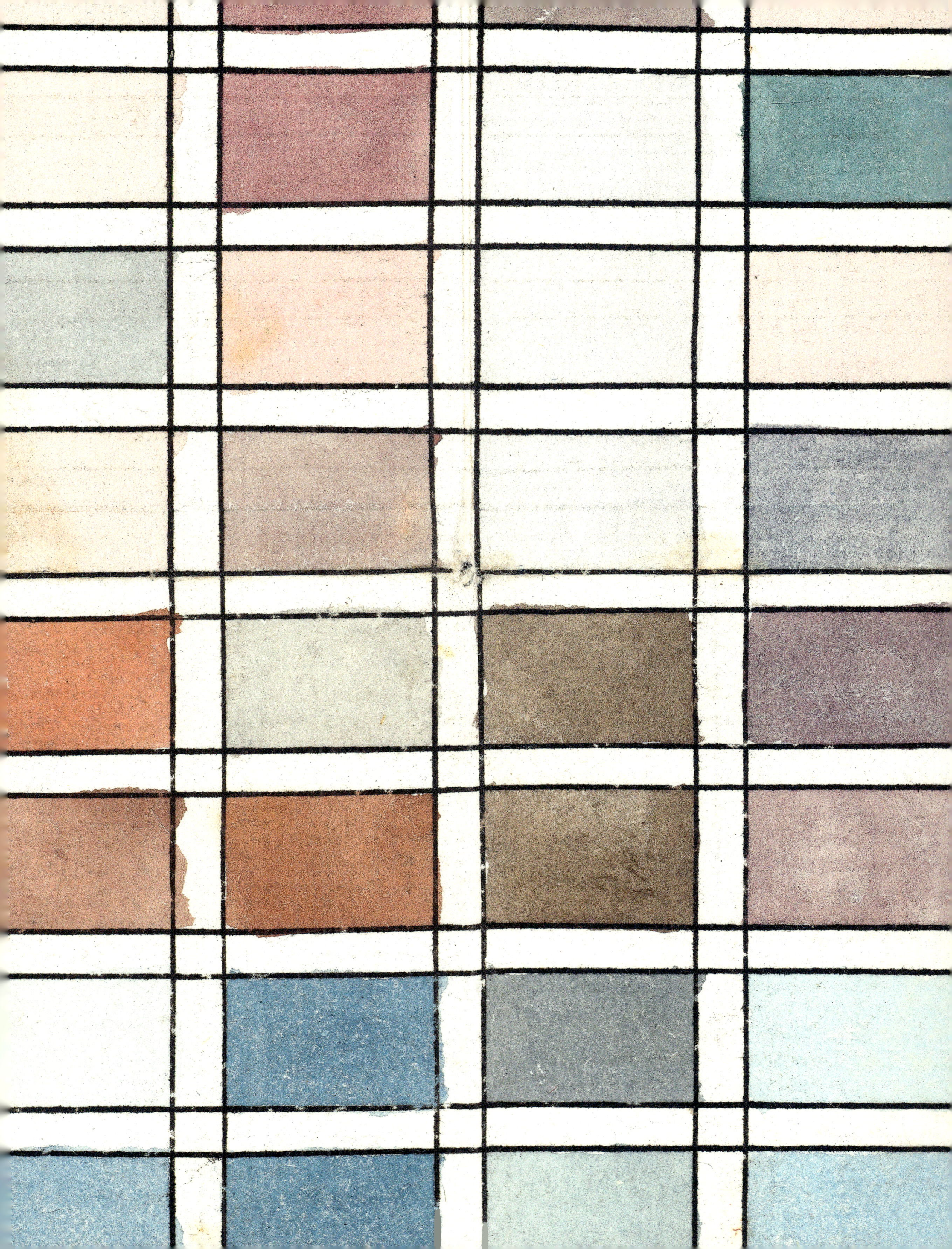

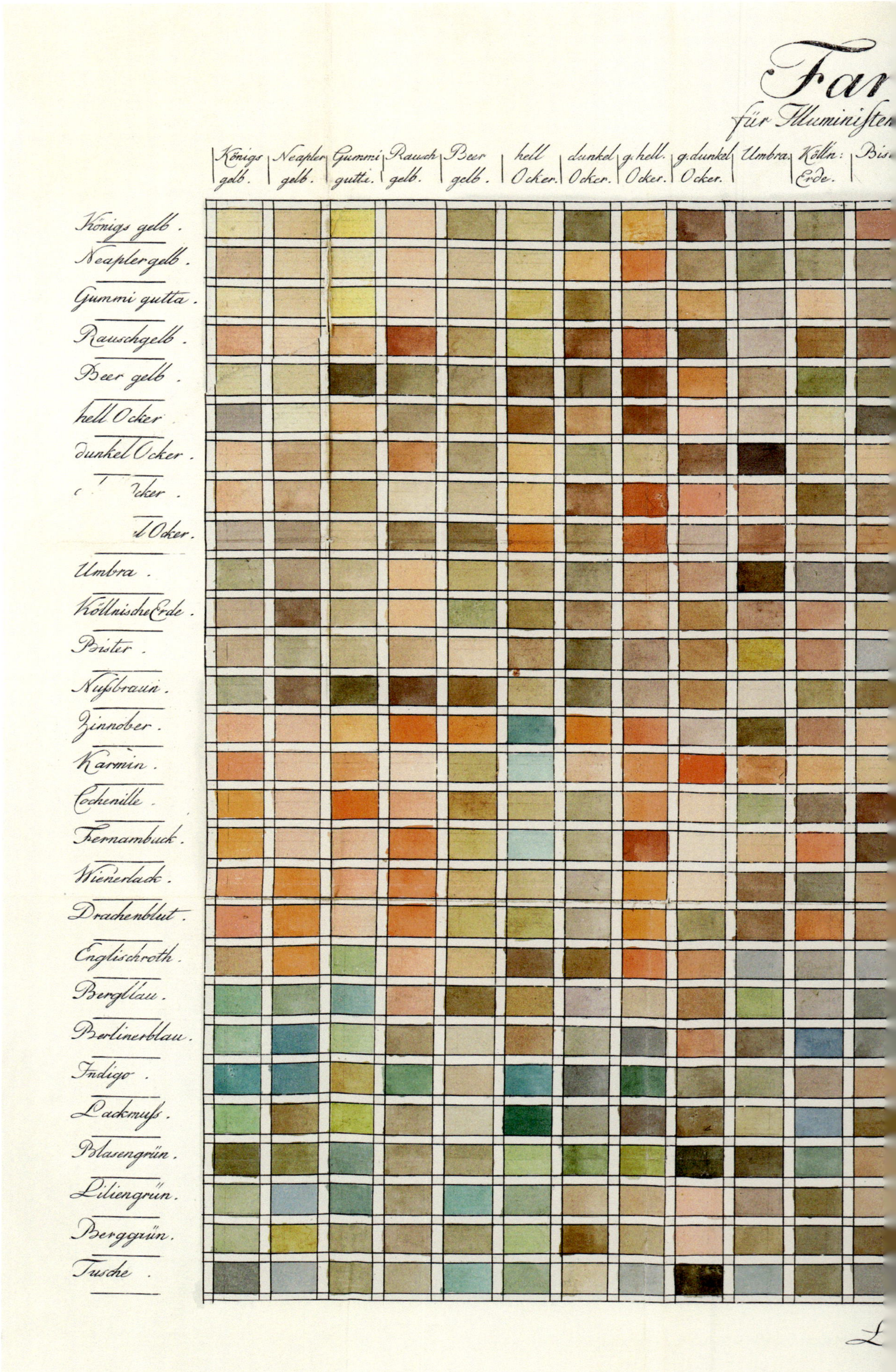
Königs gelb.
Neapler gelb.
Gummi gutti.
Rauch gelb.
Beer gelb.
hell Ocker.
dunkel Ocker.
g. hell. Ocker.
g. dunkel Ocker.
Umbra.
Köln: Erde.
Königs gelb.
Neaplergelb.
Gummi gutta.
Rauschgelb.
Beer gelb.
hell Ocker.
dunkel Ocker.
Umbra.
Köllnische Erde.
Bister.
Nußbraun.
Zinnober.
Karmin.
Cochenille.
Fernambuck.
Wienerlack.
Drachenblut.
Englischroth.
Bergblau.
Berlinerblau.
Indigo.
Lackmuß.
Blasengrün.
Liliengrün.
Berggrün.
Tusche.

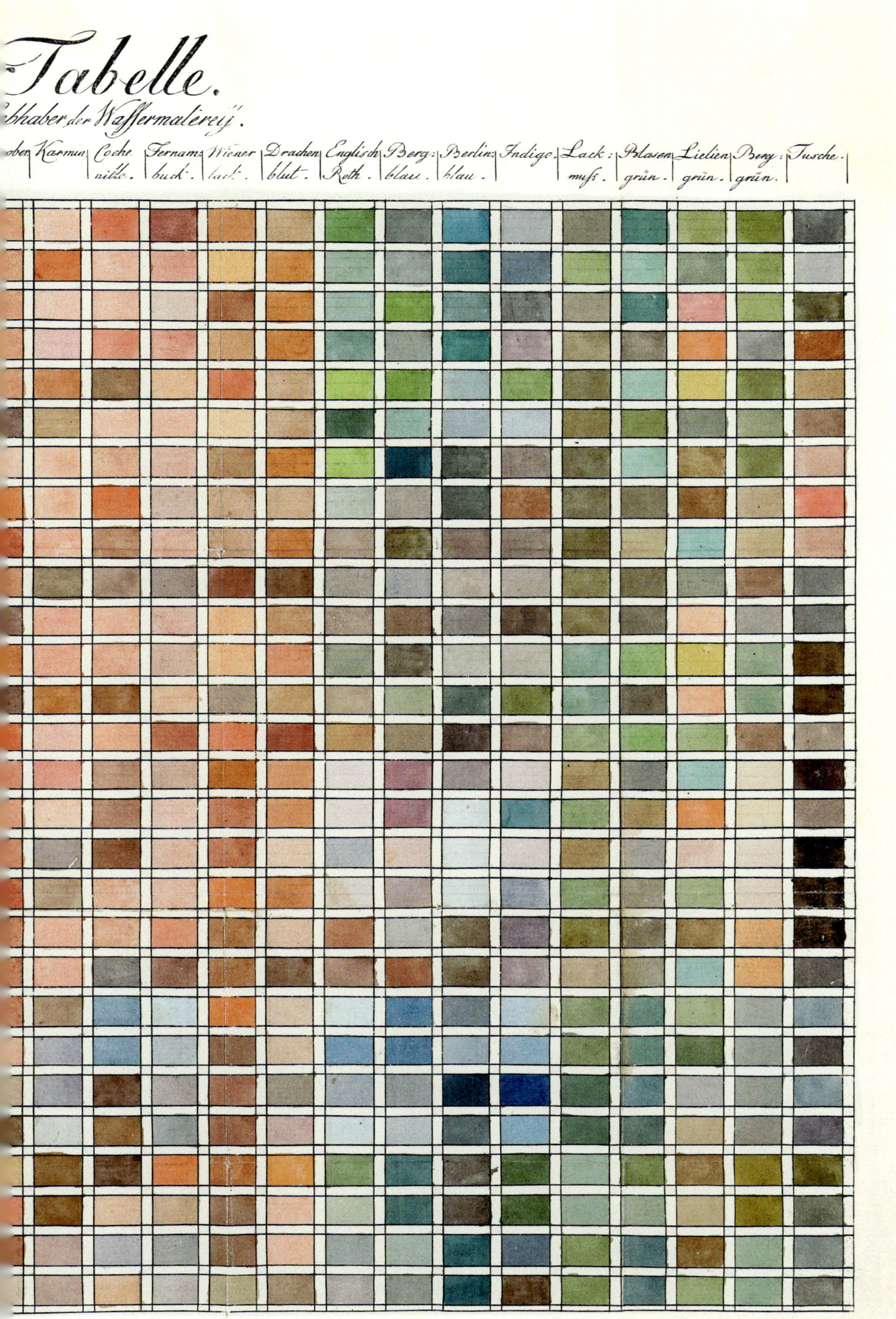
Tabelle.
bhaber der Waßermalereÿ.
Karmin
Coche nille.
Fernam buck.
Wiener lack.
Drachen blut.
Englisch Roth.
Berg: blau.
Berlin: blau.
Indigo.
Lack: muß.
Blasen grün.
Lielien grün.
Berg: grün.
Tusche.

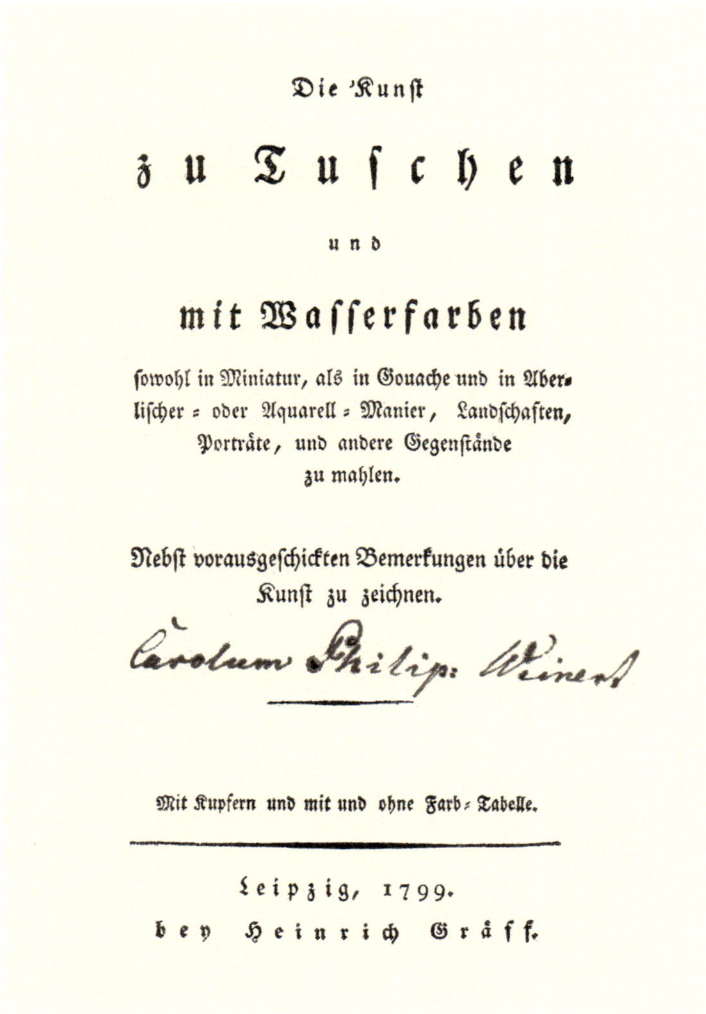

Die Kunst

zu Tuschen

und

mit Wasserfarben

sowohl in Miniatur, als in Gouache und in Aberlischer = oder Aquarell = Manier, Landschaften, Porträte, und andere Gegenstände zu mahlen.

Nebst vorausgeschickten Bemerkungen über die Kunst zu zeichnen.

Mit Kupfern und mit und ohne Farb = Tabelle.

Leipzig, 1799.
bey Heinrich Gräff.

Farb-Tabelle. Für Illuministen und Liebhaber der Wassermalerey

Die letzte Farbtabelle dieses Kapitels stammt aus einem Buch, das im Kontext der Farbgeschichte selten besprochen wurde, weil es womöglich nie als ernstzunehmendes Werk über Farbe galt. Gleichwohl war es charakteristisch für seine Zeit und wirft ein praktisches Licht auf die Maler des späten 18. Jahrhunderts, ihre Malinhalte und die Materialien, die sie zur Verfügung hatten. Johann Heinrich Meynier war ein deutscher Schriftsteller und Pädagoge und übte weitere Berufe aus, doch er war offenbar kein Experte für Farbe oder Malerei. Bekannt wurde der hochgebildete Mann mit hugenottischen Wurzeln vor allem durch seine Bücher für Kinder und Jugendliche und seine Sachbücher über Naturkunde und Geschichte. Die enorme Anzahl seiner Veröffentlichungen lässt sich nur schwer ermessen, da er unterschiedliche Schreibweisen seines Namens und mehr als ein Dutzend Pseudonyme verwendete.

Meyniers *Die Kunst zu Tuschen und mit Wasserfarben (...) zu mahlen* war an die wachsende Leserschaft von Amateurmalern mit ihrem neuerdings erschwinglichen Künstlermaterial gerichtet. Die Aquarellmalerei hatte in vielen europäischen Ländern seit den 1760er-Jahren zunehmend an Beliebtheit gewonnen. Damals waren Farben in vorgefertigten befeuchteten Blöcken auf den Markt gekommen, die man in Schachteln und Dosen mühelos mit sich herumtragen konnte. Im vollständigen Titel des Buches ist auch von der „Aberlischen Manier" in der Malerei die Rede, die der Schweizer Maler Johann Ludwig Aberli (1723–1786) propagiert hatte. Dabei handelte es sich um eine schnelle Methode, Radierungen mit Wasserfarben auszumalen, um sie wie Originalgemälde erscheinen zu lassen. Meynier verfasste sein recht umfangreiches Handbuch über die Malerei als Zeitvertreib für ein breites Laienpublikum, was ihm die Kritik eines frühen Rezensenten eintrug. Dieser beklagte den damals anscheinend verbreiteten Gedanken, dass man nur Grundkenntnisse in Materialkunde und Technik besitzen musste, um Kunst zu verstehen und zu produzieren.

Die große ausfaltbare Tafel des Buches mit der Überschrift „Farb-Tabelle. Für Illuministen und Liebhaber der Wassermalerey" war ein Anreiz für Amateurkünstler, ihre eigenen Farben zu mischen und mit diesen Mischungen individuelle Referenztafeln zu erstellen. Die im Buch veröffentlichte Tabelle war nicht koloriert und ist in diesem Zustand in einigen Exemplaren erhalten. Meyniers System sah, ohne die Mischungsverhältnisse genau anzugeben, insgesamt 784 Farbmischungen vor, die sich von der hellsten Farbe (Königsgelb) oben links in einer Diagonale bis zur schwarzen Tusche unten rechts erstreckten. Der Begleittext listet jede Mischung erneut auf, ergänzt um Hinweise auf „schmutzige", für die Malerei ungeeignete Mixturen wie „Gebrannter Dunkelocker und Drachenblut" oder „Indigo und Bister".

Tableau des couleurs. À l'usage des illustrateurs et adeptes de l'aquarelle

Le dernier graphique chromatique de ce chapitre provient d'un livre rarement évoqué du point de vue de l'histoire de la couleur, peut-être parce qu'il n'a jamais été considéré comme une œuvre sérieuse en la matière, alors qu'il est caractéristique de son temps et qu'il met en lumière les peintres de la fin du XVIIIe siècle et les matériaux dont ils disposaient. Entre autres métiers, l'Allemand Johann Heinrich Meynier exerçait ceux d'écrivain et de pédagogue, sans être, semble-t-il, un spécialiste de la couleur ou de la peinture. D'origine huguenote, très instruit, il est surtout connu pour ses œuvres de fiction destinées à la jeunesse et des ouvrages généraux sur la nature et l'histoire. Il est difficile d'évaluer la quantité de livres que cet auteur prolifique a signés car celui-ci orthographiait son nom de plusieurs manières et a employé plus d'une douzaine de pseudonymes.

Son ouvrage intitulé *Die Kunst zu Tuschen und mit Wasserfarben (...) zu mahlen* est destiné aux peintres amateurs de plus en plus nombreux en son temps, susceptibles de s'offrir des fournitures de peinture devenues bon marché. L'aquarelle s'est largement popularisée dans de nombreux pays d'Europe depuis les années 1760, période à laquelle les peintures sont vendues sous forme de tablettes humidifiées prêtes à l'emploi, facilement transportables dans des boîtes en bois ou métalliques. Le titre complet du livre de Meynier fait aussi allusion à une « manière aberlienne » de peindre, telle que la défend le peintre suisse Johann Ludwig Aberli (1723–1786), méthode rapide de coloriage d'eaux-fortes avec des lavis pour que les gravures ressemblent à des peintures originales. Destinant au grand public ce long manuel de peinture récréative, Meynier s'attire la critique dans un des premiers comptes rendus de son ouvrage, lequel déplore qu'à cette époque, beaucoup de gens pensent qu'il suffit de s'y connaître un peu en matériaux et en technique pour comprendre et créer des œuvres d'art.

Le grand tableau dépliable de *Farb-Tabelle. Für Illuministen und Liebhaber der Wassermalerey* incite les peintres amateurs à réaliser leurs propres mélanges de couleurs et à personnaliser leurs tableaux de combinaisons chromatiques. Ce tableau était fourni vierge de toute couleur et il en subsiste quelques exemplaires non remplis. Le système de Meynier propose en tout 784 mélanges, sans en préciser les proportions, disposés en diagonale du coloris le plus clair (orpiment vif), dans l'angle supérieur gauche, jusqu'à l'encre noire, dans l'angle inférieur droit. Le texte répertorie chaque combinaison et comprend des notes sur les associations « sales » et inconvenantes, comme l'ocre foncé avec le sang-de-dragon, ou l'indigo avec le bistre.

Tabla de colores. Para ilustradores y entusiastas de la acuarela

La última tabla de colores de este capítulo proviene de un libro que rara vez se ha tratado en el contexto de la historia del color, posiblemente porque nunca se ha considerado un trabajo serio sobre el color; sin embargo, es típico de su época y arroja luz sobre el tipo de personas que pintaban a finales del siglo XVIII y qué materiales tenían a su disposición. Johann Heinrich Meynier fue un escritor y educador alemán, entre otras profesiones, pero aparentemente no un especialista en el color o la pintura. De origen hugonote y con una gran formación, es más conocido como escritor de novelas populares para jóvenes y manuales sobre naturaleza e historia. Su producción fue enorme y no se puede evaluar completamente ya que escribió su nombre con varias grafías diferentes y empleó más de una docena de seudónimos.

Die Kunst zu Tuschen und mit Wasserfarben (...) zu mahlen iba dirigido al creciente número de pintores aficionados que había surgido con la aparición de nuevos materiales más asequibles. La pintura en acuarela se había popularizado en muchos países de Europa desde la década de 1760, cuando las pinturas pasaron a estar disponibles en pastillas húmedas listas para usar que podían transportarse fácilmente en cajas y latas. El título completo del libro de Meynier también se refiere a la «manera aberliana» de pintar, promovida por el pintor suizo Johann Ludwig Aberli (1723–1786), que equivalía a un método rápido de colorear aguafuertes con aguadas para que parecieran pinturas originales. Meynier escribió su extenso manual sobre la pintura para el ocio pensando en el público general, lo que le valió el reproche de uno de sus primeros críticos, quien lamentó que muchas personas creyeran que bastaba un conocimiento básico de los materiales y la técnica para comprender el arte y producirlo.

El gran gráfico desplegable del libro, titulado *Farb-Tabelle. Für Illuministen und Liebhaber der Wassermalerey*, era un incentivo para que los artistas aficionados practicaran mezclando sus propios colores y crearan tablas de referencia personales de esas combinaciones. Como tal, no estaba coloreado, y aún sobreviven algunos ejemplos de gráficos en blanco. El sistema de Meynier proponía un total de 784 mezclas de colores, cuyas proporciones no precisa, que iban en diagonal desde el color más claro (amarillo del rey) en la esquina superior izquierda hasta la tinta negra en la esquina inferior derecha. El texto adjunto vuelve a enumerar cada mezcla e incluye notas sobre combinaciones «sucias» inadecuadas, como ocre oscuro y sangre de drago, o índigo y bistre.

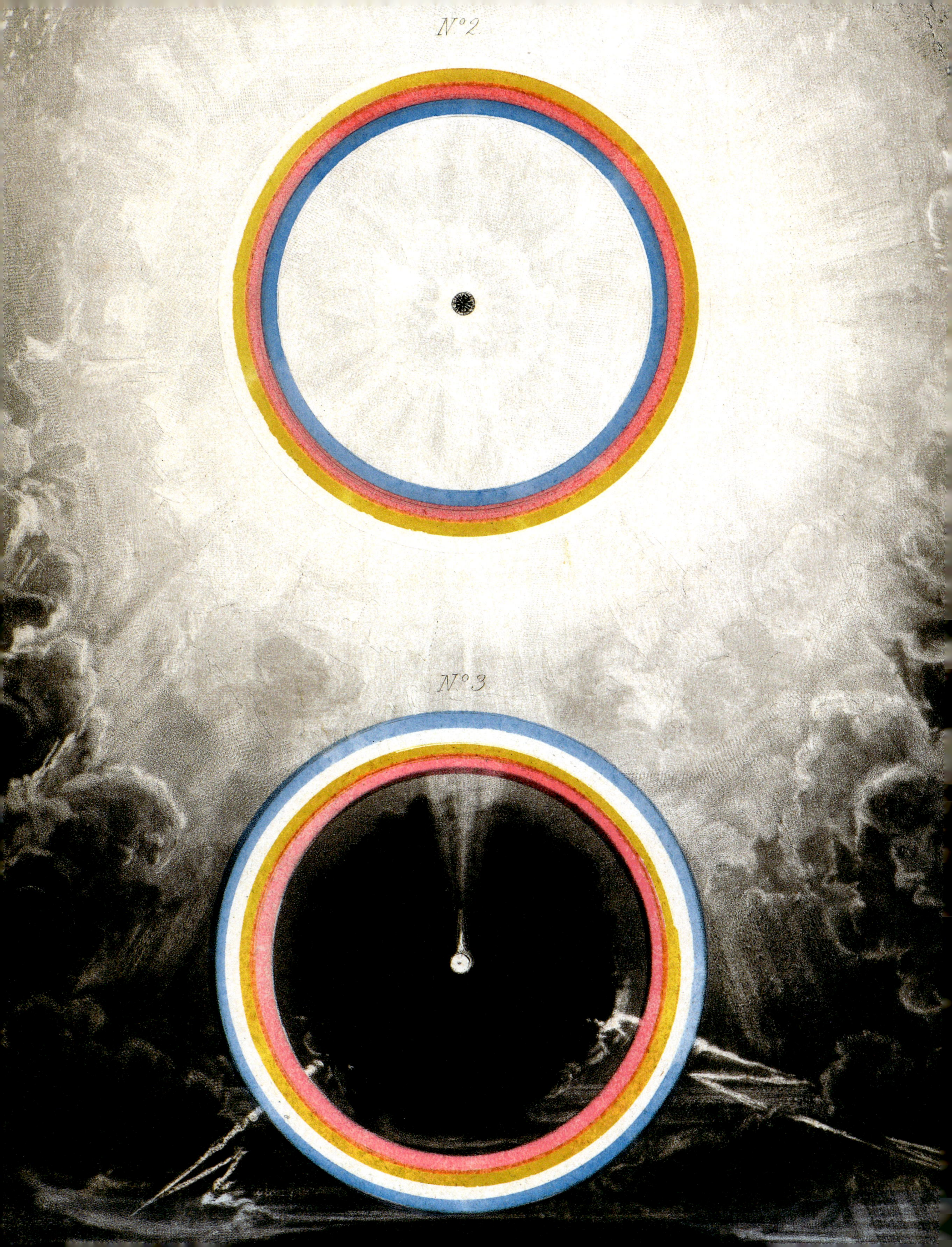
N° 2
N° 3

Chapter 2

The Shape of Colour: CIRCLES, WHEELS AND GLOBES

1708–1896

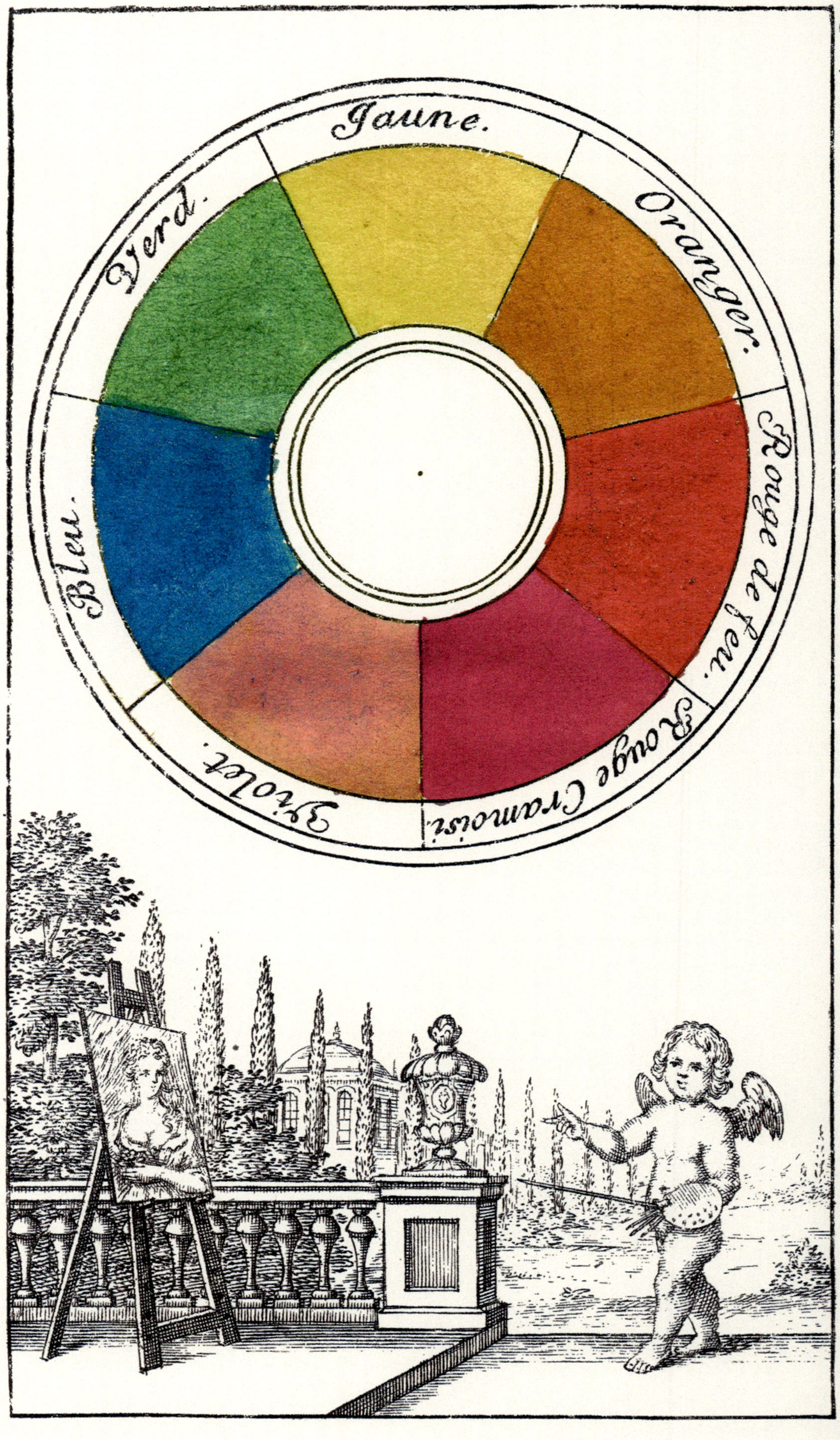
Jaune.
Oranger.
Rouge de feu.
Rouge Cramoisi.
Violet.
Bleu.
Verd.

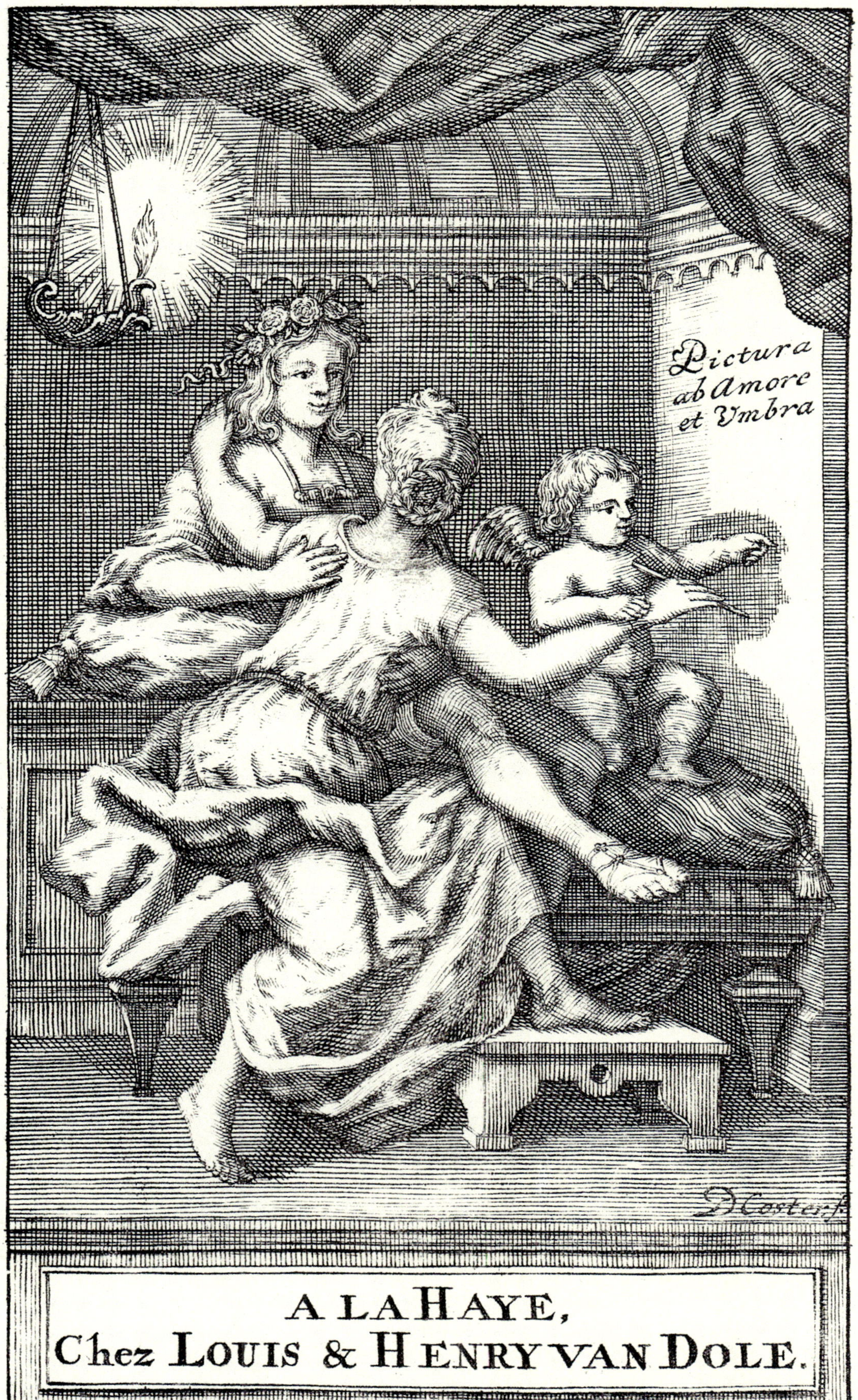
Pictura
ab Amore
et Vmbra
D Coster f.
A LA HAYE,
Chez LOUIS & HENRY VAN DOLE.

Attributed to Claude Boutet (dates unknown)

TRAITÉ DE LA PEINTURE EN MIGNATURE

A Treatise on Painting in Miniature

2 plates, 15.5 x 8.6 cm / 6 ⅛ x 3 ⅜ in., The Hague, 1708

Winterthur, Stadtbibliothek, Farbsammlung Werner Spillmann

The pair of diagrams shown here are perhaps the first colour wheels actually to appear in colour in western print culture. They were published just four years after Isaac Newton's uncoloured wheel and were likely influenced by that form on the basis of their circular design, although they are also something of a mystery. They appear in some copies of a substantial practical handbook on colour, *Traité de la peinture en mignature*, which included when it was published in 1708 a small section on painting in pastel, "Traité de la peinture au pastel". The main part of the text was first published in around 1673 and is often attributed to the French painter Claude Boutet, but this is not certain, and furthermore it may not have been Boutet who created these plates, which appear only in some of the later editions. Aside from such questions of authorship, the book contains detailed recipes and advice on how to prepare pigments and use colours when painting portraits, landscapes, flowers and other subjects. In that respect it doesn't differ greatly from other 17th- and 18th-century artists' manuals, but the carefully designed and hand-coloured engravings depicting the colour wheels are very special. Their presence in a handbook for artists, and the added figurative scenes beneath the wheels showing allegorical putti as portrait painters, makes clear that their purpose is to illustrate material colour, in contrast to Newton's colour wheel which was concerned in the main with the optical spectrum. A pair of colour wheels would usually represent an attempt to visualise secondary or tertiary colour mixtures, effectively creating another layer of colour, but in this case there is no recognisable arrangement of primaries and secondaries based on the pure, unmixed colours of red, yellow and blue. The circle on the left contains seven colours, named as violet, blue, green, yellow, orange, fire red and crimson, while the circle on the right extends this scheme to a 12-colour arrangement, adding golden yellow, red, purple, sea green and yellow-green. These additional five colours are not all mixtures of the colours found in the first wheel, and indeed they include a simple red, one of the pure "primitives" mentioned in the accompanying text. The hand-colouring in the few surviving examples of this book varies considerably, which in turn adds further mystery to this pair of charming early colour wheels for painters.

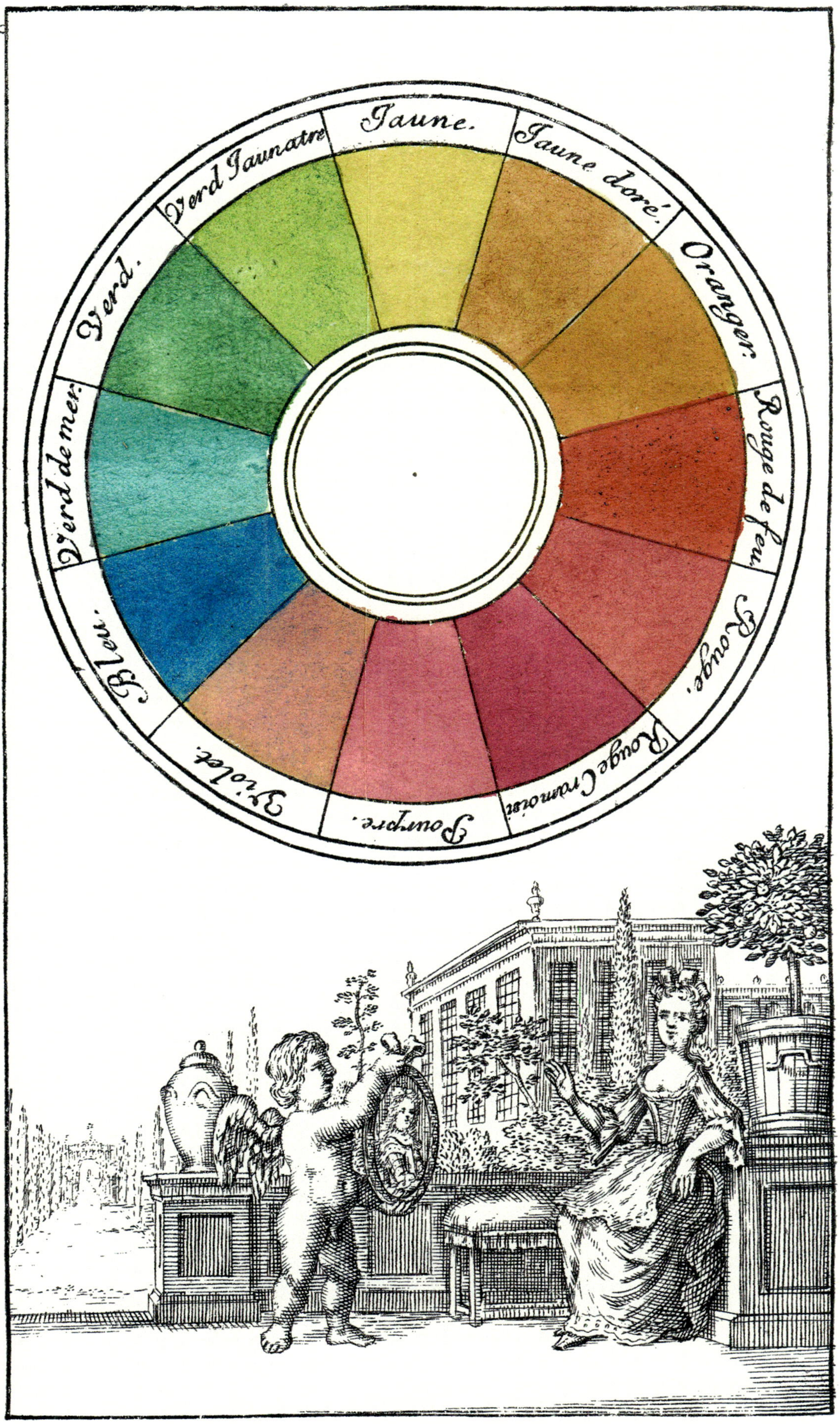
Jaune.
Jaune doré.
Oranger.
Rouge de feu.
Rouge.
Rouge Cramoisi
Pourpre.
Violet.
Bleu.
Verd de mer.
Verd.
Verd Jaunatre

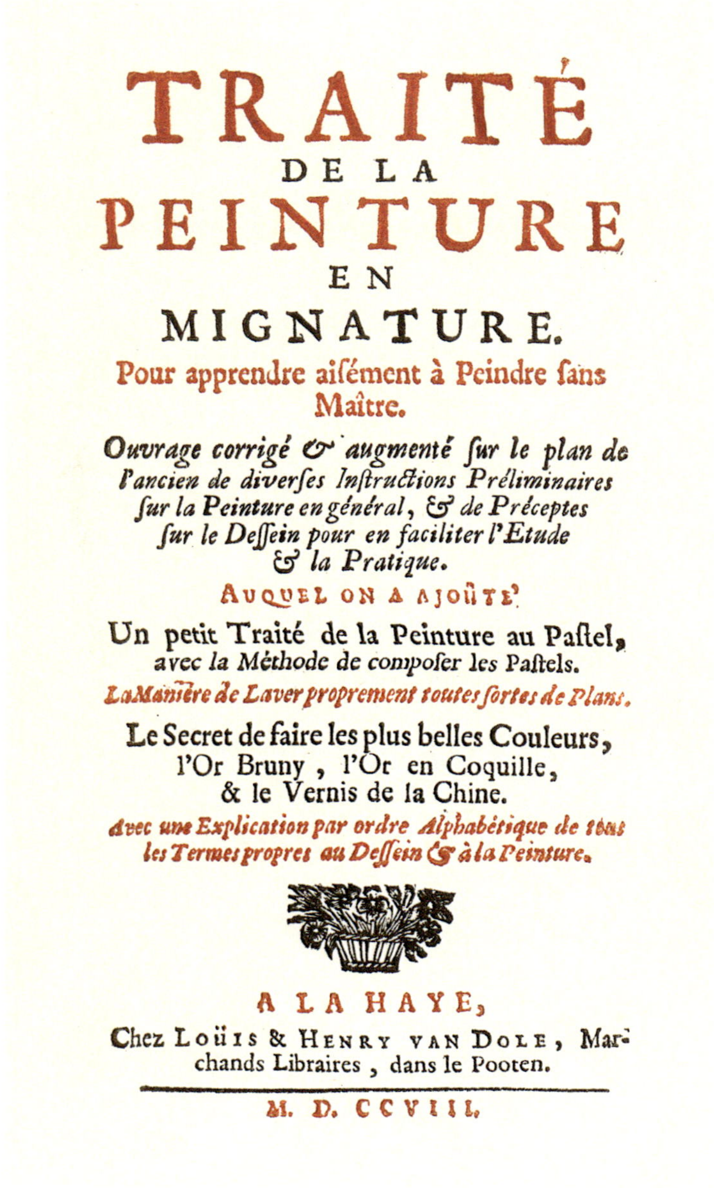

TRAITÉ
DE LA
PEINTURE
EN
MIGNATURE.
Pour apprendre aiſément à Peindre ſans Maître.
Ouvrage corrigé & augmenté ſur le plan de l'ancien de diverſes Inſtructions Préliminaires ſur la Peinture en général, & de Préceptes ſur le Deſſein pour en faciliter l'Etude & la Pratique.
AUQUEL ON A AJOÛTÉ
Un petit Traité de la Peinture au Paſtel, avec la Méthode de compoſer les Paſtels.
La Maniére de Laver proprement toutes ſortes de Plans.
Le Secret de faire les plus belles Couleurs, l'Or Bruny, l'Or en Coquille, & le Vernis de la Chine.
Avec une Explication par ordre Alphabétique de tous les Termes propres au Deſſein & à la Peinture.
A LA HAYE,
Chez LOÜIS & HENRY VAN DOLE, Marchands Libraires, dans le Pooten.
M. D. CCVIII.

Die Kunst der Miniaturmalerei

Bei dem hier gezeigten Diagrammpaar handelt es sich um die womöglich ersten Farbkreise, die in einer westlichen Publikation tatsächlich in Farbe erschienen. Sie kamen nur vier Jahre nach Isaac Newtons nicht koloriertem Farbkreis heraus, wurden in ihrer Kreisform vermutlich von diesem beeinflusst – und geben doch auch Rätsel auf. Abgedruckt waren sie in einigen Exemplaren des praktischen Farbhandbuchs *Traité de la peinture en mignature*, das bei seiner Veröffentlichung 1708 mit „Traité de la peinture au pastel" auch einen kurzen Abschnitt über Pastellmalerei enthielt. Der Hauptteil des Textes wurde erstmals um 1673 veröffentlicht und wird häufig, wenn auch nicht mit letzter Sicherheit, dem französischen Maler Claude Boutet zugeschrieben. Auch die Tafeln, die nur in einigen späteren Auflagen vorkommen, stammen möglicherweise nicht von ihm. Abgesehen von solchen Urheberschaftsfragen sind die im Buch enthaltenen detaillierten Rezepte interessant, ebenso die Erläuterungen, wie Pigmente zuzubereiten und Farben einzusetzen sind, um Porträts, Landschaften, Blumen und andere Motive zu malen. In dieser Hinsicht unterscheidet es sich nicht sehr von anderen Künstlerhandbüchern des 17. und 18. Jahrhunderts, wären da nicht die in Kupfer gestochenen, ungewöhnlichen Abbildungen der beiden sorgfältig gestalteten und handkolorierten Farbkreise. Diese Darstellung in einem Handbuch für Künstler und die darunter eingefügten Figurenszenen, in denen allegorische Putti als Porträtmaler auftreten, machen deutlich, dass die Kreise, anders als Newtons Farbkreis, bei dem das prismatische Spektrum im Vordergrund stand, stoffliche Farben verbildlichen sollten. Ein Bild von zwei Farbkreisen stellte üblicherweise den Versuch dar, sekundäre oder tertiäre Farbmischungen abzubilden und dadurch eine weitere Farbebene zu erzeugen, doch in diesem Fall ist keine Anordnung von Primär- und Sekundärfarben auf Basis der reinen, unvermischten Farben Rot, Gelb und Blau erkennbar. Der linke Kreis enthält sieben Farben, die mit Violett, Blau, Grün, Gelb, Orange, Feuerrot und Karminrot bezeichnet sind, der rechte erweitert dieses Schema unter Hinzunahme von Goldgelb, Rot, Purpur, Meergrün und Gelbgrün auf zwölf Farben. Nicht alle fünf zusätzlichen Töne wurden aus den Farben des ersten Kreises gemischt, sogar ein einfaches Rot ist dabei, eine der im Begleittext erwähnten reinen „Urfarben". Dass die Handkoloration in den wenigen noch erhaltenen Exemplaren des Buches erheblich variiert, lässt dieses Paar früher Farbkreise für Maler noch rätselhafter erscheinen.

Traité de la peinture en mignature

Les deux graphiques montrés ici sont peut-être les premiers cercles chromatiques à avoir paru en couleurs dans l'histoire de l'édition occidentale. Publiés à peine quatre ans après le disque sans couleurs d'Isaac Newton, ils sont probablement influencés par cette forme, qui fonde leur conception circulaire, même s'ils conservent aussi un certain mystère. Ils apparaissent dans certains exemplaires d'un copieux manuel pratique sur la couleur, le *Traité de la peinture en mignature*, qui comprenait, à sa publication en 1708, un petit *Traité de la peinture au pastel*. La majeure partie du texte est d'abord publiée vers 1673 et souvent attribuée au peintre français Claude Boutet, mais sans certitude. De plus, Boutet ne semble pas avoir été le créateur de ces gravures, qui n'apparaissent que dans certains tirages plus tardifs. Mis à part cette question de paternité, l'ouvrage contient des recettes et conseils détaillés pour la préparation des pigments et l'utilisation des couleurs pour peindre des portraits, des paysages, des fleurs ou d'autres sujets. De ce point de vue, il ressemble beaucoup à d'autres traités artistiques des XVIIe et XVIIIe siècles, mais les délicates gravures coloriées à la main pour représenter les cercles chromatiques y sont particulièrement remarquables. Leur présence dans un manuel à l'usage des artistes, et les scènes figuratives sous les cercles, où des *putti* allégoriques se font portraitistes, mettent en évidence que le propos est bien ici d'illustrer la couleur matérielle, par contraste avec le cercle chromatique de Newton, qui représente globalement le spectre visible. Les cercles chromatiques allant par deux constituent habituellement une tentative de visualisation des mélanges de couleurs secondaires et tertiaires, créant de fait une autre couche colorée; or dans ce cas-ci, nul assemblage visible de couleurs primaires et secondaires basées sur les couleurs pures et non mélangées rouge, jaune et bleu. Le cercle de gauche contient sept couleurs («violet, bleu, verd [*sic*], jaune, orange, rouge de feu, rouge cramoisi»), tandis que le cercle de droite enrichit la représentation jusqu'à disposer 12 teintes, en ajoutant «jaune doré, rouge, pourpre, verd [*sic*] de mer et verd-jaunatre». Ces cinq couleurs supplémentaires ne sont pas du tout présentées comme des mélanges des couleurs appartenant au premier cercle. Elles comprennent d'ailleurs un rouge simple, l'une des couleurs «primitives» mentionnées dans le texte d'accompagnement. La coloration manuelle varie énormément d'un exemplaire à l'autre, parmi ceux qui nous sont parvenus, ce qui ajoute une touche de mystère à ces deux charmants cercles chromatiques destinés aux peintres.

Tratado de la pintura en miniatura

Los dos diagramas que aquí se muestran son quizás las primeras ruedas de colores que aparecen en color en la cultura impresa occidental. Se publicaron solo cuatro años después de la rueda sin colores de Isaac Newton, diseño que probablemente influyó en su forma circular, aunque eso no puede afirmarse a ciencia cierta. Aparecen en algunos ejemplares de un importante manual práctico sobre el color, *Traité de la peinture en mignature* (1708), que en el momento de su publicación incluía una pequeña sección sobre la pintura al pastel, *Traité de la peinture au pastel*. La parte principal del texto, publicada por primera vez hacia 1673, suele atribuirse al pintor francés Claude Boutet, pero eso tampoco puede asegurarse; asimismo, es posible que Boutet no fuera el creador de las láminas, que solo aparecen en algunas de las últimas ediciones. Más allá de estas cuestiones de autoría, el libro contiene recetas detalladas y consejos sobre la preparación de pigmentos y el uso de colores en la pintura de retrato, paisaje, flores y otros temas. En ese aspecto, no difiere mucho de otros manuales para artistas de los siglos XVII y XVIII, pero los grabados cuidadosamente diseñados y coloreados a mano que representan las ruedas de colores son muy especiales. Su presencia en un manual para artistas, y las escenas figurativas añadidas debajo de las ruedas que muestran *putti* alegóricos como retratistas, dejan claro que su propósito es ilustrar el color material, en contraste con la rueda de colores de Newton, que se ocupaba principalmente del espectro óptico. La inclusión de dos ruedas de colores solía ser un intento de visualizar mezclas de colores secundarios o terciarios con la creación de otra capa de color, pero en este caso no hay una disposición reconocible de primarios y secundarios basados en el rojo, el amarillo y el azul puros y sin mezclar. El círculo de la izquierda contiene siete colores, denominados violeta, azul, verde, amarillo, naranja, rojo fuego y carmesí, mientras que el de la derecha amplía este esquema a un conjunto de 12 colores al que se han añadido el amarillo dorado, el rojo, el púrpura, el verde mar y el verde amarillo. Estos cinco colores adicionales no son todos mezclas de los colores que se encuentran en la primera rueda y, de hecho, incluyen un rojo simple, uno de los «primitivos» puros mencionados en el texto adjunto. El coloreado a mano en los pocos ejemplares de este libro que se han conservado varía considerablemente, lo que a su vez añade misterio a esta pareja de encantadoras primeras ruedas de colores para pintores.

Moses Harris (1730–*c.* 1788)

THE NATURAL SYSTEM OF COLOURS

3 plates, 35.3 x 25.5 cm | 13 ⅞ x 10 in., London, c. 1769–1776
Winterthur, Stadtbibliothek, Farbsammlung Werner Spillmann

Moses Harris was an English entomologist who wrote and illustrated this short early book with a colour system designed for painters and illustrators, while acknowledging the influence of Isaac Newton. As the continuation of his title explains, his system was intended to show the "*Regular and Beautiful Order and Arrangement, Arising from the Three Premitives* [sic], *Red, Blue, and Yellow*". No date is given in the first edition, but it was probably published between 1769 and 1776. It was dedicated to the President of the Royal Academy of Arts in London, Sir Joshua Reynolds (1723–1792), and was the first in a series of important works on colour that originated from the circle of people associated with the Royal Academy in the decades after it was founded in 1768, a fact that underlines the changing attitudes to colour at this time in art and art education.

The weighting of text (just eight pages, including the dedication) to images (two plates showing magnificent colour wheels together with charts in some copies showing examples of colour mixtures) and the aspirational design of Harris's illustrations are a clear indication of the importance he assigned to the accurate depiction of colours in the natural world. The two wheels are copperplate engravings, comprising respectively 18 sectors of "prismatic" and "compound" colour mixtures divided and numbered according to 20 degrees of intensity, with ten degrees delineated in the images. Each wheel carries a total of 360 colours. The book's subtitle however states only 660 different colours in total, as the mixtures green, orange and purple are shown in both circles. In the center of each circle are overlapping triangles of the "primitive" colours red, yellow and blue, and the "mediates" orange, green and purple, each having a black centre representing the subtractive colour mixture. A brief accompanying list of the primaries and secondaries provides equivalents of these colours from the worlds of pigments and flora.

A second, posthumous edition was published in 1811, with newly cut plates. Harris's work was widely discussed, taught and reproduced by a number of artists, teachers and scholars in the early 19th century, including Turner, Thomas Phillips and William Benjamin Sarsfield Taylor (1781–1850). In 1963 the American writer Faber Birren produced a facsimile edition of this extraordinary book and called it "perhaps the rarest known book in the literature of color". Of the orginal editions (this one and the 1811 reprint) fewer than 10 copies are currrently recorded in libraries worldwide.

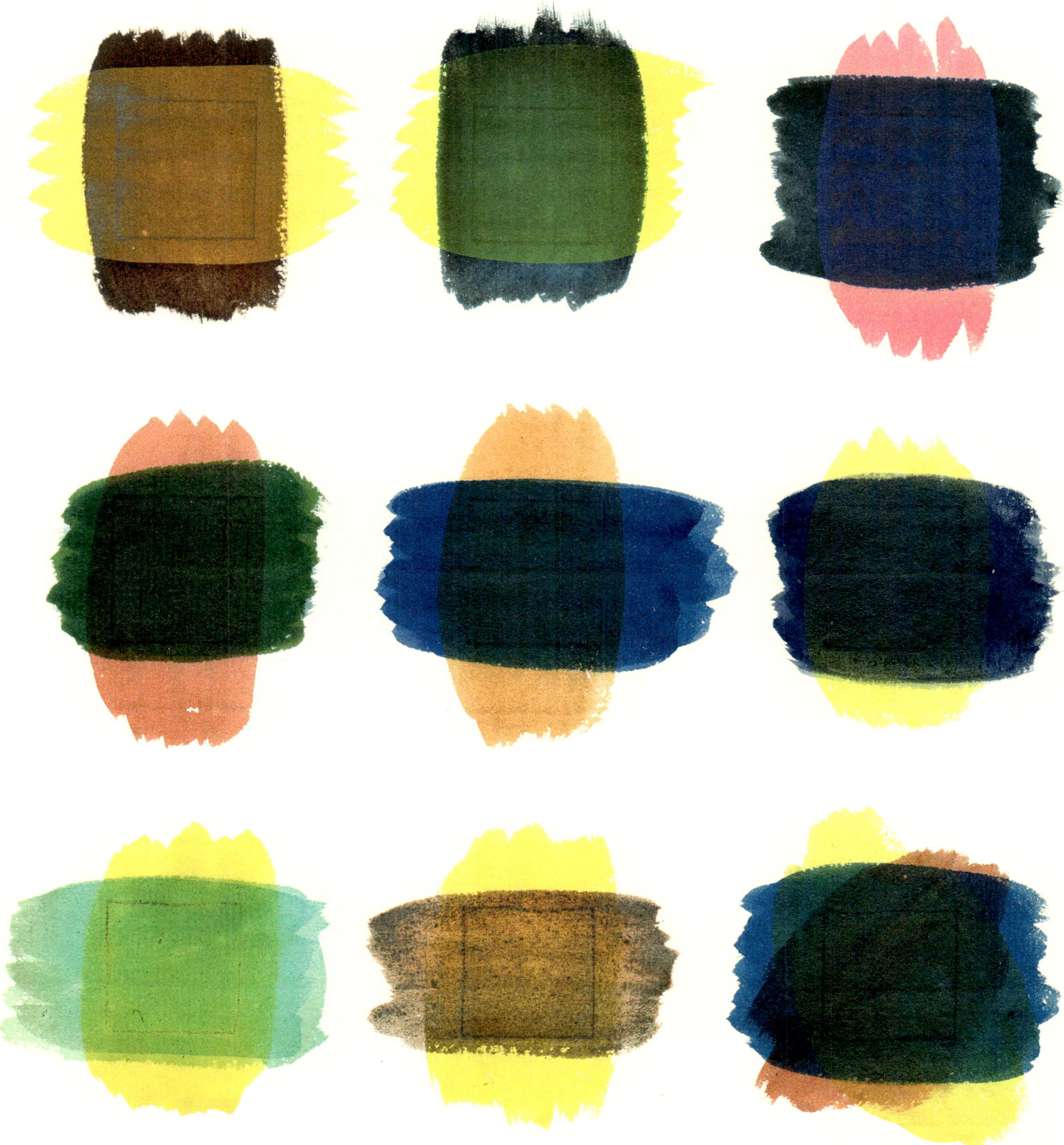

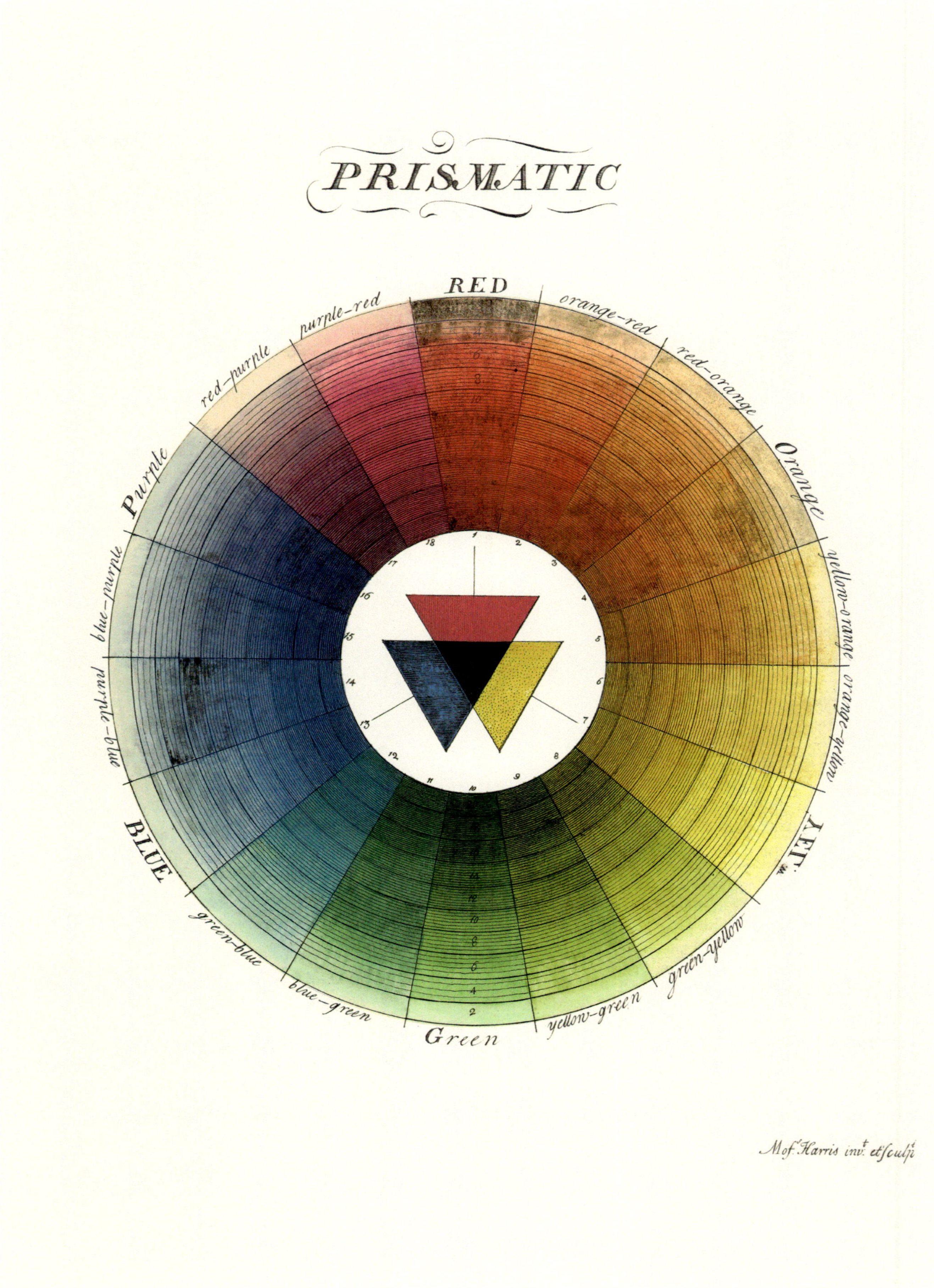
PRISMATIC
RED
orange-red
red-orange
Orange
yellow-orange
orange-yellow
YEL.
green-yellow
yellow-green
Green
blue-green
green-blue
BLUE
purple-blue
blue-purple
Purple
red-purple
purple-red
Mos. Harris inv. et sculp.

THE NATURAL

SYSTEM of COLOURS,

Wherein is difplayed the regular and beautiful Order and Arrangement,

Arifing from the Three Premitives, *Red*, *Blue*, and *Yellow*,

The manner in which each Colour is formed, and its Compofition,

The dependance they have on each other, and by their

HARMONIOUS CONNECTIONS

Are produced the Teints, or Colours, of every Object in the Creation,

And thofe Teints, tho' fo numerous as 660, are all comprifed in Thirty Three Terms, only

By MOSES HARRIS,

AUTHOR of the AURELIAN, *&c. &c.*

Printed at LAIDLER's OFFICE, Princes-ftreet, Licefter-Fields.

Das natürliche System der Farben

Schon früh verfasste der englische Entomologe Moses Harris dieses kurze Büchlein und bebilderte es mit einem Farbsystem, das sich an Maler und Illustratoren richtete, wobei er den Einfluss Isaac Newtons durchaus eingestand. Wie die Fortsetzung des Titels seiner Abhandlung erklärt, sollte sein System die „regelmäßige und schöne Ordnung und Gliederung" zeigen, „die aus den drei Urfarben Rot, Blau und Gelb erwächst". Die erste Auflage ist nicht datiert, erschien jedoch vermutlich zwischen 1769 und 1776. Sie war Sir Joshua Reynolds (1723–1792) gewidmet, dem Präsidenten der Royal Academy of Arts in London, und bildete den Auftakt zu einer Reihe von bedeutenden Farbtraktaten aus dem Umkreis der Akademie in den Jahrzehnten nach ihrer Gründung 1768. Dies unterstreicht den Wandel in den Einstellungen zur Farbe, der sich damals im Bereich der Kunst und Kunsterziehung vollzog.

Die Gewichtung von Text (nur acht Seiten einschließlich Widmung) und Bildern (zwei Tafeln mit prächtigen Farbkreisen, in einigen Exemplaren ergänzt um Tabellen mit Beispielen für Farbmischungen) und die anspruchsvolle Gestaltung der Illustrationen sind ein deutlicher Hinweis auf die Bedeutung, die Harris der exakten Wiedergabe von natürlichen Farben zuschrieb. Die in Kupfer gestochenen Farbkreise bestehen aus jeweils 18 Sektoren mit „prismatischen" und „gemischten" Farben, unterteilt und nummeriert in zwanzig Intensitätsgrade, die sich in den Abbildungen auf zehn Stufen verteilen. Jeder Kreis enthält insgesamt 360 Farben. Im Untertitel des Buches ist jedoch nur von 660 verschiedenen Farben die Rede, da die Mischfarben Grün, Orange und Violett in beiden Kreisen vertreten sind. Das Zentrum der Kreise bilden sich überlappende Dreiecke aus den „Urfarben" Rot, Gelb und Blau und den „Zwischenfarben" Orange, Grün und Violett. Das schwarze Dreieck in der Mitte steht in beiden Fällen stellvertretend für die subtraktive Farbmischung. Eine kurze Begleitliste der Primär- und Sekundärfarben im Text steuert Entsprechungen aus der Pigment- und Pflanzenwelt bei.

1811 erschien postum eine zweite Auflage des Buches mit neu gestochenen Kupfertafeln. Harris' Werk wurde im frühen 19. Jahrhundert von zahlreichen Künstlern, Pädagogen und Wissenschaftlern wie William Turner, Thomas Phillips oder William Benjamin Sarsfield Taylor (1781–1850) umfassend diskutiert, in der Lehre verwendet und reproduziert. 1963 gab der amerikanische Schriftsteller Faber Birren eine Faksimile-Ausgabe dieses außergewöhnlichen Buches heraus und bezeichnete es als „das vielleicht seltenste bekannte Buch in der Literatur über Farben". Von der Originalausgabe (und dem Nachdruck von 1811) sind derzeit weniger als 10 Exemplare in Bibliotheken weltweit verzeichnet.

Le système naturel des couleurs

Moses Harris est un entomologiste anglais, auteur et illustrateur de ce bref ouvrage, l'un des tout premiers à proposer un système chromatique destiné aux peintres et illustrateurs, tout en reconnaissant l'influence d'Isaac Newton. Comme l'explique la suite du titre, son système a pour objectif de montrer « L'ordre et la disposition réguliers et beaux découlant des trois premitives [*sic*], rouge, bleu et jaune ». Aucune date ne figure dans la première édition, mais elle a probablement été publiée entre 1769 et 1776. Dédicacé au président de la Royal Academy of Arts à Londres, Sir Joshua Reynolds (1723–1792), l'ouvrage est le premier d'une série de titres importants consacrés à la couleur, tous signés par des personnes associées à la Royal Academy dans les décennies qui suivent sa fondation en 1768, indice du changement d'attitude vis-à-vis de la couleur à cette époque, en peinture comme dans l'enseignement artistique.

La proportion du texte (huit pages tout juste, dédicace comprise) par rapport aux images (deux gravures figurant deux splendides cercles chromatiques, accompagnés, dans certains exemplaires, de tableaux donnant des exemples de mélanges de couleurs) et l'ambitieuse qualité des illustrations d'Harris indiquent clairement l'importance qu'il attribue à une représentation précise des couleurs du monde naturel. Les deux cercles, gravés au burin sur cuivre, présentent les couleurs « prismatiques » et les couleurs « composées ». Chacun est constitué de 18 segments, divisés en 20 degrés d'intensité numérotés, et présente donc 360 couleurs. Le sous-titre du livre n'évoque quant à lui que 660 couleurs différentes au total, puisque les mélanges violet, orange et vert apparaissent dans les deux cercles. Au centre du premier cercle sont superposés trois triangles des couleurs « primitives » rouge, jaune et bleu ; au centre du second, trois triangles des couleurs « médiates » violet, orange et vert. Dans les deux cas, le centre noir représente le mélange soustractif des couleurs. Une brève liste des couleurs primaires et secondaires en indique les équivalents dans l'univers des pigments et de la flore.

Une seconde édition, posthume, paraît en 1811 avec de nouvelles gravures. Les travaux d'Harris ont été largement discutés, enseignés et reproduits par de nombreux artistes, professeurs et élèves au début du XIXe siècle, dont Turner, Thomas Phillips ou encore William Benjamin Sarsfield Taylor (1781–1850). En 1963, l'écrivain états-unien Faber Birren a réalisé une édition en fac-similé de cet ouvrage extraordinaire, dont il disait que c'était « peut-être l'ouvrage le plus rare dans l'histoire des livres consacrés à la couleur ». Aujourd'hui, il ne reste des éditions originales (qu'il s'agisse de cette édition ou de la réimpression de 1811) que moins de 10 exemplaires recensés dans des bibliothèques de par le monde.

El sistema natural de los colores

Moses Harris fue un entomólogo inglés que escribió e ilustró este breve libro con un sistema de color diseñado para pintores e ilustradores, y lo hizo, como él mismo reconoció, influido por Isaac Newton. La continuación de su título explica la intención de su sistema: mostrar *«el orden y el arreglo hermosos y regulares que surgen de los tres premitivos* [sic], *rojo, azul y amarillo»*. La primera edición no está fechada, pero probablemente se publicó entre 1769 y 1776. Estaba dedicada al presidente de la Royal Academy of Arts de Londres, sir Joshua Reynolds (1723–1792), y fue la primera de una serie de importantes obras sobre el color surgidas del círculo de personas asociadas con la Royal Academy en las décadas posteriores a su fundación en 1768, un hecho que subraya las actitudes cambiantes hacia el color en ese momento en el arte y la educación artística.

El peso del texto (apenas ocho páginas, incluida la dedicatoria) en relación con las imágenes (dos láminas con magníficas ruedas de colores junto con gráficos en algunas copias que muestran ejemplos de mezclas de colores) y el calculado diseño de las ilustraciones de Harris son un claro indicio de la importancia que le asignó a la representación precisa de los colores en el mundo natural. Las dos ruedas son grabados realizados con planchas de cobre. Cada una consta de 18 partes con mezclas de colores «prismáticos» y «compuestos» divididos y numerados en 20 grados de intensidad repartidos en diez casillas. Así, cada rueda incluye un total de 360 colores. Sin embargo, el subtítulo del libro establece la existencia de tan solo 660 colores diferentes; esto se explica porque las mezclas de verde, naranja y morado aparecen en ambos círculos, lo que da como resultado ese total de 660. En el centro de cada círculo hay triángulos superpuestos de los colores «primitivos» rojo, amarillo y azul, y los «medios» naranja, verde y morado, cada uno con un centro negro que representa la síntesis sustractiva de colores. Una breve lista adjunta de colores primarios y secundarios proporciona equivalentes de estos colores del mundo de los pigmentos y la flora.

En 1811 se publicó una segunda edición póstuma con nuevas láminas. El trabajo de Harris fue muy discutido, enseñado y reproducido por artistas, maestros y académicos a principios del siglo XIX, incluidos Turner, Thomas Phillips y William Benjamin Sarsfield Taylor (1781–1850). En 1963, el escritor estadounidense Faber Birren produjo una edición facsímil de este extraordinario libro, que describió como «quizás el libro más raro que se conoce sobre el color». En la actualidad, se conservan menos de 10 copias de las ediciones originales (esta y la reimpresión de 1811) registradas en bibliotecas de todo el mundo.

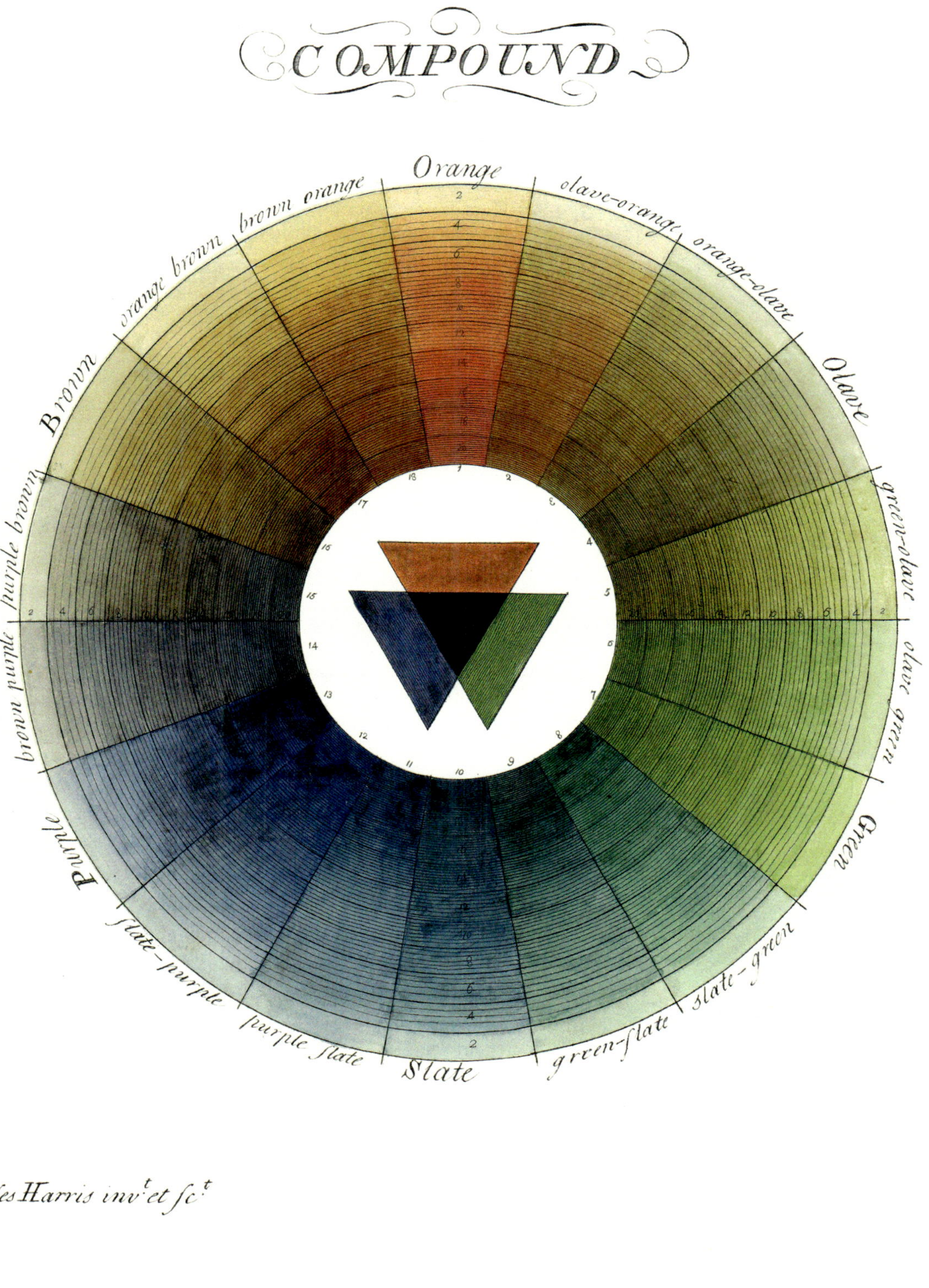
COMPOUND
Orange
olave-orange
orange-olave
Olave
green-olave
olave green
Green
slate-green
green-slate
Slate
purple slate
slate-purple
Purple
brown purple
purple brown
Brown
orange brown
brown orange
Moses Harris inv^t et sc^t

Moses Harris

AN EXPOSITION OF ENGLISH INSECTS

2 plates, 29 x 23 cm | 11 3/8 x 9 in., London, 1782
Sussex, collection of the author;
Los Angeles, Getty Research Institute (frontispiece)

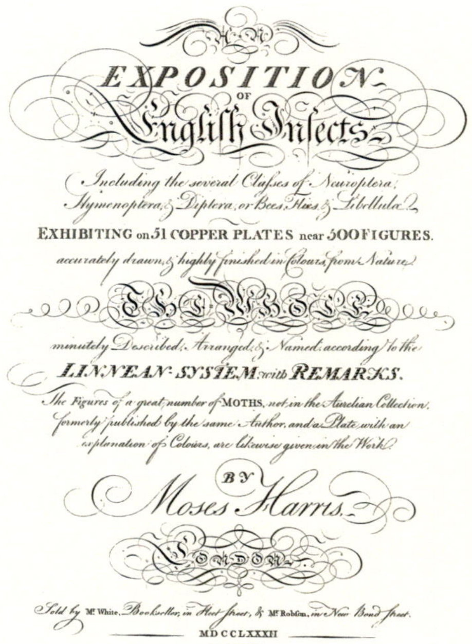

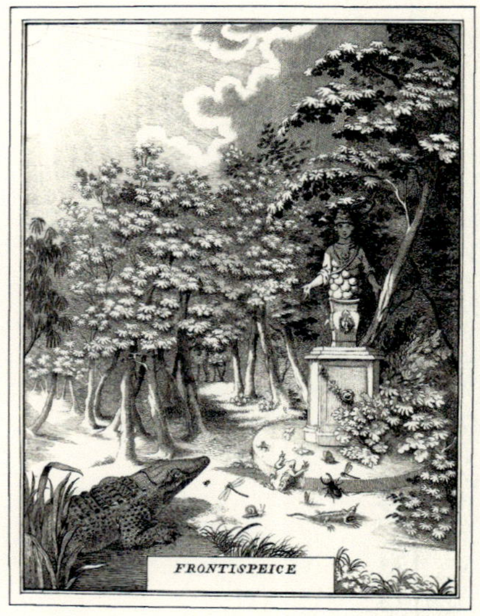

As a specialist in the study of insects Harris had previously published in 1766 *The Aurelian: or, Natural History of English Insects, Namely, Moths and Butterflies* (a book that was also mentioned on the title-page of *The Natural System of Colours*). Since he worked as an illustrator of books on the natural world Harris had an acute awareness of the subtleties of colour, especially the darker, compound shades, and of the difficulties faced by artists when using pigments. In the 18th century, the accurate representation of aspects of the natural world was of the utmost importance, especially in scientific illustration, and while Harris hoped that the colour circles published in *The Natural System of Colours* would provide guidance and assistance he went a step further in his next entomological work. In *An Exposition of English Insects* (first published in 1776), illustrated with dozens of hand-coloured etchings depicting moths, butterflies, bees, dragonflies etc., he added a modified version of his own colour wheels. Entitled "Scheme of Colours", it was designed to represent the numerous shades of brown, grey, black and chromatic colours present in the many varieties of insects. In the centre once again are the overlapping primitives forming black in the middle. Surrounding this, the first, innermost circle shows 18 colours in full intensity, derived from red, orange, yellow, green, blue and purple, while the second circle moving outwards shows lighter versions of these colours. The third circle shows tertiary mixtures, with lighter versions again in the last, outer circle. Provided with an accompanying text (in English and French) and an explanatory key for the colours in the circle, this was effectively a colour guide for the figurative illustrations in the rest of the book. The Scheme of Colours is often missing in surviving copies, suggesting that readers may have cut it out in order to place it next to the insect plates when examining the book.

SCHEME of COLOURS

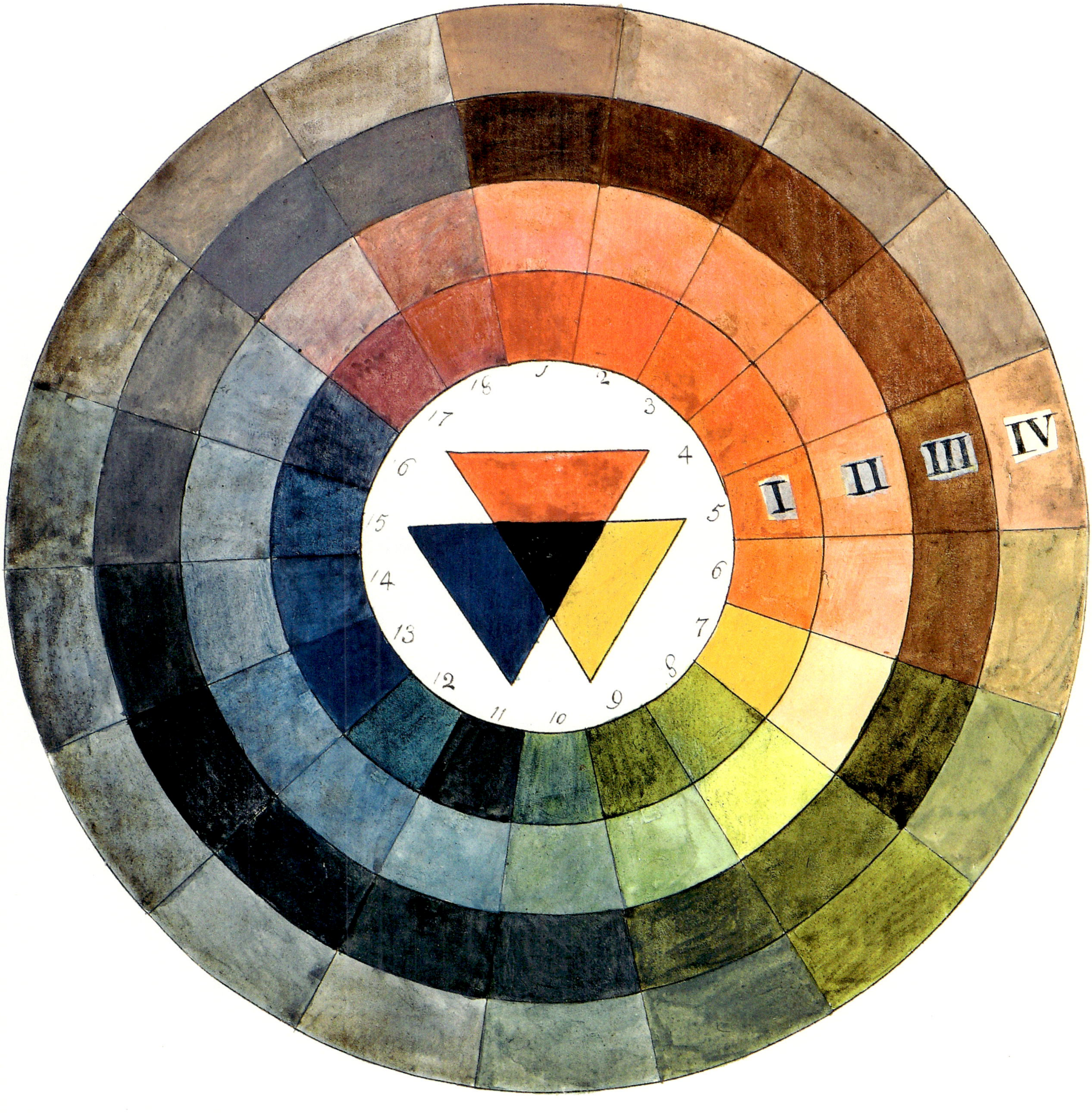

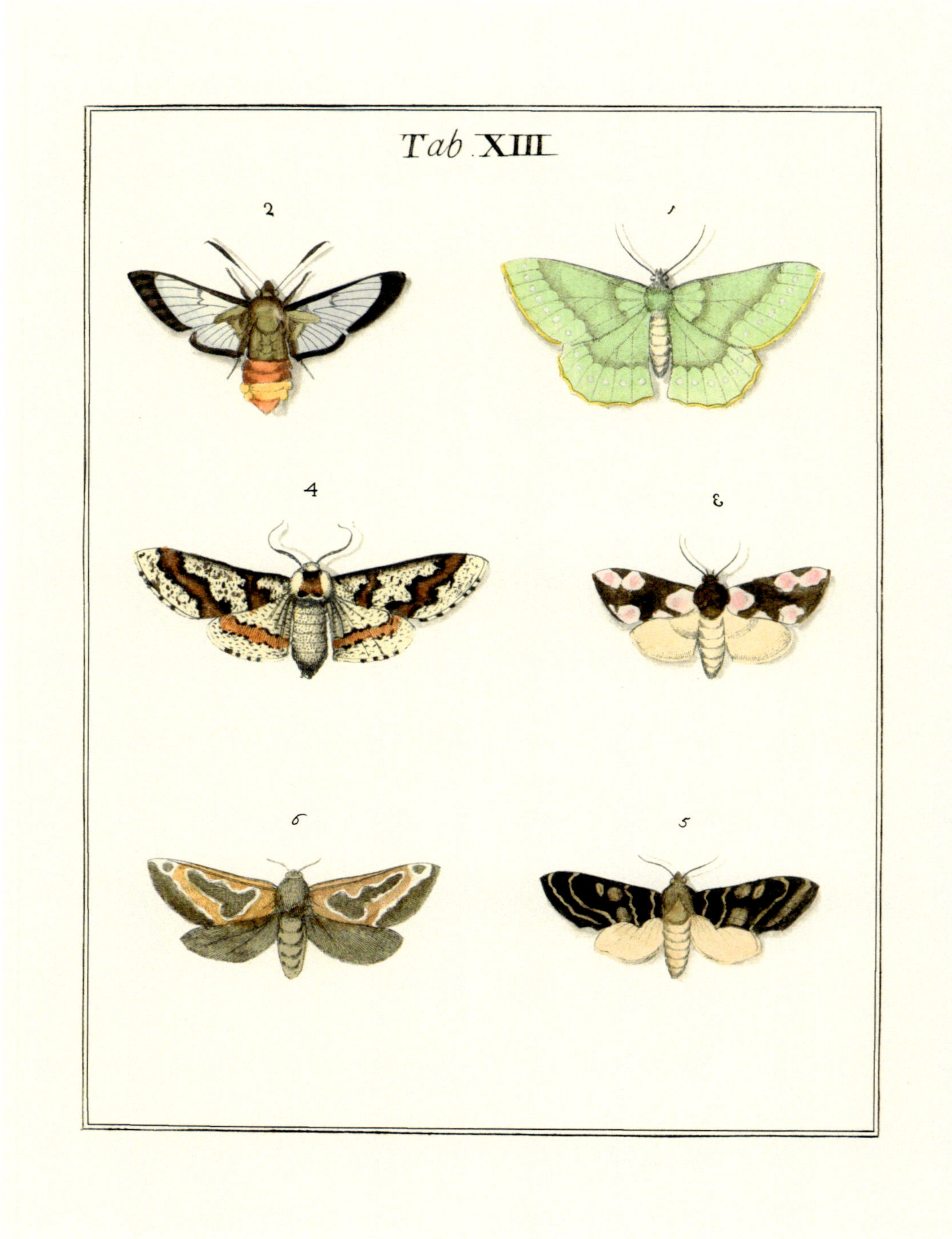

In *An Exposition of English Insects* (1776 erstmals veröffentlicht) fügte er Dutzenden von handkolorierten Radierungen mit Abbildungen von Faltern, Schmetterlingen, Bienen, Libellen und anderen Insekten eine modifizierte Version seiner Farbkreise hinzu. Dieses „Farbschema" sollte die unzähligen Schattierungen von Braun, Grau, Schwarz und sämtlichen Buntfarben abbilden, die in der artenreichen Welt der Insekten zu finden sind. Im Zentrum bilden die einander überlappenden Urfarben erneut ein schwarzes Dreieck. Daran angrenzend besteht der erste, innere Kreis aus 18 von Rot, Orange, Gelb, Grün, Blau und Violett abgeleiteten Farben in voller Intensität. Der zweite, weiter außen liegende Kreis zeigt die gleichen Farben in helleren Versionen. Der dritte Kreis zeigt tertiäre Farbmischungen, deren abermals hellere Versionen den letzten, äußeren Kreis bilden. Mit dem Begleittext (in englischer und französischer Sprache) und einer Legende zu den im Kreis vertretenen Farben war damit praktisch ein Farbführer durch die figurativen Abbildungen im Rest des Buches entstanden. Dass das Farbschema in noch erhaltenen Exemplaren oft fehlt, legt den Schluss nahe, dass es womöglich herausgeschnitten wurde, um es beim Studium des Buches neben die Insektentafeln zu legen.

Eine Darstellung englischer Insekten

Als Experte für Insektenforschung hatte Harris bereits 1766 *The Aurelian: or, Natural History of English Insects, Namely, Moths and Butterflies* (ein Buch, das auch auf der Titelseite von *The Natural System of Colours* erwähnt wurde) veröffentlicht. Als Illustrator naturkundlicher Bücher besaß Harris ein ausgeprägtes Bewusstsein für die Feinheiten von Farbe, zumal für die dunkleren, gemischten Töne, und für die Schwierigkeiten, mit denen sich Künstler im Gebrauch von Pigmenten konfrontiert sahen. Im 18. Jahrhundert waren präzise Naturdarstellungen insbesondere für wissenschaftliche Illustrationen von allergrößter Bedeutung. Während Harris hoffte, dass die in *The Natural System of Colours* veröffentlichten Farbkreise Orientierung und Hilfestellung bieten würden, ging er in seinem nächsten entomologischen Werk einen Schritt weiter.

Exposition des insectes anglois

Spécialiste de l'étude des insectes, Harris a déjà publié, en 1766, *The Aurelian: or, Natural History of English Insects, Namely, Moths and Butterflies*, ouvrage également mentionné sur la page de titre de *The Natural System of Colours*. En tant qu'illustrateur de livres consacrés à la nature, Harris a pleinement conscience des subtilités des couleurs, notamment des nuances composées les plus sombres, et des difficultés que rencontrent les artistes quand ils utilisent des pigments. Au XVIIIe siècle, il est tout à fait capital de fournir des représentations précises des éléments du monde naturel, tout particulièrement dans le domaine de l'illustration scientifique. Harris, qui espère que les cercles chromatiques publiés dans son *Natural System of Colours* puissent servir de guides et d'aides, va encore plus loin dans son ouvrage d'entomologiste suivant. Dans cette *Exposition of English Insects*, publié pour la première fois en 1776, et illustré de dizaines d'aquarelles coloriées à la main décrivant phalènes, papillons, abeilles, libellules, etc., il ajoute une version modifiée de ses propres cercles chromatiques. Intitulé « Scheme of Colours » (« Combinaison des couleurs »), ce dessin est conçu pour représenter les nombreuses nuances de brun, de gris et de noir, et les couleurs du cercle chromatique que présentent les variétés d'insectes. Au milieu, on retrouve les couleurs primitives superposées et formant un triangle central noir. Un premier cercle entourant cette forme est composé de 18 couleurs dans leur expression la plus intense, dérivées du rouge, de l'orange, du jaune, du vert, du bleu et du violet, tandis que le deuxième cercle, en allant vers l'extérieur, montre des nuances plus claires de ces mêmes couleurs. Le troisième cercle montre des mélanges de couleurs tertiaires, dont les versions plus claires forment le dernier cercle extérieur. Accompagné d'un texte en anglais et en français et d'une légende explicative des couleurs du cercle, l'ensemble constitue un véritable guide chromatique pour aborder les illustrations figuratives du reste de l'ouvrage. Cette « combinaison des couleurs » est souvent manquante dans les exemplaires parvenus jusqu'à nous, ce qui suggère que les lecteurs la détachaient peut-être pour la placer à côté des planches d'insectes lorsqu'ils consultaient l'ouvrage.

Exposición de insectos ingleses

Como especialista en el estudio de los insectos, Harris había publicado en 1766 *The Aurelian: or, Natural History of English Insects, Namely, Moths and Butterflies* (un libro que también se menciona en la portada de *The Natural System of Colours*). Desde que trabajó como ilustrador de libros sobre el mundo natural, Harris tenía una profunda conciencia de las sutilezas del color, en especial de los tonos más oscuros y compuestos, y de las dificultades a las que se enfrentan los artistas al usar pigmentos. En el siglo XVIII, la representación precisa de los aspectos del mundo natural era de suma importancia, sobre todo en la ilustración científica, y aunque Harris esperaba que los círculos de color publicados en *The Natural System of Colours* brindaran orientación y ayuda, fue un paso más allá en su siguiente trabajo entomológico. En *Exposition of English Insects* (publicado por primera vez en 1776), que está ilustrado con docenas de aguafuertes coloreados a mano que representan polillas, mariposas, abejas, libélulas, etc., añadió una versión modificada de sus propias ruedas de colores. Titulada «Esquema de colores», la diseñó para representar los numerosos tonos de marrón, gris, negro y colores cromáticos presentes en las muchas variedades de insectos. En el centro, una vez más están los primitivos superpuestos formando el color negro. A su alrededor, el círculo más interior muestra 18 colores en toda su intensidad derivados del rojo, del naranja, del amarillo, del verde, del azul y del morado, mientras que el segundo círculo muestra versiones más claras de esos colores. El tercer círculo recoge mezclas terciarias, y así hasta llegar al último círculo exterior, que contiene las versiones más claras. La rueda, a la que acompañaba un texto adjunto (en inglés y francés) y una leyenda explicativa de los colores, no era otra cosa que una guía de colores de las ilustraciones figurativas contenidas en el libro. Las copias supervivientes suelen carecer de este esquema, lo que sugiere que los lectores quizá lo recortaron para colocarlo junto a las láminas de insectos al examinar el libro.

Ignaz Schiffermüller (1727–1806)

VERSUCH EINES FARBENSYSTEMS

An Attempt at a Colour System
2 plates, 29 x 20.5 cm | 11 3/8 x 8 1/8 in., Vienna, 1772
Winterthur, Stadtbibliothek, Farbsammlung Werner Spillmann

Moses Harris and the Austrian Jesuit priest Ignaz Schiffermüller had several things in common. Both were entomologists, and therefore particularly attuned to the accurate representation of colour in their illustrative and descriptive work. They also both saw the need to provide a useful and reliable colour system that acknowledged Newton's achievements while making the necessary adaptations for artists, designers and natural scientists. In their choice of visual representation they each followed Newton's conceptual model of a circle, although their designs are subtly different. It is not known whether they were aware of each other's work.

Schiffermüller chose to use a 12-part wheel that comprised three primaries, three secondaries and six tertiaries. The circle is evenly divided into 30° segments, in contrast to Newton's asymmetrical wheel which had assigned uneven proportions to the seven named colours. Even so, the decorative pictorial additions to the diagram reveal how Schiffermüller was keen to refer to his predecessor's optical experiments, so that a rainbow in Newtonian colour sequence is depicted in three of the corners, while in the other corner there is a vignette of someone directing a ray of light through a prism. The two scenes on the right illustrate the application of Newton's theory in design and painting, while in the centre of the composition an anthropomorphised sun, with human facial features, is shown breaking through the clouds. The result is a beautiful and deceptively simple colour diagram that combines pictorial, abstract, symbolic and linguistic elements, although Schiffermüller also realised that he needed to elaborate on the range of colours it presented. In the same way as Jacob Christian Schäffer (pp. 104–107), whose book he is likely to have known, he added examples of colour charts which break down each of the 12 colours in the diagram into 12 types, labelled with the capital letters A to M and each represented in three degrees of lightness labelled a, b and c, which theoretically results in a total of 432 colours.

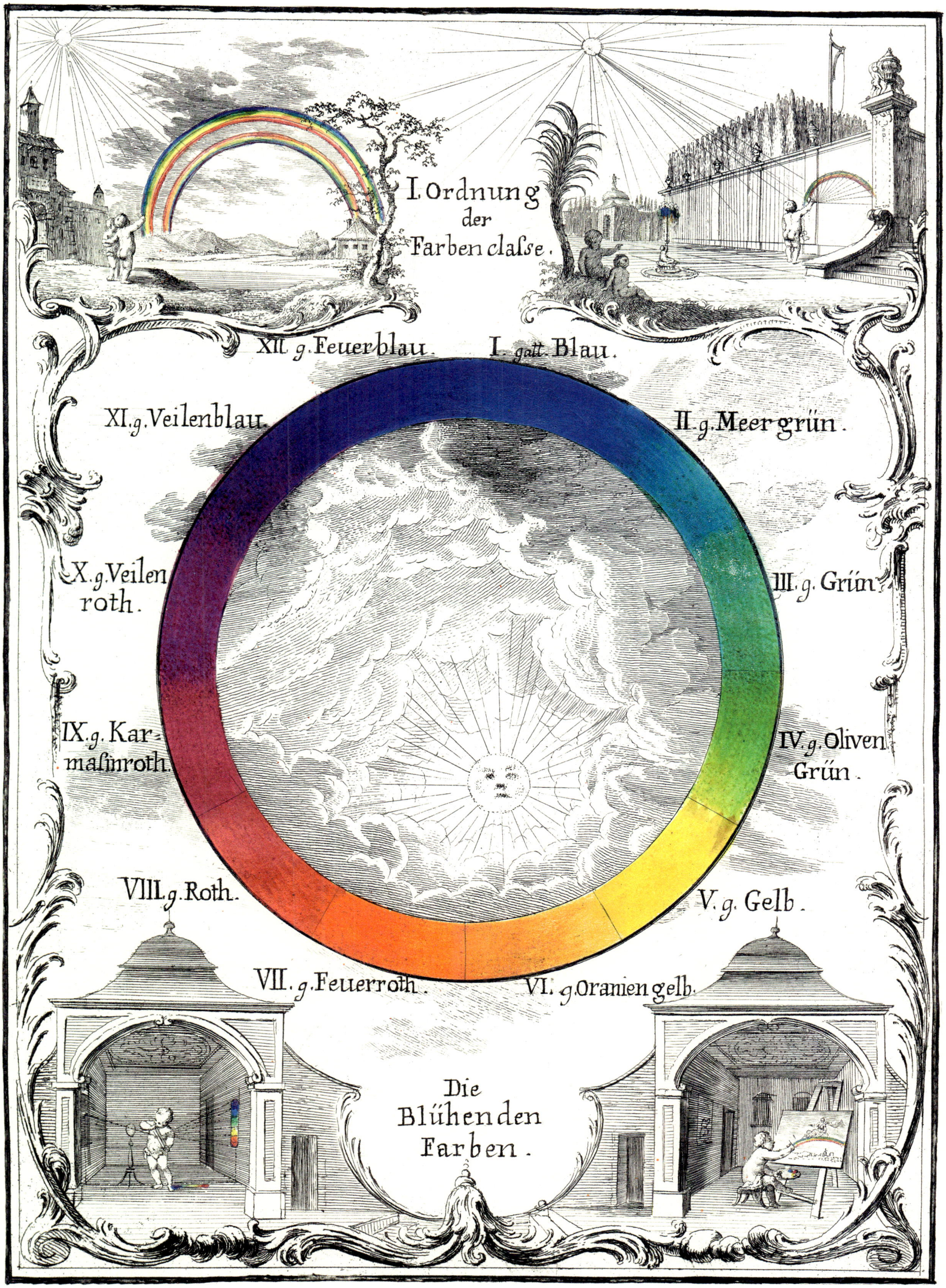
I. Ordnung
der
Farben classe.
XII. g. Feuerblau.
I. gatt. Blau.
XI. g. Veilenblau.
II. g. Meergrün.
X. g. Veilen roth.
III. g. Grün.
IX. g. Karmasinroth.
IV. g. Oliven Grün.
VIII. g. Roth.
V. g. Gelb.
VII. g. Feuerroth.
VI. g. Oranien gelb.
Die
Blühenden
Farben.

Pag. 25

I. Gatt.	Blaue Schattirung.	Color Caeruleus.	Les Nuances du Bleu.
A	a. Weiß. (Schneeweiß.) b. Silberweiß. Perlenweiß. ** c. Milchweiß. Milchfarbe. Weißlicht.	a. Albus. (Niveus. Candidus. *) b. Argenteus. c. Lacteus. Lactescens.	a. Blanc. Blanc pur. b. Blanc d'argent. Blanc argenté. Blanc de perle. c. Blanc de lait. Couleur laiteuse.
B	a. Blauweiß. Blaulicht. Porcellanfarbe.* b. Perlenfarbe. c. Blaulichtweiß. Wasserfarbe. Diamantfarbe.	a. Lacteocaeruleus. b. Margaritaceus. c. Albocaeruleus. Aqueus.	a. Bleu blanc. Bleuâtre. b. Couleur de perles. Couleur de nacre. c. Couleur d'eau.
C	a. Weißlichtblau. b. Silberblau. c. Wasserblau.	Caerulescens. Subcaeruleus.	Bleu naissant.
D	a. Mattblau. Magerblau. Schwachblau. Glockenblumfarbe. b. Graublau. Hechtenblau. * c. Blaßblau. Bleumourant. (Blümerant.) Bleichblau.	a. c. Pallide caeruleus. (Caeruleus pallens, tabescens, languidus, evanidus. b. Caesius. ** Dilute caeruleus.	Bleu mourant. Bleu pâle.
E	a. Feinblau. Sächsischblau. Zartblau. b. Aschenblau. c. Lieblichblau. Sanftblau.	Blande caeruleus. Suaviter caeruleus.	Bleu mignon. Azurin.
F	a. Himmelblau. Lebhaftblau. (Himmelschön.) b. Lichtblau. c. Saphirblau. Hellblau.	Caeruleus. Belle, laete caeruleus. (Caelestinus. Aereus.) Saphirinus.	Bleu celeste. Couleur du Ciel.

G	a. Lazurblau. (Lasurblau.) marinblau. * b. Blau. Ganzblau. Mittelb[…] Gemeinblau[…] c. Hochblau. Kornblumenblau[…]
H	a. Vollblau. Kräftigblau. Sta[…] b. Blankblau. (Blaublank *[…] zer[…] c. Türkisblau. ** (Türchino.[…]
I	a. Dunkelblau. Französischb[…] (Franzblau.) b. Stahlblau. c. Indigblau. (Endichblau. In[…]
K	a. Beerblau. * Tiefindigblau. […]kelfranz[…] b. Schlackenblau. Schmelzblau[…] kelst[…] c. Braunblau. (Blau von S[…] Schwarz und Blau.) Bleich[…]
L	a. Eisenblau. Schwarzblaulicht. b. Wolkenschwarz. c. Schwarzblau.
M	a. Blauschwarz. Blaulichtsch[…] Kohlenschwarz. * b. Sammetschwarz. Rabens[…] c. Gagatschwarz. Glänzendschw[…]

ıeus. ... uleus Cyaneus.)	*a.* Bleu d'azur. (Azur.) Bleu d'outremer. *b.* Bleu. Bleu complet. *c.* Bleu-beau. Bleu-Reine.
ıſe caeruleus. ıte caeruleus.	Bleu couvert. Bleu turquin. Couleur de turquoiſe.
ıde caeruleus. aneus ſaturus.) Indicus.	Bleu de Roi. Bleu foncé. Bleu brun clair. L'indico.
ure caeruleus. uleus *obſcurus.*) nde Cyaneus. leus ſubnigricans.) lus.	*a.* Bleu pers. Bleu brun. Fleur de Guéde. *b.* Bleu de forge. *c.* Livide. Bleuâtrebrun.
leſcens. Caeruleo- . Nigro caeruleus. ruleus. Atrocyaneus.	*a.* Bleu noirâtre. *b.* Bleu d'enfer. *c.* Bleu noir.
ruleoniger. leſcentiniger. ter. *	*a.* Noirbleu. Noirbleuâtre. Noir de Charbon. *b.* Noir. Noir velouté. *c.* Noir de jais.

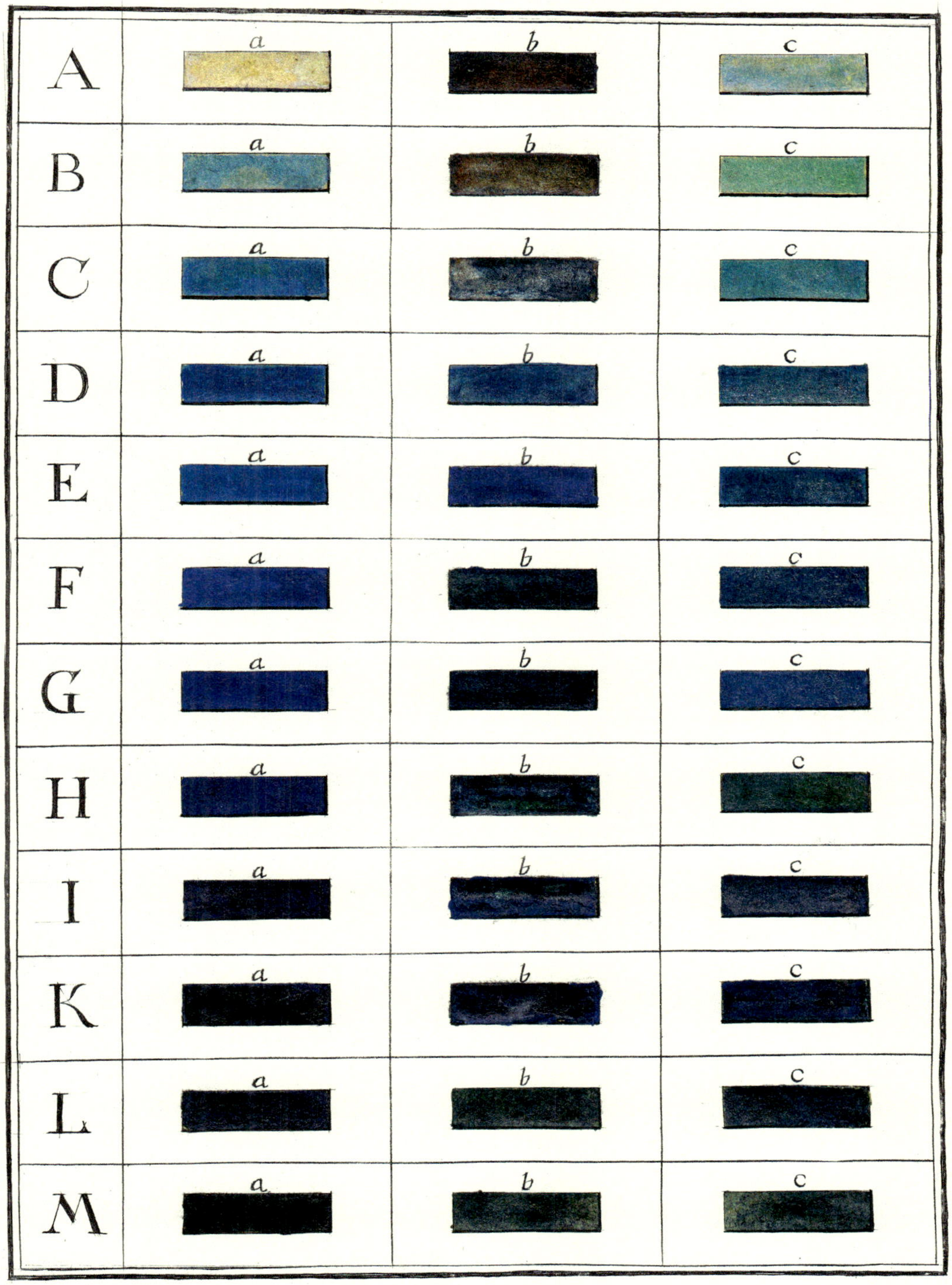

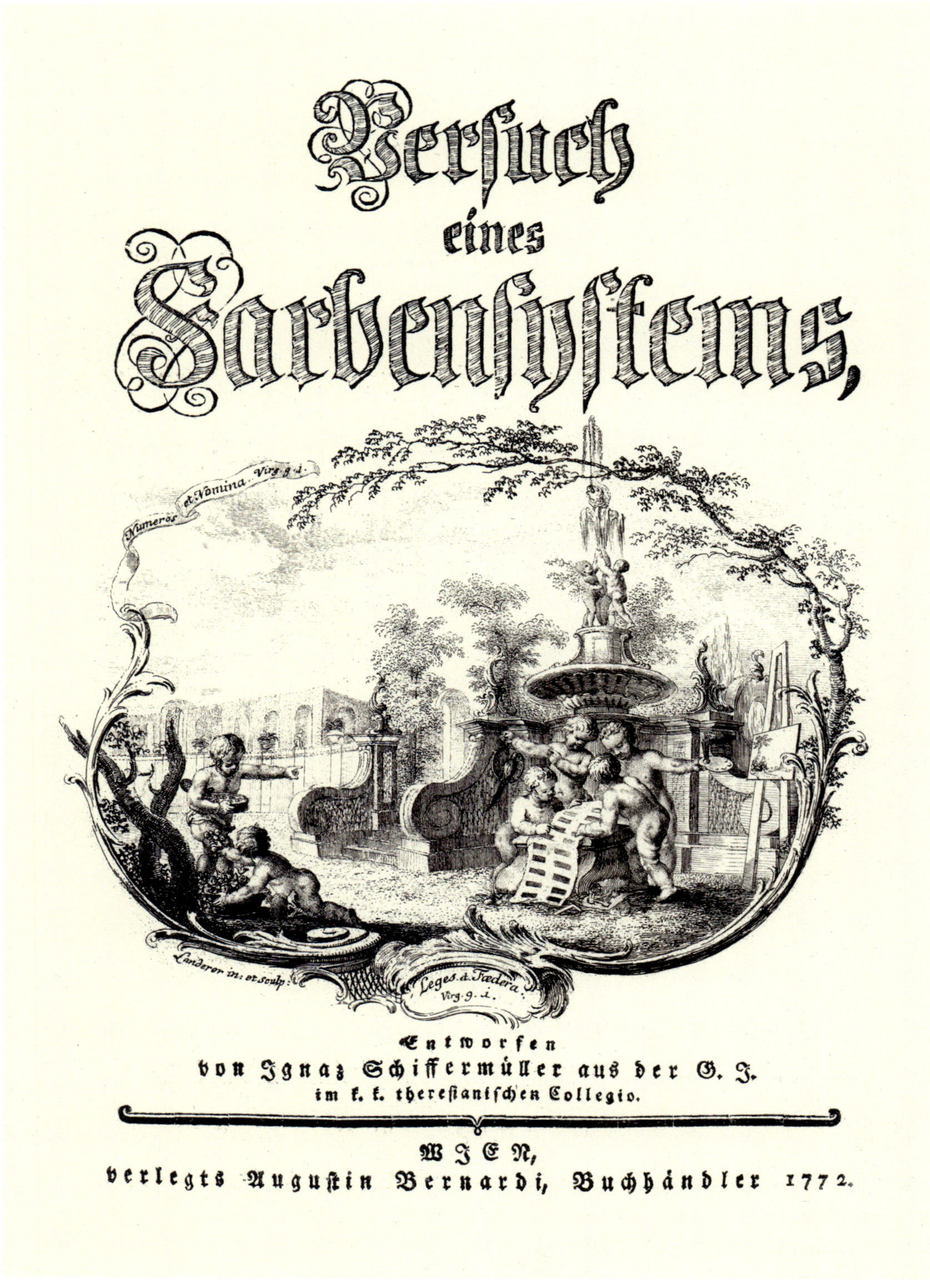

Versuch eines Farbensystems

Moses Harris und der österreichische Jesuitenpater Ignaz Schiffermüller verfolgten ähnliche Interessen und Absichten. Beide waren Entomologen und daher in ihren Illustrationen und Beschreibungen besonders auf eine akkurate Farbwiedergabe fixiert. Beide sahen zudem die Notwendigkeit, ein nützliches und verlässliches Farbsystem zu entwickeln, das die Leistungen Newtons anerkannte, aber auch notwendige Anpassungen für Künstler, Gestalter und Naturwissenschaftler vornahm. In ihrer Wahl der visuellen Mittel folgten beide dem Newton'schen Kreismodell, gestalteten es jedoch geringfügig unterschiedlich aus. Ob sie um die Arbeit des jeweils anderen wussten, ist nicht bekannt.

Schiffermüller entschied sich für einen zwölfteiligen Ring, der drei Primär-, drei Sekundär- und sechs Tertiärfarben umfasste. Anders als Newtons asymmetrisch aufgeteilter Kreis, der den sieben benannten Farben ungleiche Anteile zugewiesen hatte, ist der Ring gleichmäßig in 30-Grad-Segmente unterteilt. Und doch lassen die schmückenden Motive des Diagramms Schiffermüllers Bestreben erkennen, einen Bezug zu den optischen Experimenten seines Vorgängers herzustellen. In drei Ecken bildete er einen Regenbogen mit der von Newton entdeckten Farbfolge ab, in der vierten lenkt eine Gestalt einen Lichtstrahl durch ein Prisma. Die beiden Szenen rechts veranschaulichen die Anwendung der Newton'schen Theorie auf Gestaltung und Malerei, während im Zentrum der Komposition eine Sonne mit menschlichen Gesichtszügen durch die Wolken bricht. Das Ergebnis ist ein schönes und täuschend einfaches Farbdiagramm, das malerische, abstrakte, symbolische und sprachliche Elemente verbindet, wenngleich Schiffermüller auch erkannte, dass er das darin präsentierte Farbspektrum weiter ausführen musste. Auf die gleiche Weise wie Jacob Christian Schäffer (S. 104–107), dessen Buch er wahrscheinlich kannte, fügte er seinem Werk Beispiele für Farbtabellen hinzu, die jede der zwölf Farben des Diagramms in zwölf Typen untergliedern. Die mit den Großbuchstaben A bis M gekennzeichneten Typen stellte er in jeweils drei Helligkeitsstufen mit den Bezeichnungen a, b und c dar, was theoretisch insgesamt 432 Farben ergab.

Essai de système chromatique

Moses Harris et le prêtre jésuite autrichien Ignaz Schiffermüller ont plusieurs choses en commun. Tous deux entomologistes, ils sont particulièrement sensibles à la précision des couleurs dans leurs travaux de description et d'illustration. Ils comprennent également le besoin d'établir un système chromatique fiable et pratique, qui à la fois prenne en compte les découvertes d'Isaac Newton et soit adapté aux besoins des artistes, dessinateurs et naturalistes. Ils font tous deux le choix d'une représentation circulaire, suivant le concept de Newton, même si leur graphisme est légèrement différent. On ignore s'ils ont eu connaissance de leurs travaux respectifs.

Schiffermüller conçoit un cercle en 12 parties, comprenant trois couleurs primaires, trois secondaires et six tertiaires, et divisé en segments égaux de 30 degrés, contrairement au cercle asymétrique de Newton qui avait attribué des proportions inégales aux sept couleurs qu'il avait nommées. Malgré cela, les éléments ornementaux tout autour du cercle montrent bien à quel point Schiffermüller se réfère aux expériences optiques de son prédécesseur : un arc-en-ciel issu de la séquence chromatique newtonienne orne trois des angles de la page, tandis que dans le dernier angle, une vignette représente un personnage faisant passer un rayon de lumière à travers un prisme. Les deux scènes de droite illustrent l'application de la théorie de Newton au dessin et à la peinture, tandis qu'au centre de la composition, un soleil anthropomorphe, doté d'un visage aux traits humains, perce les nuages. L'ensemble, plus complexe qu'il n'en a l'air, forme un splendide graphique chromatique qui combine des éléments picturaux, abstraits, symboliques et linguistiques. Schiffermüller a pourtant compris qu'il lui fallait également détailler l'éventail de couleurs qu'il présentait. À l'instar de Jacob Christian Schäffer (p. 104–107), dont il connaissait probablement le livre, il ajoute donc des exemples de tableaux chromatiques qui décomposent chacune des 12 couleurs du graphique en 12 types, désignés par les majuscules A à M, chacun des types étant divisé à son tour en trois degrés de luminosité désignés par les minuscules a, b et c, soit un total théorique de 432 couleurs.

Tentativa de sistema de color

Moses Harris y el sacerdote jesuita austríaco Ignaz Schiffermüller tenían varias cosas en común. Ambos eran entomólogos y, por lo tanto, coincidían en la necesidad de una representación precisa del color en su trabajo ilustrativo y descriptivo. También veían la necesidad de proporcionar un sistema de color útil y fiable, que reconociera los logros de Isaac Newton e incorporara las adaptaciones necesarias para su uso por parte de artistas, diseñadores y científicos naturales. Para la representación visual de los colores, ambos siguieron el modelo conceptual de Newton y eligieron el círculo, aunque sus diseños son sutilmente diferentes. No se sabe si alguno estaba al corriente del trabajo del otro.

Schiffermüller optó por utilizar una rueda dividida en 12 segmentos con tres colores primarios, tres secundarios y seis terciarios. Cada segmento medía 30°, en contraste con la rueda asimétrica de Newton, que había asignado proporciones distintas a los siete colores incluidos. Aun así, la decoración pictórica que acompaña al diagrama revela el deseo de Schiffermüller de referirse a los experimentos ópticos de su predecesor, de modo que en tres esquinas aparece representado un arcoíris con la secuencia de colores newtoniana, mientras que en la otra esquina hay una viñeta de una persona dirigiendo un rayo de luz a través de un prisma. Las dos escenas de la derecha ilustran la aplicación de la teoría de Newton en el diseño y la pintura, mientras que en el centro de la composición se muestra un sol con rasgos faciales humanos que se abre paso entre las nubes. El resultado es un diagrama cromático hermoso y engañosamente simple que combina elementos pictóricos, abstractos, simbólicos y lingüísticos, aunque Schiffermüller también se dio cuenta de que necesitaba desarrollar la gama de colores que presentaba. Al igual que Jacob Christian Schäffer (págs. 104–107), cuyo libro probablemente conocía, agregó ejemplos de tablas de colores que dividen cada uno de los 12 colores del diagrama en 12 tipos distintos etiquetados con letras mayúsculas de la A a la M, cada uno de los cuales se representa en tres grados de luminosidad rotulados a, b y c, lo que en teoría da como resultado un total de 432 colores.

Johann Wolfgang von Goethe (1749–1832),
Friedrich Schiller (1759–1805)

DIE TEMPERAMENTENROSE

A Rose of Temperaments
1 watercolour drawing, ø 15.3 cm / 6 in., 1798/99
Klassik Stiftung Weimar, Goethe-Nationalmuseum

The German writer, amateur artist and scientist Johann Wolfgang von Goethe had a passion for many aspects of colour theory. His interests included the field of prismatic studies and were to a large extent fuelled by his highly critical response to Newton's *Opticks* from 1704, which resulted in his first published work on colour in 1791/92, *Beyträge zur Optik* (Contributions to Optics), whose title included an overt reference to Newton's earlier groundbreaking work (pp. 210–217). Between the date of this publication and his fully developed *Zur Farbenlehre* (*On Colour Theory*) of 1810–1812 (pp. 236–249), Goethe produced several sketches relating to his colour studies, often in response to written exchanges in which he was involved with other artists and writers. Some of these designs were later incorporated in the visual material for the plates in *Zur Farbenlehre*, while others survive as records of intellectual exchange, and examples of the ways in which Goethe developed his theories and concepts. The evocative *Temperamentenrose* was drawn in the course of an excited discussion about colour between Goethe and his fellow poet and playwright Friedrich Schiller. While the sketch itself is rather crude, it was based on extended conversations between the two men over several months with the aim of depicting connections between colours and the temperaments of the human mind. From this there emerged the idea of a kind of weathervane in the form of a 12-tone colour wheel with a large and ominous black centre. It was clearly based on Goethe's principles of polarity and contrast, in combination with the ancient concept of the four bodily humours, which could be taken to resemble four temperaments, namely choleric, sanguine, melancholic and phlegmatic. These names are indicated on the outer ring of the rose, each character type being associated with three specific occupations. For example, teachers, historians and orators are located in the green-blue phlegmatic group, while despots, heroes and adventurers appear on the opposite, choleric section, occupying the red-orange to yellow range. The colours themselves are not named.

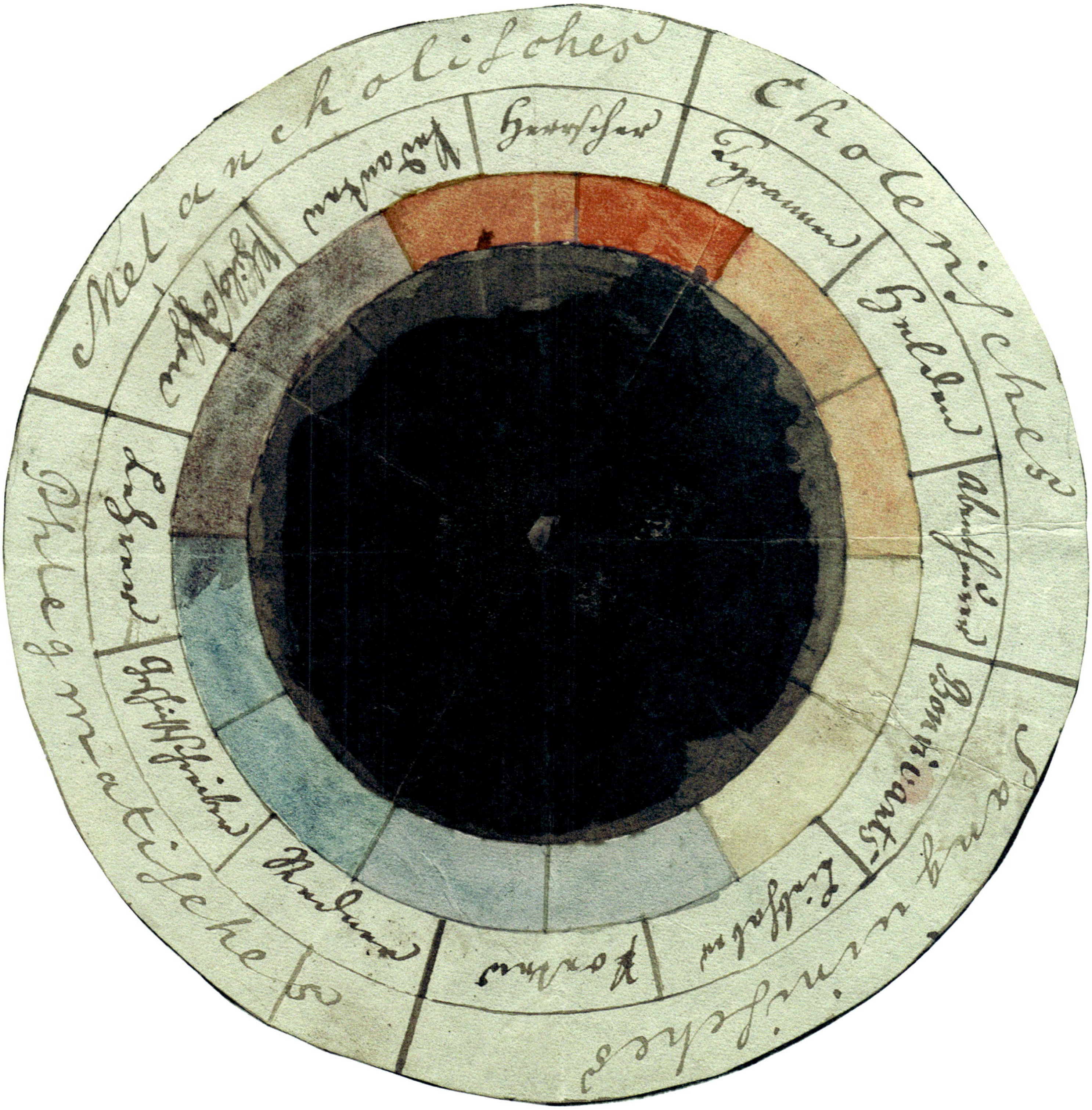
Melancholisches
Cholerisches
Sanguinisches
Phlegmatisches
Herrscher
Helden
Bonvivants

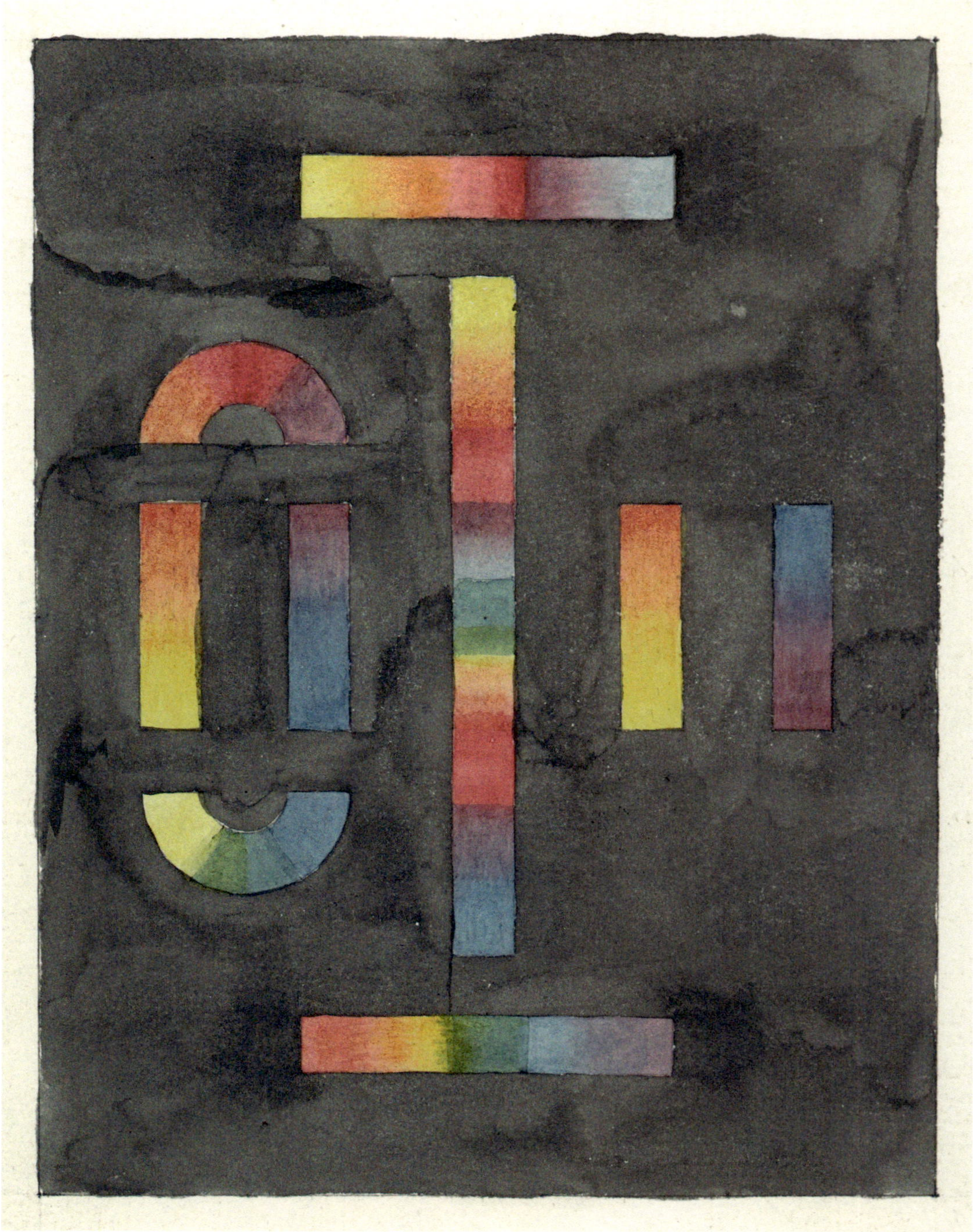

Johann Wolfgang von Goethe
A Symbolic Approach to Magnets, 1798
Symbolische Annäherung zum Magneten
Approche symbolique de l'aimant
Acercamiento simbólico al imán
Watercolour on white ribbed card, 19.7 x 16.1 cm (7¾ x 6⅜ in.)
Klassik Stiftung Weimar

Die Temperamentenrose

Goethe, der sich auch als Maler und Naturforscher betätigte, hatte eine Leidenschaft für viele Aspekte der Farbtheorie. Zu seinen Interessen gehörte auch der Bereich der prismatischen Studien, die zu einem Gutteil von seiner äußerst kritischen Reaktion auf Newtons *Opticks* von 1704 genährt wurden. Aus dieser Haltung ging 1791/92 Goethes erste Veröffentlichung über Farbe hervor, deren Titel, *Beyträge zur Optik*, sich offenkundig auf Newtons bahnbrechendes Werk bezog (S. 210–217). Zwischen dieser Publikation und seiner ausgearbeiteten Schrift *Zur Farbenlehre* aus den Jahren 1810 bis 1812 (S. 236–249) fertigte Goethe im Zusammenhang mit seinen Farbuntersuchungen – und oft als Reaktion auf Korrespondenzen, die er mit anderen Künstlern und Schriftstellern führte – mehrere Zeichnungen an. Einige davon wurden später in das Bildmaterial für die Tafeln von *Zur Farbenlehre* aufgenommen, andere haben sich als Dokumente des geistigen Austauschs und als Beispiele dafür erhalten, wie Goethe seine Theorien und Konzepte entwickelte. Die sinnträchtige Temperamentenrose entstand im Verlauf einer hitzigen Diskussion über Farbe zwischen Goethe und seinem Dichter- und Dramatikerkollegen Friedrich Schiller. Die Zeichnung wirkt zwar eher plump, beruhte jedoch auf ausgedehnten Gesprächen, die die beiden Männer über mehrere Monate hinweg mit dem Ziel führten, Beziehungen zwischen Farben und den Temperamenten des menschlichen Geistes abzubilden. Daraus entstand die Idee einer Art Wetterfahne in Form eines Farbkreises mit zwölf Farben und einem so großen wie ominösen schwarzen Zentrum. Das Modell basierte eindeutig auf Goethes Grundsätzen von Polarität und Kontrast und auf der antiken Vorstellung von den vier Körpersäften, denen vier Temperamente zugeordnet wurden: das Cholerische, das Sanguinische, das Melancholische und das Phlegmatische. Anhand dieser vier Bezeichnungen auf dem äußeren Ring der Rose wird jeder Charaktertyp mit drei speziellen Berufen assoziiert. So finden sich Lehrer, Historiker und Redner in der grün-blauen Gruppe der Phlegmatiker, während Despoten, Helden und Abenteurer im gegenüberliegenden Abschnitt der Choleriker das rot-orangefarbene bis gelbe Spektrum besetzen. Die Farben selbst sind nicht benannt.

Rose des tempéraments

L'écrivain et scientifique allemand Johann Wolfgang von Goethe, également peintre amateur, se passionne pour de nombreux aspects de la théorie des couleurs. Il s'intéresse notamment à l'étude des prismes et nourrit, dans ses réflexions, une réfutation des propos de Newton dans *Opticks* (1704). Il en sortira un premier ouvrage consacré à la couleur, en 1791–1792, *Beyträge zur Optik* (« Contributions à l'optique »), dont le titre fait ouvertement référence à l'ouvrage novateur de Newton (p. 210–217). Entre la date de cette publication et l'aboutissement de son *Zur Farbenlehre* (*Traité des couleurs*) de 1810–1812 (p. 236–249), Goethe produit plusieurs esquisses, fruits de ses études sur les couleurs, souvent dans les échanges épistolaires qu'il entretient avec d'autres artistes et écrivains. Certains de ces dessins trouveront plus tard leur place dans les éléments visuels accompagnant les planches de son *Zur Farbenlehre*; d'autres constituent simplement des traces d'échanges intellectuels et illustrent l'élaboration par Goethe de ses théories et concepts. L'évocatrice *Temperamentenrose* est dessinée pendant une discussion passionnée à propos de la couleur avec le poète et dramaturge Friedrich Schiller. Le dessin lui-même a beau être plutôt rudimentaire, il est issu de conversations nourries entre les deux hommes pendant plusieurs mois, dans le but de dépeindre les relations entre les couleurs et les tempéraments de l'esprit humain. C'est de là que naît l'idée d'une sorte de girouette prenant la forme d'un cercle chromatique de 12 tons, avec un centre noir, large et menaçant. Il faut y voir une évidente application des principes de polarité et de contraste développés par Goethe, combinés avec l'ancien concept des quatre humeurs du corps humain, vues comme l'équivalent de quatre tempéraments : colérique, sanguin, mélancolique et flegmatique. Ces termes sont inscrits sur l'anneau extérieur de la rose, et chaque tempérament est associé à trois occupations particulières. Par exemple, les enseignants, historiens et orateurs sont situés dans le groupe vert-bleu des flegmatiques, tandis que les despotes, héros et aventuriers apparaissent dans la section opposée, celle des colériques, qui couvre la gamme des rouge-orangé jusqu'au jaune. Les couleurs elles-mêmes ne sont pas nommées.

Rosa de temperamentos

El escritor, artista aficionado y científico alemán Johann Wolfgang von Goethe sentía pasión por muchos aspectos de la teoría del color. Sus intereses, que incluían el campo de los estudios prismáticos, estuvieron alimentados en gran medida por su respuesta altamente crítica a la *Opticks* de Newton de 1704, que dio como resultado su primer trabajo publicado sobre el color en 1791/1792, *Beyträge zur Optik* (Contribuciones a la óptica), cuyo título incluía una referencia abierta al trabajo pionero de Newton (págs. 210–217). Entre la fecha de publicación de esta obra y su *Zur Farbenlehre* (Teoría de los colores) completamente desarrollada de 1810–1812 (págs. 236–249), Goethe realizó varios bocetos relacionados con sus estudios del color, a menudo como parte de la correspondencia que mantenía con otros artistas y escritores. Más adelante, se incorporaron láminas con algunos de esos diseños en *Zur Farbenlehre*, mientras que otros sobreviven como registros de intercambio intelectual y ejemplos de las formas en que Goethe desarrolló sus teorías y conceptos. La sugerente *Temperamentenrose* se dibujó en el curso de una animada discusión sobre el color entre Goethe y su colega poeta y dramaturgo Friedrich Schiller. Si bien el boceto en sí es bastante tosco, se basó en las largas conversaciones que ambos hombres mantuvieron durante varios meses con el objetivo de representar las conexiones entre los colores y los temperamentos de la mente humana. De ahí surgió la idea de una especie de veleta con forma de rueda de color de 12 tonos con un gran y siniestro centro negro. Se basaba claramente en los principios de polaridad y contraste de Goethe, en combinación con el antiguo concepto de los cuatro humores corporales, que podrían considerarse como cuatro temperamentos, a saber: colérico, sanguíneo, melancólico y flemático. Estas palabras pueden leerse en el anillo exterior de la rosa, y cada tipo de carácter se asocia a tres ocupaciones específicas. Por ejemplo, los maestros, historiadores y oradores se ubican en el grupo flemático verde-azul, mientras que los déspotas, héroes y aventureros aparecen en la sección opuesta, colérica, ocupando el espectro que va del rojo-naranja al amarillo. Los colores en sí no se nombran.

Johann Wolfgang von Goethe

FARBENKREIS ZUR SYMBOLISIERUNG DES „MENSCHLICHEN GEISTES- UND SEELENLEBENS“

A Colour Wheel to Symbolise the "Life of the Human Mind and Soul"
1 pen and ink drawing, with watercolour, ø 6.9 cm / 2 ¾ in., 1809
Frankfurt am Main, Freies Deutsches Hochstift, Goethe-Haus

This small but important diagram by Goethe can be viewed in relation to the *Temperamentenrose* since it is also concerned with analogies, associations and symbolic meanings of colour and colour order, in connection with the human mind and emotions. It was created in the intense month of activity before the completion and publication of what Goethe considered one of his most important works, the three-volume *Zur Farbenlehre* (pp. 236–249). Although the diagram was not used, and differs in several respects from the colour wheels that were included in the book, this small ink and watercolour sketch is a further indication of Goethe's intensive research. It also reflects the Romantic Age, when great importance was assigned to the powers of the human soul and subjective experience. The colour order in this circle follows the familiar pattern of the primaries and their mixtures, although they are not named as such. As with the *Temperamentenrose* Goethe divided the wheel into four sections, which here symbolise moral and spiritual aspects: *Vernunft* (reason) appears over the red/yellow-red range, *Phantasie* (imagination) over the red/blue-red, *Sinnlichkeit* (sensuality) over the green/blue and *Verstand* (understanding) over the yellow/green. In addition, an inner ring is divided into compartments labelled with the attributes *schön, edel, gut, nützlich, gemein* and *unnöthig* (beautiful, noble, good, useful, ordinary and superfluous). Goethe later re-examined these aspects in the concluding part of the didactic section of *Zur Farbenlehre,* "Sinnlich-sittliche Wirkung der Farbe" (The Effect of Colour with Reference to Moral Associations).

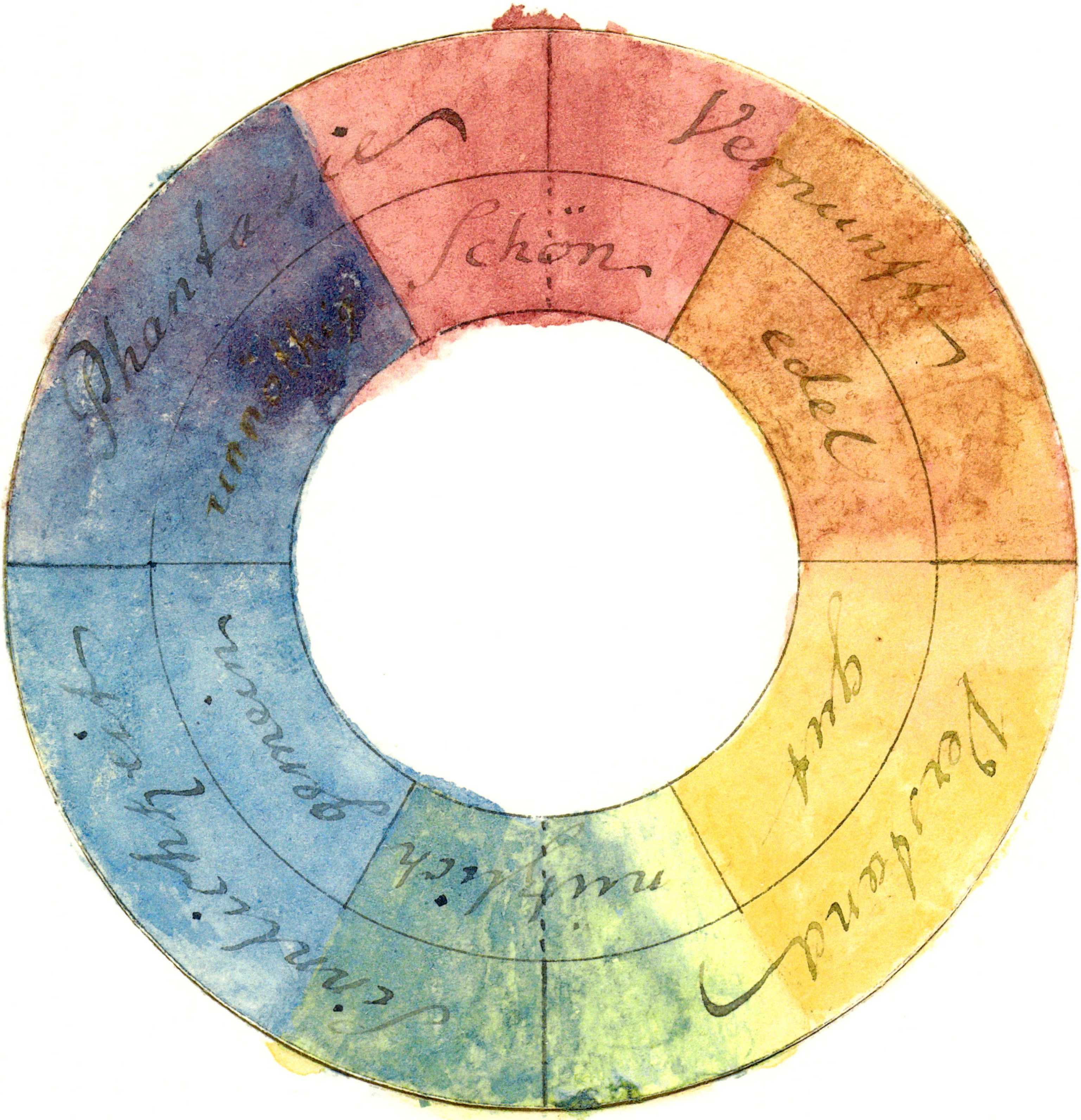
Phantasie
Vernunft
Verstand
Sinnlichkeit
schön
edel
gut
nützlich
gemein
unnöthig

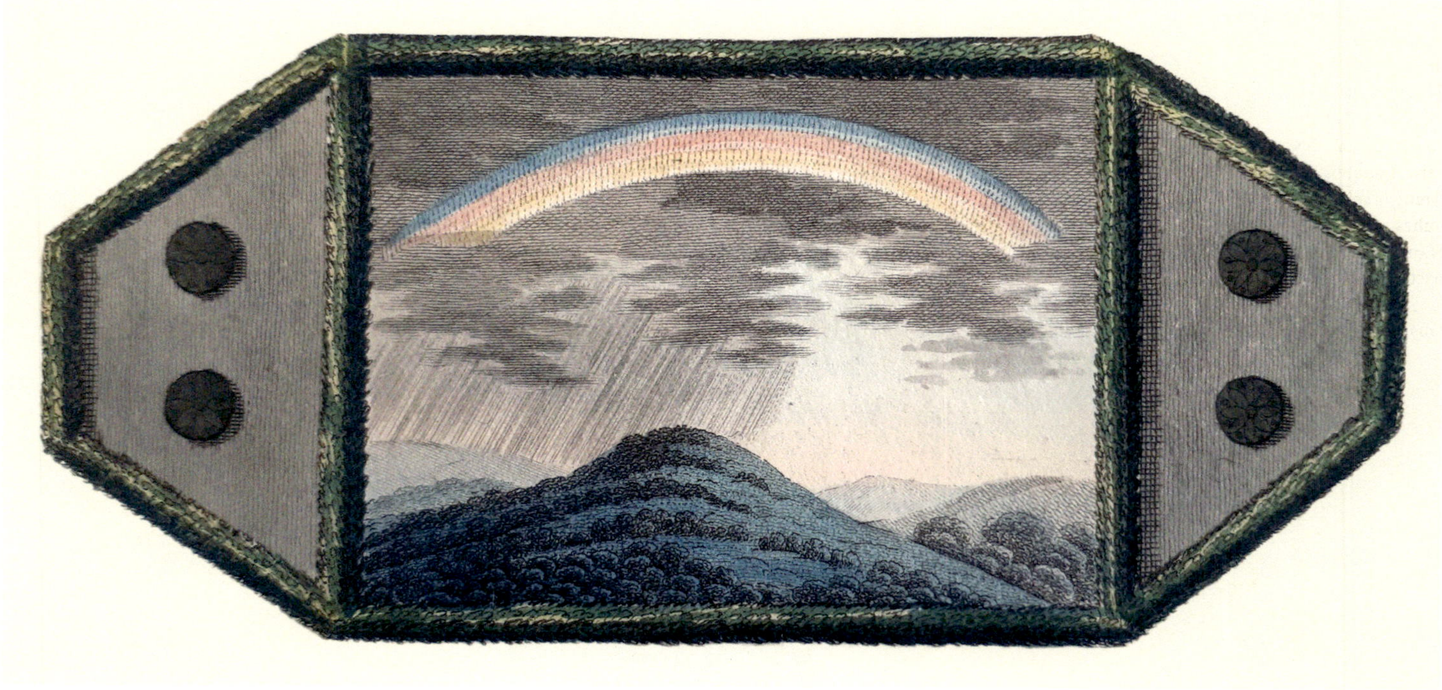

Farbenkreis zur Symbolisierung des „menschlichen Geistes- und Seelenlebens“

Dieses kleine, aber bedeutende Diagramm Goethes kann im Zusammenhang mit der *Temperamentenrose* betrachtet werden. Wie sie beschäftigt es sich mit Analogien, Assoziationen und der symbolischen Bedeutung von Farben und Farbordnungen im Hinblick auf den Geist des Menschen und seine Gefühle. Es entstand in einem Monat intensiver Beschäftigung mit Farbe, die in die Vollendung und Veröffentlichung der dreibändigen Abhandlung *Zur Farbenlehre* (S. 236–249) mündete, für Goethe eines seiner wichtigsten Werke. Wenngleich das Diagramm nicht verwendet wurde und in mehrerlei Hinsicht von den im Buch vertretenen Farbkreisen abweicht, ist die kleine aquarellierte Federzeichnung ein weiteres Indiz für Goethes aufwendige Forschungstätigkeit. Und sie ist ein Spiegel der Romantik, in der den Kräften der menschlichen Seele und dem subjektiven Erleben große Bedeutung zugemessen wurde. Die Farbordnung in diesem Kreis folgt dem vertrauten Muster der Primärfarben und ihrer Mischungen, auch wenn sie nicht als solche benannt sind. Wie bei der *Temperamentenrose* unterteilte Goethe den Kreis in vier Abschnitte, die hier moralische und geistige Aspekte symbolisieren: Dem Farbverlauf von Rot nach Gelbrot ist die „Vernunft“ zugeordnet, dem von Rot nach Blaurot die „Phantasie“, dem von Grün nach Blau die „Sinnlichkeit“ und dem von Gelb nach Grün der „Verstand“. Ein zweiter, innerer Ring besteht aus sechs Feldern, die mit den Eigenschaften „schön“, „edel“, „gut“, „nützlich“, „gemein“ und „unnöthig“ beschriftet sind. Im letzten Kapitel des didaktischen Teils von *Zur Farbenlehre* kam Goethe unter der Überschrift „Sinnlich-sittliche Wirkung der Farbe“ ausführlich auf diese Aspekte zurück.

Cercle chromatique pour symboliser « la vie intellectuelle et spirituelle de l’être humain »

Minuscule mais capital, ce graphique de Goethe peut être examiné au regard de la *Temperamentenrose* puisqu’il se base également sur des analogies, associations et significations symboliques de la couleur et de l’ordre chromatique, en relation avec l’esprit et les émotions. Goethe le crée pendant l’intense mois qui précède l’achèvement et la publication de ce qu’il tient pour l’un de ses ouvrages majeurs, *Zur Farbenlehre* (p. 236–249), en trois volumes. Bien qu’absent de l’ouvrage et différent à plusieurs titres des cercles chromatiques qui y figurent, ce petit dessin à l’encre rehaussé d’aquarelle est un indice supplémentaire des recherches approfondies de Goethe. Il est aussi un reflet de la période romantique, où l’on conférait la plus haute importance aux pouvoirs de l’âme humaine et de l’expérience subjective. L’ordre chromatique de ce petit cercle respecte le schéma habituel des couleurs primaires et leurs mélanges, bien qu’elles ne soient pas nommées.

Johann Wolfgang von Goethe
Rainbow over Green Mountains, in Grey Clouds, 1827
Regenbogen über grünendem Gebirg, in grauen Wolken
Arc-en-ciel au-dessus d'une montagne verdoyante, parmi des nuages gris
Arcoíris sobre montaña verde, entre nubes grises
Hand-coloured copperplate engraving, pasted on paper; 6.4 x 11.9 cm (2½ x 4¾ in.)
Düsseldorf, Goethe-Museum

Comme pour la *Temperamentenrose*, Goethe divise le cercle en quatre sections, qui symbolisent ici des facettes morales et spirituelles : *Vernunft* (la raison) apparaît sur la gamme rouge/jaune-rouge, *Phantasie* (l'imagination) sur la gamme rouge/bleu-rouge, *Sinnlichkeit* (la sensualité) sur le vert/bleu et *Verstand* (la compréhension) sur le jaune/vert. De plus, le cercle intérieur est partagé en compartiments désignés par les adjectifs *schön, edel, gut, nützlich, gemein* et *unnöthig* (beau, noble, bon, utile, ordinaire et superflu). Goethe révisera par la suite ces aspects dans la conclusion de la partie didactique du *Traité des couleurs*, « Sinnlich-sittliche Wirkung der Farbe » (« L'effet de la couleur en référence aux associations morales »).

Círculo de color para simbolizar «la vida de la mente y el alma humana»

Este pequeño pero importante diagrama de Goethe se puede ver en relación con la *Temperamentenrose*, ya que también se ocupa de las analogías, asociaciones y significados simbólicos del color y del orden del color en relación con la mente y las emociones humanas. Goethe lo creó durante el intenso mes de actividad anterior a la finalización y publicación de lo que consideraba una de sus obras más importantes, los tres volúmenes de *Zur Farbenlehre* (págs. 236–249). Aunque el diagrama no se usó y difiere en varios aspectos de las ruedas de colores que se incluyeron en el libro, este pequeño boceto en tinta y acuarela es una muestra más de la intensa investigación de su autor. También es reflejo de la época romántica, en la que se atribuyó gran importancia a los poderes del alma humana y la experiencia subjetiva. El orden de los colores en este círculo sigue el patrón familiar de los primarios y sus mezclas, aunque no se nombran como tales. Al igual que en la *Temperamentenrose*, Goethe dividió la rueda en cuatro secciones, que aquí simbolizan aspectos morales y espirituales: *Vernunft* (razón) aparece sobre el espectro rojo/amarillo-rojo, *Phantasie* (imaginación) sobre el rojo/azul-rojo, *Sinnlichkeit* (sensualidad) sobre el verde/azul y *Verstand* (comprensión) sobre el amarillo/verde. Además, un anillo interior está dividido en segmentos etiquetados con los atributos *schön, edel, gut, nützlich, gemein* y *unnöthig* (hermoso, noble, bueno, útil, corriente y superfluo). Goethe volvió a examinar estos aspectos en la parte final de la sección didáctica de *Zur Farbenlehre*, «Sinnlich-sittliche Wirkung der Farbe» (El efecto del color con referencia a las asociaciones morales).

Philipp Otto Runge (1777–1810)

FARBEN-KUGEL

oder Construction des Verhältnisses aller Mischungen der Farben zu einander, und ihrer vollständigen Affinität

Colour Globe
1 plate, 21.6 x 18.7 cm | 8 ½ x 7 ⅜ in., Hamburg, 1810
Hamburger Kunsthalle

It is to the German Romantic painter Philipp Otto Runge that credit is due for the first visually convincing three-dimensional concept of colour. The multi-dimensionality of colour order had of course been considered earlier, as in 1611 when the Finnish-Swedish astronomer Aron Sigfrid Forsius (1550–1624) presented a colour diagram that seems to have been intended to represent a three-dimensional shape, while a similar notion is evident in the pyramidal forms devised by Tobias Mayer (pp. 112–115), but when Runge gave his diagram the form of a sphere and presented it in the style of a terrestrial globe his model gained a more familiar, and seemingly tactile quality. The development of this spherical form, which was still limited to the two-dimensional format of a flat print, occupied much of Runge's time in the last few years of his life. His research was informed from 1806 onwards by lengthy written exchanges on colour with Goethe, who was sent a copy of the manuscript of *Farben-Kugel* prior to its publication. By coincidence, Runge's *Farben-Kugel* and the first part of Goethe's *Zur Farbenlehre* were both published in 1810, within a few months of each other, and Goethe praised the brilliance of Runge's concept in personal letters and in the later historical part of *Zur Farbenlehre*. It was, however, to be Runge's last work as he died from tuberculosis the same year.

Runge's hand-coloured etching shows the colour globe in four views, with one perspective view aligned so that the white region at the top (equivalent to the North Pole on a terrestrial globe) marks the pure white culmination of the greyscale that runs vertically through the globe, while a second perspective view shows the opposite black region (equivalent to the South Pole) that represents the darkening of the greyscale. A cross-section through the poles depicts the complementaries red and green on opposite sides of the globe as a series of concentric rings, while another cross-section represented as a 12-part colour wheel slices through the equator, around which run the pure colours, without any addition of white or black. While in practical terms this was far from being a perfect three-dimensional colour model, it worked well conceptually because it made it easy to perceive the range of colour mixtures, shades and tints within the form of the globe. The perspective views meanwhile add a sense of movement and dynamism to the model. Runge also added coloured plates in this work showing examples of colour harmonies and contrasts in the form of simple squares.

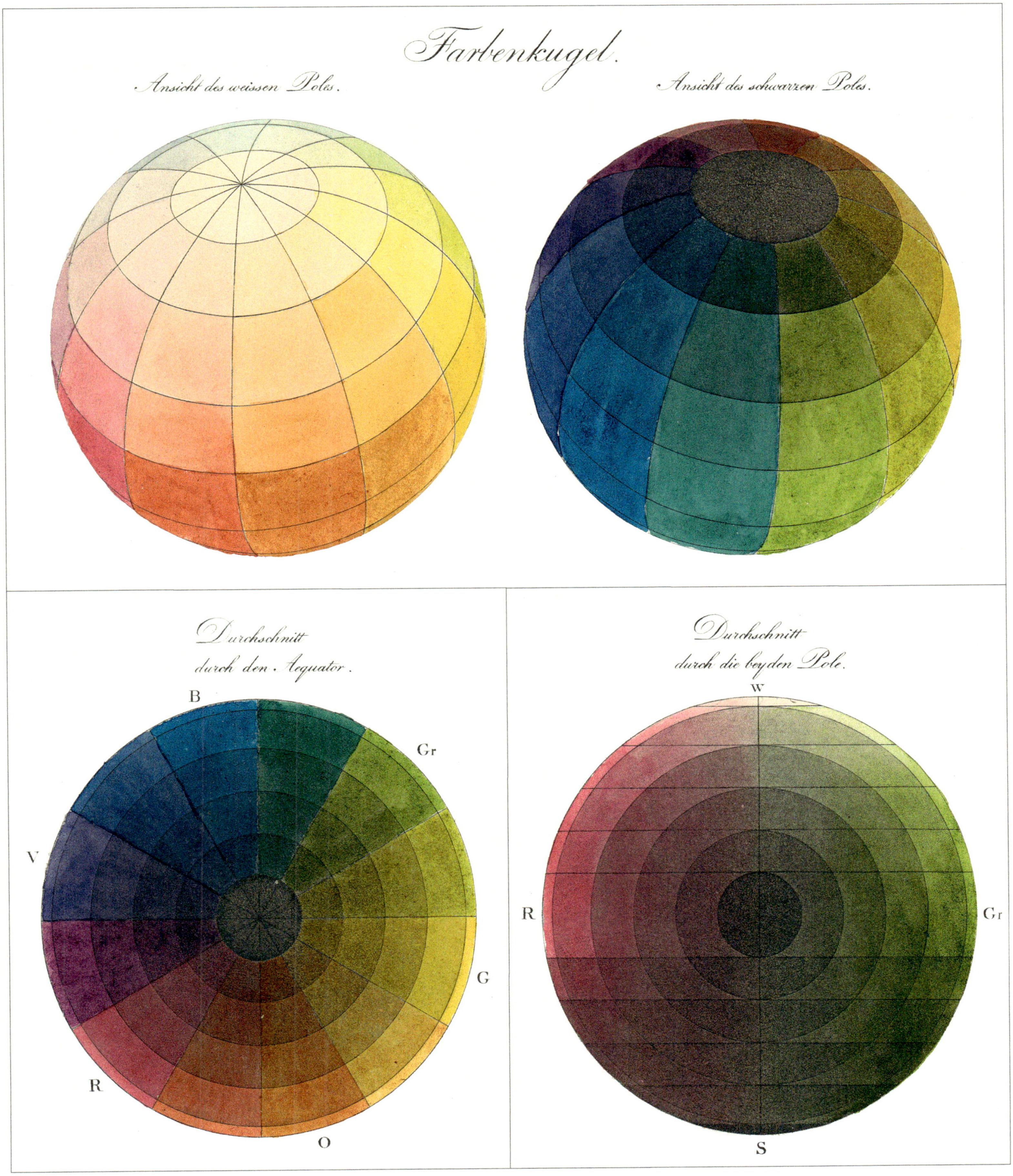
Farbenkugel.
Ansicht des weissen Poles.
Ansicht des schwarzen Poles.
Durchschnitt
durch den Aequator.
Durchschnitt
durch die beyden Pole.
B
Gr
V
G
R
O
W
R
Gr
S

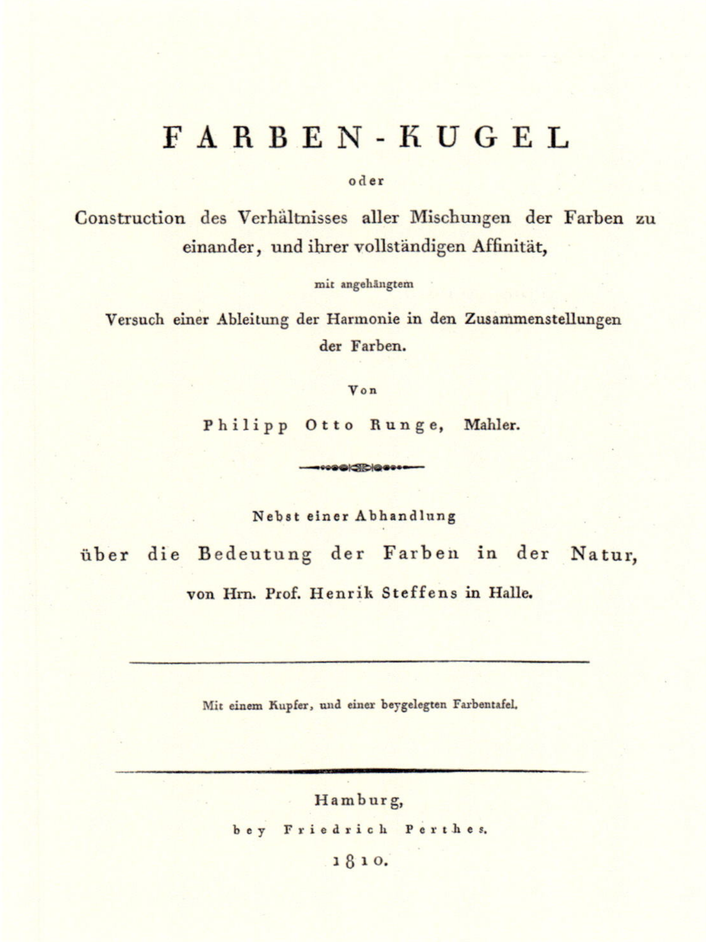
FARBEN-KUGEL

oder

Construction des Verhältnisses aller Mischungen der Farben zu einander, und ihrer vollständigen Affinität,

mit angehängtem

Versuch einer Ableitung der Harmonie in den Zusammenstellungen der Farben.

Von

Philipp Otto Runge, Mahler.

Nebst einer Abhandlung

über die Bedeutung der Farben in der Natur,

von Hrn. Prof. Henrik Steffens in Halle.

Mit einem Kupfer, und einer beygelegten Farbentafel.

Hamburg,

bey Friedrich Perthes.

1810.

Farben-Kugel

Dem deutschen Maler der Romantik Philipp Otto Runge ist das erste visuell überzeugende dreidimensionale Farbsystem zu verdanken.
Die Mehrdimensionalität der Farbenordnung wurde natürlich bereits früher erwogen, etwa als 1611 der finnisch-schwedische Astronom Aron Sigfrid Forsius (1550–1624) ein Farbdiagramm vorlegte, das allem Anschein nach eine dreidimensionale Form darstellen sollte. Eine ähnliche Vorstellung spricht aus den von Tobias Mayer entworfenen pyramidalen Figuren (S. 112–115). Doch als Runge seinem Diagramm die Form einer Kugel gab und diese im Stil eines Erdballs präsentierte, erlangte sein Modell eine vertrautere und scheinbar haptische Qualität. Die Entwicklung dieser Kugelform, die jedoch auf das zweidimensionale Format der Druckseite beschränkt war, nahm Runge in den letzten Jahren seines Lebens stark in Anspruch. Seine Forschung war seit 1806 vom Austausch längerer Briefe über Farbe zwischen ihm und Goethe geprägt, dem er eine Kopie des Manuskripts von *Farben-Kugel* schickte, bevor das Buch erschien. Durch Zufall lagen zwischen der Veröffentlichung von Runges *Farben-Kugel* und dem ersten Teil von Goethes *Zur Farbenlehre* im Jahr 1810 nur wenige Monate. Goethe lobte die Brillanz der Runge'schen Theorie in persönlichen Briefen und später im historischen Teil von *Zur Farbenlehre*. Sie sollte Runges letztes Werk sein, denn noch im selben Jahr starb der Künstler an Tuberkulose.

Runges handkolorierte Radierung zeigt seine Farben-Kugel in vier Ansichten, von denen eine perspektivische Aufsicht so ausgerichtet ist, dass die weiße Region oben (dem Nordpol eines Erdglobus entsprechend) den reinweißen Endpunkt der Grauskala markiert, die senkrecht durch die Kugel verläuft. Eine zweite Perspektive zeigt die gegenüberliegende (dem Südpol entsprechende) schwarze Region, die den dunkler werdenden Teil der Grauskala repräsentiert. Ein Schnitt durch die Pole bildet die Komplementärfarben Rot und Grün auf entgegengesetzten Seiten der Kugel in konzentrischen Ringen ab, während ein weiterer Schnitt in Form eines zwölfteiligen Farbkreises durch den Äquator verläuft. Seinen äußeren Ring bilden die reinen Farben, denen weder Weiß noch Schwarz beigemischt wurde. War die Kugel in praktischer Hinsicht auch bei Weitem kein perfektes dreidimensionales Farbmodell, so funktionierte sie doch als Konzept, weil sie es einem erleichterte, sich das Spektrum der Farbmischungen, Töne und Schattierungen in der Kugelform vorzustellen. Die perspektivischen Ansichten des Modells erzeugen derweil einen Eindruck von Bewegung und Dynamik. Darüber hinaus fügte Runge seinem Buch auch Farbtafeln mit Beispielen für Farbharmonien und -kontraste in Form einfacher Quadrate hinzu.

Sphère chromatique

C'est au peintre romantique allemand Philipp Otto Runge que l'on doit la première représentation convaincante d'une conception chromatique en trois dimensions. La multidimensionnalité de l'ordre chromatique avait bien sûr été envisagée avant : dès 1611, l'astronome finno-suédois Aron Sigfrid Forsius (1550–1624) a proposé un graphique de couleurs qui semble avoir été conçu pour représenter une forme tridimensionnelle, tandis que cette même notion est évidente dans les formes pyramidales imaginées par Tobias Mayer (p. 112–115). Mais lorsque Runge dessine une forme sphérique à la façon d'un globe terrestre, sa représentation prend une tournure familière et devient presque possible à toucher. L'élaboration de cette forme sphérique, qui était à l'époque limitée par l'impression en deux dimensions, a beaucoup occupé Runge dans les dernières années de sa vie. Ses recherches se nourrissent, à partir de 1806, par son abondante correspondance avec Goethe, à qui il fait parvenir un exemplaire manuscrit de sa *Farben-Kugel* avant publication. Le hasard fera que le livre de Runge et la première partie du *Zur Farbenlehre* de Goethe paraissent tous les deux en 1810, à quelques mois d'intervalle. Goethe fera l'éloge de la remarquable conception de Runge dans des lettres personnelles et, ultérieurement, dans la partie historique de son *Zur Farbenlehre*. Runge publie là son dernier ouvrage, et meurt de la tuberculose la même année.

Coloriée à la main, la gravure à l'eau-forte de Runge présente une sphère chromatique sous quatre angles. L'une des vues en perspective est orientée de manière que la zone blanche au sommet (l'équivalent du pôle Nord d'un globe terrestre) indique le point culminant, d'un blanc pur, de l'échelle des gris qui traverse verticalement le centre de la sphère. Une deuxième perspective montre la zone opposée (l'équivalent du pôle Sud), noire, qui représente l'obscurcissement de l'échelle des gris. En bas à droite, une coupe longitudinale par les pôles montre le rouge et le vert, complémentaires et diamétralement opposés, et déclinés sur une série d'anneaux concentriques ; en bas à gauche, une coupe transversale au niveau de l'équateur représente un disque chromatique en 12 parties, où sont figurées les couleurs pures, sans aucune adjonction de blanc ou de noir. Bien qu'en termes pratiques cette modélisation chromatique tridimensionnelle soit loin d'être parfaite, elle fonctionne bien sur le plan conceptuel car elle rend facilement perceptibles, au sein d'une même sphère, les gammes de mélanges de couleurs, les nuances et les teintes. L'inclinaison des axes des sphères, quant à elle, insuffle du mouvement et du dynamisme à la représentation. Runge a également intégré des planches coloriées à son ouvrage, montrant des exemples d'harmonies et de contrastes chromatiques sous la forme de simples carrés.

Esfera de color

El primer concepto tridimensional de color visualmente convincente se lo debemos al pintor romántico alemán Philipp Otto Runge. Por supuesto, la multidimensionalidad del orden de los colores se había considerado con anterioridad: en 1611, el astrónomo finlandés-sueco Aron Sigfrid Forsius (1550–1624) presentó un diagrama cromático que parece haber tenido la intención de representar una forma tridimensional, mientras que una noción similar es evidente en las formas piramidales ideadas por Tobias Mayer (págs. 112–115), pero cuando Runge le dio a su diagrama la forma de una esfera y lo presentó al estilo de un globo terrestre, su modelo adquirió un aspecto más familiar y una calidad aparentemente táctil. Runge ocupó gran parte de los últimos años de su vida en el desarrollo de esta forma esférica, todavía limitada por el formato bidimensional de una lámina impresa. A partir de 1806 en adelante, su investigación se apoyó en la extensa correspondencia sobre el color que mantuvo con Goethe, a quien envió una copia del manuscrito del *Farben-Kugel* antes de su publicación. La casualidad quiso que el *Farben-Kugel* de Runge y la primera parte del *Zur Farbenlehre* de Goethe se publicaran en 1810, con pocos meses de diferencia. Goethe elogió la brillantez del concepto de Runge en cartas personales y en la posterior parte histórica de *Zur Farbenlehre*. Fue el último trabajo de Runge, quien murió de tuberculosis ese mismo año.

El aguafuerte coloreado a mano de Runge muestra cuatro vistas de su esfera de color. Una de ellas ofrece una perspectiva en la que la región blanca de la parte superior (equivalente al Polo Norte en un globo terrestre) marca la culminación en blanco puro de la escala de grises que corre en vertical a lo largo del globo, mientras que una segunda vista en perspectiva muestra la región negra opuesta (equivalente al Polo Sur), que representa el oscurecimiento de la escala de grises. Una sección transversal a través de los polos representa los complementarios rojo y verde en lados opuestos del globo como una serie de anillos concéntricos, mientras que otra sección transversal representada como una rueda de color de 12 segmentos corta el ecuador, alrededor del cual corren los colores puros, sin ninguna adición de blanco o negro. Si bien en términos prácticos estaba lejos de ser un modelo de color tridimensional perfecto, funcionó bien conceptualmente porque facilitó la percepción de la gama de mezclas de colores, sombras y matices dentro de la forma del globo. Por su parte, las vistas en perspectiva añaden una sensación de movimiento y dinamismo al modelo. Runge también incluyó láminas coloreadas en este trabajo que muestran ejemplos de armonías y contrastes de colores en forma de sencillos cuadrados.

John Howard Kyan (1774–1850)

ON THE ELEMENTS OF LIGHT, AND THEIR IDENTITY WITH THOSE OF MATTER, RADIANT AND FIXED

1 plate, 24 x 15 cm | 9 ½ x 5 ⅞ in., London, 1838
Sussex, collection of the author

The frontispiece of this book presents one of the most intriguing and complex plates illustrating colour concepts in the early 19th century. Technically it was the work of two colour researchers, and it serves as a good example of the way in which existing concepts might be developed further through intellectual and artistic exchange. John Howard Kyan was an inventor, the son of an Irish copper mine owner, and in 1832 he patented a method for preserving ("kyanising") materials such as wood, paper, canvas and cloth, effectively preventing them from rotting prematurely. He was also interested in the early use of electricity, and in 1837 he became a founding member of the London Electrical Society. *On the Elements of Light, and Their Identity with Those of Matter* was published the following year, and the book is a fascinating document relating to light and colour science in the early Victorian age. A closer look at the frontispiece diagram reveals that it was based on designs and concepts by George Field (pp. 258–273), who had encouraged Kyan to continue his research into magnetism, electricity and optics and to expand an earlier essay he had written into this full-length book. A long introductory portion of Kyan's book discussed the basics of the theory of light, with reference to Newton's prismatic studies, followed by a description of Field's experiments and the different apparatus he had designed, including the Chromascope and Metrochrome, which were also illustrated. Significantly, the mezzotint frontispiece, engraved by David Lucas (1802–1881), was based on a drawing by Field, which makes it a rare surviving example of his art. The background presents an image of the creation of colours from darkness and light, amidst a dramatic pictorial display of thunder, lightning and sublime illumination, with clear religious undertones. Over this are placed three abstract figures, explained in detail by Kyan and which represent: figure 1, a strip of colours flanked by magnetic fields, "the Spectrum produced by the Newtonian Prism with the prismatic Colours, and the emblematical representation of the Magnetic forces"; figure 2, an "Aureola, obtained by the action of the Lenticular Prism of Field, on a black spot, on the white ground"; figure 3, "the Aureola from a white spot, on a dark or black ground". Field himself reworked and reused elements of this plate in several of his own subsequent books.

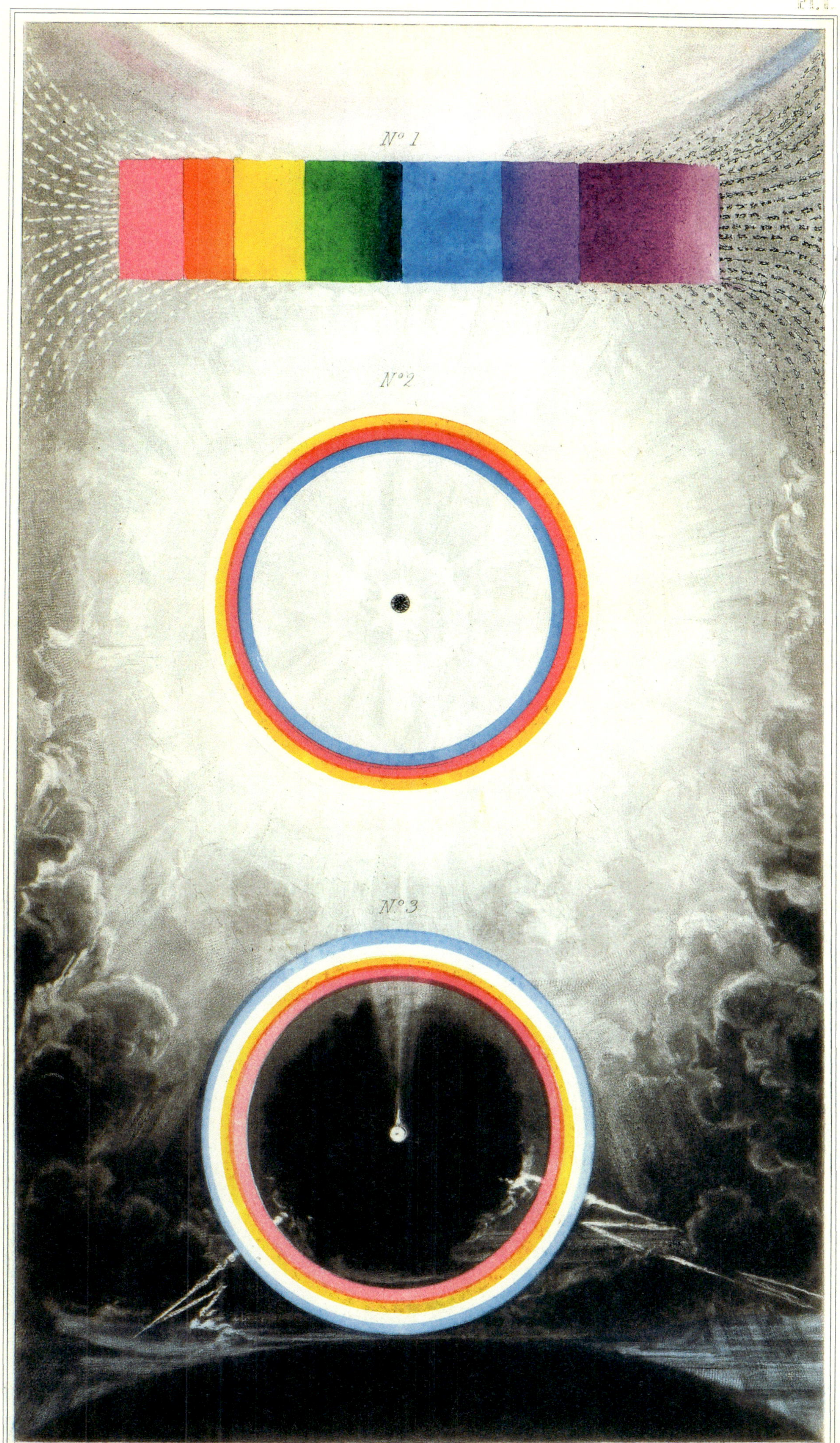

Drawn by Geo. Field. Engraved by David Lucas.

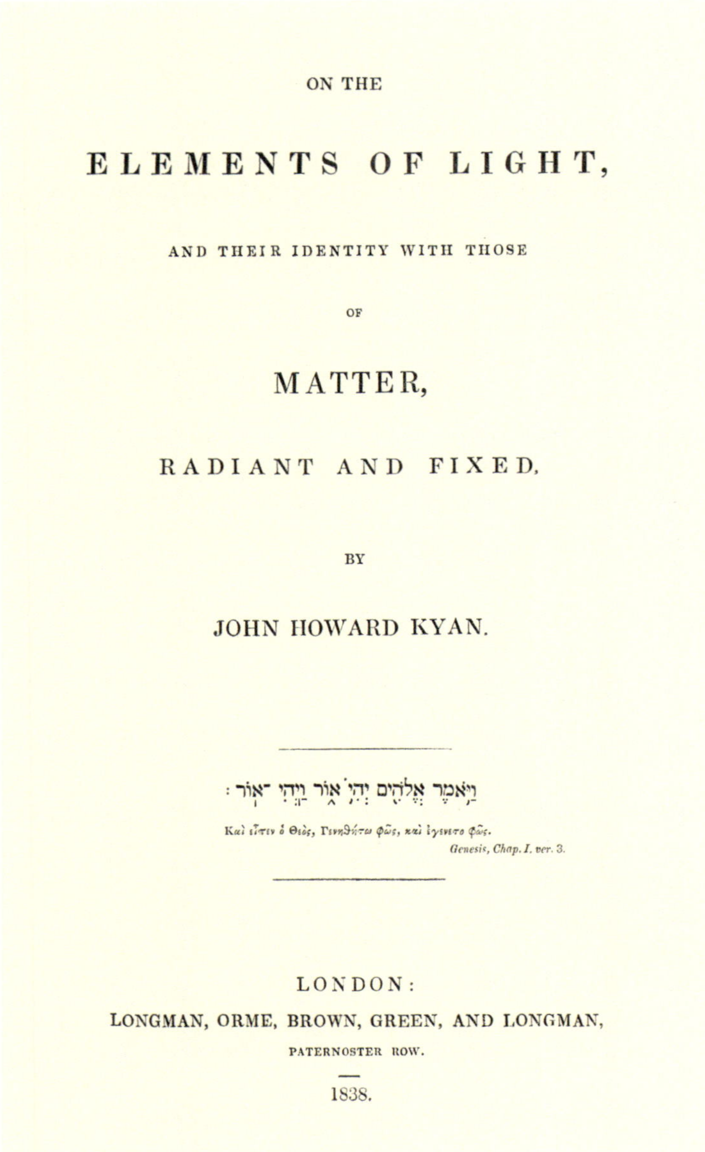

ON THE

ELEMENTS OF LIGHT,

AND THEIR IDENTITY WITH THOSE

OF

MATTER,

RADIANT AND FIXED.

BY

JOHN HOWARD KYAN.

וַיֹּאמֶר אֱלֹהִים יְהִי אוֹר וַיְהִי־אוֹר׃

Καὶ εἶπεν ὁ Θεὸς, Γενηθήτω φῶς, καὶ ἐγένετο φῶς.
Genesis, Chap. I. ver. 3.

LONDON:
LONGMAN, ORME, BROWN, GREEN, AND LONGMAN,
PATERNOSTER ROW.
1838.

Von den Bestandteilen des Lichts, und ihrer Übereinstimmung mit denen der Materie, strahlend und fest

Das Frontispiz dieses Buches zeigt eine der unwiderstehlichsten und komplexesten Tafeln zur Illustration von Farbtheorien des frühen 19. Jahrhunderts. Technisch war es das Werk zweier Farbforscher, das beispielhaft demonstriert, wie sich bestehende Theorien durch intellektuellen und künstlerischen Austausch weiterentwickeln ließen. John Howard Kyan war Erfinder. 1832 ließ der Sohn eines irischen Kupferminenbesitzers ein Verfahren zur Konservierung von Materialien wie Holz, Papier, Leinwand und Tuch („Kyanisierung") patentieren, das eine vorzeitige Fäulnis effektiv verhinderte. Kyan interessierte sich auch für die frühe Anwendung von Elektrizität und wurde 1837 Gründungsmitglied der London Electrical Society. Im Jahr darauf erschien *On the Elements of Light, and Their Identity with Those of Matter*. Das Buch ist ein faszinierendes Dokument der Licht- und Farbwissenschaft aus frühviktorianischer Zeit. Ein genauerer Blick auf das Diagramm verrät, dass es auf Entwürfen und Ideen von George Field (S. 258–273) beruhte, der Kyan ermutigt hatte, seine Magnetismus-, Elektrizitäts- und Optikforschung fortzusetzen und einen früher verfassten Aufsatz zu diesem Buch auszuarbeiten. In einer ausführlichen Einleitung diskutiert Kyan die Grundlagen der Theorie des Lichts in Bezug auf Newtons prismatische Studien. Dann folgt eine Beschreibung der Experimente Fields und der von ihm entworfenen Apparate, darunter das „Chromaskop" und das „Metrochrom", die auch abgebildet sind. Bezeichnenderweise stammte die Vorlage für das von David Lucas (1802–1881) in Mezzotinto ausgeführte Frontispiz ebenfalls von Field und macht es so zu einem der seltenen erhaltenen Beispiele seiner Kunst. Das Hintergrundbild zeigt die Erschaffung der Farben aus Dunkelheit und Licht inmitten einer dramatischen Inszenierung von Donner, Blitz und erhabener Beleuchtung mit deutlichen religiösen Untertönen. Darüber wurden drei abstrakte Figuren gelegt, die Kyan im Detail erläuterte: Figur 1, ein von Magnetfeldern flankierter mehrfarbiger Streifen, zeigt „das von Newtons Prisma erzeugte Spektrum mit den prismatischen Farben und die symbolische Darstellung der magnetischen Kräfte"; Figur 2 verkörpert eine „durch die Wirkung des Field'schen Linsenprismas auf einen schwarzen Punkt erzielte Aureole auf weißem Hintergrund"; Figur 3 schließlich stellt „die Aureole eines weißen Punkts auf einem dunklen oder schwarzen Hintergrund" dar. Field selbst überarbeitete später Elemente aus dieser Tafel und verwendete sie in mehreren seiner Bücher weiter.

Des élements de la lumière et de leurs similitudes avec ceux de la matière luminescente et solide

Le frontispice de ce livre propose l'une des planches d'illustration des concepts chromatiques parmi les plus intrigantes et complexes du début du XIXe siècle. Il est techniquement l'œuvre de deux spécialistes de la couleur et constitue un bon exemple de la façon dont des conceptions existantes peuvent être développées grâce à l'échange intellectuel et artistique. John Howard Kyan, fils d'un propriétaire de mine de cuivre irlandais, est inventeur. En 1832, il fait breveter une méthode de conservation (« procédé dit de Kyan ») de matériaux tels que le bois, le papier, la toile et le tissu, visant à éviter leur pourrissement prématuré. Il s'intéresse également à l'utilisation de l'électricité et fait partie, en 1837, des membres fondateurs de la London Electrical Society. *Des éléments de la lumière et de leurs similitudes avec ceux de la matière luminescente et solide*, publié l'année suivante, est un état des connaissances passionnant concernant la lumière et la couleur au début de l'ère victorienne. L'observation du dessin de cette première page illustrée montre qu'il se fonde sur les représentations et les concepts de George Field (p. 258–273), qui a encouragé Kyan à poursuivre ses recherches sur le magnétisme, l'électricité et l'optique, et à développer un essai écrit précédemment pour en faire un livre. Dans un long propos introductif, Kyan discute les fondements de la théorie de la lumière, en citant les travaux de Newton sur le prisme, puis il décrit les expériences de Field et les différents appareils que celui-ci a conçus, dont le « chromascope » et le « métrochrome », qui sont également illustrés. Il est à noter que la manière noire du frontispice, réalisée par David Lucas (1802–1881), s'inspire d'un dessin de Field, ce qui en fait l'une des rares traces de la production artistique de ce dernier. L'arrière-plan donne à voir la création des couleurs naissant de l'obscurité et de la lumière, au milieu d'une scénographie picturale représentant le tonnerre, l'éclair et une sublime illumination, à la connotation religieuse évidente. Au-dessus de ce décor sont disposées trois figures abstraites, dont Kyan explique en détail ce qu'elles représentent comme suit : figure 1, une bande de couleurs entre deux champs magnétiques, c'est-à-dire le « spectre produit par le prisme newtonien avec les couleurs prismatiques, et la représentation emblématique des forces magnétiques » ; figure 2, une « auréole obtenue par l'action du prisme lenticulaire de Field, subie par un point noir sur un fond blanc » ; figure 3, « l'auréole émanant d'un point blanc, sur un fond sombre ou noir ». Field lui-même retravaillera et réutilisera des éléments de cette planche dans plusieurs de ses ouvrages ultérieurs.

Los elementos de la luz y sus similitudes con los de la materia luminiscente y sólida

El frontispicio de este libro presenta una de las láminas más intrigantes y complejas de principios del siglo XIX que ilustran un concepto de color. Técnicamente, fue el trabajo de dos investigadores del color, y es un buen ejemplo de la forma en que los conceptos existentes pueden desarrollarse todavía más por medio del intercambio intelectual y artístico. John Howard Kyan fue un inventor, hijo del propietario de una mina de cobre irlandés, que en 1832 patentó un método para preservar («kyanizar») materiales como la madera, el papel, el lienzo y la tela y evitar que se pudrieran prematuramente. También se interesó en los primeros usos de la electricidad, y en 1837 se convirtió en miembro fundador de la London Electrical Society. *On the Elements of Light, and Their Identity with Those of Matter* se publicó al año siguiente. El libro es un documento fascinante sobre la ciencia de la luz y del color de principios de la época victoriana. Una mirada más atenta al diagrama del frontispicio revela que este se basó en diseños y conceptos de George Field (págs. 258–273), quien había animado a Kyan a continuar su investigación sobre el magnetismo, la electricidad y la óptica y a ampliar un ensayo anterior que había escrito y que terminó convirtiéndose en este libro. Una larga introducción al libro de Kyan aborda los conceptos básicos de la teoría de la luz, con referencia a los estudios prismáticos de Newton, a los que sigue una descripción de los experimentos de Field y de los diferentes aparatos que había diseñado, incluidos el chromascope y el metrochrome, que también aparecen ilustrados. De forma significativa, el frontispicio a media tinta, grabado por David Lucas (1802–1881), se basó en un dibujo de Field, hecho que lo convierte en uno de los pocos ejemplos de su arte que han sobrevivido. El fondo presenta una imagen de la creación de los colores a partir de la oscuridad y la luz en medio de una espectacular exhibición de truenos, relámpagos e iluminación sublime con un claro trasfondo religioso. Sobre este se colocan tres figuras abstractas, explicadas en detalle por Kyan y que representan: la figura 1, una franja de colores flanqueada por campos magnéticos, «el Espectro producido por el Prisma Newtoniano con los Colores prismáticos, y la representación emblemática de las fuerzas Magnéticas»; la figura 2, una «Aureola, obtenida por la acción del Prisma Lenticular de Field, sobre un punto negro, sobre fondo blanco»; y la figura 3, «la Aureola de un punto blanco, sobre un fondo oscuro o negro». El mismo Field reelaboró y reutilizó elementos de esta lámina en varios de sus libros.

Jean-François-Léonor Mérimée (1757–1836)

THE ART OF PAINTING IN OIL, AND IN FRESCO

2 plates, 20 x 13 cm | 7 ⅞ x 5 ⅛ in., London, 1839
Sussex, collection of the author

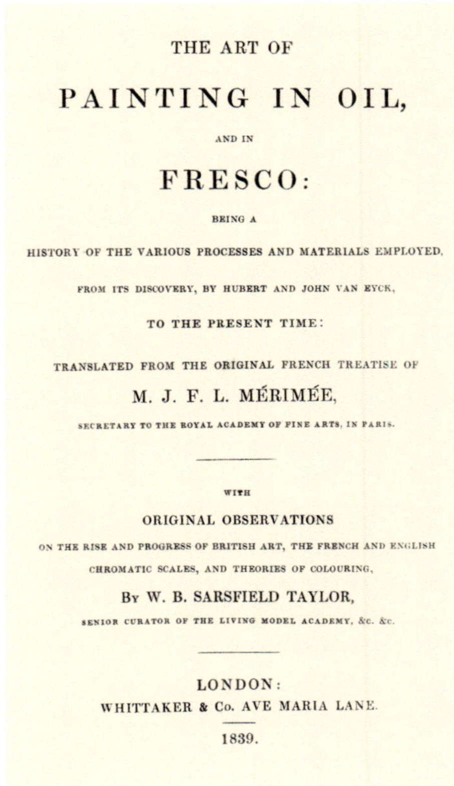
THE ART OF

PAINTING IN OIL,

AND IN

FRESCO:

BEING A

HISTORY OF THE VARIOUS PROCESSES AND MATERIALS EMPLOYED,

FROM ITS DISCOVERY, BY HUBERT AND JOHN VAN EYCK,

TO THE PRESENT TIME:

TRANSLATED FROM THE ORIGINAL FRENCH TREATISE OF

M. J. F. L. MÉRIMÉE,

SECRETARY TO THE ROYAL ACADEMY OF FINE ARTS, IN PARIS.

WITH

ORIGINAL OBSERVATIONS

ON THE RISE AND PROGRESS OF BRITISH ART, THE FRENCH AND ENGLISH

CHROMATIC SCALES, AND THEORIES OF COLOURING,

BY W. B. SARSFIELD TAYLOR,

SENIOR CURATOR OF THE LIVING MODEL ACADEMY, &c. &c.

LONDON:

WHITTAKER & Co. AVE MARIA LANE.

1839.

In 1830, a colour diagram resembling a cart wheel with spokes was included in a substantial book about the preparation and use of colours by the French artist and colour technician Jean-François-Léonor Mérimée, *De la peinture à l'huile, ou Des procédés matériels employés dans ce genre de peinture, depuis Hubert et Jean Van-Eyck jusqu'à nos jours*. In scale and range, Mérimée's work foreshadowed both George Field's *Chromatography* of 1835 (pp. 270–273) and Chevreul's writings on colour contrast (pp. 282–295). Mérimée brought together various (art) historical developments in the field of colour in conjunction with the latest material technology, while also devoting an important chapter to the harmonious use of colour. The visualisation of colour order, harmony and contrast he employed was his simple *échelle chromatique* (colour scale), in which six hand-coloured discs were arranged in a circle with the red, yellow and blue (in the original French edition) depicted slightly larger than the orange, green and purple. The discs are connected to each other by narrow bands showing the gradations between the six colours, and further bands extend inwards towards a neutral greyish-black disc in the centre. Mérimée's diagram is a highly effective visual concept and an efficient means of depicting colour connections and their positions on a colour wheel, but it is still somewhat limited and is strictly two-dimensional. In the posthumous English edition of 1839, from which the plates reproduced here are taken, *The Art of Painting in Oil, and in Fresco*, the translator (and artist) William Benjamin Sarsfield Taylor added a much-simplified version of Moses Harris's colour wheel to Mérimée's conceptual model, partly to highlight the contribution of English publications on colour theory in the preceding decades, and partly to extend Mérimée's system. In most extant copies of both editions of this work the hand-colouring differs greatly, and tends to lack subtlety and fine nuancing, certainly when compared to Harris's *Natural System of Colours* (pp. 138–143), and in a number of copies the pigments have also deteriorated significantly.

French Chromatic Scale (M. Merimée's)

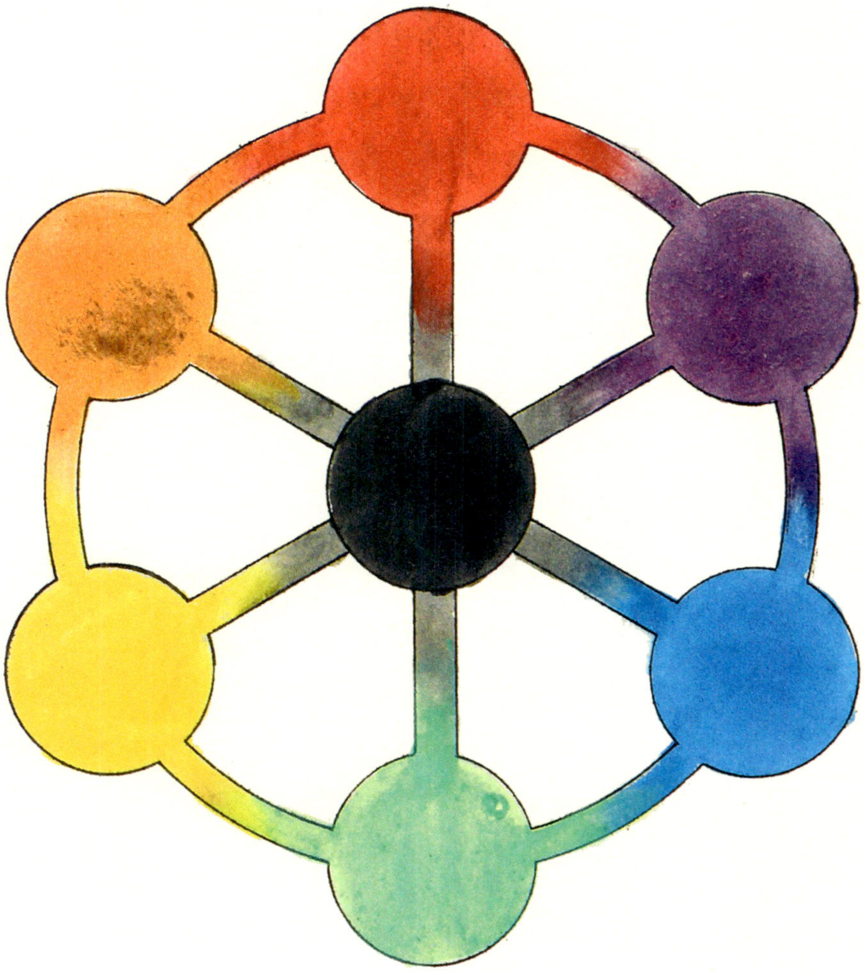

London Published by Whittaker & C° Ave Maria Lane.

A. Ducôte's Lithog.ʳ 70, Sᵗ Mart.ˢ Lane

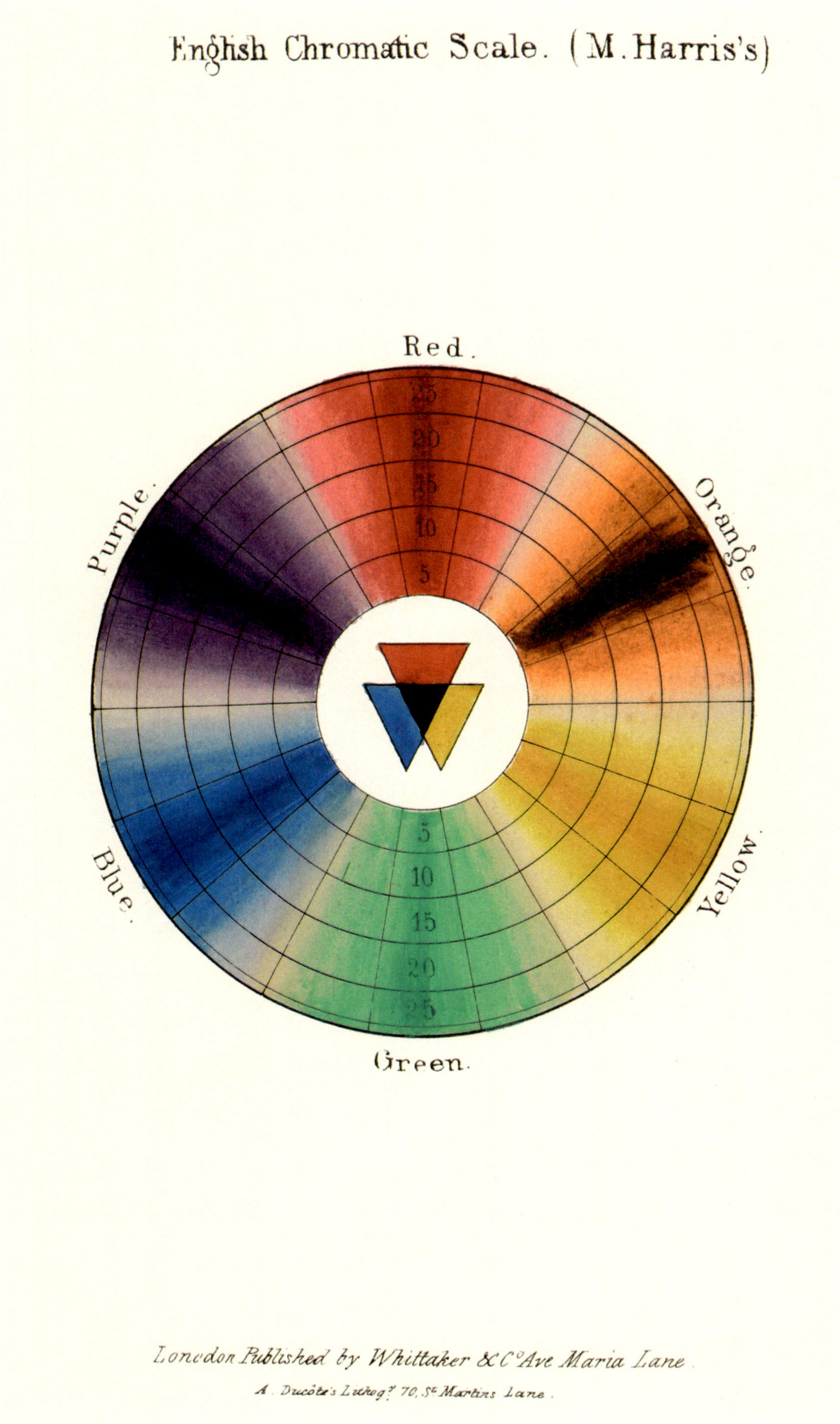

Die Kunst der Öl- und Freskenmalerei

Im Jahr 1830 wurde ein Farbdiagramm, das einem Wagenrad mit Speichen ähnelte, in ein ausführliches Buch über die Zubereitung und Verwendung von Farben aufgenommen. Verfasst hatte *De la peinture à l'huile, ou Des procédés matériels employés dans ce genre de peinture, depuis Hubert et Jean Van-Eyck jusqu'à nos jours* der französische Maler und auf Farben spezialisierte Chemiker Jean-François-Léonor Mérimée. In Umfang und Bandbreite nahm das Werk sowohl George Fields *Chromatography* von 1835 (S. 270–273) als auch Chevreuls Schriften zu den Farbkontrasten vorweg (S. 282–295). Mérimée führte verschiedene historische und kunsthistorische Entwicklungen aus dem Bereich der Farbe mit der neuesten Materialtechnologie zusammen und widmete ein wichtiges Kapitel dem harmonischen Farbgebrauch. Um Ordnung, Harmonie und Kontraste der Farben zu veranschaulichen, verwendete er eine einfache „échelle chromatique" (Farbleiter). Darin bilden sechs handkolorierte Farbscheiben einen Kreis, dessen rote, gelbe und blaue Scheibe (in der französischen Originalausgabe) etwas größer ausfallen als die orangefarbene, die grüne und die violette. Die schmalen Streifen, durch die sie verbunden sind, zeigen die Abstufungen zwischen den Farben. Weitere Streifen erstrecken sich nach innen zu einer neutralen, gräulich schwarzen Scheibe im Zentrum der Abbildung. Als visuelles Modell, um Farbverbindungen und ihre Lage in einem Farbkreis erfolgreich darzustellen, ist Mérimées Diagramm überaus effektiv, in seinen Mitteln jedoch begrenzt und überdies streng zweidimensional gegliedert. In *The Art of Painting in Oil, and in Fresco*, der 1839 postum erschienenen englischen Ausgabe, aus der die hier reproduzierten Tafeln stammen, stellte der Übersetzer (und Künstler) William Benjamin Sarsfield Taylor dem konzeptuellen Entwurf Mérimées eine stark vereinfachte Version des Farbkreises von Moses Harris an die Seite, um den Beitrag englischer Publikationen zur Farbtheorie in den vorausgegangenen Jahrzehnten hervorzuheben und Mérimées System zu erweitern. In den meisten erhaltenen Exemplaren beider Ausgaben dieses Werks weichen die Handkolorationen nicht nur erheblich voneinander ab, es fehlt ihnen auch – und ganz gewiss im Vergleich zu Harris' *Natural System of Colours* (S. 138–143) – an Feinheit und subtiler Nuancierung. In einer Reihe von Exemplaren haben zudem die Farben stark gelitten.

De la peinture à l'huile

En 1830, un graphique chromatique ressemblant à une roue de charrette à rayons figure dans un volumineux ouvrage consacré à la préparation et à l'utilisation des couleurs, signé par le peintre et coloriste français Jean-François-Léonor Mérimée, peintre et spécialiste français des pigments, *De la peinture à l'huile, ou Des procédés matériels employés dans ce genre de peinture, depuis Hubert et Jean Van-Eyck jusqu'à nos jours*. De par son envergure, l'ouvrage de Mérimée annonce à la fois le *Chromatography* de George Field en 1835 (p. 270–273) et les écrits de Chevreul sur les contrastes de couleurs (p. 282–295). Mérimée réunit plusieurs avancées dans l'histoire de l'art en matière de couleur et de progrès technologiques, tout en consacrant un important chapitre à l'utilisation harmonieuse des couleurs. Pour visualiser l'ordre des couleurs, l'harmonie et le contraste, il invente une simple « échelle chromatique » : six disques coloriés à la main sont disposés en cercle, avec le rouge, le jaune et le bleu (dans l'édition originale) légèrement plus grands que l'orange, le vert et le violet. Les disques sont reliés entre eux par d'étroites bandes montrant les gradations entre les six couleurs, et d'autres bandes s'étirent vers le disque gris-noir qui occupe le centre. Le graphique de Mérimée est un concept visuel fort efficace et très pratique pour décrire les liens entre les couleurs et leurs positions sur le cercle chromatique, mais il s'avère encore quelque peu limité et reste au stade bidimensionnel. Dans une édition anglaise posthume de 1839, d'où sont extraites les planches reproduites dans le présent ouvrage, *The Art of Painting in Oil, and in Fresco*, le traducteur (et peintre) William Benjamin Sarsfield Taylor ajoute une version très simplifiée du cercle chromatique de Moses Harris au modèle conceptuel de Mérimée. Il cherche, par là, à souligner la contribution à la théorie chromatique des publications anglaises des décennies précédentes, et également à perfectionner le système de Mérimée. Dans la plupart des exemplaires subsistants des deux éditions de cet ouvrage, la mise en couleurs diffère beaucoup et les nuances manquent de subtilité et de finesse, notamment en comparaison avec le *Natural System of Colours* d'Harris (p. 138–143). Les pigments de plusieurs exemplaires se sont grandement détériorés.

El arte de la pintura al óleo y al fresco

En 1830, un diagrama cromático que se parece a una rueda de carreta con radios se incluyó en un importante libro sobre la preparación y el uso de los colores del artista y especialista en pigmentos francés Jean-François-Léonor Mérimée, *De la peinture à l'huile, ou Des procédés matériels employés dans ce genre de peinture, depuis Hubert et Jean Van-Eyck jusqu'à nos jours*. En escala y alcance, el trabajo de Mérimée prefiguró tanto la *Chromatography* de George Field de 1835 (págs. 270–273) como los escritos de Chevreul sobre el contraste de colores (págs. 282–295). Mérimée juntó varios desarrollos históricos (artísticos) en el campo del color con la última tecnología de materiales y dedicó un capítulo importante al uso armonioso del color. La visualización del orden, de la armonía y del contraste de los colores que empleó fue su sencilla «échelle chromatique» (escala de colores), en la que seis discos coloreados a mano están dispuestos a lo largo de un círculo, con el rojo, el amarillo y el azul (en la edición original en francés) ocupando espacios ligeramente más grandes que el naranja, el verde y el morado. Los discos están conectados entre sí por bandas estrechas que muestran las gradaciones entre los seis colores, y otras bandas se extienden hacia el interior, donde hay un disco de un color neutro negro grisáceo. El diagrama de Mérimée es un concepto visual muy efectivo y un medio eficiente para representar las conexiones entre los colores y sus posiciones en una rueda, pero todavía es algo limitado y estrictamente bidimensional. En la edición póstuma en inglés de 1839, de donde se han tomado las láminas que aquí se reproducen, de *The Art of Painting in Oil, and in Fresco*, el traductor (y artista) William Benjamin Sarsfield Taylor agregó una versión muy simplificada de la rueda de colores de Moses Harris al modelo conceptual de Mérimée, en parte para resaltar la contribución de las publicaciones inglesas a la teoría del color en las décadas anteriores y en parte para ampliar el sistema de Mérimée. En la mayoría de las copias existentes de ambas ediciones de este trabajo, el coloreado a mano difiere mucho y tiende a carecer de sutileza y matices finos, en especial cuando se comparan con el *Natural System of Colours* de Harris (págs. 138–143). En varias copias, los pigmentos también se han deteriorado de forma significativa.

Guido Schreiber (1799–1871)

DIE FARBENLEHRE

part 3 of *Das technische Zeichnen*

Colour Theory, part 3 of *Technical Drawing*
8 plates, 18.8 x 13 cm | 7 ⅜ x 5 ⅛ in., Leipzig, 1868
Dresden, Sächsische Landesbibliothek –
Staats- und Universitätsbibliothek

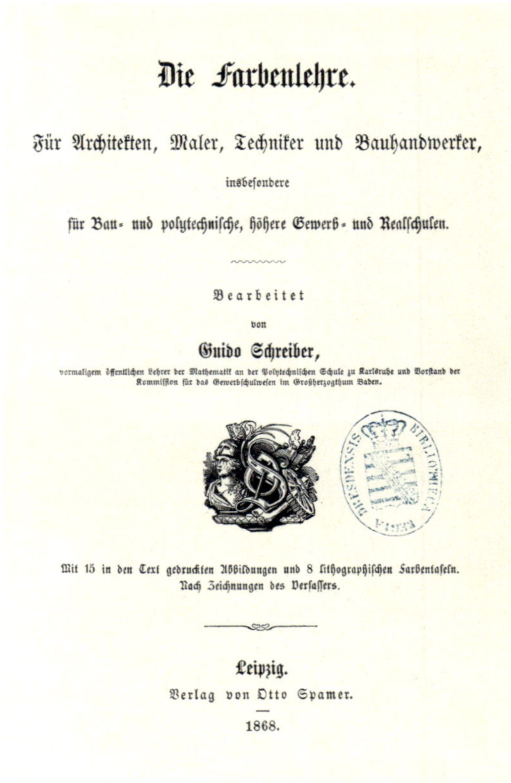
Die Farbenlehre.

Für Architekten, Maler, Techniker und Bauhandwerker,
insbesondere
für Bau- und polytechnische, höhere Gewerb- und Realschulen.

Bearbeitet
von
Guido Schreiber,
vormaligem öffentlichen Lehrer der Mathematik an der Polytechnischen Schule zu Karlsruhe und Vorstand der Kommission für das Gewerbschulwesen im Großherzogthum Baden.

Mit 15 in den Text gedruckten Abbildungen und 8 lithographischen Farbentafeln.
Nach Zeichnungen des Verfassers.

Leipzig.
Verlag von Otto Spamer.
1868.

Guido Schreiber's contribution to colour literature is rarely discussed, perhaps because it falls into the category of practical and educational handbooks, without claiming to offer any new theories or concepts. However, this small-format publication with its many chromolithographic plates deserves closer inspection as it succinctly presents colour theory to a readership intended to include architects, painters, technicians and builders. It formed the third part of a series of text-books on architecture for use in schools and colleges, as well as independent study.

Schreiber was a German mathematician and geometry teacher with a military background, and in his teaching and research he showed a particular interest in pictorial composition, light diffraction, the aesthetics of colour and coloured shadows. Schreiber's analytical approach and didactic tone are evident in this and his other published works, and his *Farbenlehre* is partly an interpretation and simplified presentation of Goethe's works on colour (he even copied the wording from Goethe's well-known title). In consequence, Schreiber's book can be seen as an interesting case study concerning the reception given to Goethe's colour theory and its later application just over half a century after it was published. The colour plates are of surprisingly high quality, especially considering that this would have been a book issued as a relatively cheap edition. The images showing ornamental designs were clearly influenced by the magnificent publication on colour in the decorative arts by Owen Jones (1809–1874), *The Grammar of Ornament* (1856), and indeed Jones is duly discussed in a chapter on colour and aesthetics. One plate simply and very successfully illustrates the phenomenon of coloured shadows (p. 178), while Schreiber's depictions of colour wheels and chromatic order are impressive in their clear design, making the best possible use of the small print format and lithography for the colour reproduction. Plate 2 is particularly interesting for as well as a fairly standard 12-part colour wheel it shows a simplified version of Runge's colour sphere (p. 163), segmented like an orange and as if flattened into a two-dimensional shape so as to fit into the book.

Tafel I.

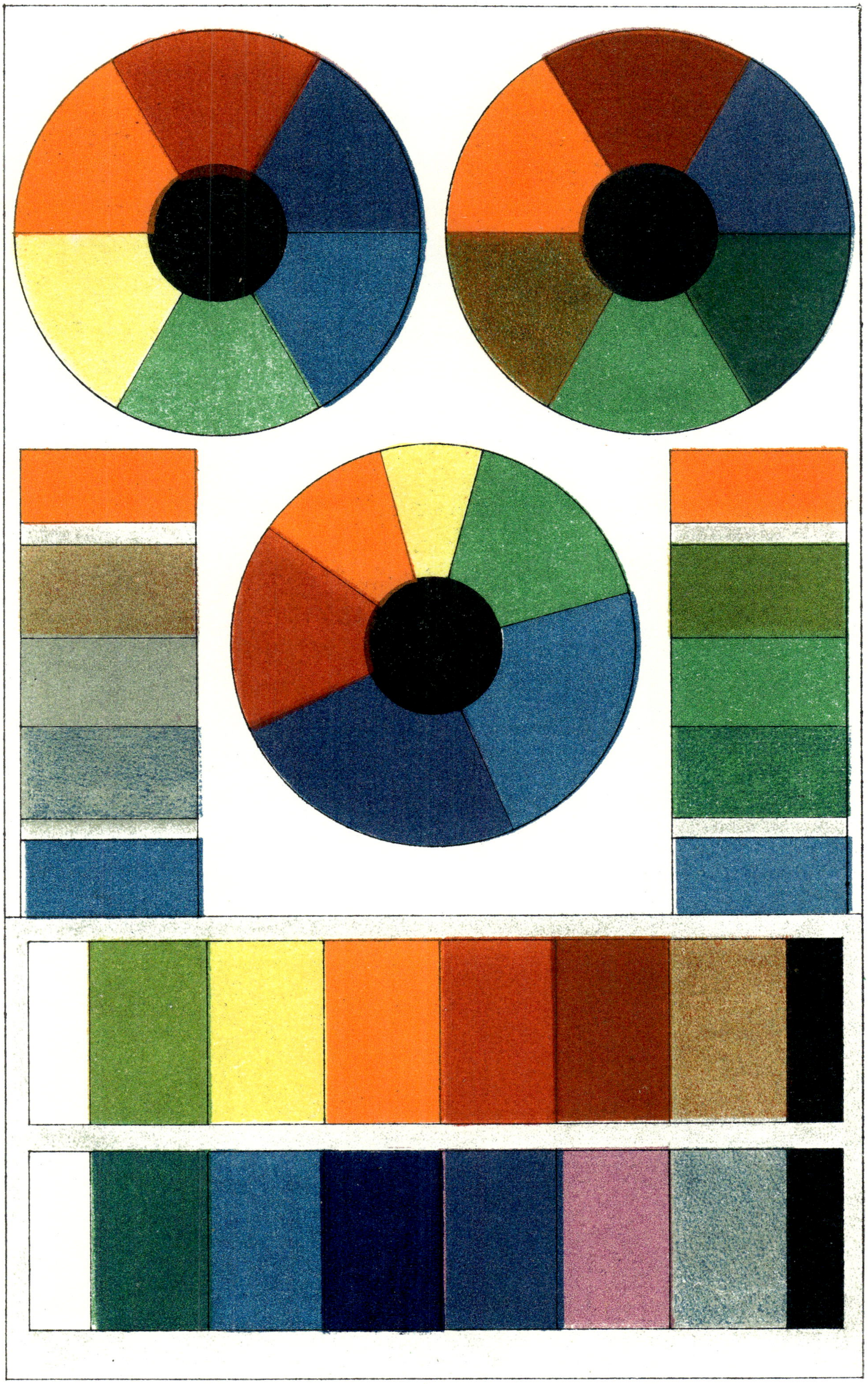

G. Schreibers Farbenlehre

Leipzig: Verlag von Otto Spamer.

F. A. Brockhaus' Geogr.-artist. Anstalt, Leipzig

Tafel II.

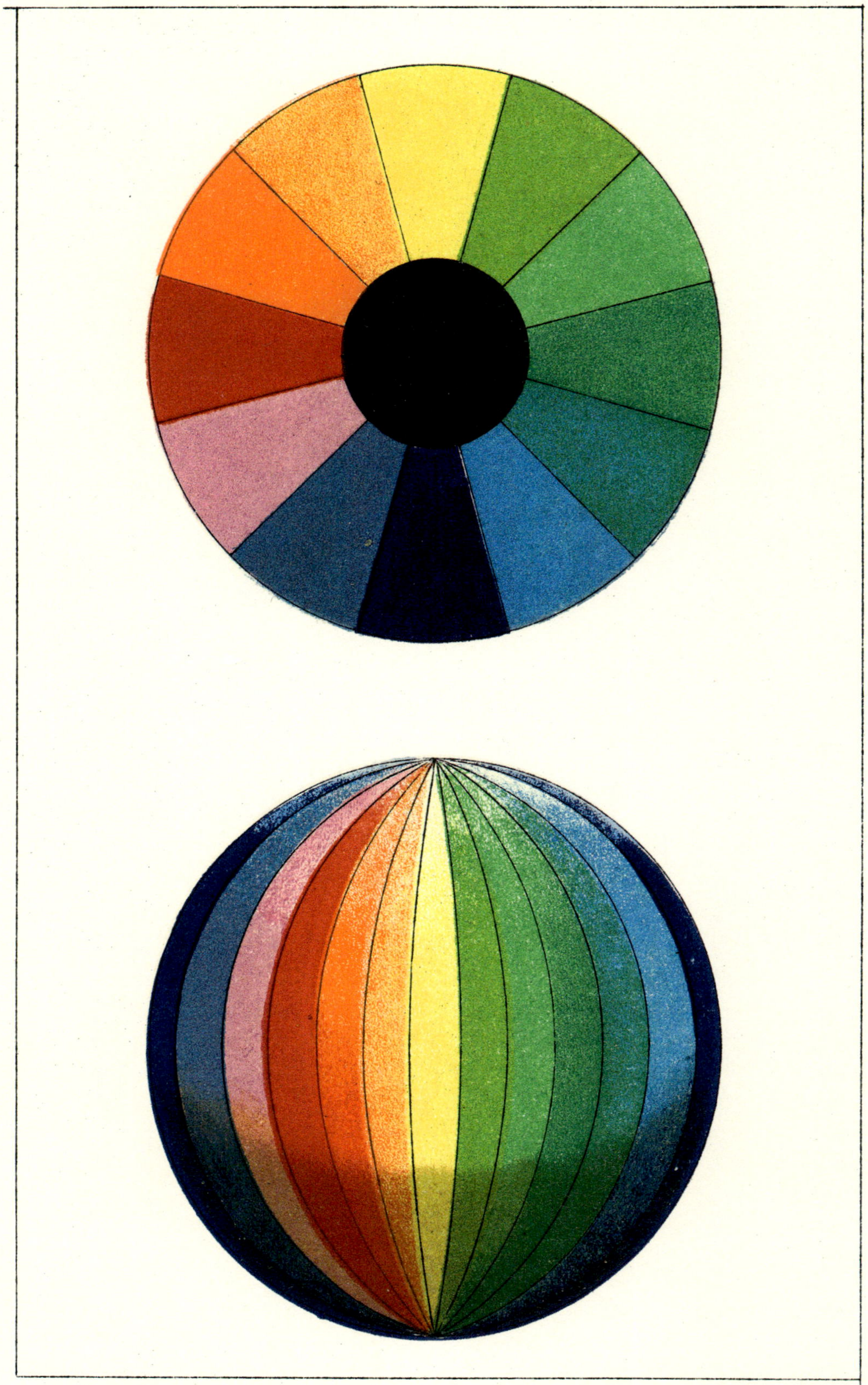

G. Schreibers Farbenlehre Leipzig Verlag von Otto Spamer.

F. A. Brockhaus Geogr.-artist. Anstalt Leipzig

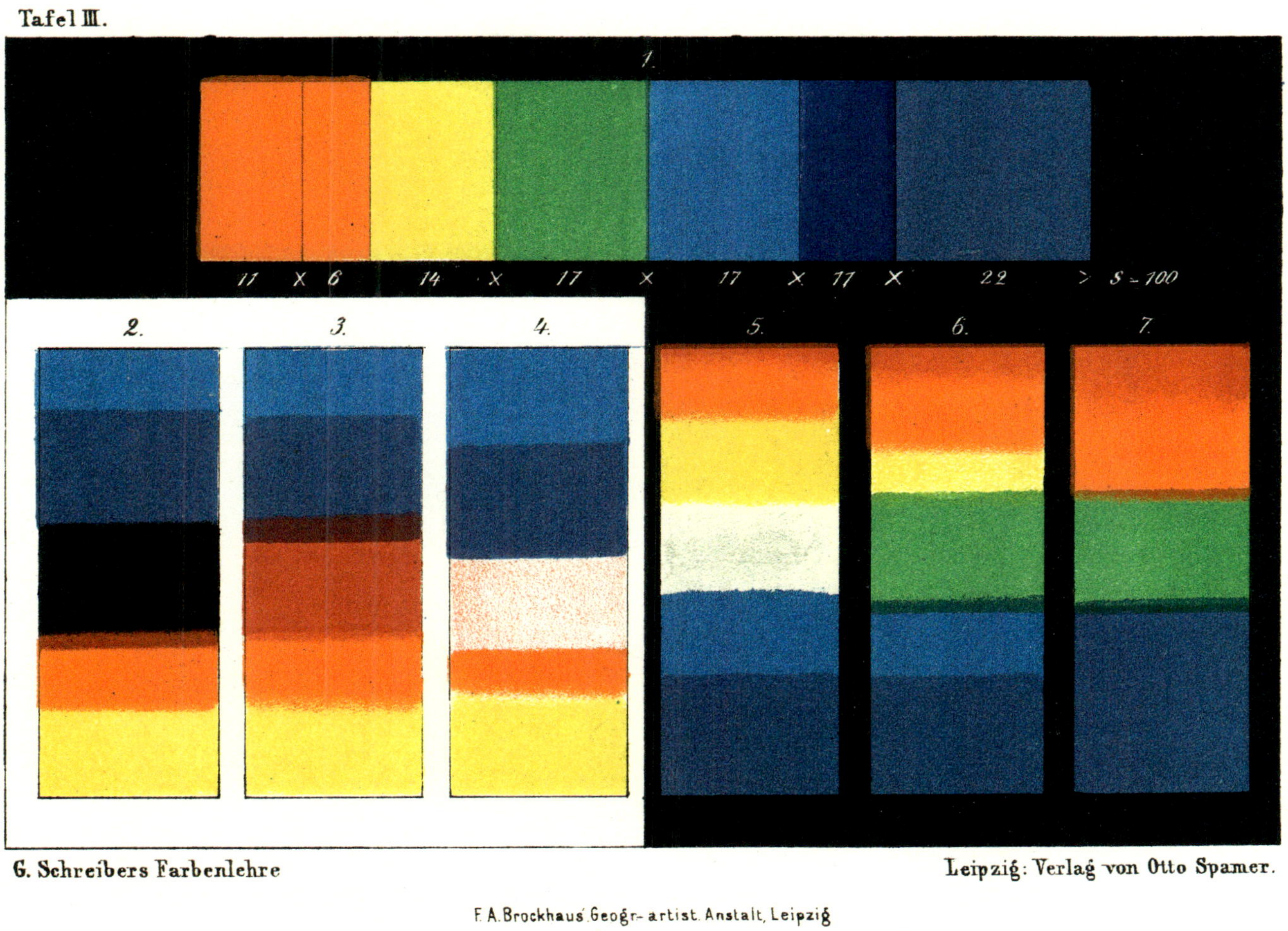

Die Farbenlehre, *Das technische Zeichnen*, Teil 3

Guido Schreibers Beitrag zur Farbliteratur wird selten besprochen, weil er womöglich in die Kategorie der Praxis- und Lehrhandbücher fällt, ohne Anspruch auf neue Theorien oder Konzepte zu erheben. Doch diese kleinformatige Publikation mit ihren vielen farblithografischen Tafeln verdient eine genauere Untersuchung, weil sie die Farbtheorie für eine angestrebte Leserschaft von Architekten, Malern, Technikern und Bauingenieuren in prägnanter Form aufbereitet. Sie bildete den dritten Teil einer Reihe von Architekturlehrbüchern für Schulen, Hochschulen und das Selbststudium.

Schreiber war ein deutscher Mathematiker und Geometrielehrer mit militärischem Hintergrund. In seiner Lehre und Forschung zeigte er ein besonderes Interesse für den Aufbau von Bildern, für die Lichtbrechung und die Ästhetik von Farben und farbigen Schatten. Schreibers analytischer Ansatz und didaktischer Ton sind in dieser und anderen seiner Veröffentlichungen offensichtlich, und seine *Farbenlehre* ist zum Teil eine Interpretation und vereinfachte Darstellung der Werke Goethes über Farbe (er übernahm sogar den Begriff aus Goethes berühmtem Titel). Sein Buch kann daher als interessantes Beispiel für die Rezeption von Goethes Farbtheorie und ihrer Anwendung gut ein halbes Jahrhundert nach ihrem Erscheinen betrachtet werden. Die Farbtafeln sind von erstaunlich hoher Qualität, was besonders bemerkenswert ist, da dieses Buch als vergleichsweise preiswerte Ausgabe herausgegeben wurde. Die Abbildungen von Ornamenten waren eindeutig beeinflusst von der prächtigen kunstgewerblichen Publikation *The Grammar of Ornament* (1856) des Architekten Owen Jones (1809–1874), der in einem Kapitel über Farbe und Ästhetik gebührend gewürdigt wird. Während eine Tafel auf einfache und überzeugende Weise das Phänomen der farbigen Schatten veranschaulicht (S. 178), beeindrucken Schreibers Darstellungen von Farbkreisen und der Ordnung der Farben mit ihrer klaren Gestaltung, die aus dem kleinen Druckformat und dem lithografischen Reproduktionsverfahren das Beste herausholt. Tafel 2 ist besonders interessant, weil sie neben einem recht gewöhnlichen zwölfteiligen Farbkreis eine vereinfachte Version der Runge'schen Farben-Kugel (S. 163) zeigt, die mit ihren orangenartigen Spalten aussieht, als sei sie zu einer zweidimensionalen Form abgeflacht worden, um in das Buch hineinzupassen.

Tafel IV.

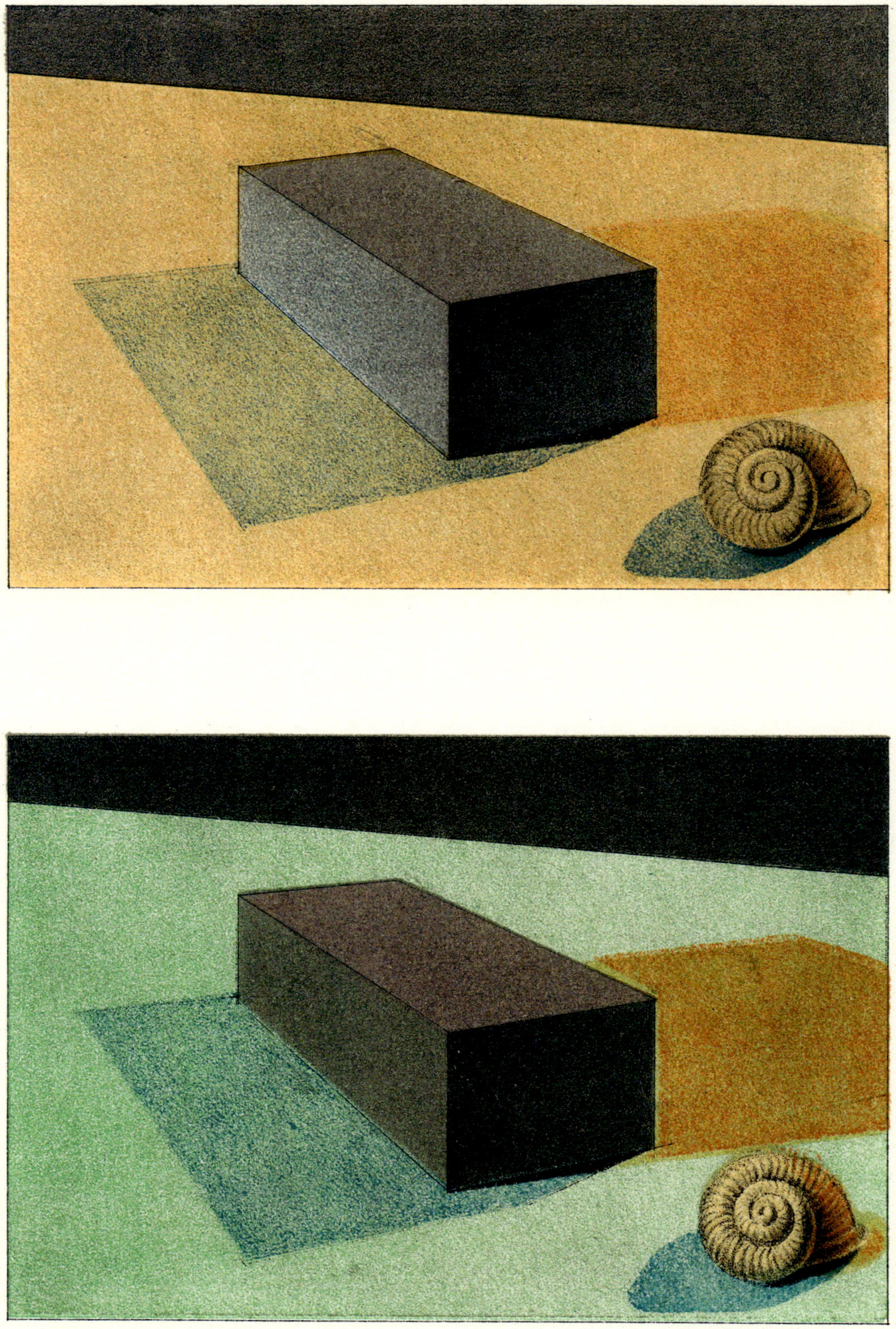

G. Schreibers Farbenlehre　　Leipzig: Verlag von Otto Spamer.

F. A. Brockhaus' Geogr.-artist. Anstalt, Leipzig

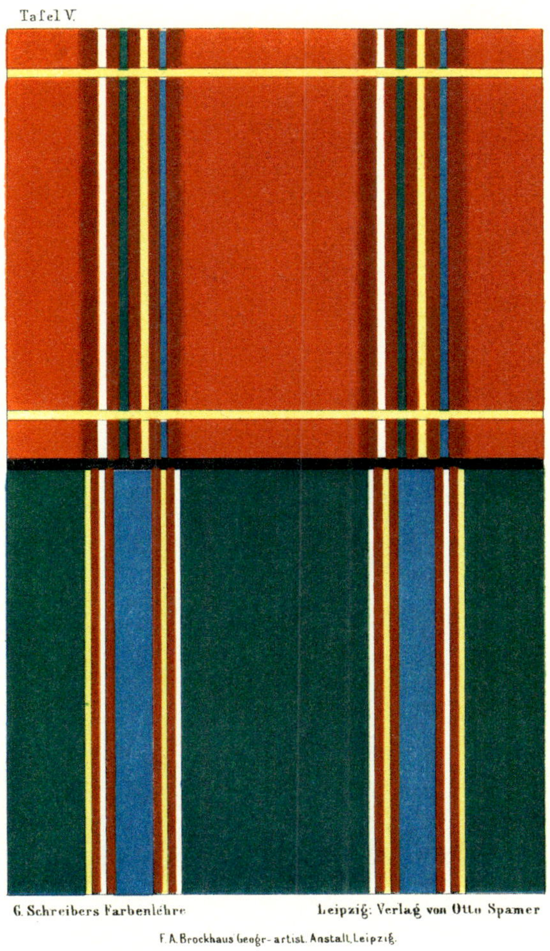

Théorie des couleurs, troisième volume du *Dessin technique*

La contribution de Guido Schreiber à la littérature sur la couleur est rarement examinée, peut-être parce qu'elle se range dans la catégorie des manuels pratiques et pédagogiques, sans revendiquer l'apport de théories ou concepts nouveaux. Pourtant, cette publication de petit format et ses nombreuses chromolithographies mérite d'être regardée de plus près. Elle présente succinctement la théorie des couleurs à un lectorat d'architectes, de peintres, de techniciens et de bâtisseurs, et constitue le troisième volume d'une série de manuels sur l'architecture à usage des élèves et des étudiants, ainsi que pour l'étude à titre privé.

Schreiber est un mathématicien allemand, devenu professeur de géométrie après une carrière militaire. Dans son enseignement et ses recherches, il montre un intérêt tout particulier pour la composition picturale, la diffraction de la lumière, l'esthétique chromatique et les ombres colorées. Tous ses ouvrages, dont celui-ci, sont clairement marqués par l'approche analytique et le ton didactique de Schreiber, et *Farbenlehre* est pour partie une interprétation et une présentation simplifiée des travaux de Goethe sur la couleur (l'auteur copie même la formulation du titre du célèbre ouvrage de Goethe). Par conséquent, le livre de Schreiber peut être vu comme une intéressante étude de cas sur la réception du *Zur Farbenlehre* de Goethe et les applications de celui-ci, à peine plus d'un siècle après sa parution. Les planches de couleurs sont d'une qualité surprenante, surtout si l'on tient compte du fait qu'il s'agit probablement d'une édition relativement bon marché. Les dessins ornementaux sont de toute évidence influencés par le splendide ouvrage sur l'emploi de la couleur dans les arts décoratifs d'Owen Jones (1809–1874), *The Grammar of Ornament* (1856). Jones est d'ailleurs dûment étudié dans un chapitre consacré à la couleur et l'esthétique. Une planche en particulier illustre de façon à la fois simple et remarquable le phénomène des ombres colorées (p. 178). La description visuelle que propose Schreiber des cercles chromatiques et de l'ordre chromatique est impressionnante de clarté, et tire le meilleur parti d'un petit format d'impression et de la reproduction des couleurs par lithographie. La planche 2 est à ce titre particulièrement intéressante car, outre un cercle chromatique assez classique en 12 sections, elle montre une version simplifiée de la sphère des couleurs de Runge (p. 163), segmentée comme une orange et comme aplatie en une forme bidimensionnelle, afin de tenir dans ce petit volume.

Tafel VII.

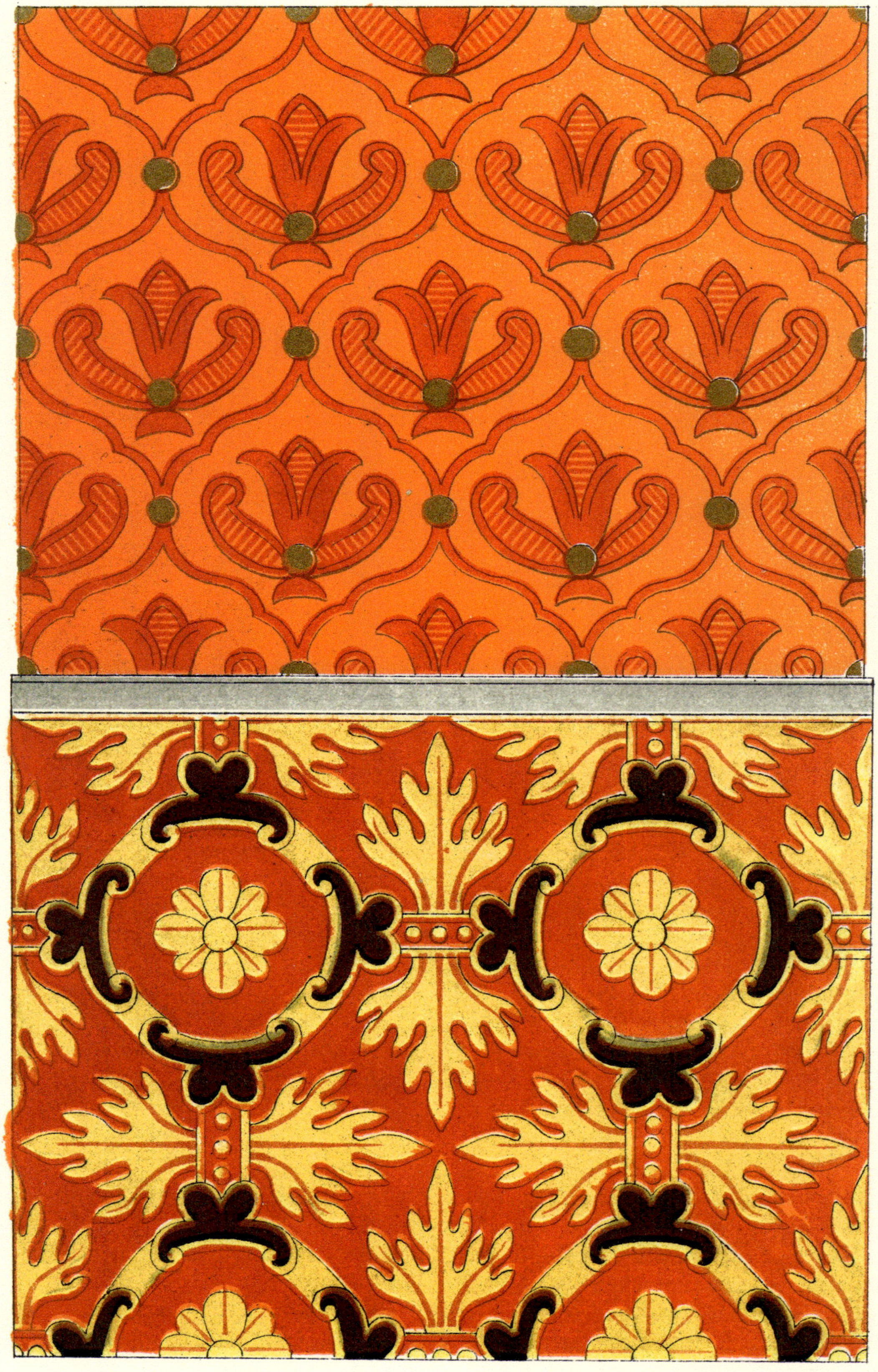

G. Schreibers Farbenlehre   Leipzig: Verlag von Otto Spamer.

F. A. Brockhaus' Geogr.-artist. Anstalt Leipzig

Tafel VIII.

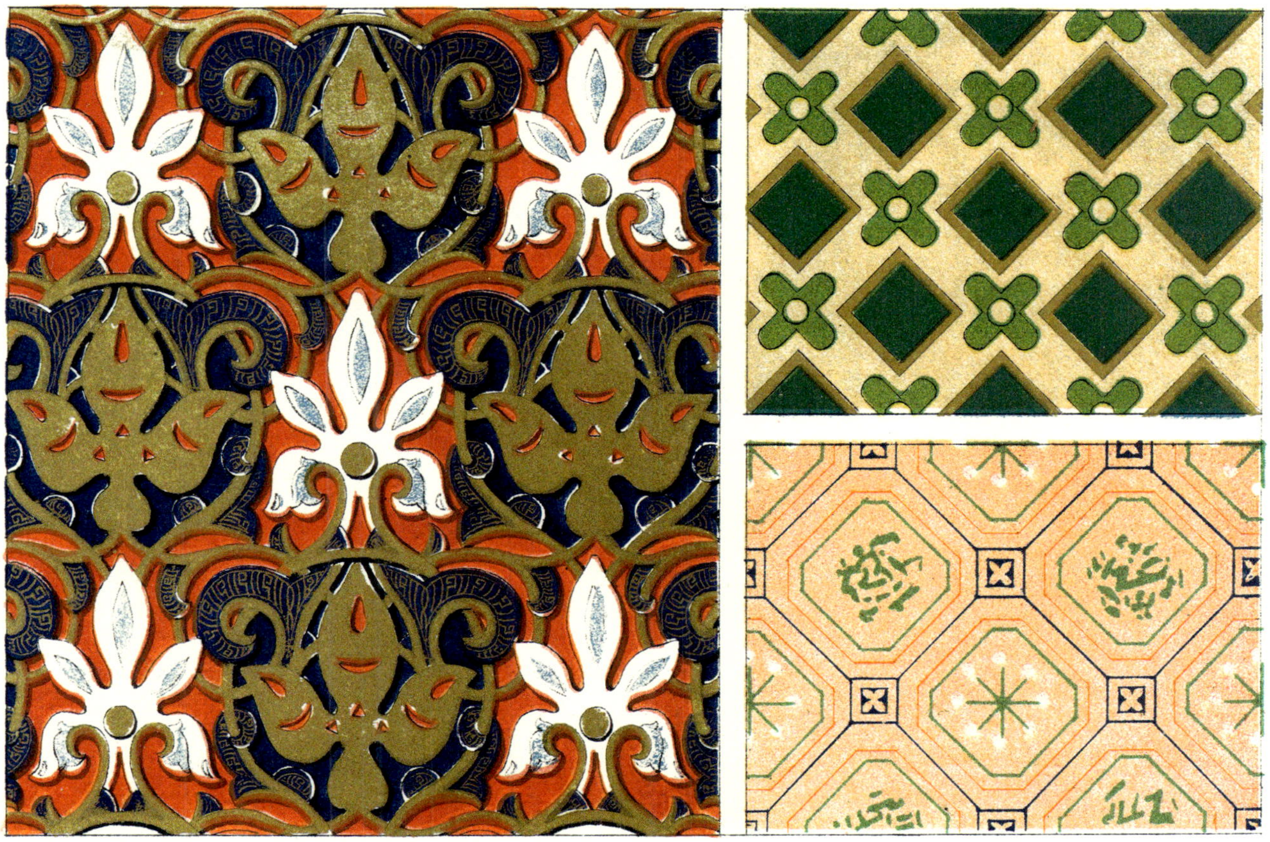

G. Schreibers Farbenlehre

Leipzig: Verlag von Otto Spamer.

F. A. Brockhaus' Geogr.-artist. Anstalt, Leipzig

Teoría del color, parte 3 de *Dibujo técnico*

La contribución de Guido Schreiber a los trabajos sobre el color rara vez se aborda, quizás porque se engloba en la categoría de manuales prácticos y educativos que no pretenden ofrecer nuevas teorías o conceptos. Sin embargo, esta publicación de pequeño formato con muchas láminas cromolitográficas merece una mayor atención, ya que presentaba de manera sucinta la teoría del color a un público que buscaba incluir a arquitectos, pintores, técnicos y constructores. Fue la tercera parte de una serie de libros de texto sobre arquitectura para uso en escuelas y universidades, así como para el estudio independiente.

Schreiber fue un matemático y profesor de geometría alemán con formación militar que en su trabajo como docente e investigador mostró un especial interés por la composición pictórica, la difracción de la luz, la estética del color y las sombras coloreadas. El enfoque analítico y el tono didáctico de Schreiber son evidentes en esta y otras obras publicadas, y su *Farbenlehre* es en parte una interpretación y una presentación simplificada de las obras de Goethe sobre el color (incluso copió el término principal del conocido título de Goethe). En consecuencia, el libro de Schreiber puede verse como un interesante caso de estudio de la recepción que tuvo la teoría del color de Goethe y su aplicación medio siglo después de su publicación. Las láminas en color sorprenden por su alta calidad, en especial si se tiene en cuenta que este libro se publicó como una edición relativamente barata. Las imágenes que muestran diseños ornamentales estaban claramente influenciadas por la magnífica publicación sobre el color en las artes decorativas de Owen Jones (1809–1874), *The Grammar of Ornament* (1856); de hecho, Jones se analiza debidamente en un capítulo sobre el color y la estética. Una lámina ilustra de manera sencilla y muy efectiva el fenómeno de las sombras de colores (pág. 178), mientras que las representaciones de Schreiber de las ruedas de colores y del orden cromático son impresionantes por la claridad de su diseño, que aprovecha del mejor modo posible el pequeño formato de la lámina y la litografía para la reproducción de los colores. La lámina 2 tiene un especial interés porque, aparte de ser una rueda de color con 12 partes bastante estándar, muestra una versión simplificada de la esfera de color de Runge (pág. 163), segmentada como una naranja y aplanada en una forma bidimensional para que encaje en el libro.

Wilhelm von Bezold (1837–1907)

DIE FARBENLEHRE IM HINBLICK AUF KUNST UND KUNSTGEWERBE

Colour Theory in Relation to Art and Art Industry
3 plates, 22.1 x 13.7 cm | 8 ¾ x 5 ⅜ in., Braunschweig, 1874
Munich, Bayerische Staatsbibliothek

Wilhelm von Bezold was a physics professor at the Royal Bavarian Polytechnic School in Munich, whose range of interests included technology, science, architecture and the decorative arts. He is perhaps best known for his work with Ernst Wilhelm von Brücke (1819–1892) during which they discovered the 'Bezold–Brücke effect', the perceptual change in the hue of a colour relative to the intensity of light.

Bezold's substantial work on colour theory presented here reflects the developments in colour and optical science in the later 19th century, in particular the research of James Clerk Maxwell (1831–1879) and Hermann von Helmholtz (1821–1894) who had both recently made significant contributions in the study of trichromatic vision. The chromolithographic plates in Bezold's *Die Farbenlehre* are of a superior quality, as may be seen with plate I which shows the spectrum of sunlight accompanied by the respective spectra of yellow (gamboge) and (Prussian) blue followed by mixtures of the two, as observed through a prism and against a black ground. This explains the black background colour of the plates, which also offsets the colours to dramatic effect. Plate II features a pair of saturated colour wheels, also placed on a black background. In an age before colour photography (Maxwell was working on it at the time, but the world was still many decades away from it being widely available), these plates must have looked quite spectacular. Bezold noted in the book that he himself considered the colour plates to be a success, but at the same time he remarked on the challenges and limitations of colour printing. His colour wheels relate to an innovative three-dimensional colour concept in the form of a cone, based on 10 colours with a greyscale running through its centre, and culminating in black at the top of the cone. A translation of Bezold's book was published just two years later in Boston, with the title *The Theory of Color in Its Relation to Art and Art-Industry*, while a new and revised edition of the original work appeared in Germany in 1921, a testament to the high regard in which Bezold's work on colour was still held at that date.

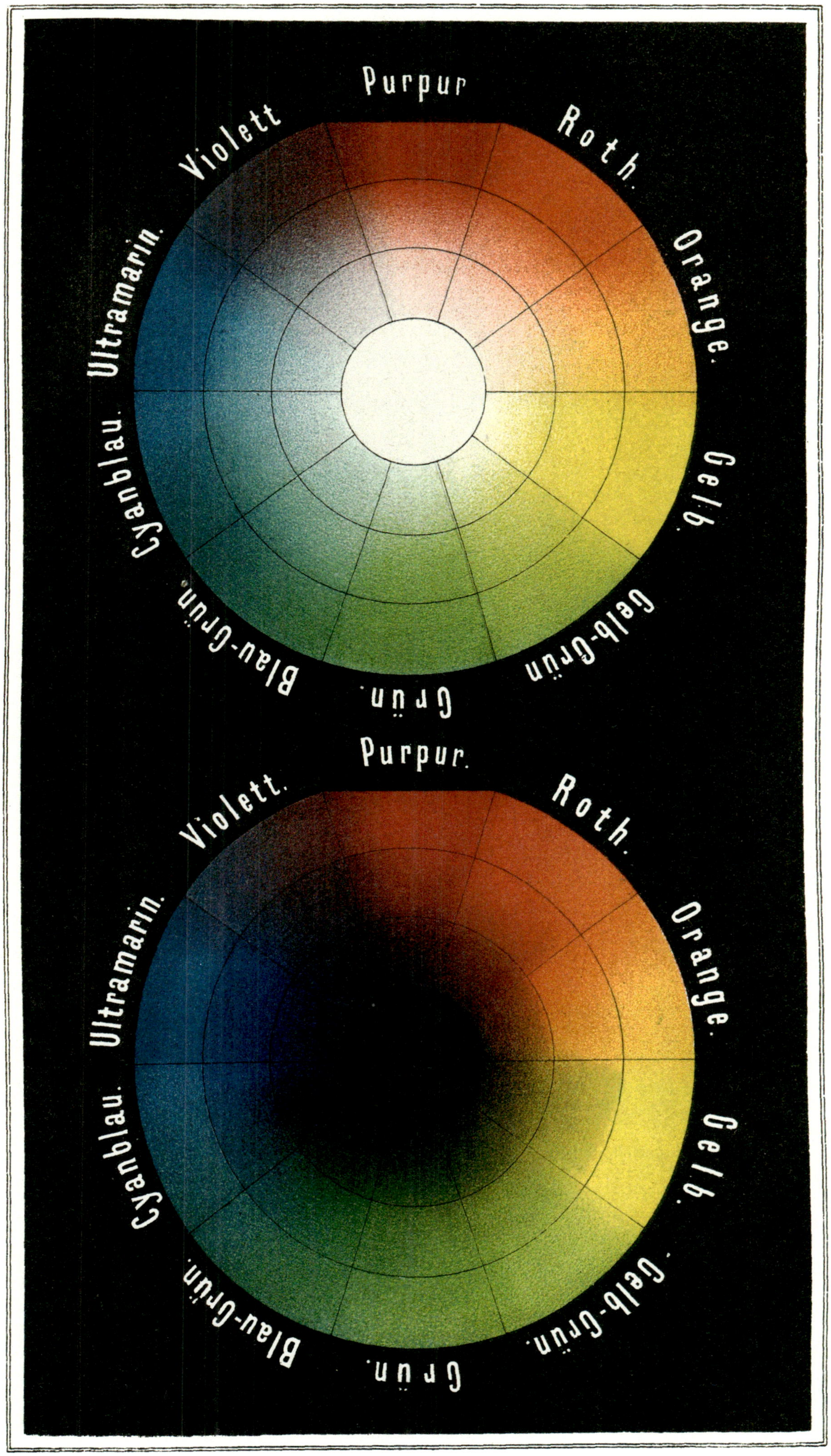

Chromolith. v. A. Schütze, Berlin. Braunschweig, Verlag von G. Westermann.

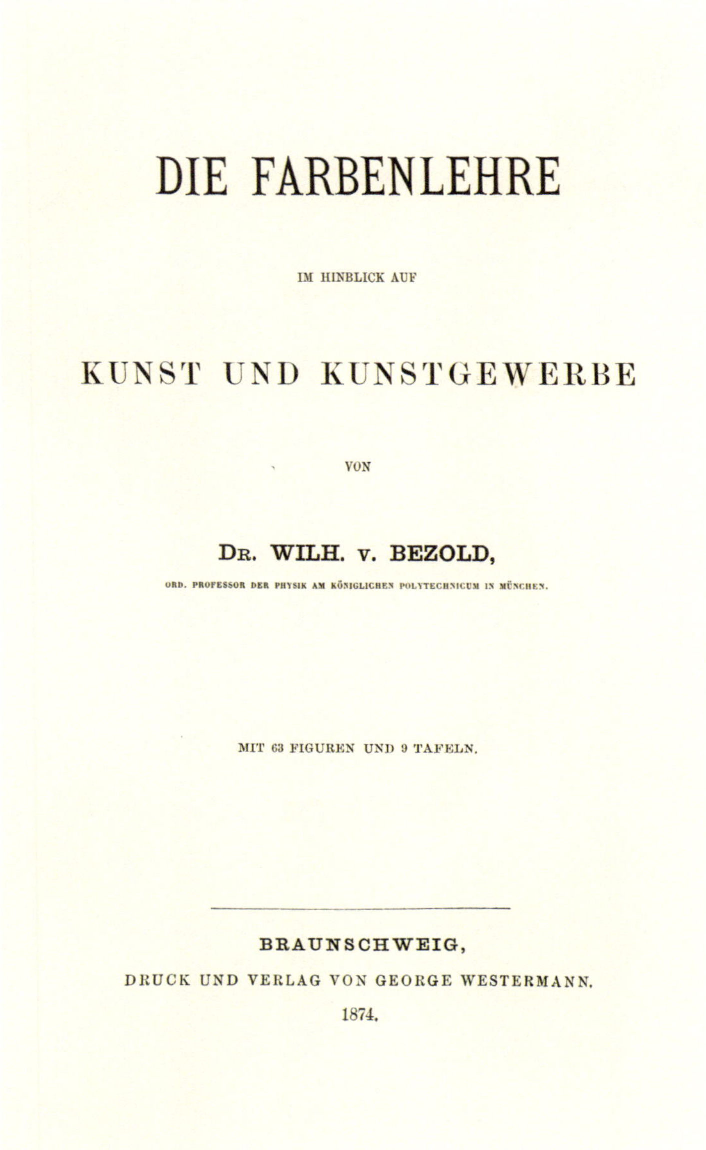

DIE FARBENLEHRE

IM HINBLICK AUF

KUNST UND KUNSTGEWERBE

VON

DR. WILH. v. BEZOLD,

ORD. PROFESSOR DER PHYSIK AM KÖNIGLICHEN POLYTECHNICUM IN MÜNCHEN.

MIT 63 FIGUREN UND 9 TAFELN.

BRAUNSCHWEIG,

DRUCK UND VERLAG VON GEORGE WESTERMANN.

1874.

Die Farbenlehre im Hinblick auf Kunst und Kunstgewerbe

Wilhelm von Bezold war Physikprofessor an der Polytechnischen Schule München und interessierte sich neben Technik und Wissenschaft auch für Architektur und Kunstgewerbe. Bekannt ist er wohl vor allem für seine Zusammenarbeit mit Ernst Wilhelm von Brücke (1819–1892). Gemeinsam entdeckten die beiden Wissenschaftler den nach ihnen benannten Bezold-Brücke-Effekt, die Veränderung eines empfundenen Farbtons in Abhängigkeit von der Intensität des Lichts.

Bezolds hier vorgestelltes bedeutendes Werk zur Farbtheorie spiegelt die Entwicklungen von Farbwissenschaft und Optik im ausgehenden 19. Jahrhundert wider, insbesondere die Arbeiten von James Clerk Maxwell (1831–1879) und Hermann von Helmholtz (1821–1894), die beide kurz zuvor wesentliche Beiträge zur Erforschung des Dreifarbensehens geleistet hatten. Die hohe Qualität der farblithografischen Tafeln in *Die Farbenlehre* zeigt sich gleich in Tafel I, auf der das Sonnenspektrum und die entsprechenden Spektren der Farben Gummigutt (Gelb) und Preußischblau zu sehen sind, gefolgt von Mischungen der beiden Farben, jeweils betrachtet durch ein Prisma vor schwarzem Hintergrund. Dies erklärt die schwarze Hintergrundfarbe der Tafeln und setzt zugleich die Buntfarben auf dramatische Weise dagegen ab. Tafel II zeigt ein Paar gesättigte Farbkreise, ebenfalls auf schwarzem Grund. In der Zeit vor der Farbfotografie (sie wurde etwa zeitgleich von Maxwell entwickelt, doch die Welt war noch Jahrzehnte von ihrer allgemeinen Verfügbarkeit entfernt) müssen diese Farbtafeln spektakulär gewirkt haben. Im Buch merkte Bezold an, dass er selbst sie für einen Erfolg hielt, äußerte sich jedoch auch über die Herausforderungen und Grenzen des Farbdrucks. Seine Farbkreise beziehen sich auf ein innovatives dreidimensionales Farbsystem in Kegelform, basierend auf zehn Farben und einer Grauachse im Zentrum, die an der Kegelspitze in Schwarz gipfelt. Während nur zwei Jahre später in Boston eine englische Übersetzung des Buches mit dem Titel *The Theory of Color in Its Relation to Art and Art-Industry* herauskam, zeugt die 1921 in Deutschland erschienene überarbeitete Auflage des Originalwerks davon, wie angesehen Bezolds Farbtheorie damals immer noch war.

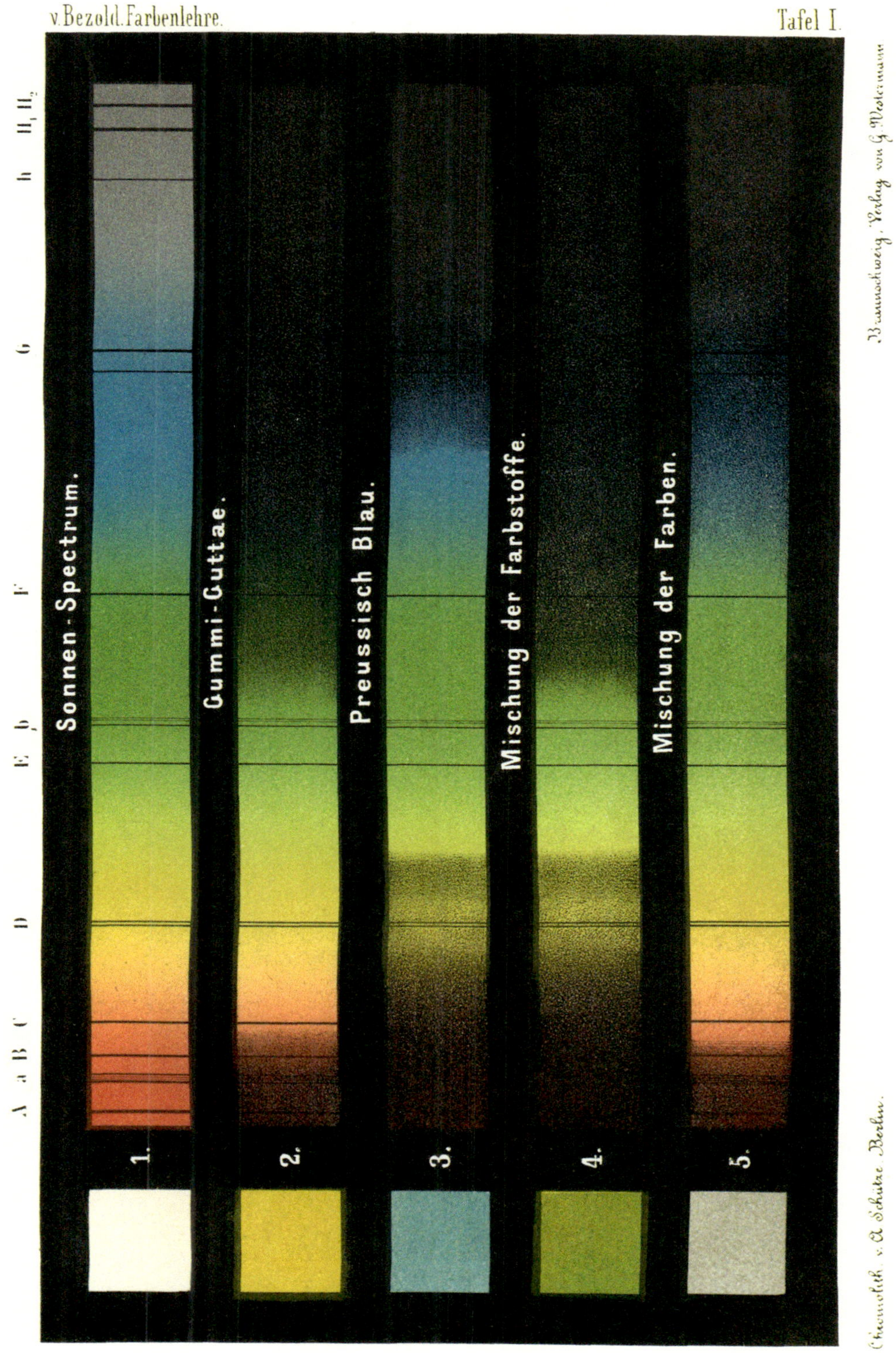
v. Bezold. Farbenlehre.
Tafel I.
Sonnen-Spectrum.
Gummi-Guttae.
Preussisch Blau.
Mischung der Farbstoffe.
Mischung der Farben.
1.
2.
3.
4.
5.

La théorie de la couleur en relation avec l'art et les métiers de l'art

Wilhelm von Bezold est professeur de physique à l'École polytechnique royale de Bavière, à Munich, qui forme tout à la fois à la technologie, à l'architecture, aux sciences et aux arts décoratifs. C'est sans doute pour ses travaux avec Ernst Wilhelm von Brücke (1819–1892) qu'il est le plus connu : ils ont ensemble découvert ce qui sera appelé l'effet Bezold-Brücke, qui désigne le changement de perception des nuances d'une couleur selon l'intensité de la lumière.

Ce qui est présenté ici du travail approfondi de Bezold sur la théorie des couleurs reflète les avancées scientifiques en matière chromatique et optique de la fin du XIXe siècle. Deux autres éminents chercheurs, James Clerk Maxwell et Hermann von Helmholtz (1821–1894), viennent à l'époque d'apporter leurs importantes contributions dans l'étude de la vision trichromatique. Les planches chromolithographiques de *Die Farbenlehre* sont de qualité supérieure, comme on peut l'observer sur la planche I : le spectre de la lumière du soleil y est accompagné des spectres respectifs du jaune (gomme-gutte) et du bleu (de Prusse) qui sont suivis du mélange des deux, tels qu'observés à travers un prisme et sur un fond noir. Ainsi s'explique le noir employé pour le fond, qui renforce l'effet spectaculaire des couleurs des planches. La planche II présente deux cercles de couleurs saturées, également placés sur un fond noir. Ces planches devaient être très impressionnantes à voir en ces temps antérieurs à la photographie en couleurs (Maxwell était déjà en train d'en développer la technique, mais elle ne se répandrait largement dans le monde que bien des décennies plus tard). Bezold note dans son ouvrage qu'il est satisfait des planches en couleurs, mais il souligne dans le même temps les limites et le défi que représente l'impression en couleurs. Ses cercles découlent d'un concept chromatique tridimensionnel novateur : une forme conique comprenant 10 couleurs, avec une échelle de gris structurant le centre, et qui culmine au sommet du cône avec le noir. Une traduction en anglais du livre de Bezold est publiée à peine deux ans plus tard à Boston, sous le titre *The Theory of Color in Its Relation to Art and Art-Industry*. Une nouvelle édition revue et corrigée de l'ouvrage original paraît en Allemagne en 1921, ce qui témoigne de la haute considération dans laquelle le travail de Bezold est tenu à l'époque.

La teoría del color en relación con el arte y la industria del arte

Wilhelm von Bezold era profesor de Física en la Real Escuela Politécnica de Baviera en Múnich. Sus intereses abarcaban la tecnología, la ciencia, la arquitectura y las artes decorativas. Quizás sea más conocido por su trabajo con Ernst Wilhelm von Brücke (1819–1892), con quien descubrió el «efecto Bezold-Brücke», el cambio perceptible en el matiz de un color en función de la intensidad de la luz.

El importante trabajo de Bezold sobre la teoría del color que se presenta aquí refleja los desarrollos en la ciencia óptica y del color a finales del siglo XIX, en particular la investigación de James Clerk Maxwell y Hermann von Helmholtz (1821–1894), quienes recientemente habían hecho contribuciones significativas en el estudio de la visión tricromática. Las láminas cromolitográficas de *Die Farbenlehre* son de una calidad superior, como puede verse en la lámina I, que muestra el espectro de la luz solar acompañado de los respectivos espectros del amarillo (gutagamba) y del azul (prusiano) seguidos de mezclas de ambos, observados a través de un prisma y sobre un fondo negro que hace que los colores contrasten con un efecto dramático. La lámina II presenta un par de ruedas de colores saturados, también colocadas sobre un fondo negro. En una era anterior a la fotografía en color (Maxwell la estaba desarrollando en ese momento, pero el mundo todavía estaba a muchas décadas de que estuviera ampliamente disponible), estas láminas debieron de tener un aspecto bastante espectacular. Bezold señaló en el libro que él mismo las consideraba un éxito, pero también comentó los desafíos y las limitaciones de la impresión en color. Sus ruedas se relacionan con un innovador concepto de color tridimensional en forma de cono, basado en 10 colores y con una escala de grises que recorre su centro y culmina en el negro de la cúspide. Una traducción del libro de Bezold se publicó solo dos años después en Boston con el título *The Theory of Color in Its Relation to Art and Art-Industry*, mientras que una edición nueva y revisada del trabajo original apareció en Alemania en 1921, prueba de la alta estima que todavía se tenía de él en esas fechas.

v. Bezold, Farbenlehre. Tafel III.

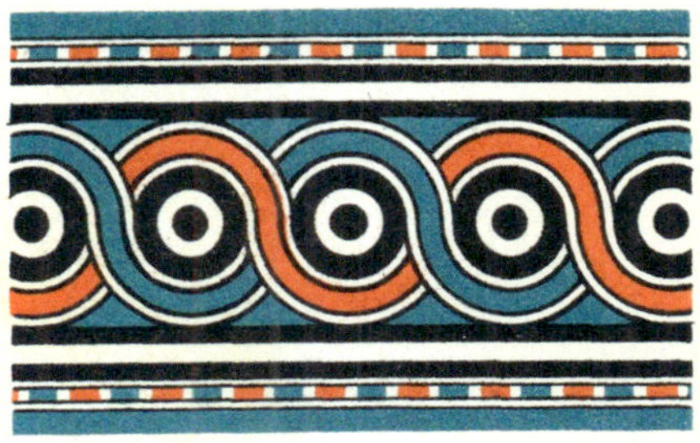

Fig. 1.

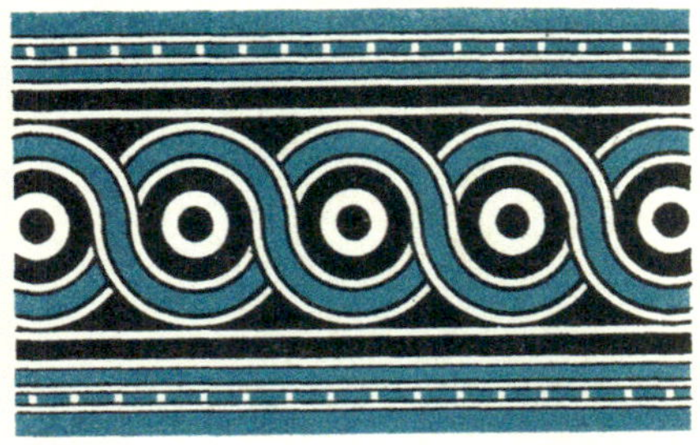

Fig. 2.

Fig. 3.

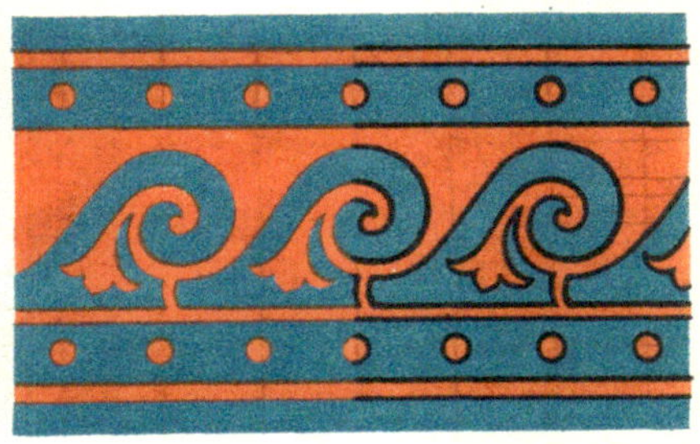

Fig. 4.

Fig. 5.

Fig. 6.

Fig. 7.

Fig. 8.

Chromolith. v. A. Schütze Berlin. Braunschweig, Verlag von G. Westermann.

Edwin D. Babbitt (1828–1905)

THE PRINCIPLES OF LIGHT AND COLOR

4 plates, 24.2 x 16.5 cm | 9 ½ x 6 ½ in., New York, 1878
London, Royal College of Art, Colour Reference Library

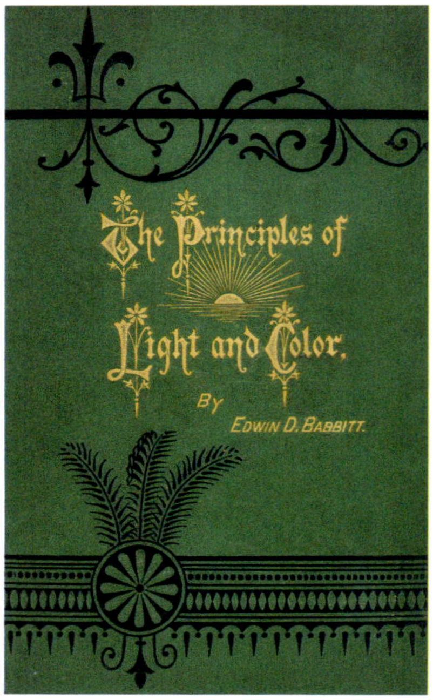

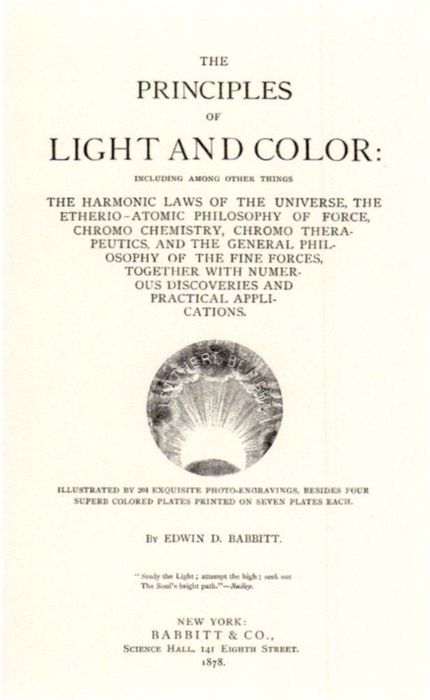

Edwin Dwight Babbitt was a New York-born teacher and stationery maker, who decided to pursue a radical career change in mid-life. Seemingly without any formal training he became a self-declared physician and mystic healer, being listed in a directory from the 1870s as a "magnetist and author of a health guide". In 1878 he published this lengthy work on the healing qualities of colour and light, which marked him as one of the pioneers of the alternative medicine method of chromotherapy. Now largely discredited as pseudoscience, the purported healing powers of colour (always in connection with light) had been investigated and discussed since earlier historical times, and found a receptive audience in the late 19th century, an age characterised by quackery and various misconceptions in the world of medicine. Babbitt was strongly criticised by several figures during his later career, but the fascination with the "elaborate and elegant (book by) this bold speculator", as one critic put it, was evident, not least when it was translated into a number of other languages.

The 200,000 words of Babbitt's text were embellished with 204 photoengravings and four coloured plates prepared by the New York printer John Fahnestock, who received a special acknowledgement from the author for creating illustrations "which for beauty I have not seen surpassed on either side of the ocean". One of the plates features a 14-part colour wheel in the shape of a stylised flower-head that identifies the qualities of the different colours in categories marked around the circumference. In keeping with Babbitt's belief in the therapeutic effect of individual colours he chose a pale sky-blue paper, since he considered it to be handsome and "soothing to the nerves of the eye". Among the objects he devised to assist with colour therapy were Chromo-lenses, double convex lenses made of different coloured glass that were hollow and could be filled with water, which would then focus the colour into a ray and transfer its healing powers to the individual; and the Chromolume, a panel of stained-glass with 16 sections of coloured panes that could be placed in sunlight so that the patient could bathe in the healing, coloured light. Babbitt was also a businessman, and his book included ordering information and price lists for these and other items. For all its contested content, as, for example, Babbitt's diagram of the imagined shape of an atom or his claim that coloured light could be an alternative for anaesthetics, the book is a fascinating snapshot of early ideas about colour psychology, psychotherapy and the effects of colour on mood, especially in architectural spaces.

BABBITT'S PRINCIPLES OF LIGHT AND COLOR.—PLATE I.

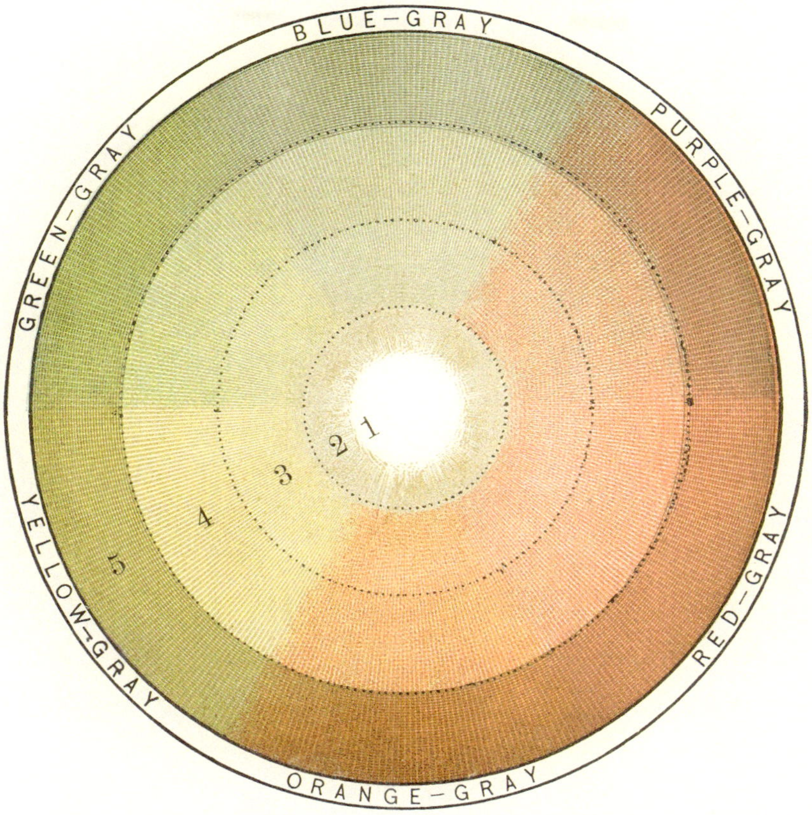

VARIOUS SHADES OF GRAY ARRANGED IN ANALOGICAL HARMONY.

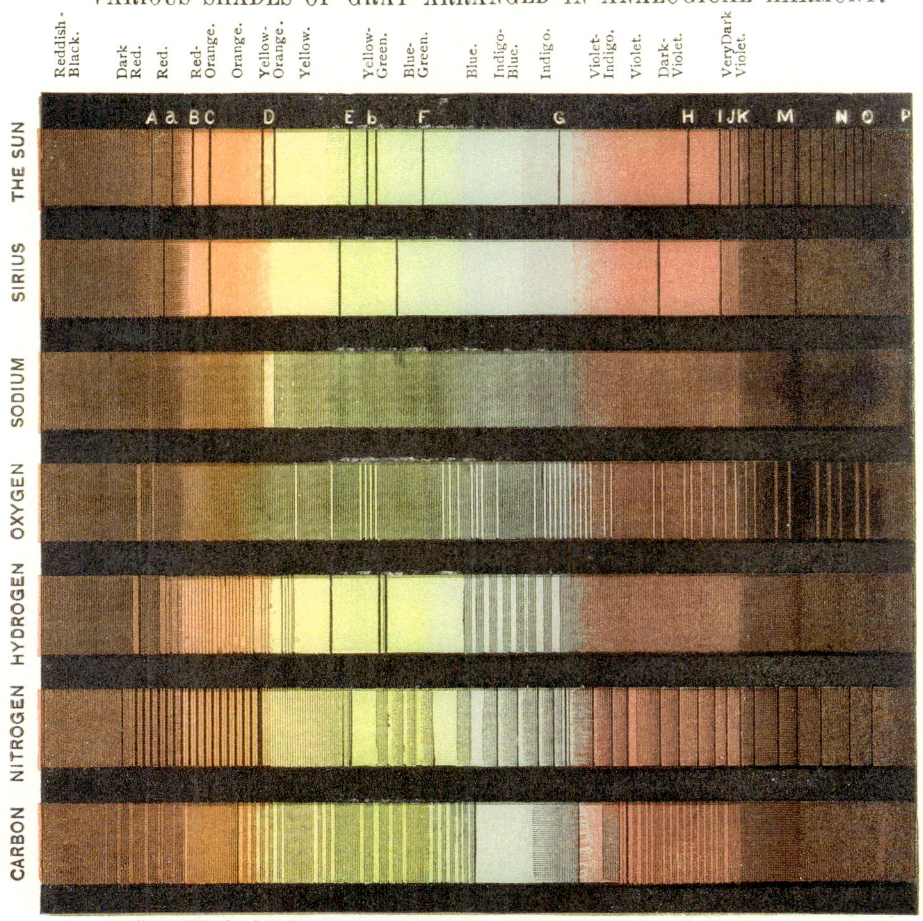

SPECTRA OF THE SUN, SIRIUS, AND SEVERAL ELEMENTS.

Die Grundsätze von Licht und Farbe

In der Mitte seines Lebens beschloss der in New York geborene Lehrer und Schreibwarenhersteller Edwin Dwight Babbitt, sich beruflich radikal zu verändern, und erklärte sich, anscheinend ohne jede formale Ausbildung, zum Arzt und mystischen Heiler. In einem Verzeichnis aus den 1870er-Jahren wird er als „Magnetist und Autor eines Gesundheitsratgebers" geführt. 1878 veröffentlichte Babbitt dieses seitenstarke Werk über die heilenden Eigenschaften von Farbe und Licht, das ihn in der alternativen Medizin zum Pionier der Farbtherapie machte. Die angeblichen Heilkräfte von Farbe (stets im Zusammenhang mit Licht), die heute weithin als pseudowissenschaftlich angezweifelt werden, waren in der Geschichte schon seit Langem erforscht und diskutiert worden. Im späten 19. Jahrhundert, einer Zeit der Quacksalberei und allerlei falscher Vorstellungen in der Welt der Medizin, fanden sie ein aufnahmebereites Publikum. Babbitt wurde im weiteren Verlauf seiner Karriere von mehreren Seiten scharf kritisiert, doch die Begeisterung für das „ausführliche und elegante [Buch] dieses kühnen Spekulanten", wie ein Kritiker es ausdrückte, war offensichtlich – nicht zuletzt, als es in zahlreiche andere Sprachen übersetzt wurde.

Die zweihunderttausend Worte des Buchtextes waren mit 204 Fotogravüren und vier Farbtafeln ausgeschmückt. Letztere hatte der New Yorker Drucker John Fahnestock produziert, der von Babbitt mit einem besonderen Dank bedacht wurde; er habe Abbildungen geschaffen, „die ich an Schönheit auf beiden Seiten des Ozeans niemals übertroffen sah". Eine Tafel zeigt einen 14-teiligen Farbkreis in Form einer stilisierten Blüte, auf deren äußerem Ring die Eigenschaften der unterschiedlichen Farben vermerkt sind. Seinem Glauben an die therapeutische Wirkung einzelner Farben entsprechend, wählte Babbitt ein blasshimmelblaues Papier, das aus seiner Sicht attraktiv und „beruhigend für die Nerven des Auges" war. Als Hilfsmittel für die Farbtherapie entwickelte er unter anderem die „Chromo-Lenses" (Farblinse), eine bikonvexe hohle Linse aus farbigem Glas, die sich mit Wasser füllen ließ, um die Farbe zu einem Strahl zu bündeln und ihre Heilkräfte auf den Anwender zu übertragen. Eine weitere Erfindung von ihm war das „Chromolume" (Farblicht), eine Art Buntglasfenster im Holzrahmen mit 16 verschiedenen Farbzonen, das ins Sonnenlicht gestellt werden konnte, um die Patienten im heilenden Farblicht baden zu lassen. Da Babbitt auch Geschäftsmann war, enthielt sein Buch Bestellangaben und Preislisten zu diesen und anderen Artikeln. Trotz aller umstrittenen Inhalte, etwa Babbitts Diagramm vom vermeintlichen Aussehen eines Atoms oder seiner Behauptung, farbiges Licht könnte eine Alternative zu Betäubungsmitteln darstellen, ist das Buch eine faszinierende Momentaufnahme früher Vorstellungen von Farbpsychologie, Farbpsychotherapie und der Wirkung von Farben auf die menschliche Stimmung, insbesondere im architektonischen Raum.

Les principes de la lumière et de la couleur

Edwin Dwight Babbitt, enseignant natif de New York et papetier, prend un tournant radical au milieu de sa vie. Apparemment sans la moindre formation, il s'autoproclame médecin et guérisseur mystique, et figure dans l'annuaire, à partir des années 1870, en tant que « magnétiseur et auteur d'un guide pour la santé ». En 1878, il fait paraître ce long ouvrage sur les propriétés curatives de la couleur et de la lumière, qui le place parmi les pionniers de la méthode de soin alternative appelée chromothérapie, depuis qualifiée de pseudoscience et donc discréditée. Les prétendus pouvoirs guérisseurs de la couleur (toujours en relation avec la lumière), qui faisaient déjà de très longue date l'objet de maints recherches et débats, trouvent un fort écho auprès du public de la fin du XIXe siècle, une époque qui voit croître le charlatanisme et nombre d'idées farfelues dans le domaine médical. Babbitt est sévèrement critiqué par plusieurs figures importantes à la fin de sa carrière, mais l'ouvrage « élaboré et élégant [de] cet impudent spéculateur », selon les mots d'un contempteur, opère une évidente fascination, y compris lorsqu'il est traduit dans de nombreuses langues.

Le texte de Babbitt, fort de 200 000 mots, est orné de 204 photogravures et quatre planches en couleurs, préparées par l'imprimeur new-yorkais John Fahnestock. Pour l'occasion, l'auteur remercie tout particulièrement celui-ci pour la création des illustrations « dont [il] n'a rien vu de comparable en beauté d'un côté comme de l'autre de l'océan ». L'une des planches représente un cercle chromatique en 14 parties, sous forme de fleur stylisée, et identifie les qualités des différentes couleurs par des catégories inscrites sur tout le pourtour. Croyant fermement à l'effet thérapeutique de chaque couleur prise individuellement, Babbitt a choisi un papier bleu ciel clair, qu'il tient pour beau et « apaisant pour les nerfs des yeux ». Il est également le concepteur de plusieurs objets employés pour pratiquer la thérapie par la couleur : ses « Chromo-lenses » (lunettes chromatiques) sont des lunettes à doubles verres convexes, taillés dans du verre coloré et creux, et que l'on remplit d'eau. Ainsi, la couleur se focalise en un rayon qui transmet son pouvoir guérisseur à la personne. Son « Chromolume » est un vitrail composé de 16 sections de verre coloré et que l'on place face aux rayons du soleil, de façon que la personne soit baignée par la lumière colorée et curative. Babbitt est par ailleurs un homme d'affaires : son volume inclut les renseignements permettant de passer commande et la liste des tarifs pour tous les objets et articles. L'ouvrage comprend encore d'autres éléments fortement contestés, comme le schéma imaginé par Babbitt pour représenter la forme d'un atome, ou encore l'affirmation selon laquelle la lumière colorée serait une alternative possible aux anesthésiques. Cependant, il constitue un aperçu fascinant des premières idées en matière de psychologie des couleurs,

BABBITT'S PRINCIPLES OF LIGHT AND COLOR.—Plate II

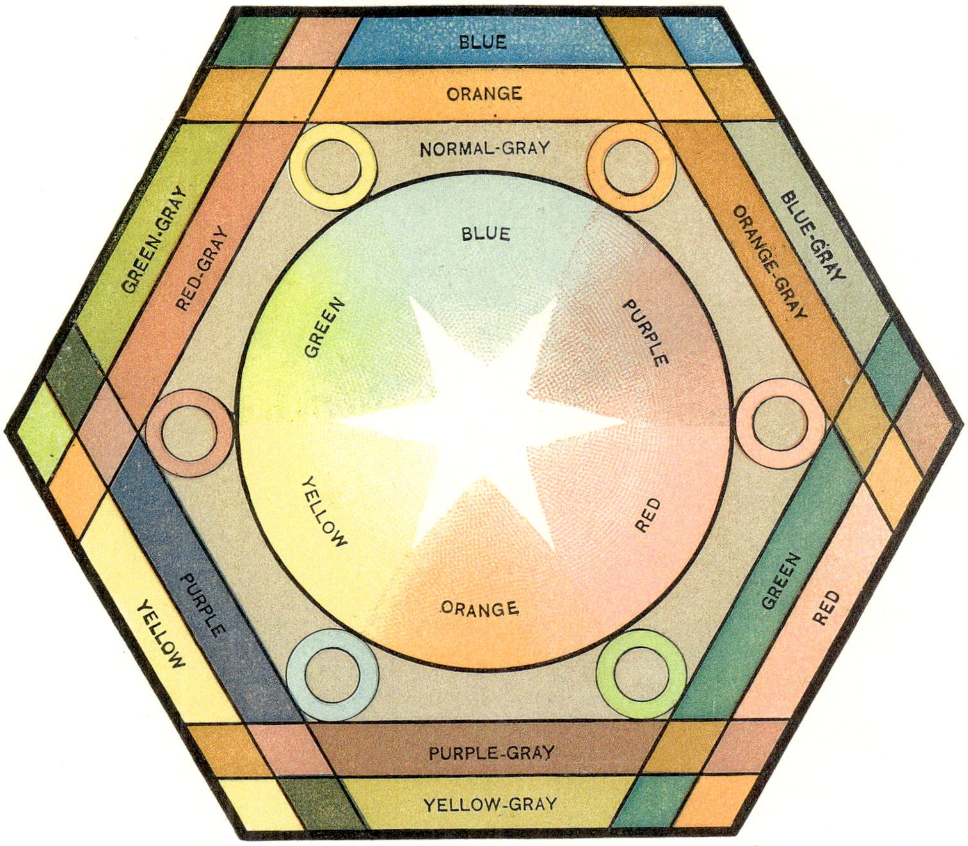

CHROMATIC HARMONY OF GRADATION AND CONTRAST.

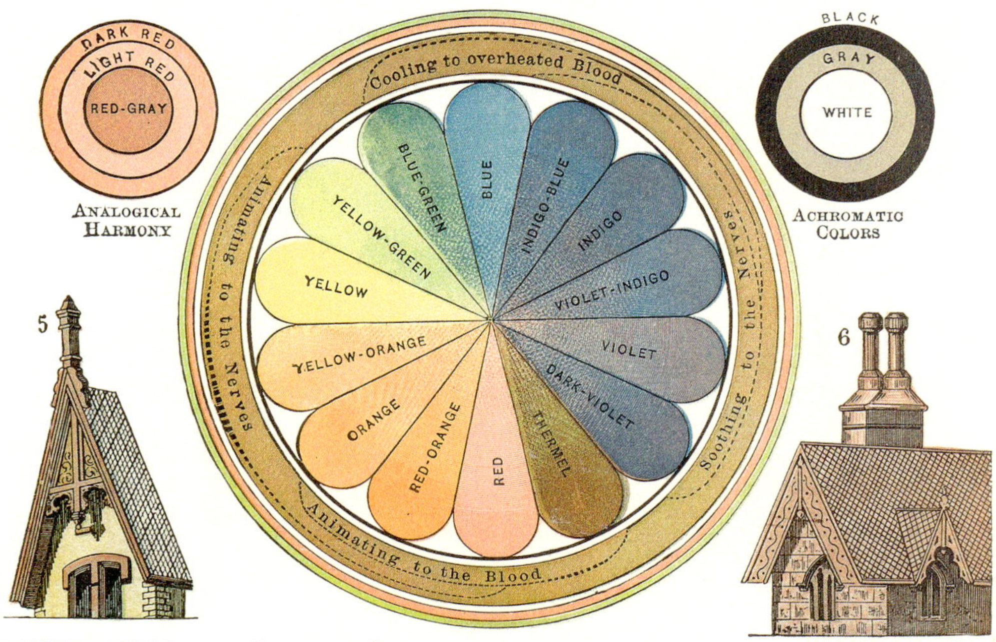

CONTRASTING HARMONY

Chromatic Colors placed opposite those which form a Chemical Affinity with them

ANALOGICAL HARMONY

In the above elaborate combination of colors the artist has found it impossible to get every feature accurate, although he has many beautiful and pure tints. The grays on both plates I. and II. are not sufficiently subdued, the chromatic colors standing out too brilliantly, the red, for instance, in the house, fig. 6, being too strong, etc. For description of plates see pp. 63, 65, 66, 69, 71, etc. The spectra are described on p. 217.

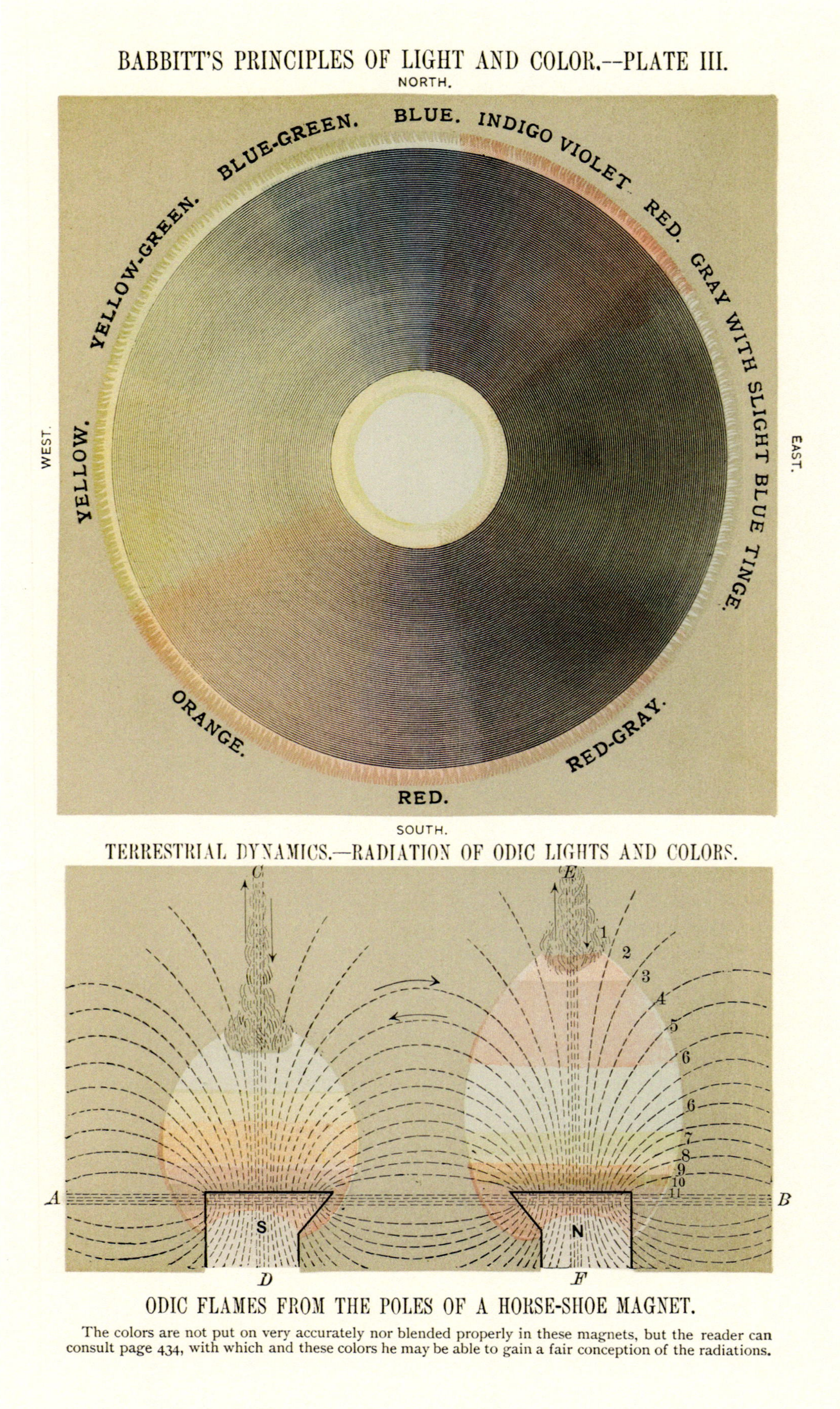

de psychothérapie par les couleurs et d'effets de la couleur sur l'humeur, notamment dans la conception des espaces architecturaux.

Los principios de la luz y del color

Edwin Dwight Babbitt fue un maestro y papelero nacido en Nueva York que optó por un cambio radical de carrera a mediana edad, cuando se autoproclamó médico y sanador místico pese a no haber seguido aparentemente ningún tipo de educación formal. En un directorio de la década de 1870 aparece como «magnetista y autor de una guía de salud». En 1878 publicó este extenso trabajo sobre las cualidades curativas del color y la luz, que lo convirtió en uno de los pioneros del método de la cromoterapia, un tratamiento de la medicina alternativa hoy desacreditado en gran medida como seudociencia. Los supuestos poderes curativos del color (siempre en conexión con la luz) ya se habían investigado y discutido en el pasado, y encontraron un público receptivo a finales del siglo XIX, una época caracterizada por la charlatanería y los malentendidos en el mundo de la medicina. Babbitt fue muy criticado por varias figuras durante esta etapa de su carrera, pero la fascinación por el «elaborado y elegante [libro de] este audaz especulador», como lo describió un crítico, fue evidente, sobre todo tras su traducción a varios idiomas.

Las 200 000 palabras del texto de Babbitt iban acompañadas de 204 fotograbados y cuatro láminas en color elaboradas por el impresor neoyorquino John Fahnestock, quien recibió un reconocimiento especial del autor por crear unas ilustraciones «cuya belleza no he visto superada a ningún lado del océano». Una de las láminas muestra una rueda de color de 14 partes con forma de flor estilizada que identifica las cualidades ligadas a los diferentes colores por medio de unas categorías que se indican a su alrededor. En línea con su creencia sobre el efecto terapéutico de los colores, Babbitt eligió un papel azul cielo pálido, ya que lo consideraba hermoso y «relajante para los nervios de la vista». Entre los objetos que ideó para ayudar con la terapia del color estaban las «Chromolenses», lentes dobles convexas hechas de vidrios de diferentes colores que estaban huecas y podían llenarse con agua para concentrar la luz en un rayo y transferir los poderes curativos del color; y el «Chromolume», un panel con 16 secciones de vidrios de colores que se exponían al sol para que los pacientes tomaran baños de luz coloreada curativa. Babbitt también era un hombre de negocios y su libro incluía información sobre pedidos y listas de precios para estos y otros artículos. A pesar de su contenido tan cuestionado (por ejemplo, el diagrama con la forma imaginada de un átomo o la afirmación de que la luz coloreada podía ser una alternativa a los anestésicos), el libro es una instantánea fascinante de las primeras ideas sobre la psicología del color, la psicoterapia y los efectos del color en el estado de ánimo, en especial en espacios arquitectónicos.

BABBITT'S PRINCIPLES OF LIGHT AND COLOR.—PLATE IV.

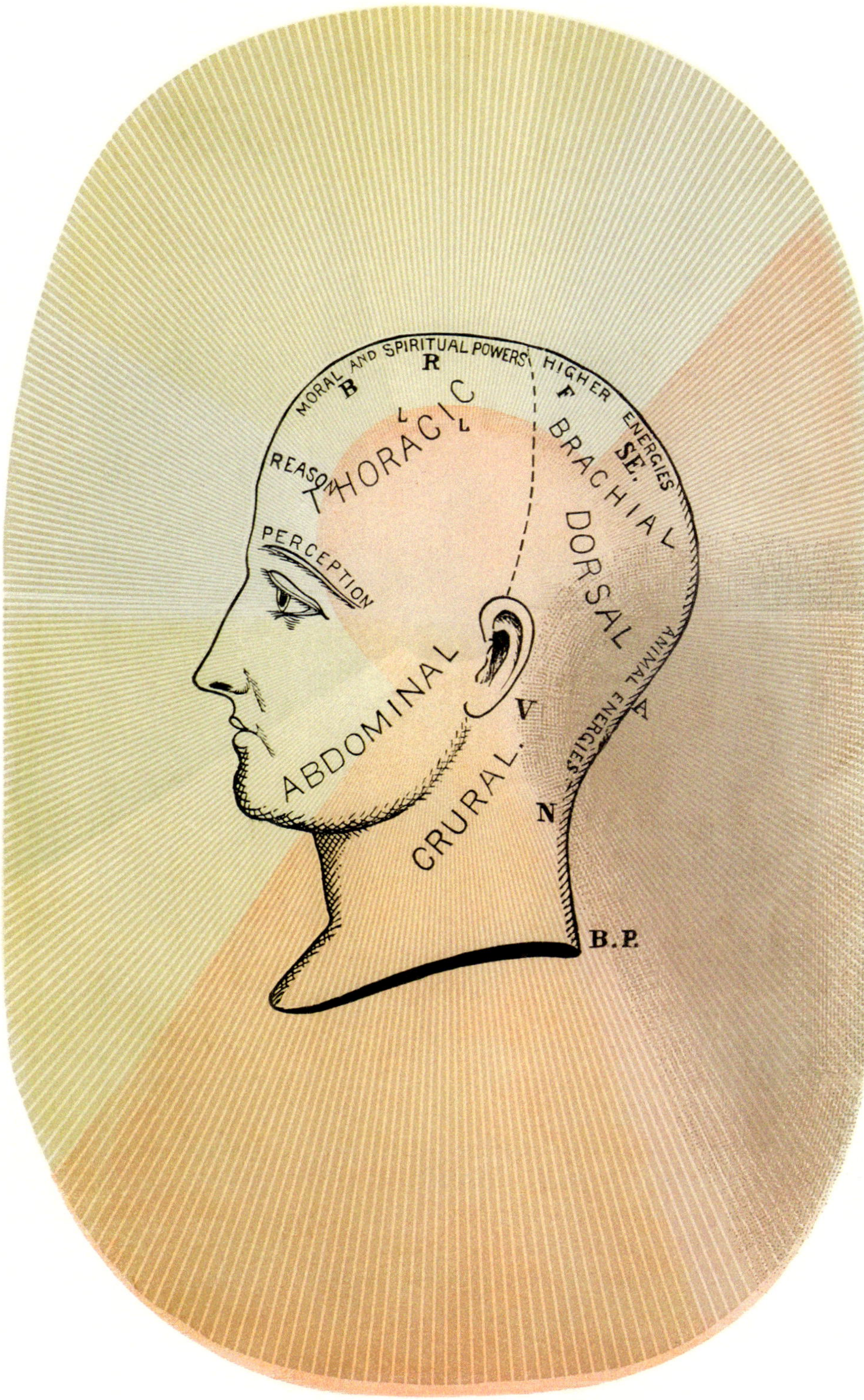

PSYCHIC LIGHTS AND COLORS.

On the lower face of the above the artist has placed the green and yellow too low. The yellow should come over the mouth, then a slight orange merging into a red at the chin, which continues all the way to the occiput, at which last point it assumes a more muddy cast.

Robert Steinheil (1863–1944)

LA REPRODUCTION DES COULEURS

par la superposition des trois couleurs simples

The Reproduction of Colours

19 plates, 36 x 29 cm | 14 ⅛ x 11 ⅜ in., Paris, 1896
Los Angeles, Getty Research Institute

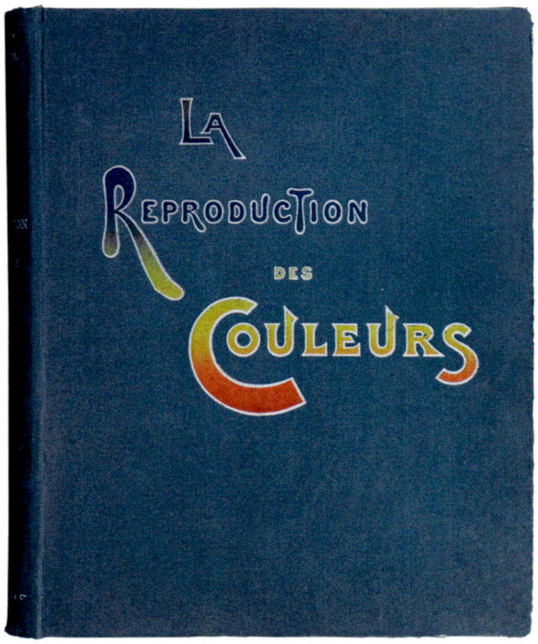

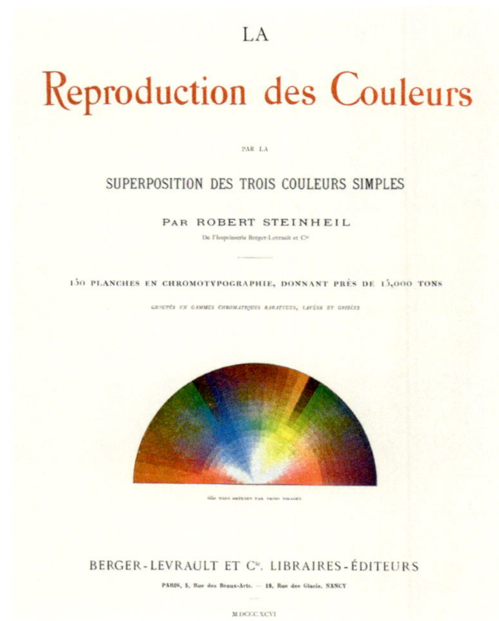

Robert Steinheil's book is a masterpiece of late 19th-century decorative book design and illustration, and as such it is also self-referential, for this work is a celebration of the technical possibilities offered by halftone colour printing in the 1890s. The full title, *La Reproduction des Couleurs par la superposition des trois couleurs simples*, makes clear that the book's purpose is to demonstrate how to reproduce colour by overlaying the "simple" chromatic primaries yellow, red and blue. The cover is typical of aesthetic *fin-de-siècle* style, with its elaborate title graphics on the front board coloured in prismatic shading. The title-page gives a further taste of the quality of the illustrations that follow, with its decorative luminous semi-circle of 660 individual tints, illustrating the principles of simultaneous colour contrasts.

The book was printed by the historic French firm Berger-Levrault et Cie, and comprises 43 pages of letterpress main text, including a glossary of colour terms, a taxonomy of colour (with reference to Chevreul, Lacouture and others as influences) together with alphabetical and numerical tables of all the "nuances" of colour shown in the illustrations. The text is followed by an "Atlas" of 150 colour plates, printed on three types of paper in different batches of the copies, showing the series of colour combinations. These mostly take the form of colour wheels with sectors divided into 10 gradations that illustrate the printing process and provide exact proportions for each nuance or shade, identified by three numbers. There are 16 plates showing colour overlays in numbered 10 x 10 grid systems, while the final colour plate is an assemblage of several overlays in different shapes and patterns, including the semi-circular image from the title-page repeated at the centre. In total, the plates show almost 15,000 individual colours. This is a highly ambitious book about colour printing, produced at the highest level of expertise, and is a magnificent testament to the technological achievements of the industrial age. The print run must have been small, as it is now one of the rarest of historical books on colour.

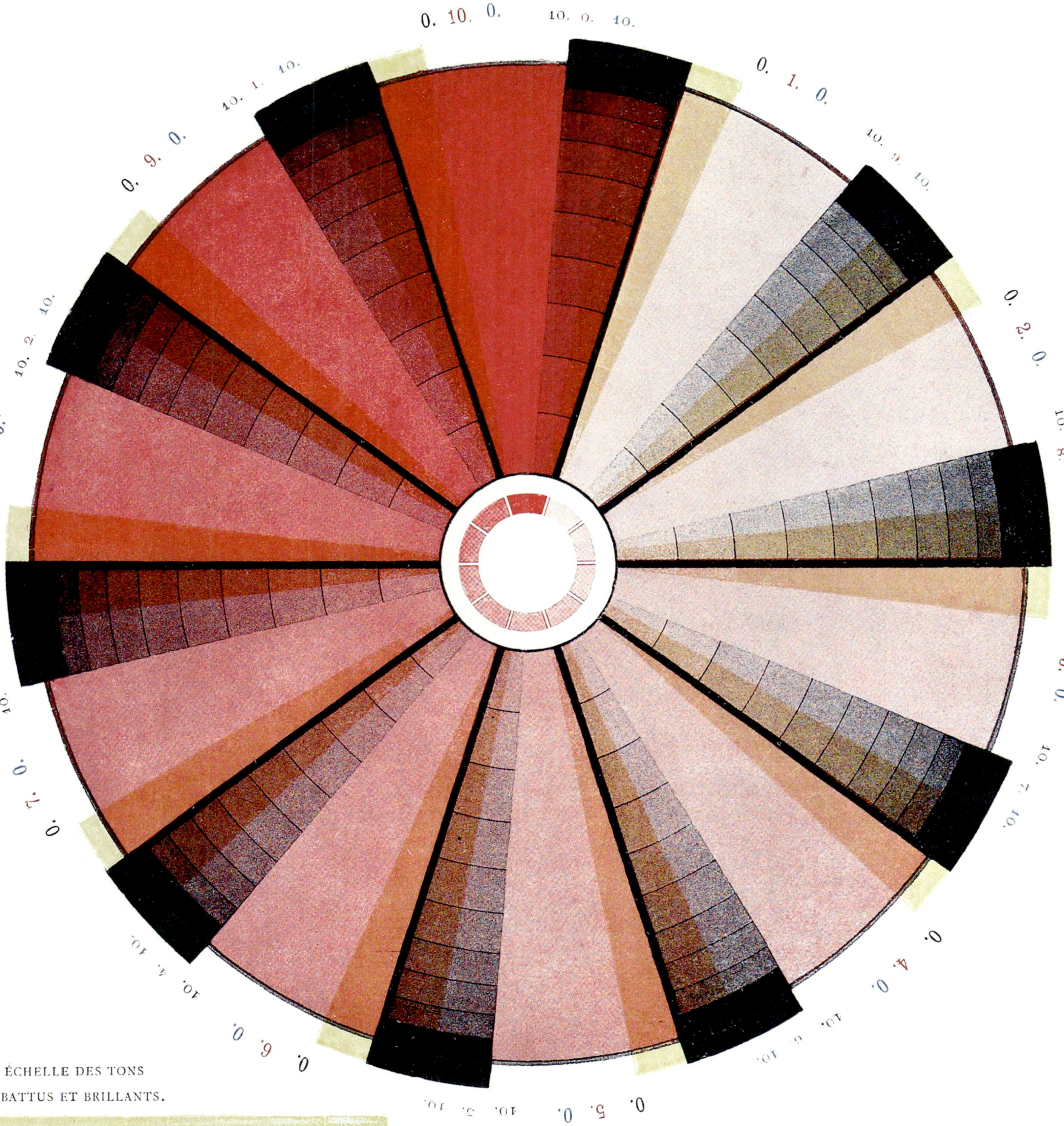

ÉCHELLE DES TONS
RABATTUS ET BRILLANTS.

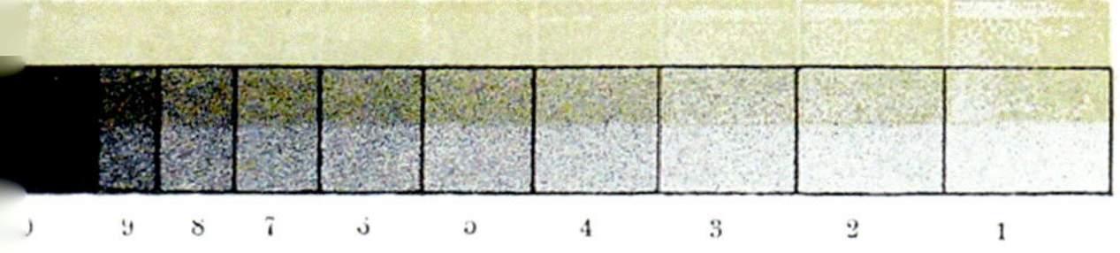

N. B. Les chiffres en noir se rapportent aux tons jaunes.

LE ROUGE.

Ses tons lavés et leurs gammes.

Pl. 3

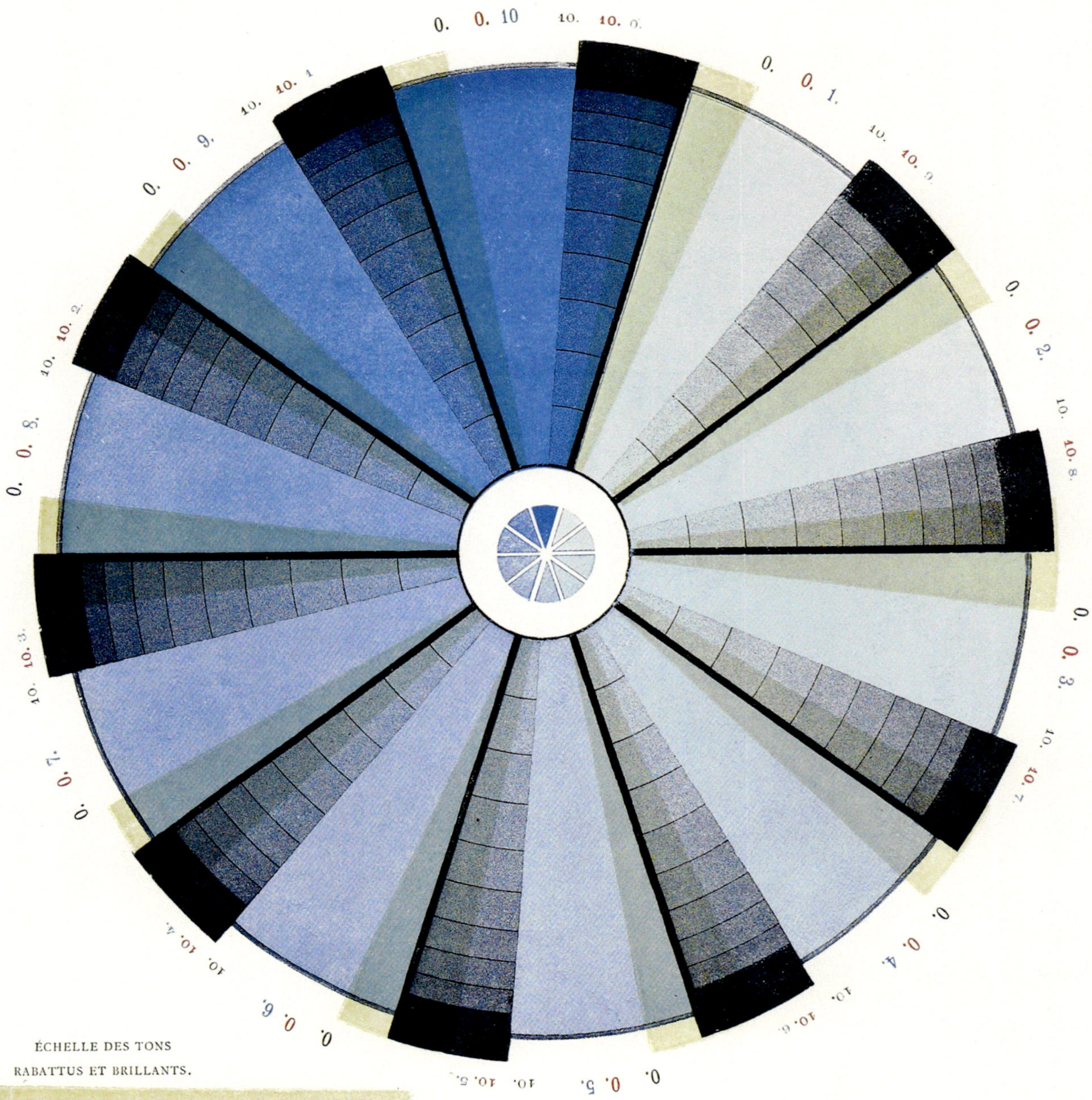

ÉCHELLE DES TONS
RABATTUS ET BRILLANTS.

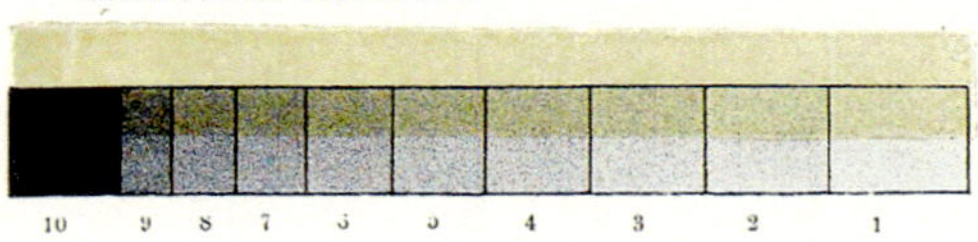

N. B. Les chiffres en noir se rapportent aux tons jaunes.

LE BLEU.

Ses tons lavés et leurs gammes.

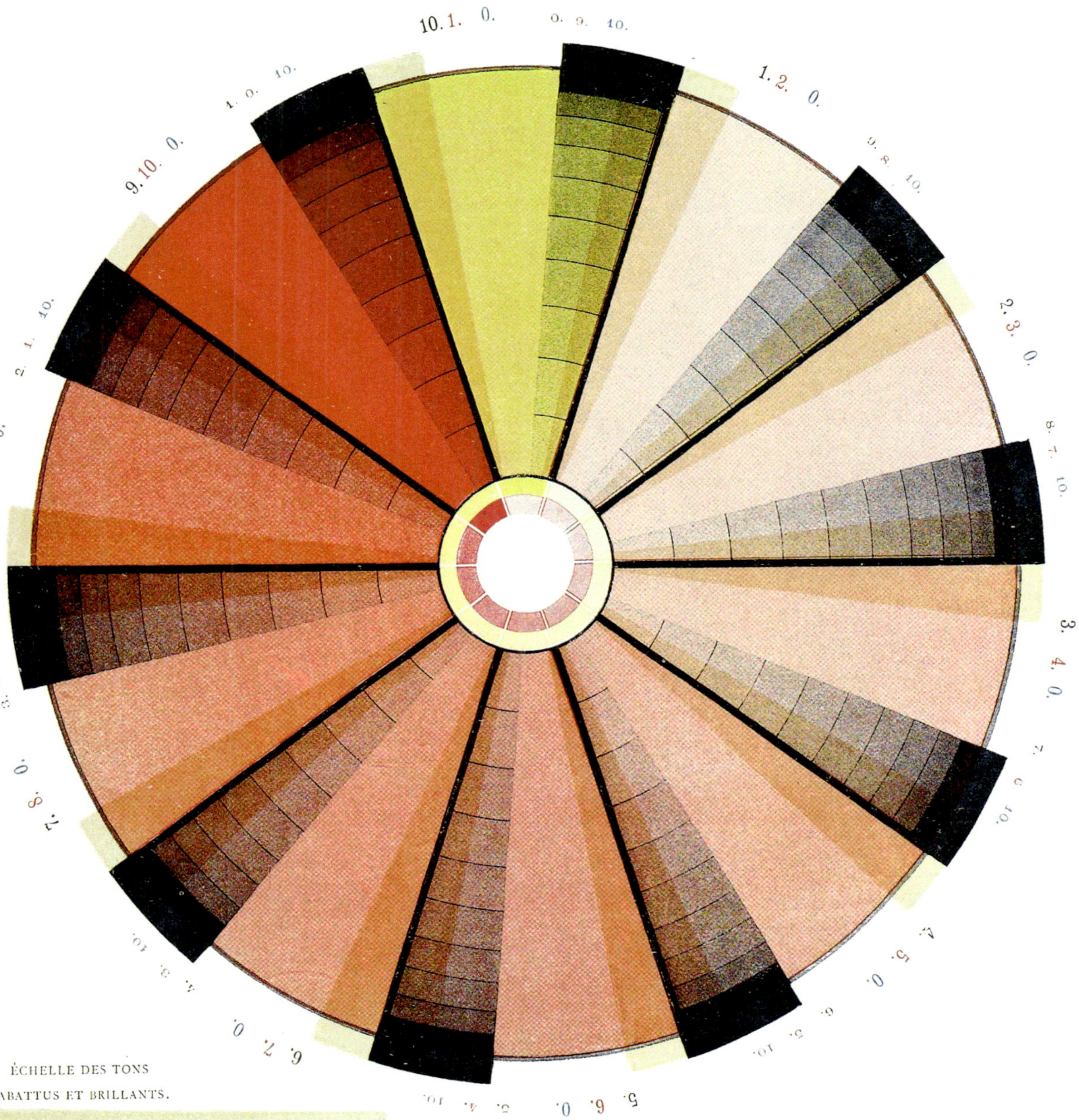

ÉCHELLE DES TONS
RABATTUS ET BRILLANTS.

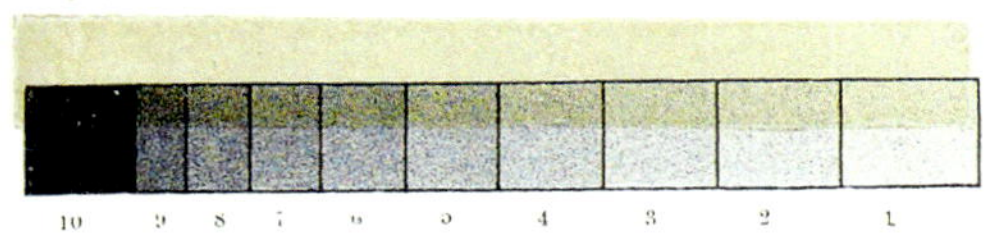

N. B. Les chiffres en noir se rapportent aux tons jaunes.

LES NUANCES 10.1.0. et 9.10.0.

Leurs tons lavés et leurs gammes.

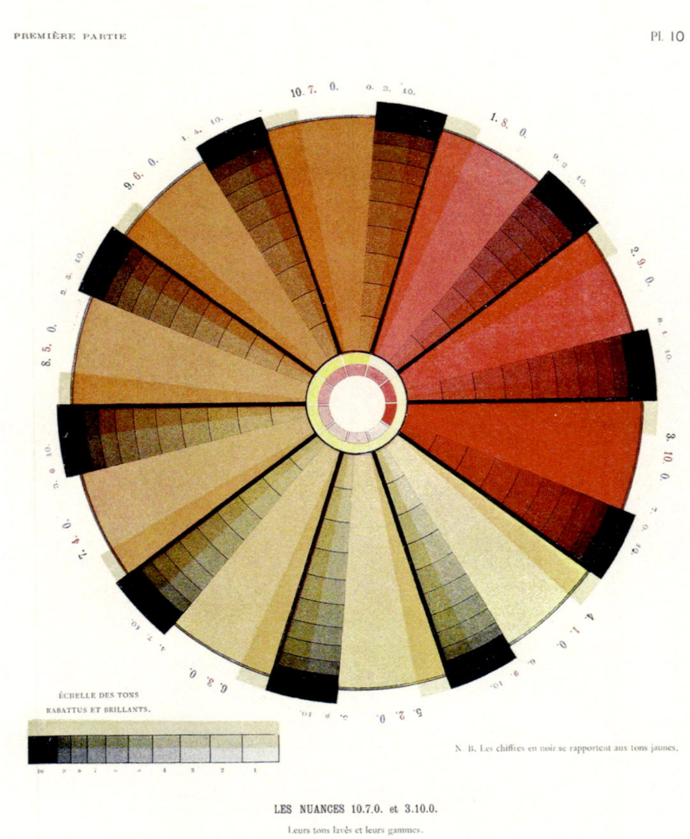

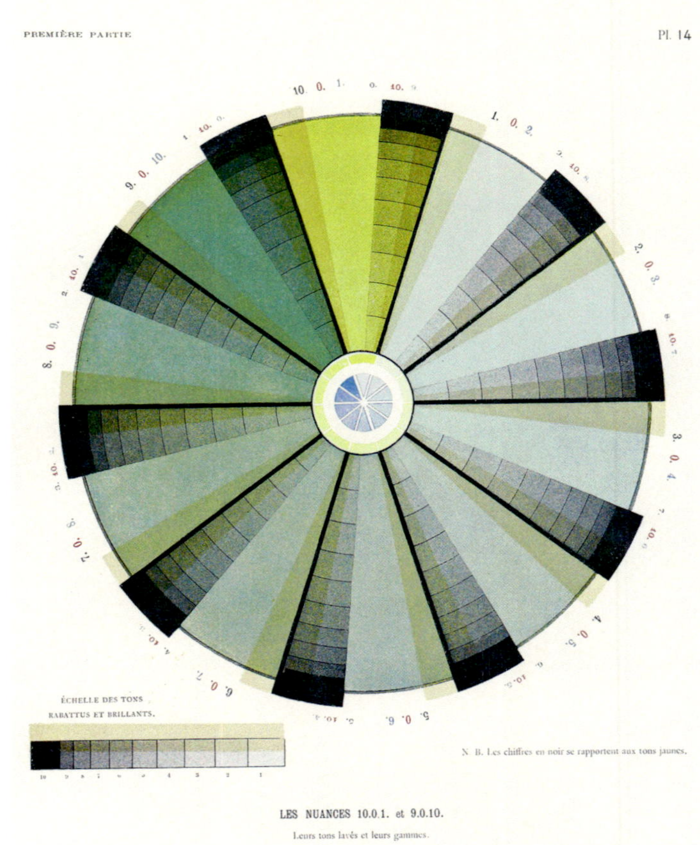

Die Reproduktion von Farben

Robert Steinheils Buch ist ein Meisterwerk der dekorativen Buchgestaltung und -illustration des späten 19. Jahrhunderts und als solches auch selbstreferenziell, da es die technischen Möglichkeiten des Farbrasterdrucks in den 1890er-Jahren zelebrierte. Der vollständige Titel, *La Reproduction des Couleurs par la superposition des trois couleurs simples*, macht deutlich, dass das Buch demonstrieren sollte, wie Farbe reproduziert wird, indem man die „einfachen" Primärfarben Gelb, Rot und Blau übereinanderlegt. Der Bucheinband mit seiner kunstvollen, in prismatischer Abstufung gedruckten Titelgrafik auf dem Vorderdeckel ist typisch für die Ästhetik des Fin de Siècle. Die Titelseite liefert einen weiteren Vorgeschmack auf die folgenden Abbildungen. Ihr schmückender, leuchtkräftiger Halbkreis aus 660 einzelnen Farbtönen veranschaulicht die Grundsätze der simultanen Farbkontraste.

Das von dem alteingesessenen französischen Unternehmen Berger-Levrault et Cie gedruckte Werk enthält 43 Buchdruckseiten Haupttext, darunter ein Glossar der Farbbegriffe und eine Farbtaxonomie (mit Hinweisen auf Einflüsse von Chevreul, Lacouture und anderen) nebst alphabetischen und numerischen Tabellen aller in den Abbildungen vorkommenden farblichen „Nuancen". Auf den Text folgt ein „Atlas" von 150 Farbtafeln, der je nach Charge des Buches auf drei verschiedenen Papiertypen gedruckt wurde und die Serie der Farbkombinationen zeigt. Diese nehmen größtenteils die Form von Farbkreisen an, deren in zehn Abstufungen unterteilte Sektoren das Druckverfahren abbilden und für jede Nuance oder Schattierung genaue, durch drei Zahlen bestimmte Mischverhältnisse angeben. Auf 16 Tafeln sind Farbüberlagerungen in nummerierten Rastern aus jeweils zehn mal zehn Feldern zu sehen, während die letzte Farbtafel aus einer Ansammlung mehrerer Überlagerungen in verschiedenen Formen und Mustern besteht, darunter als Wiederholung im Zentrum die halbkreisförmige Abbildung von der Titelseite. Insgesamt zeigen die Tafeln fast fünfzehntausend Einzelfarben. Das äußerst ambitionierte, mit einem Höchstmaß an Fachwissen produzierte Buch über den Farbdruck zeugt auf prachtvolle Weise von den technologischen Errungenschaften des Industriezeitalters. Seine Auflage muss klein gewesen sein, denn es zählt heute zu den seltensten historischen Büchern über Farbe.

PREMIÈRE PARTIE

Pl. 34

Pour les Tons voisins
plus et moins jaunes
voir les planches 45 et 133.

Pour les Tons voisins
plus et moins rouges
voir les planches 124 et 44.

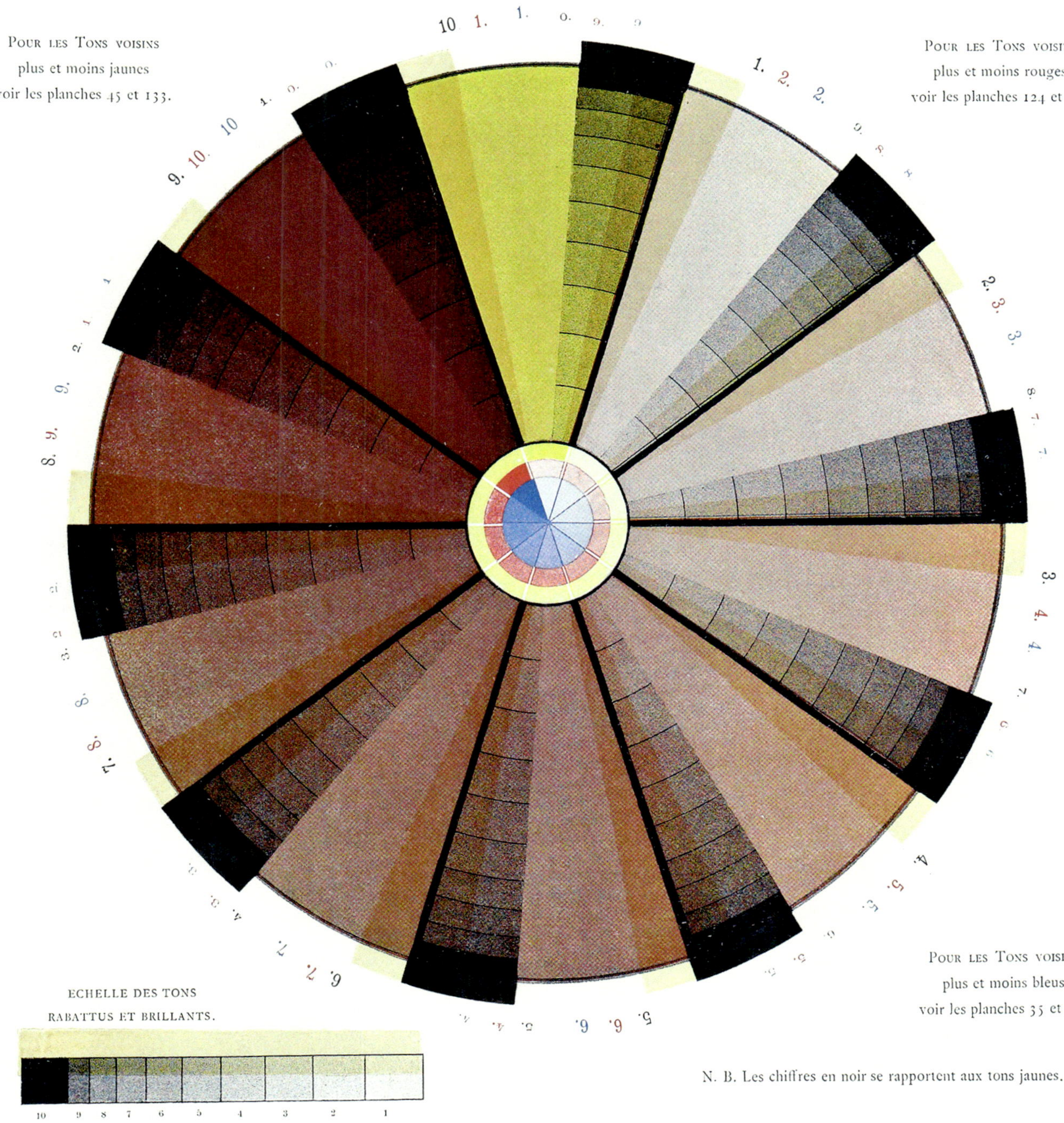

Pour les Tons voisins
plus et moins bleus
voir les planches 35 et 43.

N. B. Les chiffres en noir se rapportent aux tons jaunes.

LES NUANCES 10.1.1 et 9.10.10

Leurs tons lavés et leurs gammes.

PREMIÈRE PARTIE

Pl. 52

Pour les Tons voisins
plus et moins jaunes
voir les planches 63 et 41

Pour les Tons voisins
plus et moins rouges
voir les planches 42 et 62.

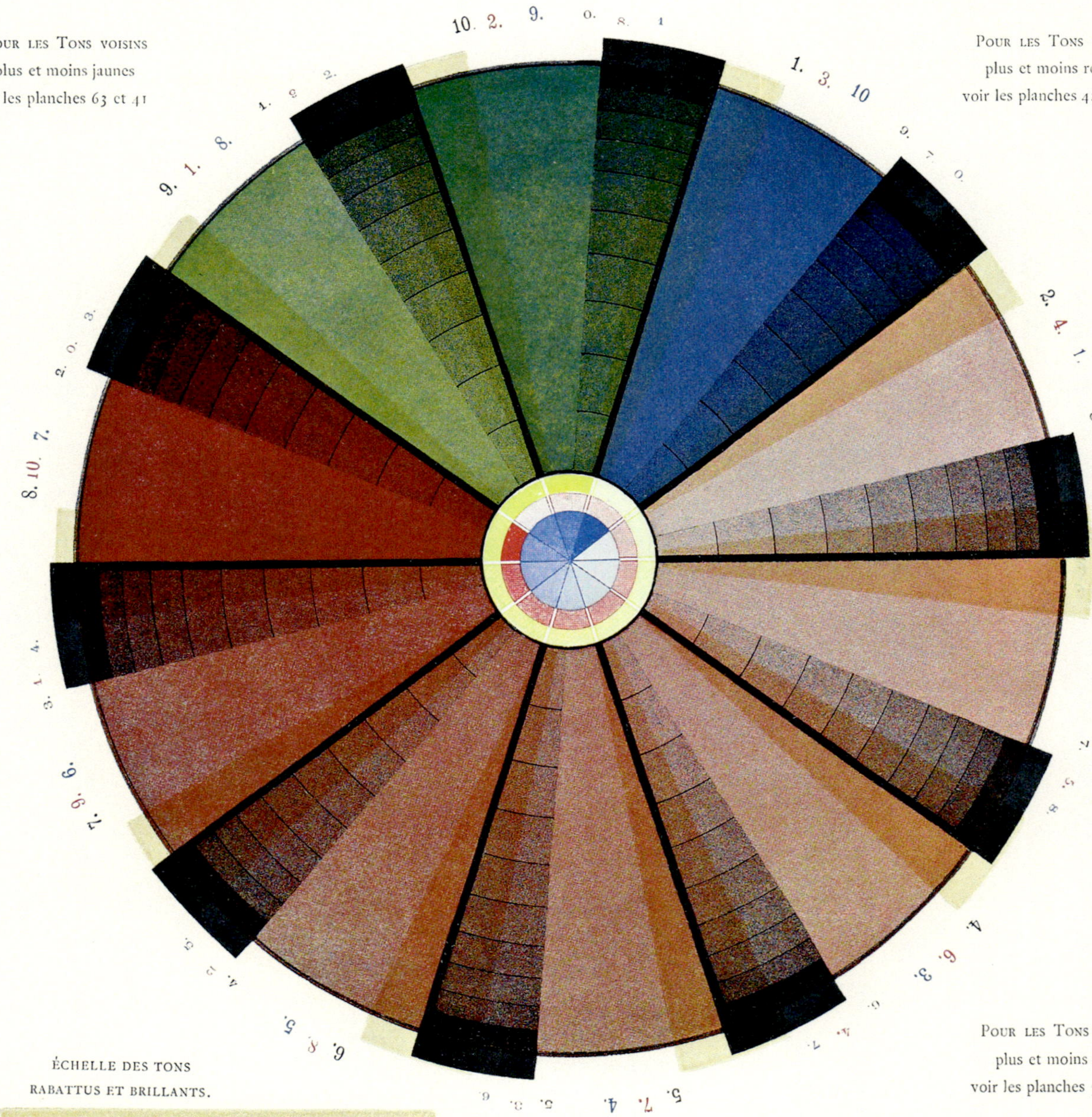

Pour les Tons voisins
plus et moins bleus
voir les planches 53 et 51.

ÉCHELLE DES TONS
RABATTUS ET BRILLANTS.

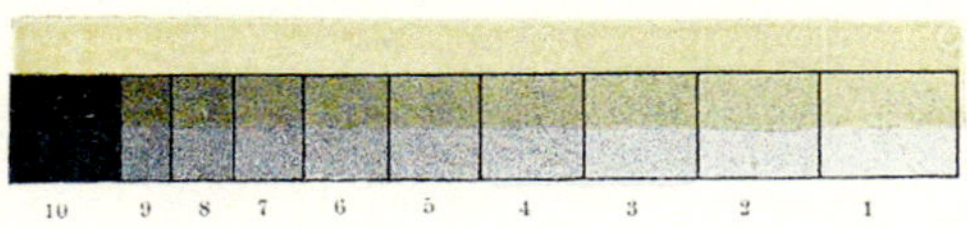

N. B. Les chiffres en noir se rapportent aux tons jaunes.

LES NUANCES 10.2.9, 8.10.7 et 1.3.10.

Leurs tons lavés et leurs gammes.

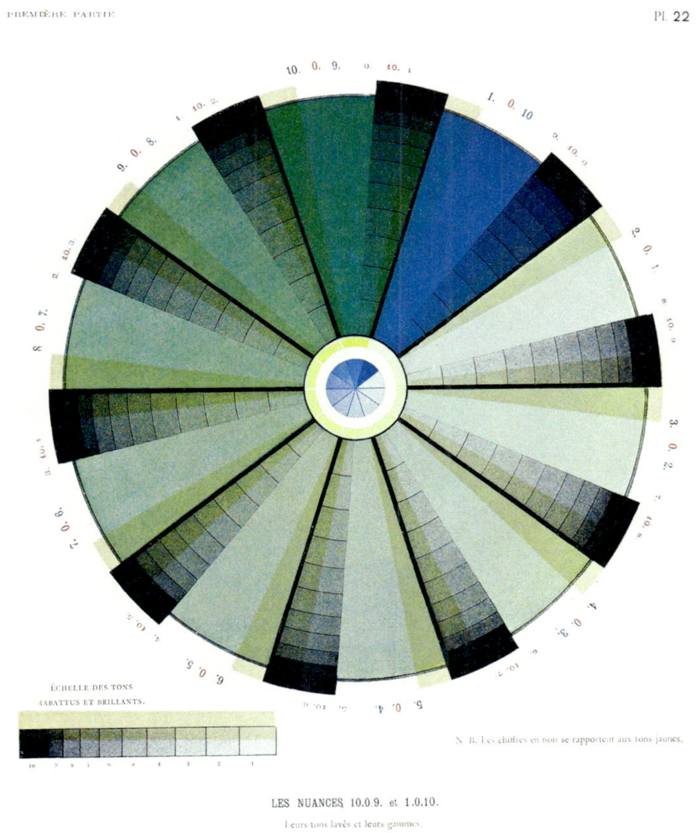

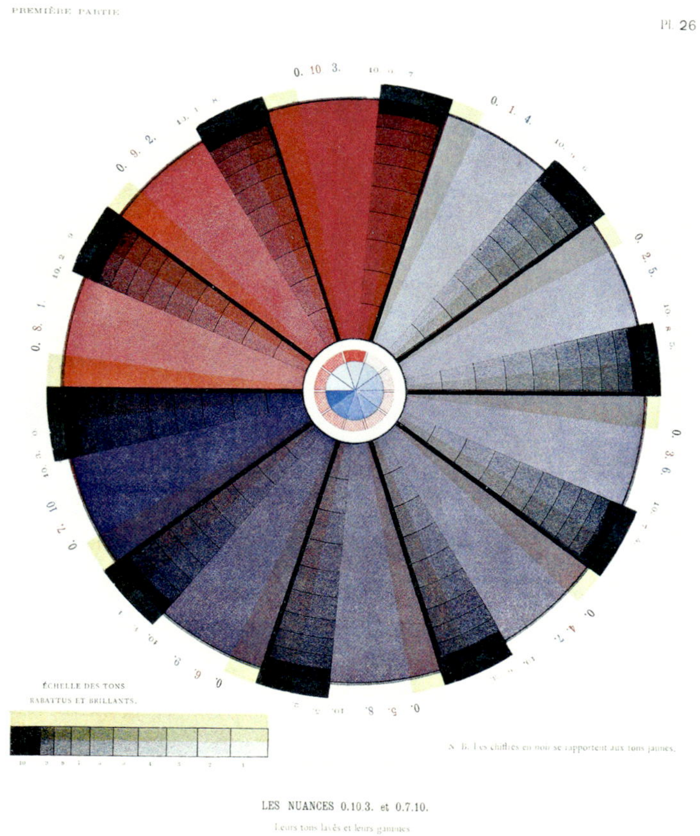

La Reproduction des Couleurs

Le livre de Robert Steinheil est un chef-d'œuvre consacré au dessin et à l'illustration éditoriale de la fin du XIXe siècle. À ce titre, on peut dire qu'il est autoréférentiel, dans la mesure où il célèbre les possibilités techniques offertes par l'impression en couleurs demi-teintes des années 1890. Le titre complet, *La Reproduction des Couleurs par la superposition des trois couleurs simples*, pose clairement que le propos du livre est de démontrer comment reproduire des couleurs en superposant celles appelées « simples », autrement dit les couleurs primaires, jaune, rouge et bleu. La couverture est typique du style esthétique fin-de-siècle, avec son lettrage recherché aux nuances prismatiques. La page de titre donne un avant-goût de la qualité des illustrations qui suivent : elle figure un lumineux demi-cercle décoratif de 660 teintes différentes, qui représente les principes des contrastes simultanés des couleurs.

Imprimé par la très ancienne maison française Berger-Levrault et Cie, le livre contient 43 pages de texte typographié, dont un glossaire chromatique, une taxonomie de la couleur (qui reconnaît l'influence de Chevreul, Lacouture et d'autres), ainsi que des tables alphabétique et numérique de toutes les nuances de couleurs utilisées dans les illustrations. Le texte est suivi d'un « Atlas » de 150 planches en couleurs qui figurent les séries de combinaisons chromatiques et sont imprimées sur trois types de papier différents selon les lots d'exemplaires. La plupart des planches se présentent sous la forme de cercles chromatiques dont les portions sont divisées en 10 gradations, qui illustrent le processus d'impression et indiquent les proportions exactes pour chaque nuance, identifiée par trois numéros. Il y a par ailleurs 16 planches qui montrent les superpositions de couleurs par un système de grilles numérotées 10 x 10, la dernière étant un assemblage de plusieurs superpositions de formes et motifs différents, dont la figure en demi-cercle de la page de titre, reproduite ici au centre. Au total, les planches rassemblent près de 15 000 couleurs différentes. Un ouvrage extrêmement ambitieux sur l'impression en couleurs, produit avec un très haut niveau de savoir-faire, qui reste un magnifique témoignage des réussites technologiques de l'ère industrielle. En raison d'un tirage probablement réduit, l'ouvrage est devenu l'un des livres historiques sur la couleur les plus rares.

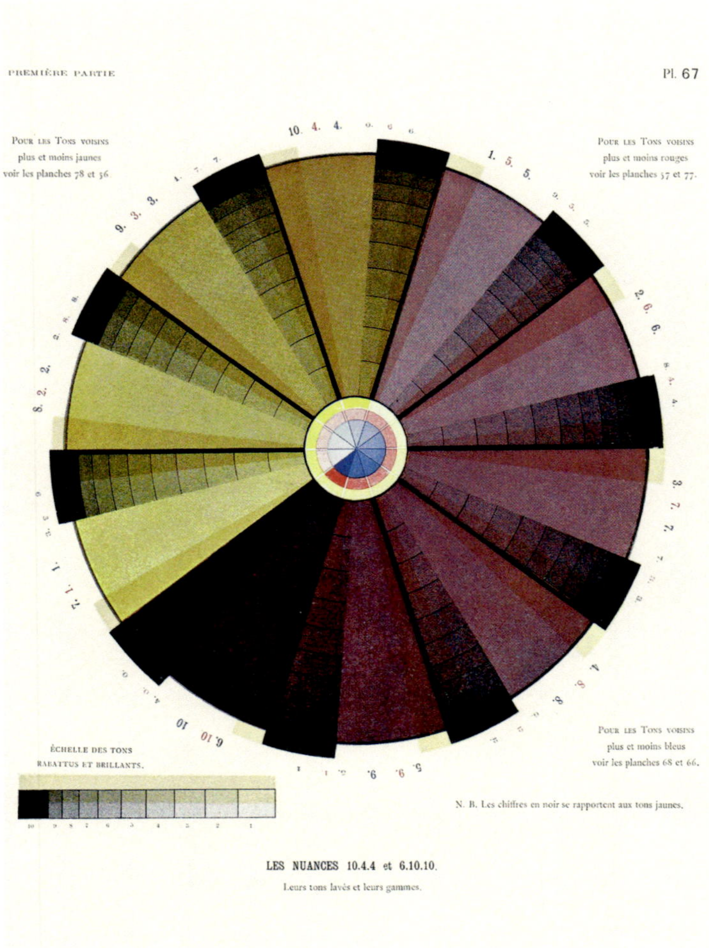

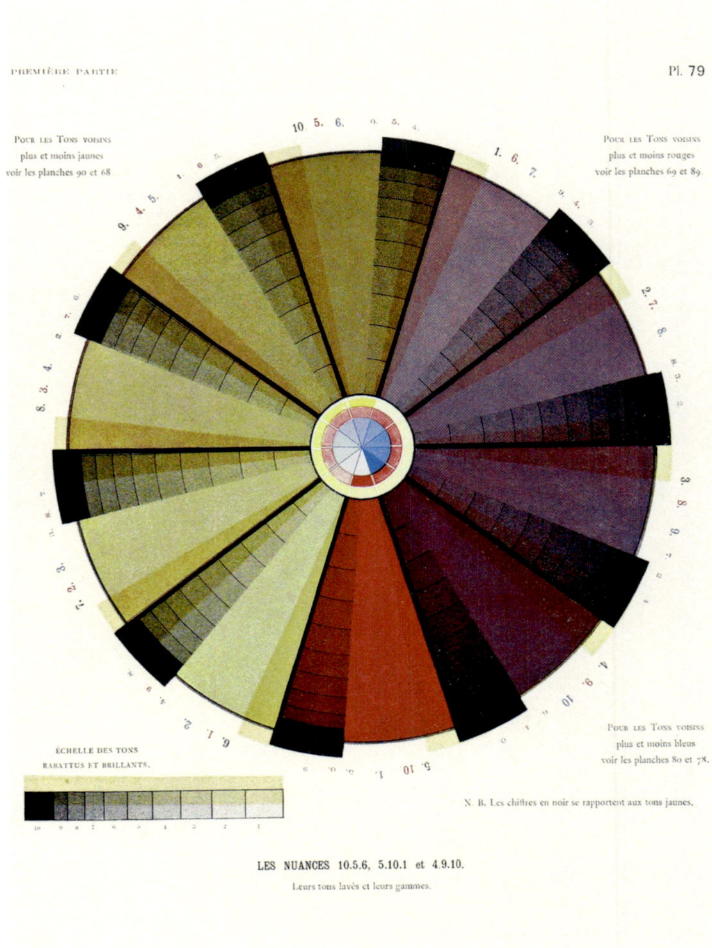

La reproducción de los colores

El libro de Robert Steinheil es una obra maestra del diseño de libros ilustrados de finales del siglo XIX. Como tal, también es autorreferencial, ya que es una celebración de las posibilidades técnicas que ofrecía la impresión en colores de tonos medios en la década de 1890. El título completo, *La Reproduction des Couleurs par la superposition des trois couleurs simples*, deja claro que su propósito es mostrar de qué modo se pueden reproducir colores mediante la superposición de los colores primarios «simples»: amarillo, rojo y azul. El diseño es típico de la estética *fin-de-siècle*. El elaborado título de la cubierta está coloreado con una escala de colores prismáticos, mientras que la portada da una muestra más de la calidad de las ilustraciones que vienen a continuación con la inclusión de un semicírculo luminoso decorativo con 660 tonos distintos que ilustra los principios del contraste de colores simultáneos.

La histórica empresa francesa Berger-Levrault et Cie imprimió el libro. Este consta de 43 páginas de texto principal, que incluyen un glosario de términos relacionados con el color, una taxonomía del color (con referencia a Chevreul, Lacouture y otras influencias) y tablas alfabéticas y numéricas de todos los «matices» de color mostrados en las ilustraciones. Al texto lo sigue un «Atlas» de 150 láminas en color, que se imprimieron en tres tipos de papel diferentes a lo largo de toda la tirada, con la serie de combinaciones de colores. La mayoría presentan la forma de una rueda de color con sectores divididos en 10 gradaciones que ilustran el proceso de impresión y detallan las proporciones exactas para cada matiz o sombra, que se identifica con tres números. Hay 16 láminas que muestran superposiciones de color en sistemas de cuadrículas numeradas de 10 × 10, mientras que la lámina de color final es un conjunto de varias superposiciones en diferentes formas y patrones, e incluye en el centro la imagen semicircular que ya aparece en la portada. En total, las láminas recogen casi 15 000 colores distintos. Este es un libro muy ambicioso sobre la impresión en color, producido con el más alto nivel de experiencia, y es un testimonio magnífico de los logros tecnológicos de la era industrial. La tirada debió de haber sido pequeña, ya que ahora es uno de los libros históricos en color más raros.

PREMIÈRE PARTIE

Pl. 58

Pour les Tons voisins
plus et moins jaunes
voir les planches 69 et 47

Pour les Tons voisins
plus et moins rouges
voir les planches 48 et 68.

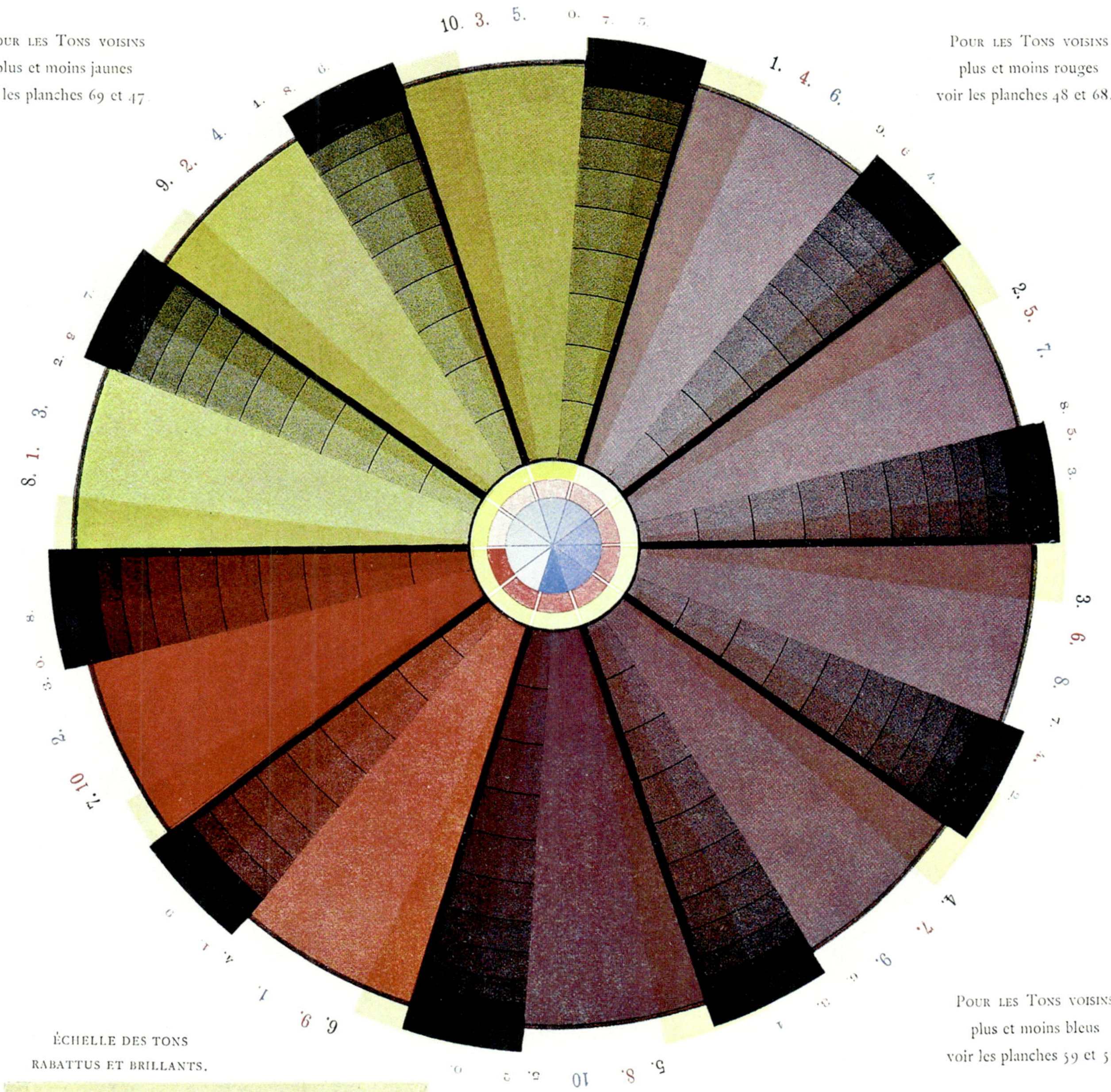

Pour les Tons voisins
plus et moins bleus
voir les planches 59 et 57.

ÉCHELLE DES TONS
RABATTUS ET BRILLANTS.

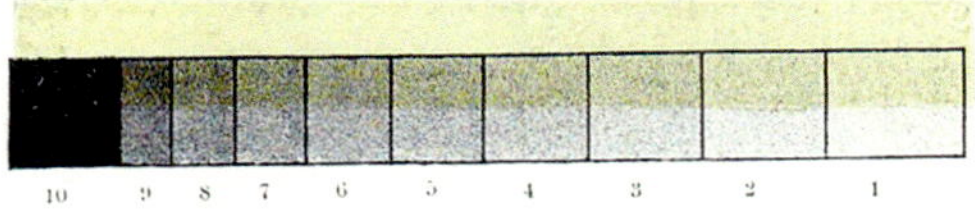

N. B. Les chiffres en noir se rapportent aux tons jaunes.

LES NUANCES 10.3.5, 7.10.2 et 5.8.10.

Leurs tons lavés et leurs gammes.

DEUXIÈME PARTIE. Pl. 134

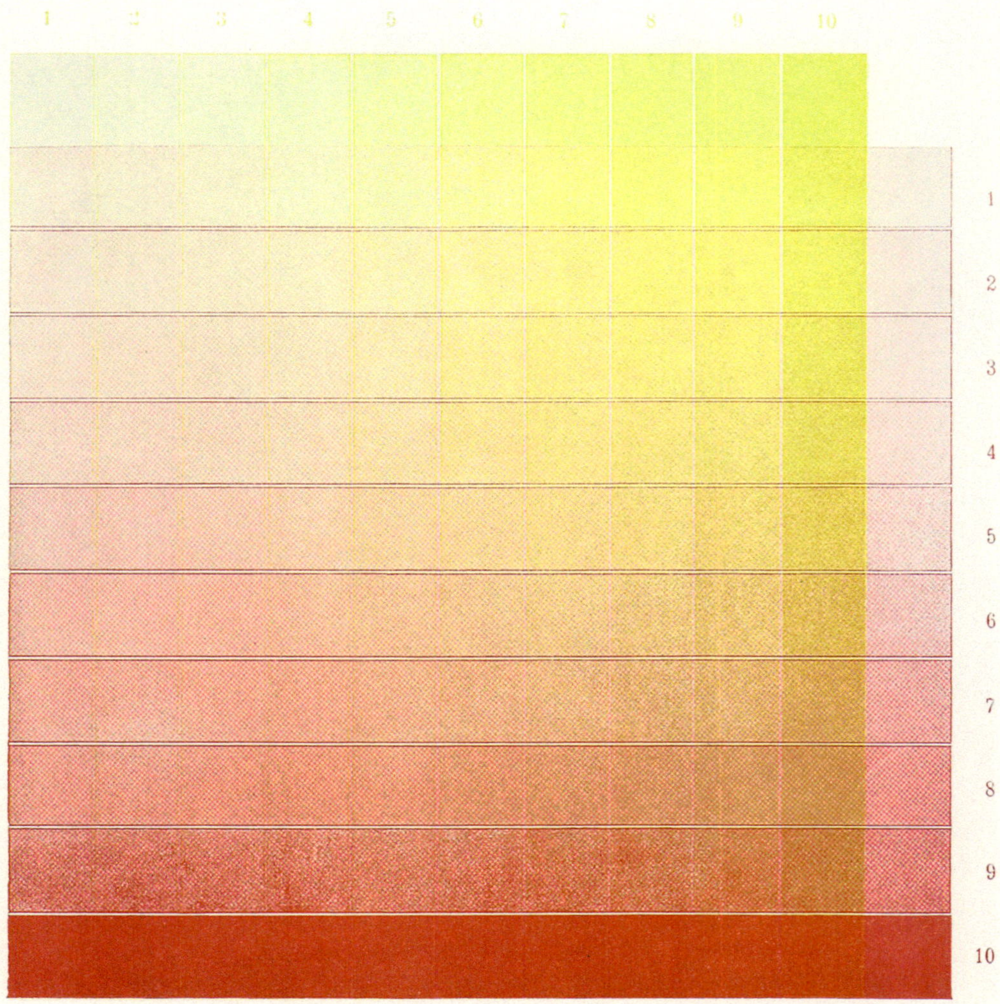

LES CENT COMBINAISONS DU ROUGE SUR LE JAUNE

Chaque ton occupe le carré d'intersection d'une bande verticale et d'une bande horizontale.

Pour avoir la formule numérique du ton d'un carré quelconque, ajouter à la droite du chiffre de la bande jaune le chiffre de la bande rouge, suivi du zéro représentant le bleu absent.

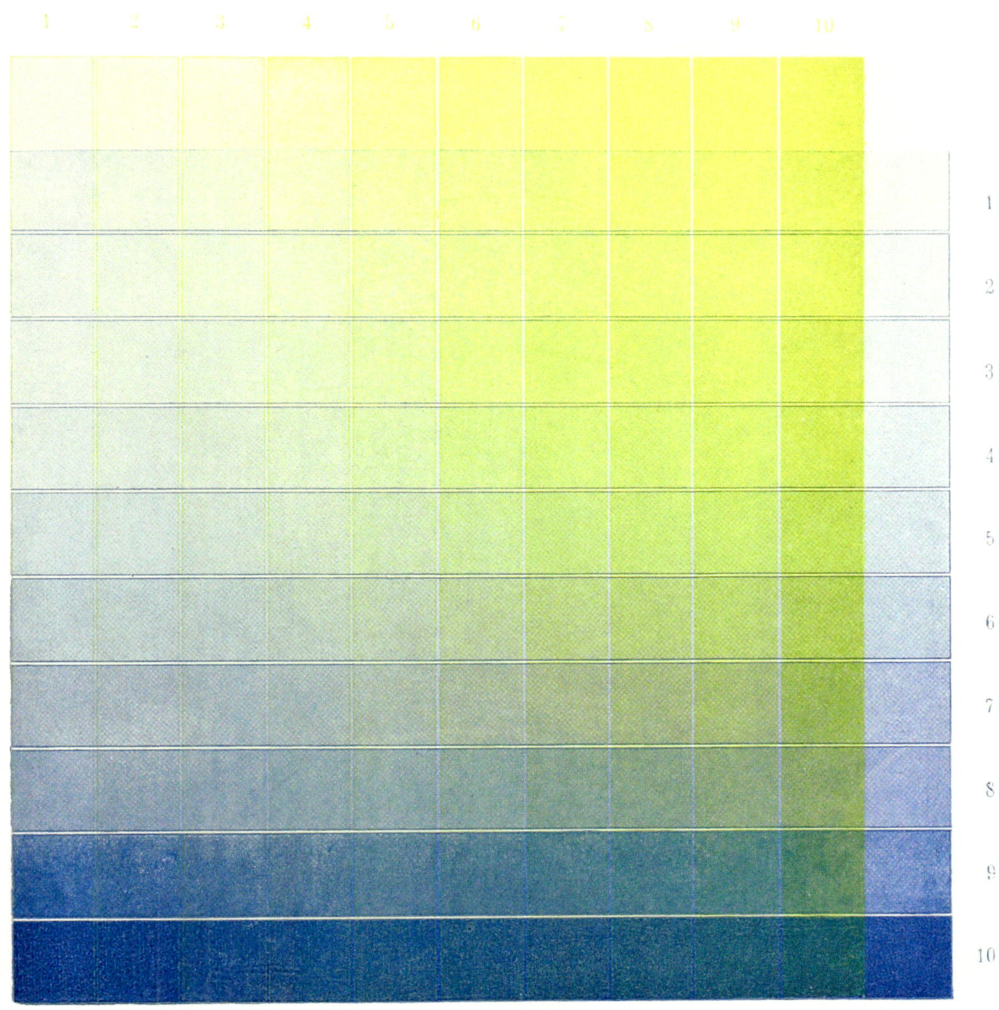

LES CENT COMBINAISONS DU BLEU SUR LE JAUNE

Chaque ton occupe le carré d'intersection d'une bande verticale et d'une bande horizontale.

Pour avoir la formule numérique du ton d'un carré quelconque, ajouter à la droite du chiffre de la bande jaune un zéro représentant le rouge absent, suivi du chiffre de la bande bleue.

DEUXIÈME PARTIE. Pl. 138

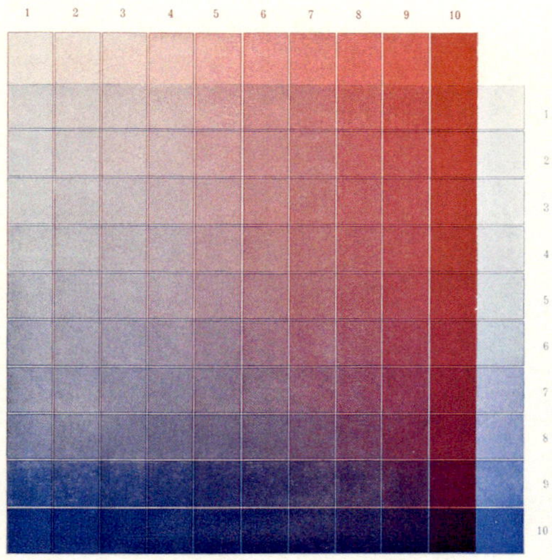

LES CENT COMBINAISONS DU BLEU SUR LE ROUGE

Chaque ton occupe le carré d'intersection d'une bande verticale et d'une bande horizontale.

Pour avoir la formule numérique du ton d'un carré quelconque, ajouter à la droite du zéro représentant le jaune absent, le chiffre de la bande rouge, suivi du chiffre de la bande bleue.

DEUXIÈME PARTIE. Pl. 139

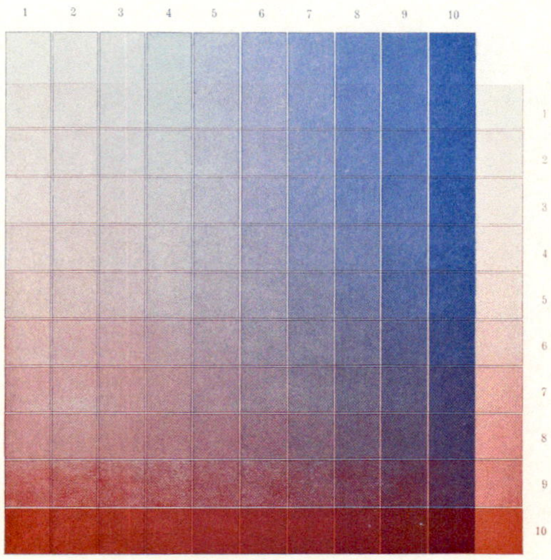

LES CENT COMBINAISONS DU ROUGE SUR LE BLEU

Chaque ton occupe le carré d'intersection d'une bande verticale et d'une bande horizontale.
L'ordre du tirage étant interverti, la formule conventionnelle ne peut être appliquée à cette planche.

DEUXIÈME PARTIE. Pl. 143

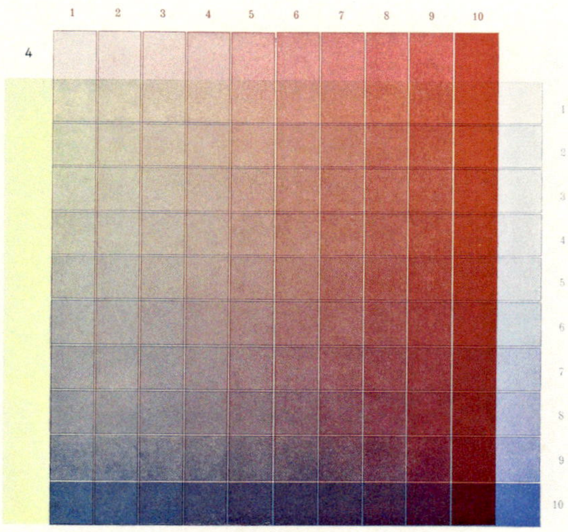

LES CENT COMBINAISONS DU ROUGE ET DU BLEU SUR LE JAUNE N° 4

N. B. Chaque ton occupe le carré d'intersection d'une bande verticale rouge et d'une bande horizontale bleue.

Pour avoir la formule numérique du ton d'un carré quelconque, ajouter à la droite du chiffre d'intensité du jaune, commun à tous les carrés, le chiffre de la bande rouge, suivi du chiffre de la bande bleue.

Ex.: Voir à la planche 141 le ton marqué 2. 3. 6.

DEUXIÈME PARTIE. Pl. 149

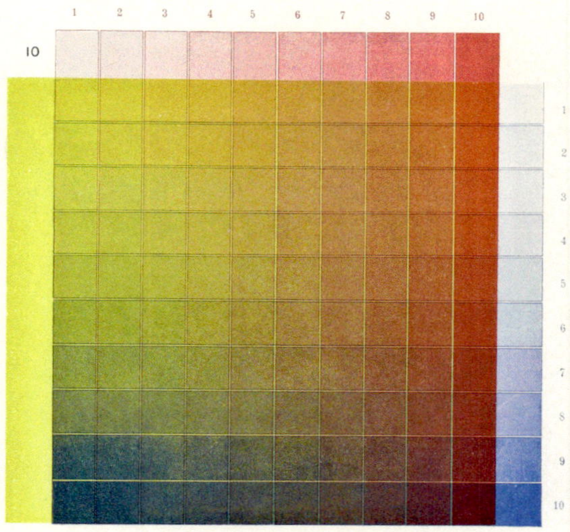

LES CENT COMBINAISONS DU ROUGE ET DU BLEU SUR LE JAUNE N° 10

N. B. Chaque ton occupe le carré d'intersection d'une bande verticale rouge et d'une bande horizontale bleue.

Pour avoir la formule numérique du ton d'un carré quelconque, ajouter à la droite du chiffre d'intensité du jaune, commun à tous les carrés, le chiffre de la bande rouge, suivi du chiffre de la bande bleue.

Ex.: Voir à la planche 141 le ton marqué 2. 3. 6.

TROISIÈME PARTIE

Pl. 150

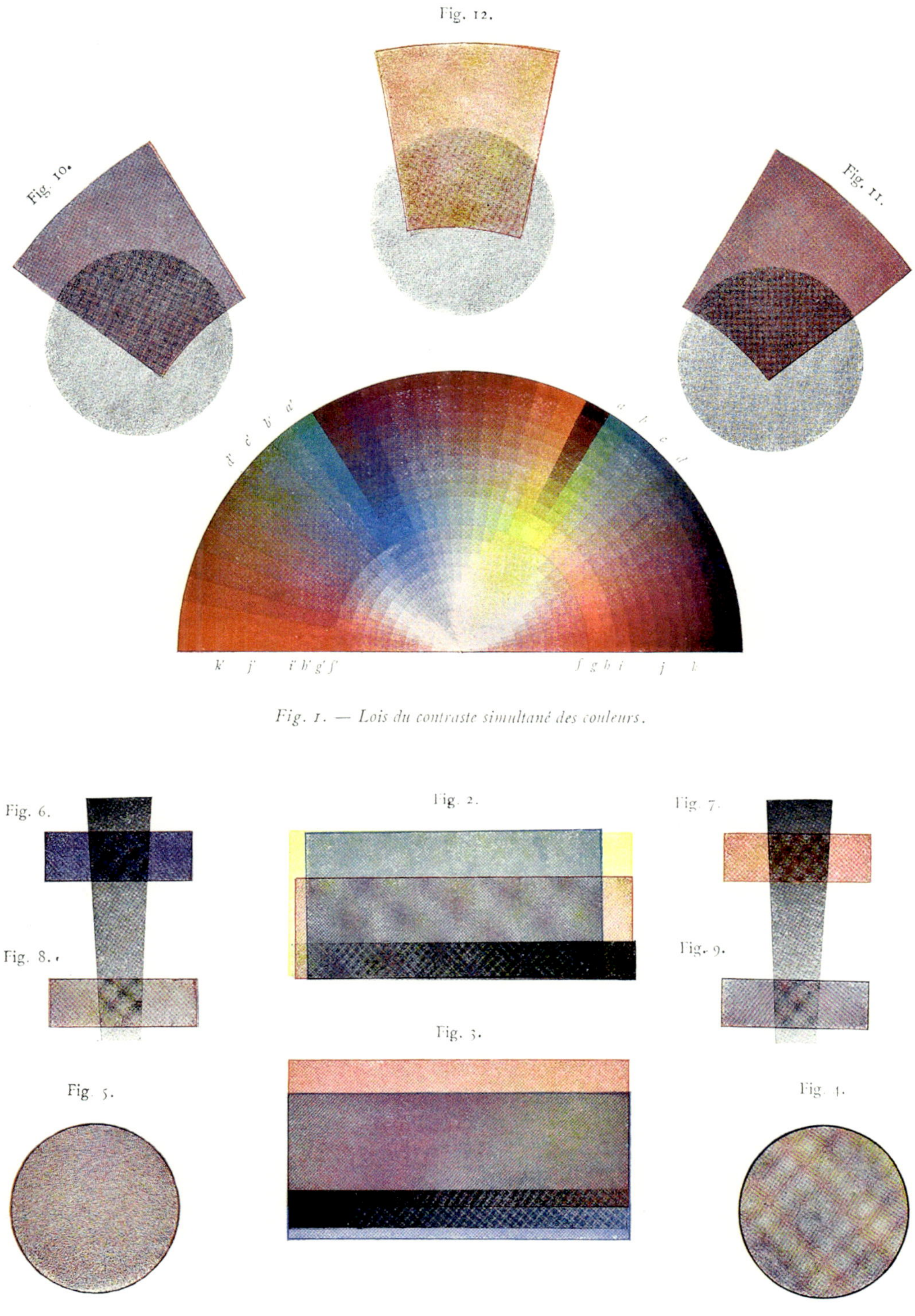

QUELQUES RÉSULTATS DES SUPERPOSITIONS DE COULEURS.

Fig. 19."

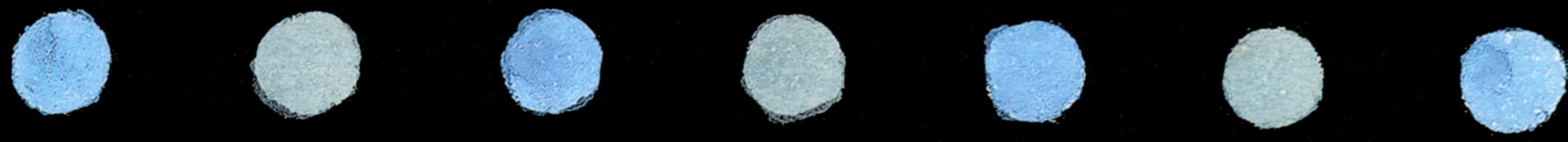

Fig. 14."

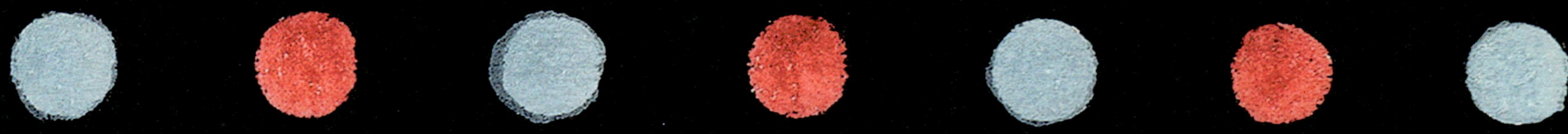

Fig. 16."

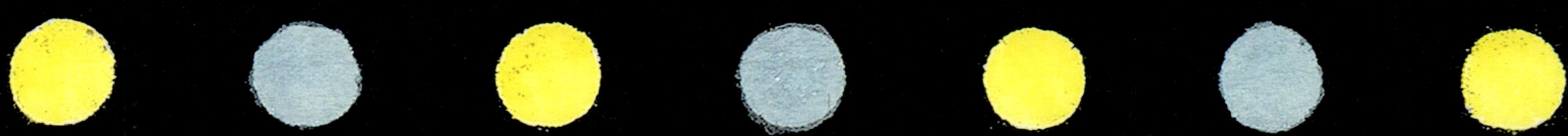

Fig. 17."

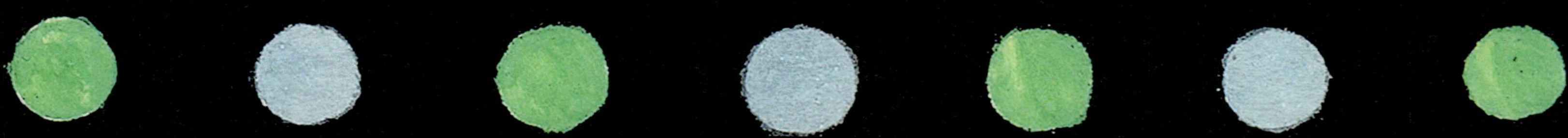

Fig. 18."

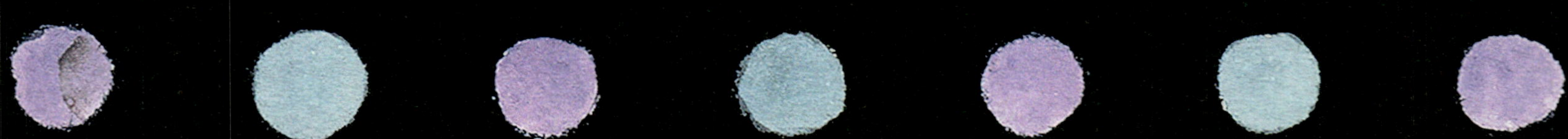

Chapter 3

New Elucidations: THE RISE OF COLOUR THEORY

1704–1864

Isaac Newton (1643–1727)

OPTICKS

or, a Treatise of the Reflexions, Refractions, Inflexions and Colours of Light

13 plates, 25 x 21 cm | 9 ⅞ x 8 ¼ in., London, 1704
Washington, D.C., Library of Congress

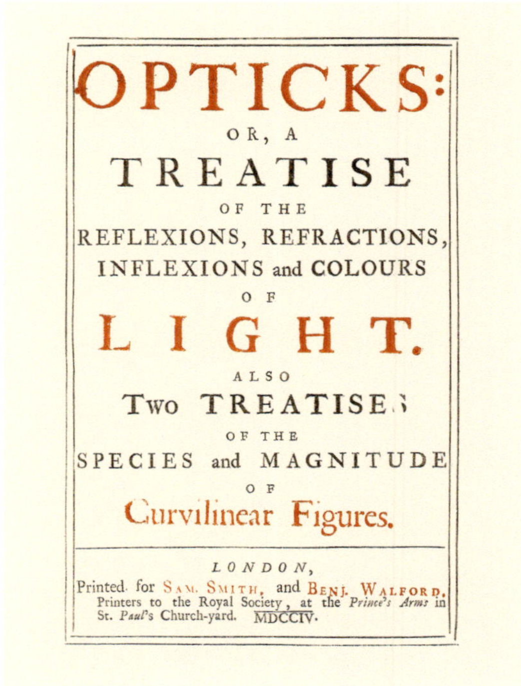

Newton's *Opticks* of 1704 is regularly named as one of the milestones in colour history. It made an immediate impact on the arts and sciences in the 18th century, although it is worth noting that Newton had been occupied with the subject of colour for decades before this book was published, and that he was himself part of a long line of other figures who worked with colour concepts and research. He was certainly not the first to have carried out optical studies, but in terms of scale and thoroughness his work stands out by some distance from that of both his contemporaries and his predecessors. Newton's status and reputation, together with advances in the print culture of his day, also helped to disseminate and popularise his work, with this book being translated into several languages soon after the first edition in English. It was discussed, developed and adapted from a number of critical perspectives by several other writers in the 18th and 19th centuries, and even today there is hardly a text, programme or talk on colour that does not mention Newton. That first small prism he bought in the 1660s when still a young man has thrown a very long ray of colour across the centuries, and his colour circle is one of the most important and recognisable diagrams in the history of science.

Newton began experimenting with prisms at Woolsthorpe Manor, his home in Lincolnshire, in 1666. He first presented the results of his "crucial experiment" of splitting pure sunlight into the prismatic colour spectrum in the *Philosophical Transactions of the Royal Society* in London in 1671. With this experiment he proved that the colours of the spectrum visible to the human eye cannot be split further into more colours, and in fact form white light again when redirected through a second prism, thus laying out the principles of additive, or immaterial, colour mixture. At first, Newton noted five identifiable colours but later settled on seven, which he understood to be analogous to musical scales, namely red, orange, yellow, green, blue, indigo and violet. Despite the linear appearance of colour order in his experiments he envisioned a circular shape to represent it from an early stage, eventually creating the (uncoloured) colour wheel that was printed in *Opticks*. There are hundreds of engravings in *Opticks*, illustrating the many stages and aspects of Newton's decades of research, but the surprisingly small colour wheel, with its uneven proportional divisions, is the most potent and influential of his diagrams. Within a few years it was adapted by other researchers, in an attempt to meet the needs of artists working with material colour.

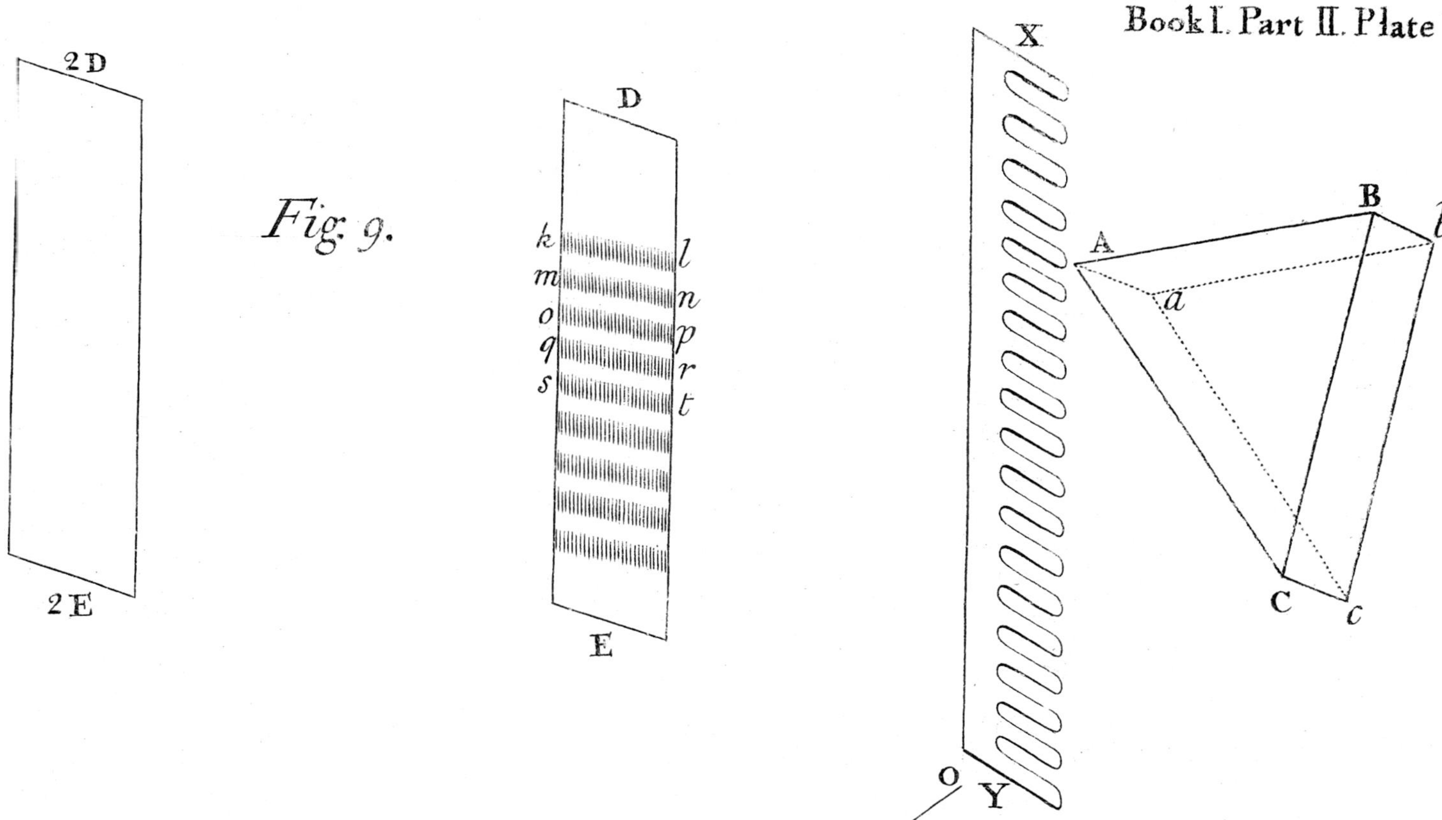

Fig. 9.

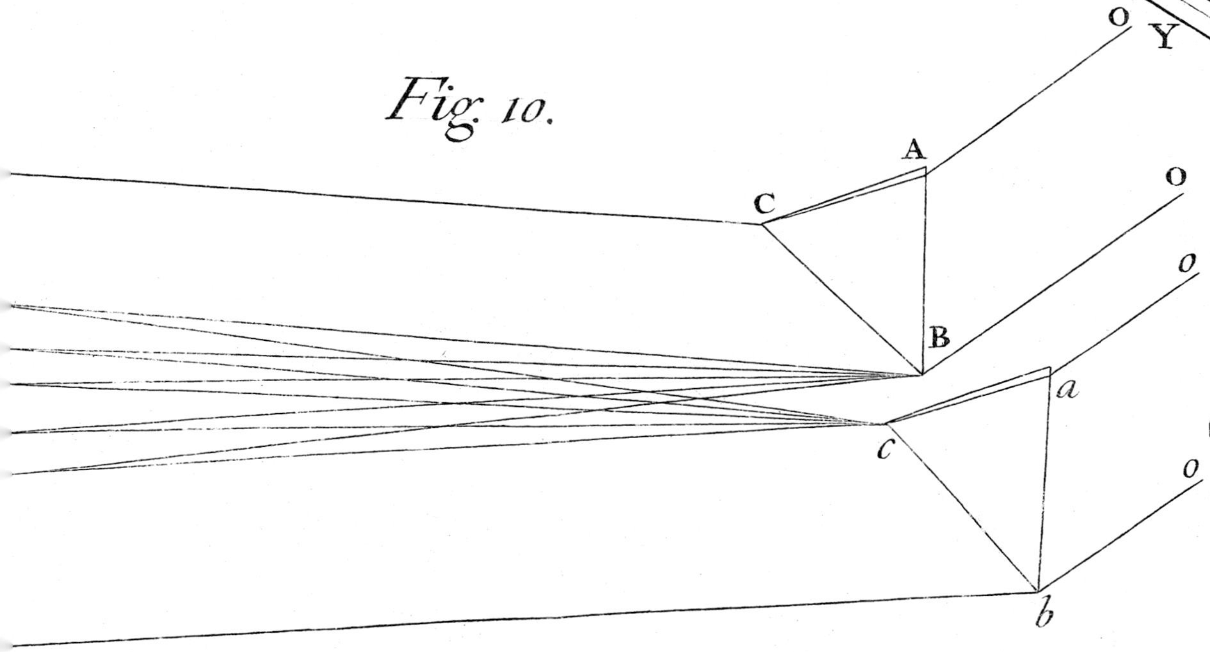

Fig. 10.

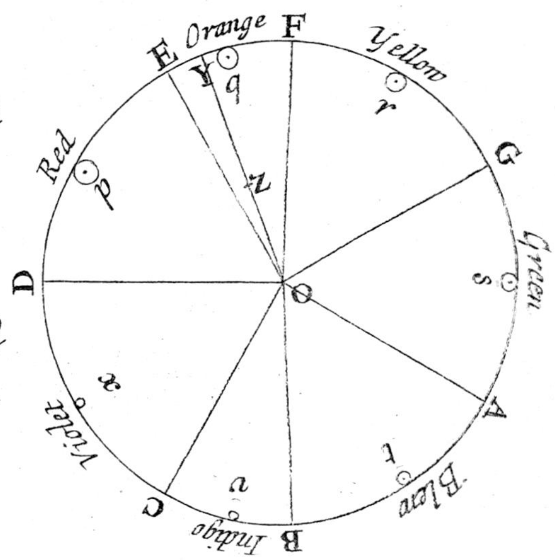

Fig. 11.

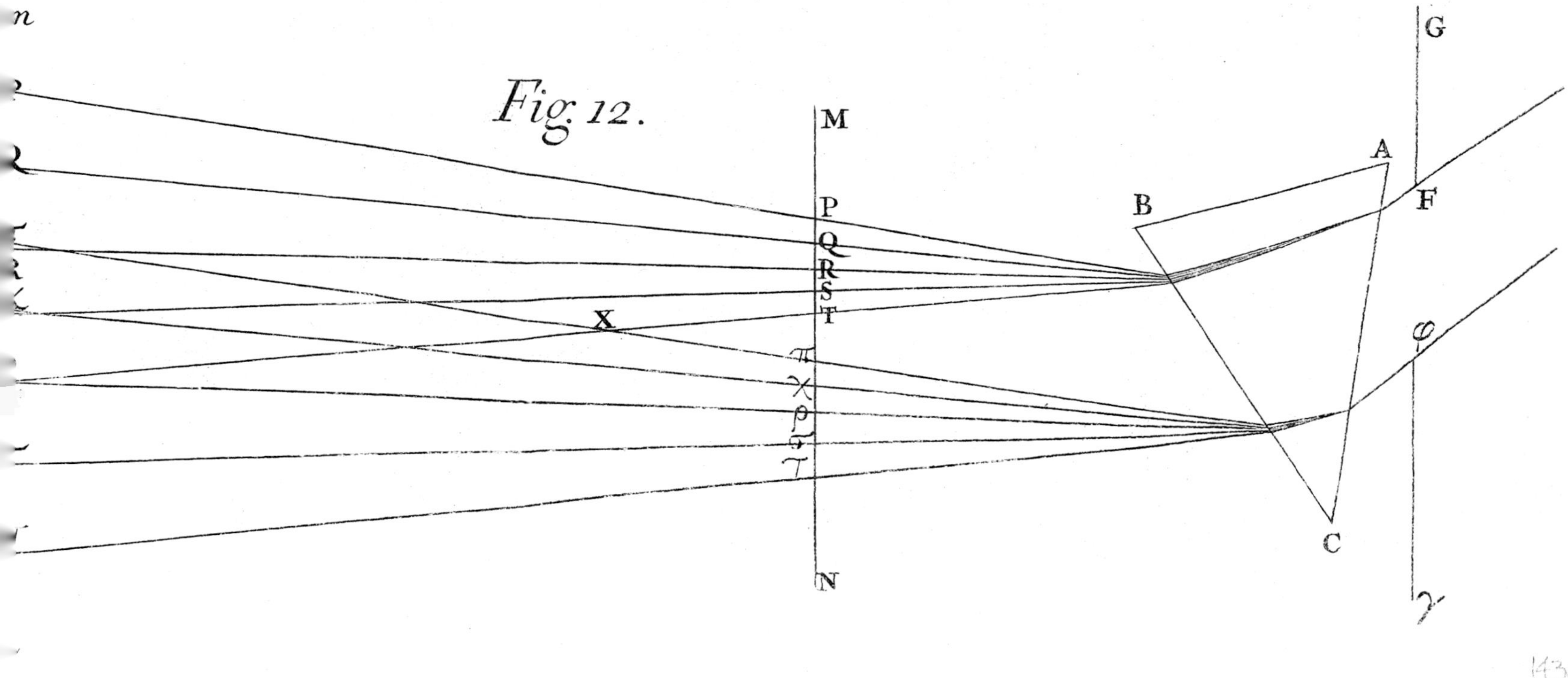

Fig. 12.

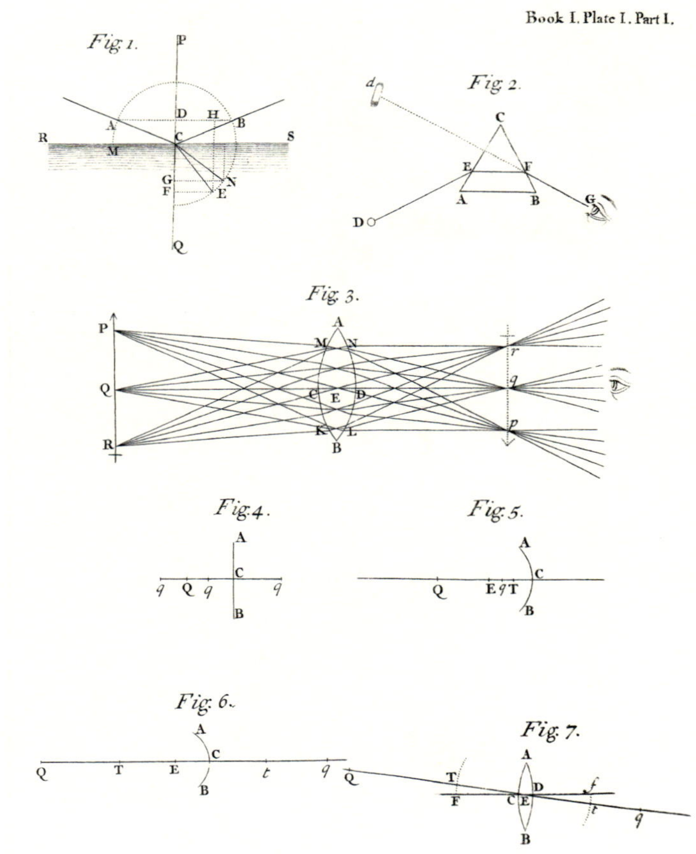

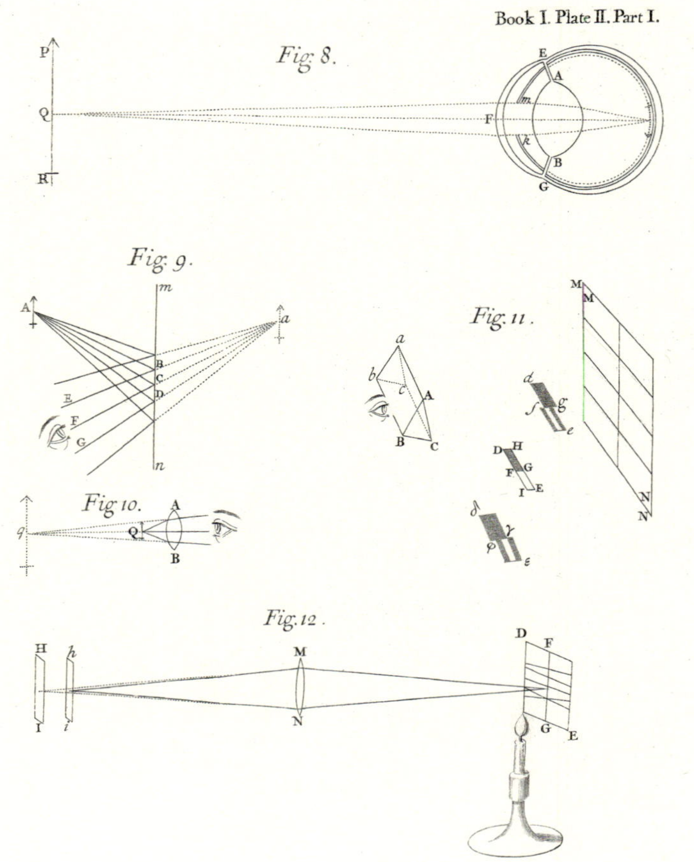

Optik oder Abhandlung über Spiegelungen, Brechungen, Beugungen und Farben des Lichts

Newtons *Opticks* aus dem Jahr 1704 wird regelmäßig als Meilenstein der Farbgeschichte bezeichnet. Das Buch hatte unmittelbare Auswirkungen auf die Kunst und Wissenschaft des 18. Jahrhunderts. Nicht zu vergessen ist jedoch, dass sich Newton bereits Jahrzehnte vor dessen Veröffentlichung mit dem Thema Farbe befasst hatte und selbst zu einer Vielzahl von Personen gehörte, die sich mit Farbsystemen und Farbforschung beschäftigten. Er war mit Sicherheit nicht der Erste, der optische Studien durchführte, doch im Hinblick auf Umfang und Gründlichkeit hebt sich seine Arbeit deutlich von der seiner Zeitgenossen und Vorgänger ab. Newtons Stellung und Reputation, aber auch die Fortschritte im Druckwesen seiner Zeit, halfen sein Werk zu verbreiten und bekannt zu machen. Kurz nach Erscheinen der englischen Erstauflage wurde das Buch in mehrere Sprachen übersetzt. Im 18. und 19. Jahrhundert wurde es von anderen Autoren aus unterschiedlichen Perspektiven besprochen, weiterentwickelt und angepasst, und selbst heutzutage gibt es kaum einen Text, ein Programm oder eine Diskussion über Farbe, in denen Newton nicht erwähnt wird. Das erste kleine Prisma, das er in den 1660er-Jahren als junger Mann erwarb, hat einen sehr langen Farbstrahl durch die Jahrhunderte geworfen, und sein Farbkreis zählt zu den wohl wichtigsten Diagrammen der Wissenschaftsgeschichte mit besonderem Wiedererkennungswert.

Newton begann 1666 in seinem Elternhaus Woolsthorpe Manor in der englischen Grafschaft Lincolnshire mit Prismen zu experimentieren. In den *Philosophical Transactions of the Royal Society* veröffentlichte er 1671 erstmals die Ergebnisse seines „entscheidenden Versuchs“, das Sonnenlicht in die Farben des prismatischen Spektrums aufzuspalten. Mit diesem Versuch bewies er, dass sich die für das menschliche Auge sichtbaren Spektralfarben nicht in weitere Farben aufspalten lassen und bei Umlenkung durch ein zweites Prisma tatsächlich wieder weißes Licht ergeben. Damit hatte er die Grundsätze der additiven oder immateriellen Farbmischung dargelegt. Anfangs notierte Newton fünf identifizierbare Farben, entschied sich dann aber für sieben, die er mit den sieben Stufen der Tonleiter gleichsetzte: Rot, Orange, Gelb, Grün, Blau, Indigo und Violett. Trotz der linearen Ordnung der Farben in seinen Experimenten malte er sich für ihre Darstellung schon früh eine Kreisform aus und schuf schließlich den in *Opticks* abgedruckten (nicht kolorierten) Farbkreis. In Hunderten von Stichen veranschaulicht das Buch die vielen Stadien und Aspekte der jahrzehntelangen Forschungen Newtons, doch der überraschend kleine Farbkreis mit seinen verschieden großen Unterteilungen ist das wirkungsvollste und einflussreichste seiner Diagramme. In dem Versuch, die Bedürfnisse von Künstlern zu erfüllen, die mit stofflichen Farben arbeiteten, wurde der Kreis innerhalb

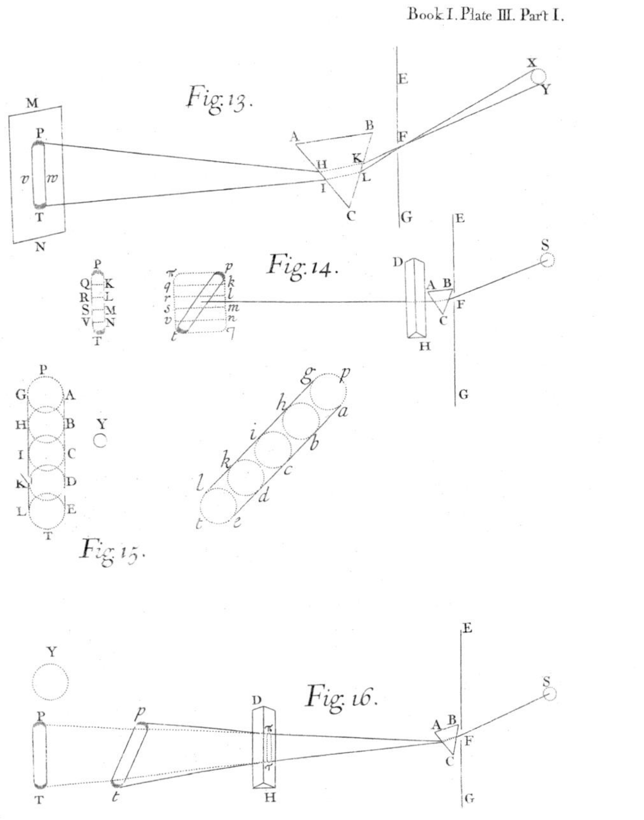

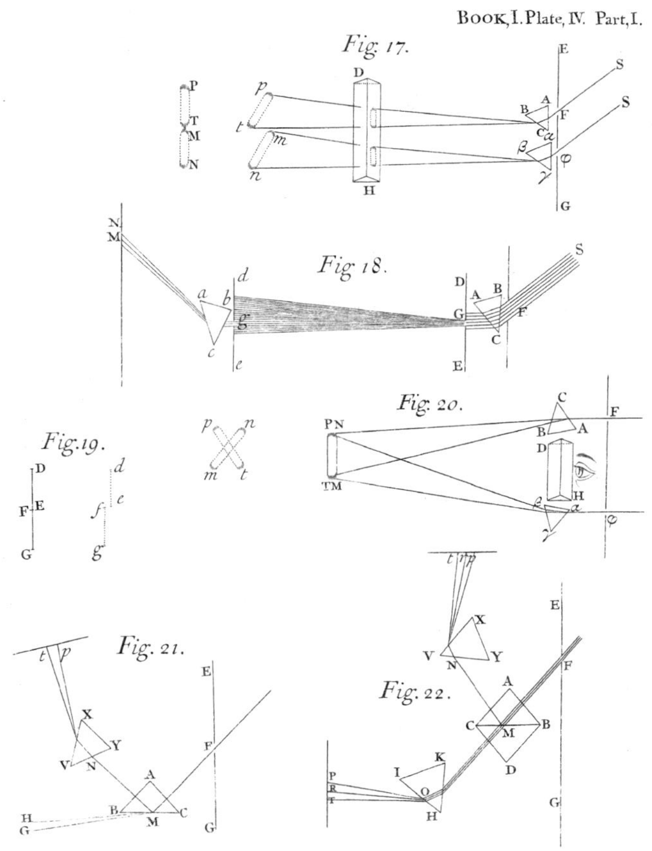

weniger Jahre von anderen Forschern übernommen und weiterentwickelt.

Traité d'optique sur les réflexions, réfractions, inflexions et couleurs de la lumière

Couramment considéré comme l'un des ouvrages majeurs de l'histoire de la couleur, l'*Opticks* de Newton a paru en 1704 et a imprimé sa marque sur les arts et les sciences du XVIIIe siècle. Quand paraît son traité, Newton s'intéresse au sujet depuis plusieurs décennies déjà. Il s'inscrit dans une longue lignée d'auteurs ayant étudié la couleur et les manières de la concevoir. S'il n'est pas le premier à avoir réalisé des recherches en optique, il se distingue grandement de ses prédécesseurs comme de ses contemporains par l'ampleur et la profondeur de ses travaux. Le statut et la notoriété de Newton, ainsi que les progrès de l'imprimerie, lui permettent de diffuser auprès d'un vaste public cet ouvrage qui, peu après sa première édition en anglais, sera traduit dans plusieurs langues. Son traité est analysé, enrichi et adapté grâce aux divers points de vue critiques de plusieurs auteurs des XVIIIe et XIXe siècles. Aujourd'hui encore, rares sont les textes, programmes de recherche ou conférences sur la couleur qui ne citent pas Newton. Le petit prisme qu'il achète au cours de sa jeunesse dans les années 1660 diffusera à travers les siècles un immense rayon de couleur. Son cercle chromatique est l'un des graphiques les plus marquants et reconnaissables de l'histoire des sciences.

C'est en 1666 que Newton réalise ses premières expérimentations avec des prismes dans son manoir de Woolsthorpe, dans le Lincolnshire. Il présente les résultats de son « expérience capitale » – la division de la lumière solaire pure en un spectre de couleurs prismatiques – dans les *Philosophical Transactions of the Royal Society*, à Londres en 1671. Cette manipulation lui permet de démontrer que les couleurs du spectre visible par l'œil humain se limitent à un nombre donné de teintes et qu'elles se recomposent sous la forme de lumière blanche lorsqu'on les dirige vers un second prisme. Il établit ainsi les principes du mélange additif, ou immatériel, des couleurs. Newton identifie d'abord cinq couleurs, puis sept, qu'il estime analogues à des gammes musicales : rouge, orange, jaune, vert, bleu, indigo et violet. Malgré l'aspect linéaire de l'ordre chromatique de ses expériences, il conçoit très tôt l'idée d'une forme circulaire pour représenter celui-ci et créera le cercle chromatique (mais incolore) qui figure dans *Opticks*. Ce traité contient des centaines de gravures illustrant les nombreux aspects et étapes des longues recherches de Newton. Pourtant, c'est un petit cercle chromatique, étonnamment modeste avec ses divisions inégalement proportionnées, qui constitue son graphique le plus frappant et dont l'influence sera gigantesque. En quelques années, il est remanié par plusieurs chercheurs qui tentent de le mettre au service des peintres travaillant avec la couleur matérielle.

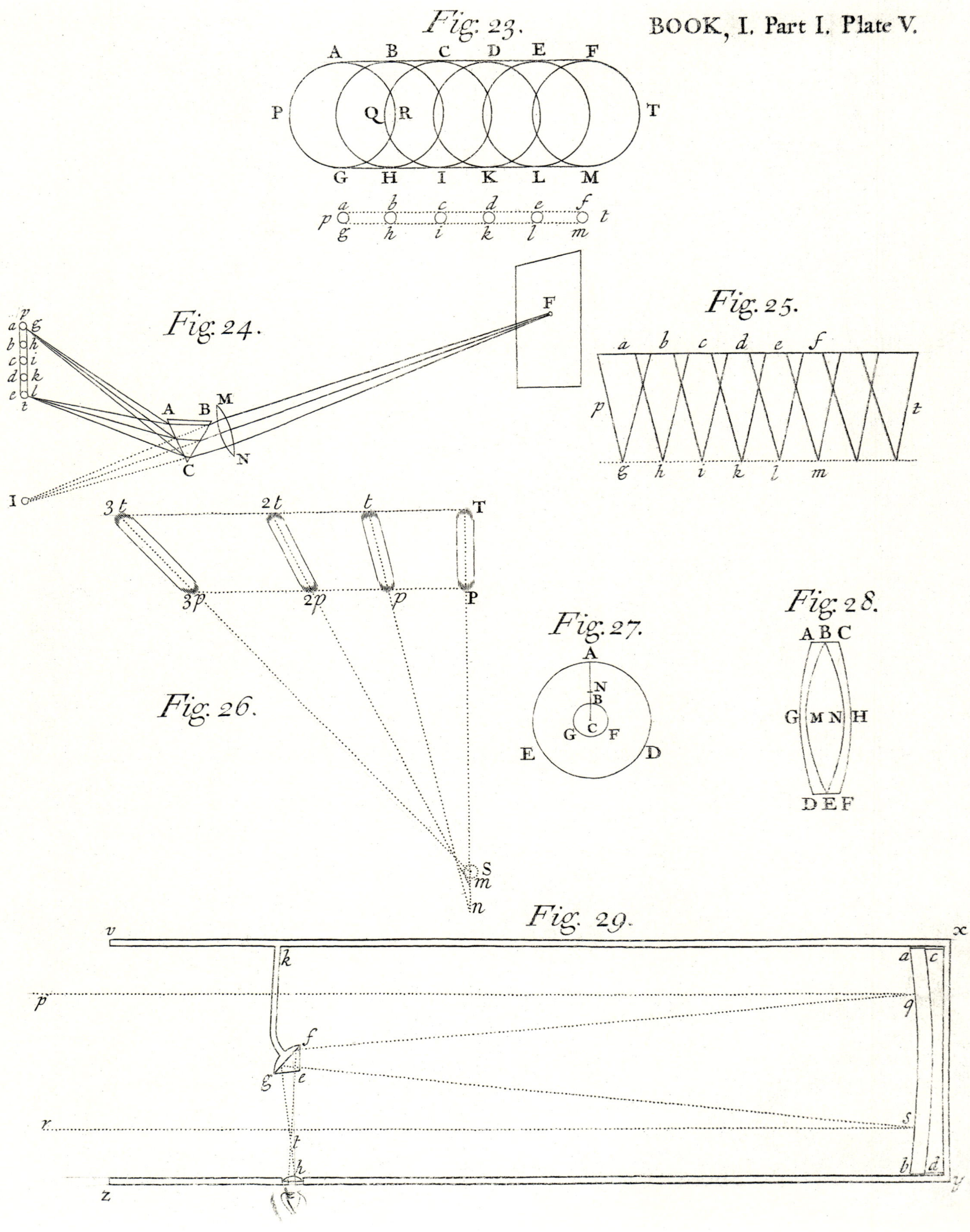
BOOK, I. Part I. Plate V.
Fig: 23.
A B C D E F
P Q R T
G H I K L M
a b c d e f
p t
g h i k l m
Fig: 24.
F
p a g b h c i d k e l t
A B M
C N
I
Fig: 25.
a b c d e f
p t
g h i k l m
3t 2t t T
3p 2p p P
Fig: 26.
S m n
Fig: 27.
A N B C G F E D
Fig: 28.
ABC
G M N H
DEF
Fig: 29.
v x
k a c
p q
f
g e
r s
t
h b d
z y

Book I. Part II. Plate I.

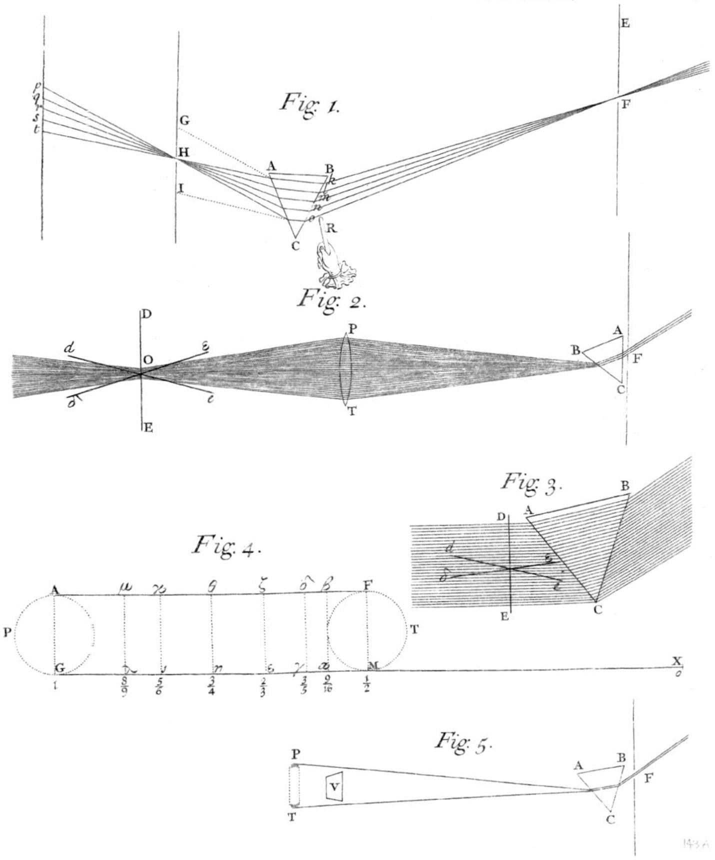

Book I. Part II. Plate II.

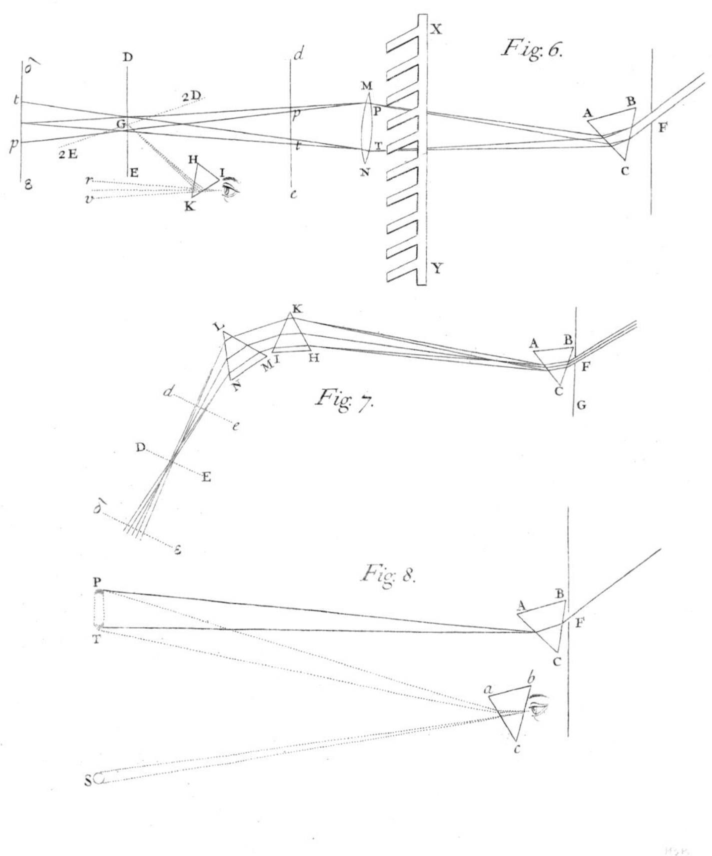

Book I. Part II. Plate

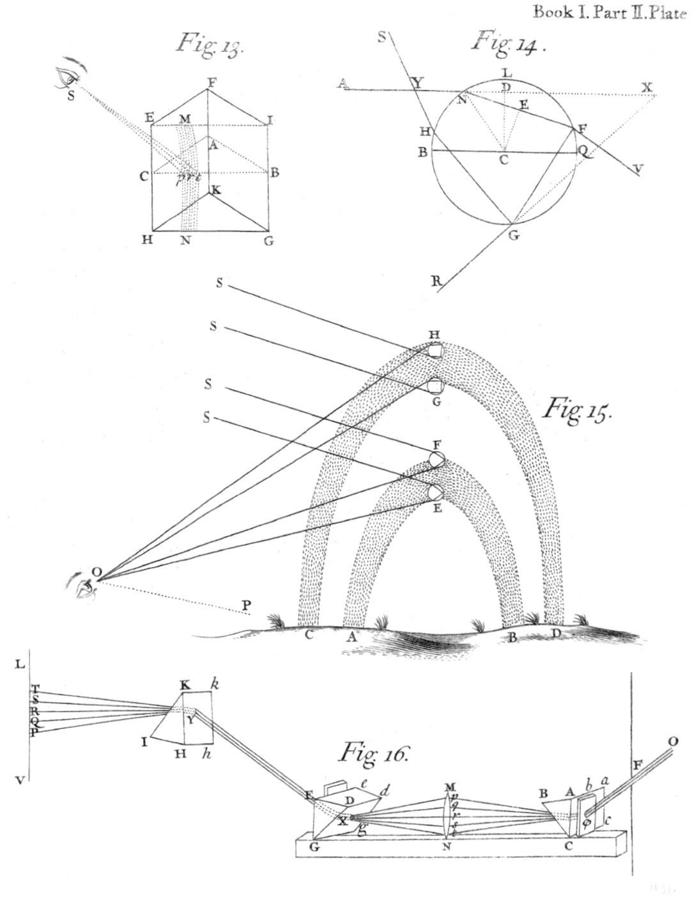

Book III. Plate I.

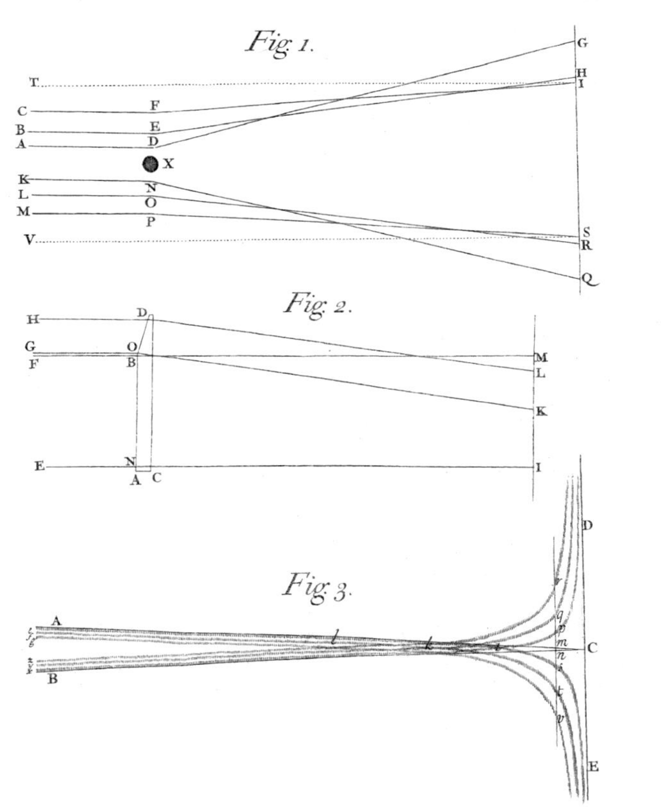

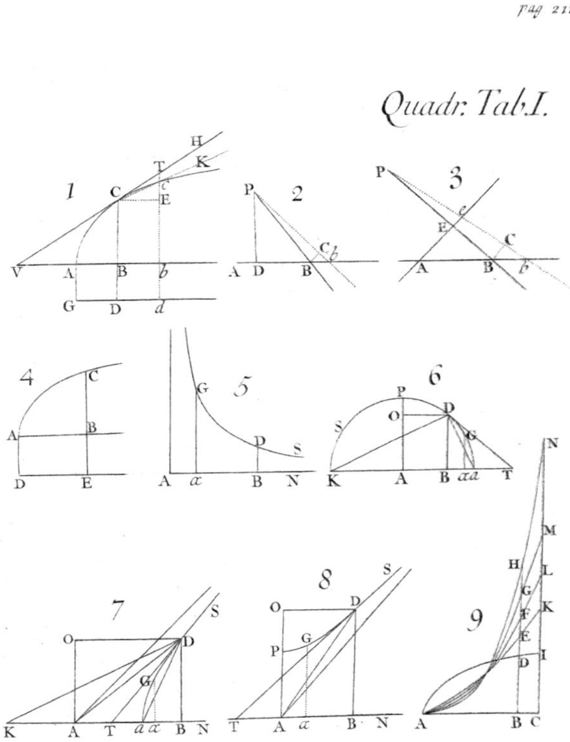

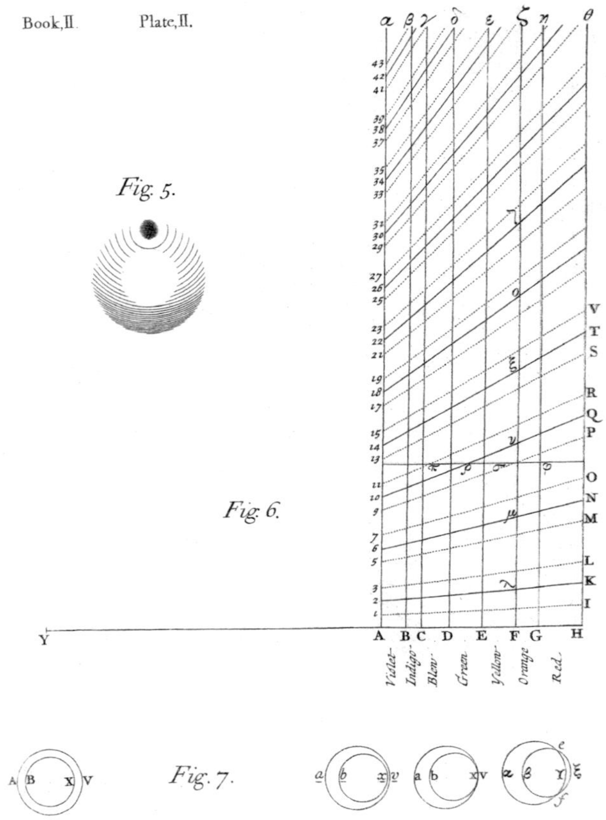

Optica, o Tratado de las reflexiones, refracciones, inflexiones y colores de la luz

La *Opticks* de Newton de 1704 tuvo un impacto inmediato en las artes y las ciencias en el siglo XVIII, y suele citarse como un hito de la historia del estudio del color. Newton se había ocupado de la materia durante décadas antes de que se publicara este libro, y era parte de una larga lista de figuras que habían investigado y trabajado los conceptos de color. Es cierto que no fue el primero en realizar estudios ópticos, pero su trabajo destaca sobre el de sus contemporáneos y predecesores en escala y minuciosidad. El estatus y la reputación de Newton, junto con los avances en la cultura de la impresión de su época, también ayudaron a difundir y popularizar esta obra, que se tradujo a varios idiomas poco después de la primera edición en inglés. En los siglos XVIII y XIX, varios autores la discutieron, desarrollaron y adaptaron desde una serie de perspectivas críticas, e incluso hoy en día es difícil encontrar un texto, programa o charla sobre el color que no mencione a Newton. El pequeño prisma que compró en la década de 1660 cuando todavía era un hombre joven ha proyectado un rayo de color deslumbrante a través de los siglos, y su círculo de color es uno de los diagramas más importantes y reconocibles de la historia de la ciencia.

Newton comenzó a experimentar con prismas en Woolsthorpe Manor, la casa de Lincolnshire donde nació, en 1666. Presentó por primera vez los resultados de su «experimento crucial» de dividir la luz solar pura en el espectro de colores prismáticos en las *Philosophical Transactions of the Royal Society* en Londres en 1671. Con ese experimento, demostró que los colores del espectro visible para el ojo humano no pueden dividirse en más colores y que, de hecho, vuelven a formar luz blanca cuando se redirigen a través de un segundo prisma, lo que estableció los principios de la mezcla de colores aditivos o inmateriales. Al principio, Newton señaló cinco colores identificables, pero luego se decidió por siete, que entendió como análogos a las escalas musicales, a saber: rojo, naranja, amarillo, verde, azul, índigo y violeta. A pesar del orden lineal de los colores en sus experimentos, desde el principio imaginó una forma circular, y finalmente creó la rueda de colores (sin colorear) que se imprimió en *Opticks*. Hay cientos de grabados en *Opticks* que ilustran las etapas y aspectos de las décadas de investigación de Newton, pero esa rueda de colores sorprendentemente pequeña, con sus divisiones de proporciones desiguales, es su diagrama más potente e influyente. A los pocos años, otros investigadores lo adaptaron en un intento por satisfacer las necesidades de los artistas que trabajaban con colores materiales.

Book, II. Plate, I.

Fig. 1.

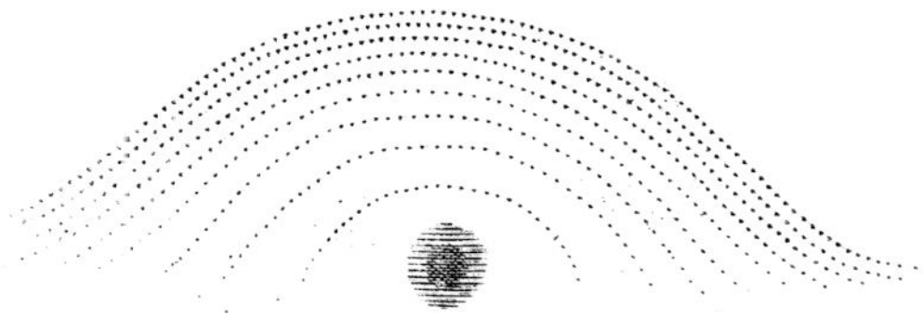

Fig. 2.

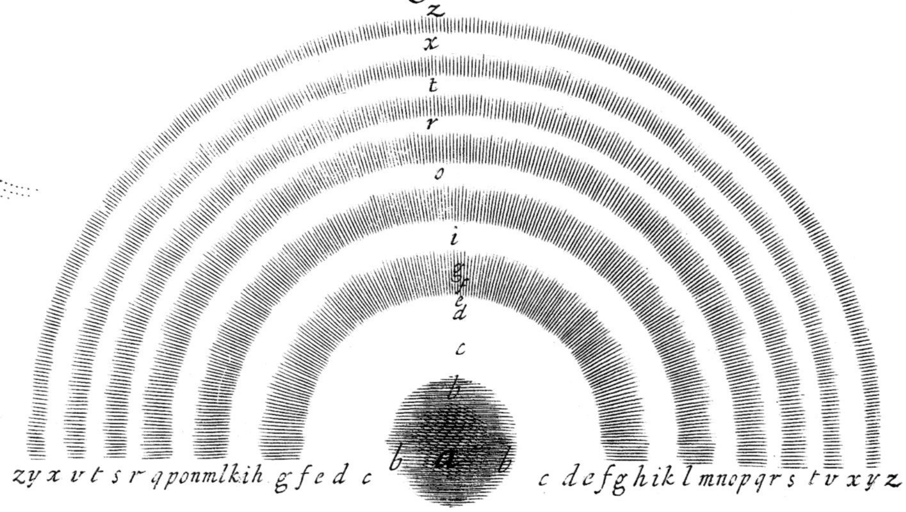

Fig. 3.

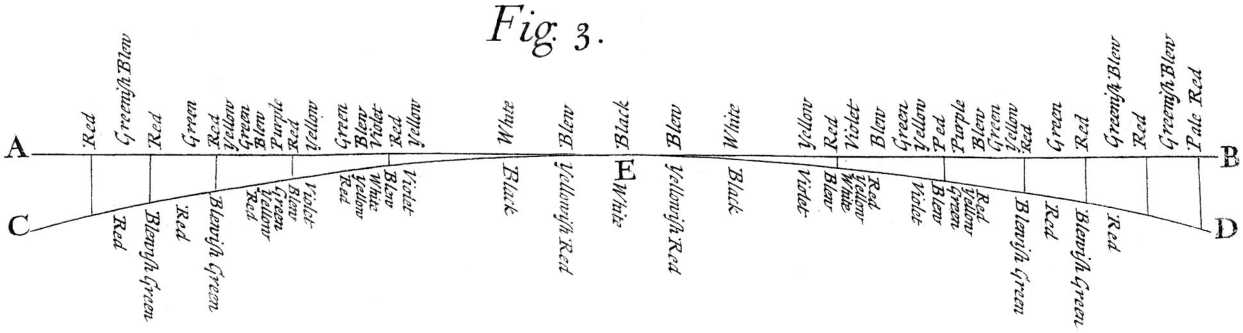

Fig. 4.

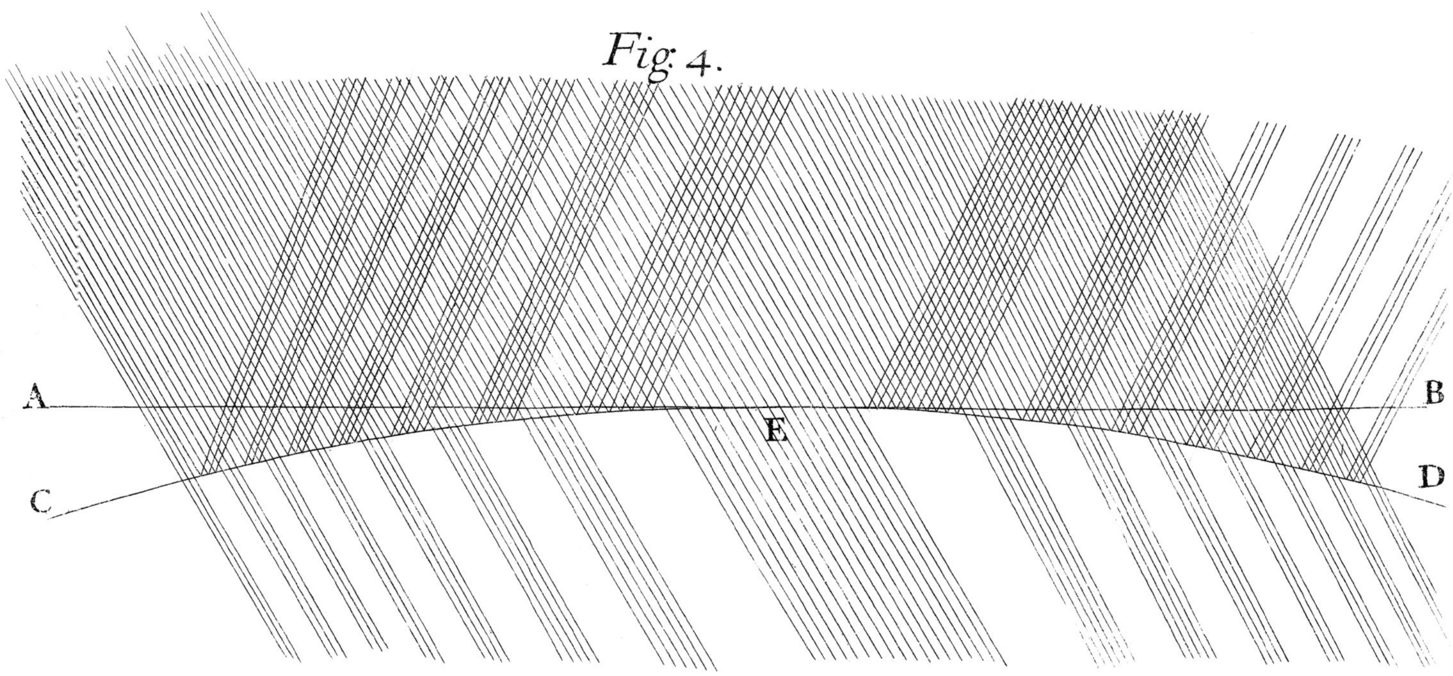

Mary Gartside (*c.* 1755–1819)

AN ESSAY ON LIGHT AND SHADE, ON COLOURS, AND ON COMPOSITION IN GENERAL

10 plates, 3 illustrated pages, 29 x 22.5 cm / 11 ⅜ x 8 ⅞ in., London, 1805
Sussex, collection of the author

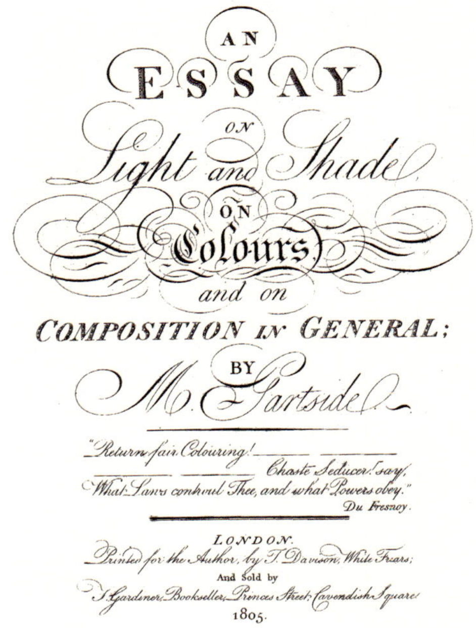
AN ESSAY ON Light and Shade ON Colours and on COMPOSITION IN GENERAL; BY M. Gartside.

"Return fair Colouring! ——— Chaste Seducer! say, What Laws control Thee, and what Powers obey."
Du Fresnoy.

LONDON. Printed for the Author, by T. Davison, White Friars; And Sold by J. Gardiner, Bookseller, Princes Street, Cavendish Square. 1805.

The English flower painter Mary Gartside was very likely the first woman to have published illustrated books on colour. Women were not excluded from colour history, and had played a part as allegorical figures, colourists, assistants, illuminators, translators, teachers and in other roles, but rarely had they been able to participate as authors or theorists. One early exception is Catherine Perrot (1620–169?), who taught art at the French court and published a treatise on painting miniatures in 1686. In some ways, Gartside's road to publishing books on colour was similar, since she too worked as a painting teacher, which was an acceptable occupation for a woman at the time and also justified writing about it. Very little is known about her life, but she secured the patronage of Lady Sophia Grey of Dunham Massey, in Cheshire, and managed to carve out a lifelong career in art which included publishing a pamphlet and three books on colour between around 1804 and 1808. Although she could not enrol at the Royal Academy as a student, she exhibited some of her botanical drawings there in 1781 and at other venues in London until 1808, thereby establishing a personal network of artists, teachers, botanists and colour researchers, while always carefully operating within what was socially acceptable for an unmarried woman in Georgian England. But there is more to the story of Mary Gartside.

Her books, of which *An Essay on Light and Shade* was the first, were overtly conceived as watercolour painting manuals for young women, but beyond that Gartside succeeded in placing herself within a tradition of colour theorists, adding the words "new theory" to the title of her second book. Her work mentioned other researchers, among them Isaac Newton and Moses Harris, making it clear that she had a sound knowledge of colour discourse, and it is possible that she knew James Sowerby (pp. 230–235) in person.

An Essay on Light and Shade is illustrated with two pictorial plates (soft-ground etchings), two coloured tables, a further coloured engraving showing a colour circle described as a "ball" and eight plates of colour "blots". It is these last eight plates that have given Gartside a special place in colour history. While essentially examples of her colour theory applied to composition and the harmonious arrangement of colours (white, yellow, orange, green, scarlet, blue, violet and crimson), they are not coloured engravings but freely painted masses of colour that vaguely resemble flowers seen through wet glass, and because they were hand-painted the images are different in each copy of the book. The concept was very clearly Gartside's own, but she could not have painted all the blots herself for the entire print run, and it is likely that she used some of her students to help out. Gartside's colour blots were inventive, radical, ambitious and suffused with an abstract beauty that has only recently begun to be recognised and appreciated.

Yellow. 2.

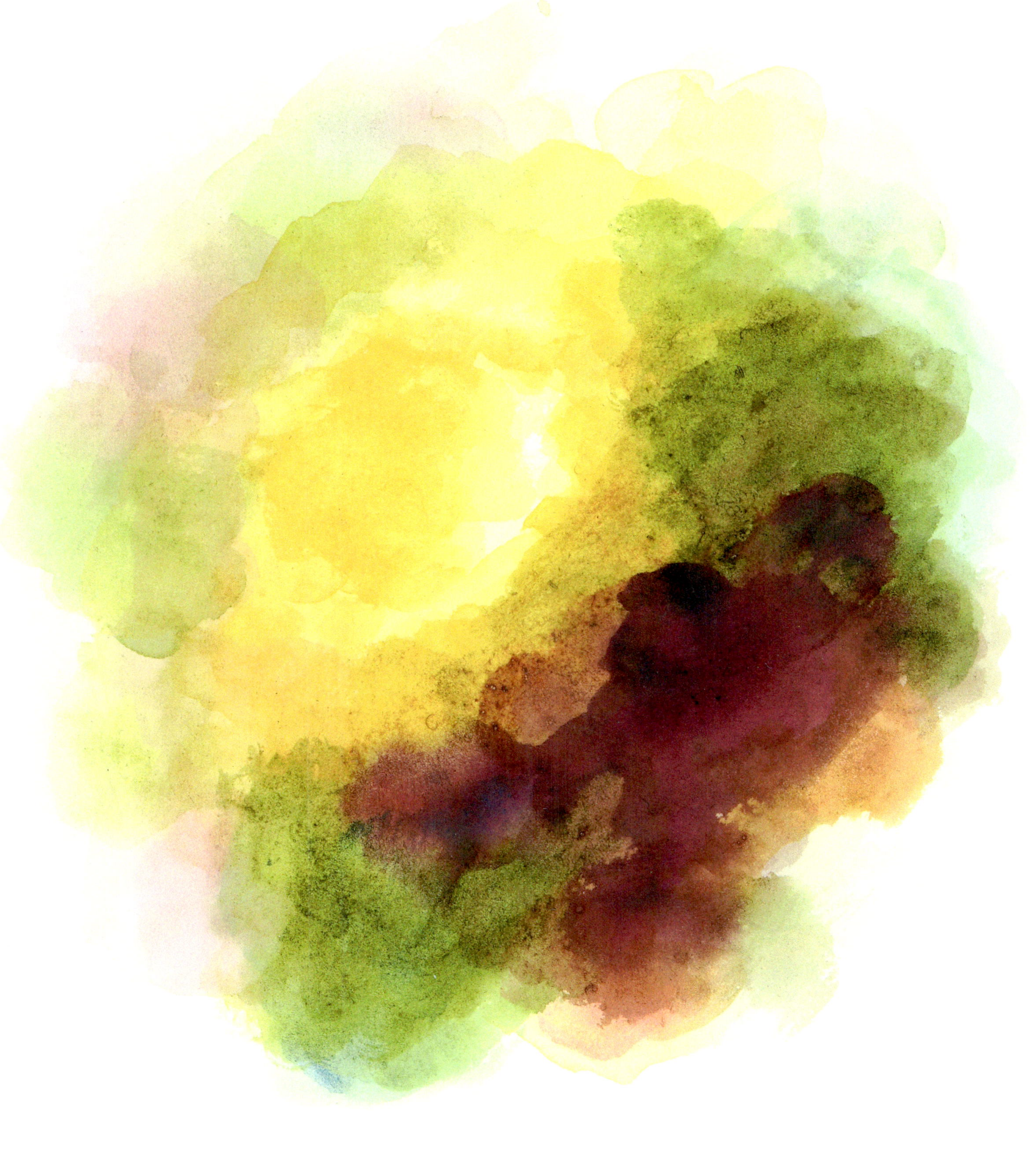

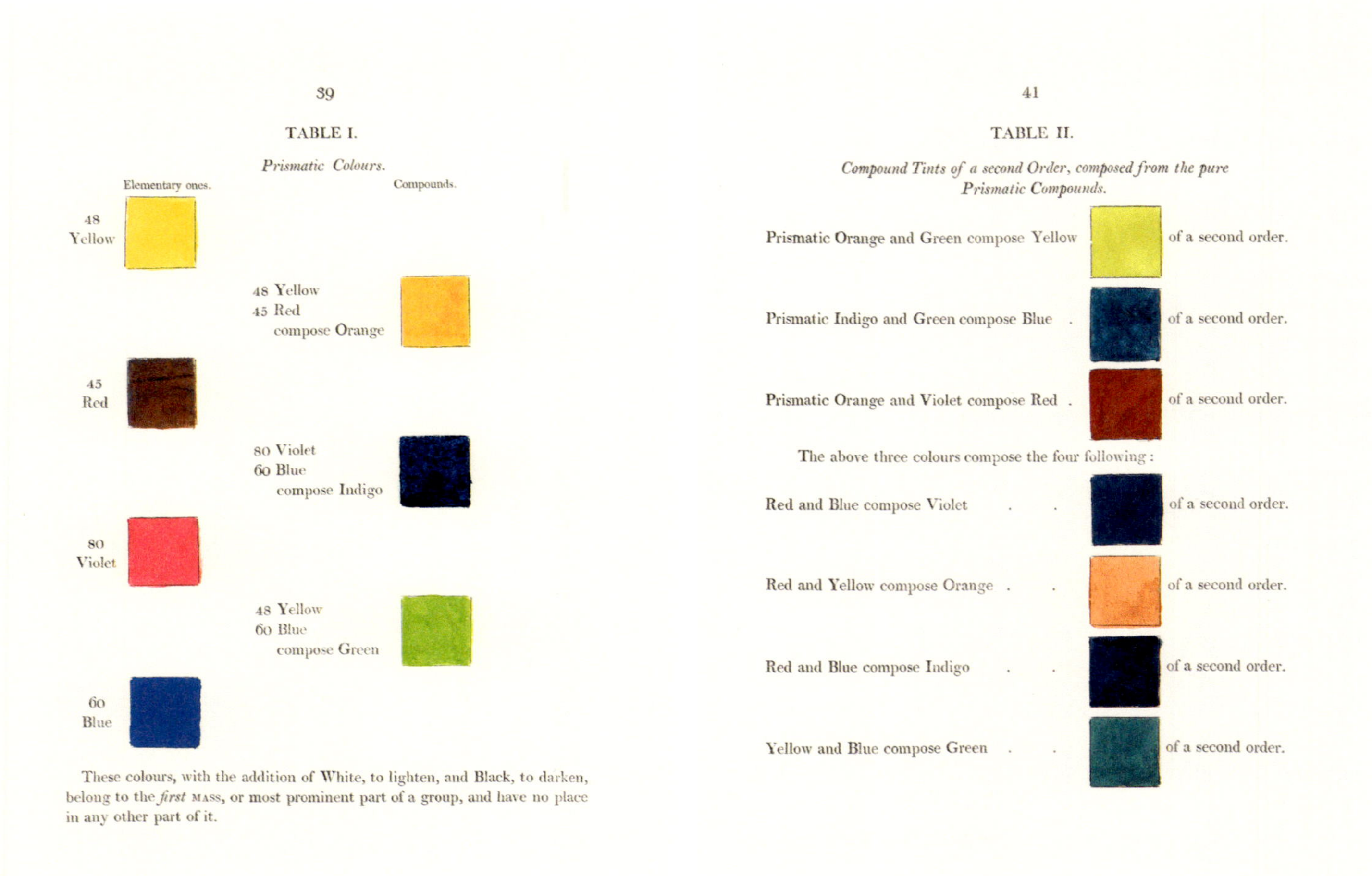

Ein Aufsatz über Licht und Schatten, über Farben und über Komposition im Allgemeinen

Die englische Blumenmalerin Mary Gartside war die höchstwahrscheinlich erste Frau, die ein illustriertes Buch über Farbe veröffentlichte. Auch Frauen haben in der Geschichte der Farbtheorie eine Rolle gespielt, zum Beispiel als allegorische Figuren, Koloristinnen, Helferinnen, Illuminatorinnen, Übersetzerinnen oder Lehrerinnen, doch selten waren sie in der Lage gewesen, einen Beitrag als Autorinnen oder Theoretikerinnen zu leisten. Eine frühe Ausnahme bildete Catherine Perrot (1620–169?), die am französischen Hof Kunst unterrichtete und 1686 eine Abhandlung über Miniaturmalerei herausgab. In mancherlei Hinsicht verlief Gartsides Weg zur Veröffentlichung von Farbenbüchern ähnlich, denn auch sie unterrichtete Malerei, was als Beschäftigung für Frauen damals akzeptiert wurde und auch rechtfertigte, dass sie darüber schrieben. Über Gartsides Leben ist sehr wenig bekannt. Man weiß, dass sie sich die Unterstützung von Lady Sophia Grey aus Dunham Massey in der Grafschaft Cheshire erwarb und es ihr gelang, ein Leben lang im Bereich der Kunst zu arbeiten und um 1804 bis 1808 eine Broschüre und drei Bücher über Farbe zu publizieren. Obwohl sie sich nicht als Studentin an der Royal Academy of Arts einschreiben konnte, stellte Gartside dort 1781 und an anderen Orten in London bis 1808 ihre botanischen Zeichnungen aus und baute sich auf diese Weise ein persönliches Netzwerk aus Künstlerinnen und Künstlern, Lehrern, Botanikern und Farbforschern auf. Gleichzeitig achtete sie sorgsam darauf, nur in einem Rahmen zu agieren, der für eine unverheiratete Frau im georgianischen England gesellschaftlich anerkannt war. Doch das ist nicht die ganze Geschichte von Mary Gartside.

Ihre Bücher, von denen *An Essay on Light and Shade* das erste war, waren offenkundig als Aquarellmalhandbücher für junge Frauen angelegt. Darüber hinaus jedoch stellte sich Gartside erfolgreich in eine Tradition von Farbtheoretikern, indem sie dem Titel ihres zweiten Buches die Worte „neue Farbenlehre“ hinzufügte. In ihrem Werk erwähnte sie andere Forscher, darunter Isaac Newton und Moses Harris, und machte deutlich, dass sie sich im Farbdiskurs auskannte. Persönlich bekannt war sie möglicherweise mit James Sowerby (S. 230–235).

An Essay on Light and Shade ist mit zwei malerischen Tafeln (Weichgrundätzungen), zwei farbigen Tabellen, einem weiteren kolorierten Stich eines als „Ball“ bezeichneten Farbkreises und acht Tafeln mit farbigen „Flecken“ bebildert. Diese letzten acht Tafeln brachten Gartside eine Sonderstellung in der Geschichte der Farben ein. Im Wesentlichen handelt es sich dabei um Beispiele für ihre Farbtheorie, angewandt auf den

Bildaufbau und die harmonische Anordnung von Farben (Weiß, Gelb, Orange, Grün, Scharlachrot, Blau, Violett und Karminrot). Doch anstelle von kolorierten Stichen sind frei aufgetragene Farben zu sehen, die entfernt an Blüten hinter nassen Glasscheiben erinnern, und weil sie von Hand gemalt wurden, unterscheiden sich die Bilder von Buch zu Buch. Das Konzept stammte zweifellos von Gartside selbst. Da sie jedoch unmöglich alle Flecken der gesamten Auflage eigenhändig gemalt haben kann, ließ sie sich wahrscheinlich von ihren Schülerinnen helfen. Gartsides Farbflecken waren originell, radikal, ambitioniert und sind zudem von einer abstrakten Schönheit, die erst seit Kurzem Anerkennung und Würdigung erfahren hat.

Essai sur la lumière et l'ombre, sur les couleurs et sur la composition en général

Peintre de fleurs, l'Anglaise Mary Gartside est très probablement la première femme à avoir publié des ouvrages illustrés ayant pour thème la couleur. Si les femmes ont joué un rôle dans l'histoire de la couleur, en y contribuant notamment comme figures allégoriques, coloristes, assistantes, enlumineuses, traductrices, pédagogues, rares sont celles qui ont pu accéder au statut d'autrices ou de théoriciennes. L'une des toutes premières exceptions est Catherine Perrot (1620–169?), qui enseigna la peinture à la cour de France et publia un traité sur l'art de la miniature en 1686. À certains égards, Gartside suit une voie semblable avant de faire paraître des livres traitant de la couleur, puisqu'elle enseigne aussi la peinture, activité acceptable pour une femme de son époque et prétexte à publier. On sait fort peu de choses de sa biographie, si ce n'est qu'elle a obtenu le patronage de Lady Sophia Grey de Dunham Massey, dans le Cheshire, et parviendra à mener une carrière artistique tout au long de sa vie. Elle est l'autrice d'un opuscule et de trois ouvrages consacrés à la couleur, parus entre 1804 et 1808. Bien qu'elle ne puisse s'inscrire à la Royal Academy, elle y expose une partie de ses dessins botaniques en 1781 et dans d'autres salons londoniens jusqu'en 1808. Elle se crée ainsi un réseau de peintres, enseignants, botanistes et chercheurs sur la couleur, en veillant constamment à rester dans les limites de ce que la société anglaise permet à une femme célibataire à l'époque georgienne. Mais l'histoire de Mary Gartside ne s'arrête pas là.

Ses ouvrages, dont le premier est cet « Essai sur la lumière et l'ombre », prennent la forme délibérée de manuels d'aquarelle destinés aux jeunes femmes. Mais leur autrice parvient à s'inscrire dans une lignée de théoriciens de la couleur en ajoutant au titre de la nouvelle édition les mots « nouvelle théorie ». Elle y cite d'autres chercheurs, comme Isaac Newton et Moses Harris et montre, par là même, qu'elle a une connaissance solide du discours sur la couleur. Il est même possible qu'elle ait rencontré James Sowerby (p. 230–235).

An Essay on Light and Shade est illustré de deux planches

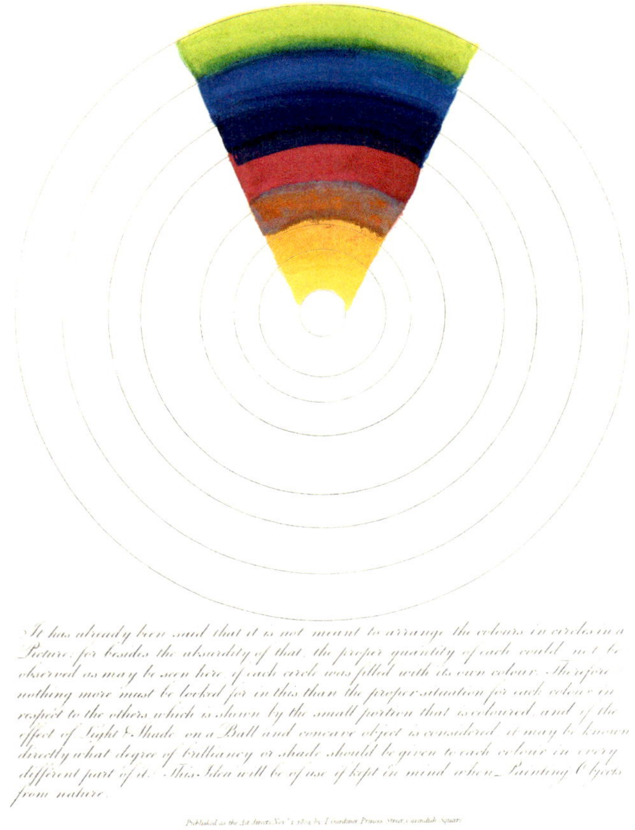

descriptives (eaux-fortes au vernis mou), de deux tableaux de couleurs, d'une autre gravure en couleurs représentant un cercle chromatique qualifié de « boule » et de huit planches de « taches » de couleurs. C'est grâce à ces dernières planches que Gartside occupe une place particulière dans l'histoire de la couleur. Il s'agit avant tout d'exemples de sa théorie chromatique appliquée à la composition et à l'organisation harmonieuse des teintes (blanc, jaune, orange, vert, écarlate, bleu, violet et cramoisi). Ce ne sont pas des gravures colorées, mais des taches de couleurs qui ressemblent un peu à des fleurs vues à travers du verre humide. Peintes à la main, ces images diffèrent d'un exemplaire à l'autre de l'ouvrage. Si Gartside est bien l'autrice de cette conception, il est impossible qu'elle ait peint elle-même les taches de tous les exemplaires du tirage et elle a probablement bénéficié de l'aide de ses disciples. Tout à la fois inventives, radicales et ambitieuses, ses taches de couleurs se caractérisent par une beauté abstraite qui n'est reconnue et appréciée que depuis peu.

Ensayo sobre la luz y la sombra, sobre los colores y sobre la composición en general

La pintora de flores inglesa Mary Gartside fue muy probablemente la primera mujer que publicó libros ilustrados sobre el color. Las mujeres habían participado en la historia del color como figuras alegóricas, coloristas, ayudantes,

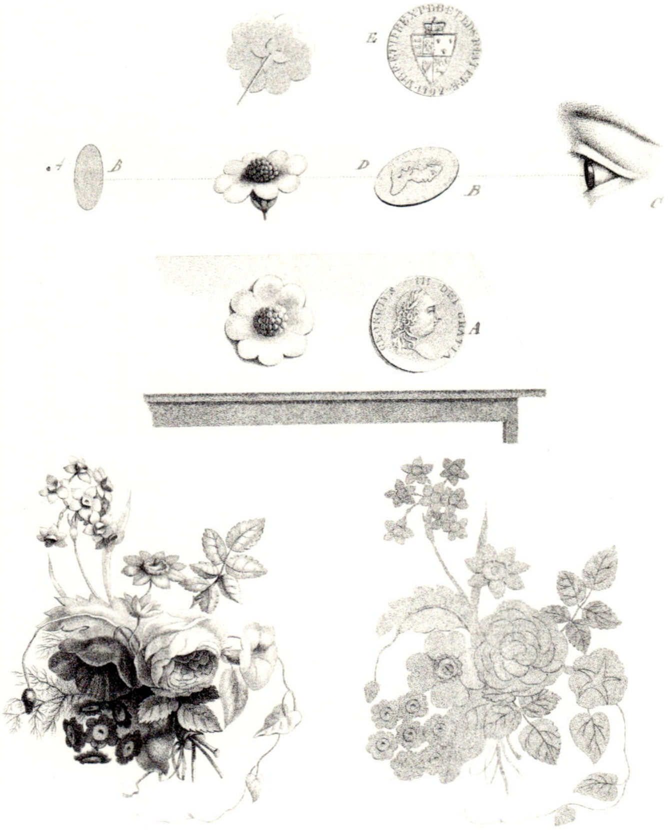

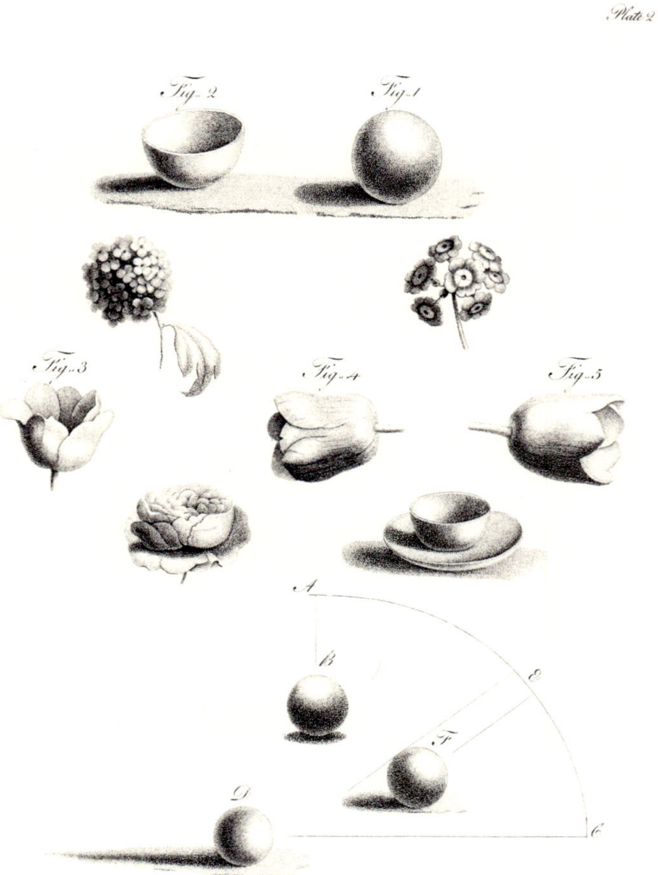

iluminadoras, traductoras y docentes, entre otros roles, pero rara vez habían podido participar como autoras o teóricas. Una de las primeras excepciones es Catherine Perrot (1620–169?), quien enseñó arte en la corte francesa y publicó un tratado sobre la pintura de miniaturas en 1686. En cierto modo, el camino de Gartside hacia la publicación de libros sobre el color fue similar, ya que ella también trabajó como profesora de pintura, una ocupación aceptable para una mujer en ese momento que también justificaba escribir sobre la materia. Se sabe muy poco acerca de su vida, pero consiguió el mecenazgo de lady Sophia Grey de Dunham Massey, en Cheshire, y logró labrarse una carrera de por vida en el arte que incluyó la publicación de un folleto y tres libros sobre el color entre 1804 y 1808. Aunque no pudo matricularse en la Royal Academy como estudiante, expuso allí algunos de sus dibujos botánicos en 1781 y en otros lugares de Londres hasta 1808, lo que le permitió establecer una red de contactos personal formada por artistas, profesores, botánicos e investigadores del color, aunque siempre actuando dentro de lo que era socialmente aceptable para una mujer soltera en la Inglaterra georgiana. Pero la historia de Mary Gartside trasciende ese marco.

Sus libros, de los cuales *An Essay on Light and Shade* fue el primero, se presentaron como manuales de pintura en acuarela para mujeres jóvenes. Más allá de eso, Gartside logró ubicarse dentro de una tradición de teóricos del color al añadir el término «nueva teoría» al título de su segundo libro. Su trabajo menciona a otros investigadores, con Isaac Newton y Moses Harris entre ellos, lo que deja claro que tenía un sólido conocimiento del discurso del color, y es posible que conociera a James Sowerby (págs. 230–235) en persona.

An Essay on Light and Shade está ilustrado con dos láminas grabadas al aguafuerte, dos tablas de colores, otro grabado en color que muestra un círculo de color descrito como una «bola» y ocho láminas de «manchas» de colores. Estas últimas ocho láminas son las que le han dado a Gartside un lugar especial en la historia de la disciplina. Si bien son en esencia ejemplos de su teoría del color aplicada a la composición y la disposición armoniosa de los colores (blanco, amarillo, naranja, verde, escarlata, azul, violeta y carmesí), no son grabados coloreados sino masas de color pintadas libremente que se asemejan vagamente a flores vistas a través de un vidrio húmedo, y debido a que fueron pintadas a mano, las imágenes son diferentes en cada copia del libro. El concepto era claramente propio de Gartside, pero esta no pudo haber pintado ella misma todas las manchas de toda la tirada, y es probable que se valiera de la ayuda de algunos de sus alumnos. Las manchas de color de Gartside son inventivas, radicales, ambiciosas y están impregnadas de una belleza abstracta que solo recientemente se ha comenzado a reconocer y apreciar.

White: 1.

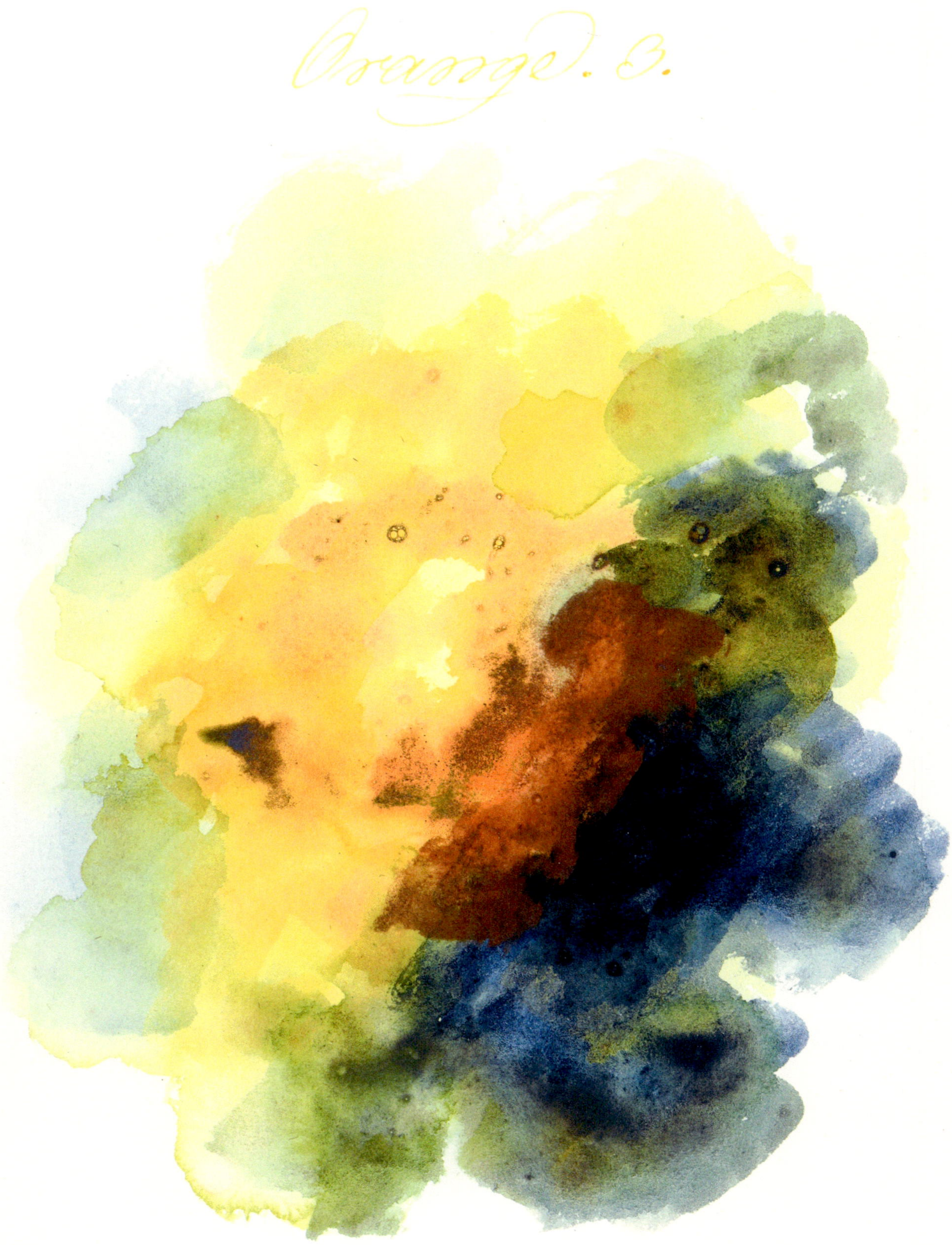
Orange. 3.

Green. 4.

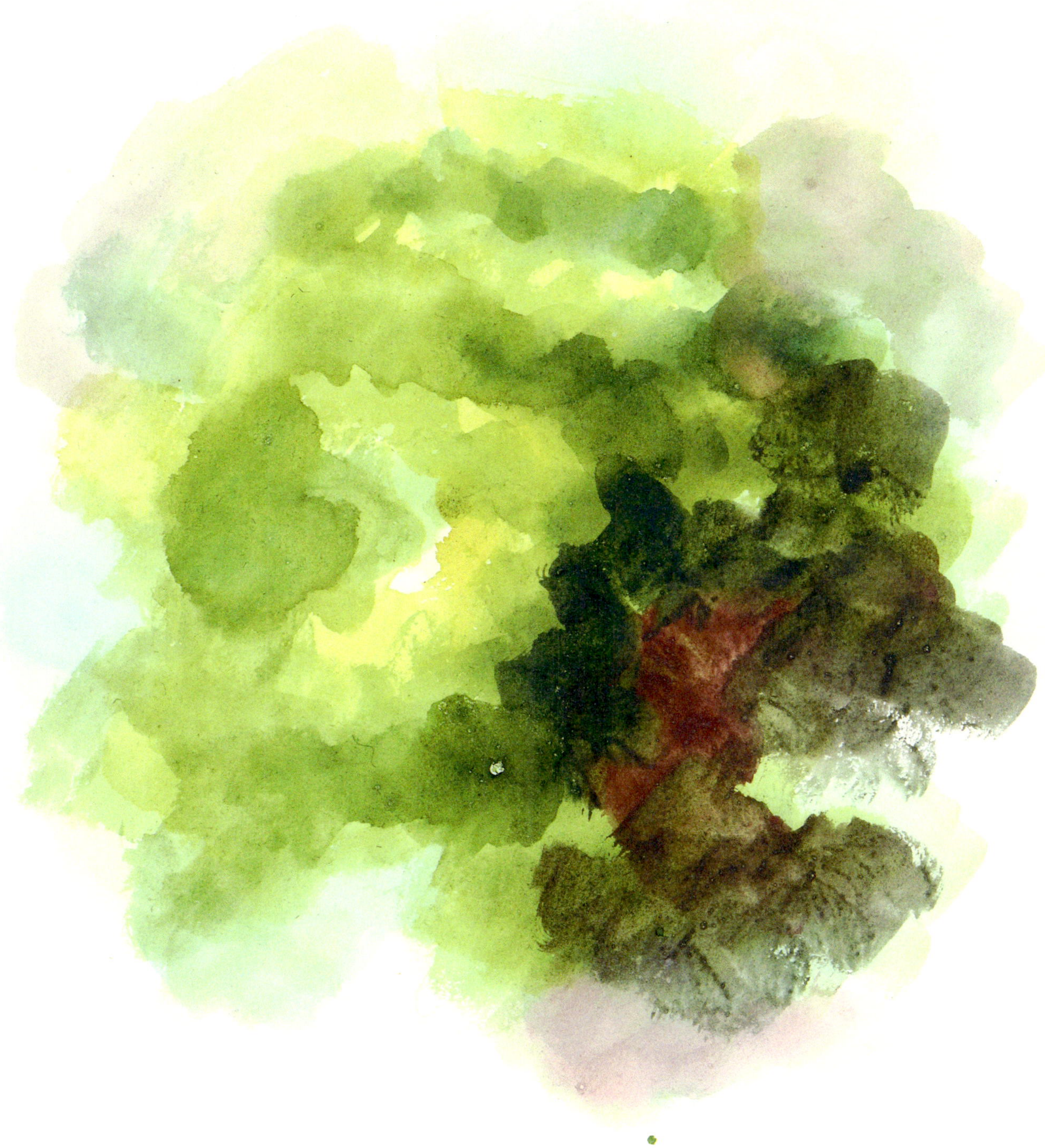

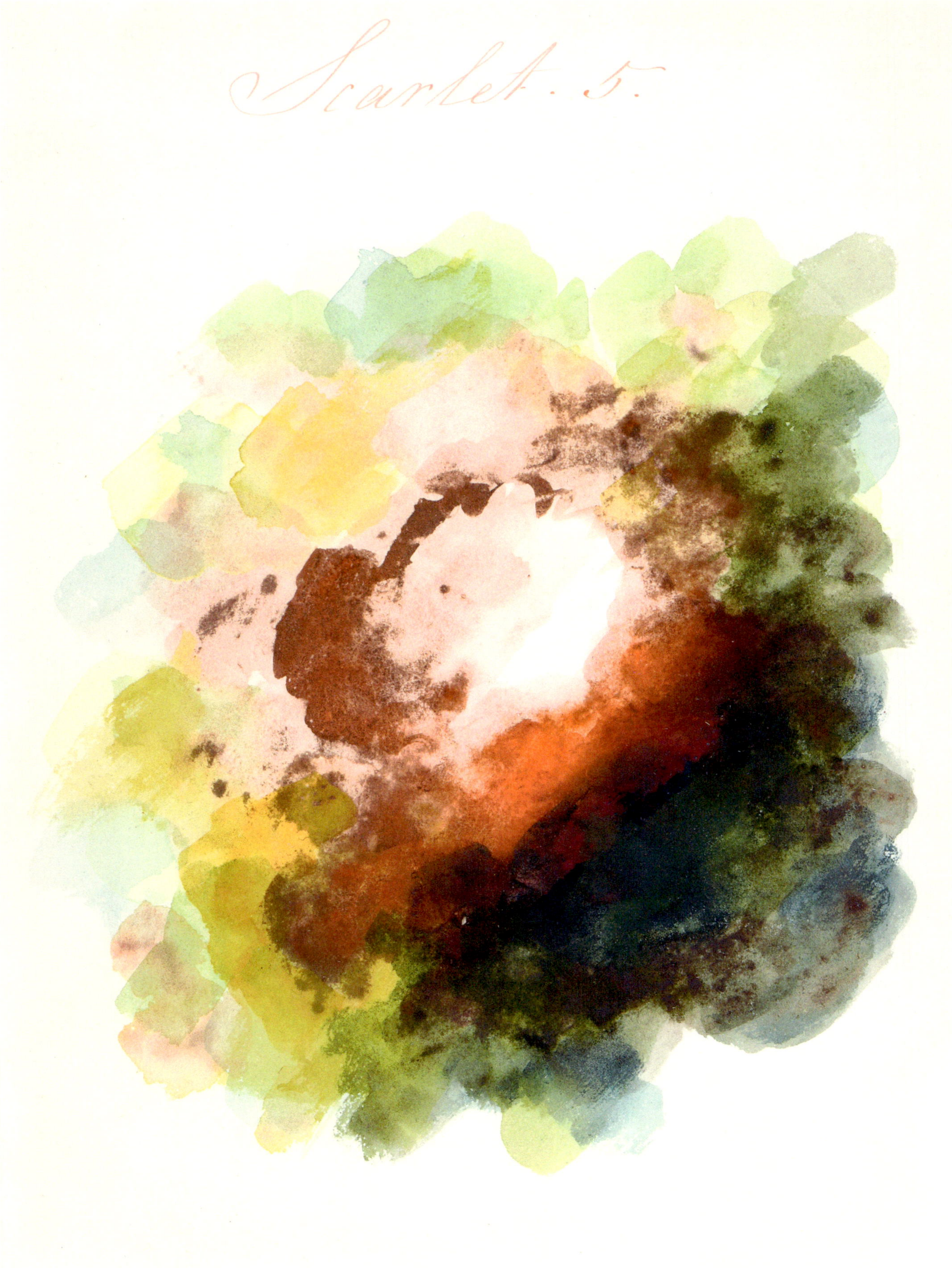
Scarlet. 5.

Blue 6.

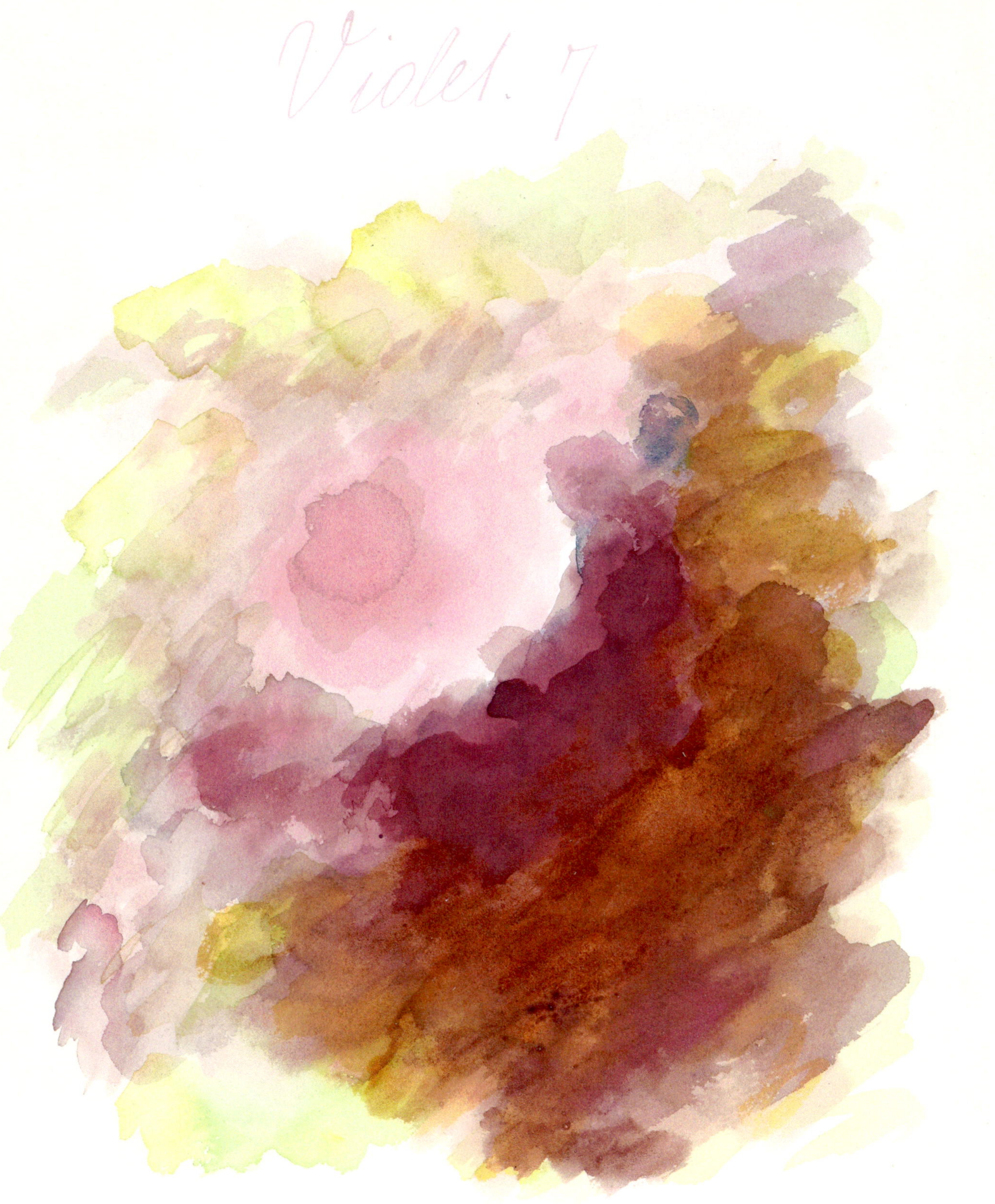
Violet 7

Crimson.

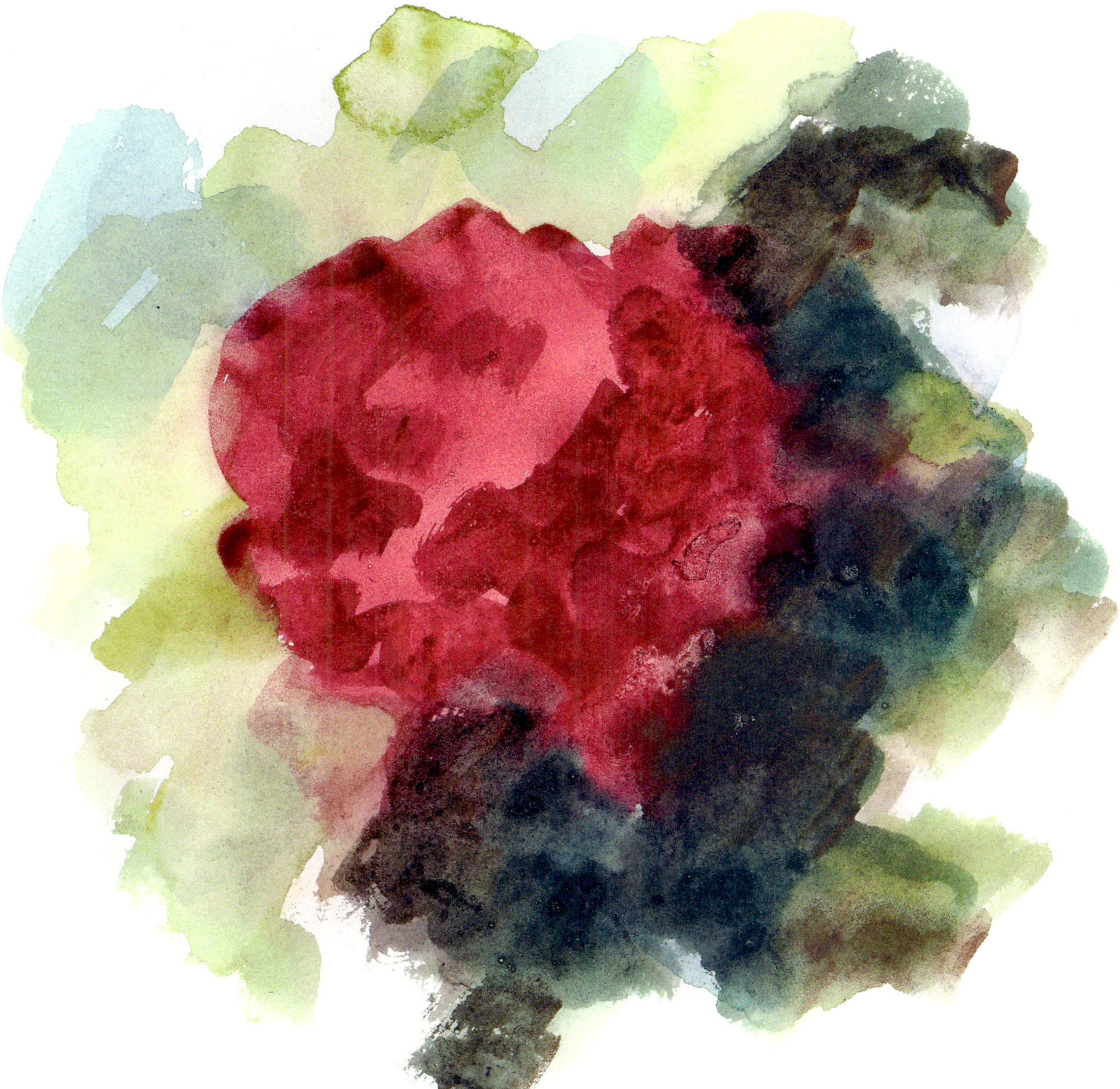

James Sowerby (1757–1822)

A NEW ELUCIDATION OF COLOURS, ORIGINAL PRISMATIC, AND MATERIAL

Showing Their Concordance in Three Primitives, Yellow, Red, and Blue

7 plates, 31 x 25.5 cm | 12 ¼ x 10 in., London, 1809
London, Wellcome Collection

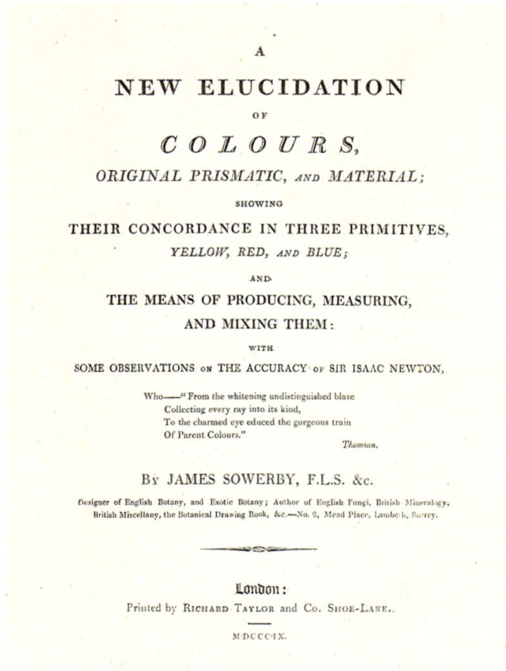
A

NEW ELUCIDATION

OF

COLOURS,

ORIGINAL PRISMATIC, AND MATERIAL;

SHOWING

THEIR CONCORDANCE IN THREE PRIMITIVES,

YELLOW, RED, AND BLUE;

AND

THE MEANS OF PRODUCING, MEASURING,

AND MIXING THEM:

WITH

SOME OBSERVATIONS ON THE ACCURACY OF SIR ISAAC NEWTON,

Who—"From the whitening undistinguished blaze
Collecting every ray into its kind,
To the charmed eye educed the gorgeous train
Of Parent Colours."

Thomson.

BY JAMES SOWERBY, F.L.S. &c.

Designer of English Botany, and Exotic Botany; Author of English Fungi, British Mineralogy, British Miscellany, the Botanical Drawing Book, &c.—No. 2, Mead Place, Lambeth, Surrey.

London:

Printed by RICHARD TAYLOR and Co. SHOE-LANE.

M·DCCC·IX.

The naturalist James Sowerby published this book at an exciting time for anyone interested in this field. A few years earlier, the polymath Thomas Young (1773–1829) had presented his theory of trichromatic colour vision, while Runge and Goethe were both busy at this time working on their colour concepts and imminent publications. It is probable that Sowerby would have heard that Mary Gartside's wondrously illustrated book on colour and flower painting had gone into a second edition in 1808 (now titled *An Essay on a New Theory of Colours*), and to this wealth of major new works on colour he added his own, *A New Elucidation of Colours*, which he dedicated to Newton, whose *Opticks* had been published just over 100 years earlier. A "new" approach then, an attempt to redirect the emphasis to light and shade and the differences between colours, but an approach that still paid homage to Newton's achievements. Sowerby also attempted to remove confusion and explain the difference between additive (immaterial) and subtractive (material) colour mixtures, while as a naturalist and prolific illustrator he shared with Moses Harris a concern with the subtleties of colour in the natural world, concentrating in particular on the painters' primaries red, yellow and blue and choosing as their "most perfect" pigment equivalents carmine, gamboge and Prussian blue. Sowerby was the inventor of a Chromatometer, a type of prism to measure the amount of dark and light in colours and which was presented and described in *A New Elucidation*. The device was meant to be used with the plates in the book that show bands of colour or black (Goethe had done something similar in 1791; see pp. 28–29, 459). Sowerby's main colour diagram, the "Chromatic Scale" (Table 5), is an elegant variation on a colour wheel, overlaid with a three-banded triangle in which the primaries in their purest form are located on the outer bands in the middle portions between the vertices, with the tints becoming brighter towards the centre. The three additional diamonds added to the bottom of the triangle contain further mixtures, in a similar way to Harris's second colour wheel, resulting in a total of 63 colours, plus the white central triangle. It is a beautifully executed diagram, using layers of transparent watercolour paint, albeit one that remained firmly two-dimensional.

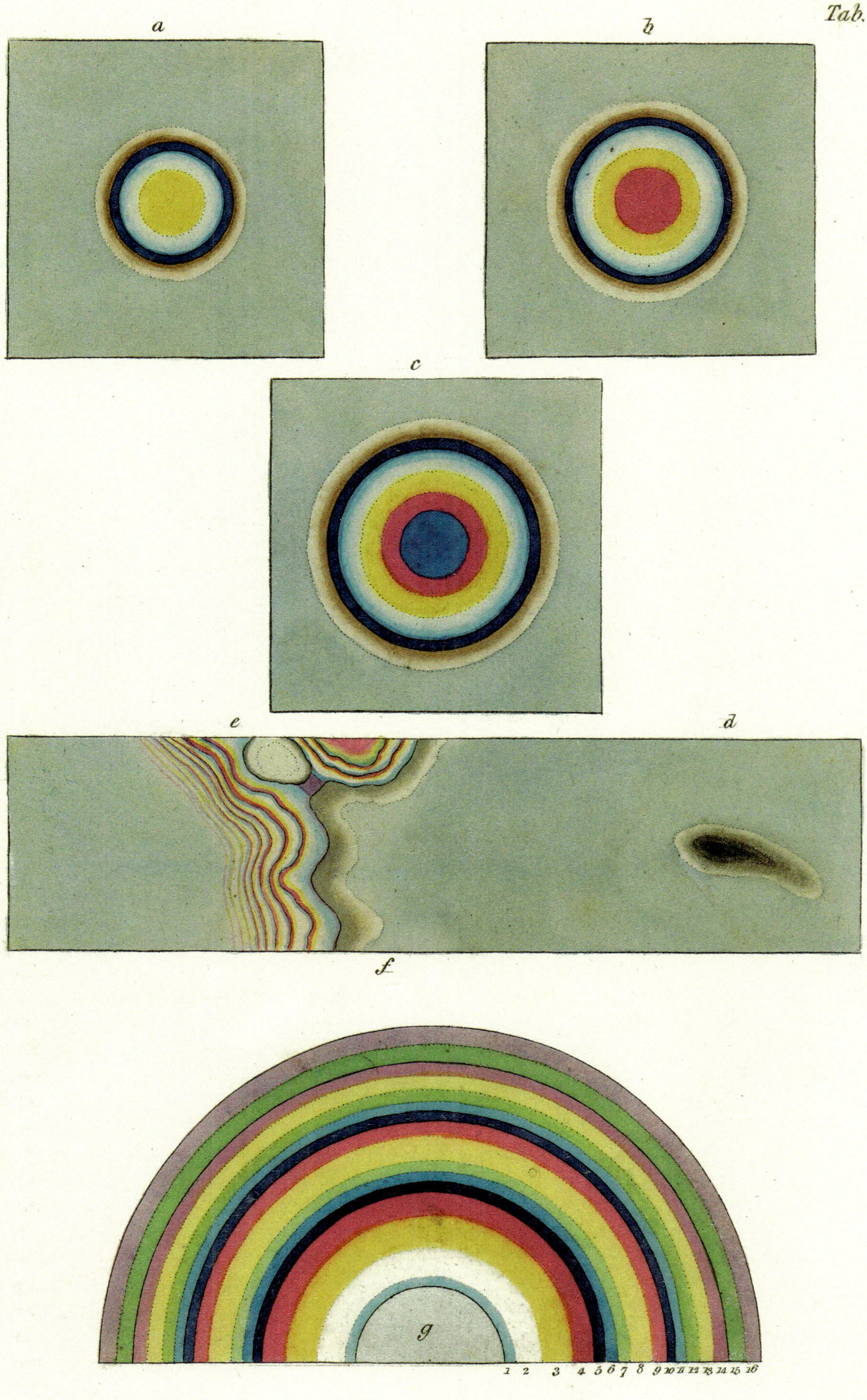

Oct. 1. 1807. Publish'd by Jas. Sowerby London.

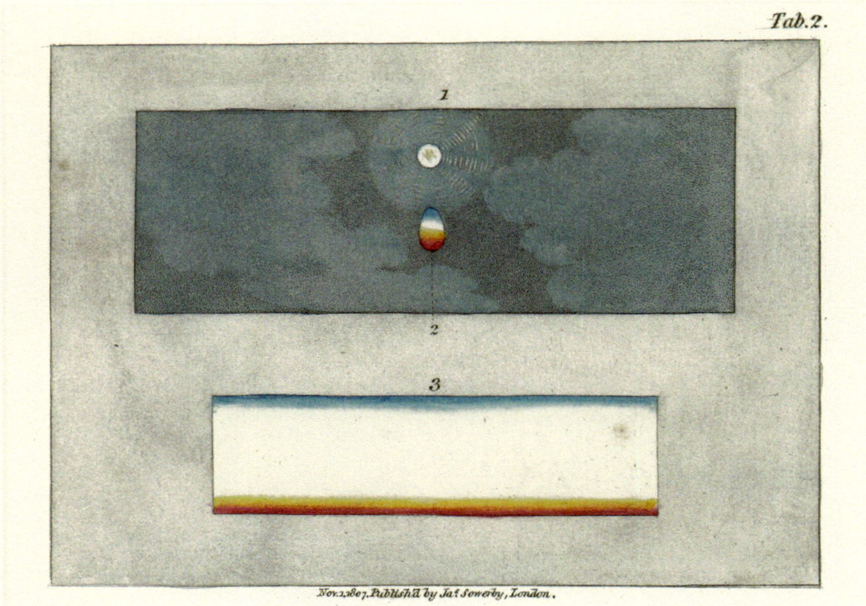

Neue Erläuterung der Farben, original prismatisch und stofflich; mit einer Demonstration ihrer Übereinstimmung in den drei Grundfarben Gelb, Rot und Blau

Der Naturforscher James Sowerby veröffentlichte dieses Buch in einer für alle Farbinteressierten aufregenden Zeit. Wenige Jahre zuvor hatte der Universalgelehrte Thomas Young (1773–1829) seine Theorie des Dreifarbensehens vorgelegt, während Runge und Goethe mit der Arbeit an ihren Farbsystemen und bevorstehenden Publikationen beschäftigt waren. Sowerby dürfte auch davon gehört haben, dass Mary Gartsides wundersam illustriertes Buch über Farbe und Blumenmalerei 1808 in zweiter Auflage erschienen war (jetzt mit dem Titel *An Essay on a New Theory of Colours*). Zu dieser Fülle bedeutender neuer Werke über Farbe steuerte er mit *A New Elucidation of Colours* sein eigenes Buch bei und widmete es Newton, dessen *Opticks* gut hundert Jahre zuvor erschienen war. Ein „neuer" Ansatz also, ein Versuch, die Betonung auf Licht und Schatten und auf die Unterschiede zwischen den Farben umzulenken, der jedoch gleichwohl den Errungenschaften Newtons verpflichtet war. Sowerby versuchte auch, die Verwirrung zu beseitigen und den Unterschied zwischen additiven (immateriellen) und subtraktiven (stofflichen) Farbmischungen zu erklären, während er als Naturforscher und produktiver Illustrator mit Moses Harris ein Interesse für die farblichen Feinheiten in der Natur teilte. Dabei konzentrierte er sich mit Rot, Gelb und Blau speziell auf die Primärfarben der Malerei und wählte als ihre „perfektesten" Pigmententsprechungen Karmin, Gummigutt und Preußischblau. Sowerby war der Erfinder des „Chromatometers", einer Vorrichtung, die ein Prisma und einen schwarzen Keil auf Papier umfasste und von ihm mit einem Barometer zum Messen von Farben verglichen wurde. Es bildete einen wichtigen Teil von *A New Elucidation* und war mit den Tafeln im Buch zu verwenden, die farbige oder schwarze Streifen zeigten (Goethe hatte 1791 etwas Ähnliches vorgelegt; siehe S. 28–29, 459).

Sowerbys Hauptfarbdiagramm ist die „Chromatische Skala" (Tafel 5), eine elegante Variante eines Farbkreises mit einem dreistreifigen Dreieck darin, auf dem die Primärfarben in ihrer reinsten Form in den mittleren Abschnitten der drei äußeren Streifen platziert sind. Zur Mitte hin werden die Farbtöne heller. Die drei zusätzlichen Rauten an der unteren Spitze des Dreiecks enthalten ähnlich wie Harris' zweiter Farbkreis weitere Farbmischungen, sodass sich neben dem zentralen weißen Dreieck insgesamt 63 Farbtöne ergeben. Für das schön ausgeführte, aber streng zweidimensionale Diagramm wurden transparente Wasserfarben schichtweise übereinander aufgetragen.

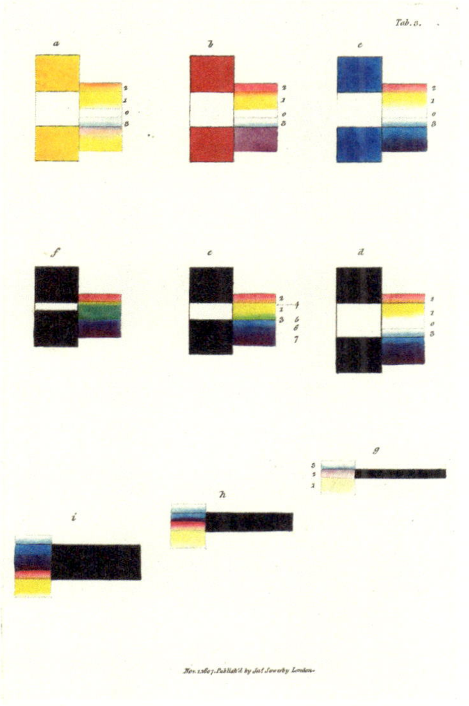

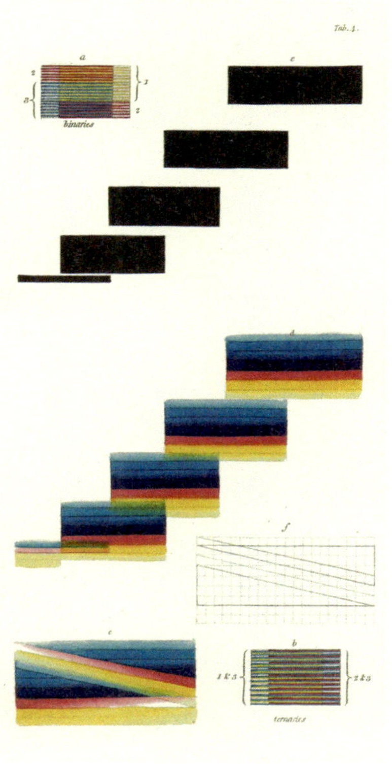

Nouvelle élucidation des couleurs, prismatiques et matérielles, et montrant leur correspondance avec les trois couleurs primaires, jaune, rouge et bleu

Naturaliste anglais, James Sowerby publie cet ouvrage à un moment passionnant pour quiconque s'intéresse au sujet. Quelques années plus tôt, l'érudit Thomas Young (1773–1829) a présenté sa théorie de la vision trichromatique, tandis que Runge et Goethe travaillaient à leurs conceptions respectives et à leurs publications imminentes. Il est probable que Sowerby ait entendu parler de la parution de la seconde édition du livre merveilleusement illustré que Mary Gartside consacre à la couleur et à la peinture de fleurs, paru en 1808 sous le titre *An Essay on a New Theory of Colours* (« Essai sur une nouvelle théorie des couleurs »). À cet ensemble d'œuvres nouvelles et importantes, Sowerby ajoute sa propre contribution – « Nouvelle élucidation des couleurs » – qu'il dédie à Newton, dont l'*Opticks* a paru plus d'un siècle plus tôt. Il s'agit donc d'une approche « nouvelle », d'une tentative de revenir à l'examen de la lumière et de l'ombre et des différences entre les couleurs, qui rend néanmoins hommage aux avancées de Newton. Sowerby cherche également à clarifier les choses et à expliquer la différence entre mélanges de couleurs additif (immatériel) et soustractif (matériel). Naturaliste et illustrateur prolifique, il partage aussi l'attention que porte Moses Harris aux subtilités des couleurs du monde naturel, s'intéresse en particulier aux couleurs primaires des peintres – rouge, jaune et bleu – et choisit comme équivalents « les plus parfaits » le carmin, la gomme-gutte et le bleu de Prusse. Sowerby est aussi l'inventeur d'un « chromatomètre », dispositif composé d'un prisme et d'un coin noir, destiné à la mesure des couleurs et qu'il compare à un « baromètre » chromatique, auquel est consacrée une partie importante de son ouvrage. Cet appareil devait être utilisé avec les planches du livre montrant des bandes de couleurs ou de noir (Goethe avait réalisé un dispositif identique en 1791 ; voir p. 28–29, 459).

L'« échelle chromatique » (Table 5), le principal graphique de couleurs de Sowerby, est une variante élégante du cercle chromatique, auquel se superpose un triangle à trois bandes, où les couleurs primaires les plus pures se situent dans les bandes périphériques des portions centrales, entre les sommets du triangle, et s'éclaircissent vers le centre. Les trois losanges ajoutés au bas du triangle contiennent d'autres mélanges de couleurs, à la manière du cercle des couleurs secondaires de Harris, et rassemblent en tout 63 teintes, en plus du triangle blanc situé au centre. Ce graphique magnifiquement réalisé superpose des couches d'aquarelle translucide, mais conserve un aspect résolument bidimensionnel.

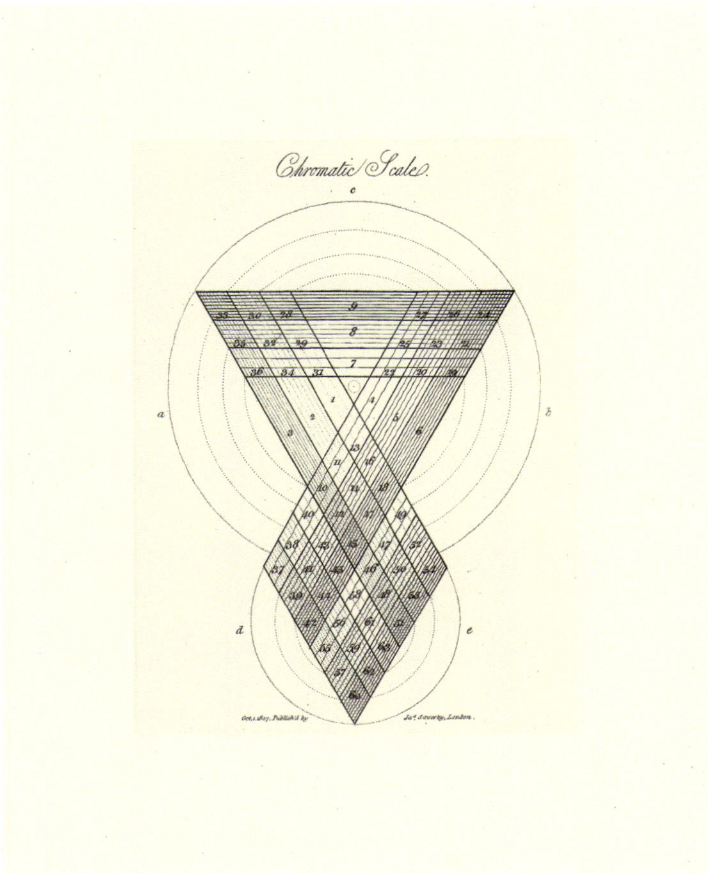

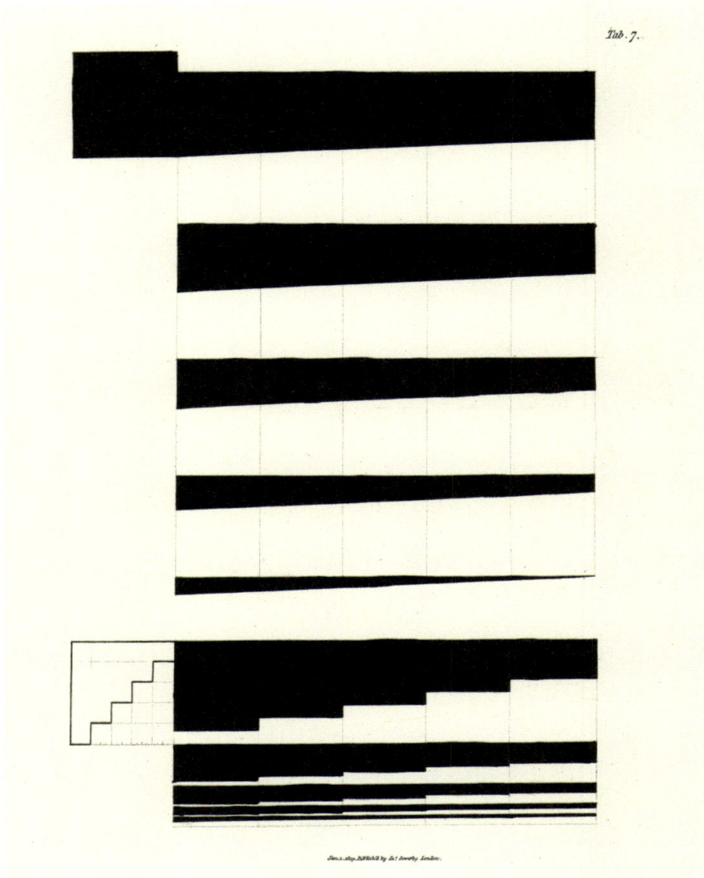

Nueva elucidación de los colores prismáticos originales y materiales y su concordancia con los tres primarios: amarillo, rojo y azul

El naturalista James Sowerby publicó este libro en un momento emocionante para cualquier persona interesada en el color. Unos años antes, el erudito Thomas Young (1773–1829) había presentado su teoría de la visión del color tricromática, mientras que Runge y Goethe trabajaban por aquel entonces en sus conceptos de color y publicaciones inminentes. Es probable que Sowerby supiera que el libro maravillosamente ilustrado de Mary Gartside sobre el color y la pintura de flores había alcanzado una segunda edición en 1808 (ahora titulada *An Essay on a New Theory of Colours*), y a esa gran cantidad de importantes obras nuevas sobre el color añadió la suya propia, *A New Elucidation of Colours*, que dedicó a Newton, cuya *Optica* se había publicado hacía poco más de 100 años. El suyo era, por lo tanto, un enfoque «nuevo», un intento de redirigir la atención hacia la luz y la sombra y las diferencias entre los colores, pero un enfoque que todavía rendía homenaje a los logros de Newton. Sowerby también intentó aclarar la diferencia entre las mezclas de colores aditivos (inmateriales) y sustractivos (materiales), mientras que como naturalista y prolífico ilustrador compartió con Moses Harris la preocupación por las sutilezas del color en el mundo natural, y dedicó especial atención a los colores primarios de los pintores (el rojo, el amarillo y el azul), para los que eligió como pigmentos equivalentes «más perfectos» el carmín, la gutagamba y el azul de Prusia. Sowerby también fue el inventor de un «chromatometer», un dispositivo que cuenta con un prisma y una cuña negra que comparó con un «barómetro» de color, y que estaba destinado a la medición de los colores. El dispositivo, al que dedicó una parte importante de *A New Elucidation*, estaba pensado para su uso con las láminas del libro que muestran bandas de color o negras (Goethe había hecho algo similar en 1791; véanse págs. 28–29, 459).

El diagrama cromático principal de Sowerby, la «escala cromática» (tabla 5), es una versión elegante de una rueda de colores con un triángulo de tres bandas superpuesto en el que las formas más puras de los colores primarios ocupan las partes medias situadas entre los vértices de las bandas exteriores, con los tonos haciéndose más claros a medida que se aproximan al centro. Los tres rombos añadidos a la parte inferior del triángulo contienen más mezclas, de manera similar a la segunda rueda de colores de Harris, lo que da como resultado un total de 63 colores, más el triángulo central blanco. Es un diagrama bellamente ejecutado en el que se emplearon capas de acuarela transparente, aunque estrictamente bidimensional.

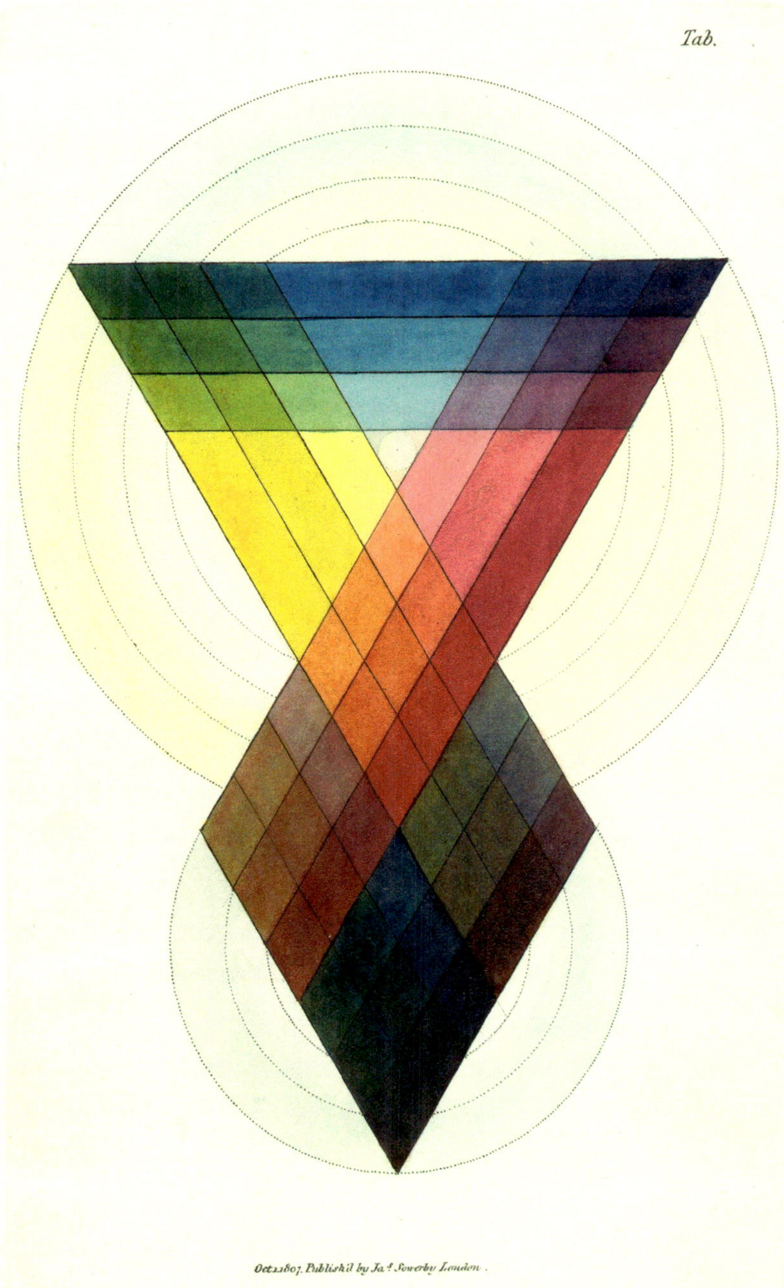
Tab.
Oct 1 1807. Publish'd by Jas Sowerby London.

Johann Wolfgang von Goethe

ZUR FARBENLEHRE

On Colour Theory
16 plates, 23 x 19.3 cm / 9 x 7⅝ in., Vienna, 1810–1812
Berlin, Universitätsbibliothek der Humboldt-Universität

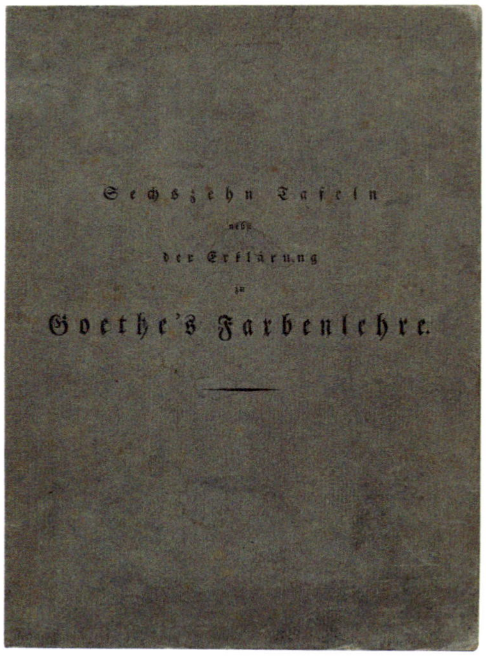
Sechszehn Tafeln
nebst
der Erklärung
zu
Goethe's Farbenlehre.

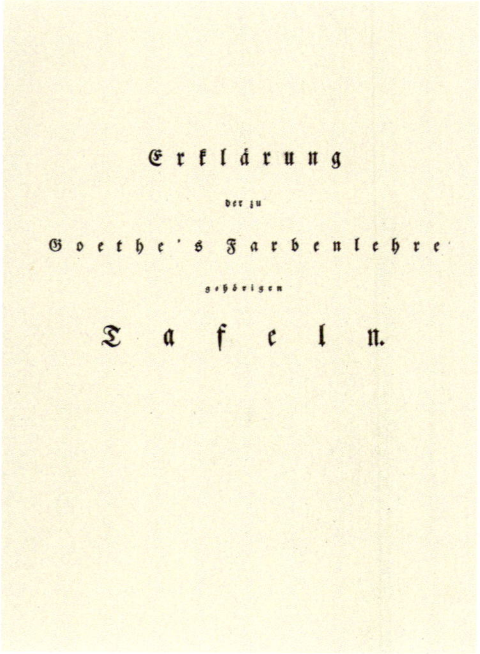
Erklärung
der zu
Goethe's Farbenlehre
gehörigen
Tafeln.

It is hard to overestimate the significance of Goethe's *Zur Farbenlehre*, which was the result of more than 20 years of research. It was first published in Tübingen in 1810 in three volumes, as two volumes of text and one volume, an "Atlas", of 17 engraved plates, 12 of them hand-coloured. Goethe himself, in characteristically boastful style, considered that he was "the only person who knows the truth in the difficult science of colours", and was immensely proud of this work.

Two decades earlier he had published a short treatise on colour, *Beyträge zur Optik*, which was in part a critical attack on Newton's *Opticks* and in which Goethe proposed a dualistic colour system based on the polarity of darkness and light, represented by the "pure" colours blue and yellow from which all other colours can be produced, including the "impure" red. Goethe argued further that black and white are not colours, but representations of total darkness and the brightest light. Readers were encouraged to recreate certain experiments with a prism, using a set of printed coloured playing-cards that accompanied the book (pp. 28–29, 459). Over the following 20 years Goethe extended and developed his theory, often as a result of discussions with other artists and writers, and this culminated in *Zur Farbenlehre*, one of the most comprehensive texts on colour written since Newton's great work. However, where Newton had sought to be as objective and rational as possible, Goethe's approach reflected the different times in which he lived, and indeed in the Romantic Age subjective experience and emotions associated with colour were often considered more important than hard scientific facts and experimental predictability. As such Goethe paved the way for new attitudes in relation to colour, for example in psychology, while his legacy was and remains considerable, far beyond the limits of the German-speaking world. The first English edition of the main "didactic" part of his *Farbenlehre* was translated by Charles Lock Eastlake and published in London in 1840 as *Goethe's Theory of Colours*, but the manuscript may have been in circulation in the British art world from as early as 1820, since it would seem to have directly inspired Turner (p. 47) and several other artists.

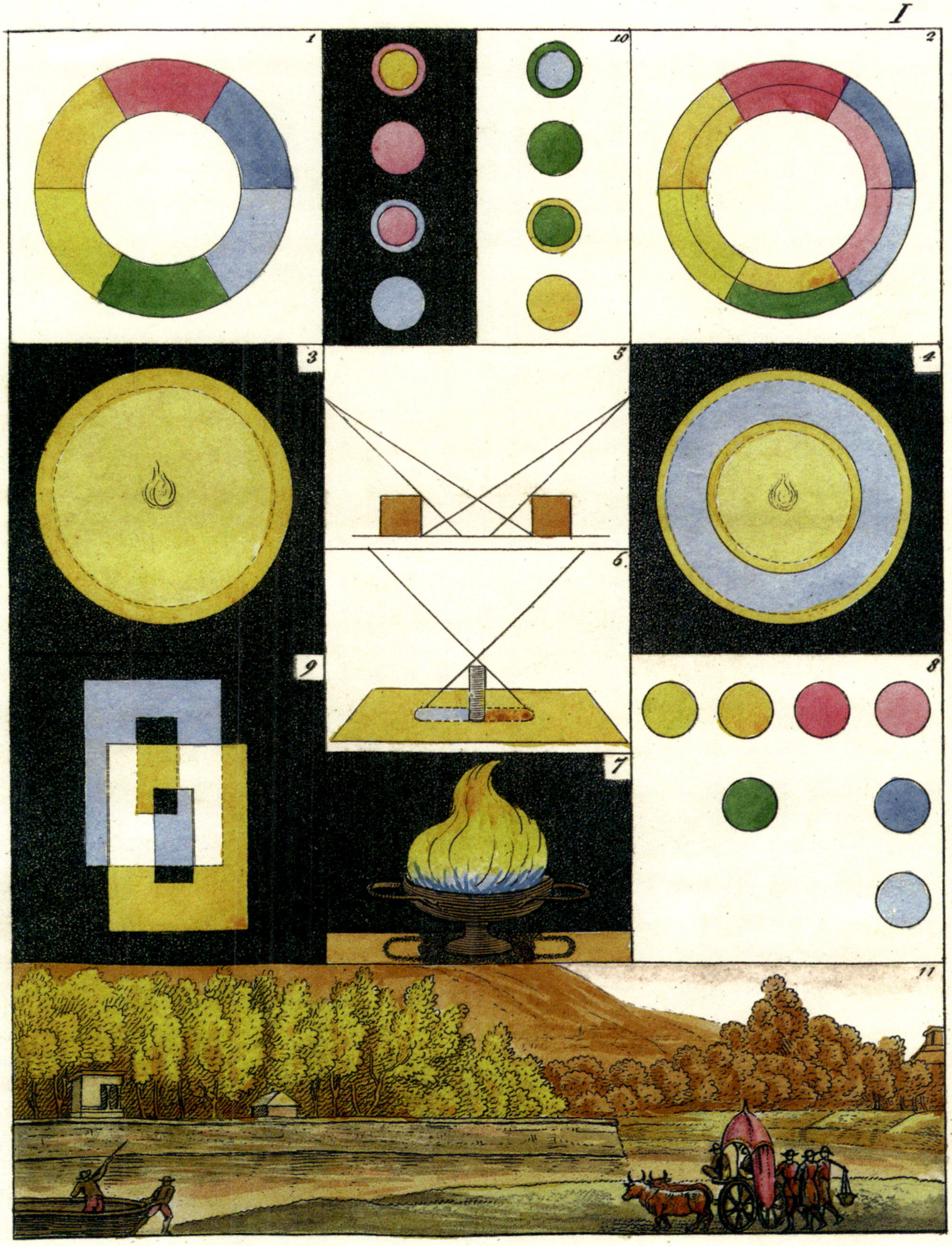
I
1
10
2
3
5
4
6.
9
8
7
11

Zur Farbenlehre

Die Bedeutung des von Goethe verfassten Werks *Zur Farbenlehre*, dem mehr als zwanzig Jahre Forschungsarbeit vorausgingen, ist nicht hoch genug einzuschätzen. Es erschien 1810 in Tübingen in zwei Textbänden und einem „Atlas" aus 17 Kupfertafeln, von denen zwölf handkoloriert waren. Goethe selbst befand, er sei „in der schwierigen Wissenschaft der Farbenlehre der Einzige [...], der das Rechte weiß", und war enorm stolz auf sein Werk.

Zwei Jahrzehnte zuvor hatte er mit *Beyträge zur Optik* (1791/92) eine kurze zweiteilige Abhandlung über Farbe veröffentlicht, die in Teilen kritisch mit Newtons *Opticks* ins Gericht ging und in der Goethe ein dualistisches Farbsystem vorschlug, das auf der Polarität von Dunkelheit und Licht beruhte. Diese wurden von den „reinen" Farben Blau und Gelb repräsentiert, aus denen sich alle anderen Farben herstellen ließen, auch das „nie ganz reine" Rot. Goethe argumentierte weiter, dass Schwarz und Weiß keine Farben seien, sondern Repräsentanten der völligen Finsternis und des hellsten Lichts. Die Leserschaft wurde ermuntert, bestimmte Experimente mit einem Prisma nachzuvollziehen und hierfür die gedruckten und kolorierten Spielkarten zu verwenden, die dem ersten Teil der *Beyträge* beilagen (S. 28–29, 459). In den folgenden zwanzig Jahren baute Goethe diese Theorie aus und entwickelte sie weiter, oft aufgrund von Diskussionen mit anderen Künstlern und Schriftstellern. Am Ende dieser Entwicklung stand *Zur Farbenlehre*, eine der umfänglichsten Schriften über Farbe seit Newtons bedeutendem Werk. Doch während Newton so objektiv und rational wie möglich hatte sein wollen, spiegelte Goethes Ansatz die anderen Zeiten wider, in denen er lebte – und tatsächlich wurden subjektives Erleben und mit Farbe in Verbindung gebrachte Gefühle in der Romantik oft für wichtiger gehalten als harte wissenschaftliche Fakten und experimentelle Berechenbarkeit. Insofern bereitete Goethe den Weg für neue Betrachtungsweisen von Farbe, etwa in der Psychologie, und sein Vermächtnis weit über die Grenzen des deutschsprachigen Raums hinaus ist bis heute beträchtlich. Die erste englischsprachige Ausgabe des „didaktischen" Hauptteils von *Zur Farbenlehre* wurde von Charles Lock Eastlake übersetzt und kam 1840 unter dem Titel *Goethe's Theory of Colours* in London heraus. Das Manuskript könnte in der britischen Kunstwelt jedoch schon seit 1820 in Umlauf gewesen sein, da es Turner und mehrere andere Künstler unmittelbar beeinflusst zu haben scheint (S. 47).

Traité des couleurs

L'importance du *Zur Farbenlehre* de Goethe, résultat de plus de deux décennies de recherches, est inestimable. Paru pour la première fois en 1810 à Tübingen, il se compose de trois volumes, deux tomes de texte et un « Atlas » de 17 planches gravées, dont 12 coloriées à la main. Extrêmement fier de son œuvre, Goethe se proclame « seule personne connaissant la vérité dans la science difficile des couleurs ».

Vingt ans auparavant, il avait publié un petit traité en deux parties sur le même sujet, *Beyträge zur Optik* (1791–1792), qui se voulait une critique sévère de l'*Opticks* de Newton, dans laquelle il proposait en outre un système chromatique duel, fondé sur l'opposition entre obscurité et lumière, laquelle était représentée par les couleurs « pures » bleu et jaune, d'où peuvent naître toutes les autres couleurs, y compris le rouge « impur ». Goethe allait plus loin en affirmant que le noir et le blanc ne sont pas des couleurs mais des représentations de l'obscurité totale et de la lumière la plus vive. Il encourageait ses lecteurs à refaire certaines de ses expériences à l'aide d'un prisme et d'un jeu de cartes à jouer imprimées en couleurs, qui accompagne la première partie de *Beyträge zur Optik* (p. 28–29, 459). C'est souvent à la suite de discussions avec d'autres artistes et écrivains que Goethe approfondira sa théorie et aboutira au *Zur Farbenlehre*, l'un des textes les plus détaillés sur le sujet depuis la grande œuvre de Newton. Tandis que le scientifique anglais visait avant tout l'objectivité et la rationalité, Goethe est le produit de son temps. À l'époque romantique, l'expérience subjective et les émotions en rapport avec les couleurs sont fréquemment considérées comme plus importantes que les faits scientifiques tangibles et le caractère prévisible de l'expérimentation. Ainsi Goethe a-t-il ouvert la voie à de nouvelles manières d'appréhender la couleur, en psychologie par exemple. Son héritage demeure considérable et dépasse largement les frontières du monde germanophone. Due à Charles Lock Eastlake, la première édition traduite en anglais de la partie principale, « didactique », du *Zur Farbenlehre* a paru à Londres en 1840 sous le titre *Goethe's Theory of Colours*. Toutefois, le manuscrit de cette édition pourrait avoir circulé parmi les peintres britanniques dès 1820, car il semble avoir directement inspiré Turner et d'autres artistes (p. 47).

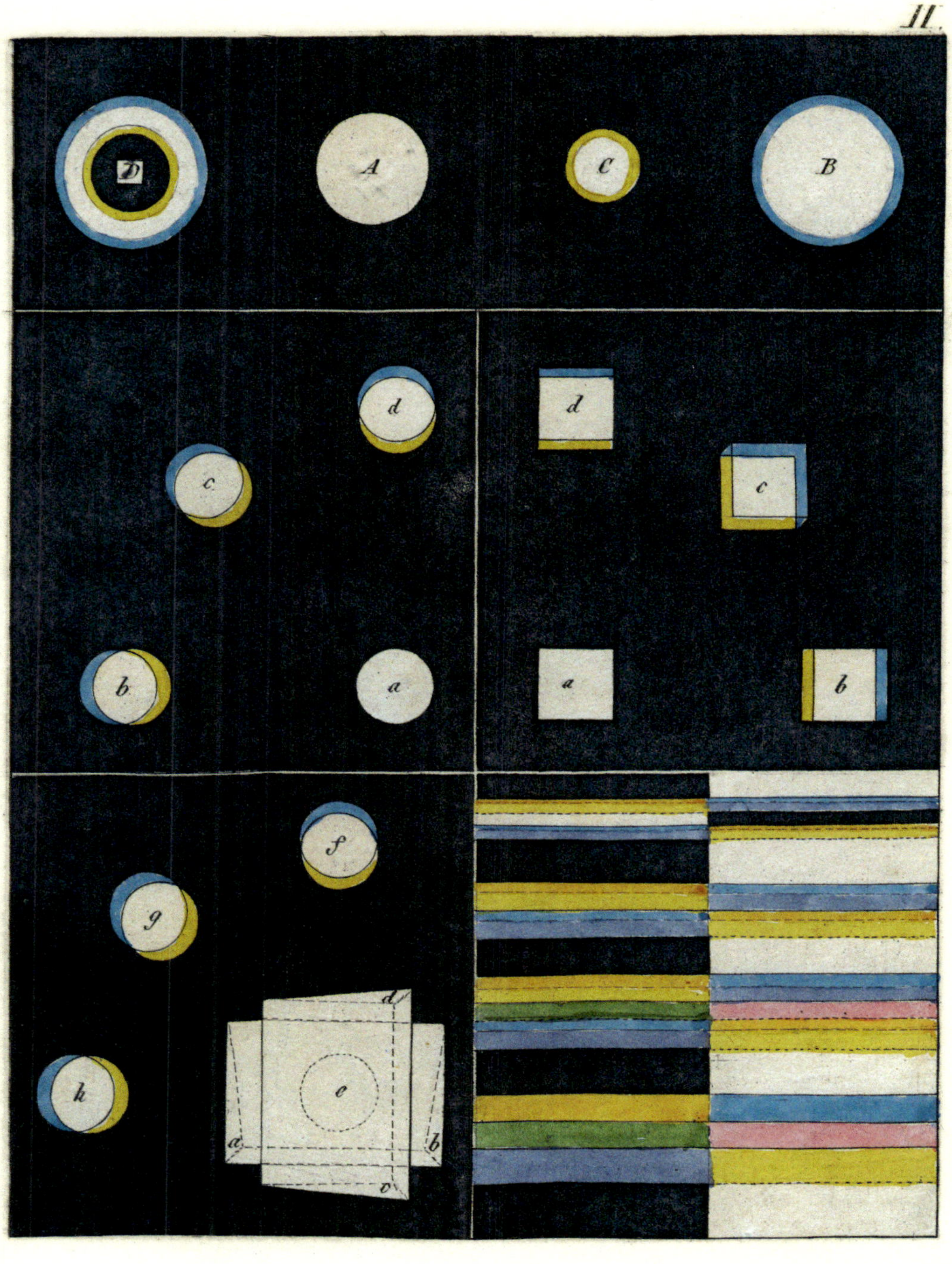
II.
D
A
C
B
d
c
b
a
d
c
a
b
f
g
h
e
a
b
c
d

II a

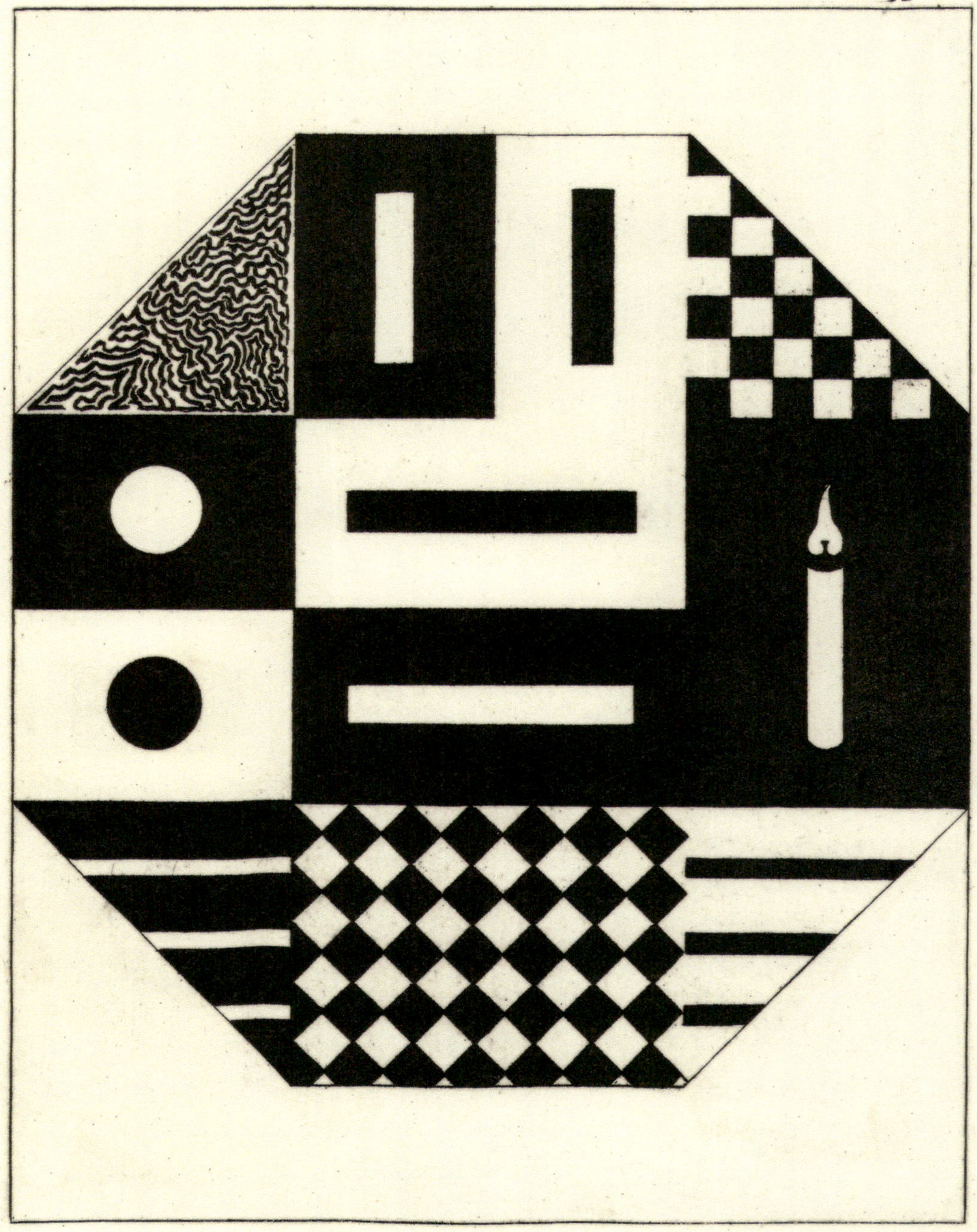

III.

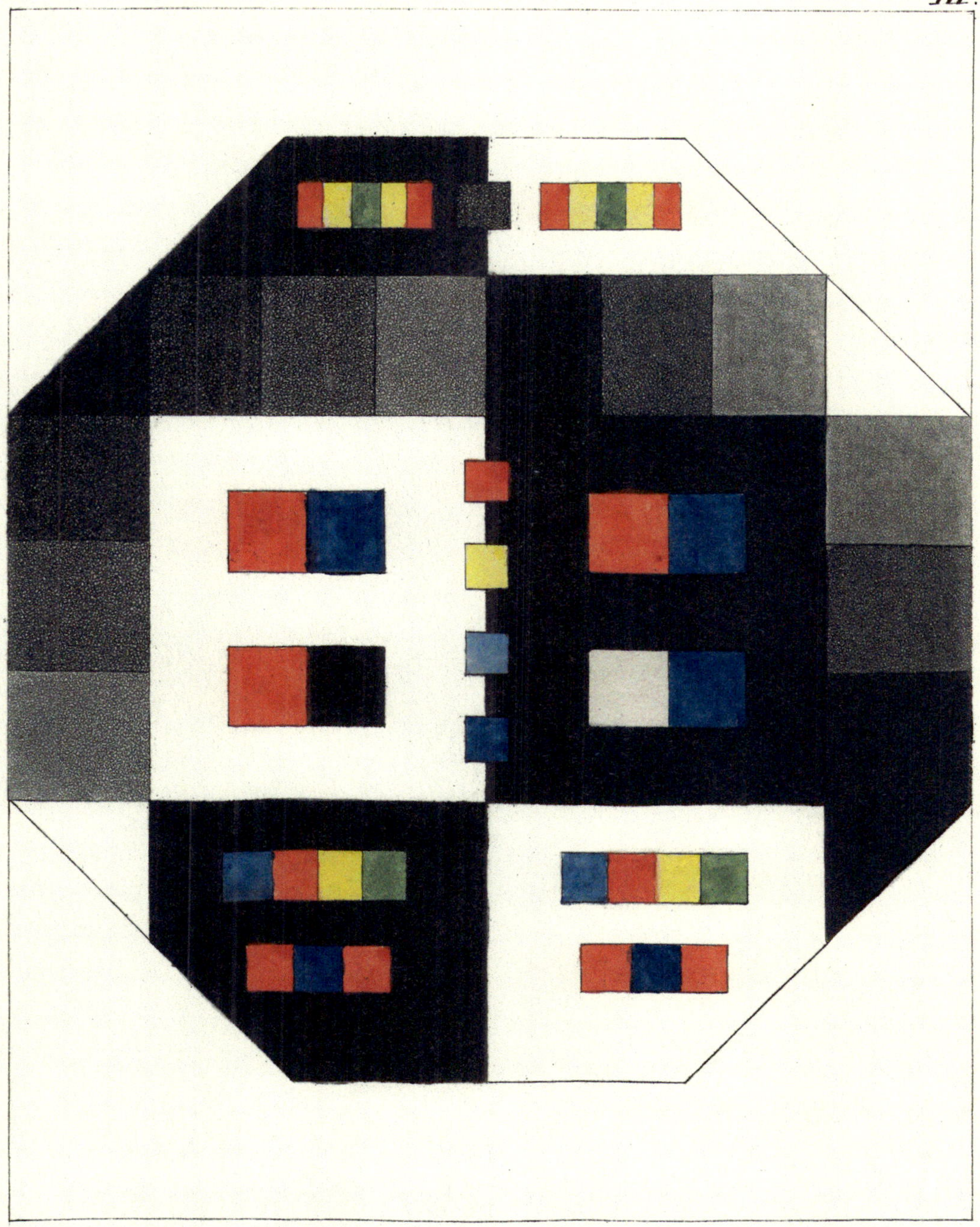

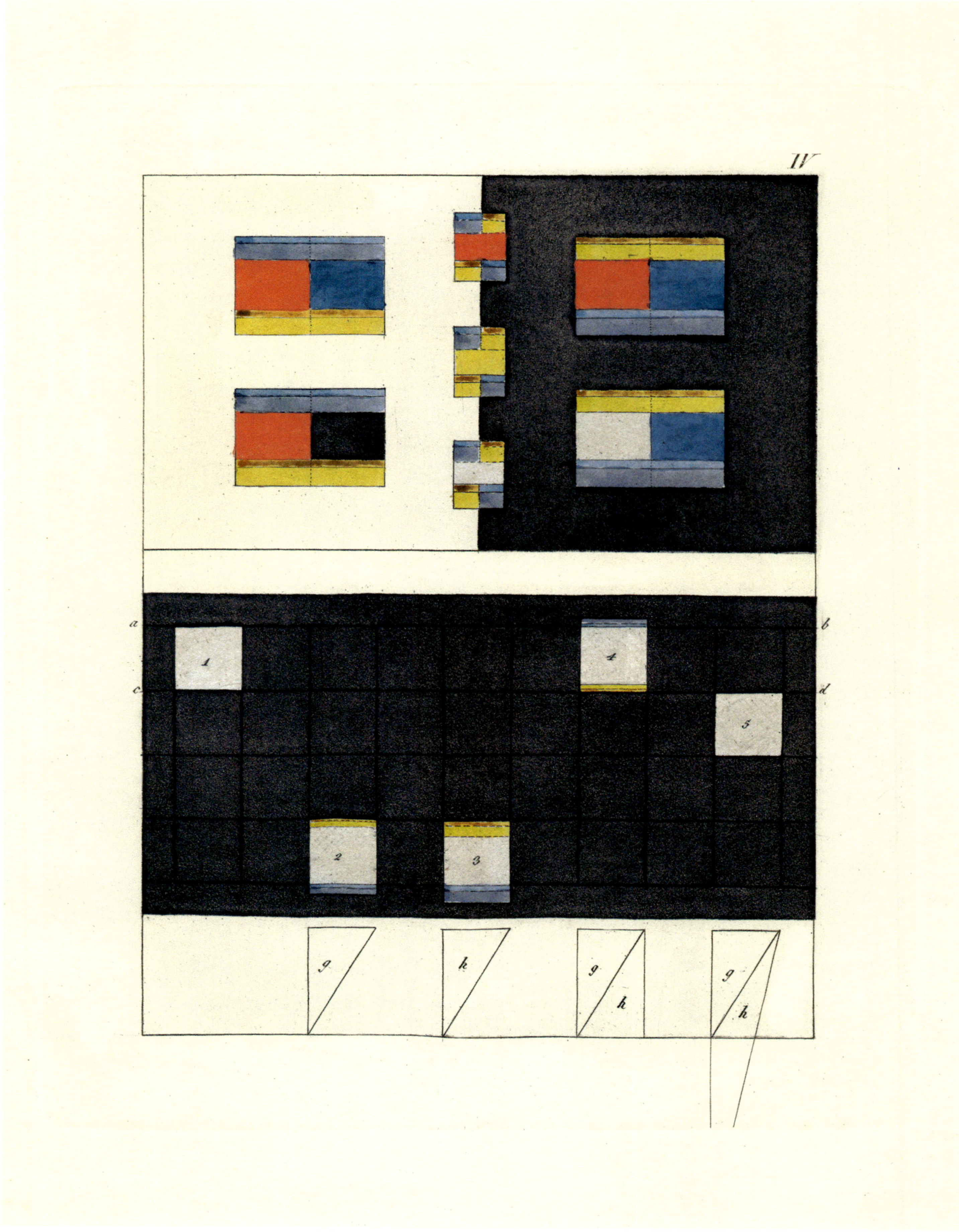
IV
a
b
c
d
1
4
5
2
3
g
h
g
h
g
h

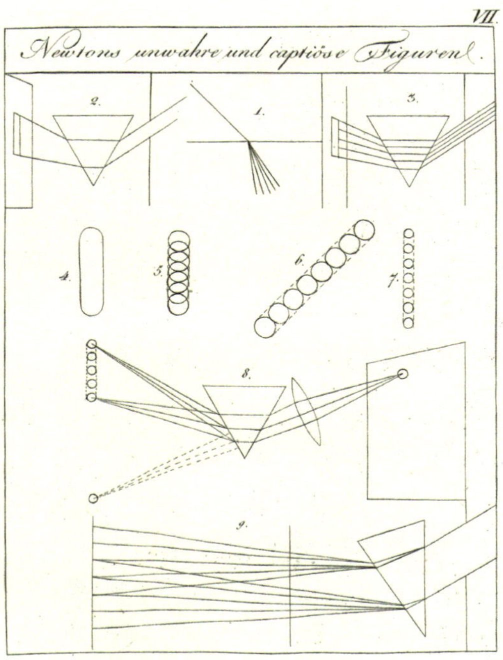

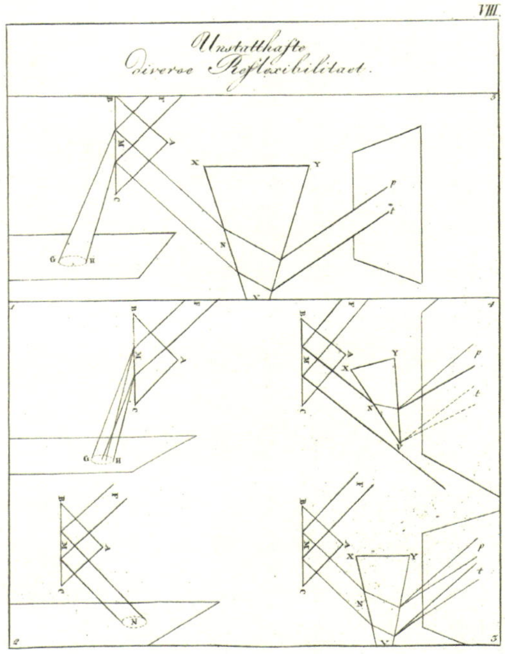

Teoría del color

Es difícil estimar la importancia de *Zur Farbenlehre* de Goethe, que fue el resultado de más de 20 años de investigación. Se publicó por primera vez en Tubinga en 1810 en tres volúmenes, dos de texto y un «Atlas» con 17 láminas grabadas, 12 de ellas coloreadas a mano. El propio Goethe se consideraba «la única persona que conoce la verdad en la difícil ciencia de los colores», y se sintió inmensamente orgulloso de esta obra.

Dos décadas antes había publicado un breve tratado en dos partes sobre el color, *Beyträge zur Optik* (1791/1792), que era en parte un ataque crítico a la *Opticks* de Newton y en el que Goethe proponía un sistema de color dualista basado en la polaridad de la oscuridad y la luz, representada por los colores «puros» azul y amarillo, a partir de los cuales se pueden producir el resto de los colores, incluido el rojo «impuro». Goethe argumentó asimismo que el blanco y el negro no son colores, sino representaciones de la oscuridad total y la luz más brillante. El tratado animaba a sus lectores a recrear ciertos experimentos con un prisma y un juego de naipes de colores impresos que acompañaba a la primera parte de *Beyträge* (págs. 28–29, 459). Durante los siguientes 20 años, Goethe amplió y desarrolló su teoría, a menudo como resultado de discusiones con otros artistas y escritores, y eso culminó en *Zur Farbenlehre*, uno de los textos más completos sobre el color posteriores a la gran obra de Newton. Sin embargo, donde Newton había buscado ser lo más objetivo y racional posible, el enfoque de Goethe reflejaba las diferentes épocas en las que vivió y, de hecho, en la época romántica, la experiencia subjetiva y las emociones asociadas con el color a menudo se consideraban más importantes que los hechos científicos puros y la previsibilidad experimental. Así, Goethe allanó el camino a nuevas actitudes en relación con el color, por ejemplo, en psicología, y su legado fue y sigue siendo considerable, con un alcance que va mucho más allá de los límites del mundo de habla alemana. La primera edición en inglés de la parte principal «didáctica» de su teoría del color la tradujo Charles Lock Eastlake y se publicó en Londres en 1840 como *Goethe's Theory of Colours*, pero es posible que el manuscrito hubiera circulado en el mundo del arte británico desde una fecha tan temprana como 1820, ya que parece haber inspirado directamente a Turner y otros artistas (pág. 47).

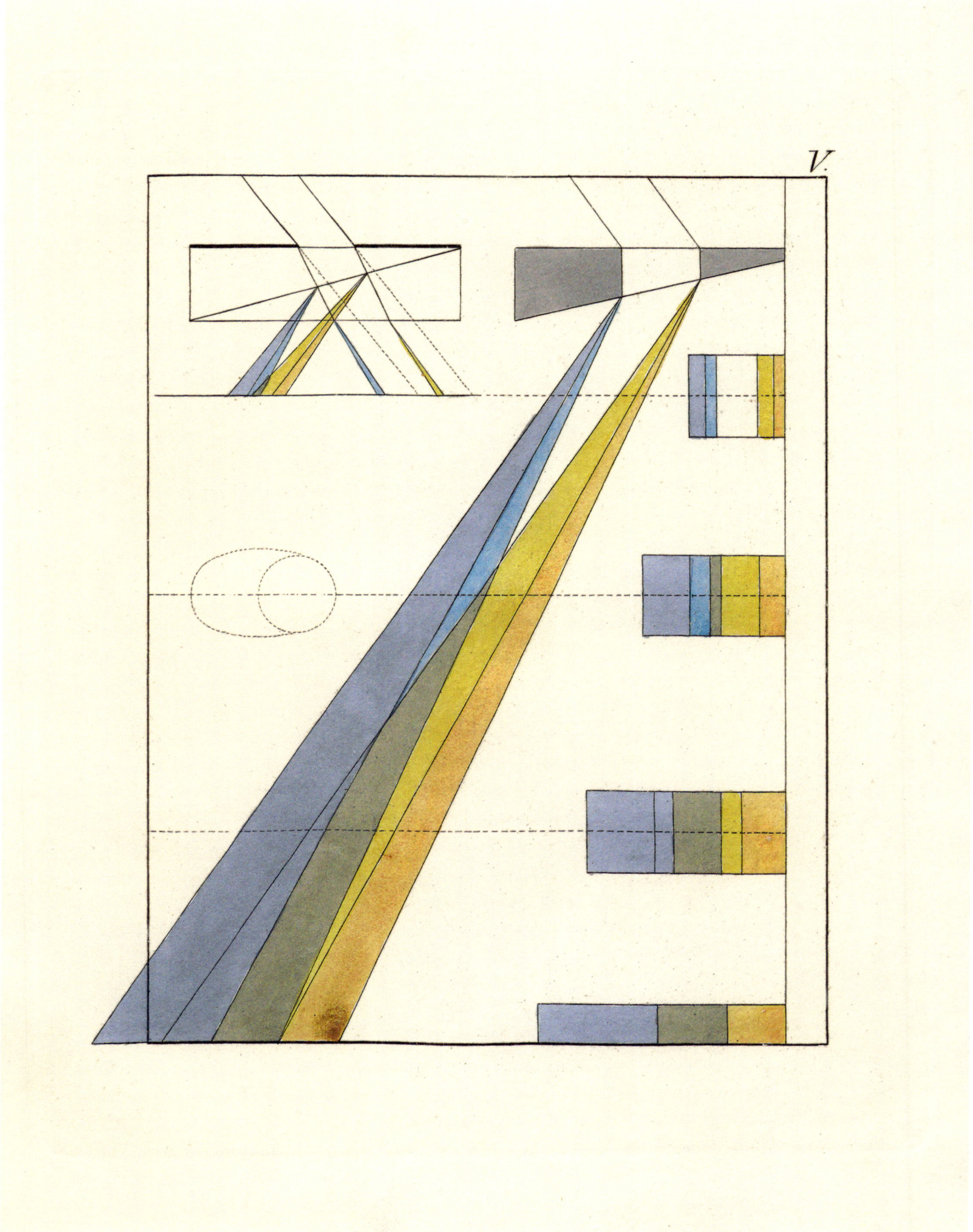
V.

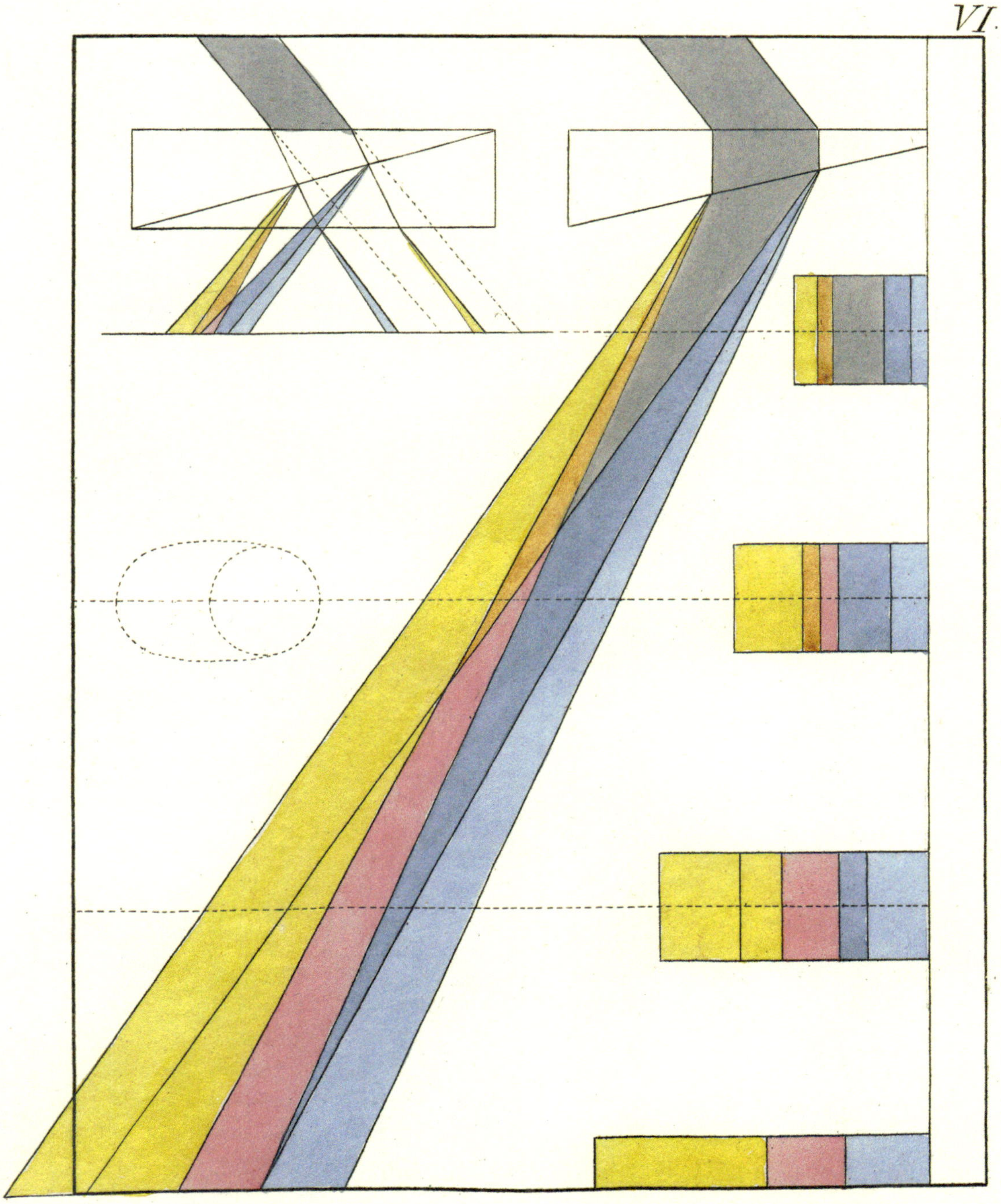
VI.

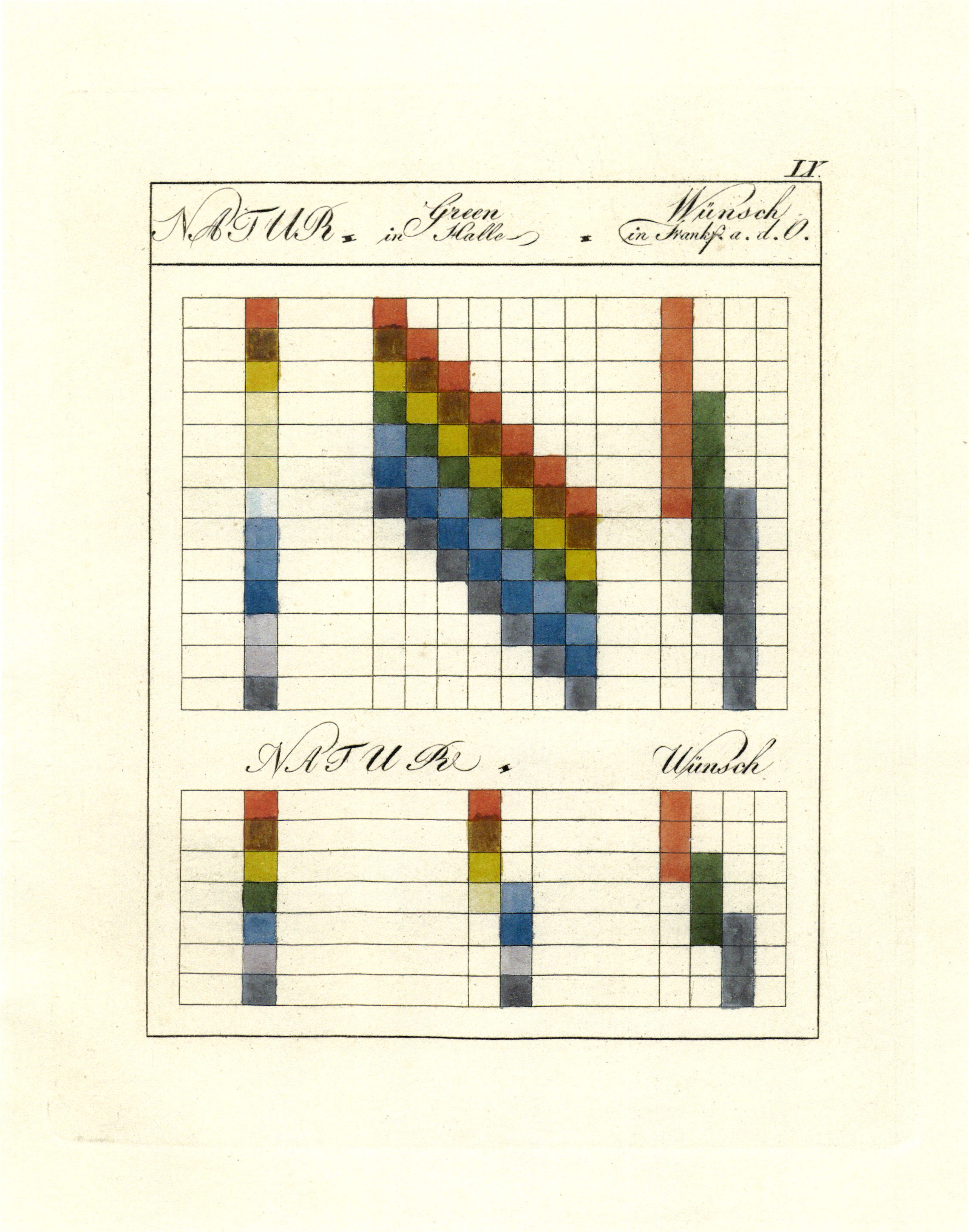
LV.
NATUR. in Green Halle. Wünsch in Frankf. a. d. O.
NATUR. Wünsch

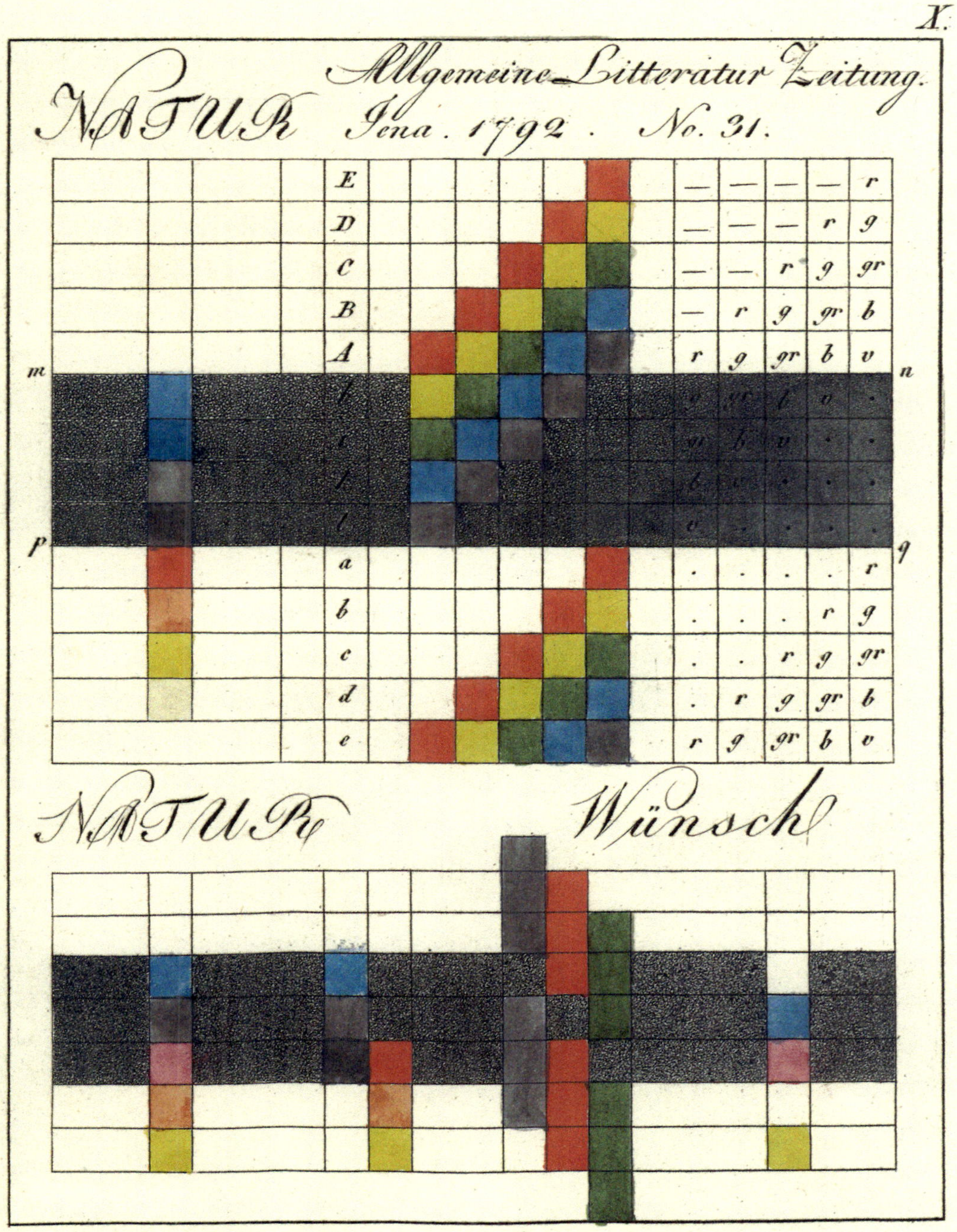
X.
Allgemeine Litteratur Zeitung.
NATUR Jena. 1792. No. 31.
E
D
C
B
A
m
n
p
q
a
b
c
d
e
NATUR
Wünsch

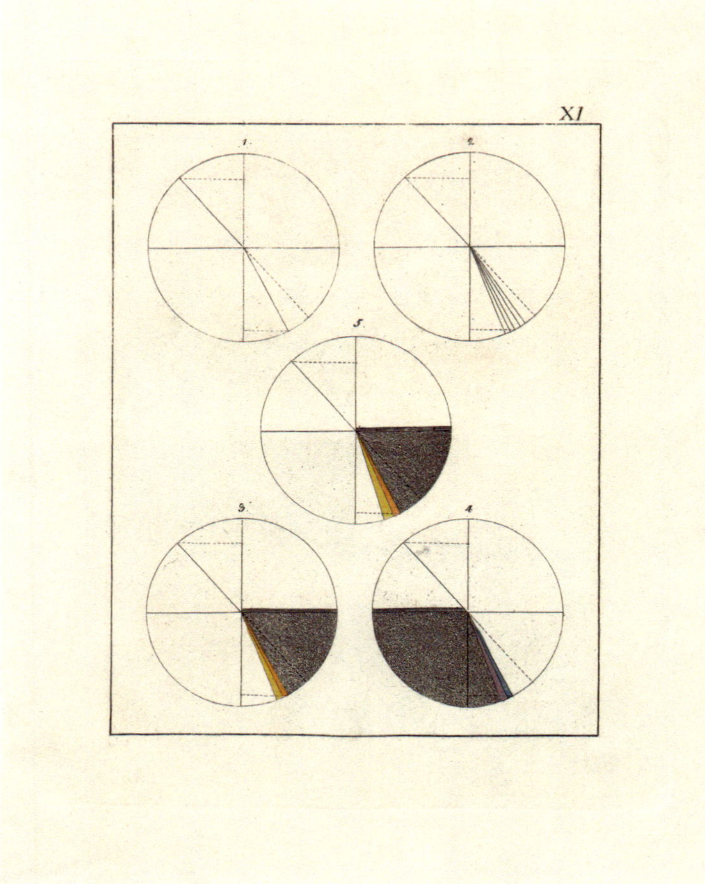
XI

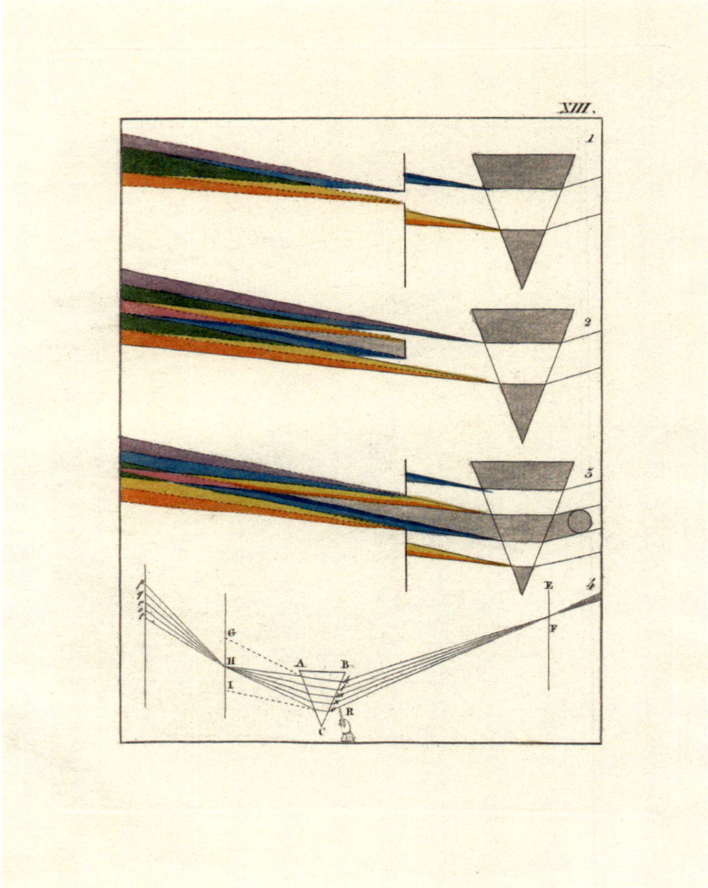
XIII.

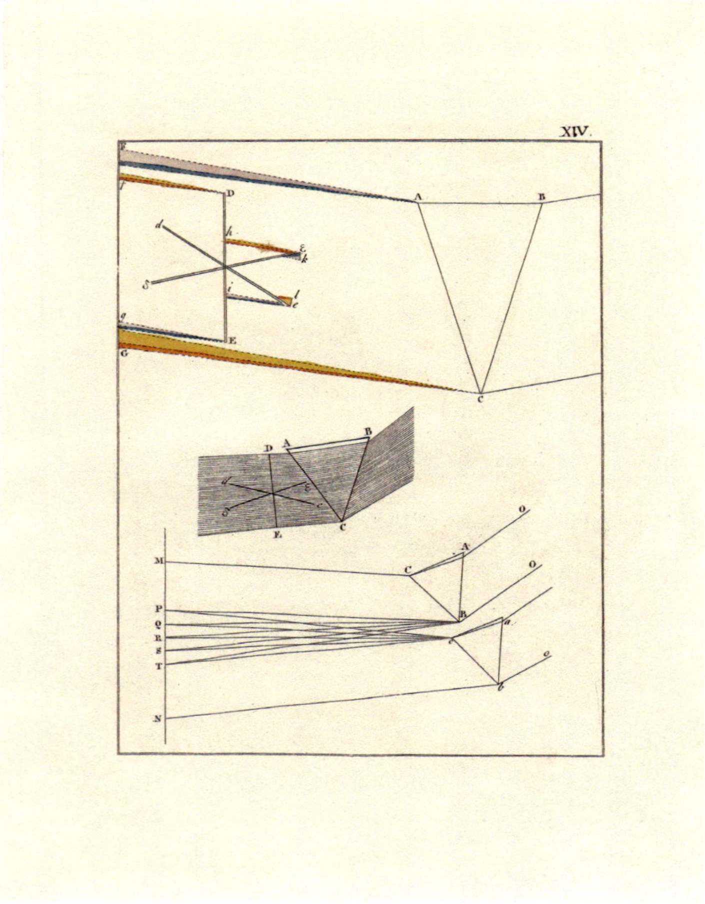
XIV.

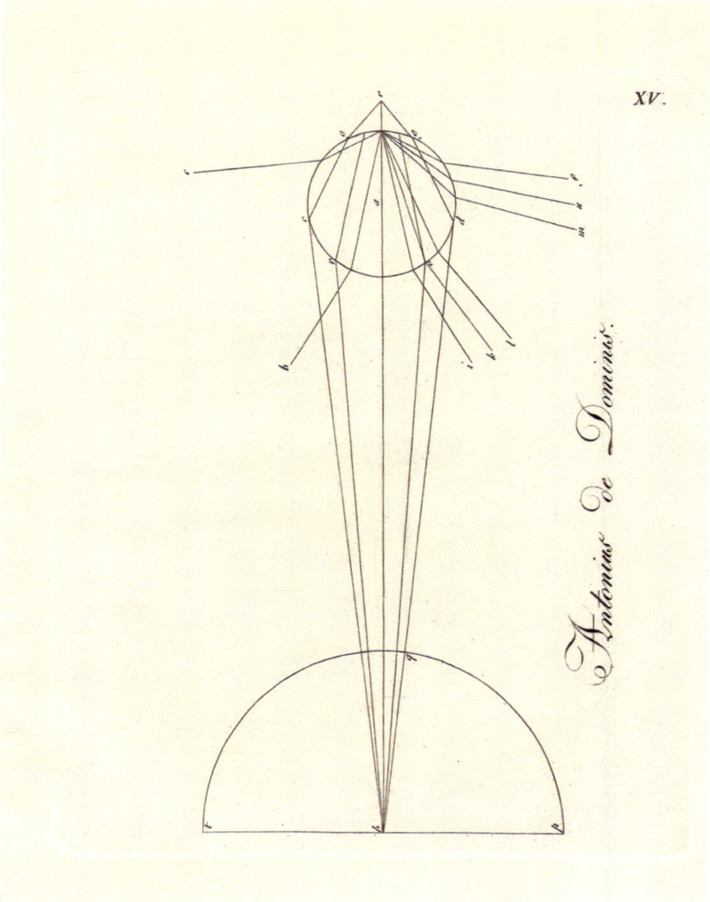
XV.
Antonius de Dominis.

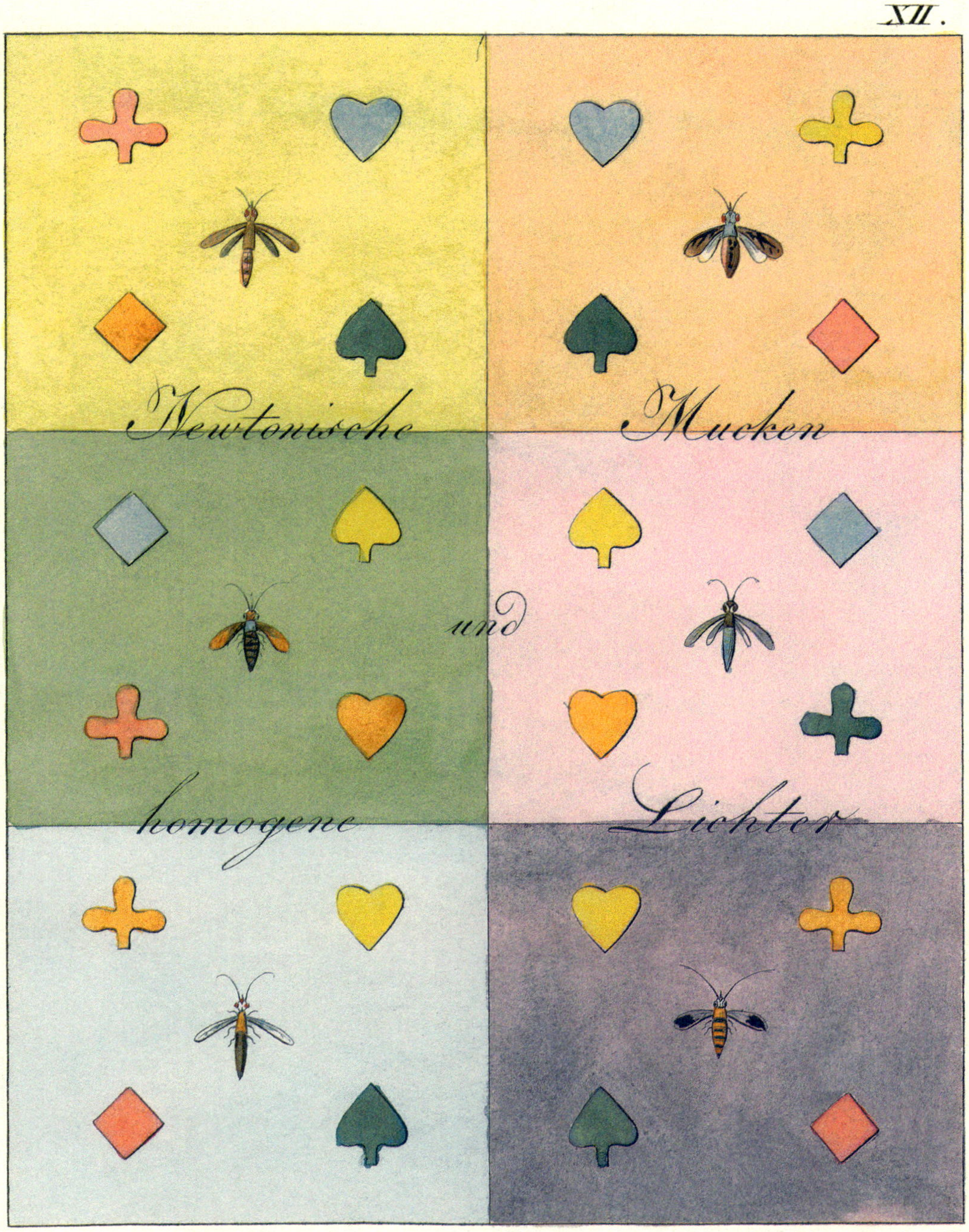
XII.
Newtonische Mucken
und
homogene Lichter

Matthias Klotz (1748–1821)

GRÜNDLICHE FARBENLEHRE

A Comprehensive Doctrine on Colour

7 plates, 41.5 x 33.5 cm | 16 ⅜ x 13 ¼ in., Munich, 1816
Munich, Bayerische Staatsbibliothek

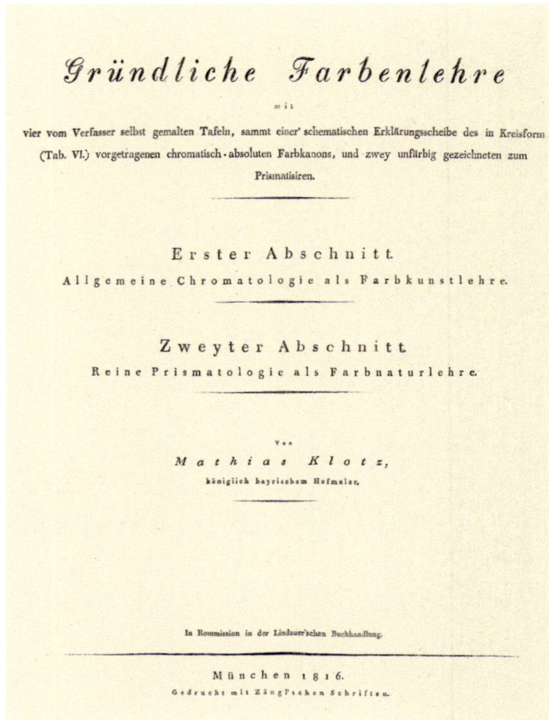

Gründliche Farbenlehre

mit

vier vom Verfasser selbst gemalten Tafeln, sammt einer schematischen Erklärungsscheibe des in Kreisform (Tab. VI.) vorgetragenen chromatisch-absoluten Farbkanons, und zwey unfärbig gezeichneten zum Prismatisiren.

Erster Abschnitt.

Allgemeine Chromatologie als Farbkunstlehre.

Zweyter Abschnitt.

Reine Prismatologie als Farbnaturlehre.

Von

Mathias Klotz,

königlich bayrischem Hofmaler.

In Kommission in der Lindauer'schen Buchhandlung.

München 1816.

Gedruckt mit Zängl'schen Schriften.

Matthias Klotz's *Gründliche Farbenlehre* was many years in the making and contains some of the most beautiful illustrations of all 19th-century books on colour, and yet this work is not very well known. The author, a portrait painter from Strasbourg, obtained a position as a scenic artist for theatre and stage sets at the Bavarian court in the 1770s, and this work would doubtless have given him a heightened sense of the interaction between colour, perspective and light. In the decades before his *Gründliche Farbenlehre* was published, Klotz wrote several pamphlets and booklets, some of them running to dozens of pages, announcing the comprehensive colour theory he was in the process of developing. He also approached Goethe about a possible collaboration but was met with silence (although Goethe did make some notes about Klotz in his diaries).

Klotz's intention was to provide a broad scientific and artistic overview of colour theory, and structured his treatise in two sections. The first part introduced *Chromatische Farbenlehre* (Chromatic colour theory) and dealt with material colour and the painter's three primaries, yellow, blue and red (he called the latter *purpur*, or purple), and their relations to black and white, or darkness and light; the second part was entitled *Prismatik* (Prismatics) and elaborated on various optical experiments similar to those Goethe had undertaken. When the book was eventually published it was met with both praise and criticism, as with the reviewer in 1817 who commented on the author's age at the time, thereby implying that his theory was old-fashioned while also deploring his lack of scientific expertise – but still writing gushingly about the book's illustrations. The seven plates, originally designed as fold-outs, are very early examples of the 'painterly' method of lithographic printing, including what is possibly the first depiction of a greyscale. Four of the plates were meant to be coloured, including the sophisticated 24-part colour wheel, which represented a three-dimensional colour solid, but when the book was published buyers had a choice between coloured and uncoloured copies (for those who might want to colour the plates in themselves). Of the 500 copies printed 45 were coloured by Klotz himself, and while his attempt to produce a work that would be equal to Goethe's was ultimately futile, the result was certainly beautiful.

Dreiquelliger Ursprung aller Färbigkeit, durch Natur und Kunst in prismatischer Richtigkeit und Energie

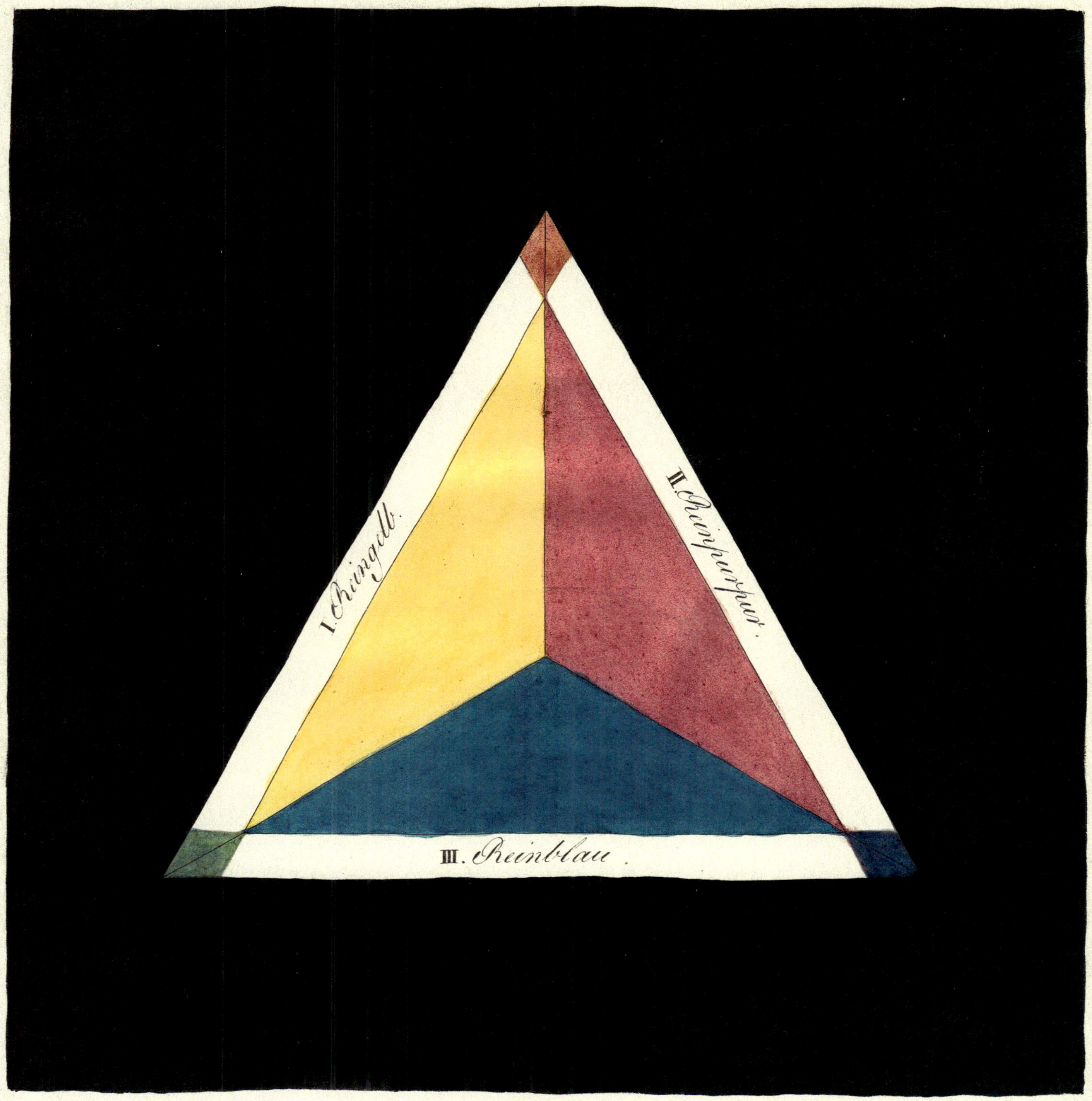

Tab. I.

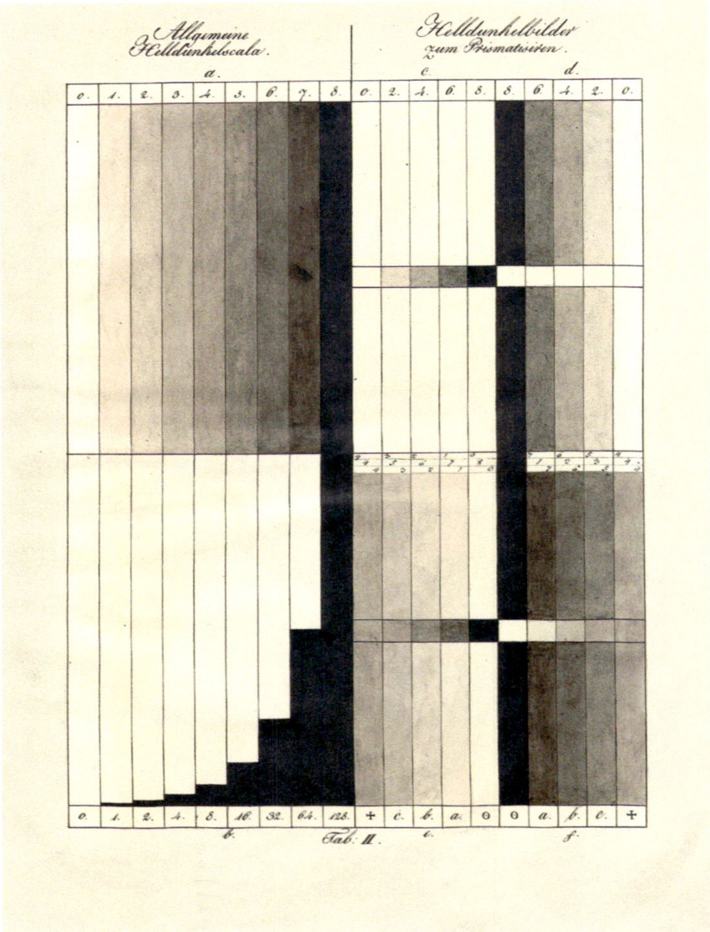

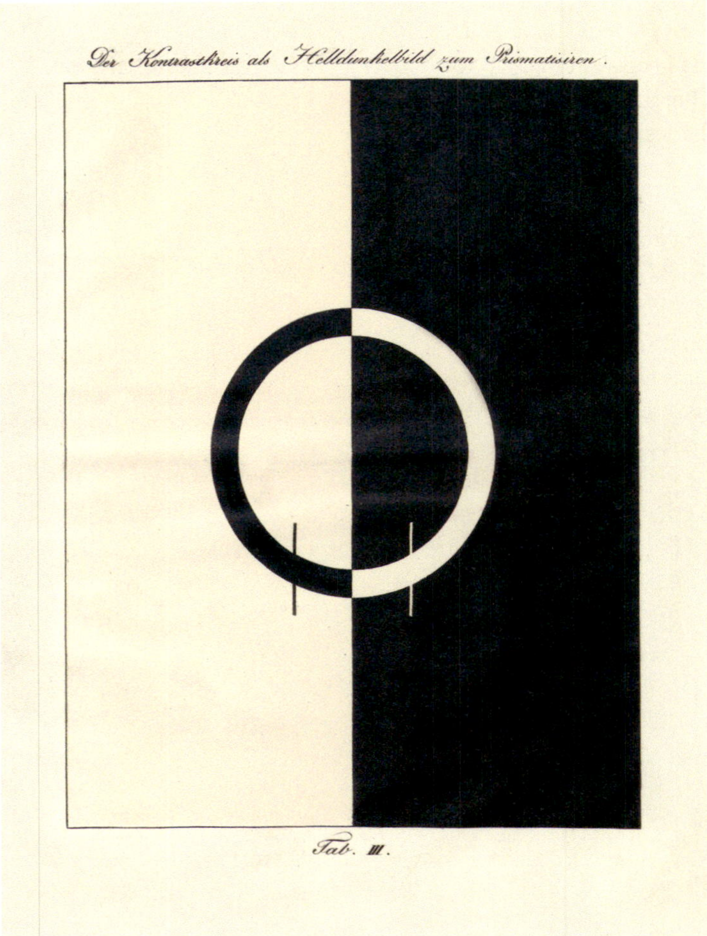

Gründliche Farbenlehre

Die Entstehung der *Gründlichen Farbenlehre* von Matthias Klotz zog sich über viele Jahre hin. Obwohl das Buch unter allen Farbenbüchern des 19. Jahrhunderts einige der schönsten Illustrationen enthält, ist es nicht sehr bekannt. Der Autor, ein Porträtmaler aus Straßburg, kam in den 1770er-Jahren als Kulissenmaler für Bühnenbilder an den bayrischen Hof. Diese Arbeit verschaffte ihm ohne Zweifel ein gesteigertes Gespür für das Zusammenspiel von Farbe, Perspektive und Licht. In den Jahrzehnten vor Erscheinen seiner *Gründlichen Farbenlehre* verfasste Klotz mehrere Aufsätze und teils mehrere Dutzend Seiten lange Broschüren, in denen er die umfassende Farbtheorie ankündigte, die gerade von ihm entwickelt wurde. Wegen einer möglichen Zusammenarbeit wandte er sich auch an Goethe, stieß dort jedoch auf Schweigen (auch wenn sich in Goethes Tagebüchern einige Notizen über Klotz finden).

Klotz' Ziel war ein breit angelegter wissenschaftlicher und künstlerischer Überblick über die Farbtheorie, den er in zwei Abschnitte gliederte. Der erste Teil stellte die „Chromatische Farbenlehre" vor und befasste sich mit stofflicher Farbe, mit Gelb, Blau und Rot (Letzteres wurde Purpur genannt) als den drei Urfarben der Malerei und mit deren Beziehungen zu Schwarz und Weiß oder Dunkelheit und Licht. Der zweite Teil trug die Überschrift „Prismatik" und ging ausführlich auf verschiedene optische Experimente ein, wie sie in ähnlicher Weise auch Goethe durchgeführt hatte. Als das Buch schließlich veröffentlicht wurde, schlugen ihm Lob und Kritik entgegen. So kommentierte ein Kritiker 1817 das damalige Alter des Autors, um anzudeuten, dass seine Theorie altmodisch war, und bemängelte Klotz' fehlende wissenschaftliche Kompetenz, äußerte sich hingegen überschwänglich zu den Illustrationen des Buches. Die sieben ursprünglich zum Ausklappen konzipierten Tafeln bilden sehr frühe Beispiele für die „malerische" Methode der Lithografie einschließlich der vielleicht ersten Abbildung einer Grauskala. Vier Tafeln sollten farbig sein, darunter der ausgeklügelte 24-teilige Farbkreis, der stellvertretend für einen dreidimensionalen Farbkörper stand. Doch als das Buch veröffentlicht wurde, hatten Käuferinnen und Käufer die Wahl zwischen Ausfertigungen in Farbe und in Schwarz-Weiß (für jene, die ihre Tafeln vielleicht selbst ausmalen wollten). Von den fünfhundert Exemplaren wurden 45 von Klotz selbst koloriert. Während sein Versuch, ein Werk auf Augenhöhe mit Goethes *Zur Farbenlehre* zu produzieren, am Ende vergeblich war, besteht an der Schönheit seines Werks sicherlich kein Zweifel.

Principes fondamentaux de la couleur

Il a fallu de nombreuses années à Matthias Klotz pour rédiger *Gründliche Farbenlehre*

(« Principes fondamentaux de la couleur »), dont les illustrations sont parmi les plus belles de tous les ouvrages du XIXe siècle sur le sujet. Pourtant, cette œuvre reste méconnue. Portraitiste strasbourgeois, Klotz est engagé comme peintre de théâtre et de décors à la cour de Bavière au cours des années 1770. Cette fonction l'a assurément doté d'une connaissance fine de l'interaction des couleurs, de la perspective et de la lumière. Au fil des décennies qui précèdent la parution de cet ouvrage, il rédige plusieurs opuscules et brochures qui comptent parfois plusieurs dizaines de pages et sont autant d'étapes annonçant sa grande théorie des couleurs. Il sollicite également Goethe pour envisager une collaboration, mais sa demande reste sans réponse (bien que Goethe mentionne Klotz dans ses journaux).

Dans son traité en deux parties, Klotz propose un vaste panorama scientifique et artistique des théories des couleurs. La première partie, « Chromatische Farbenlehre » (« Théorie chromatique »), traite de la couleur matérielle et des trois couleurs primaires du peintre, jaune, bleu et rouge (désignant cette dernière par le substantif *Purpur*, pourpre), dans leurs rapports avec le noir et le blanc, ou l'obscurité et la lumière. Intitulée « Prismatik », la deuxième partie entre dans les détails de diverses expériences optiques, semblables à celles que Goethe a réalisées. Lors de sa parution, l'ouvrage reçoit autant d'éloges que de critiques : l'auteur d'un compte rendu de 1817 souligne l'âge de l'auteur, sous-entend ainsi que sa théorie est démodée et déplore l'absence de savoir-faire scientifique, mais commente les illustrations du livre sur un ton dithyrambique. Initialement conçues pour être dépliées, les sept planches figurent parmi les tout premiers exemples de la méthode « picturale » d'impression lithographique et pourraient constituer la première description d'une échelle des gris. Quatre des planches auraient dû être en couleurs, notamment le cercle chromatique divisé en 24 sections qui représente un solide en trois dimensions. Mais le livre paraît en deux versions, avec ou sans couleurs (pour les lecteurs qui souhaiteraient colorier eux-mêmes les planches). Sur les 500 exemplaires imprimés, 45 sont coloriés à la main par Klotz en personne. Si sa tentative de réaliser une œuvre qui soit l'égale de celle de Goethe s'avère vaine, le résultat n'en est pas moins superbe.

Teoría completa del color

La *Gründliche Farbenlehre* (Teoría completa del color) de Matthias Klotz tardó muchos años en elaborarse y contiene algunas de las ilustraciones más bellas de todos los libros sobre el color del siglo XIX; pese a ello, esta obra no es muy conocida. El autor, un retratista de Estrasburgo, obtuvo un puesto como escenógrafo y artista escénico teatral en la corte bávara en la década de 1770, un trabajo que sin duda agudizó su sentido de la interacción entre el color, la perspectiva y la luz. En las décadas anteriores a la publicación de su *Gründliche Farbenlehre*, Klotz escribió varios folletos y cuadernillos, algunos de ellos con decenas de páginas, donde anunciaba la teoría completa del color que estaba desarrollando. También tanteó una posible colaboración con Goethe, pero se encontró con el silencio como respuesta (aunque Goethe escribió algunas notas sobre Klotz en sus diarios).

La intención de Klotz era proporcionar una amplia visión científica y artística de la teoría del color, y estructuró su tratado en dos secciones. La primera parte presenta la *Chromatische Farbenlehre* (Teoría del color cromático) y se ocupa del color material y los tres colores primarios del pintor (el amarillo, el azul y el rojo, este último denominado *purpur*, o púrpura, en el texto), y sus relaciones con el blanco y el negro, o la oscuridad y la luz; la segunda parte, titulada *Prismatik* (Prismática), profundiza en varios experimentos ópticos similares a los que Goethe había realizado. Cuando el libro finalmente se publicó, recibió elogios y críticas, como una reseña de 1817 que hacía referencia a la edad del autor en ese momento, lo que implicaba que su teoría estaba pasada de moda, y lamentaba asimismo su falta de experiencia científica, aunque a su vez elogiaba las ilustraciones. Las siete láminas, diseñadas originalmente como desplegables, son ejemplos de primera época del método «pictórico» de impresión litográfica, incluida la que posiblemente sea la primera representación de una escala de grises. Cuatro láminas estaban pensadas para ser coloreadas, incluida la sofisticada rueda de color de 24 partes, que representa un color sólido tridimensional, pero cuando se publicó el libro, los compradores pudieron elegir entre copias en color y sin colorear (para aquellos que quisieran colorear las láminas ellos mismos). De las 500 copias impresas, 45 fueron coloreadas por el propio Klotz, y aunque su intento de producir una obra que fuera igual a la de Goethe fue en última instancia vano, el resultado fue ciertamente hermoso.

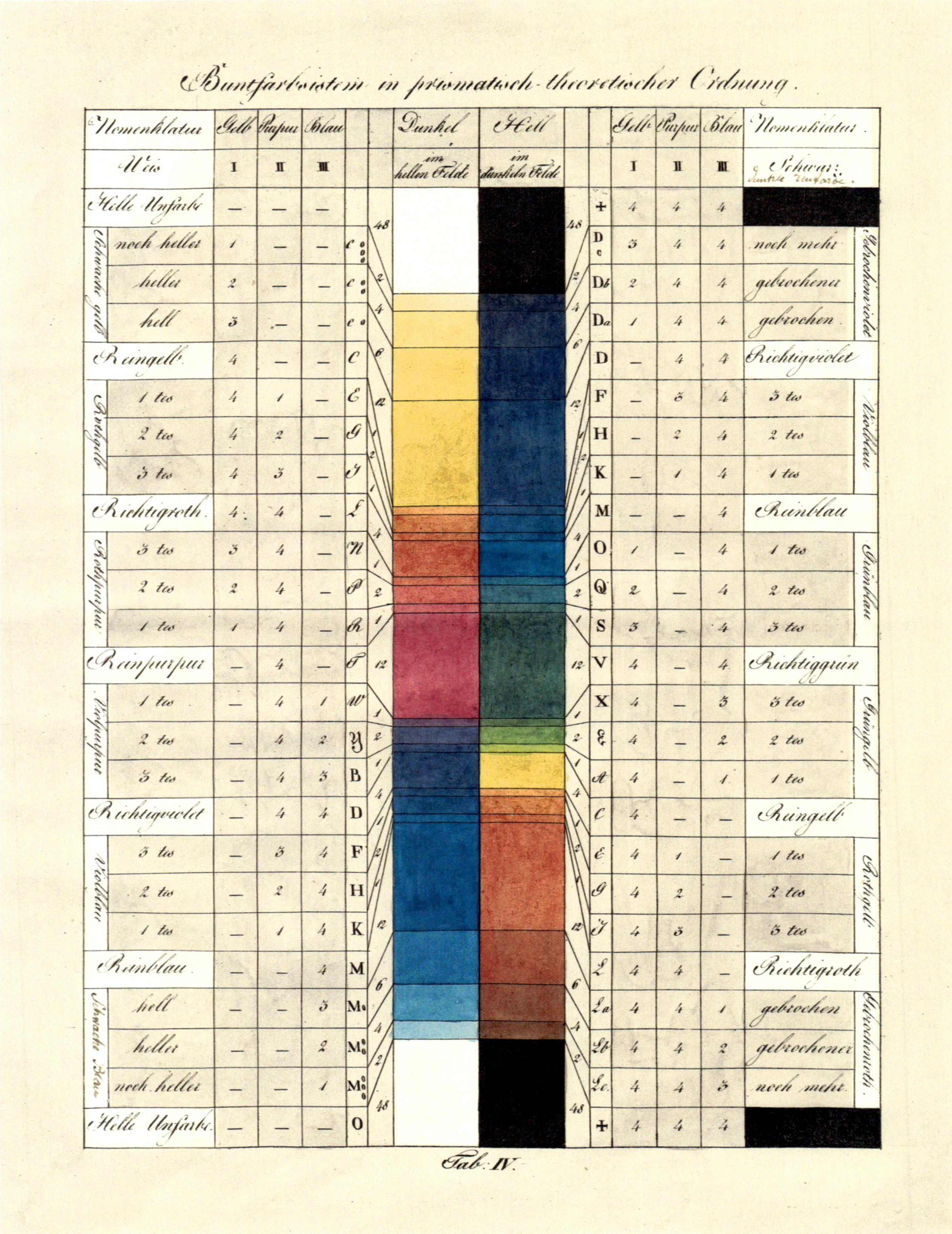
Buntfarbsystem in prismatisch-theoretischer Ordnung.
Nomenklatur | Gelb | Purpur | Blau | Dunkel | Hell | Gelb | Purpur | Blau | Nomenklatur
Weis | I | II | III | im hellen Felde | im dunklen Felde | I | II | III | Schwarz: dunkle Unfarbe
Helle Unfarbe | — | — | — | + | 4 | 4 | 4
Schwache gelb: noch heller | 1 | — | — | D c | 3 | 4 | 4 | noch mehr
heller | 2 | — | — | Db | 2 | 4 | 4 | gebrochener
hell | 3 | — | — | Da | 1 | 4 | 4 | gebrochen | Gebrochenviolet
Reingelb | 4 | — | — | C | D | — | 4 | 4 | Richtigviolet
Rothgelb: 1 tes | 4 | 1 | — | E | F | — | 3 | 4 | 3 tes
2 tes | 4 | 2 | — | G | H | — | 2 | 4 | 2 tes
3 tes | 4 | 3 | — | J | K | — | 1 | 4 | 1 tes | Violblau
Richtigroth | 4 | 4 | — | L | M | — | — | 4 | Reinblau
Rothpurpur: 3 tes | 3 | 4 | — | N | O | 1 | — | 4 | 1 tes
2 tes | 2 | 4 | — | P | Q | 2 | — | 4 | 2 tes
1 tes | 1 | 4 | — | R | S | 3 | — | 4 | 3 tes | Grünblau
Reinpurpur | — | 4 | — | T | V | 4 | — | 4 | Richtiggrün
Violpurpur: 1 tes | — | 4 | 1 | W | X | 4 | — | 3 | 3 tes
2 tes | — | 4 | 2 | Y | Z | 4 | — | 2 | 2 tes
3 tes | — | 4 | 3 | B | A | 4 | — | 1 | 1 tes | Grüngelb
Richtigviolet | — | 4 | 4 | D | C | 4 | — | — | Reingelb
Violblau: 3 tes | — | 3 | 4 | F | E | 4 | 1 | — | 1 tes
2 tes | — | 2 | 4 | H | G | 4 | 2 | — | 2 tes
1 tes | — | 1 | 4 | K | J | 4 | 3 | — | 3 tes | Rothgelb
Reinblau | — | — | 4 | M | L | 4 | 4 | — | Richtigroth
Schwache Blau: hell | — | — | 3 | Mo | La | 4 | 4 | 1 | gebrochen
heller | — | — | 2 | Mo | Lb | 4 | 4 | 2 | gebrochener
noch heller | — | — | 1 | Mo | Lc | 4 | 4 | 3 | noch mehr | Gebrochenroth
Helle Unfarbe | — | — | — | O | + | 4 | 4 | 4
48 | 2 | 4 | 6 | 12 | 1 | 2 | 1 | 4 | 1 | 2 | 1 | 12 | 1 | 2 | 1 | 4 | 1 | 2 | 12 | 6 | 4 | 2 | 48
Tab: IV.

Entstehung der Mittelfarben aus den Urfarben

durch Ausdehnung aus dem innern Dunkel ins äußere Hell: | durch Ausdehnung aus dem äußern Dunkel ins innere Hell.

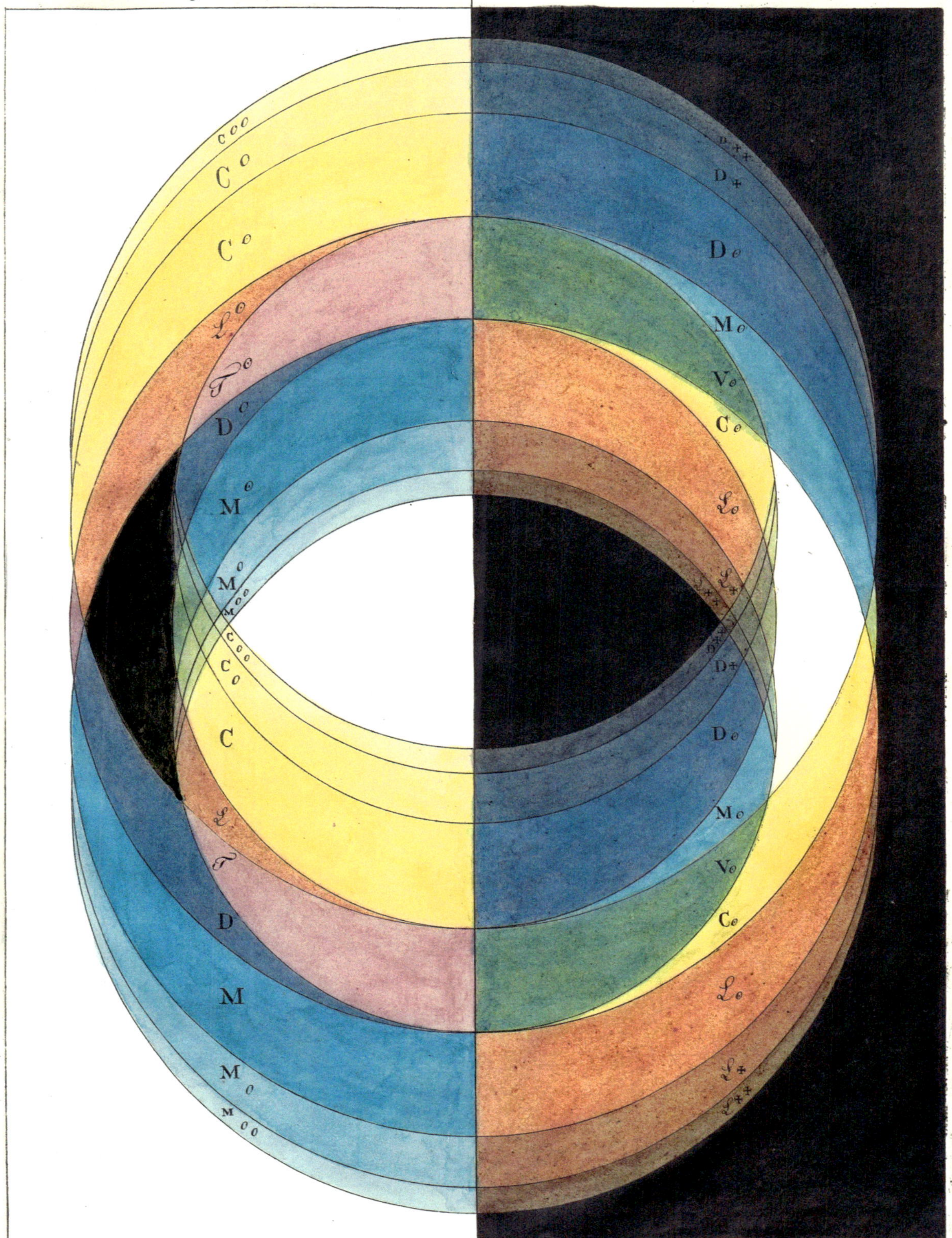

Tab. V.

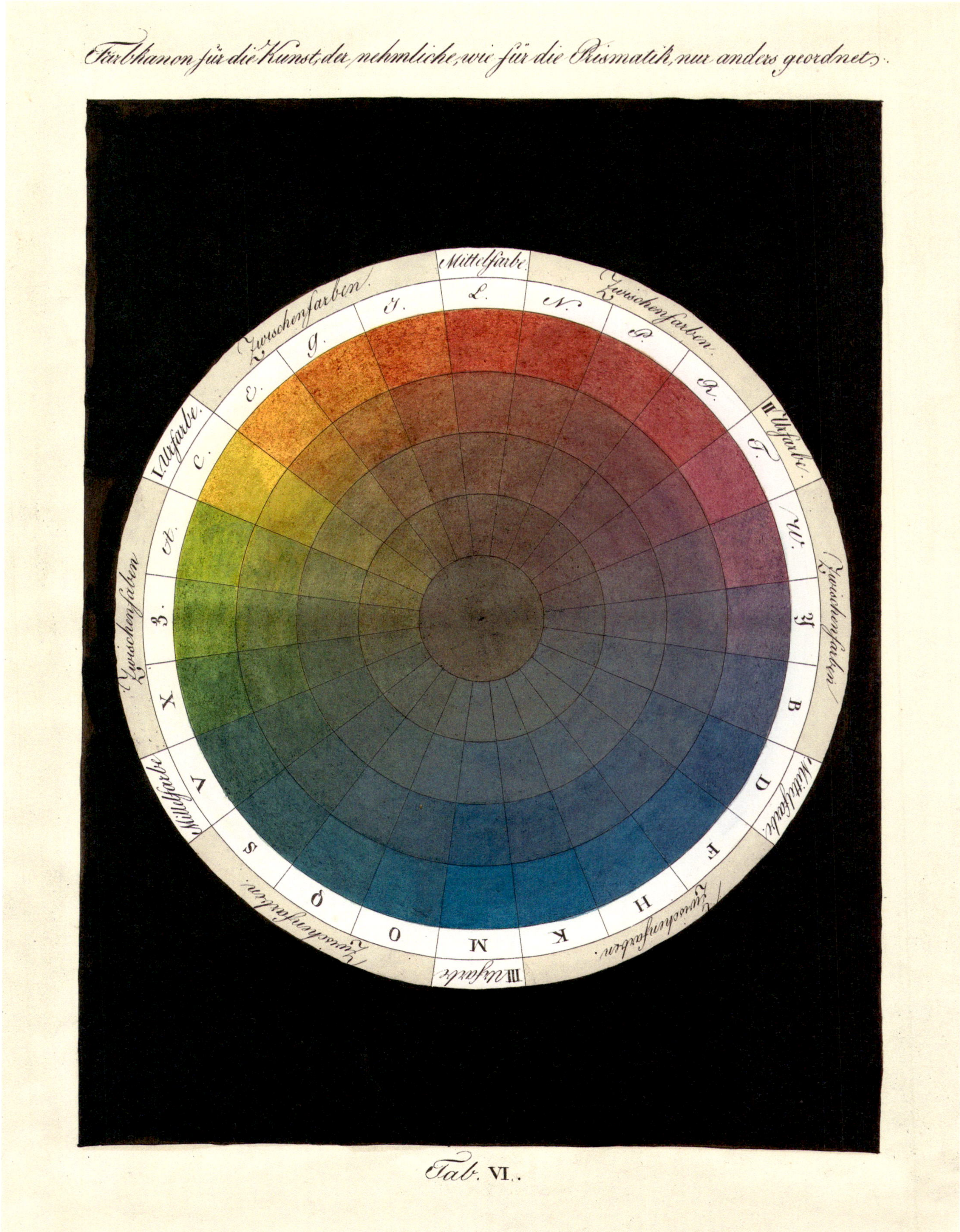
Farbkanon für die Kunst, der nehmliche, wie für die Prismatik, nur anders geordnet.
Mittelfarbe.
Zwischenfarben.
II Urfarbe.
Zwischenfarben.
Mittelfarbe.
Zwischenfarben.
III Urfarbe
Zwischenfarben.
Mittelfarbe.
Zwischenfarben
I. Urfarbe.
Zwischenfarben
L. N. P. R. T. W. Z. B D F H K M O Q S V X Z. A. C. E. G. J.
Tab. VI.

Erklärungsscheibe zum absoluten Farbkanon Tab. VI, welcher auf den prismatischen, Tab. IV. gegründet ist.

Von A und B an bis W und X bezeichnet der moderne Vorbuchstabe die Majorfarbe und der folgende antike ihre eigentliche Gegenfarbe im Minor. Das gothische Y und Z bezeichnen diese Farben als indifferent von Major und Minor.

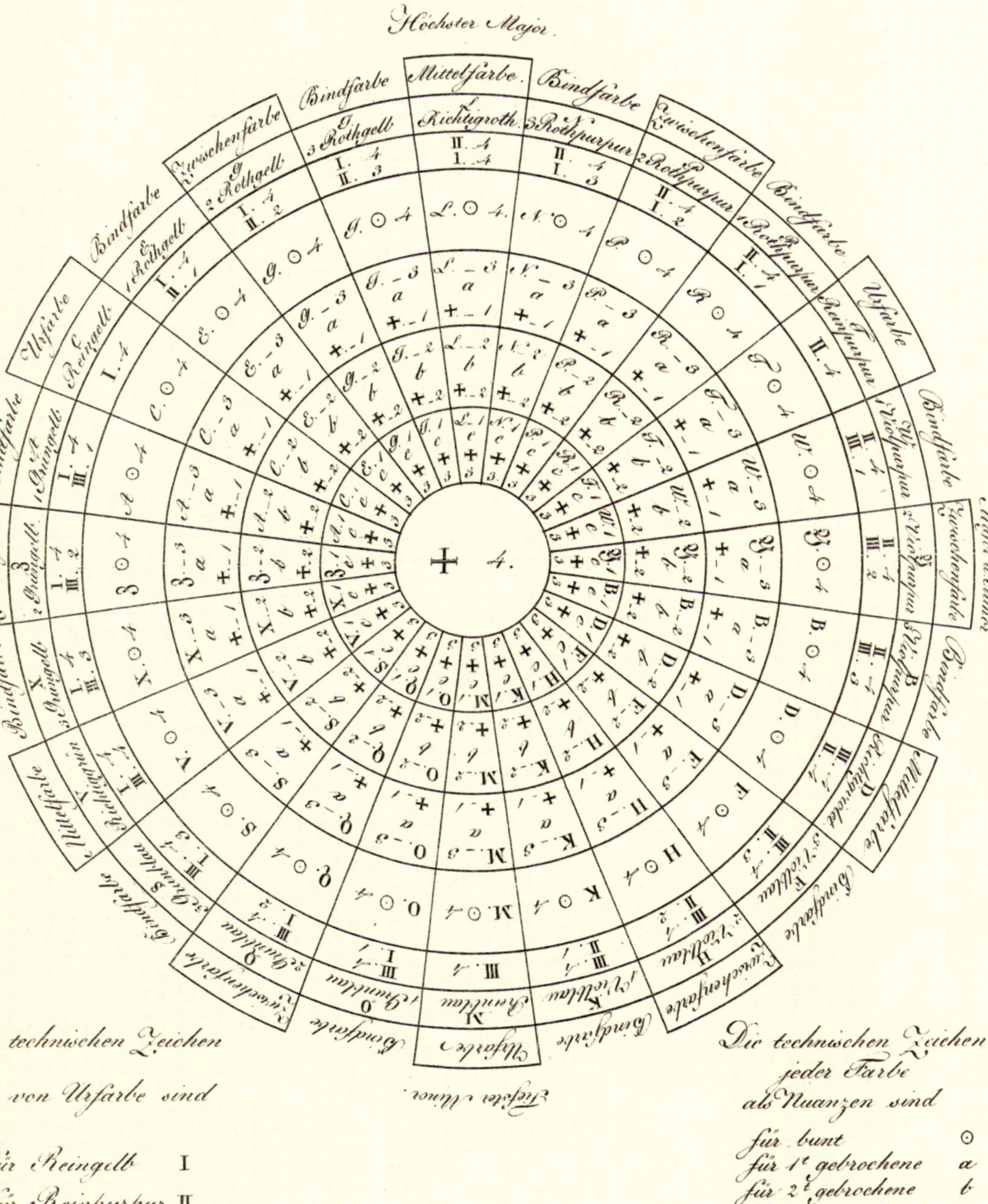

Die technischen Zeichen als von Urfarbe sind

für Reingelb I
für Reinpurpur II
für Reinblau III

Die technischen Zeichen jeder Farbe als Nuanzen sind

für bunt ⊙
für 1.te gebrochene a
für 2.te gebrochene b
für 3.te gebrochene c
für tode Unfarbe +

Tab. VII.

George Field (*c.* 1777–1854)

CHROMATICS

or, An Essay on the Analogy and Harmony of Colours

6 plates, 8 illustrated pages, 30.5 x 24.5 cm | 12 x 9 ⅝ in., London, 1817
Sussex, collection of the author

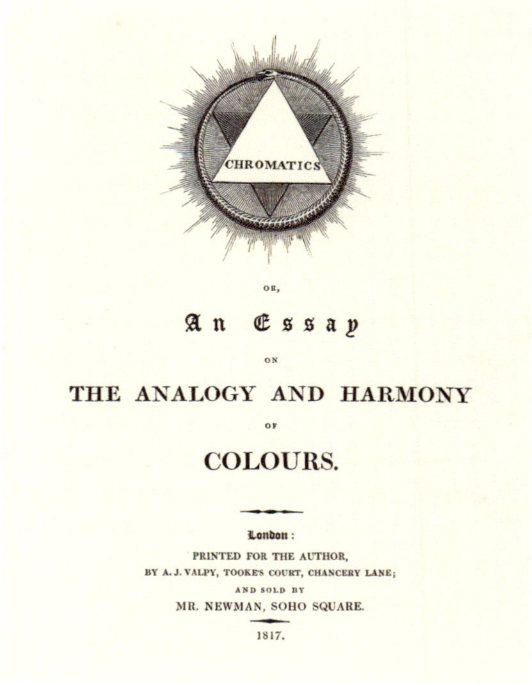
CHROMATICS

OR,

An Essay

ON

THE ANALOGY AND HARMONY

OF

COLOURS.

London:

PRINTED FOR THE AUTHOR,

BY A. J. VALPY, TOOKE'S COURT, CHANCERY LANE;

AND SOLD BY

MR. NEWMAN, SOHO SQUARE.

1817.

The entrepreneurial chemist George Field was also a pigment and dye maker and a colour researcher, as well as being extremely well connected in art circles in Britain in the early part of the 19th century. During his relatively long and successful life he provided many artists with high-quality pigments and paints, and published many books, including several on colour, which combined chemistry with theory, religious beliefs with colour order, and music with colour harmonies. Field had been experimenting with the manufacture of colours since 1804 and invented a number of tools and machines for use in pressing, filtering and producing pigments and dyes. Over the years he operated pigment factories in Bristol, London and Syon Hill Park in Middlesex, and from 1808 he began supplying pigments to artists, artists' suppliers, printers and publishers, including William Winsor and Henry Newton who later founded the art material suppliers Winsor & Newton. Field was a practising Christian and made analogies between the three primary colours and the Christian concept of the Holy Trinity. For their earthly representations he selected a trio of pure and stable pigments, namely red madder, lemon yellow and ultramarine blue.

Field's first book on colour, *Chromatics, or An Essay on the Analogy and Harmony of Colours*, was published in an edition of 250 copies. Ever keen to invent words, it seems, he explained that "*Chromatics* denotes the science of the relations of *light*, *shade*, and *colours*." Field recognised the importance of presenting colour concepts visually and took great care with the book's illustrations. The diagrams develop gradually, with the triangles that represent the trichromatic system serving as a guiding shape from which horizontal mixture charts, hexagons and eventually star shapes are derived, the stars being a variation on the familiar colour wheel. As with Newton and others, Field believed there was a strong connection between colour and music, an analogy he first discussed in *Chromatics* and which he visualised as an image of musical scales and their colour equivalents. This diagram, too, is based on a triangle. All the illustrations in *Chromatics* are hand-coloured, possibly done by Field himself or else under his close supervision. The intricate colour patterns of the sequence of five stars are especially impressive.

EXAMPLE IX.

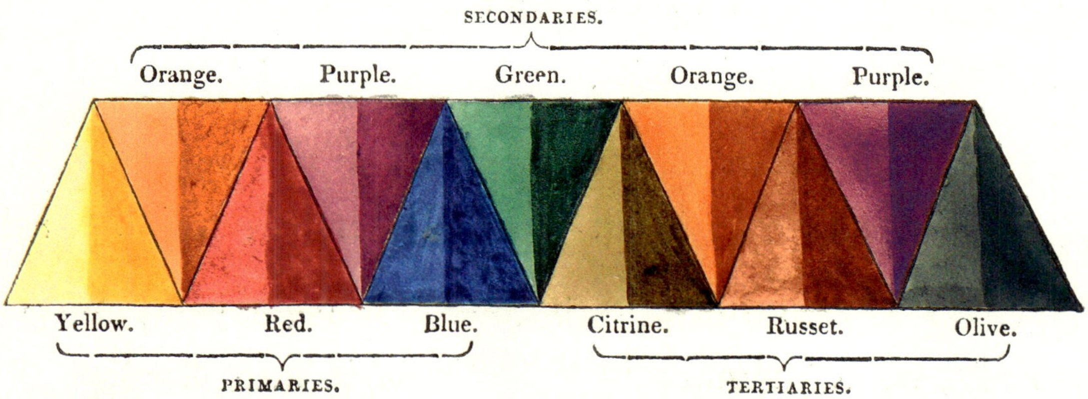

§ 20. Such are the *particular* distinctions, relations, and gradations of colours, as determined by the various predominance of their first principles through an orderly and infinite progress to the neutral grey; the position at which this predominance terminates, and the equilibrium of these first principles is re-established in unity according to a naturally perfect system.

§ 21. As the *neutralization or negation of colours* depends upon the reunion of the three primaries, (§ 8) it is evident that each of the *primary colours* is neutralized by that

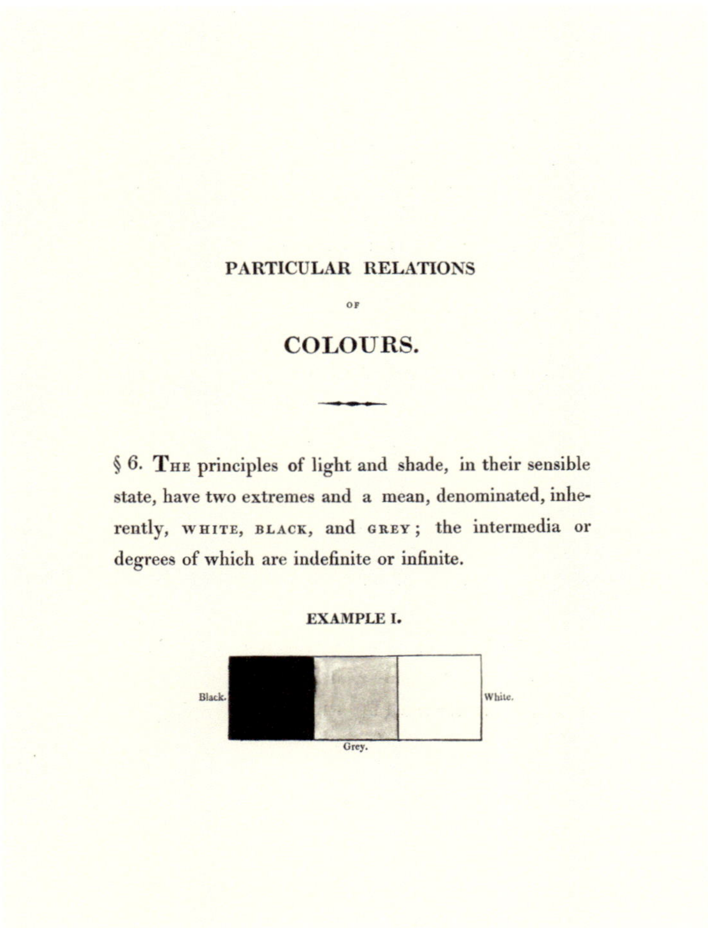

PARTICULAR RELATIONS

OF

COLOURS.

§ 6. The principles of light and shade, in their sensible state, have two extremes and a mean, denominated, inherently, WHITE, BLACK, and GREY; the intermedia or degrees of which are indefinite or infinite.

EXAMPLE I.

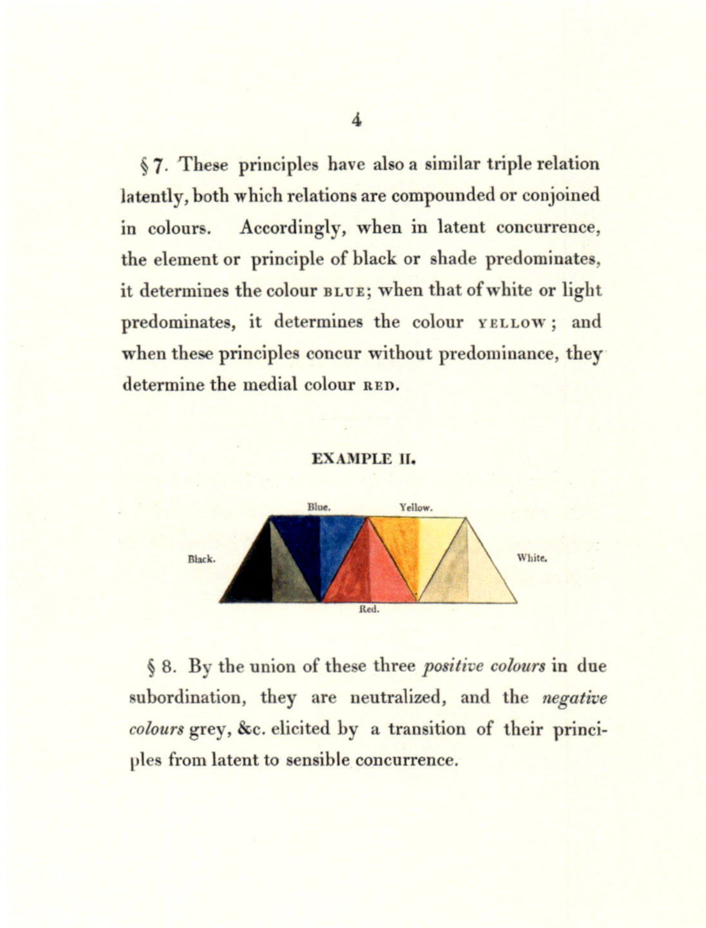

4

§ 7. These principles have also a similar triple relation latently, both which relations are compounded or conjoined in colours. Accordingly, when in latent concurrence, the element or principle of black or shade predominates, it determines the colour BLUE; when that of white or light predominates, it determines the colour YELLOW; and when these principles concur without predominance, they determine the medial colour RED.

EXAMPLE II.

§ 8. By the union of these three *positive colours* in due subordination, they are neutralized, and the *negative colours* grey, &c. elicited by a transition of their principles from latent to sensible concurrence.

Chromatik, oder: Ein Versuch über die Analogie und Harmonie der Farben

Der geschäftstüchtige Chemiker George Field war auch Pigment- und Farbstoffhersteller und als aktiver Farbforscher in den Kreisen der britischen Kunstwelt des frühen 19. Jahrhunderts außerordentlich gut vernetzt. Im Verlauf seines recht langen und erfolgreichen Lebens versorgte er zahlreiche Künstler mit hochwertigen Pigmenten und Farben und veröffentlichte viele Bücher, darunter die Farbenbücher, die Chemie mit Theorie, religiöse Überzeugungen mit Farbordnungen und Musik mit Farbharmonien verknüpften. Seit 1804 hatte Field mit Verfahren zur Farberzeugung experimentiert und eine Reihe von Werkzeugen und Maschinen zum Pressen, Filtern und Herstellen von Pigmenten und Farbstoffen erfunden. Im Laufe der Jahre betrieb er Pigmentfabriken in Bristol, London und Syon Hill Park in der Grafschaft Middlesex. Von 1808 an lieferte er Pigmente an Künstler, Künstlerbedarfsläden, Druckereien und Verleger, darunter William Winsor und Henry Newton, die später Winsor & Newton gründeten, einen Fachhandel für Künstlermaterialien. Als praktizierender Christ stellte Field Analogien zwischen den drei Primärfarben und dem christlichen Konzept der heiligen Dreifaltigkeit her. Für ihre Darstellung wählte er mit Krapprot, Zitronengelb und Ultramarinblau drei reine und stabile Pigmente aus.

Fields erstes Buch über Farbe, *Chromatics, or, An Essay on the Analogy and Harmony of Colours*, erschien in einer Auflage von 250 Exemplaren. Als begeisterter Wortschöpfer, so scheint es, erklärte er: „*Chromatik* bezeichnet die Wissenschaft von den Beziehungen zwischen *Licht*, *Schatten* und *Farben*.“ Field erkannte, wie wichtig es war, Farbtheorien in Bildern darzulegen, und verwandte große Sorgfalt auf die Illustrationen des Buches. Die Diagramme entwickeln sich Schritt für Schritt, wobei die Dreiecke, die das trichromatische System darstellen, als Leitform dienen, aus der sich waagerechte Farbmischungsgrafiken, Sechsecke und schließlich Sterne ergeben, Letztere als Variationen des vertrauten Farbkreises. Wie Newton und andere Farbforscher glaubte Field fest an eine starke Beziehung zwischen Farbe und Musik, eine Analogie, die er zuerst in *Chromatics* erörterte und als Abbildung von Tonleitern und ihren farblichen Entsprechungen anschaulich machte. Auch dieses Diagramm basiert auf der Dreiecksform. Alle Illustrationen in *Chromatics* wurden, möglicherweise von Field selbst oder unter seiner strengen Aufsicht, von Hand koloriert. Die Abfolge der fünf Sterne mit ihren komplizierten Farbmustern ist besonders beeindruckend.

5

§ 9. Thus the PRIMARY COLOURS resulting from the analysis, or concurring in the synthesis of these principles or fundamentals in union are *three;* the lowest number capable of uniting in variety, harmony, or system; and this variety of union can be only three.

§ 10. First, from the mixing of the primaries blue and red; proceeds the secondary PURPLE on the dark extreme.

EXAMPLE III.

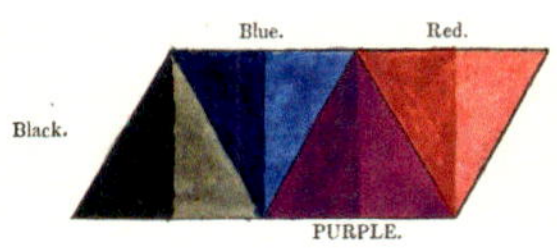

§ 11. Secondly, from that of yellow and red proceeds ORANGE on the light extreme.

6

EXAMPLE IV.

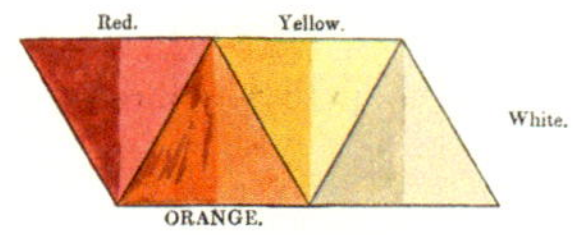

§ 12. And thirdly, from the union of blue and yellow, proceeds medially the secondary GREEN.

EXAMPLE V.

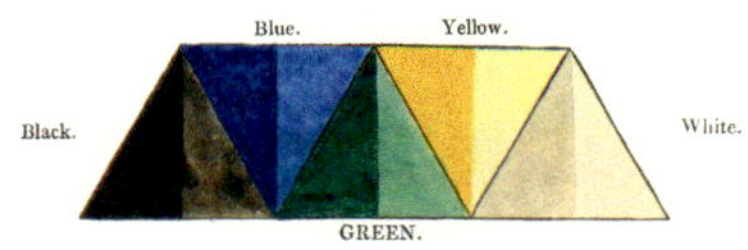

§ 13. It follows, of course, that the SECONDARY COLOURS are capable of the same variety of union as their prima-

La chromatique ou Essai sur l'analogie et l'harmonie des couleurs

Chimiste doué pour les affaires, George Field est également fabricant de pigments et de teintures et spécialiste des couleurs. En ce début du XIXe siècle, il est extrêmement bien introduit dans tous les cercles artistiques britanniques. Tout au long d'une vie relativement longue et bien remplie, il fournit de nombreux artistes en pigments et peintures de qualité et publie plusieurs ouvrages. Ceux consacrés à la couleur établissent des liens entre la chimie et la théorie, les croyances religieuses et l'ordre chromatique, ou encore la musique et les harmonies chromatiques. Field commence ses expérimentations en matière de fabrication de couleurs dès 1804 et met au point un certain nombre d'outils et de machines pour presser, filtrer et produire des pigments et des teintures. Au fil des années, il dirigera plusieurs fabriques de pigments à Bristol, Londres et Syon Hill Park dans le Middlesex. À partir de 1808, il fournit en pigments les peintres, les imprimeurs, les éditeurs et les marchands de couleurs, notamment William Winsor et Henry Newton qui fonderont la maison de matériel de beaux-arts Winsor & Newton. Chrétien pratiquant, Field établit des analogies entre les trois couleurs primaires et le concept de la Sainte Trinité. Il choisit, pour leur représentation matérielle, un trio de pigments purs et stables : garance, jaune citron et bleu outremer.

Premier opus que Field consacre à la couleur, *Chromatics, or, An Essay on the Analogy and Harmony of Colours* est tiré à 250 exemplaires. L'homme aime, semble-t-il, inventer de nouveaux mots : ainsi explique-t-il que « *la chromatique* désigne la science des relations entre la *lumière*, l'*ombre* et les *couleurs* ». Field comprend l'importance d'une représentation visuelle des concepts chromatiques et apporte le plus grand soin aux illustrations de son livre. Les graphiques se déclinent progressivement, et le système trichromatique représenté par des triangles sert de forme de référence d'où dérivent ensuite des tableaux de mélanges horizontaux, hexagonaux puis en forme d'étoile, ces derniers étant des variations du cercle chromatique classique. À l'instar de Newton et d'autres, Field croit aux rapports étroits entre la couleur et la musique et étudie cette analogie pour la première fois dans *Chromatics*. Il représente celle-ci par une gamme de notes accompagnées de leurs équivalents colorés et agencés en petits triangles. Toutes les illustrations de *Chromatics* sont peintes à la main, peut-être par Field lui-même ou sous sa supervision. Les motifs colorés complexes des cinq étoiles sont particulièrement remarquables.

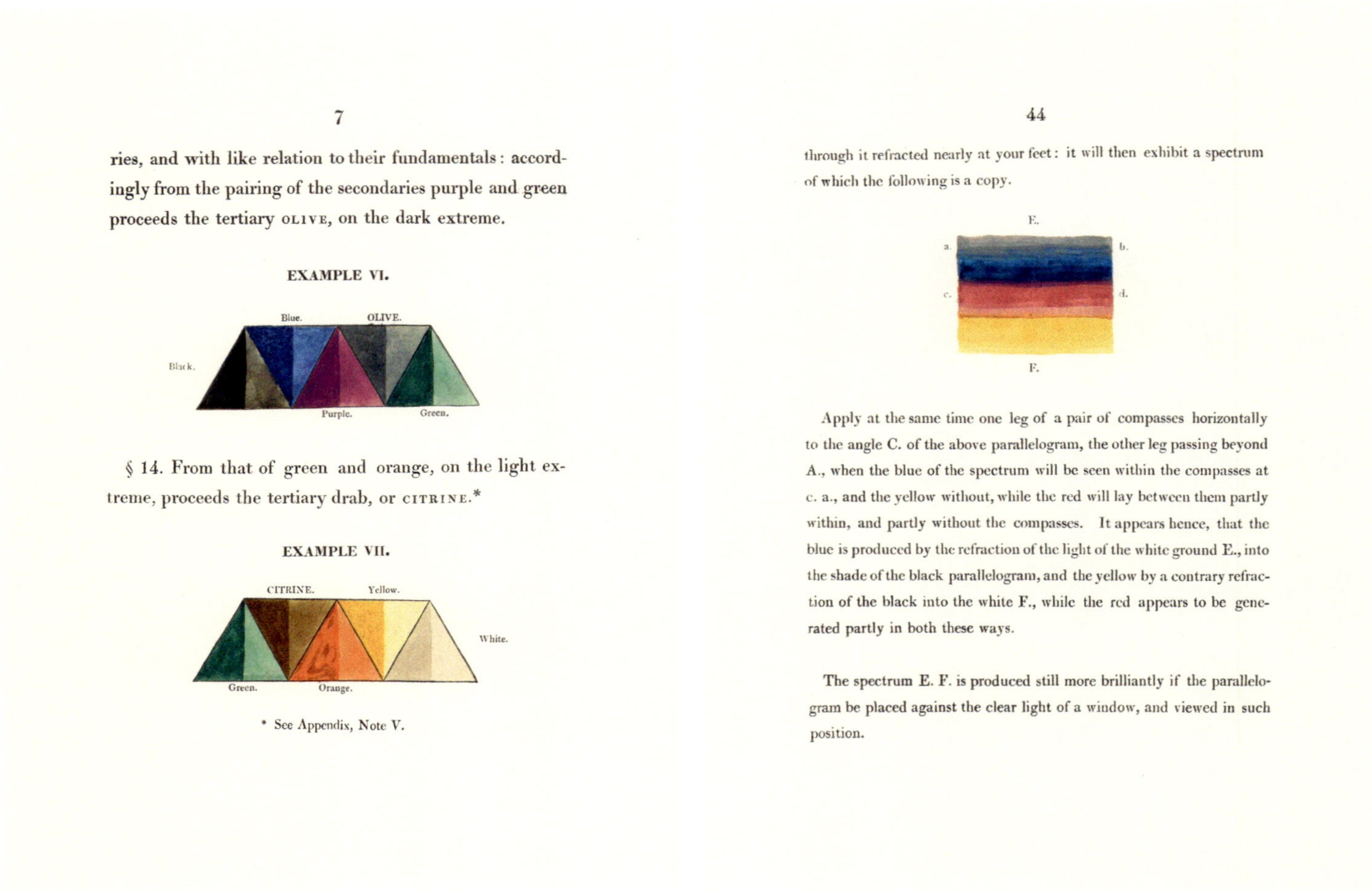

7

ries, and with like relation to their fundamentals: accordingly from the pairing of the secondaries purple and green proceeds the tertiary OLIVE, on the dark extreme.

EXAMPLE VI.

§ 14. From that of green and orange, on the light extreme, proceeds the tertiary drab, or CITRINE.*

EXAMPLE VII.

* See Appendix, Note V.

44

through it refracted nearly at your feet: it will then exhibit a spectrum of which the following is a copy.

Apply at the same time one leg of a pair of compasses horizontally to the angle C. of the above parallelogram, the other leg passing beyond A., when the blue of the spectrum will be seen within the compasses at c. a., and the yellow without, while the red will lay between them partly within, and partly without the compasses. It appears hence, that the blue is produced by the refraction of the light of the white ground E., into the shade of the black parallelogram, and the yellow by a contrary refraction of the black into the white F., while the red appears to be generated partly in both these ways.

The spectrum E. F. is produced still more brilliantly if the parallelogram be placed against the clear light of a window, and viewed in such position.

Cromática, o Ensayo sobre la analogía y la armonía de los colores

El empresario químico George Field también fue fabricante de pigmentos y tintes e investigador del color, además de estar muy bien conectado en los círculos artísticos de Gran Bretaña a principios del siglo XIX. Durante su vida relativamente larga y exitosa, proporcionó a muchos artistas pigmentos y pinturas de alta calidad y publicó muchos libros, incluidos varios sobre el color, que combinaban la química con la teoría, las creencias religiosas con el orden de los colores y la música con las armonías de los colores. Field había experimentado con la fabricación de colores desde 1804, e inventó una serie de herramientas y máquinas para prensar, filtrar y producir pigmentos y tintes. A lo largo de los años, dirigió fábricas de pigmentos en Bristol, Londres y Syon Hill Park en Middlesex, y desde 1808 comenzó a suministrar pigmentos a artistas, proveedores de artistas, impresores y editores, como William Winsor y Henry Newton, quienes más tarde fundaron la empresa proveedora de materiales artísticos Winsor & Newton. Field fue un cristiano practicante, y estableció analogías entre los tres colores primarios y el concepto cristiano de la Santísima Trinidad. Para sus representaciones terrenales, seleccionó un trío de pigmentos puros y estables, a saber: rojo de alizarina, amarillo limón y azul ultramar.

El primer libro de Field sobre el color, *Chromatics, or, An Essay on the Analogy and Harmony of Colours*, se publicó en una edición de 250 copias. Siempre dispuesto, al parecer, a inventar palabras, explicó que «*Chromatics* significa ciencia de las relaciones de la *luz*, la *sombra* y los *colores*». Field reconoció la importancia de presentar visualmente los conceptos de color y puso especial atención en las ilustraciones del libro. Los diagramas se desarrollan gradualmente, con los triángulos que representan el sistema tricromático actuando como una forma guía de la que se derivan las tablas de mezclas horizontales, los hexágonos y, finalmente, las formas de estrellas, siendo estas una variación de la familiar rueda de colores. Al igual que Newton y otros, Field creía que había una fuerte conexión entre el color y la música, una analogía que abordó por primera vez en *Chromatics* y que visualizó como una imagen de escalas musicales y sus colores equivalentes. Ese diagrama también se basa en un triángulo. Todas las ilustraciones de *Chromatics* están coloreadas a mano, posiblemente por el mismo Field o bajo su estrecha supervisión. Los elaborados patrones de color de la secuencia de cinco estrellas son especialmente memorables.

EXAMPLE X.

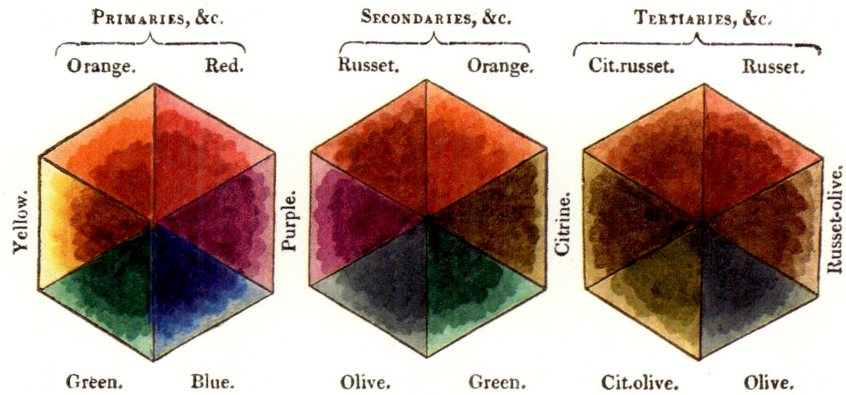

§ 24. The compounds of the latter order of colours approach near to the perfect neutral. *Perfect neutrality depends, however, upon a due subordination of the primary colours* in which blue predominates in proportion to the depth of the compound, and yellow is subordinate to red ; (§ 30) or *of the secondaries* in which purple predominates, and orange is subordinate to green, (§ 38) or finally *of the*

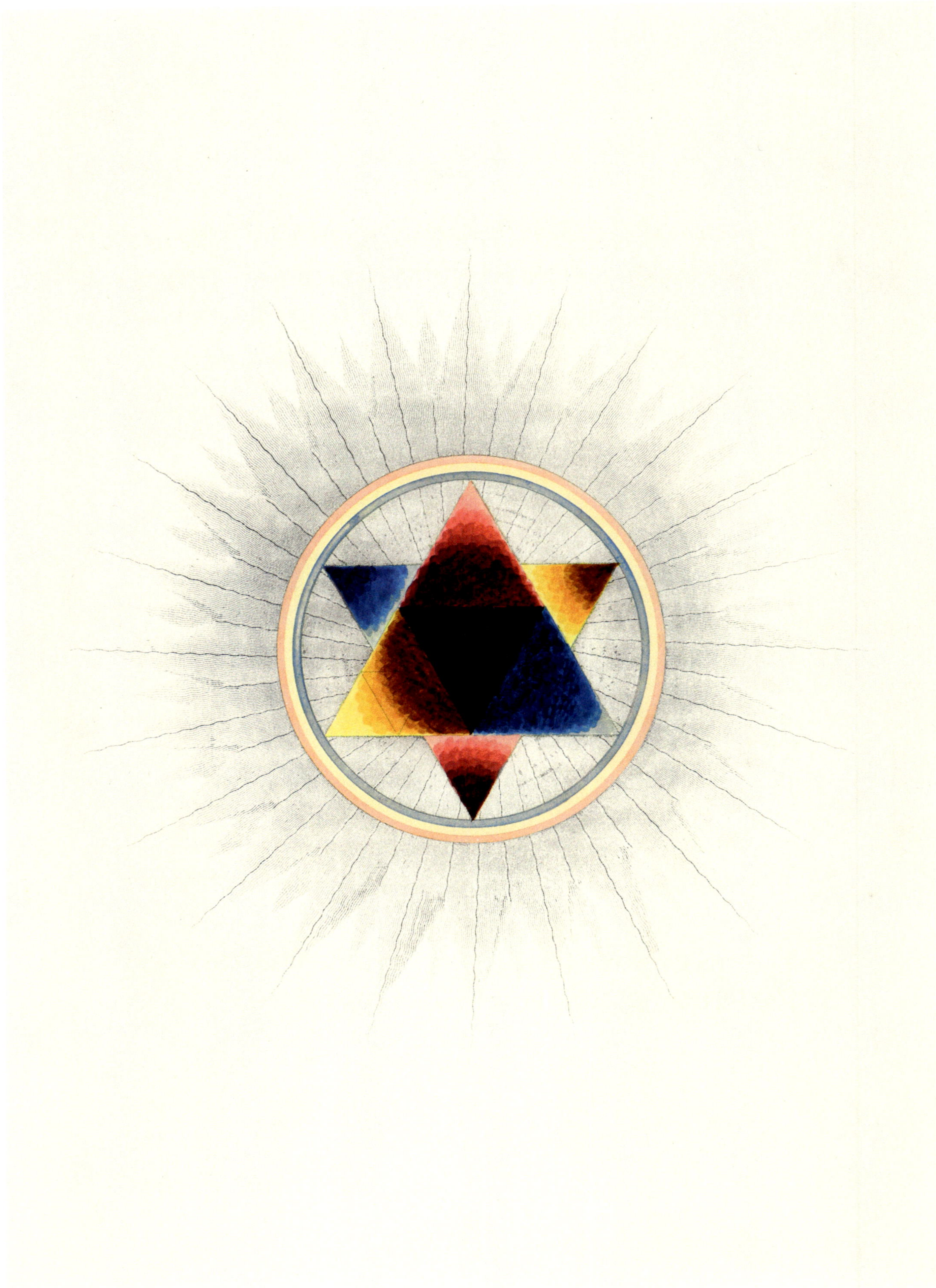

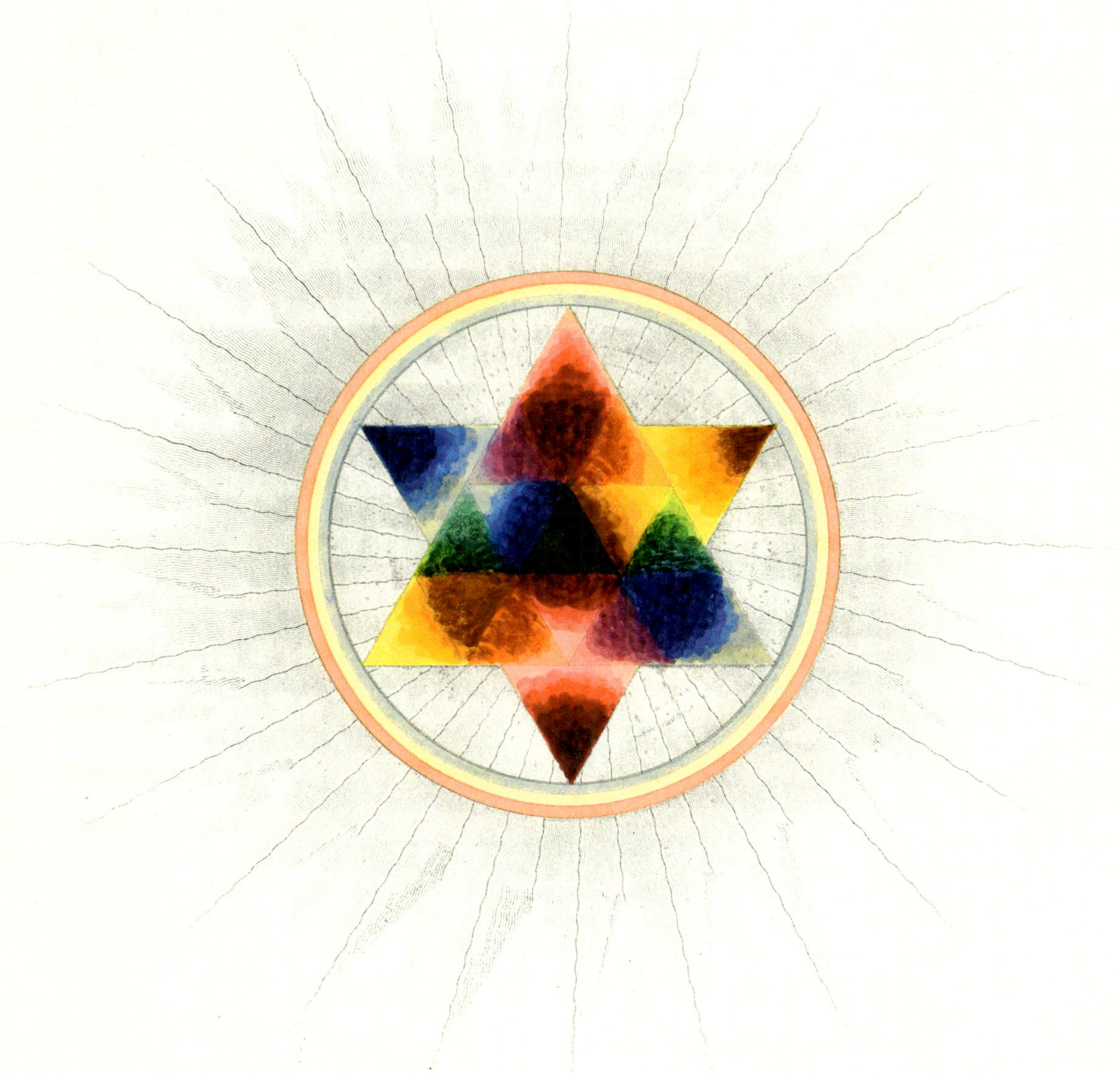

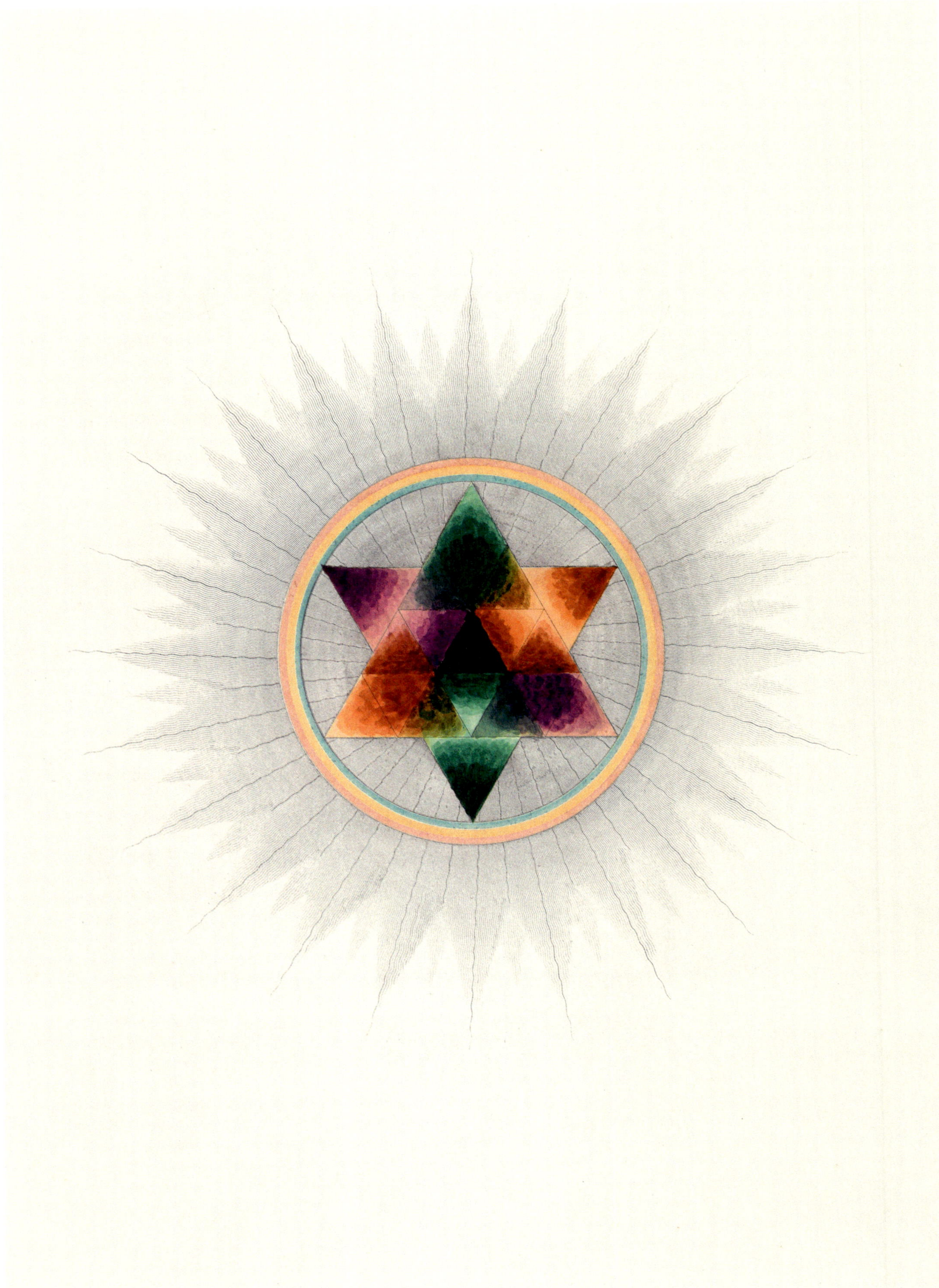

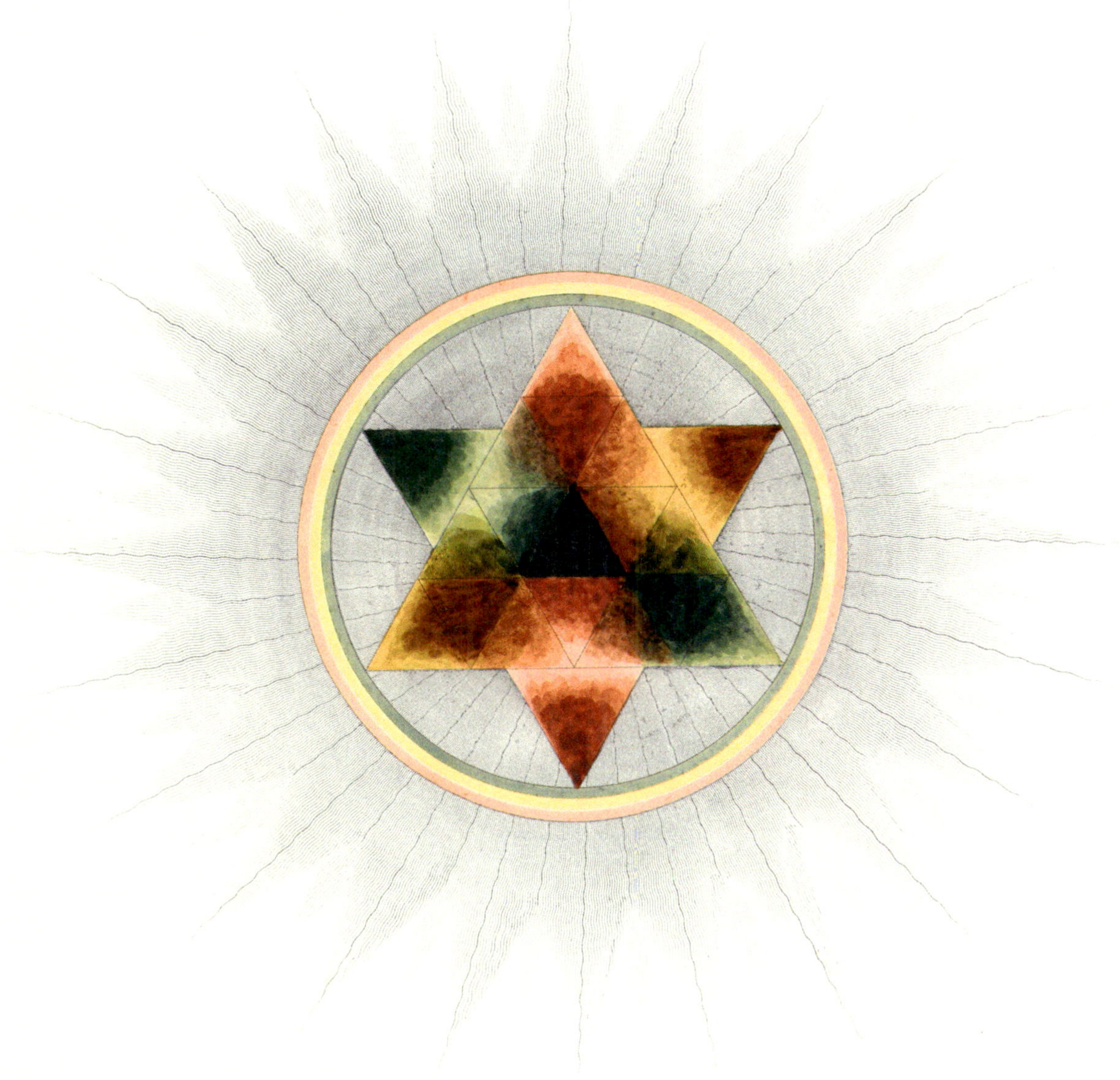

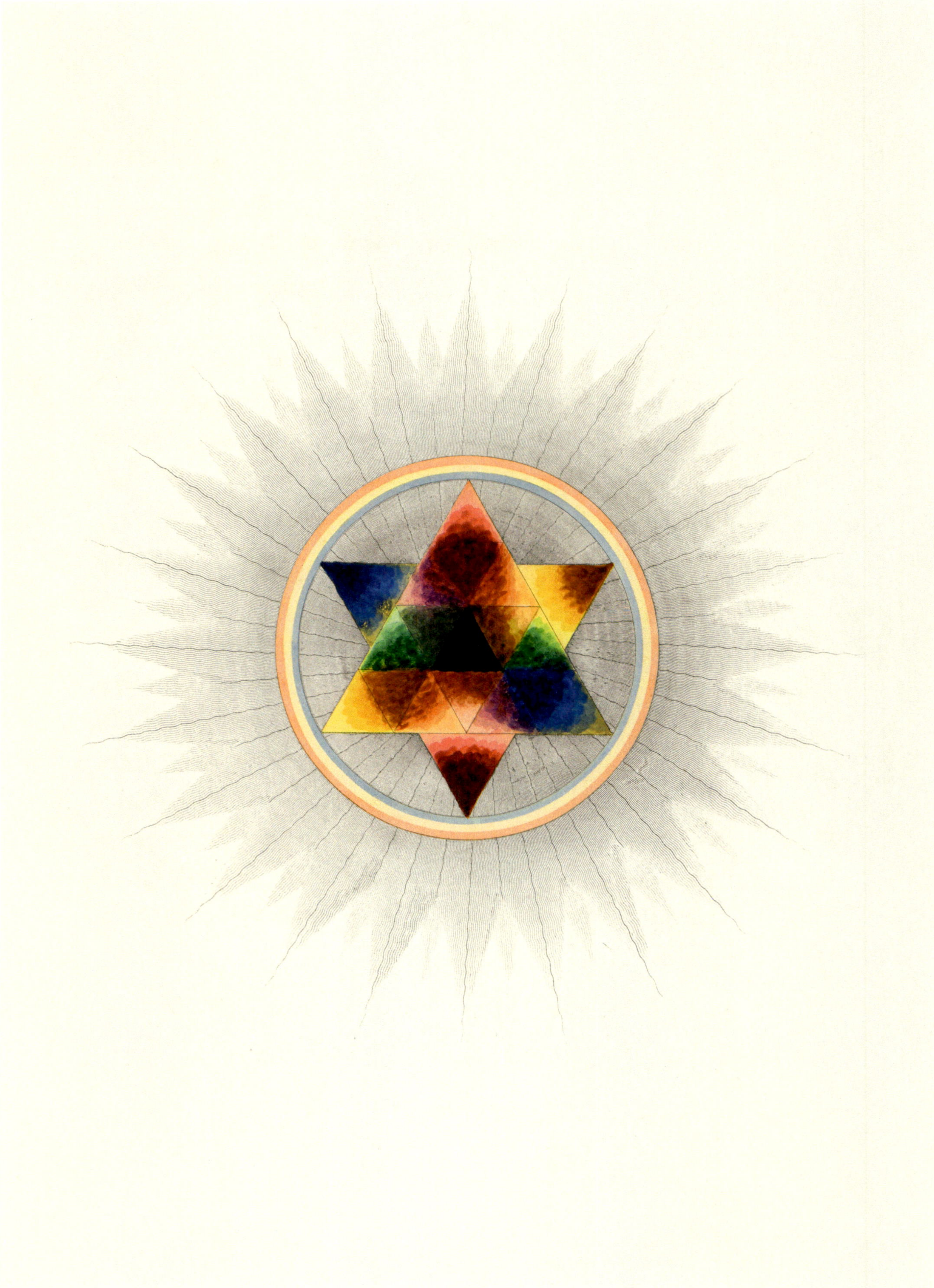

EXAMPLE XVI.

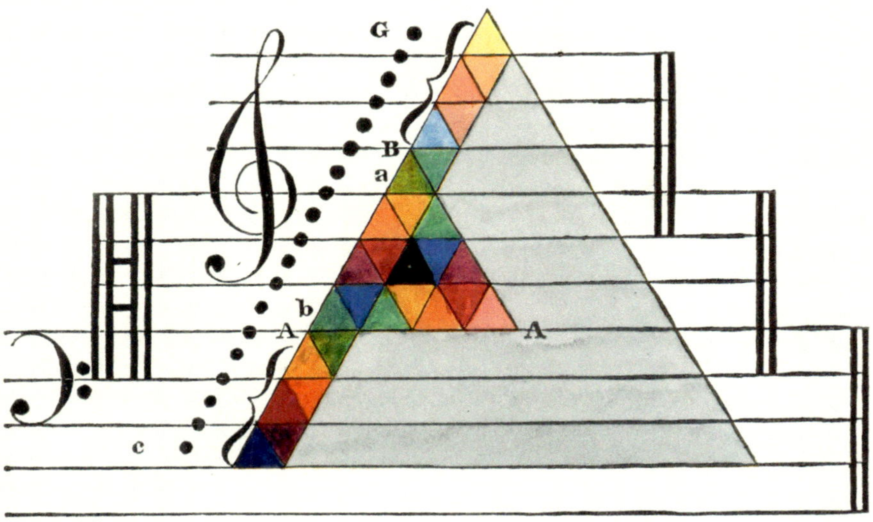

ANALOGOUS SCALE OF SOUNDS AND COLOURS.

Chr. E

George Field

CHROMATOGRAPHY

or, A Treatise on Colours and Pigments, and of Their Powers in Painting

2 plates, 34 x 25.2 cm | 13 ⅜ x 9 ⅞ in., London, 1835
Los Angeles, Getty Research Institute

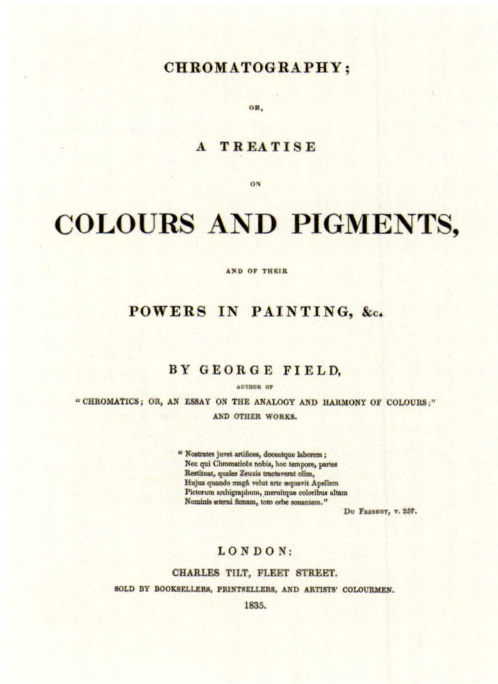

CHROMATOGRAPHY;

OR,

A TREATISE

ON

COLOURS AND PIGMENTS,

AND OF THEIR

POWERS IN PAINTING, &c.

BY GEORGE FIELD,

AUTHOR OF

"CHROMATICS; OR, AN ESSAY ON THE ANALOGY AND HARMONY OF COLOURS;"
AND OTHER WORKS.

"Nostrates juvet artifices, doceatque laborem;
Nec qui Chromaticês nobis, hoc tempore, partes
Restituat, quales Zeuxis tractaverat olim,
Hujus quando magâ velut arte æquavit Apellem
Pictorum archigraphum, meruitque coloribus altam
Nominis æterni famam, toto orbe sonantem."

DU FRESNOY, v. 257.

LONDON:

CHARLES TILT, FLEET STREET.

SOLD BY BOOKSELLERS, PRINTSELLERS, AND ARTISTS' COLOURMEN.

1835.

Field's first book on colour, the privately published *Chromatics* (1817), had brought him to the attention of the colour world both at home and abroad, but it was a short text, mostly theoretical and expensively produced, with only a small print run. His more influential and extensive work, *Chromatography*, was published 18 years later and this time on a subscription basis. The list of subscribers included many of the leading artists of the time, including Turner, Thomas Lawrence and David Wilkie. Running to 276 pages of text, including an index, it was a much more substantial and practically-oriented reference work, based on years of testing pigments for quality and stability. The book as first published contained only two illustrations, one of which is an engraved plate that precedes the chapter on "new optical instruments" and depicts some of them, including Field's own "Chromascope" and "Metrochrome", while the other illustration is an elaborate frontispiece, which appears to be a highly intelligent conflation of the many images that featured in *Chromatics*. The coloured band of triangles at the bottom of the plate shows the range of primaries, secondaries and tertiaries and their relation to black and white, while the flower-shaped diagram of "Chromatic Equivalents" acts as a compass that enables the user to calculate precise proportions of the different colour mixtures. The example shown here is uncoloured, but the same diagram appears in coloured and adapted form in later editions of Field's books.

In his works Field was responding not only to the need to establish a proper understanding of good pigments, but also the general need for scientifically sound and reliable published information on colour. This explains the continued popularity of his books, all of which were reprinted in expanded or revised editions, including posthumously. His last work, *Rudiments of the Painters' Art; or a Grammar of Colouring* (1850), was a much smaller and more affordable handbook with simpler illustrations. A radically shortened and revised edition of *Chromatography* was published by Winsor and Newton in 1869, and again in 1885, with additional texts and amendments respectively by Thomas W. Salter and J. Scott Taylor.

Fig. 1

Experiment I.

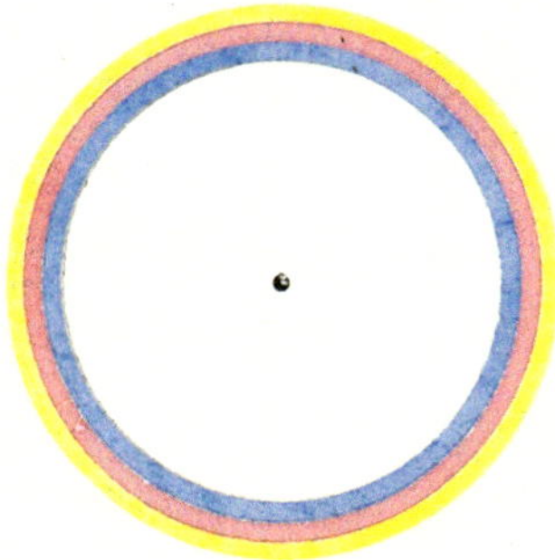

Chromatic Equivalents.

Fig. 2. Exp. XXVIII.

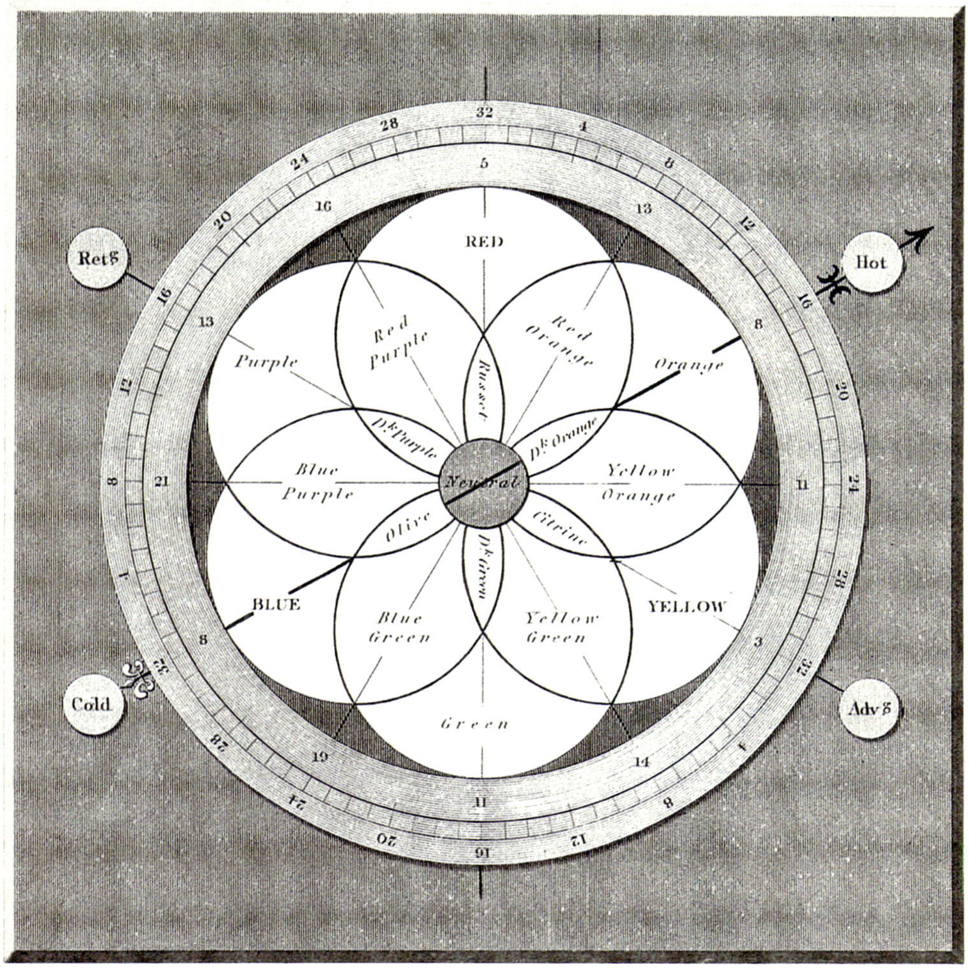

Definitive or Fundamental Scale of Colours

Fig. 3

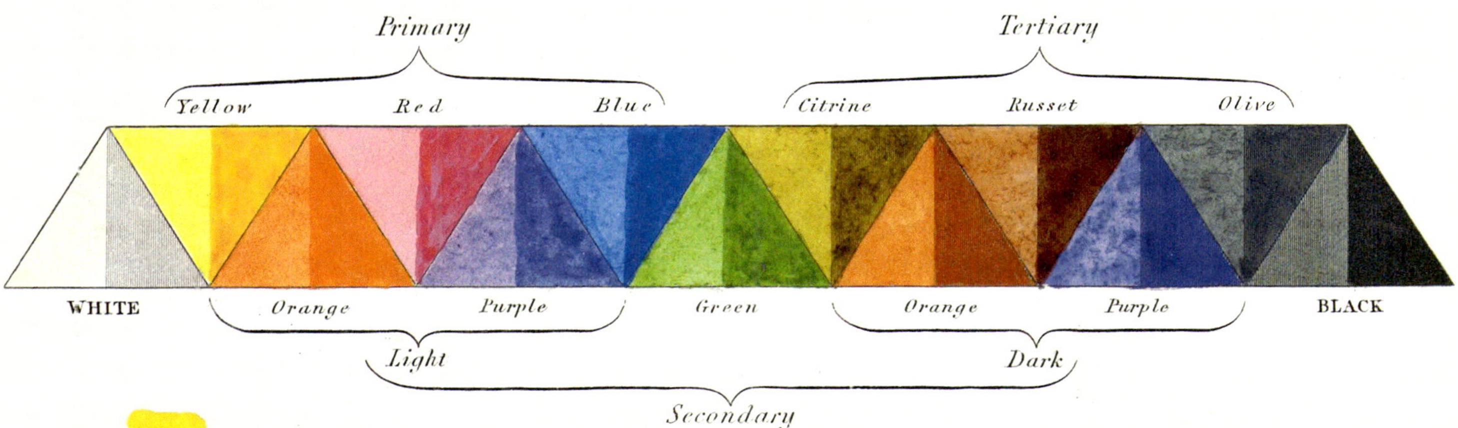

d inv.r

E. Turrell Sculp.

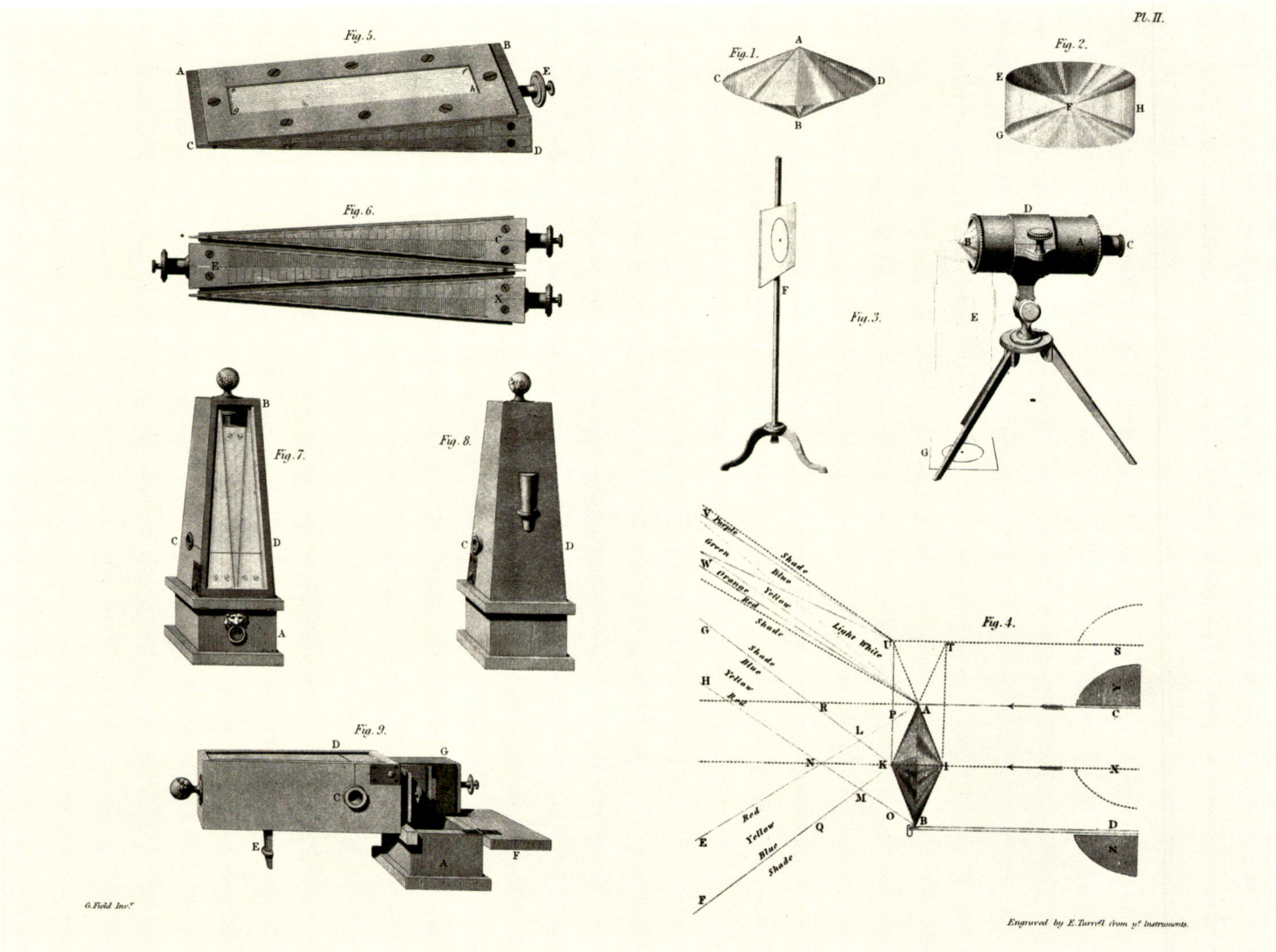

Chromatographie. Eine Abhandlung über Farben und Pigmente, so wie deren Anwendung in der Malerkunst

Fields erstes Buch über Farbe, das privat herausgegebene *Chromatics* (1817), hatte die Farbwelt im In- und Ausland auf ihn aufmerksam gemacht, war jedoch nur ein kurzer, überwiegend theoretischer und kostspielig produzierter Text mit kleiner Auflage gewesen. Sein an Einfluss und Umfang bedeutenderes Werk, *Chromatography*, erschien 18 Jahre später auf Subskriptionsbasis. Die Liste der Subskribenten umfasste zahlreiche führende Künstler der Zeit, darunter Turner, Thomas Lawrence und David Wilkie. Mit 276 Textseiten inklusive Register war das Buch ein deutlich gehaltvolleres und praxisorientierteres Nachschlagewerk, dem die jahrelange Erprobung von Pigmenten auf Qualität und Stabilität zugrunde lag. Allerdings enthielt die erste Auflage nur zwei Bildtafeln. Ein dem Kapitel über „neue optische Instrumente" vorangestellter Kupferstich zeigt unter anderem ein „Chromaskop" und ein „Metrochrom", die Field selbst entwickelt hatte. Die zweite Tafel, ein aufwendiges Frontispiz, erscheint wie eine hochintelligente Verschmelzung der vielen Abbildungen aus *Chromatics*. Der aus farbigen Dreiecken zusammengesetzte Streifen am unteren Rand der Tafel zeigt das Spektrum der Primär-, Sekundär- und Tertiärfarben und ihr Verhältnis zu Schwarz und Weiß. Das blütenförmige Diagramm der „Chromatischen Äquivalente" darüber fungiert als Kompass, der es dem Anwender ermöglicht, die exakten Verhältnisse der unterschiedlichen Farbmischungen zu berechnen. Während das hier gezeigte Beispiel nicht koloriert ist, taucht das Diagramm in späteren Ausgaben der von Field verfassten Bücher auch farbig überarbeitet auf.

Mit seinen Werken reagierte Field nicht nur auf die Notwendigkeit, ein Verständnis für gute Pigmente zu etablieren, sondern auch auf den allgemeinen Bedarf an wissenschaftlich fundierten und verlässlichen Informationen über Farbe in gedruckter Form. Das erklärt die fortgesetzte Popularität seiner Bücher, die in verbesserten und erweiterten Auflagen auch postum noch nachgedruckt wurden. Sein letztes Werk, *Rudiments of the Painters' Art; or a Grammar of Colouring* (1850), war ein wesentlich kleineres und erschwinglicheres Handbuch mit schlichteren Illustrationen. Eine radikal gekürzte und durchgesehene

Ausgabe von *Chromatography* erschien 1869 und erneut 1885 bei Winsor & Newton. Zusätzliche Texte und Änderungen hatte im ersten Fall Thomas W. Salter besorgt, im zweiten J. Scott Taylor.

La chromatographie ou Traité sur les couleurs et les pigments et leurs forces en peinture

Chromatics, le premier ouvrage de Field sur le sujet et publié à compte d'auteur en 1817, a eu un fort écho dans l'univers de la couleur, tant en Angleterre qu'à l'étranger. Cependant, c'était un texte court et principalement théorique, avec un tirage limité du fait d'un coût de fabrication élevé. Dix-huit ans plus tard, Field publie *Chromatography*, qui deviendra son ouvrage le plus marquant. Cette fois, il le finance grâce à une souscription, à laquelle participent de nombreuses grandes figures de la peinture de l'époque, dont Turner, Thomas Lawrence et David Wilkie. Avec 276 pages de texte, dont un index, ce volumineux ouvrage de référence, destiné aux applications pratiques, est le fruit d'années d'expériences pour obtenir les pigments de la meilleure qualité et les plus stables possible. Dans sa première mouture, l'ouvrage ne contient que deux illustrations. L'une est une planche gravée qui introduit le chapitre consacré aux « nouveaux instruments optiques » et en décrit certains, dont le « chromascope » et le « métrochrome », inventions de l'auteur. L'autre est un frontispice complexe, sorte d'habile compilation de nombreuses images qui figuraient dans *Chromatics*. La bande colorée de triangles au bas de la planche présente la gamme des couleurs primaires, secondaires et tertiaires et dans leurs rapports avec le noir et le blanc. Quant au graphique en forme de fleur des « Équivalents chromatiques », il permet, telle une boussole, de calculer les proportions précises des différents mélanges de couleurs. L'exemple montré ici est sans couleurs, mais une version colorée et modifiée apparaît dans des éditions ultérieures des livres de Field.

Par ses travaux, Field répond au besoin d'établir une connaissance claire des pigments de qualité, mais plus généralement de disposer d'une publication de référence scientifiquement étayée concernant la couleur. C'est ce qui explique le succès durable de ses ouvrages, tous réimprimés dans des versions augmentées et révisées, y compris à titre posthume. Son dernier livre, *Rudiments of the Painters' Art; or a Grammar of Colouring* (1850), est un manuel plus petit et moins coûteux, avec des illustrations plus simples. Une version considérablement raccourcie et révisée de *Chromatography* est publiée par Winsor et Newton en 1869. Une autre paraît en 1885, modifiée par J. Scott Taylor et avec de nouveaux textes de Thomas W. Salter.

Cromatografía, o Tratado sobre colores y pigmentos y su poder en la pintura

El primer libro de Field sobre el color, *Chromatics* (1817), que publicó de forma privada, había llamado la atención del mundo del color en su país y en el extranjero, pero era un texto corto, en su mayor parte teórico y caro de producir, del que solo se había realizado una tirada pequeña. Su trabajo más influyente y extenso, *Chromatography*, se publicó 18 años después, y esta vez por suscripción. La lista de suscriptores contaba con muchos de los principales artistas de la época, incluidos Turner, Thomas Lawrence y David Wilkie. Con 276 páginas de texto, incluido un índice, era un trabajo de referencia mucho más sustancial y orientado a la práctica, basado en años de pruebas de calidad y estabilidad de los pigmentos. El libro, tal como se publicó por primera vez, contenía solo dos ilustraciones, una de las cuales es una lámina grabada que precede al capítulo sobre «nuevos instrumentos ópticos», de los que muestra algunos, incluidos el «chromascope» y el «metrochrome» de Field, mientras que la otra ilustración es un frontispicio elaborado, que parece ser una fusión muy inteligente de las muchas imágenes que aparecen en *Chromatics*. La banda de triángulos de colores en la parte inferior de la lámina muestra el rango de primarios, secundarios y terciarios y su relación con el blanco y el negro, mientras que el diagrama con forma de flor de «equivalentes cromáticos» actúa como una brújula que permite al usuario calcular con precisión las proporciones de las diferentes mezclas de colores. El ejemplo que se muestra aquí no está coloreado, pero el mismo diagrama aparece en forma coloreada y adaptada en ediciones posteriores de los libros de Field.

En sus trabajos, Field respondía no solo a la necesidad de lograr una comprensión adecuada de los buenos pigmentos, sino también a la necesidad general de publicar información científicamente sólida y fiable sobre el color. Eso explica la continua popularidad de sus libros, todos los cuales fueron reimpresos en ediciones ampliadas o revisadas, incluso póstumamente. Su último trabajo, *Rudiments of the Painters' Art; or a Grammar of Colouring* (1850), fue un manual mucho más pequeño y asequible con ilustraciones más sencillas. Winsor & Newton publicó una edición radicalmente abreviada y revisada de *Chromatography* en 1869, y de nuevo en 1885, con textos adicionales y ediciones de Thomas W. Salter y J. Scott Taylor.

Charles Hayter (1761–1835)

A NEW PRACTICAL TREATISE ON THE THREE PRIMITIVE COLOURS

Assumed as a Perfect System of Rudimental Information

6 plates, 26.5 x 21 cm | 10 ⅜ x 8 ¼ in., London, 1826
London, Wellcome Collection;
Los Angeles, Getty Research Institute (frontispiece)

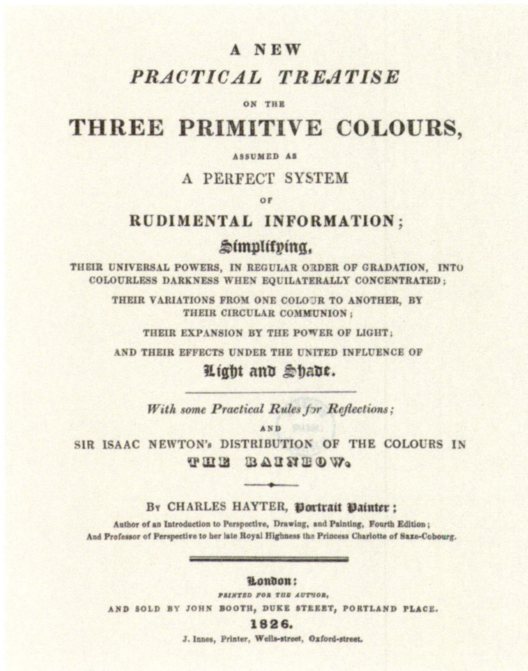
A NEW
PRACTICAL TREATISE
ON THE
THREE PRIMITIVE COLOURS,
ASSUMED AS
A PERFECT SYSTEM
OF
RUDIMENTAL INFORMATION;
Simplifying,
THEIR UNIVERSAL POWERS, IN REGULAR ORDER OF GRADATION, INTO COLOURLESS DARKNESS WHEN EQUILATERALLY CONCENTRATED;
THEIR VARIATIONS FROM ONE COLOUR TO ANOTHER, BY THEIR CIRCULAR COMMUNION;
THEIR EXPANSION BY THE POWER OF LIGHT;
AND THEIR EFFECTS UNDER THE UNITED INFLUENCE OF
Light and Shade.

With some Practical Rules for Reflections;
AND
SIR ISAAC NEWTON'S DISTRIBUTION OF THE COLOURS IN
THE RAINBOW.

BY CHARLES HAYTER, Portrait Painter;
Author of an Introduction to Perspective, Drawing, and Painting, Fourth Edition;
And Professor of Perspective to her late Royal Highness the Princess Charlotte of Saxe-Cobourg.

London:
PRINTED FOR THE AUTHOR,
AND SOLD BY JOHN BOOTH, DUKE STREET, PORTLAND PLACE.
1826.
J. Innes, Printer, Wells-street, Oxford-street.

Charles Hayter was an art teacher to members of the British royal family, some of whom, especially the sisters of George IV, showed a great interest in botanical illustration and painting technique. Hayter published his treatise at a time when painting in watercolour was becoming an increasingly popular and affordable pastime, although, while it is certainly a practical handbook, he was also keen to present his work as a serious theoretical system. The last part of the sub-title informs the reader that this book offers "some practical rules for reflections; and Sir Isaac Newton's distribution of the colours in the rainbow". A book then that helped people paint, but which also provided the intellectual and historical background to colour theory.

By the 1820s painting manuals were being published in large numbers (see Chapter 5), with some of them cheerfully embracing mechanical colour-printing methods while others still relied on the hand-colouring of engravings. Hayter's confident claim that he had produced a "perfect system of rudimental information" should perhaps be interpreted as a marketing device, conveying the message that even though it was a book about the basics of painting, it was perfect in that respect. Hayter also wanted it to be affordable, and remarked that he would have liked to include more colour illustrations but was aware of the effect this would have had on the price. His strategy seems to have worked since *A New Practical Treatise* went into a second edition the same year, followed by five further editions through until 1845. The frontispiece shown here (p. 275) was originally uncoloured, but a later owner of the copy coloured it in by hand.

Visually and conceptually, the hand-coloured illustrations in Hayter's book are clearly influenced by those of Moses Harris, whose work he had encountered in 1813 and most of all in the best-known of Hayter's plates, "The Painters Compass" (p. 279), which shows three colour wheels, each with increasing levels of brightness in three stages, from the centre to the outer edge. Where Harris's wheels have a central triangle, Hayter used overlapping discs to represent material colour, while to the right of the wheels there are two vertical scales of warm (yellow, orange, red, purple and indigo) and cold (indigo, blue, green, yellow and pale yellow) colours, both bookended by black and white.

Professor of Perspective to Her late R.H. the Princess Charlotte of Sax Cobourg.

HAYTER'S COMPENDIUM,

Exemplifying the Natural and unavoidable consequences of the equilateral Union by gradual and systematical concentration, of the Three Primitive Colours, according with Leonardo Da Vinci's proposition.

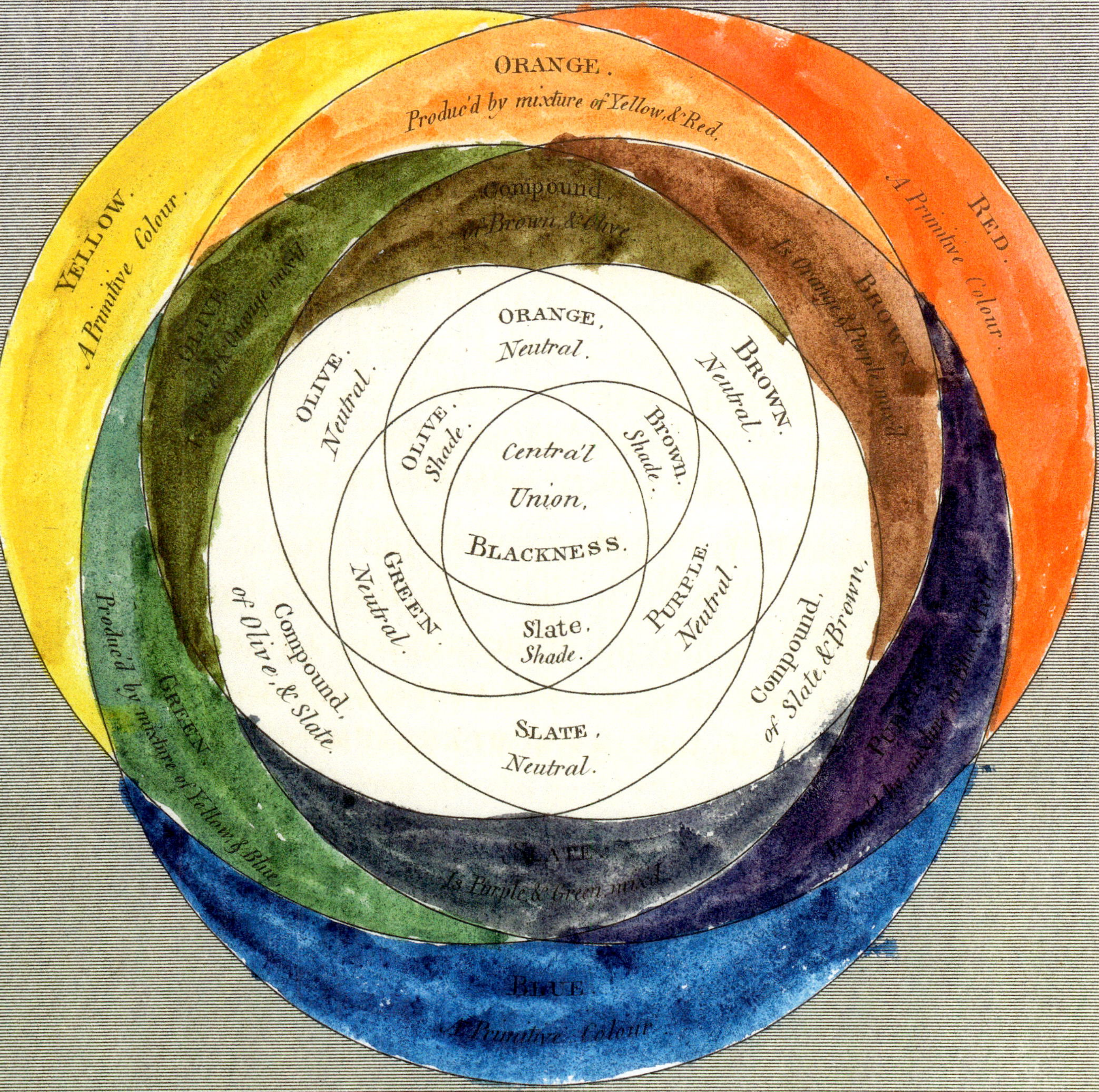

Chas. Hayter. Invent & delint. *J. Turnbull Sc.*

FRONTISPIECE.

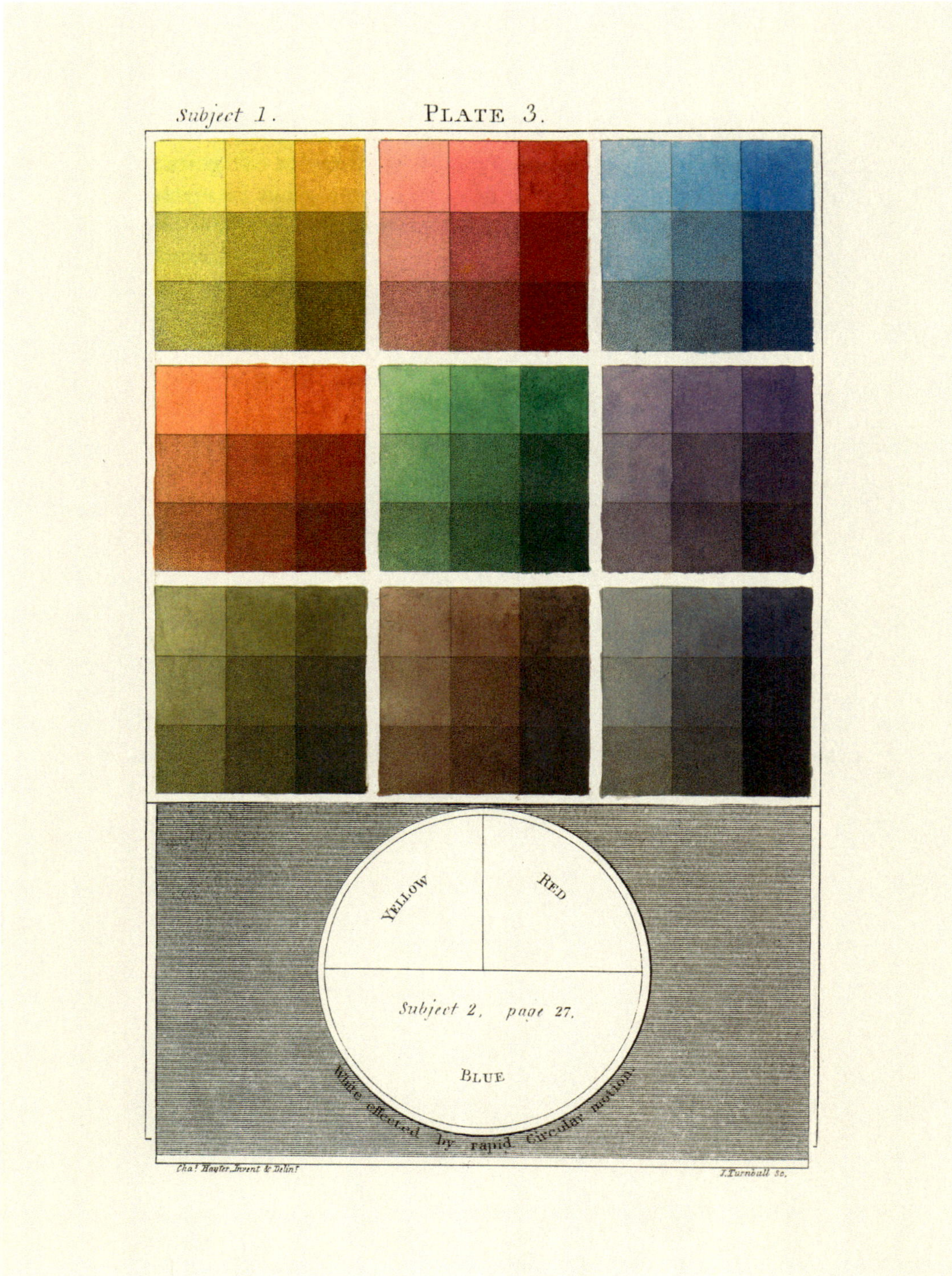

Hintergrund zur Farbtheorie mitlieferte.

In den 1820er-Jahren kamen Malhandbücher in großer Zahl auf den Markt. Einige griffen beherzt die mechanischen Farbdruckverfahren auf, andere setzten nach wie vor auf handkolorierte Kupferstiche. Hayters selbstbewusste Behauptung, er habe ein „perfektes System elementarer Informationen" geschaffen, sollte man vielleicht als Marketingtrick mit der Botschaft interpretieren, dass man es hier zwar mit einem Buch über das Einmaleins der Malerei zu tun habe, es als solches aber perfekt sei. Hayter wollte auch, dass das Buch bezahlbar blieb, und merkte an, er hätte gern mehr farbige Abbildungen eingefügt, sei sich aber der damit verbundenen Auswirkungen auf den Preis bewusst gewesen. Seine Strategie scheint aufgegangen zu sein, denn noch im selben Jahr erschien eine zweite Auflage von *A New Practical Treatise*, auf die bis 1845 fünf weitere folgten. Die hier gezeigte Frontispiz-Radierung (S. 275) war ursprünglich nicht koloriert, wurde aber von einem Besitzer des Exemplars mit der Hand koloriert.

Optisch und konzeptuell sind die handkolorierten Abbildungen in diesem Buch deutlich von denen eines Moses Harris beeinflusst, dessen Werk Hayter 1813 kennengelernt hatte. Dies gilt vor allem für die bekannteste seiner Tafeln, die den Titel „Der Malerkompass" (S. 279) trägt. Sie zeigt drei Farbkreise mit drei jeweils von innen nach außen angeordneten Helligkeitsstufen. Während bei Harris

Neue praktische Abhandlung über die drei Urfarben

Charles Hayter war Kunstlehrer am britischen Königshaus, wo sich vor allem die Schwestern von George IV. sehr für botanische Illustrationen und Maltechniken interessierten. Er veröffentlichte seine Abhandlung zu einer Zeit, in der die Aquarellmalerei als Freizeitbeschäftigung immer beliebter und erschwinglicher wurde. Zwar hatte Hayter ein praktisches Handbuch geschrieben, doch wollte er sein Werk auch als ernsthaftes theoretisches System verstanden wissen.

Der letzte Teil des Untertitels informiert die Leserschaft über „einige praktische Regeln für Spiegelungen; und Sir Isaac Newtons Verteilung der Farben im Regenbogen", die im Buch zu finden seien – einem Buch also, das Anleitungen zum Malen enthielt, aber auch den geistigen und historischen

Dreiecke das Zentrum bilden, wählte Hayter für die stofflichen Grundfarben einander überlappende Scheiben. Rechts neben die Kreise setzte er, jeweils eingerahmt von Schwarz und Weiß, zwei senkrechte Skalen mit warmen (Gelb, Orange, Rot, Violett und Indigo) und kalten Farben (Indigo, Blau, Grün, Gelb und Hellgelb).

Nouveau traité pratique sur les trois couleurs primaires

Certains membres de la famille royale britannique à qui Charles Hayter enseigne les arts, notamment les sœurs du roi George IV, s'intéressent particulièrement à l'illustration botanique et aux techniques picturales. Lorsque Hayter publie son traité, l'aquarelle est en train de devenir un passe-temps de plus en plus populaire et abordable. Pourtant, même si l'ouvrage est un manuel pratique, son auteur se propose également d'y présenter un système théorique sérieux. La dernière partie du sous-titre informe le lecteur que le livre offre des « règles pratiques permettant la réflexion ; et la distribution des couleurs dans l'arc-en-ciel telle qu'établie par sir Isaac Newton ». En somme, un ouvrage qui aide tout un chacun à peindre autant qu'il présente les bases intellectuelles et historiques de la théorie des couleurs.

Dans les années 1820, les manuels de peinture sont désormais monnaie courante (voir le chapitre 5). Certains adoptent allègrement les méthodes mécaniques d'impression polychrome, d'autres reposent toujours sur une mise en couleurs manuelle de gravures. Pour sa part, Hayter affirme avec assurance avoir mis au point un « système parfait d'informations rudimentaires », peut-être une formule publicitaire pour faire passer l'idée qu'il propose, certes, les bases de la peinture, mais sous une forme parfaite. Hayter souhaite aussi que son livre soit bon marché : il souligne qu'il aurait aimé y intégrer davantage d'illustrations en couleurs, mais qu'il est conscient de l'impact qu'elles auraient eu sur le prix. Sa stratégie semble avoir durablement fonctionné : son « Nouveau traité pratique sur les trois couleurs primaires » est réédité la même année, puis de nouveau à cinq reprises jusqu'en 1845. L'eau-forte illustrant le frontispice (p. 275) et présentée ici était initialement sans couleurs. C'est l'un des possesseurs de cet exemplaire qui l'a coloriée à la main.

En termes visuels et conceptuels, les illustrations coloriées à la main d'Hayter sont clairement influencées par celles de Moses Harris, dont il a découvert les travaux en 1813. C'est le cas notamment de sa célèbre « Boussole des peintres » (p. 279), constituée de trois cercles chromatiques présentant chacun trois degrés d'intensité, qui vont croissant depuis le centre vers l'extérieur. Là où Harris dessinait des triangles au centre, Hayter choisit de superposer des disques représentant les couleurs matérielles. À droite des cercles, deux échelles verticales présentent des gradations baptisées « effet chaud » (jaune, orange, rouge, violet et indigo) et « effet froid » (indigo, bleu, vert, jaune et jaune clair), encadrées à chaque extrémité par le noir et le blanc.

Nuevo tratado práctico sobre los tres colores primarios

Charles Hayter fue profesor de arte de miembros de la familia real británica, algunos de los cuales, especialmente las hermanas de Jorge IV, mostraron un gran interés por la ilustración botánica y la técnica pictórica. Hayter publicó su tratado en un momento en que pintar con acuarela se estaba convirtiendo en un pasatiempo cada vez más popular y asequible, aunque, si bien es un manual práctico, también tenía el deseo de presentar su trabajo como un sistema teórico serio. La última parte del subtítulo informa al lector de que el libro ofrece «algunas reglas prácticas para las reflexiones; y la distribución de los colores en el arcoíris de sir Isaac Newton». Se trata de un libro que ayudó a la gente a pintar, pero que también proporcionó los antecedentes intelectuales e históricos de la teoría del color.

En la década de 1820, los manuales de pintura se publicaban en grandes tiradas (véase el capítulo 5). En algunos de ellos, se adoptaron alegremente los métodos mecánicos de impresión en color, mientras que otros todavía confiaban en el coloreado a mano de los grabados. La atrevida afirmación de Hayter de que había producido un «sistema perfecto de información básica» tal vez debe interpretarse como una estrategia de mercadotecnia para transmitir el mensaje de que, aunque se trataba de un libro sobre los conceptos básicos de la pintura, era perfecto en ese sentido. Hayter también quería que fuera asequible, y comentó que le hubiera gustado incluir más ilustraciones en color, pero que era consciente del efecto que eso habría tenido en el precio. Su estrategia parece haber funcionado: *A New Practical Treatise* llegó a la segunda edición ese mismo año, a la que siguieron otras cinco hasta 1845. El grabado del frontispicio (pág. 275) que aquí se muestra originalmente no tenía color, pero el propietario de la copia lo coloreó a mano.

Visual y conceptualmente, las ilustraciones coloreadas a mano de su libro están claramente influenciadas por las de Moses Harris, cuya obra había conocido Hayter en 1813. Esa influencia se percibe, sobre todo, en la más conocida de sus láminas, «The Painters Compass» (pág. 279), que muestra tres ruedas de color, cada una con niveles crecientes de brillo repartidos en tres etapas, desde el centro hasta el borde exterior. Allí donde las ruedas de Harris tienen un triángulo central, Hayter empleó discos superpuestos que representan el color material, mientras que a la derecha de las ruedas hay dos escalas verticales con colores cálidos (amarillo, naranja, rojo, morado e índigo) y fríos (índigo, azul, verde, amarillo y amarillo pálido), ambas delimitadas por el blanco y el negro.

PLATE 1.

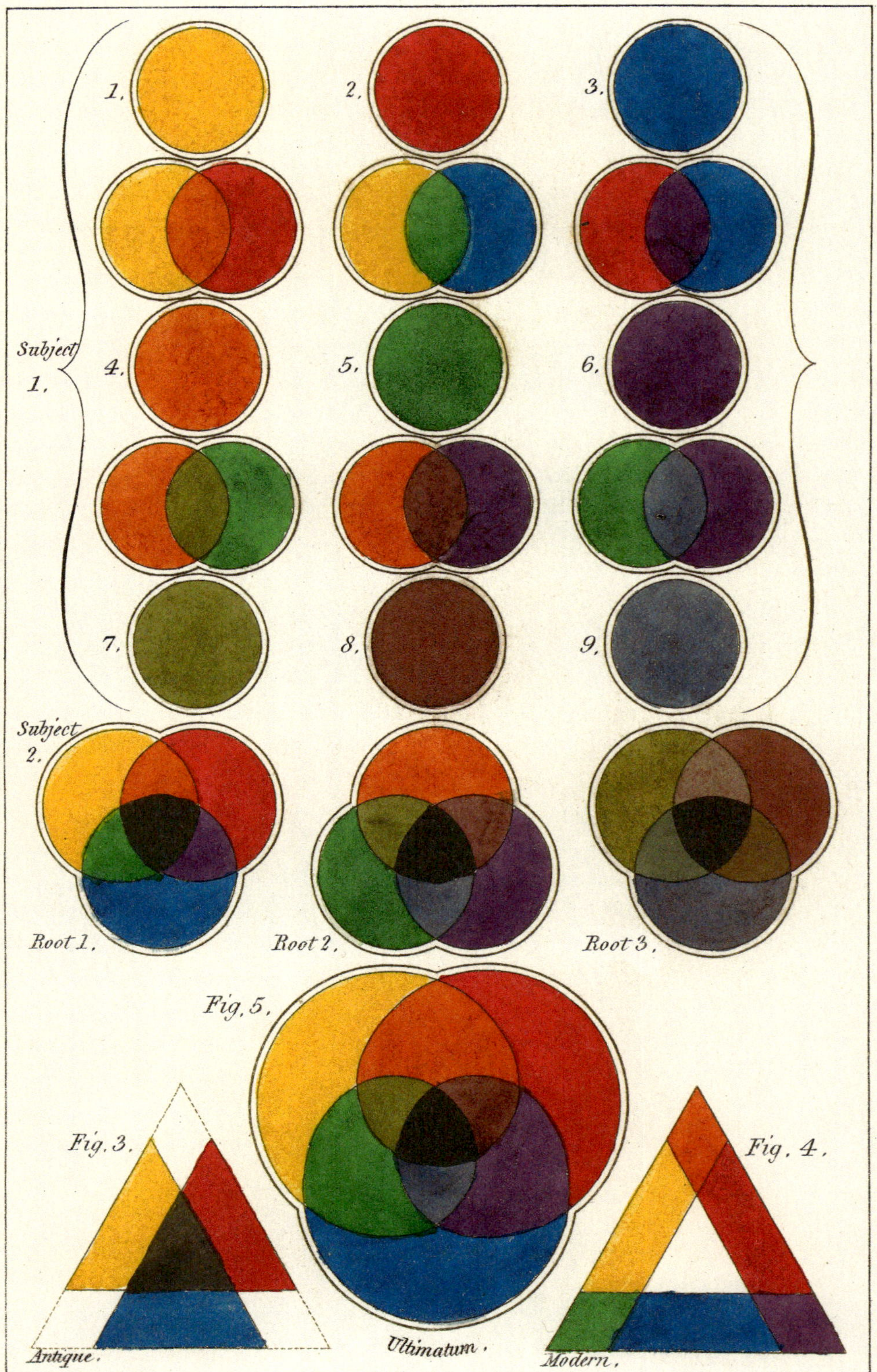

Chas. Hayter. Invent. & delint.

J. Turnbull. Sc.

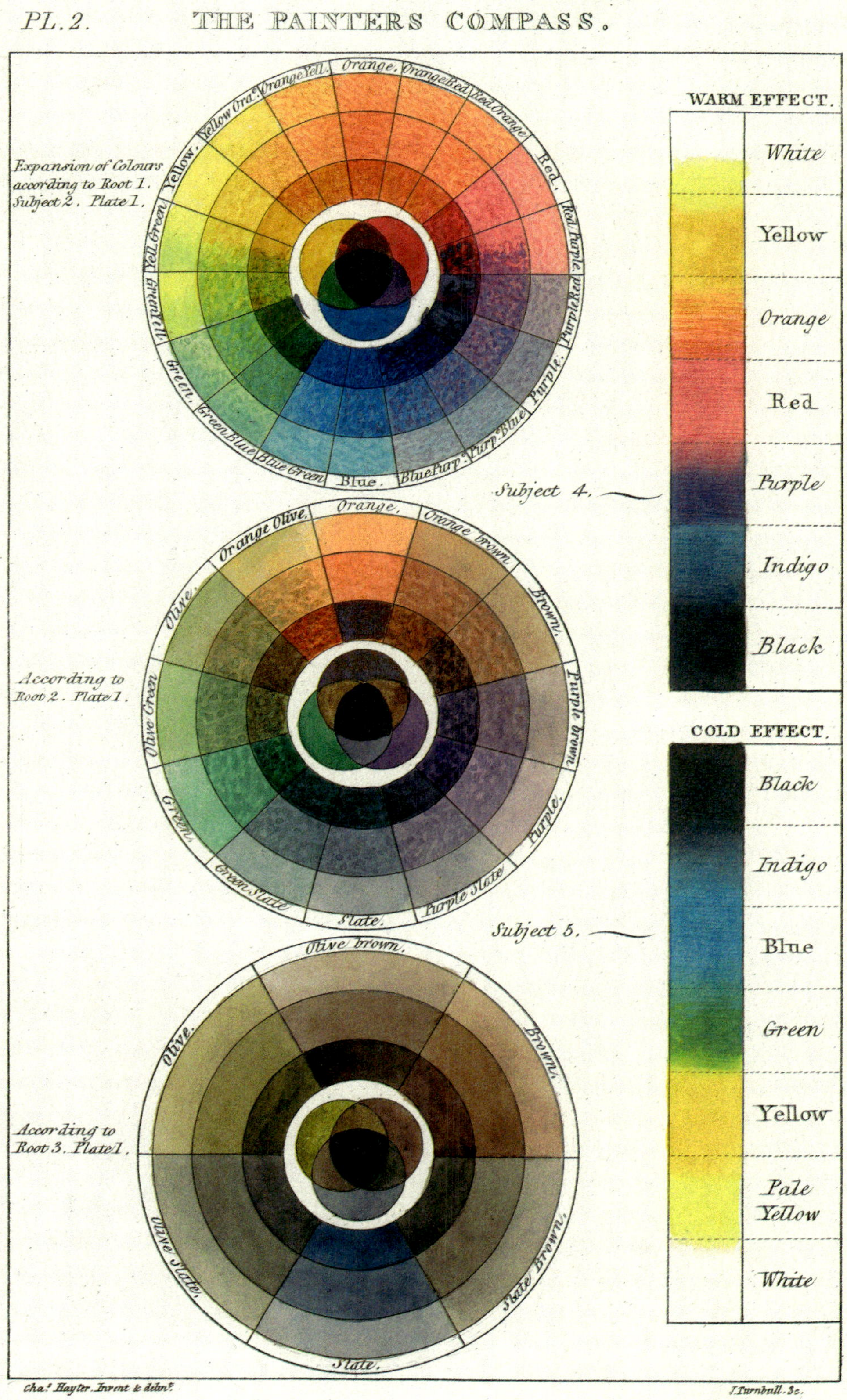
PL. 2.
THE PAINTERS COMPASS.
Expansion of Colours according to Root 1. Subject 2. Plate 1.
Yellow. Yellow Ord.e Orange Yell. Orange. Orange Red. Red Orange. Red. Red Purple. Purple Red. Purple. Purp. Blue. Blue Purp. Blue. Blue Green. Green Blue. Green. Green Yell. Yell. Green.
Subject 4.
Orange. Orange brown. Brown. Purple brown. Purple. Purple Slate. Slate. Green Slate. Green. Olive Green. Olive. Orange Olive.
According to Root 2. Plate 1.
Subject 5.
Olive brown. Brown. Slate Brown. Slate. Olive Slate. Olive.
According to Root 3. Plate 1.
WARM EFFECT.
White
Yellow
Orange
Red
Purple
Indigo
Black
COLD EFFECT.
Black
Indigo
Blue
Green
Yellow
Pale Yellow
White
Cha.s Hayter. Invent & delin.t
J. Turnbull. Sc.

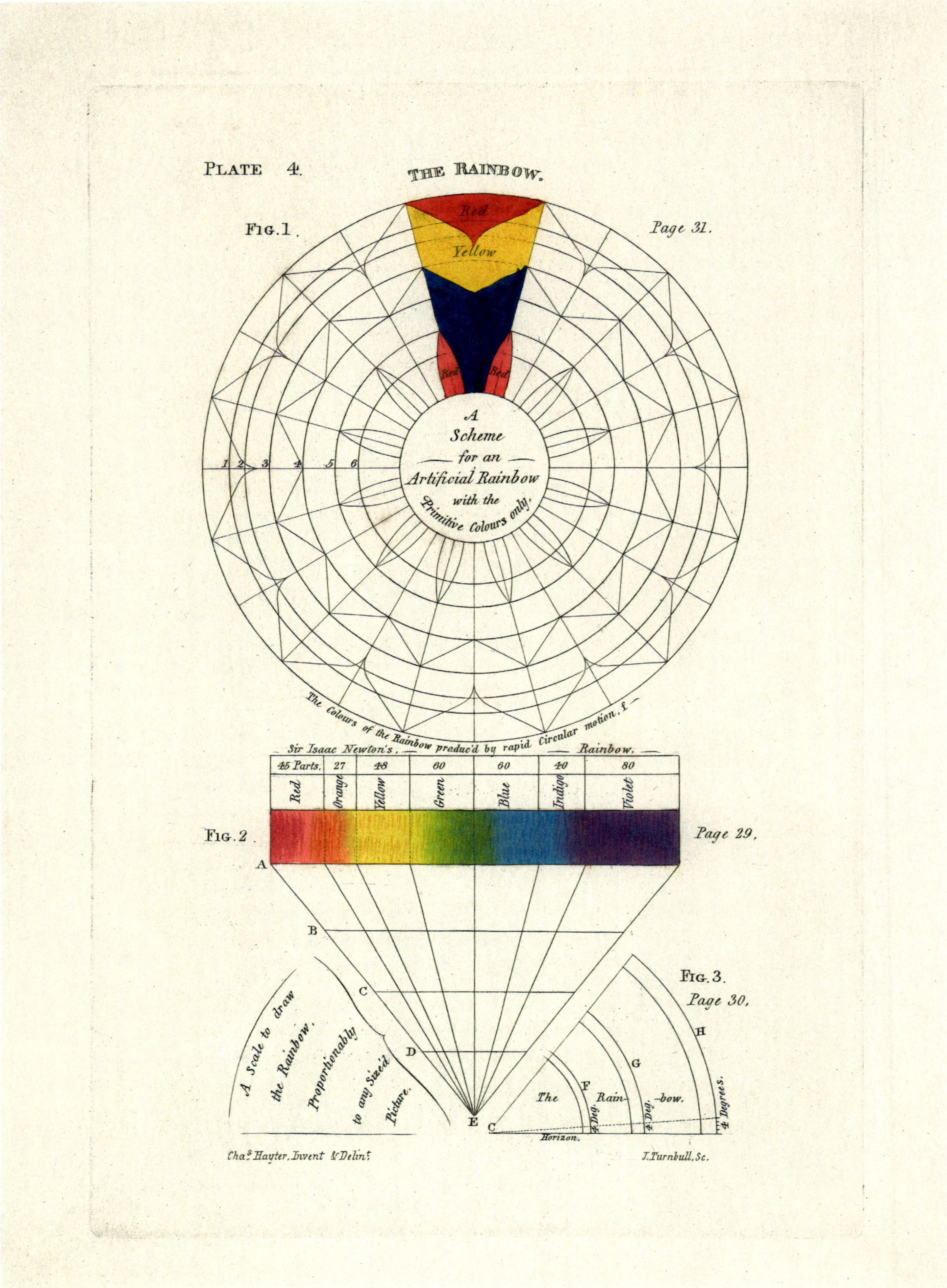
PLATE 4.
THE RAINBOW.
FIG. 1.
Page 31.
Red
Yellow
Red
Red
A Scheme for an Artificial Rainbow with the Primitive Colours only.
1 2 3 4 5 6
The Colours of the Rainbow produc'd by rapid Circular motion, f.
Sir Isaac Newton's. Rainbow.
45 Parts. 27 48 60 60 40 80
Red Orange Yellow Green Blue Indigo Violet
FIG. 2.
Page 29.
A
B
C
D
E
A Scale to draw the Rainbow, Proportionably to any Sized Picture.
FIG. 3.
Page 30.
H
G
F
The Rain- -bow.
4 Deg.
4 Deg.
4 Degrees.
C
Horizon.
Chas Hayter, Invent & Delint
J. Turnbull, Sc.

PLATE . V.

ADDENDA.

The multiplication of Tints by various Intersections may be, not only amusing, but very instructive as The following Contrivance exhibits the three Primitives — and their six Compounds pure, with XXI additional Compounds by Couplets.

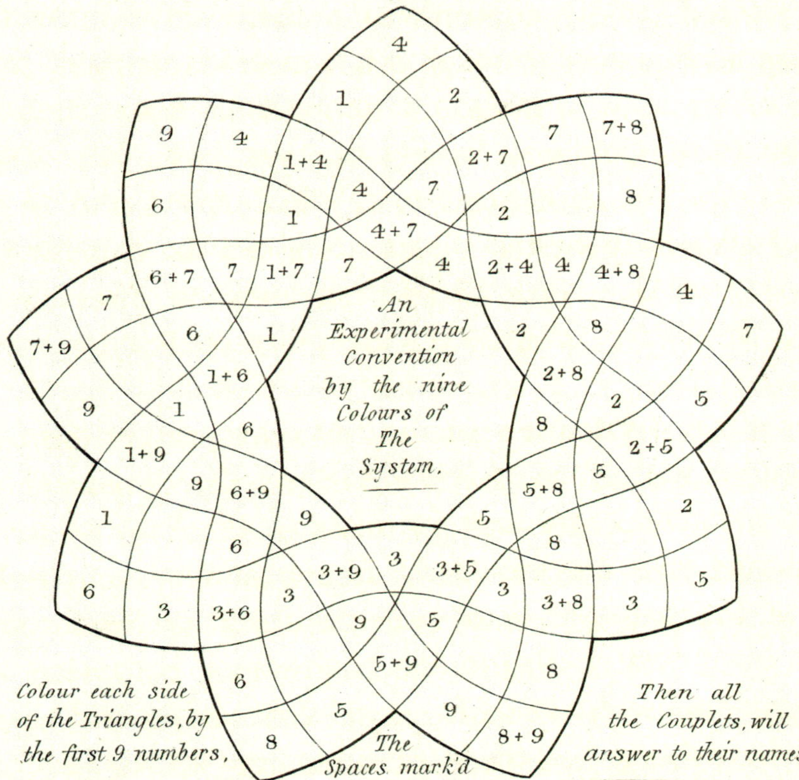

Colour each side of the Triangles, by the first 9 numbers, The Spaces mark'd Then all the Couplets, will answer to their names.

1 are	Yellow,	}	Primitive Colours,
2	Red,		
3	Blue,		
4	Orange,	}	First Compound,
5	Purple,		
6	Green,		
7	Brown,	}	Second Compound,
8	Slate,		
9	Olive,		
1+4	Yellow,	and	Orange,
1+7	Yellow,	— + —	Brown,
1+6	Yellow,	— + —	Green,
1+9	Yellow	— + —	Olive,
2+7	Red	— + —	Brown,
2+4	Red	— + —	Orange,

2 + 5	Red	+	Purple,
2 + 8	Red	+	Slate,
3 + 6	Blue	+	Green,
3 + 9	Blue	+	Olive,
3 + 5	Blue	+	Purple,
3 + 8	Blue	+	Slate,
4 + 7	Orange	+	Brown,
6 + 9	Green	+	Olive,
4 + 8	Orange	+	Slate,
5 + 8	Purple	+	Slate,
5 + 9	Purple	+	Olive,
6 + 7	Green	+	Brown,
7 + 8	Brown	+	Slate,
7 + 9	Brown	+	Olive,
8 + 9	Slate	+	Olive,

Michel-Eugène Chevreul (1786–1889)

DE LA LOI DU CONTRASTE SIMULTANÉ DES COULEURS

On the Principles of Simultaneous Contrast in Colours
33 plates, 27.4 x 24 cm | 10 ¾ x 9 ½ in., Paris, 1839
London, Royal College of Art, Colour Reference Library

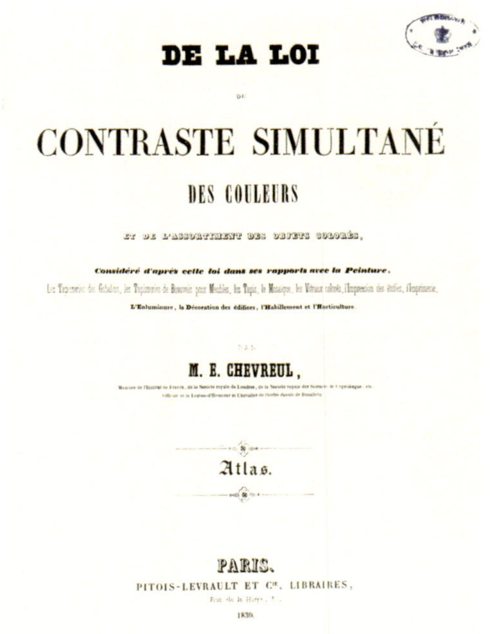
DE LA LOI
DU
CONTRASTE SIMULTANÉ
DES COULEURS
ET DE L'ASSORTIMENT DES OBJETS COLORÉS,
Considéré d'après cette loi dans ses rapports avec la Peinture,
Les Tapisseries des Gobelins, les Tapisseries de Beauvais pour Meubles, les Tapis, la Mosaïque, les Vitraux colorés, l'Impression des étoffes, l'Imprimerie,
L'Enluminure, la Décoration des édifices, l'Habillement et l'Horticulture.
PAR
M. E. CHEVREUL,
Atlas.
PARIS.
PITOIS-LEVRAULT ET Cie, LIBRAIRES,
1839.

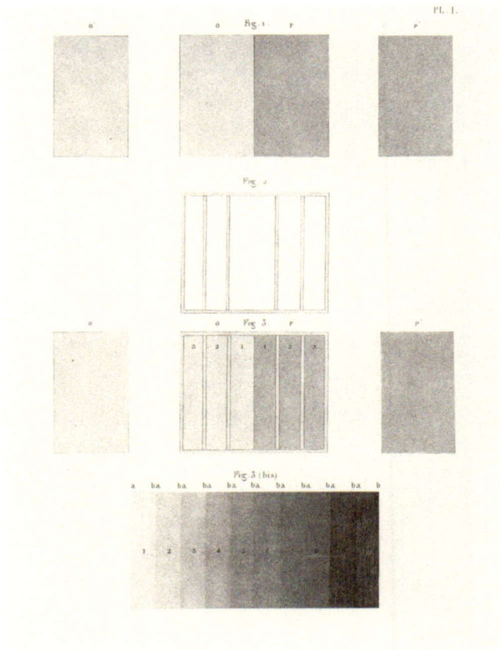

In the 19th century Michel-Eugène Chevreul was for France what Goethe was for Germany and George Field for England: the author who produced the most wide-ranging, substantial and influential work in the field of colour at this period. However, the three decades that separate Goethe's *Zur Farbenlehre* and Chevreul's *De la loi du contraste simultané des couleurs* reveal how quickly the world was changing during the early years of the 19th century.

When he was younger Chevreul had worked at the Muséum National d'Histoire Naturelle in Paris as an assistant to the chemist and pharmacist Nicolas-Louis Vauquelin, whom he succeeded as Professor of Chemistry at the museum. During his time as a consultant for the famous Gobelin tapestry factories in the 1820s, Chevreul began to formulate his laws of simultaneous colour contrast after having observed how the proximity of colours in woven fabrics affects how they are processed by the human eye. He poured his findings and years of related research into *De la loi du contraste simultané des couleurs*, which was first published in 1839 and within a year had been translated into German, with an English edition following in 1854. From the outset, the book was directed unapologetically at many different disciplines, though with a strong focus on manufacturing. This was the first peak of the industrial age, and Chevreul was well aware of the impact of new technology, such as steam power and electricity, on the arts. As such, his work differs significantly from Goethe's much more philosophical colour theory.

The book presented here, with its 735 pages of text, was accompanied by an experimental volume of around 40 lithographed plates, many of them coloured and some of them triple-folded. In order to demonstrate his notions of colour contrast, Chevreul placed hundreds of small, coloured dots against backgrounds of different colours, combining lithography and hand-colouring. He also introduced an uncoloured 72-part colour wheel, which he ingeniously turned into a three-dimensional object by attaching an overlay section along its radius. This represented the greyscale and could be lifted away from the surface of the diagram.

Figure 89

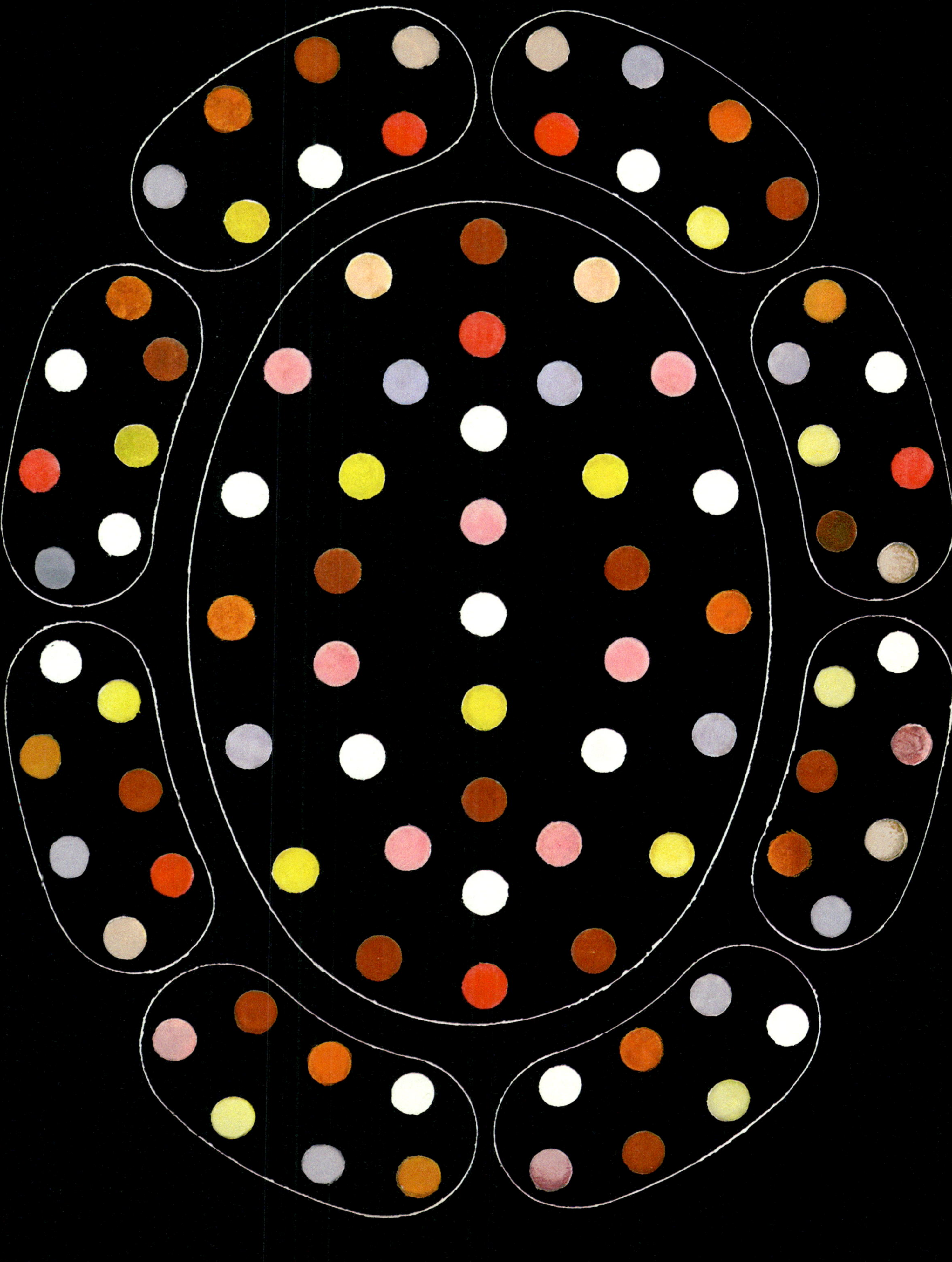

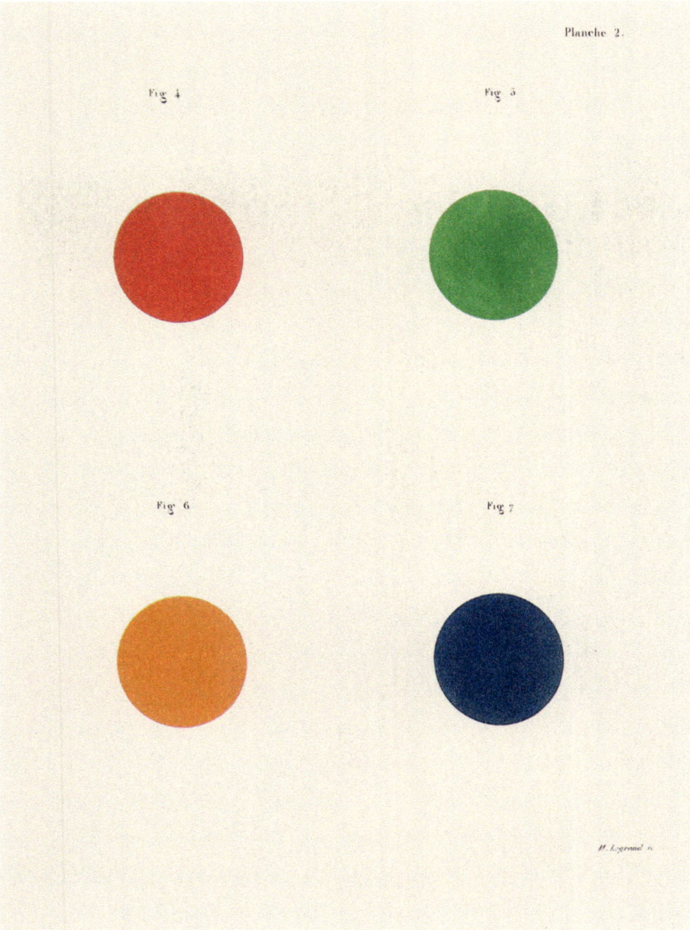
Planche 2.
Fig. 4
Fig. 5
Fig. 6
Fig. 7

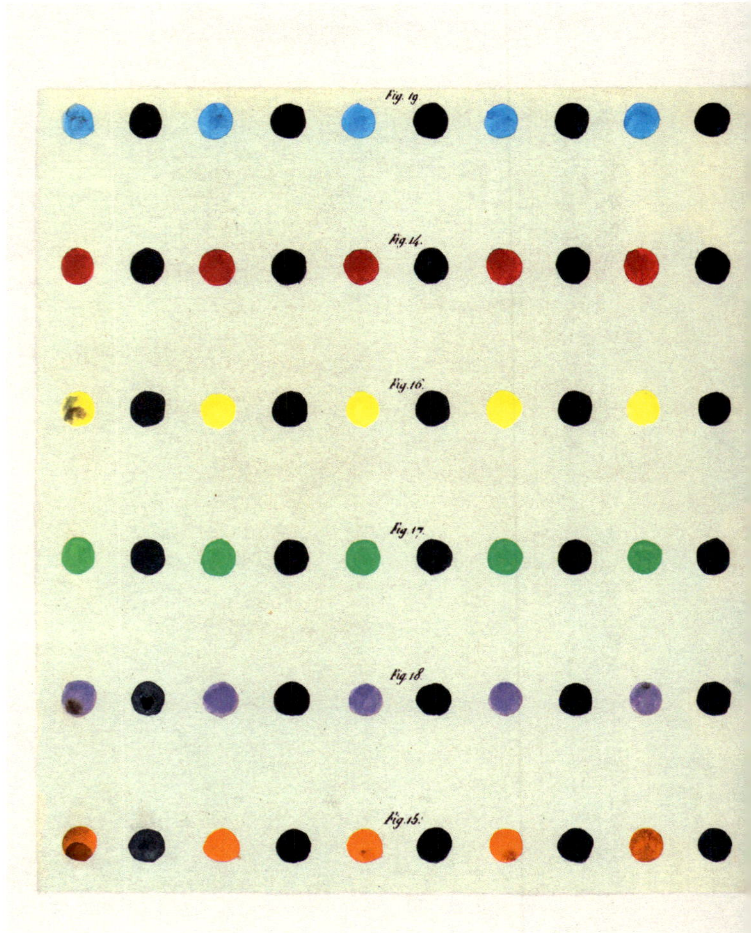
Fig. 19
Fig. 14.
Fig. 16.
Fig. 17.
Fig. 18.
Fig. 15.

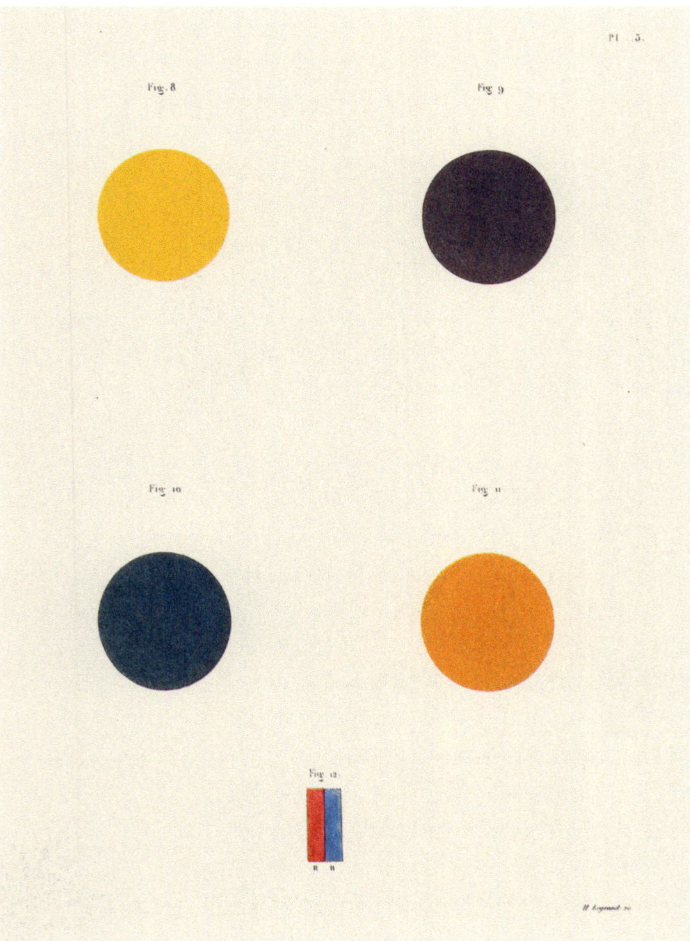
Pl. 3.
Fig. 8
Fig. 9
Fig. 10
Fig. 11
Fig. 12.

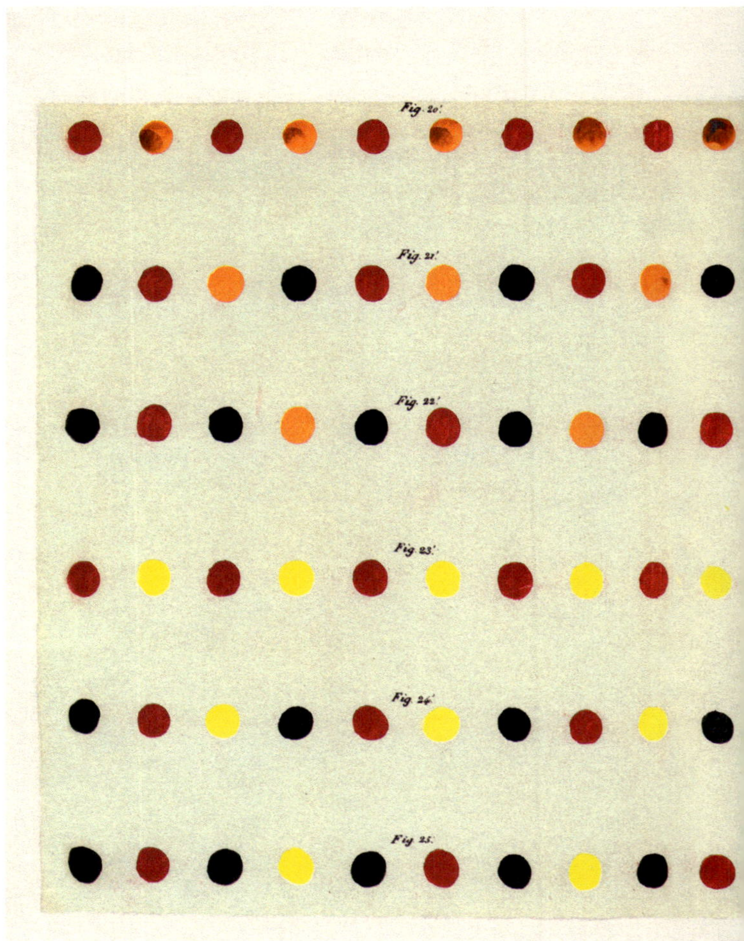
Fig. 20.
Fig. 21.
Fig. 22.
Fig. 23.
Fig. 24.
Fig. 25.

…nts des couleurs simples et binaires des artistes avec le blanc, le noir et le gris.

Pl. 5. Pl. 7.

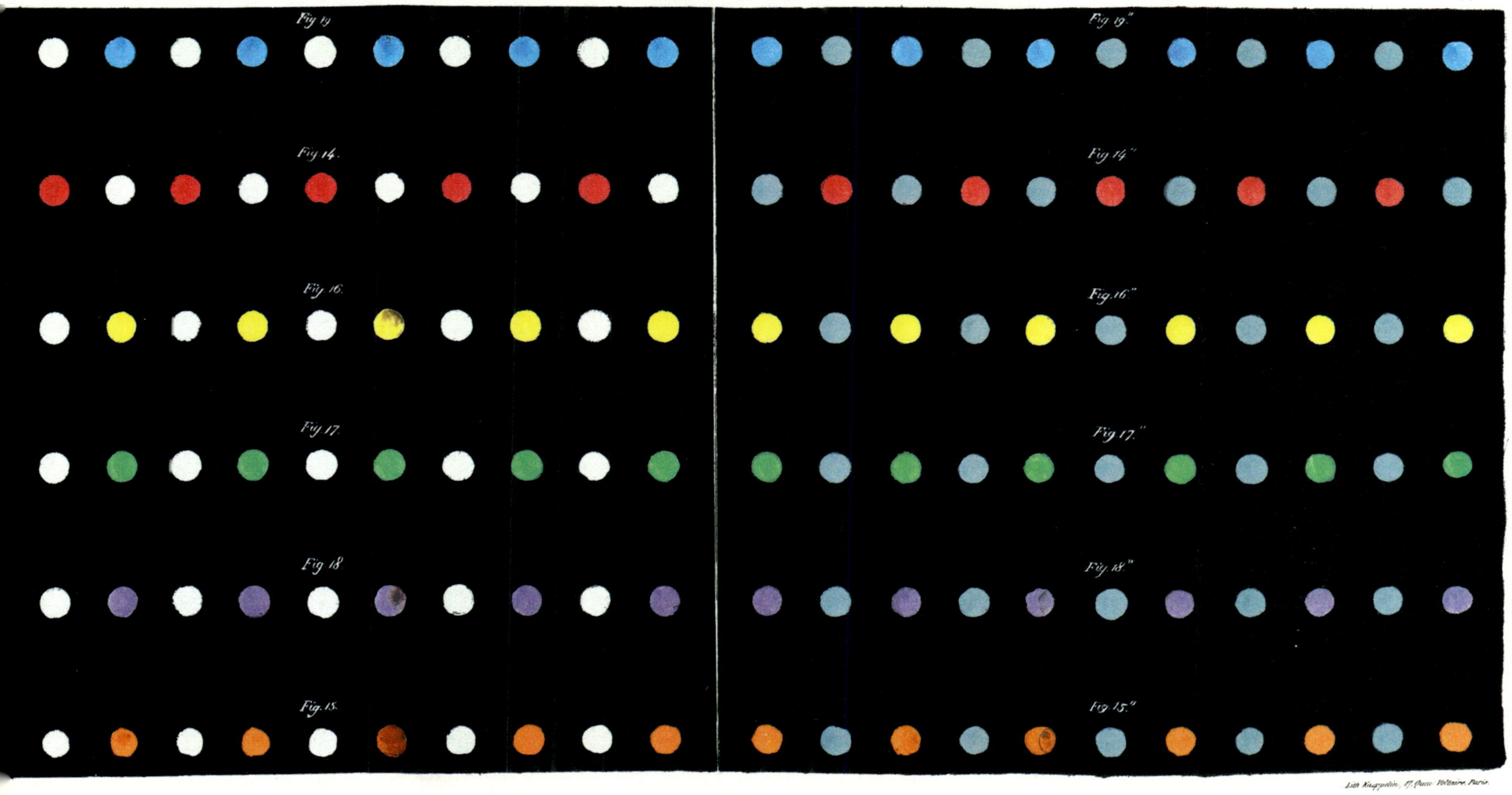

Assortiments de deux couleurs lumineuses avec le blanc, le noir et le gris.

Pl. 8. Pl. 10.

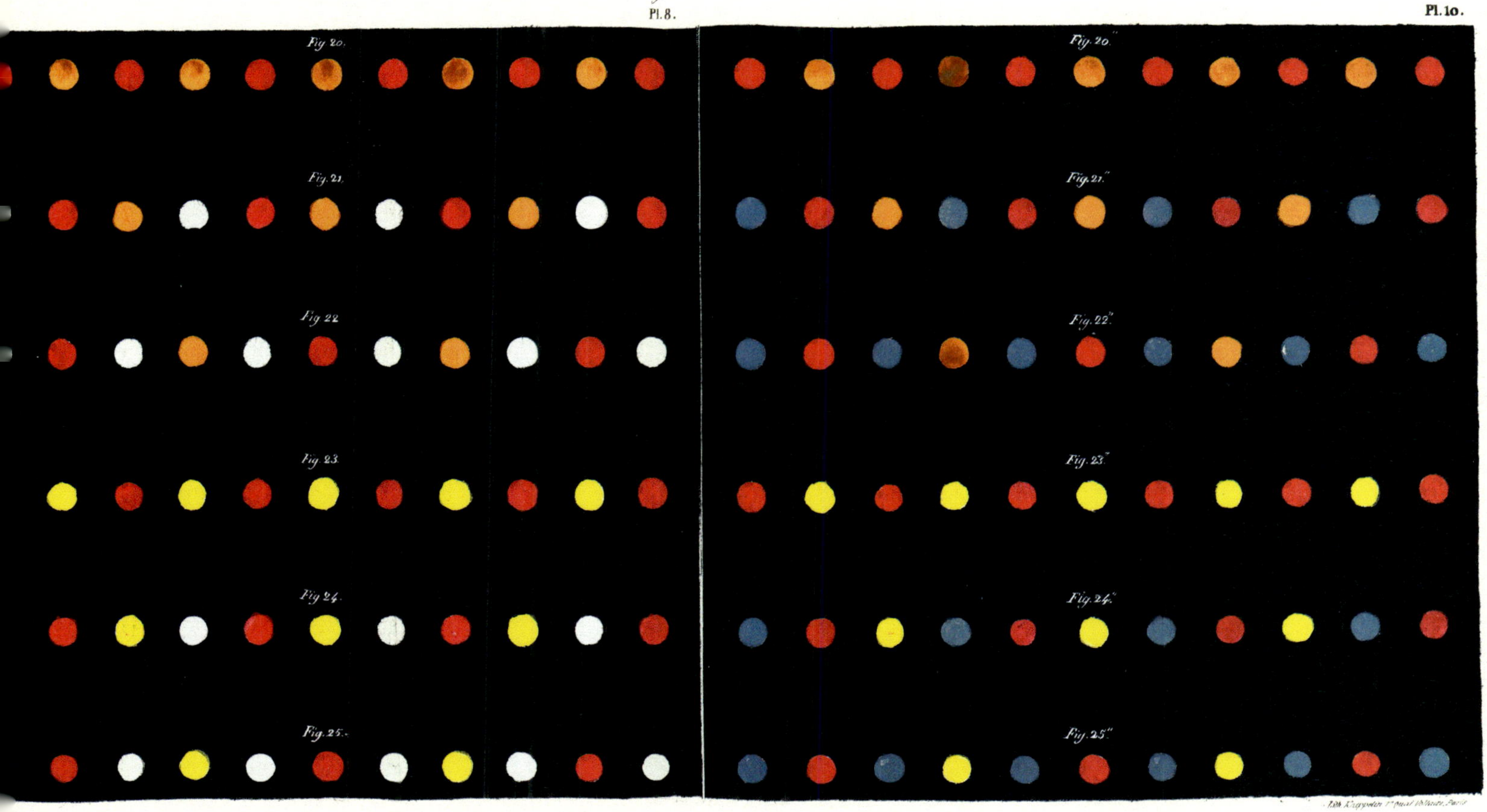

Vom Prinzip des simultanen Farbkontrasts

Im 19. Jahrhundert war Michel-Eugène Chevreul für Frankreich, was Goethe für Deutschland und George Field für England war: der Autor, der in dieser Zeit die am breitesten gefächerten, gewichtigsten und einflussreichsten Arbeiten zur Farbforschung produzierte. Dennoch legen die drei Jahrzehnte, die zwischen Goethes *Zur Farbenlehre* und Chevreuls *De la loi du contraste simultané des couleurs* liegen, auf faszinierende Weise offen, wie schnell sich die Welt im ersten Drittel des 19. Jahrhunderts veränderte.

Als junger Mann hatte Chevreul im Muséum national d'Histoire naturelle in Paris als Assistent des Chemikers und Apothekers Nicolas-Louis Vauquelin gearbeitet, dem er dort als Professor für Chemie nachfolgte. Während seiner Zeit als Berater der berühmten staatlichen Gobelinmanufaktur in den 1820er-Jahren begann Chevreul seine Gesetze des simultanen Farbkontrasts zu formulieren, nachdem er beobachtet hatte, dass die Nähe von Farben in gewebten Stoffen Einfluss darauf hat, wie das Auge sie verarbeitet. Seine Befunde und jahrelangen Forschungen zu verwandten Phänomenen ließ er in *De la loi du contraste simultané des couleurs* einfließen, das 1839 erschien und binnen eines Jahres ins Deutsche übersetzt wurde. Eine englische Ausgabe folgte 1854. Von Anfang an richtete sich das Buch unmissverständlich an viele verschiedene Disziplinen, wenn auch mit besonderem Fokus auf die Warenherstellung. Das Industriezeitalter hatte seinen ersten Höhepunkt erreicht, und Chevreul war sich über die Auswirkungen, die neue Technologien wie die Dampfkraft oder die Elektrizität auf die Kunst hatten, ganz im Klaren. Insofern unterscheidet sich sein Werk erheblich von Goethes weitaus philosophischer angelegter Farbtheorie.

Das hier vorgestellte Buch mit seinen 735 Textseiten wurde von rund vierzig lithografischen Bildtafeln in einem experimentellen „Atlas" begleitet, viele davon koloriert und einige dreimal gefaltet. Um seine Vorstellungen vom Farbkontrast zu demonstrieren, setzte Chevreul in einer Kombination aus Lithografie und manueller Kolorierung Hunderte von kleinen farbigen Punkten auf verschieden eingefärbte Hintergründe. Darüber hinaus führte er einen 72-teiligen Farbkreis ein, den er auf geniale Weise in ein dreidimensionales Objekt verwandelte, indem er auf seinem Radius eine Grauskala in Form eines aufstellbaren Viertelkreises anbrachte.

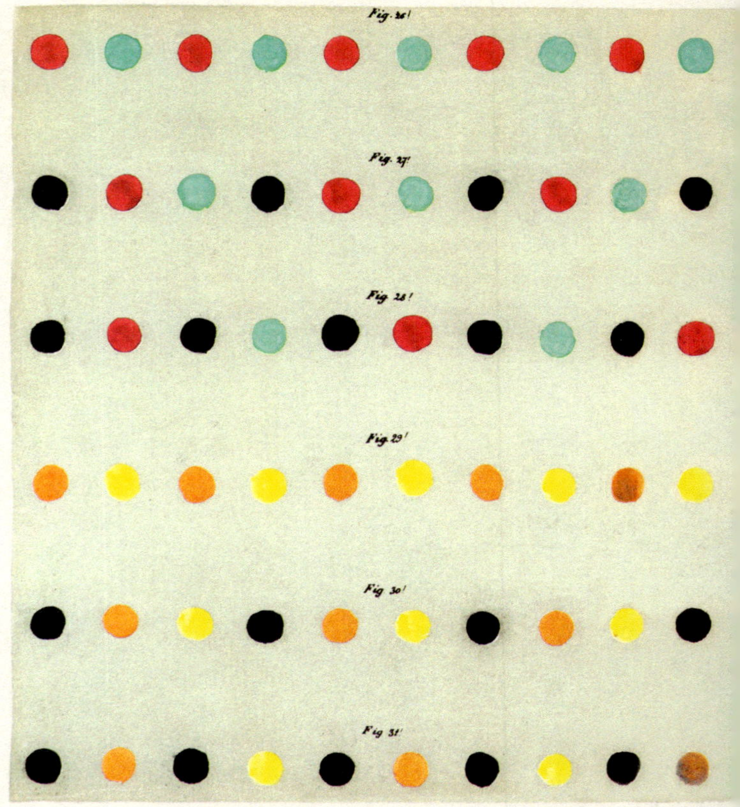

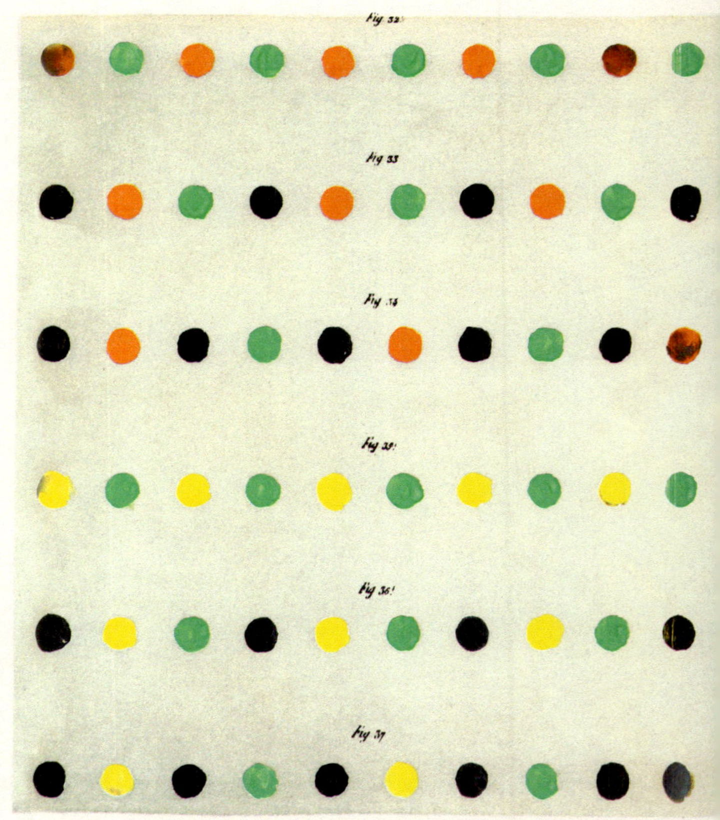

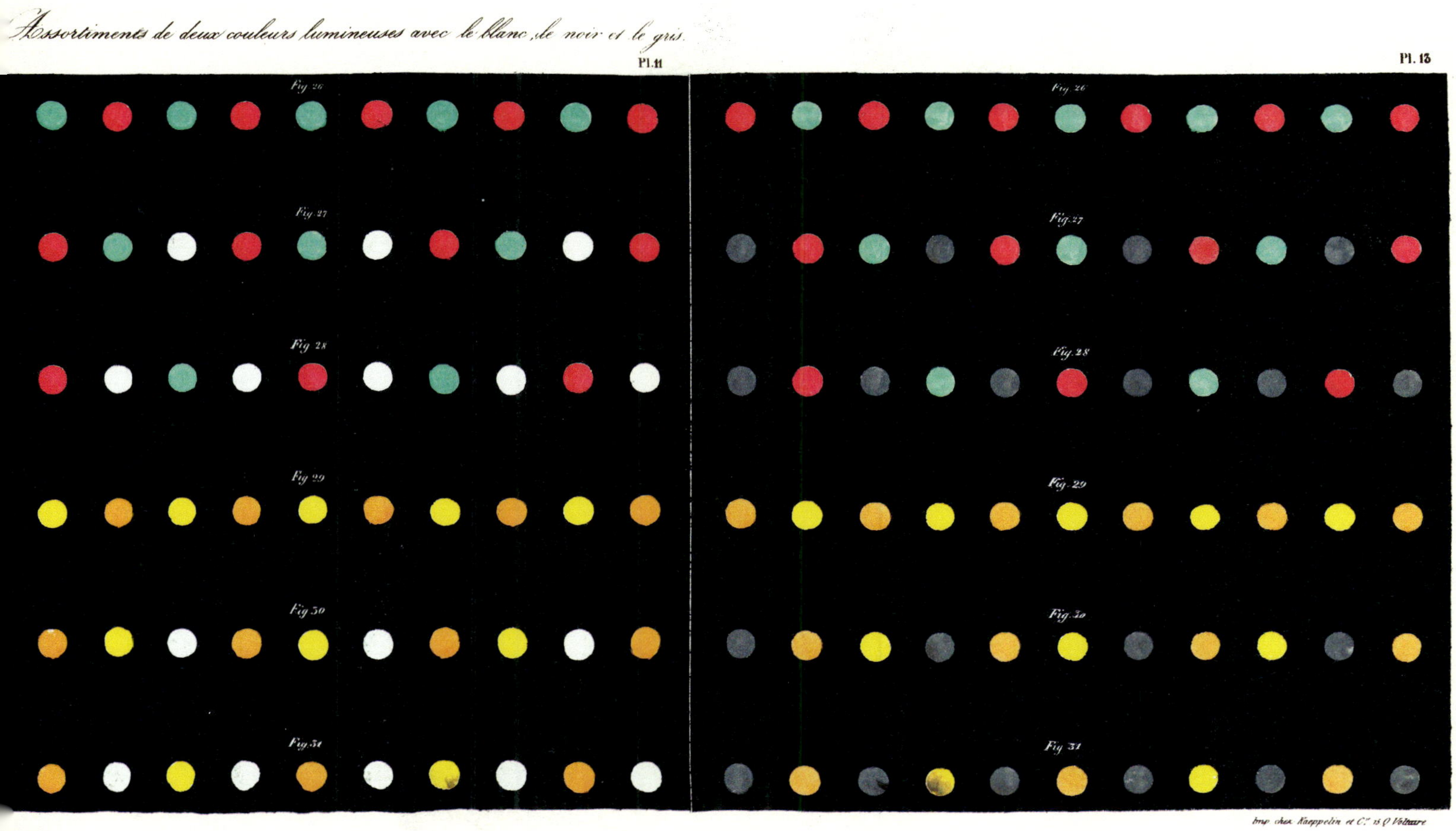
Assortiments de deux couleurs lumineuses avec le blanc, le noir et le gris.
Pl. 11
Pl. 13
Fig. 26
Fig. 27
Fig. 28
Fig. 29
Fig. 30
Fig. 31
Imp. chez Kaeppelin et Cie 15 Q. Voltaire

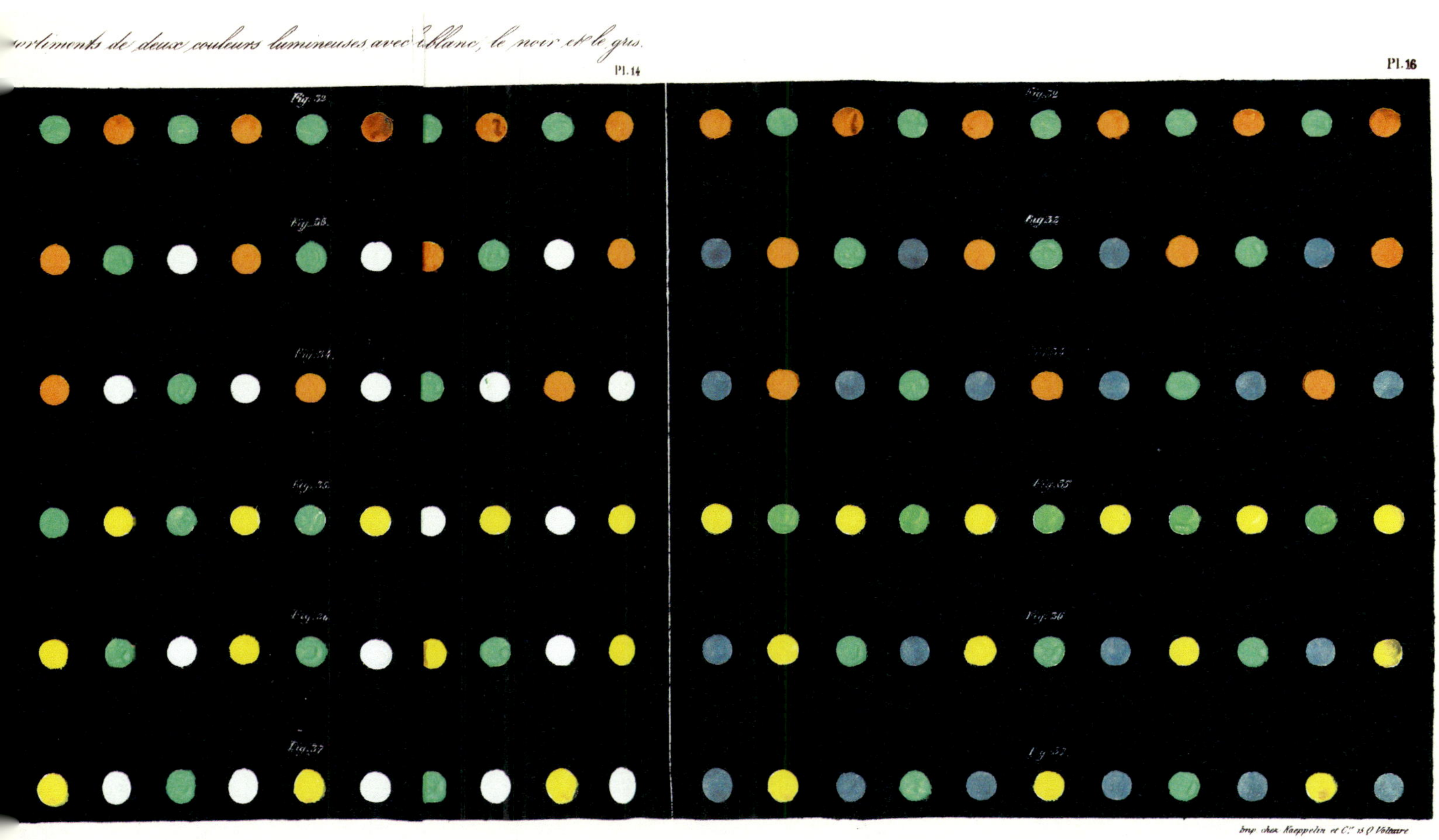
…rtiments de deux couleurs lumineuses avec le blanc, le noir et le gris.
Pl. 14
Pl. 16
Fig. 32
Fig. 33
Fig. 34
Fig. 35
Fig. 36
Fig. 37
Imp. chez Kaeppelin et Cie 15 Q. Voltaire

De la loi du contraste simultané des couleurs

Au XIXe siècle, Michel-Eugène Chevreul est à la France ce que Goethe est à l'Allemagne et George Field à l'Angleterre : l'auteur le plus prolifique et le plus influent en matière de travaux sur la couleur. Mais l'écart qui sépare *Zur Farbenlehre* (*Traité des couleurs*) de Goethe et l'ouvrage de Chevreul, qui paraît trois décennies plus tard, est révélateur de la rapidité fascinante avec laquelle le monde change en ce début de siècle.

Dans sa jeunesse, Chevreul a travaillé au Muséum national d'histoire naturelle de Paris en tant qu'assistant du chimiste et pharmacien Nicolas-Louis Vauquelin, auquel il succèdera comme professeur de chimie dans l'institution. Dans les années 1820, il œuvre comme conseiller de la célèbre Manufacture des Gobelins. C'est en observant que l'œil humain traite différemment les couleurs rapprochées dans les tissages qu'il commence à formuler ses lois du contraste simultané des couleurs. Les recherches qu'il mène au fil des années seront la matière première de cet ouvrage publié en 1839, traduit en allemand dès l'année suivante, puis en anglais en 1854. Le livre, d'emblée destiné à différentes disciplines, privilégie cependant le domaine de la fabrication. En cette première heure de gloire de l'ère industrielle, Chevreul est tout à fait conscient de l'impact des technologies nouvelles, telles que la machine à vapeur ou l'électricité, sur la production artistique. En cela, son ouvrage se distingue grandement de la théorie de la couleur, beaucoup plus philosophique, de Goethe.

Dans le livre présenté ici, les 735 pages de texte sont assorties d'une partie expérimentale contenant une quarantaine de planches, dont beaucoup sont en couleurs et certaines en trois pans dépliables. Pour démontrer la notion de contraste des couleurs, Chevreul peint des centaines de petites pastilles sur des fonds de différentes couleurs, en combinant lithographie et coloriage manuel. Il propose également, pour représenter l'échelle des gris, un cercle chromatique sans couleurs de 72 sections, qui se mue en objet en trois dimensions par l'apposition d'un rabat le long de son rayon.

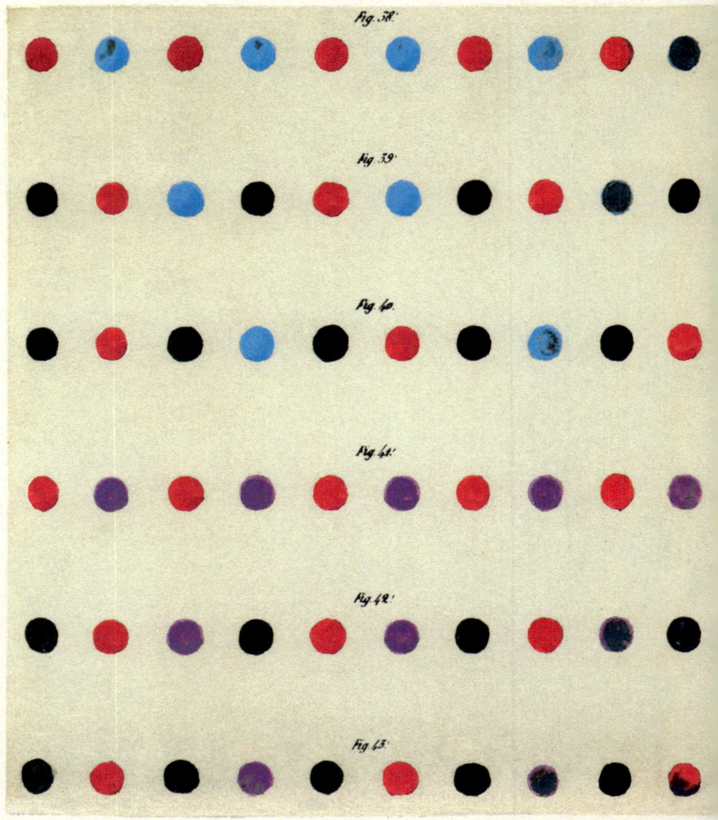

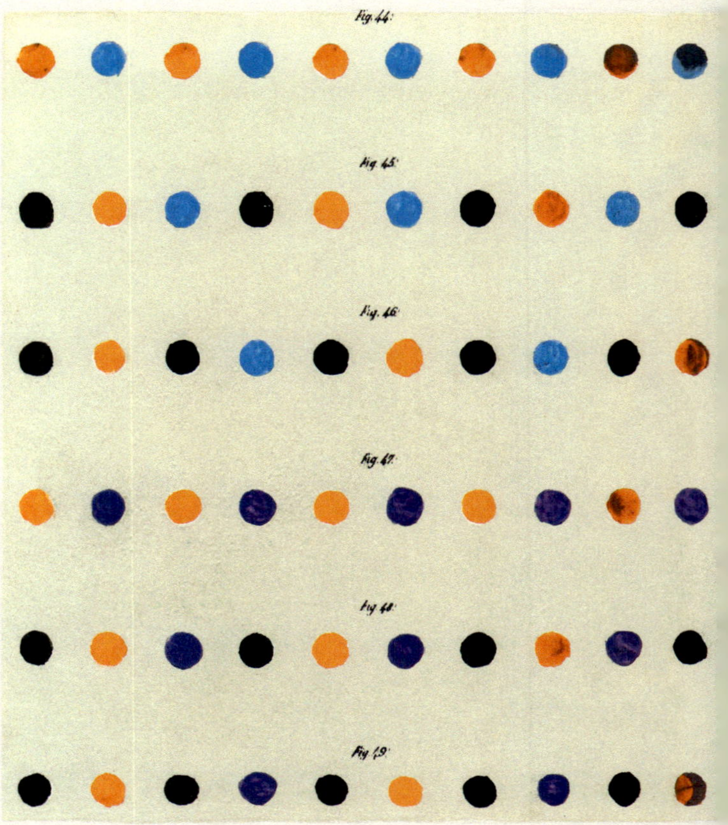

d'une couleur lumineuse et d'une couleur sombre avec le blanc, le noir et le gris

Pl. 17. Pl. 19.

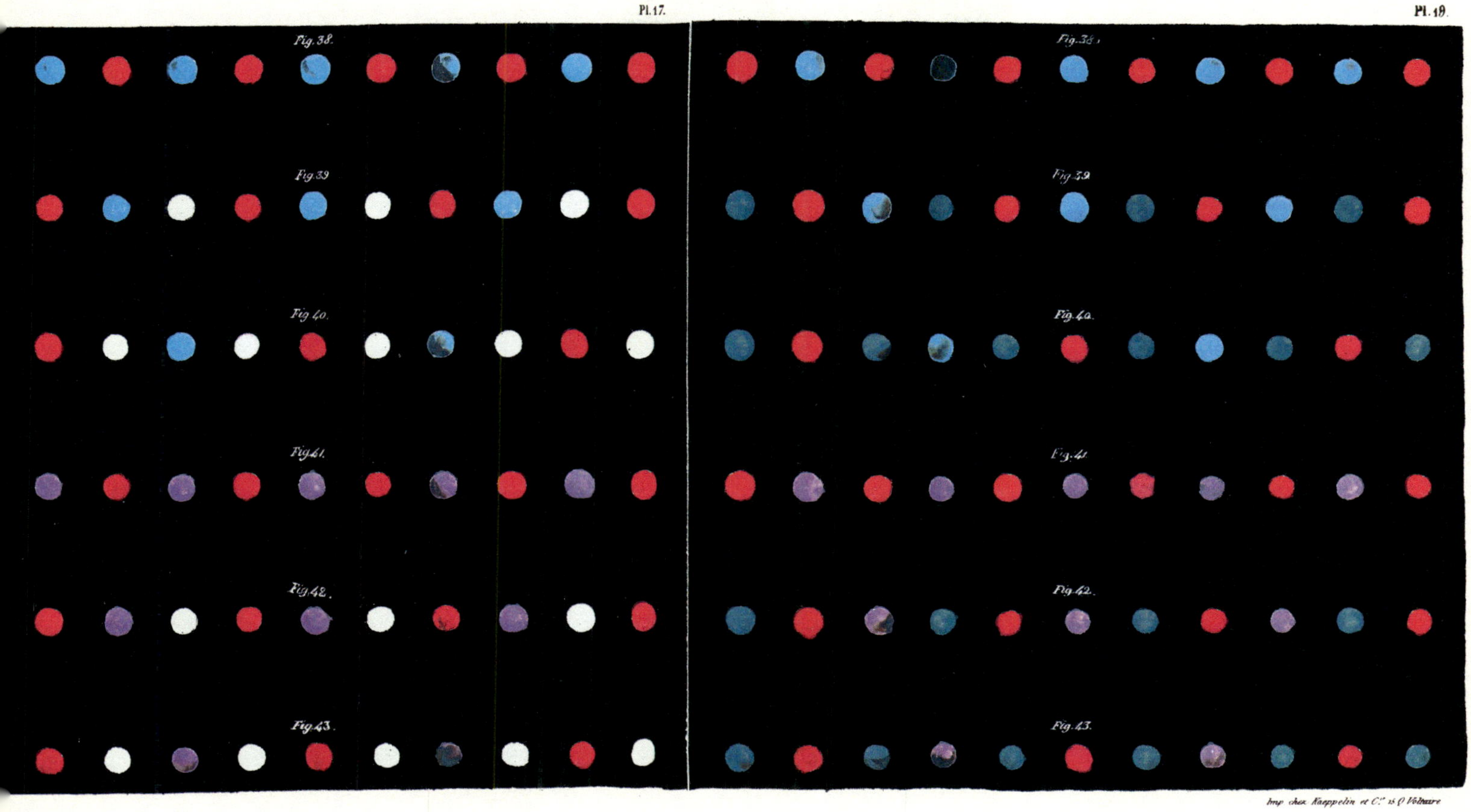

Imp. chez Kaeppelin et Cie 15 Q. Voltaire

d'une couleur lumineuse et d'une couleur sombre avec le blanc, le noir et le gris

Pl. 20. Pl. 22.

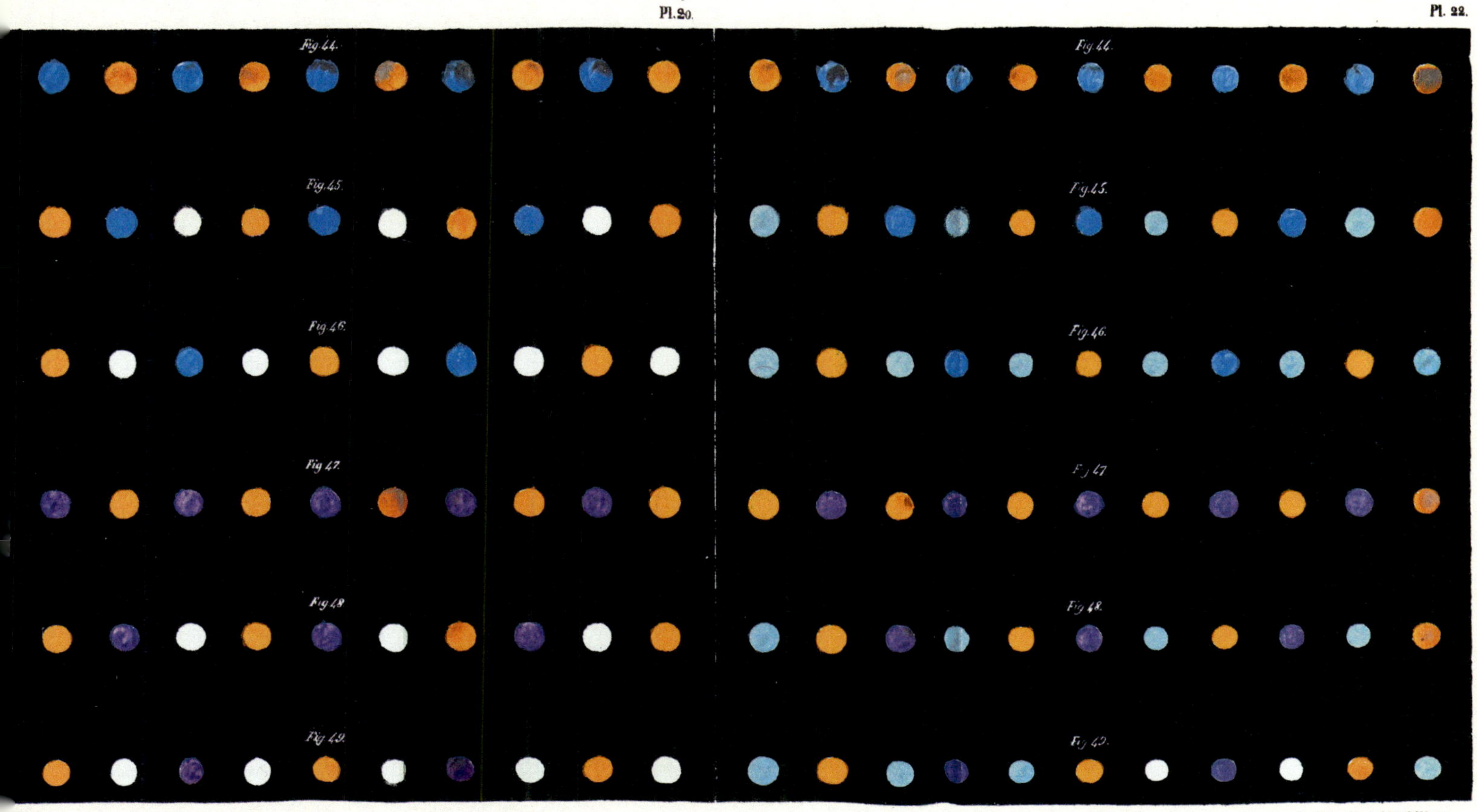

Imp. chez Kaeppelin et Cie 15 Q. Voltaire

Sobre la ley del contraste simultáneo de los colores

En el siglo XIX, Michel-Eugène Chevreul fue para Francia lo que Goethe fue para Alemania y George Field para Inglaterra: el autor que produjo la obra más amplia, sustancial e influyente en este campo en ese período. Sin embargo, las tres décadas que separan *Zur Farbenlehre* de Goethe y *De la loi du contraste simultané des couleurs* de Chevreul revelan de forma fascinante lo rápido que estaba cambiando el mundo durante los primeros años del siglo XIX.

Cuando era más joven, Chevreul había trabajado en el Muséum national d'histoire naturelle de París como ayudante del químico y farmacéutico Nicolas-Louis Vauquelin, a quien sucedió como profesor de química en el museo. Durante los años que trabajó como asesor de las famosas fábricas de tapices Gobelin en la década de 1820, Chevreul comenzó a formular sus leyes de contraste de colores simultáneos tras observar cómo la proximidad de los colores en los tejidos afecta al modo en que el ojo humano los procesa. Vertió sus hallazgos y años de investigación relacionada en *De la loi du contraste simultané des couleurs*, que se publicó por primera vez en 1839, al año siguiente se tradujo al alemán y en 1854 al inglés. Desde el principio, el libro se dirigió abiertamente a muchas disciplinas diferentes, aunque con un fuerte enfoque en la fabricación. La era industrial se encontraba entonces en su primer apogeo, y Chevreul era muy consciente del impacto de las nuevas tecnologías, como la máquina de vapor y la electricidad, en las artes. Como tal, su trabajo difiere significativamente de la teoría del color mucho más filosófica de Goethe.

El libro que aquí se presenta cuenta con 735 páginas de texto acompañadas de una sección experimental de unas 40 láminas litografiadas, muchas de ellas coloreadas y algunas en forma de tríptico desplegable. Para demostrar sus ideas, Chevreul colocó cientos de pequeños puntos de colores sobre fondos de diferentes colores, para lo que combinó litografía y coloreado a mano. También incluyó una rueda de color de 72 partes sin colorear, que ingeniosamente convirtió en un objeto tridimensional al superponer una sección a lo largo de su radio que representaba la escala de grises y podía levantarse de la superficie del diagrama.

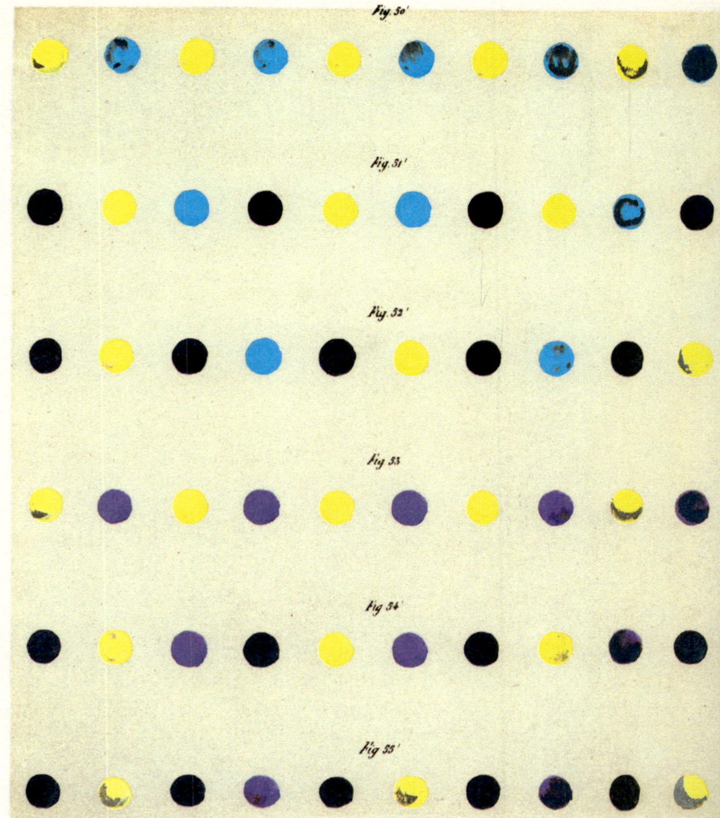

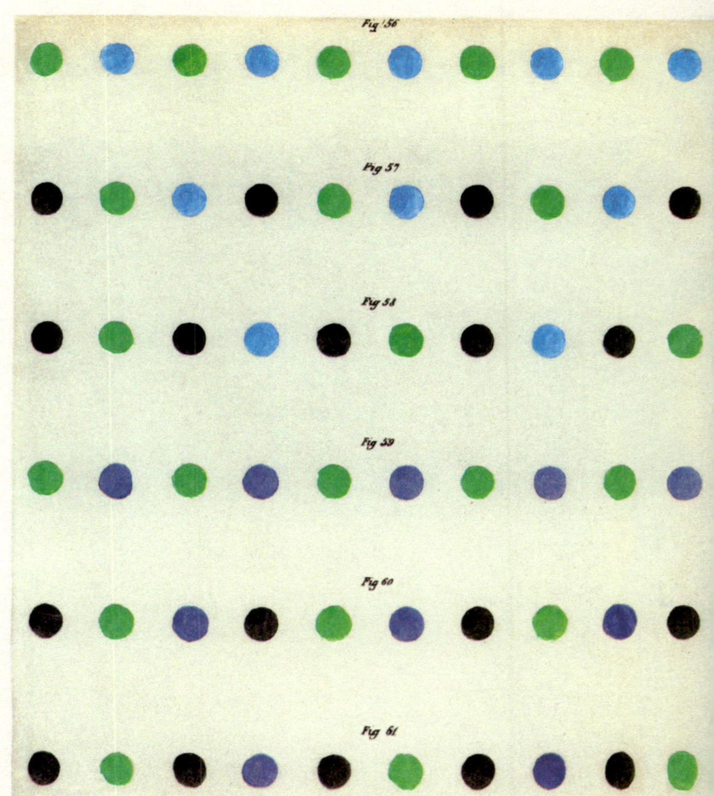

…ts d'une couleur lumineuse et d'une couleur sombre avec le blanc le noir et le gris.

Pl. 25 Pl. 25

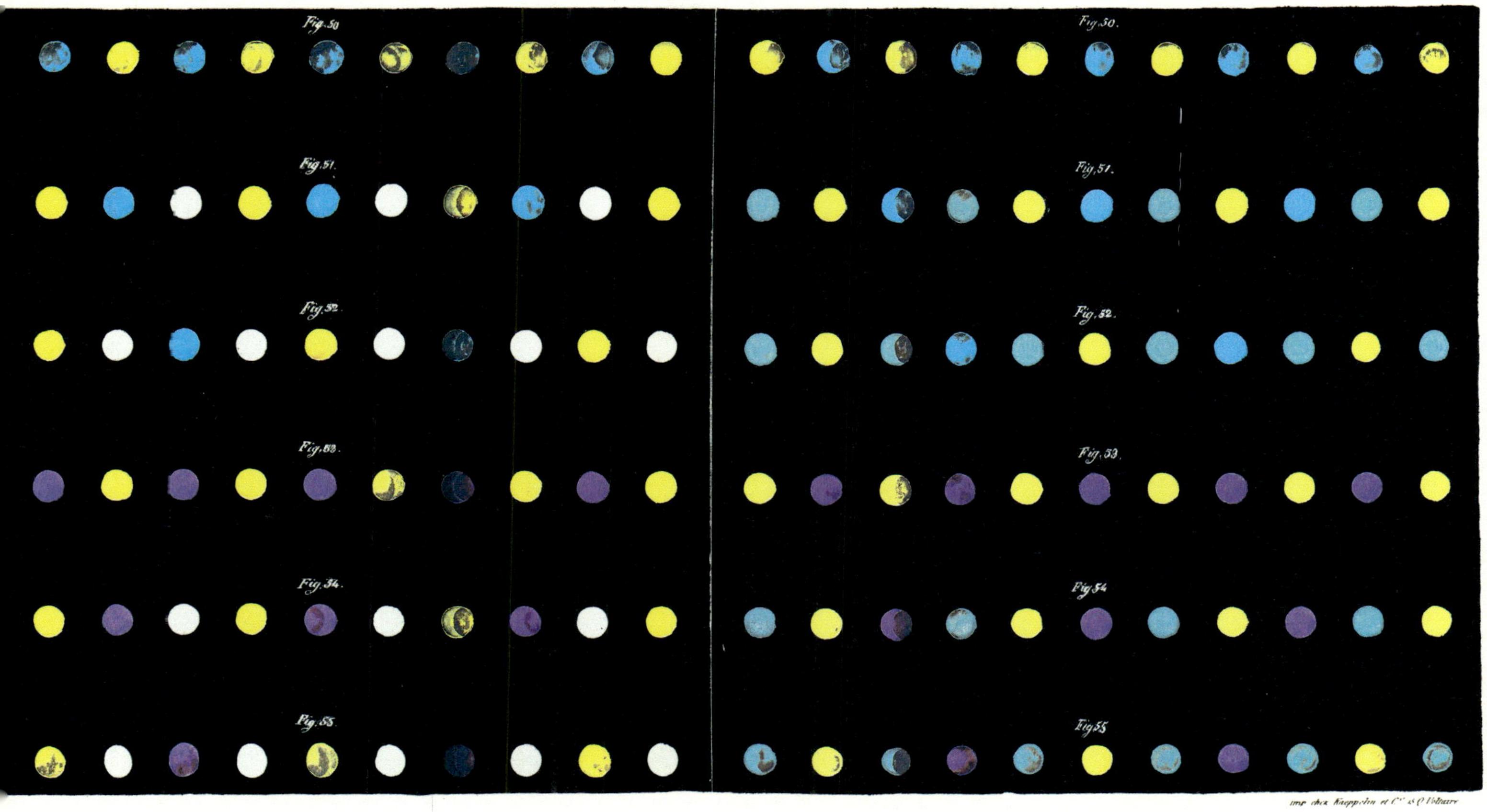

Imp. chez Kaeppelin et Cie 15 Q. Voltaire

…nts, d'une couleur lumineuse et d'une couleur sobre avec le blanc, le noir et le gris.

Pl. 26 Pl. 28

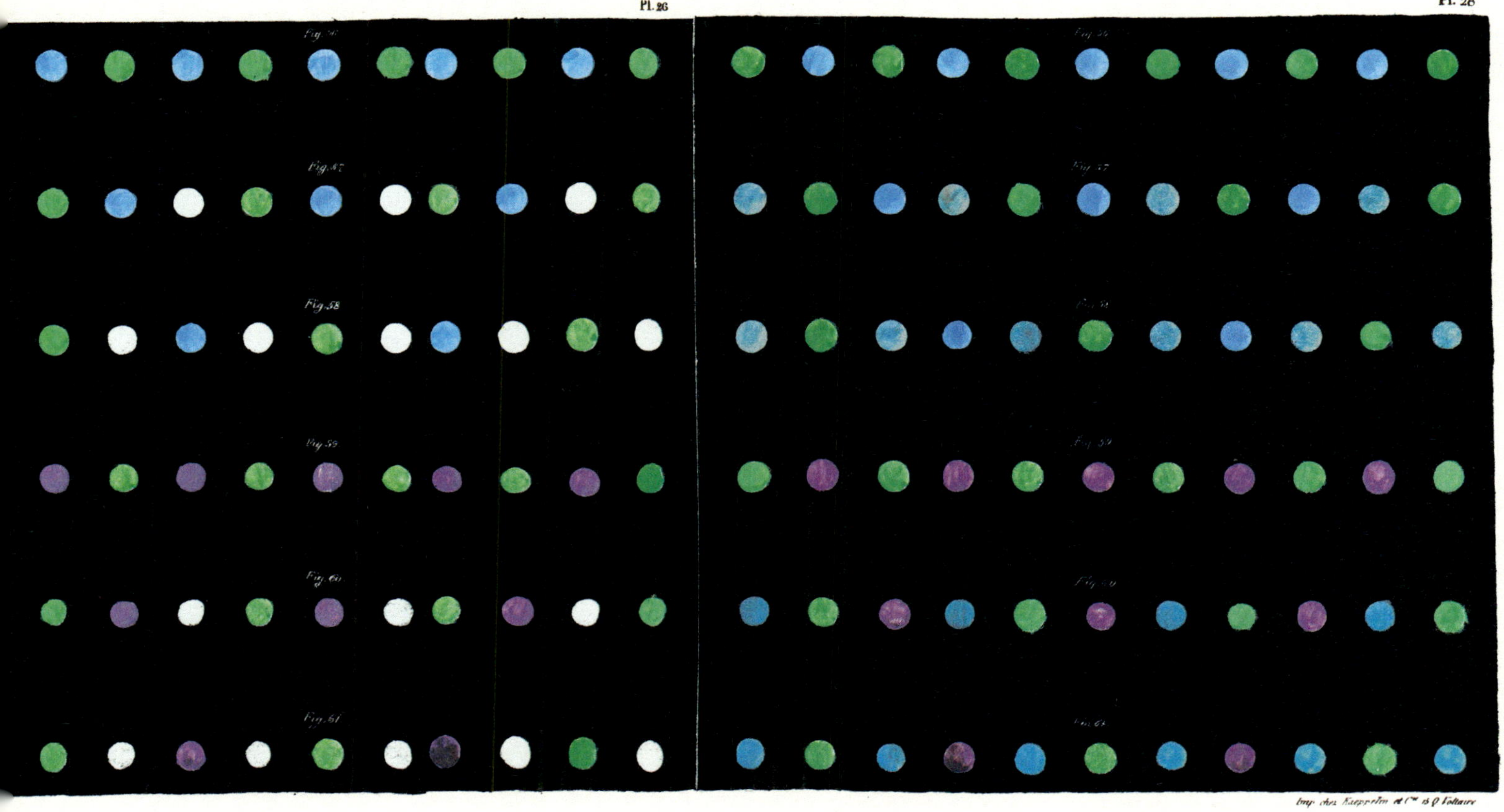

Imp. chez Kaeppelin et Cie 15 Q. Voltaire

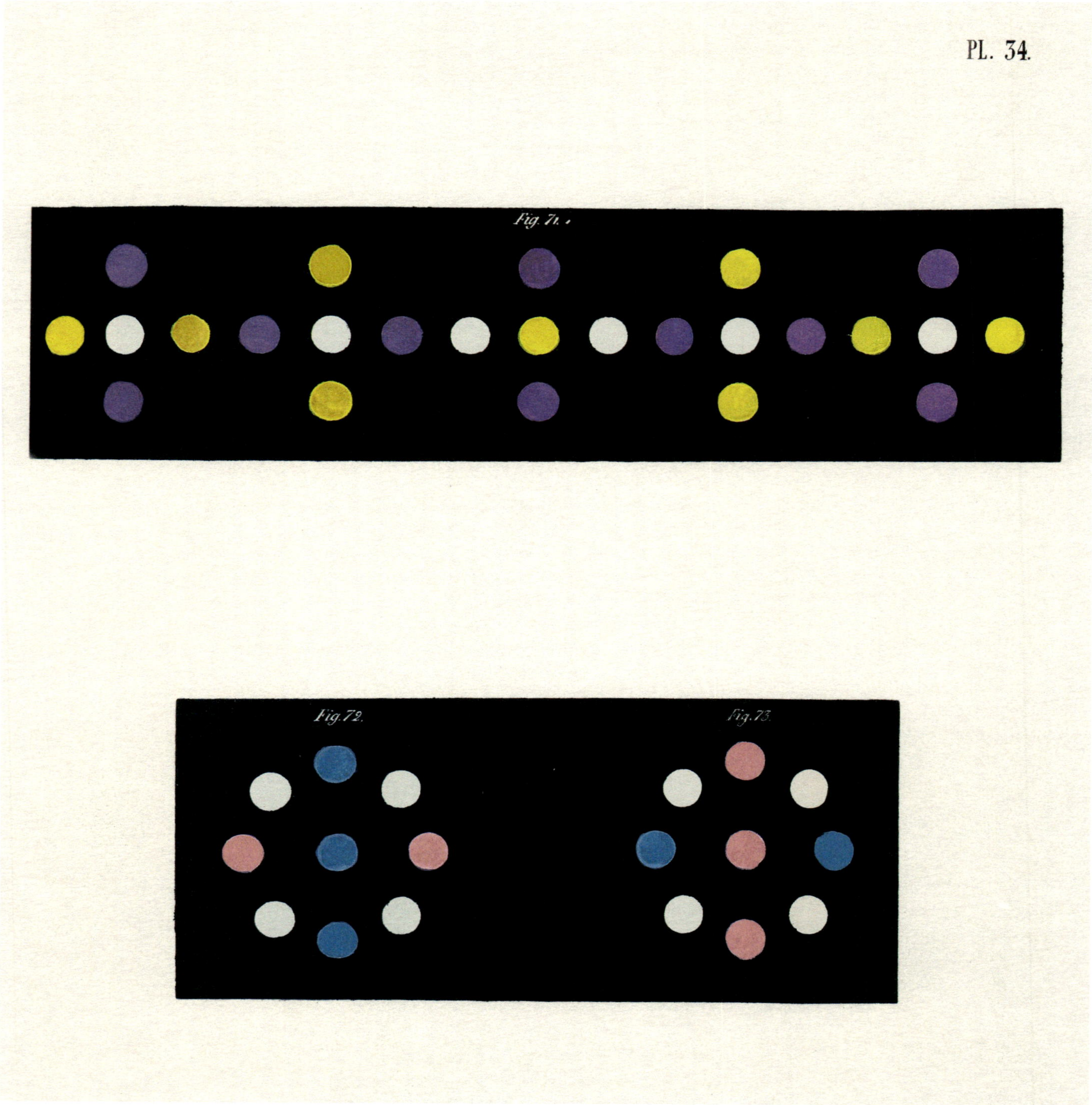
PL. 34.
Fig. 71.
Fig. 72.
Fig. 73.

Fig. 74.

PL. 35.

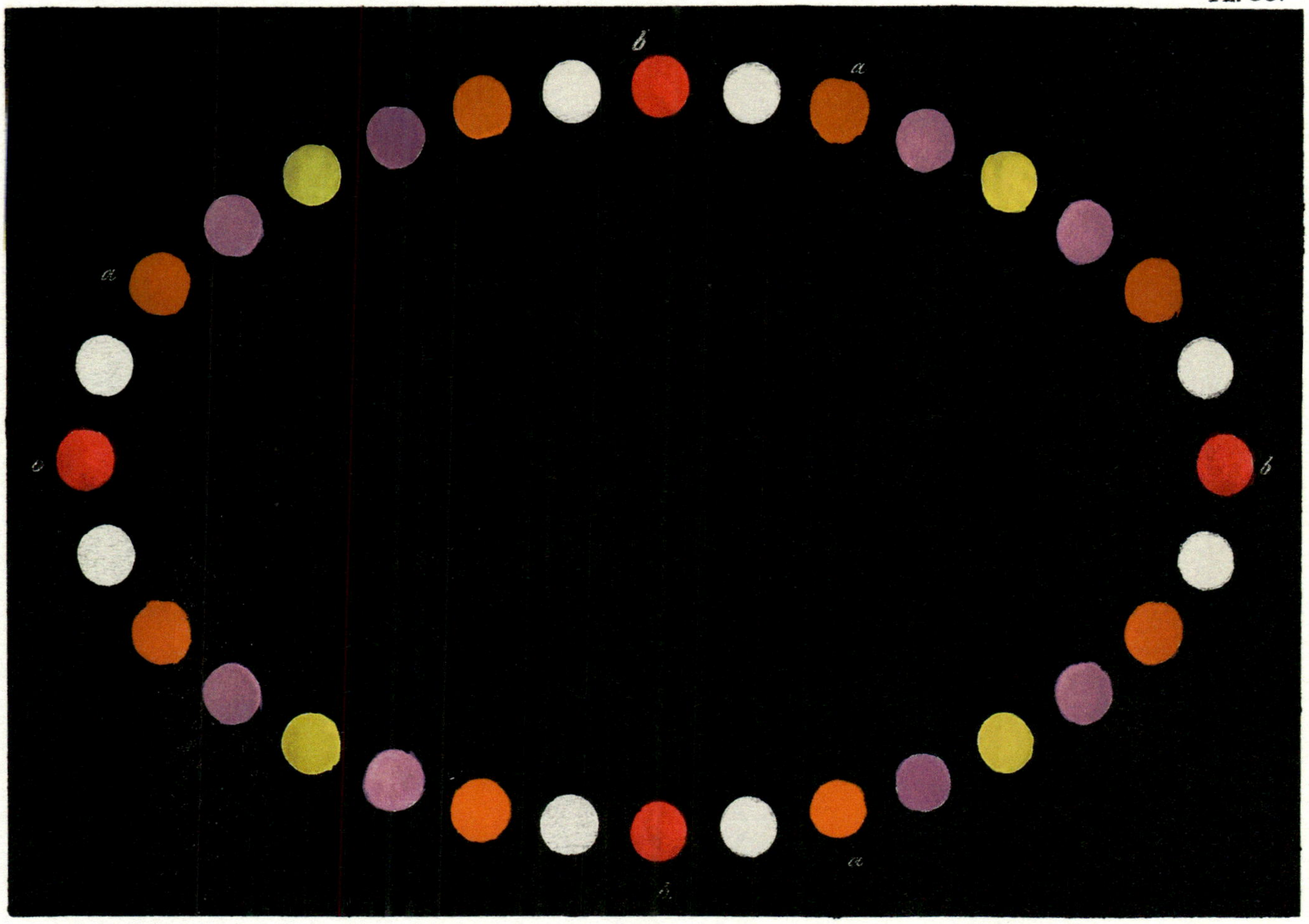

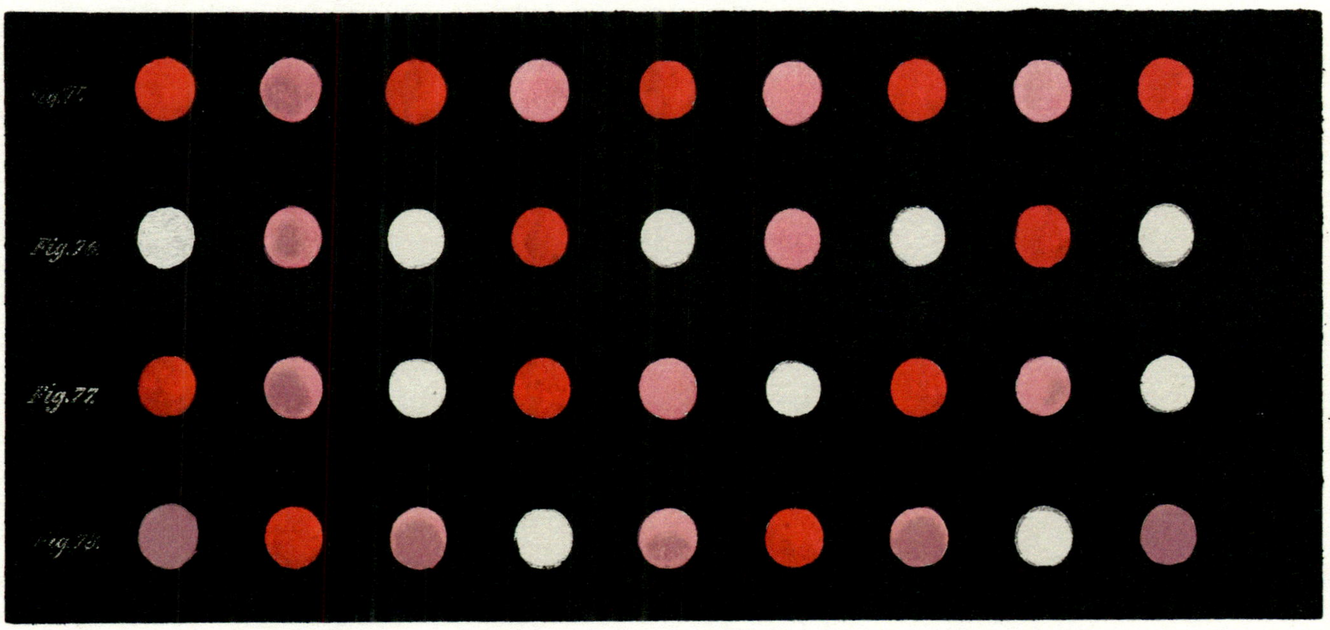

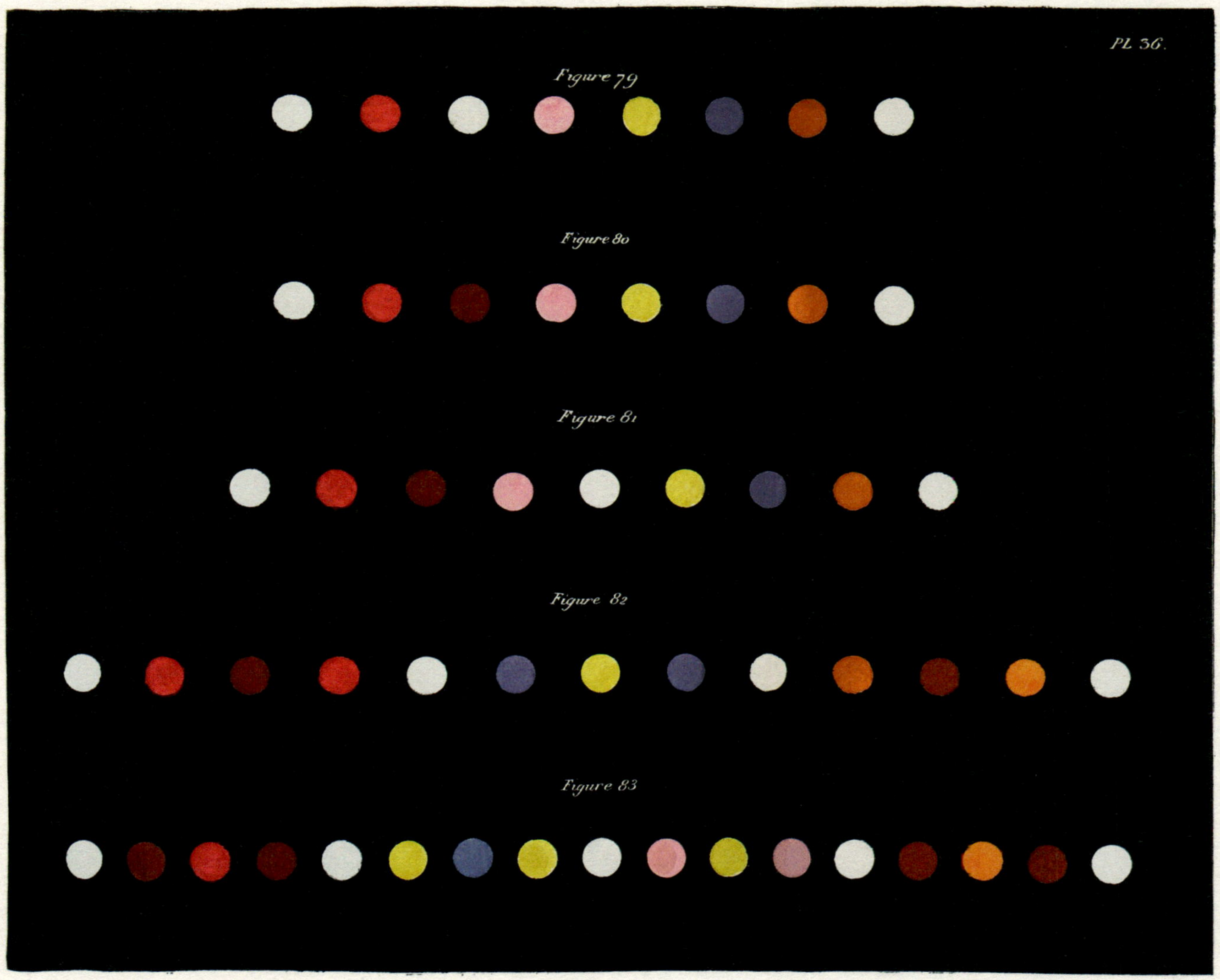
Pl. 36.
Figure 79
Figure 80
Figure 81
Figure 82
Figure 83

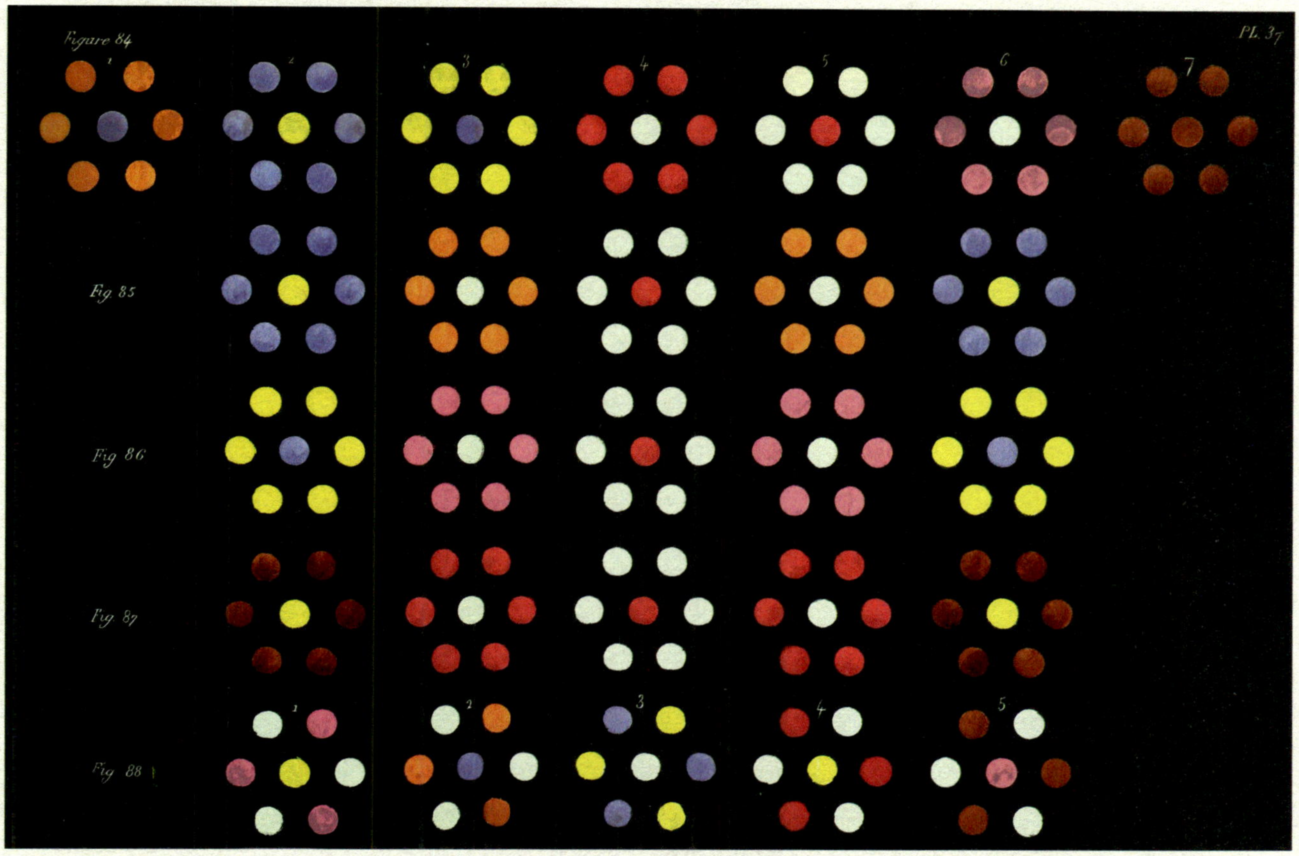
Figure 84
PL. 37
1
2
3
4
5
6
7
Fig 85
Fig 86
Fig 87
Fig 88

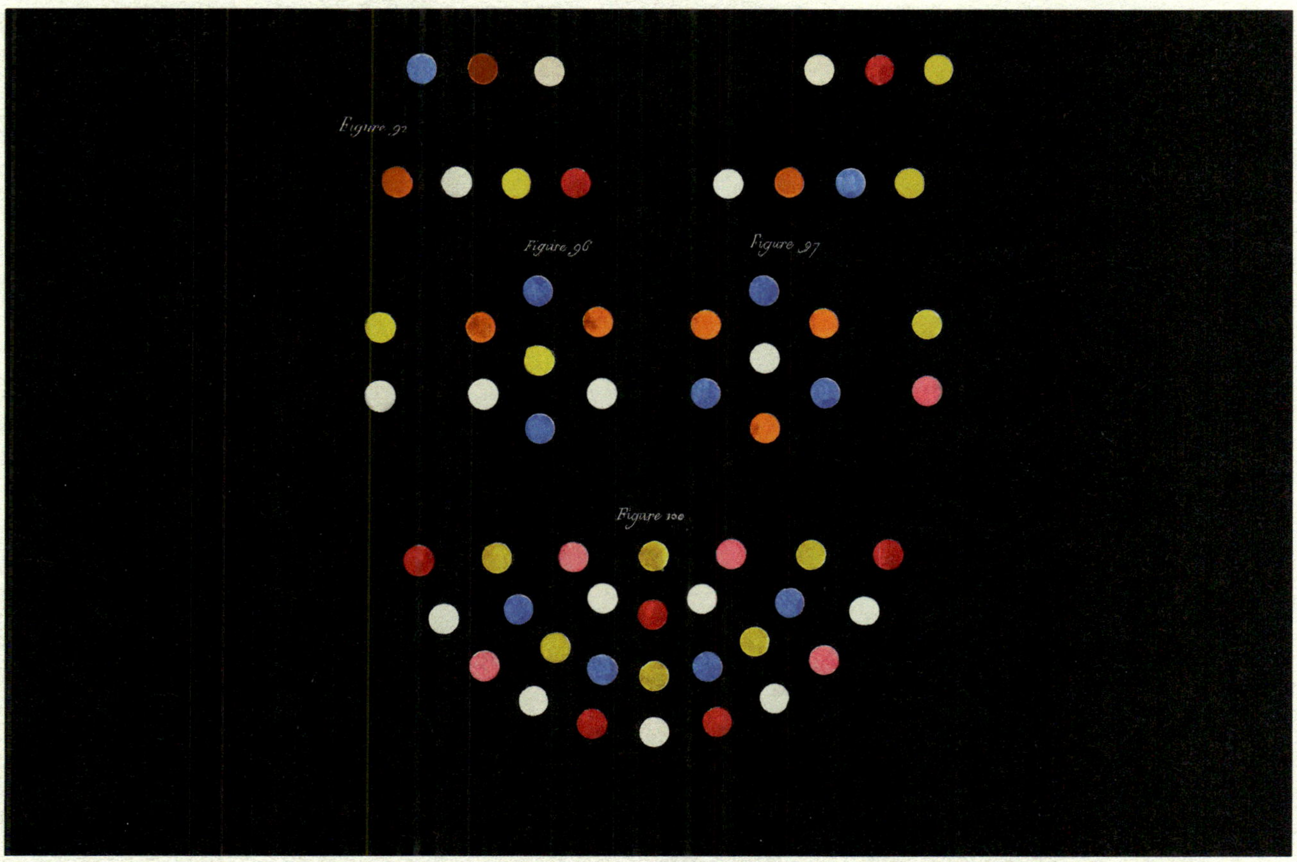
Figure 92
Figure 96
Figure 97
Figure 100

Michel-Eugène Chevreul

DES COULEURS

et de leurs applications aux arts industriels à l'aide des cercles chromatiques

On Colours, and Their Uses in the Industrial Arts as Considered with the Aid of Chromatic Circles
13 plates, 37 x 29 cm | 14 ⅝ x 11 ⅜ in., Paris, 1864
New Haven, Yale University, Robert B. Haas Family Arts Library Special Collections, Faber Birren Collection of Books on Color

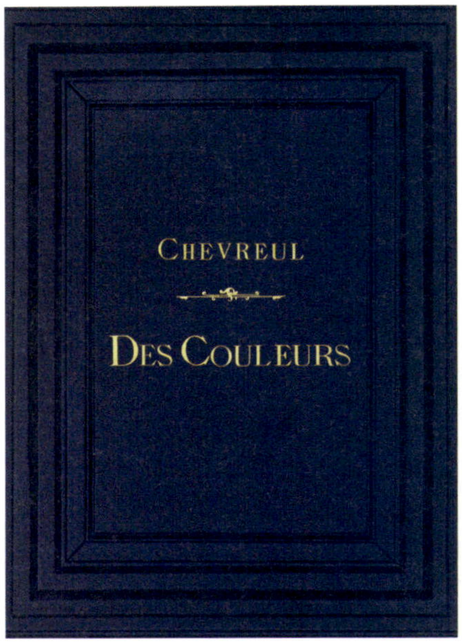

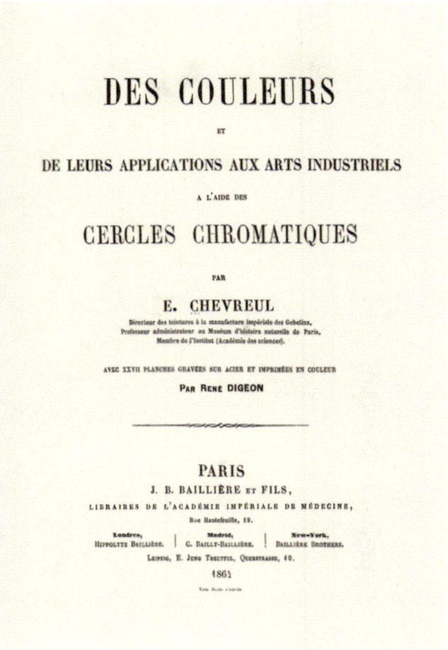
DES COULEURS

ET

DE LEURS APPLICATIONS AUX ARTS INDUSTRIELS

A L'AIDE DES

CERCLES CHROMATIQUES

PAR

E. CHEVREUL

Directeur des teintures à la manufacture impériale des Gobelins,
Professeur administrateur au Muséum d'histoire naturelle de Paris,
Membre de l'Institut (Académie des sciences).

AVEC XXVII PLANCHES GRAVÉES SUR ACIER ET IMPRIMÉES EN COULEUR
PAR RENÉ DIGEON

PARIS
J. B. BAILLIÈRE ET FILS,
LIBRAIRES DE L'ACADÉMIE IMPÉRIALE DE MÉDECINE,
Rue Hautefeuille, 19.

Londres, Hippolyte Baillière. | Madrid, C. Bailly-Baillière. | New-York, Baillière Brothers.
Leipzig, E. Jung Treuttel, Querstrasse, 10.
1864

The success of Chevreul's *De la loi du contraste simultané des couleurs* of 1839 resulted in the book being issued in several further editions, in which Chevreul constantly sought to improve the visual representations of his colour concepts. By the 1860s chromolithographic colour printing had been significantly refined, making possible the publication of a number of magnificently illustrated books that showcased an increasingly colourised world. This later work by Chevreul is one of those masterpieces.

In 1855 Chevreul had started working in conjunction with the engraver René-Henri Digeon, and together they produced a large set of colour circles, using a four-stage printing process, the stunning *Cercles chromatiques de M. E. Chevreul* (1855). This work had been based on the original uncoloured circle from 1839, and in 1861 the section of plates from *De la loi du contraste simultané des couleurs* was itself printed again, as *Exposé d'un moyen de définir et de nommer les couleurs* (A proposal for a method to be used for defining and naming colours), which was included in the 33rd volume of the *Mémoires de l'Académie des Sciences*. In 1864 it was further extended and published as a title on its own, as the edition shown here: *Des couleurs et de leurs applications aux arts industriels à l'aide des cercles chromatiques*.

A long fold-out plate, "Couleurs d'un Spectre Solaire" (Colours of a solar spectrum, pp. 298–299), displays the full linear prismatic spectrum of visible colours against a dark background and was intended as a standard for the different hues to be found in the accompanying circles. Other plates show columns of colours in subtle gradations, formed by additions of white and black (for example, "Gamme chromatique bleu", p. 300).

Digeon and Chevreul's group of 12 full-page colour circles is unrivalled in 19th-century print culture with regard to their technical and conceptual brilliance. The first circle shows the pure prismatic colours at a stage of maximum saturation. This is repeated in the second circle, but with a division into 72 sections, which is then used in all the subsequent circles. The remaining 10 circles show the development from this state of maximum brightness to almost complete darkness, with the increasing addition of black. The last few circles show astonishingly subtle shades of grey and black, with just a hint of colour in each of them.

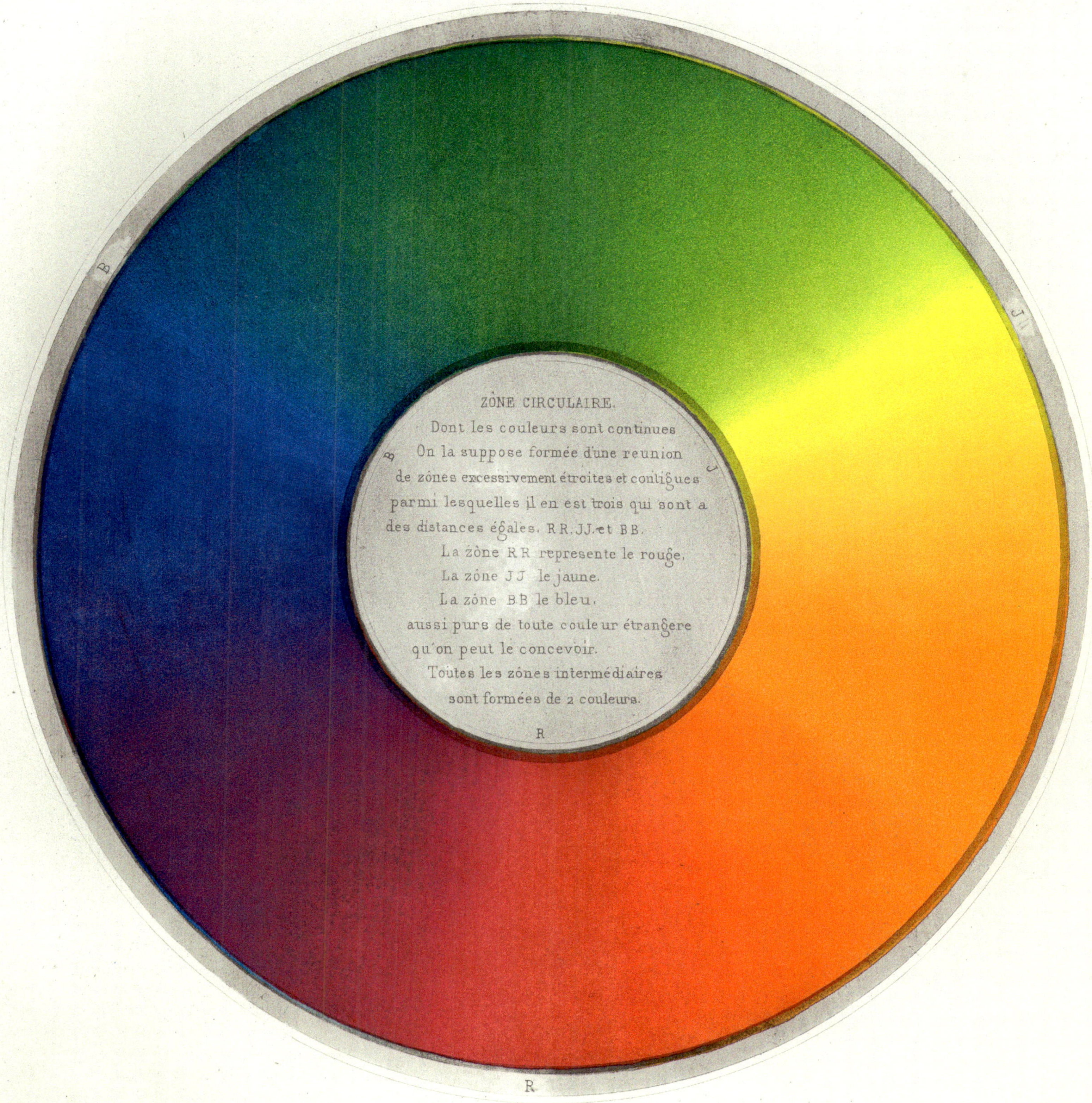
ZÔNE CIRCULAIRE.
Dont les couleurs sont continues
On la suppose formée d'une reunion
de zônes excessivement étroites et contigues
parmi lesquelles il en est trois qui sont a
des distances égales. R R. J J. et B B.
La zône R R represente le rouge,
La zône J J le jaune.
La zône B B le bleu,
aussi purs de toute couleur étrangere
qu'on peut le concevoir.
Toutes les zônes intermédiaires
sont formées de 2 couleurs.
B
J
R
B
J
R

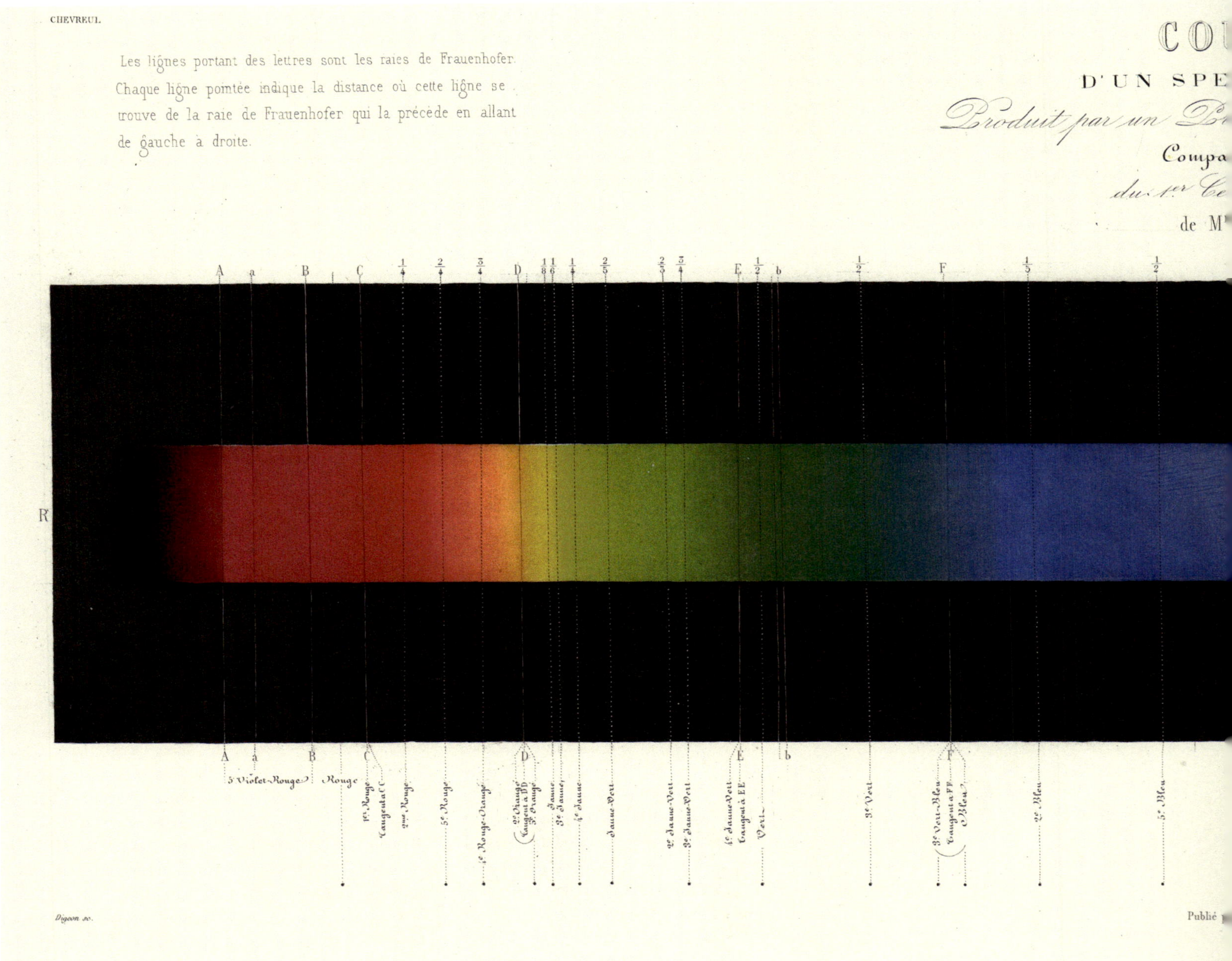
CHEVREUL.
Les lignes portant des lettres sont les raies de Frauenhofer. Chaque ligne pointée indique la distance où cette ligne se trouve de la raie de Frauenhofer qui la précède en allant de gauche à droite.
D'UN SPE
Compa
de M
R
A a B C D E b F
5 Violet-Rouge
Rouge
1er Rouge
Tangent à CC
2me Rouge
5e Rouge
4e Rouge-Orangé
2e Orangé
Tangent à DD
3e Orangé
Jaune
3e Jaune
4e Jaune
Jaune-Vert
2e Jaune-Vert
3e Jaune-Vert
4e Jaune-Vert
Tangent à EE
Vert
3e Vert
3e Vert-Bleu
Tangent à FF
Bleu
2e Bleu
5e Bleu
Digeon sc.
Publié

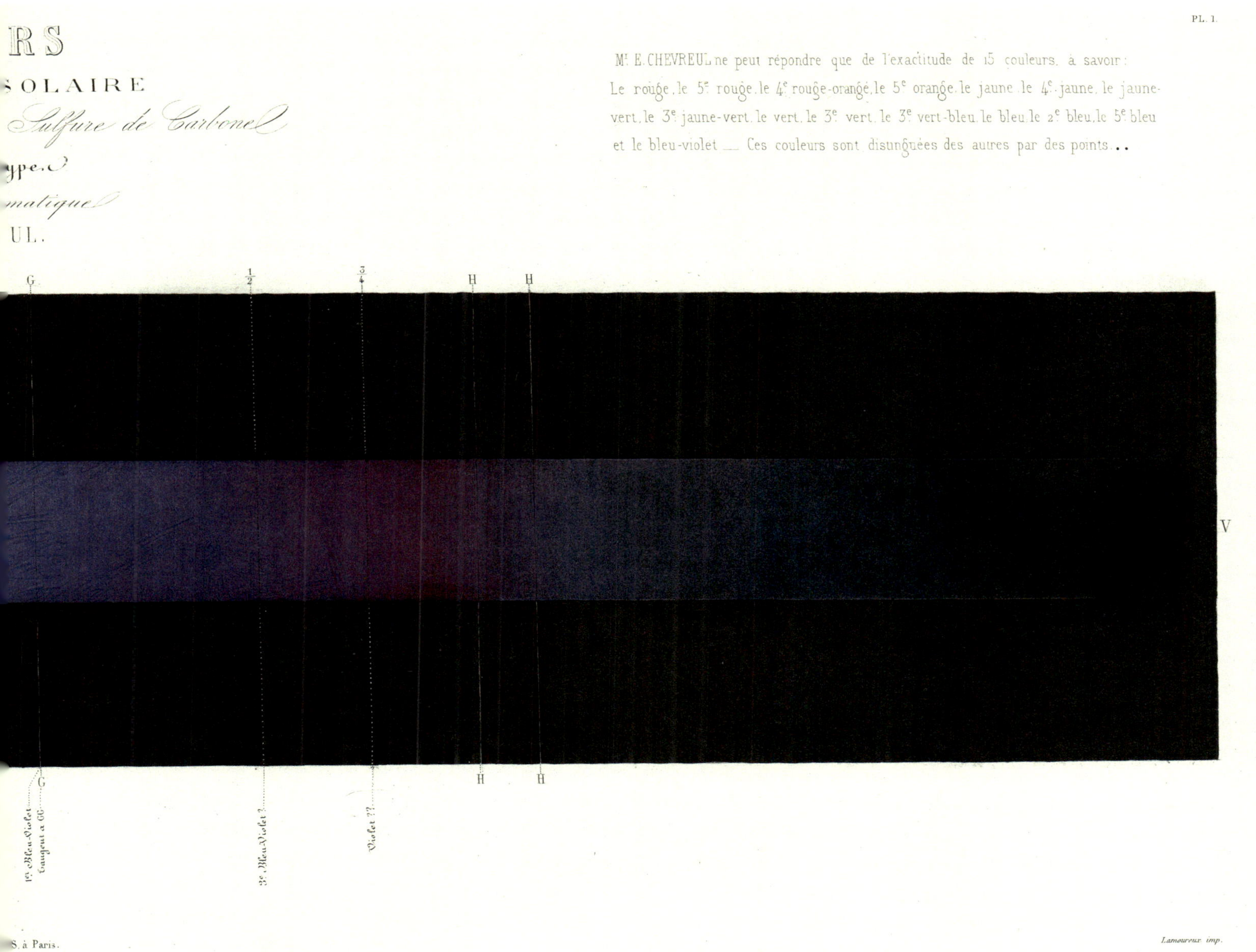
PL. 1.
RS
SOLAIRE
Sulfure de Carbone
ype.
matique
UL.
Mr. E. CHEVREUL ne peut répondre que de l'exactitude de 15 couleurs, à savoir:
Le rouge, le 5e rouge, le 4e rouge-orangé, le 5e orangé, le jaune, le 4e jaune, le jaune-vert, le 3e jaune-vert, le vert, le 3e vert, le 3e vert-bleu, le bleu, le 2e bleu, le 5e bleu et le bleu-violet __ Ces couleurs sont distinguées des autres par des points..
G
1/2
3/4
H
H
V
G
H
H
1er Bleu-Violet
tangent a GG
3e Bleu-Violet ?
Violet ??
S. à Paris.
Lamoureux imp.

MANIÈRE

Dont Mr Chevreul conçoit qu'une couleur, qui est indéfinie
en allant du blanc au noir, est distinguée en parties
définies qu'il appelle tons.

Il désigne les tons d'une même couleur par le mot :

GAMME DE CETTE COULEUR.

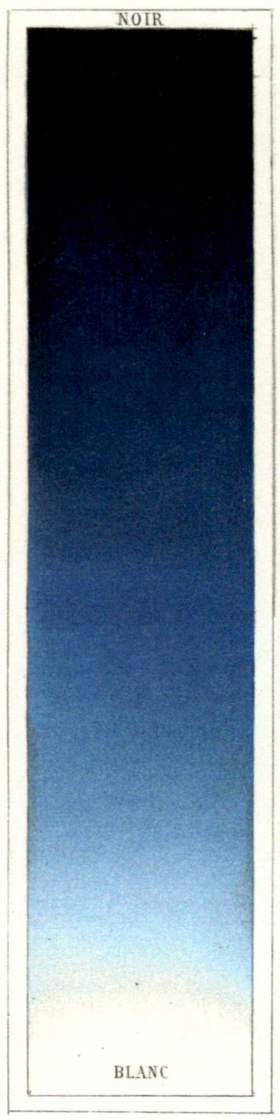

Fig. 1.

Exemple de la graduation d'une couleur.

On va du blanc au noir

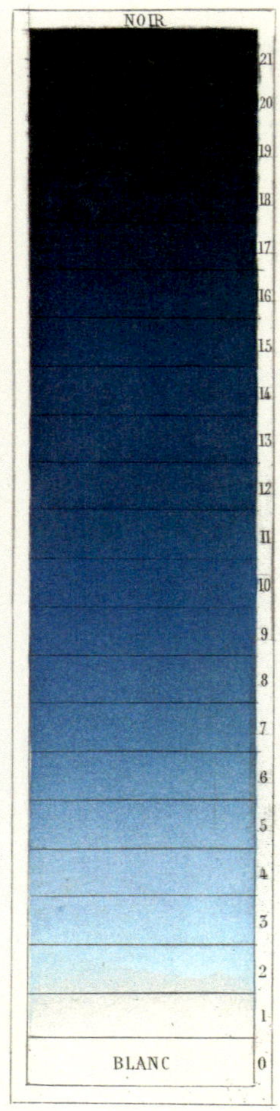

Fig. 2.

C'est la figure 1 divisée en 22 parties superficielles égales

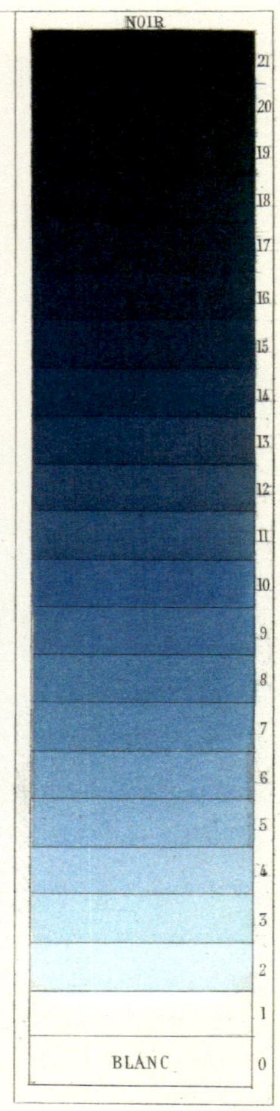

Fig. 3.

C'est la figure 2 avec la différence que la couleur qui est graduée d'une manière continue dans la figure 2 sur chaque partie superficielle en allant du blanc au noir, est repartie d'une manière uniforme sur chaque partie superficielle ; dès lors la couleur est discontinue dans l'ensemble des parties superficielles.

C'est ainsi que la fig. 3 représente ce que Mr Chevreul appelle la gamme des tons bleus au nombre de vingt.

On peut dire que le blanc est zéro et que le noir est le ton 21

Digeon sc. *Lamoureux imp.*

Publié par J.B. BAILLIÈRE et FILS, à Paris.

CHEVREUL. Pl. IV

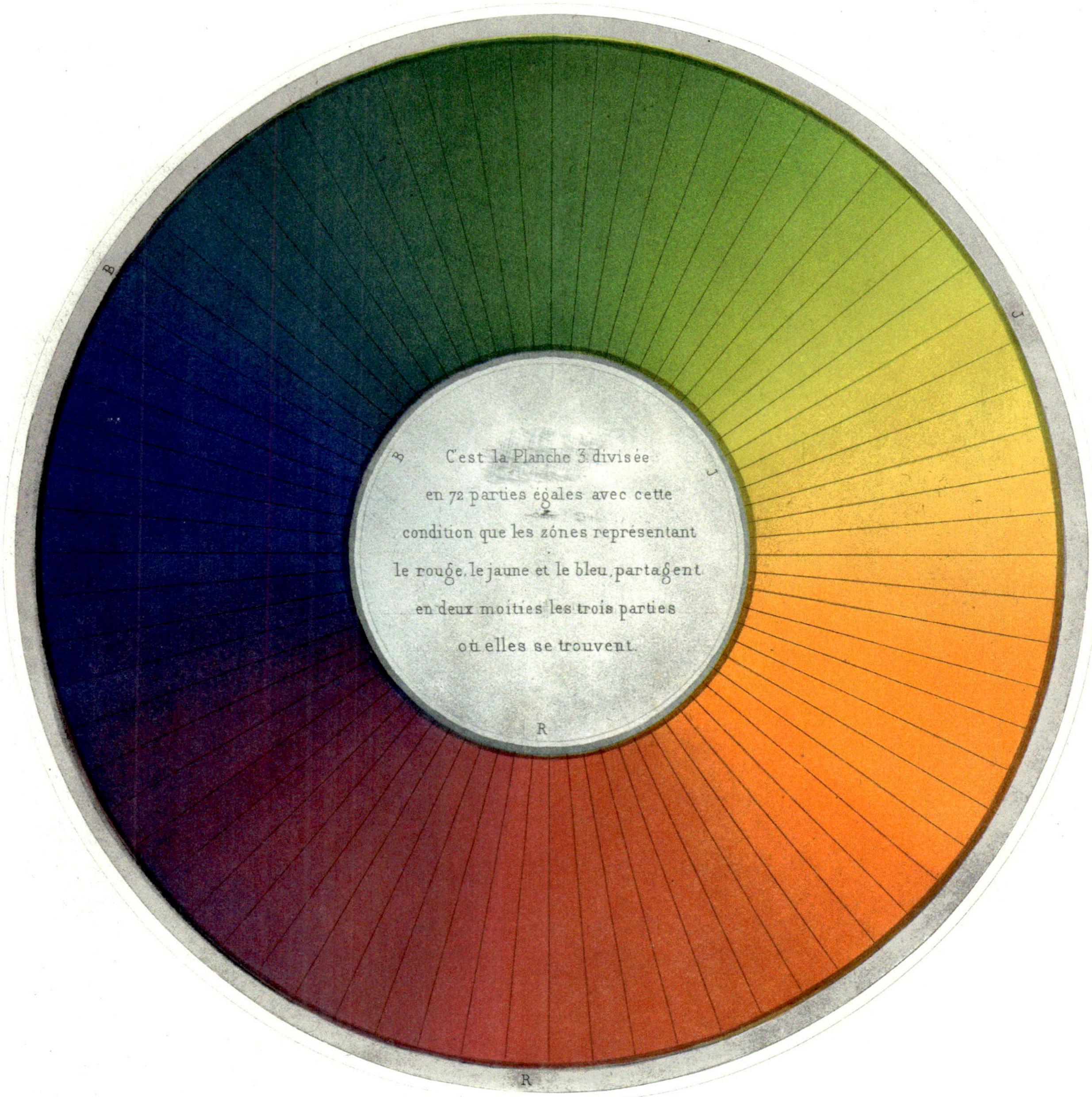

Si on se représente chaque partie de la zône Pl. 4. uniformément teinte de la *Couleur nuancée* dont elle est couverte, on aura *72 Couleurs types* suffisamment distinctes pour qu'on puisse y rapporter les couleurs franches, comme on rapporte les tons d'une même couleur à la figure 3. de la Planche 2.

Observation: La conséquence du mélange uniforme de la couleur est toute simple pour toutes les parties qui ne renferment que deux couleurs; mais les parties divisées par RR. JJ. BB. renfermant trois couleurs, doivent être l'objet d'une remarque particulière. Elles comprennent en effet chacune trois couleurs, par exemple, la moitié de la partie représentant le rouge qui est du coté du jaune contient du jaune, comme la moitié qui regarde le bleu contient du bleu. Dès lors, par le mélange, la partie rouge doit être mêlée de jaune et de bleu, à la vérité en très petite quantité, mais ce mélange ne change pas la qualité du rouge autrement qu'en l'ombrant un peu, par la raison que des couleurs matérielles complémentaires donnent du Noir par leur mélanges.

Si l'on se représentait la zône circulaire colorée par la transmission ou la réflexion des rayons colorés du prisme, le rouge au lieu d'être ombré, serait éclairci parce que les lumières colorées complémentaires reproduisent du blanc par leur mélange.

Publié par J. B. BAILLIERE et FILS, à Paris

Lamoureux Imp.

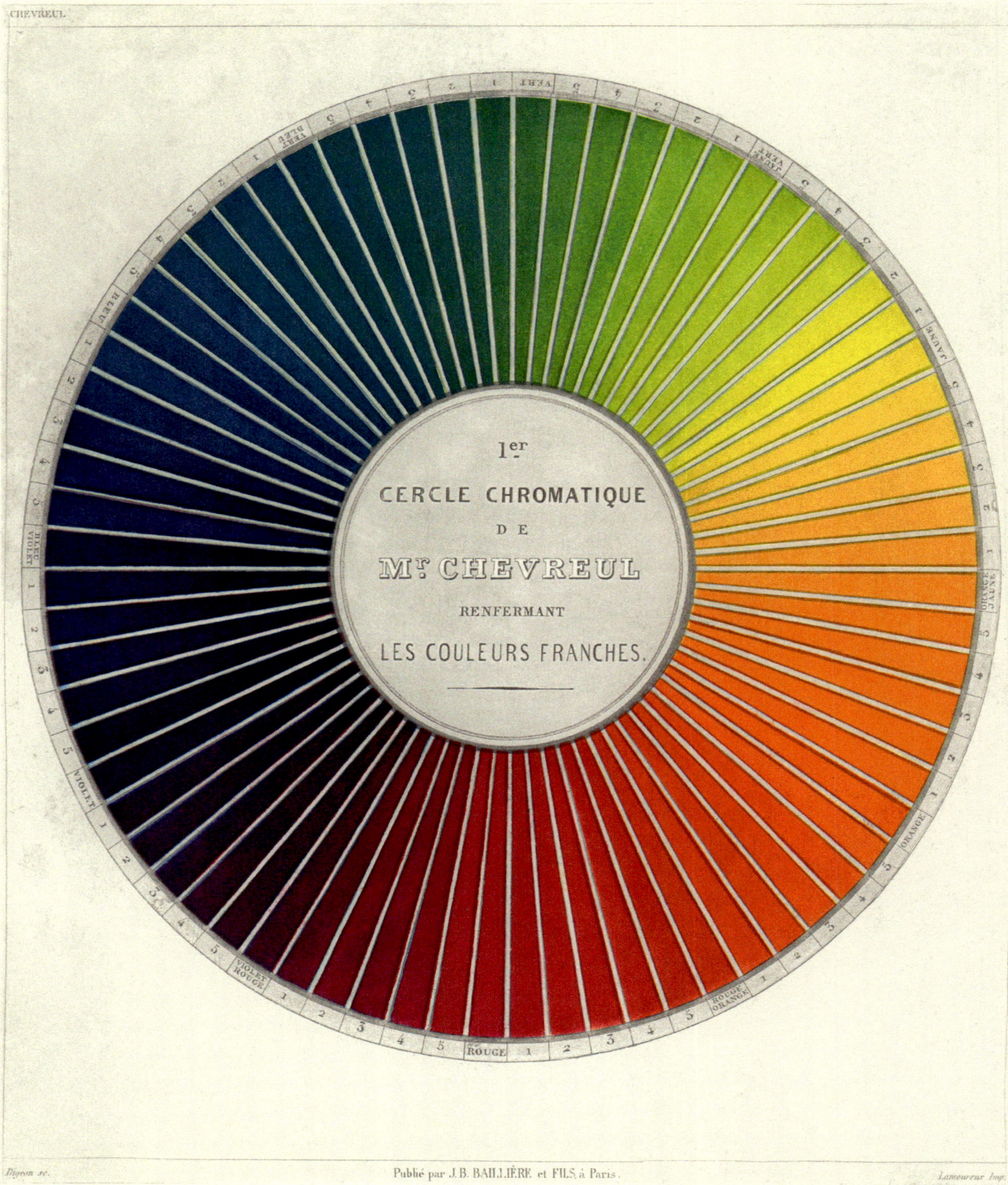
CHEVREUL
1er
CERCLE CHROMATIQUE
DE
Mr CHEVREUL
RENFERMANT
LES COULEURS FRANCHES.
ROUGE
ORANGE
JAUNE
VERT
BLEU
VIOLET
Publié par J.B. BAILLIÈRE et FILS, à Paris.

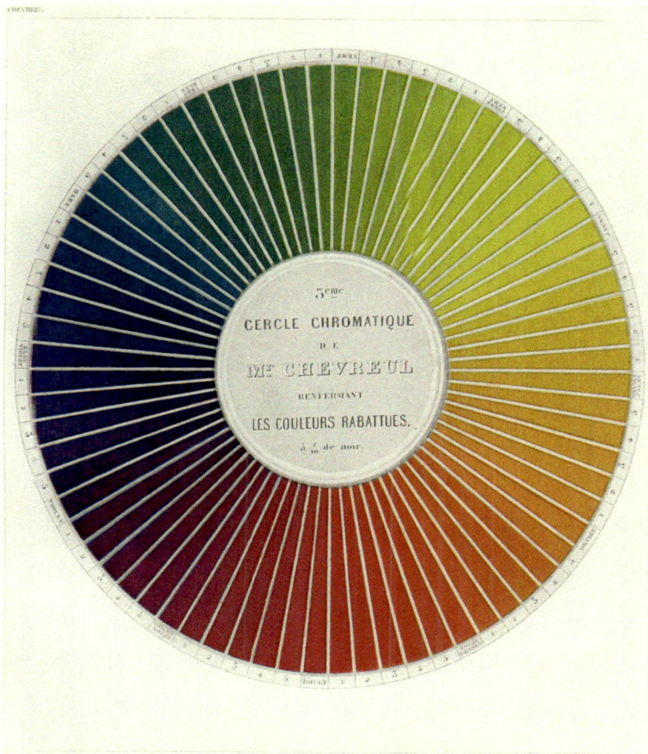

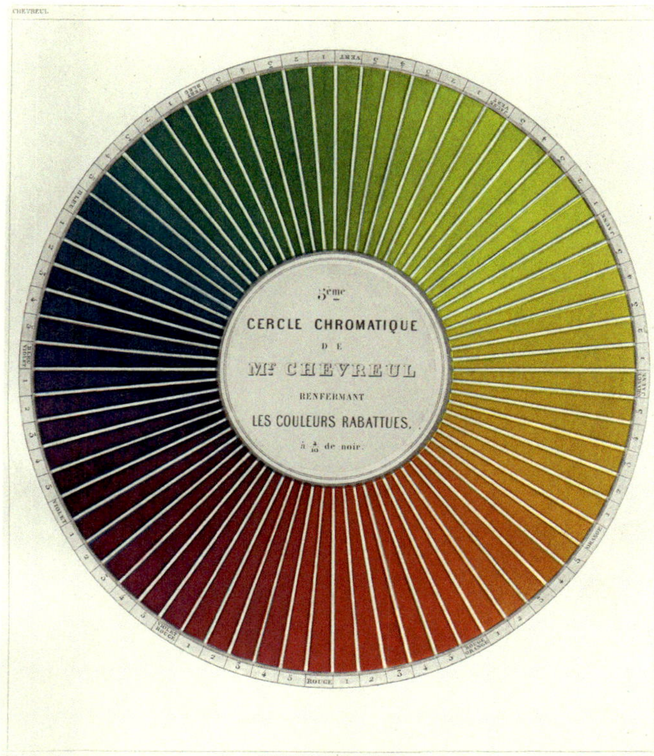

Über Farben und ihre Anwendungen in Industrie und Gewerbe mithilfe von Farbkreisen

Der Erfolg von Chevreuls *De la loi du contraste simultané des couleurs* (1839) bescherte dem Buch mehrere Neuauflagen, in denen der Autor laufend versuchte, die bildlichen Darstellungen seiner Farbsysteme zu verbessern. In den 1860er-Jahren hatte sich die Farblithografie erheblich weiterentwickelt und machte die Herausgabe prachtvoll illustrierter Bücher möglich, die eine zunehmend eingefärbte Welt darboten. Dieser spätere Band aus Chevreuls Hand ist eines jener Meisterwerke.

Chevreul hatte 1855 eine Zusammenarbeit mit dem Kupferstecher René-Henri Digeon begonnen. Gemeinsam schufen sie mithilfe eines vierstufigen Druckverfahrens die überwältigenden *Cercles chromatiques de M. E. Chevreul* (1855), eine umfangreiche Serie von Farbkreisen, die auf dem ursprünglichen, nicht kolorierten Kreis von 1839 beruhte. 1861 wurde der Farbatlas aus *De la loi du contraste simultané des couleurs* nachgedruckt und als *Exposé d'un moyen de définir et de nommer les couleurs* („Vorschlag für eine Methode der Bestimmung und Benennung von Farben") als Teil des 33. Bandes der *Mémoires de l'Académie des Sciences* veröffentlicht. Hier gezeigt werden Tafeln aus der erweiterten Fassung, die 1864 unter dem Titel *Des couleurs et de leurs applications aux arts industriels à l'aide des cercles chromatiques* („Über Farben und ihre Anwendungen in Industrie und Gewerbe mithilfe von Farbkreisen") als eigenständiges Werk erschien.

Eine lange Ausklapptafel mit dem Titel „Couleurs d'un spectre solaire" (Farben eines Sonnenspektrums, S. 298–299) zeigt das komplette prismatische Spektrum der sichtbaren Farben in linearer Form vor einem dunklen Hintergrund und sollte als Maßstab für die verschiedenen Farbtöne in den folgenden Farbkreisen dienen. Auf weiteren Tafeln sind Farbsäulen in feinen Abstufungen zu sehen, die durch Hinzugabe von Weiß oder Schwarz entstehen (zum Beispiel die „Gamme chromatique bleu" oder „Blaue Farbskala", S. 300).

Die Gruppe der zwölf ganzseitigen Farbkreise von Digeon und Chevreul ist in ihrer technischen und konzeptuellen Brillanz im Druckwesen des 19. Jahrhunderts einmalig. Der erste Kreis zeigt die reinen prismatischen Farben im Zustand maximaler Sättigung. Im zweiten Kreis wiederholt sich das Bild, nun jedoch mit einer Unterteilung in 72 Abschnitte, die in allen folgenden Kreisen beibehalten wird. Die restlichen zehn Kreise zeigen die Entwicklung von diesem Zustand maximaler Leuchtkraft bis hin zu fast völliger Dunkelheit durch ansteigendes Zusetzen

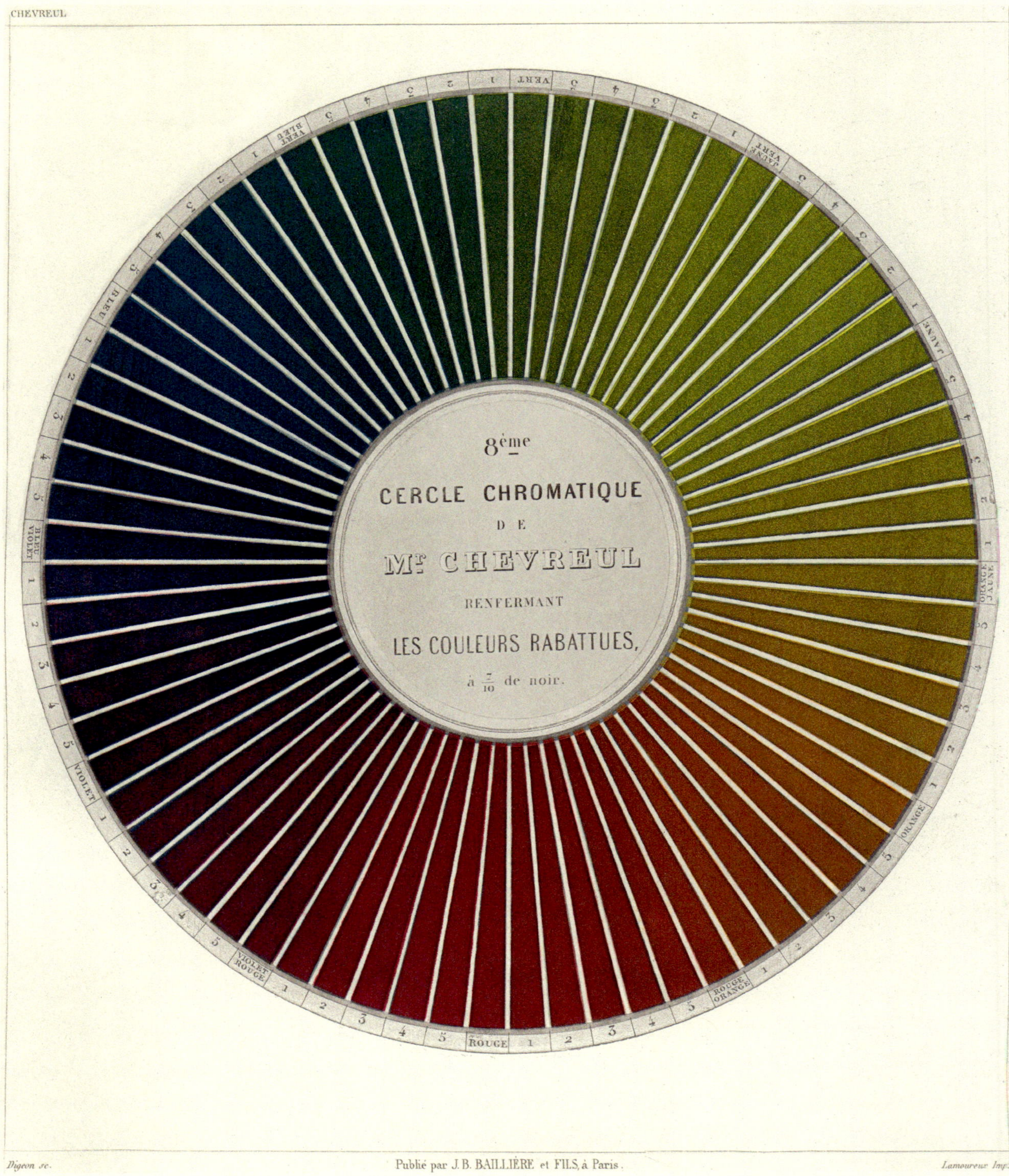
CHEVREUL
8ème
CERCLE CHROMATIQUE
DE
Mr CHEVREUL
RENFERMANT
LES COULEURS RABATTUES,
à $\frac{7}{10}$ de noir.
ROUGE
ROUGE ORANGE
ORANGE
ORANGE JAUNE
JAUNE
JAUNE VERT
VERT
VERT BLEU
BLEU
BLEU VIOLET
VIOLET
VIOLET ROUGE
Digeon sc.
Publié par J.B. BAILLIÈRE et FILS, à Paris.
Lamoureux Imp.

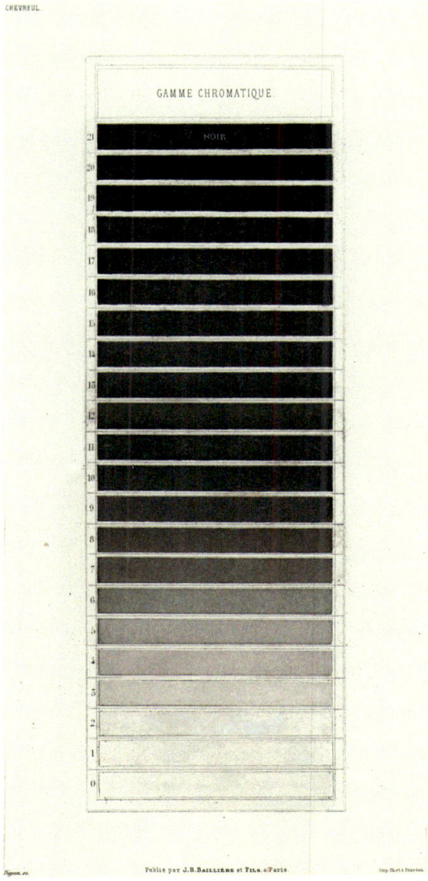

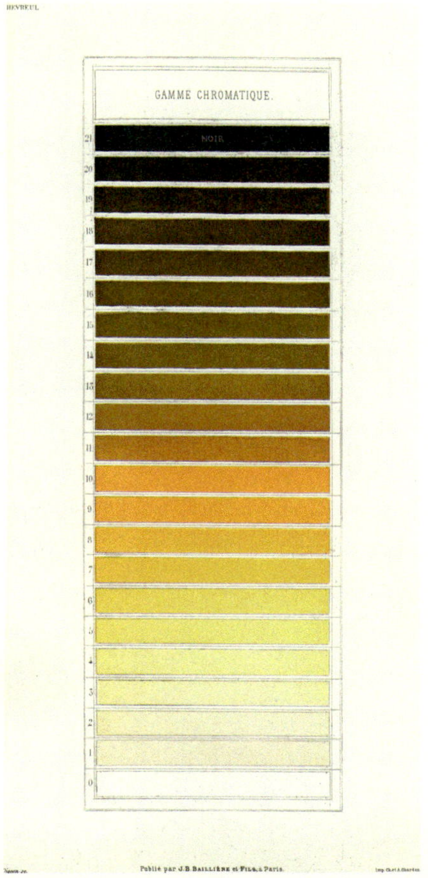

von Schwarz. Die letzten Kreise bestechen mit erstaunlich feinen, nur einen Hauch Buntfarbe enthaltenden Schattierungen von Grau und Schwarz.

Des couleurs et de leurs applications aux arts industriels à l'aide des cercles chromatiques

Le succès rencontré par *De la loi du contraste simultané des couleurs* en 1839 permet à Chevreul d'améliorer progressivement les représentations visuelles de ses conceptions chromatiques dans plusieurs rééditions. L'impression par chromolithographie s'est considérablement améliorée et, dans les années 1860, de nombreux ouvrages superbement illustrés voient le jour. Dans le monde de plus en plus coloré qui se déploie, le nouvel ouvrage de Chevreul fait figure de chef-d'œuvre.

En 1855, Chevreul entame une collaboration avec le graveur René-Henri Digeon. Ensemble, ils produisent toute une panoplie de cercles chromatiques en utilisant un processus d'impression en quatre étapes : ce sont les éblouissants *Cercles chromatiques de M. E. Chevreul* (1855), issus du cercle sans couleurs de 1839. En 1861, les planches figurant dans *De la loi du contraste simultané des couleurs* sont également réimprimées sous le titre *Exposé d'un moyen de définir et de nommer les couleurs*, qui sera inclus dans le 33e volume des *Mémoires de l'Académie des Sciences*. En 1864, cet opus est encore augmenté et doté d'un titre propre, comme sur l'édition présentée ici : *Des couleurs et de leurs applications aux arts industriels à l'aide des cercles chromatiques*.

Une longue planche dépliable intitulée « Couleurs d'un spectre solaire » (p. 298–299) montre la décomposition complète du spectre prismatique des couleurs visibles sur un fond sombre et sert de référence pour les différentes teintes que l'on trouve dans les cercles. D'autres planches montrent les couleurs disposées en colonnes dont les gradations subtiles sont obtenues par ajout de blanc et de noir (voir par exemple la gamme chromatique des bleus, p. 300).

Les 12 pleines pages de cercles chromatiques de Digeon et Chevreul restent un sommet absolu dans toute la production éditoriale du XIXe siècle, tant au plan technique que conceptuellement. Les couleurs prismatiques pures du premier cercle sont figurées dans leur saturation maximale. Cette présentation se répète dans le deuxième cercle, mais avec une division en 72 sections que l'on retrouve dans tous les cercles suivants. Les dix cercles restant montrent la déclinaison depuis le stade de luminosité maximale jusqu'à l'obscurité presque totale, au gré de l'ajout de noir. Les tout derniers cercles présentent les gris et les noirs au moyen d'infimes touches de couleur étonnamment subtiles.

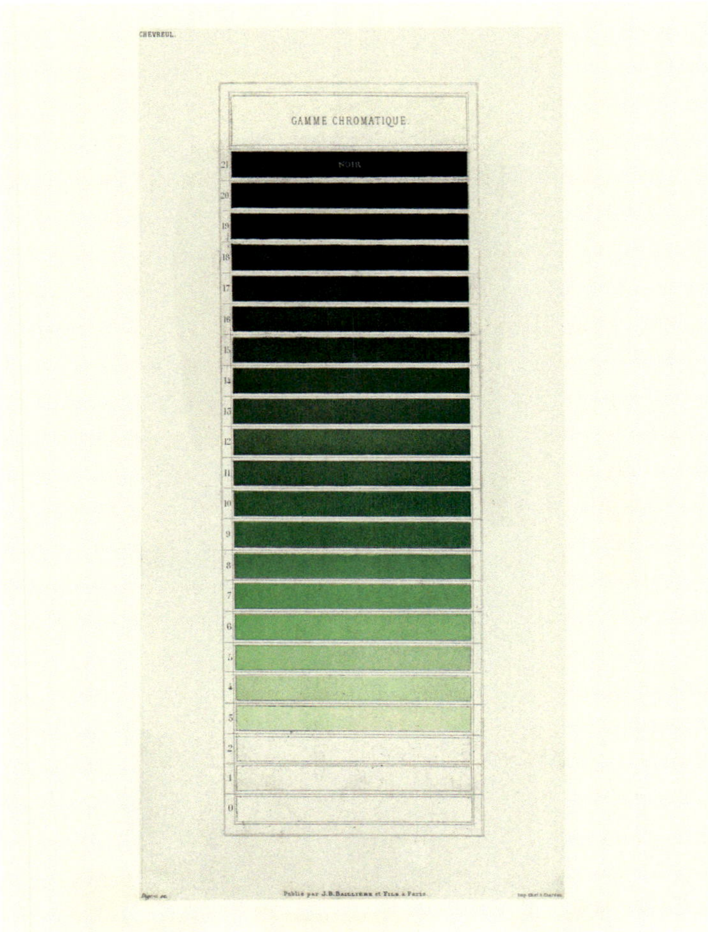

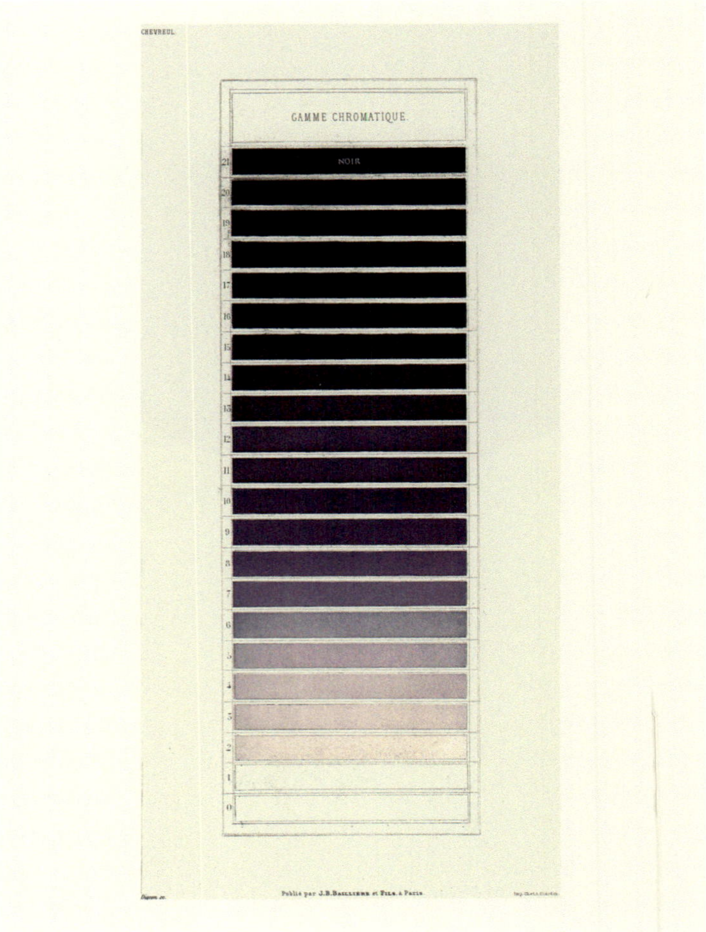

Sobre los colores y sus applicaciones en las artes industriales con la ayuda de círculos cromáticos

El éxito de *De la loi du contraste simultané des couleurs* de 1839 propició que se hicieran varias reediciones del libro, donde Chevreul buscó constantemente mejorar las representaciones visuales de sus conceptos de color. En la década de 1860, la impresión cromolitográfica en color se había perfeccionado de forma significativa, lo que hizo posible la publicación de una serie de libros magníficamente ilustrados que mostraban un mundo cada vez más colorido. Este último trabajo de Chevreul es una de esas obras maestras.

En 1855, Chevreul había comenzado a trabajar en colaboración con el grabador René-Henri Digeon, y juntos produjeron un gran conjunto de círculos de colores mediante un proceso de impresión en cuatro etapas, los impresionantes *Cercles chromatiques de M. E. Chevreul* (1855). Esta obra se había basado en el círculo sin colorear original de 1839, y en 1861 se volvió a imprimir la sección de láminas de *De la loi du contraste simultané des couleurs* como *Exposé d'un moyen de définir et de nommer les couleurs* (Propuesta de un método para definir y nombrar los colores), que se incluyó en el volumen 33 de *Mémoires de l'Académie des Sciences*. En 1864 se amplió aún más y se publicó como un título independiente, que es la edición que aquí se muestra: *Des couleurs et de leurs applications aux arts industriels à l'aide des cercles chromatiques* (Sobre los colores y sus aplicaciones en las artes industriales con la ayuda de círculos cromáticos).

Una gran lámina desplegable, «Couleurs d'un Spectre Solaire» (Colores de un espectro solar, págs. 298–299), muestra el espectro prismático lineal completo de colores visibles sobre un fondo oscuro. Se concibió como un estándar para los diferentes tonos que se encuentran en los círculos que lo acompañan. Otras láminas muestran columnas de colores en sutiles gradaciones, formadas por adiciones de blanco y negro (por ejemplo, «Gamme chromatique bleu», pág. 300).

La brillantez técnica y conceptual del conjunto de 12 círculos de color a página completa de Digeon y Chevreul no tiene rival en la cultura impresa del siglo XIX. El primer círculo muestra los colores prismáticos puros en una etapa de máxima saturación. Eso se repite en el segundo círculo, esta vez dividido en 72 secciones, estructura que se repite en los demás círculos. Los 10 círculos restantes muestran el desarrollo desde este estado de brillo máximo hasta la oscuridad casi completa con la adición creciente de negro. Los últimos círculos muestran tonos asombrosamente sutiles de grises y negros, con solo un toque de color en cada uno de ellos.

CHEVREUL

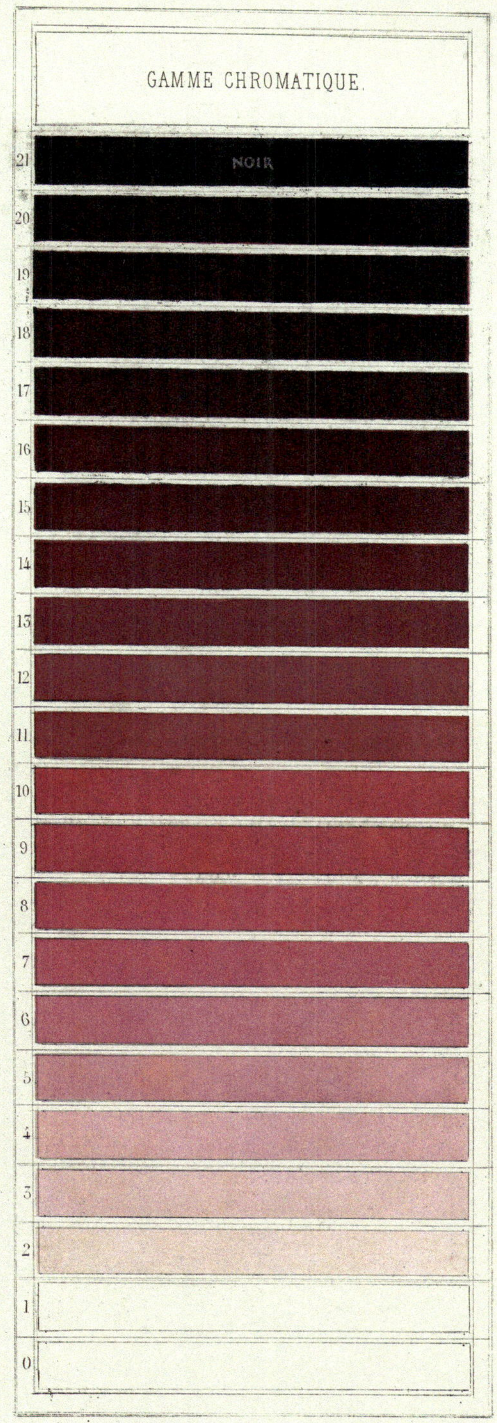

Digeon sc.

Publié par J.B. Baillière et Fils, à Paris.

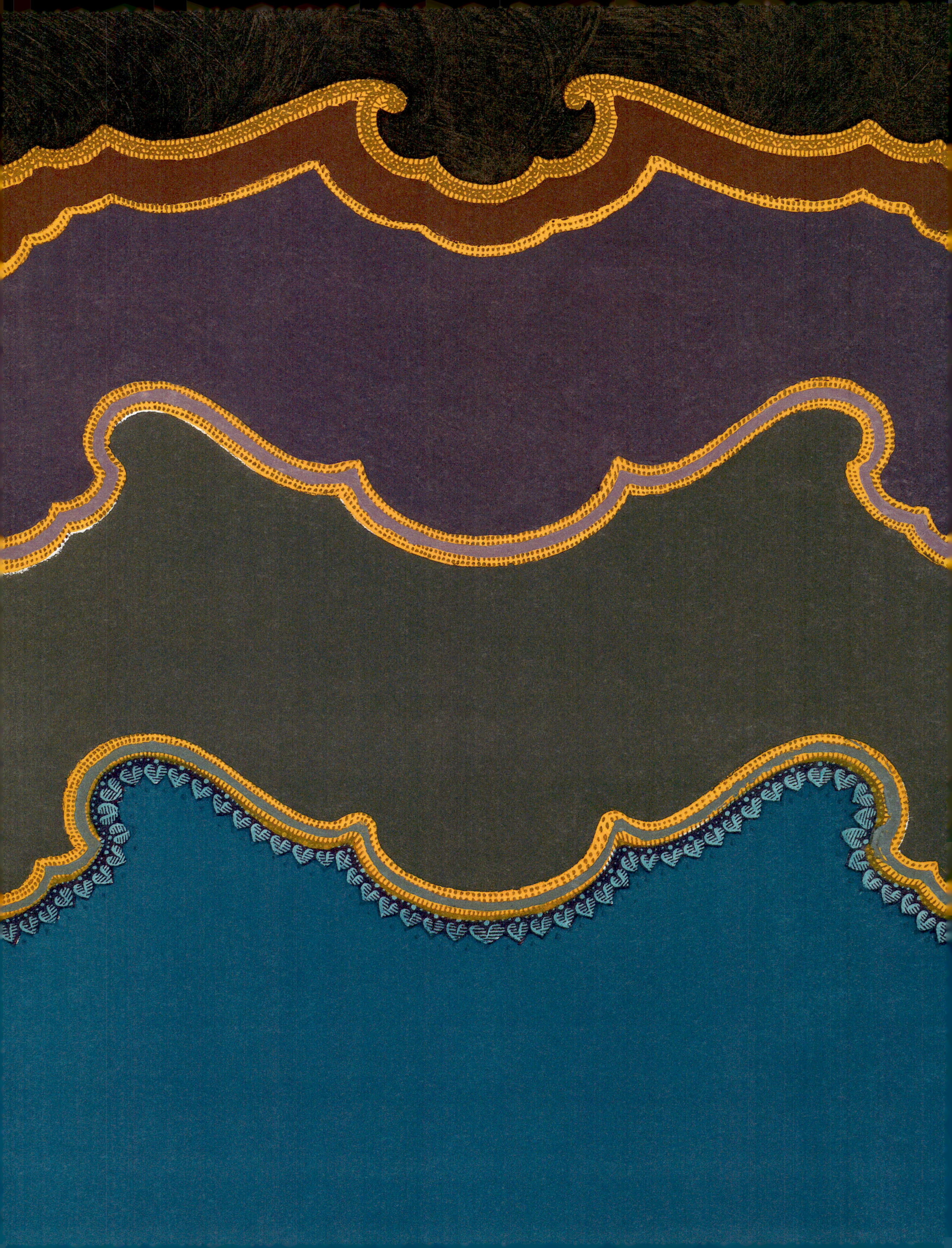

Chapter 4

Colours as Guiding Lights: NOMENCLATURES AND STANDARDS

1821–1950

Patrick Syme (1774–1845)

WERNER'S NOMENCLATURE OF COLOURS

13 illustrated pages, 24 x 15 cm | 9 ½ x 5 ⅞ in.,
Edinburgh, 1821
Fife, University of St Andrews Library,
Department of Special Collections

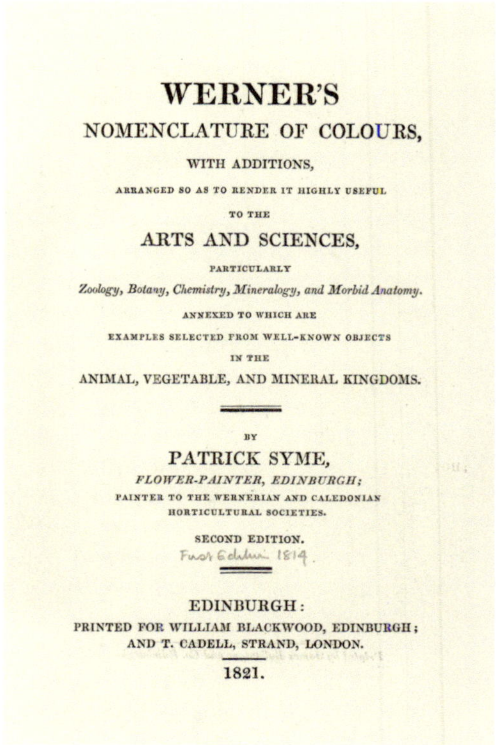
WERNER'S
NOMENCLATURE OF COLOURS,
WITH ADDITIONS,
ARRANGED SO AS TO RENDER IT HIGHLY USEFUL
TO THE
ARTS AND SCIENCES,
PARTICULARLY
Zoology, Botany, Chemistry, Mineralogy, and Morbid Anatomy.
ANNEXED TO WHICH ARE
EXAMPLES SELECTED FROM WELL-KNOWN OBJECTS
IN THE
ANIMAL, VEGETABLE, AND MINERAL KINGDOMS.

BY
PATRICK SYME,
FLOWER-PAINTER, EDINBURGH;
PAINTER TO THE WERNERIAN AND CALEDONIAN
HORTICULTURAL SOCIETIES.

SECOND EDITION.
First Edition 1814.

EDINBURGH:
PRINTED FOR WILLIAM BLACKWOOD, EDINBURGH;
AND T. CADELL, STRAND, LONDON.
1821.

A book about colour that reveals much about how 18th-century thinking, knowledge, research and print culture influenced developments in the early 19th century, *Werner's Nomenclature of Colours* is a work that manages to connect different disciplines, languages and cultures, as well as countries and people. Its own history is complex and fascinating, and benefits from the (true) story that the young Charles Darwin kept a copy of it in his coat pocket on his voyage to South America and Australia aboard HMS *Beagle* in the 1830s. Its small size may have been one reason why he took it along during his search for new species, but it had more to offer than mere portability.

The titular Werner was the German geologist Abraham Gottlob Werner, who in 1774 published a book on rocks and minerals, *Von den äusserlichen Kennzeichen der Foßilien* (On the external characteristics of mineral remains), which included a section about colour, albeit with no illustrations. Werner grouped his list of 54 individual colours into eight families (white, grey, black, blue, green, yellow, red and brown) and this list soon became an internationally influential standard for colour classification. Werner's book was translated into Hungarian, French and English before Patrick Syme, a flower painter at the Wernerian and Caledonian Horticultural Societies in Edinburgh, published his own English translation in 1814, with a second edition appearing in 1821.

Syme added two further colour groups (purple and orange) and doubled the list of individual colours to 108, adding another two in the translation's second edition. His aim was to provide a colour reference work that would be useful for "the Arts and Sciences, Particularly Zoology, Botany, Chemistry, Mineralogy, and Morbid Anatomy", and especially so in fieldwork. His most important addition, however, to Werner's original work was a set of plates depicting all the individual colours, with small hand-painted squares for each one. They are presented as charts, and Syme also noted equivalents for most of the colours using sources from the animal (including humans and insects), vegetable and mineral kingdoms. Emphasis was placed on the appearance and recognisability of each colour, to aid taxonomical identification. Within that, some of the names given to the colours are conspicuously visceral, such as *Veinous* and *Arterial Blood Red*, *Liver Brown* and *Skimmed Milk White*, the latter being found in the "white of the human eyeballs".

WHITES.

No.	Names.	Colours.	Animal.	Vegetable.	Mineral.
1	*Snow White.*		*Breast of the black headed Gull.*	*Snow-Drop.*	*Carara Marble and Calc Sinter.*
2	*Reddish White.*		*Egg of Grey Linnet.*	*Back of the Christmas Rose.*	*Porcelain Earth.*
3	*Purplish White.*		*Junction of the Neck and Back of the Kittiwake Gull.*	*White Geranium or Storks Bill.*	*Arragonite.*
4	*Yellowish White.*		*Egret.*	*Hawthorn Blossom.*	*Chalk and Tripoli.*
5	*Orange coloured White.*		*Breast of White or Screech Owl.*	*Large Wild Convolvulus.*	*French Porcelain Clay.*
6	*Greenish White.*		*Vent Coverts of Golden crested Wren.*	*Polyanthus Narcissus.*	*Calc Sinter.*
7	*Skimmed milk White.*		*White of the Human Eyeballs.*	*Back of the Petals of Blue Hepatica.*	*Common Opal.*
8	*Greyish White.*		*Inside Quill-feathers of the Kittiwake.*	*White Hamburgh Grapes.*	*Granular Limestone.*

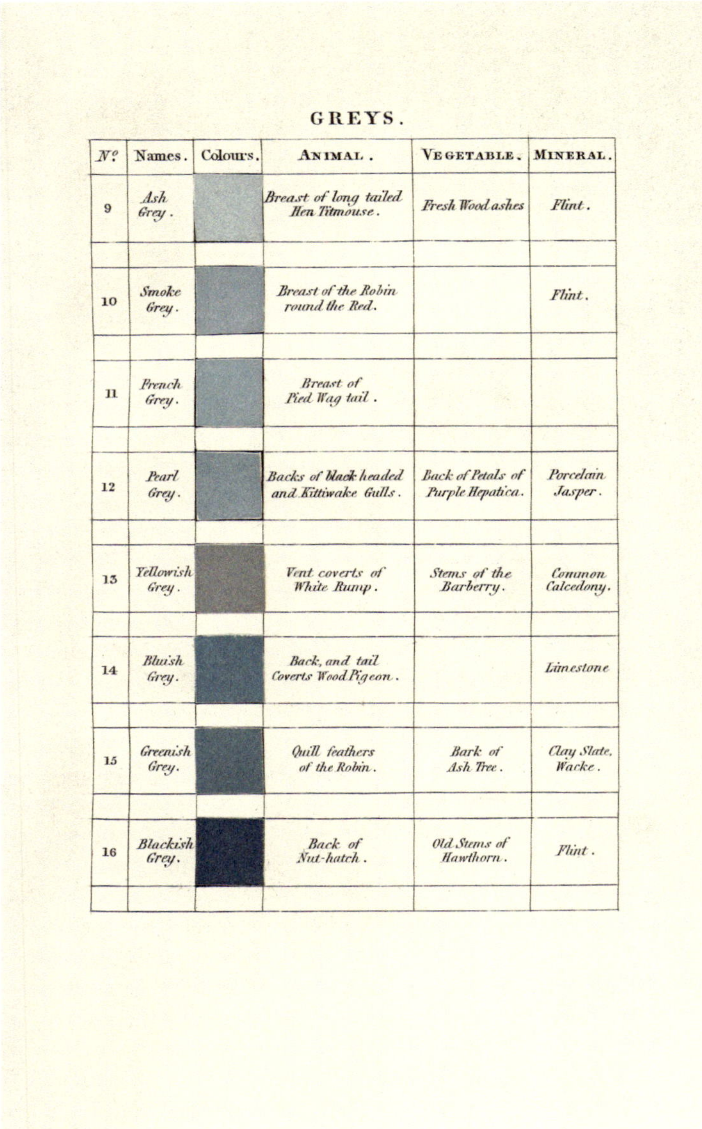

GREYS.

Nº	Names.	Colours.	Animal.	Vegetable.	Mineral.
9	Ash Grey.		Breast of long tailed Hen Titmouse.	Fresh Wood ashes	Flint.
10	Smoke Grey.		Breast of the Robin round the Red.		Flint.
11	French Grey.		Breast of Pied Wag tail.		
12	Pearl Grey.		Backs of black headed and Kittiwake Gulls.	Back of Petals of Purple Hepatica.	Porcelain Jasper.
13	Yellowish Grey.		Vent coverts of White Rump.	Stems of the Barberry.	Common Calcedony.
14	Bluish Grey.		Back, and tail Coverts Wood Pigeon.		Limestone
15	Greenish Grey.		Quill feathers of the Robin.	Bark of Ash Tree.	Clay Slate. Wacke.
16	Blackish Grey.		Back of Nut-hatch.	Old Stems of Hawthorn.	Flint.

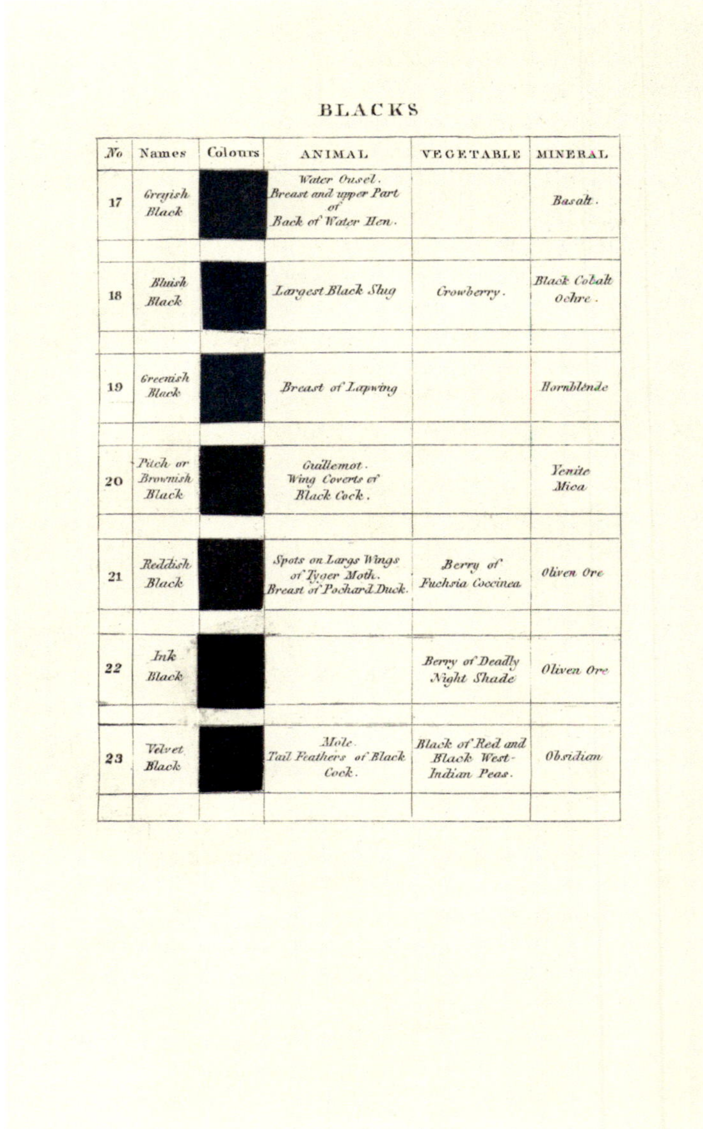

BLACKS

No	Names	Colours	ANIMAL	VEGETABLE	MINERAL
17	Greyish Black		Water Ousel. Breast and upper Part of Back of Water Hen.		Basalt.
18	Bluish Black		Largest Black Slug	Crowberry.	Black Cobalt Ochre.
19	Greenish Black		Breast of Lapwing		Hornblende
20	Pitch or Brownish Black		Guillemot. Wing Coverts of Black Cock.		Yenite Mica
21	Reddish Black		Spots on Largs Wings of Tyger Moth. Breast of Pochard Duck.	Berry of Fuchsia Coccinea	Oliven Ore
22	Ink Black			Berry of Deadly Night Shade	Oliven Ore
23	Velvet Black		Mole. Tail Feathers of Black Cock.	Black of Red and Black West-Indian Peas.	Obsidian

Werners Nomenklatur der Farben

Werner's Nomenclature of Colours ist ein Farbenbuch, das viel über den Einfluss verrät, den die Vorstellungen und das Wissen, die Forschung und das Druckwesen des 18. Jahrhunderts auf die Entwicklungen des frühen 19. Jahrhunderts ausübten. Es gelingt dem Buch, verschiedene Disziplinen, Sprachen und Kulturen, aber auch Länder und Menschen miteinander zu verknüpfen. Seine Entstehung ist komplex und faszinierend. Dahinter verbirgt sich die (wahre) Geschichte, dass ein Exemplar des Buches in der Manteltasche des jungen Charles Darwin steckte, als dieser in den 1830er-Jahren an Bord der HMS *Beagle* auf Expeditionsreise nach Südamerika und Australien ging. Unter anderem das kleine Format mag ihn bewogen haben, das Buch auf der Suche nach neuen Arten bei sich zu tragen, doch tatsächlich hatte es weit mehr zu bieten, als nur leicht transportierbar zu sein.

Der titelgebende Werner war der deutsche Geologe Abraham Gottlob Werner, der 1774 ein Buch über Steine und Minerale herausgab. *Von den äußerlichen Kennzeichen der Foßilien* enthielt auch einen Abschnitt über Farbe, allerdings ohne Abbildungen. Werner stellte eine Liste von 54 Einzelfarben auf, die er in acht Farbfamilien (Weiß, Grau, Schwarz, Blau, Grün, Gelb, Rot und Braun) zusammenfasste. Binnen kurzer Zeit entwickelte sich die Liste zur international einflussreichen Norm für Farbeinteilungen. Werners Buch wurde ins Ungarische, Französische und Englische übersetzt, bevor Patrick Syme, ein Blumenmaler der Wernerian Society und der Caledonian Horticultural Society in Edinburgh, 1814 seine eigene englische Übersetzung veröffentlichte, die 1821 in zweiter Auflage erschien.

Syme fügte Werners Liste zwei weitere Farbgruppen (Purpur und Orange) hinzu, verdoppelte die Liste der Einzelfarben auf 108 und erweiterte sie für die zweite Auflage auf 110. Ihm schwebte ein Farbhandbuch vor, das für „die Kunst und Wissenschaft, insbesondere die Zoologie, Botanik, Chemie, Mineralogie und pathologische Anatomie" nützlich wäre, vor allem in der Feldforschung.

BLUES

No.	Names	Colours	ANIMAL	VEGETABLE	MINERAL
24	Scotch Blue		Throat of Blue Titmouse.	Stamina of Single Purple Anemone.	Blue Copper Ore.
25	Prussian Blue.		Beauty Spot on Wing of Mallard Drake.	Stamina of Bluish Purple Anemone.	Blue Copper Ore
26	Indigo Blue				Blue Copper Ore.
27	China Blue		Rhynchites Nitens	Back Parts of Gentian Flower.	Blue Copper Ore from Chessy.
28	Azure Blue.		Breast of Emerald crested Manakin	Grape Hyacinth. Gentian.	Blue Copper Ore.
29	Ultra marine Blue.		Upper Side of the Wings of small blue Heath Butterfly.	Borrage.	Azure Stone or Lapis Lazuli.
30	Flax-flower Blue.		Light Parts of the Margin of the Wings of Devil's Butterfly.	Flax flower.	Blue Copper Ore
31	Berlin Blue.		Wing Feathers of Jay.	Hepatica.	Blue Sapphire.
32	Verditter Blue				Lenticular Ore.
33	Greenish Blue			Great Fennel Flower.	Turquois. Flour Spar.
34	Greyish Blue.		Back of blue Titmouse	Small Fennel Flower.	Iron Earth.

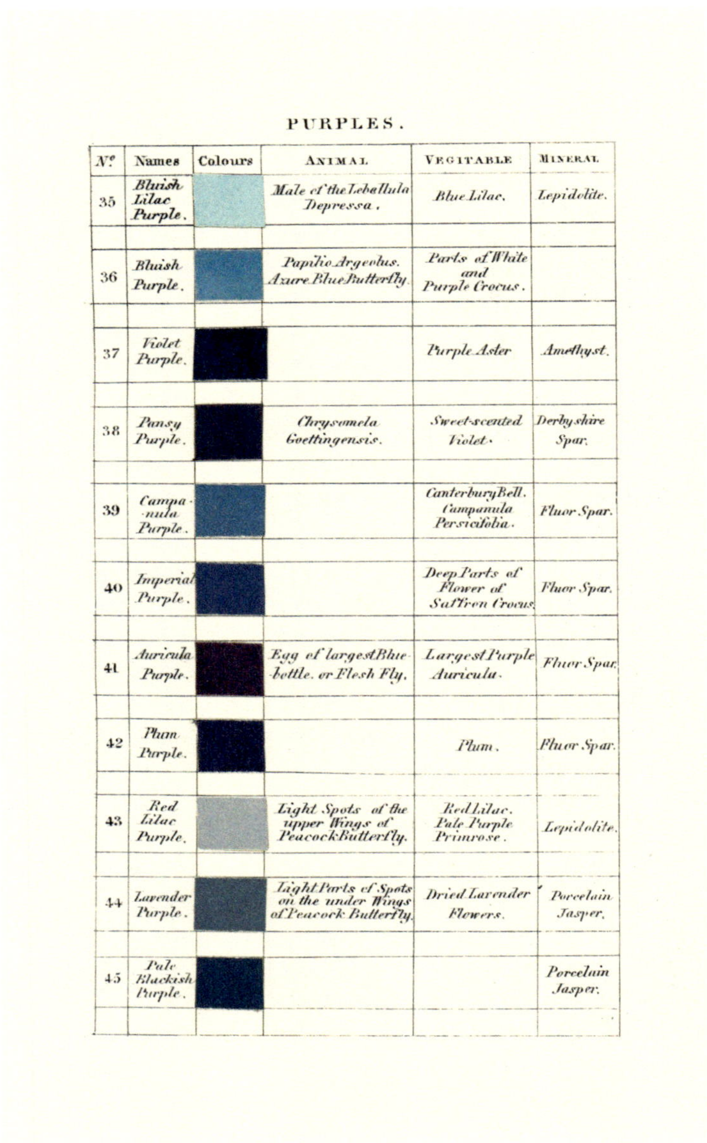

PURPLES.

Nº	Names	Colours	Animal	Vegitable	Mineral
35	Bluish Lilac Purple.		Male of the Lebellula Depressa.	Blue Lilac.	Lepidolite.
36	Bluish Purple.		Papilio Argeolus. Azure Blue Butterfly.	Parts of White and Purple Crocus.	
37	Violet Purple.			Purple Aster	Amethyst.
38	Pansy Purple.		Chrysomela Goettingensis.	Sweet-scented Violet.	Derbyshire Spar.
39	Campa-nula Purple.			Canterbury Bell. Campanula Persicifolia.	Fluor Spar.
40	Imperial Purple.			Deep Parts of Flower of Saffron Crocus.	Fluor Spar.
41	Auricula Purple.		Egg of largest Blue-bottle. or Flesh Fly.	Largest Purple Auricula.	Fluor Spar.
42	Plum Purple.			Plum.	Fluor Spar.
43	Red Lilac Purple.		Light Spots of the upper Wings of Peacock Butterfly.	Red Lilac. Pale Purple Primrose.	Lepidolite.
44	Lavender Purple.		Light Parts of Spots on the under Wings of Peacock Butterfly.	Dried Lavender Flowers.	Porcelain Jasper.
45	Pale Blackish Purple.				Porcelain Jasper.

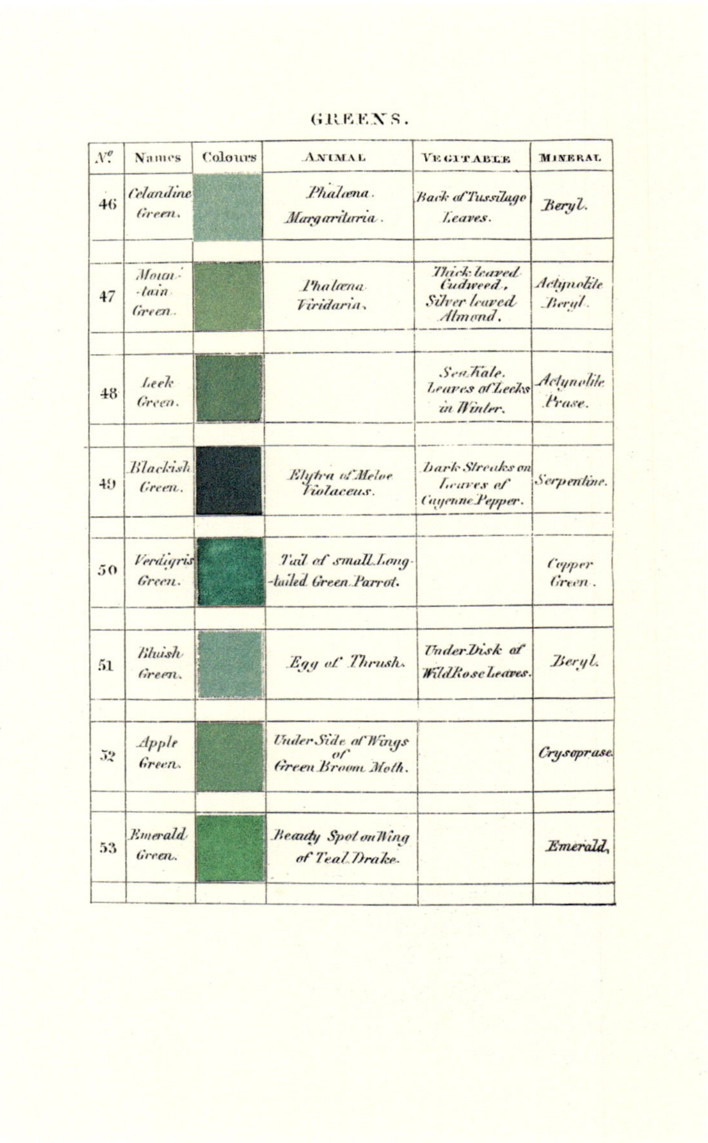

GREENS.

Nº	Names	Colours	Animal	Vegitable	Mineral
46	Celandine Green.		Phalæna. Margaritaria.	Back of Tussilago Leaves.	Beryl.
47	Moun-tain Green.		Phalæna Viridaria.	Thick leaved Cudweed. Silver leaved Almond.	Actynolite Beryl.
48	Leek Green.			Sea Kale. Leaves of Leeks in Winter.	Actynolite Prase.
49	Blackish Green.		Elytra of Meloe Violaceus.	Dark Streaks on Leaves of Cayenne Pepper.	Serpentine.
50	Verdigris Green.		Tail of small Long-tailed Green Parrot.		Copper Green.
51	Bluish Green.		Egg of Thrush.	Under Disk of Wild Rose Leaves.	Beryl.
52	Apple Green.		Under Side of Wings of Green Broom Moth.		Crysoprase.
53	Emerald Green.		Beauty Spot on Wing of Teal Drake.		Emerald.

Seine wichtigste Beigabe zu Werners Werk waren jedoch die Tafeln, auf denen sämtliche Einzelfarben in kleinen handbemalten Quadraten abgebildet waren. Syme präsentierte sie in Tabellenform und notierte für die meisten Farben Entsprechungen aus dem Reich der Tiere (einschließlich Menschen und Insekten), der Pflanzen und der Minerale. Betont wurden Erscheinung und Erkennbarkeit einer Farbe, um taxonomische Bestimmungen zu erleichtern. Einige Farbbezeichnungen sind auffallend viszeral, etwa „Venen“ und „Arterienblutrot“, „Leberbraun“ oder „Magermilchweiß“, von denen Letzteres im „Weiß des menschlichen Augapfels“ zu finden war.

Nomenclature des couleurs de Werner

Très révélatrice de la manière dont la pensée, les connaissances, les recherches et l'imprimerie du XVIIIe siècle ont influencé le début du siècle suivant, la « Nomenclature des couleurs de Werner » s'intéresse à la couleur en parvenant à couvrir toute une variété de disciplines, de langues, de cultures, de pays et de personnalités. Complexe et passionnante, l'histoire de cet ouvrage bénéficie en outre de l'affirmation (vraie) selon laquelle le jeune Charles Darwin en conservait un exemplaire dans la poche de son manteau, au cours du voyage qu'il effectua en Amérique du Sud et en Australie à bord du HMS *Beagle* dans les années 1830. C'est peut-être en raison de son petit format que le scientifique anglais l'emporta dans sa quête d'espèces nouvelles. Mais ce livre a bien d'autres atouts.

Le Werner du titre est le géologue allemand Abraham Gottlob Werner qui, en 1774, publia un ouvrage portant sur les roches et les minéraux, *Von den äußerlichen Kennzeichen der Foßilien (Traité des caractères extérieurs des fossiles)*, dont une partie est consacrée à la couleur, bien qu'elle soit dépourvue d'illustrations. Werner classe 54 teintes en huit familles de couleurs (blanc, gris, noir, bleu, vert, jaune, rouge et brun), liste qui deviendra vite une référence internationale majeure pour les catégorisations chromatiques. L'ouvrage de Werner fut traduit en hongrois, français et anglais, avant que Patrick Syme,

GREENS.

No.	Names	Colours	ANIMAL	VEGETABLE	MINERAL
54	Grass Green		Scarabæus Nobilis.	General Appearance of Grass Fields. Sweet Sugar Pear	Uran Mica.
55	Duck Green		Neck of Mallard	Upper Disk of Yew Leaves.	Ceylanite
56	Sap Green.		Under Side of lower Wings of Orange tip Butterfly.	Upper Disk of Leaves of woody Night Shade.	
57	Pistachio Green.		Neck of Eider Drake	Ripe Pound Pear. Hypnum like Saxifrage.	Crysolite.
58	Asparagus Green.		Brimstone Butterfly.	Variegated Horse-Shoe Geranium.	Beryl.
59	Olive Green.			Foliage of Lignum vitæ.	Epidote. Olvene Ore.
60	Oil Green.		Animal and Shell of common Water Snail.	Nonpareil Apple from the Wall.	Beryl
61	Siskin Green.		Siskin.	Ripe Coalmar Pear. Irish Pitcher Apple.	Uran Mica.

YELLOWS.

No.	Names	Colours.	ANIMAL	VEGETABLE	MINERAL
62	Sulphur Yellow.		Yellow Parts of large Dragon Fly.	Various Coloured Snap dragon.	Sulphur
63	Primrose Yellow.		Pale Canary Bird.	Wild Primrose	Pale coloured Sulphur.
64	Wax Yellow.		Larva of large Water Beetle.	Greenish Parts of Nonpareil Apple.	Semi Opal.
65	Lemon Yellow.		Large Wasp or Hornet.	Shrubby Goldylocks.	Yellow Orpiment.
66	Gamboge Yellow.		Wings of Goldfinch. Canary Bird.	Yellow Jasmine.	High coloured Sulphur.
67	Kings Yellow.		Head of Golden Pheasant.	Yellow Tulip. Cinque foil.	
68	Saffron Yellow.		Tail Coverts of Golden Pheasant.	Anthers of Saffron Crocus.	

YELLOWS.

Nº	Names	Colours	ANIMAL.	VEGITABLE.	MINERAL.
69	Gallstone Yellow.		Gallstones.	Marigold Apple.	
70	Honey Yellow.		Lower Parts of Neck of Bird of Paradise.		Fluor Spar.
71	Straw Yellow.		Polar Bear.	Oat Straw.	Schorlite. Calamine.
72	Wine Yellow.		Body of Silk Moth.	White Currants.	Saxon Topax.
73	Sienna Yellow.		Vent Parts of Tail of Bird of Paradise.	Stamina of Honey-suckle.	Pale Braxilian Topax.
74	Ochre Yellow.		Vent Coverts of Red Start.		Porcelain Jasper.
75	Cream Yellow.		Breast of Teal Drake		Porcelain Jasper.

ORANGE.

Nº	Names	Colours.	ANIMAL.	VEGITABLE.	MINERAL.
76	Dutch Orange.		Crest of Golden crested Wren.	Common Marigold. Seedpod of Spindle-tree.	Streak of Red Orpiment.
77	Buff Orange.		Streak from the Eye of the King Fisher.	Stamina of the large White Cistus.	Natrolite.
78	Orpiment Orange.		The Neck Ruff of the Golden Pheasant. Belly of the Warty Newt.	Indian Cress.	
79	Brownish Orange.		Eyes of the largest Flesh Fly.	Style of the Orange Lily	Dark Brazilian Topax.
80	Reddish Orange.		Lower Wings of Tyger Moth.	Hemimeris. Buff Hibiscus.	
81	Deep Reddish Orange.		Gold Fish lustre abstracted.	Scarlet Leadington Apple.	

peintre de fleurs aux Sociétés horticoles werneriennes et calédoniennes à Édimbourg, ne publie sa propre traduction anglaise en 1814, suivie d'une seconde édition en 1821.

À la liste de Werner, Syme ajoute deux groupes (violet et orange) et double le nombre de teintes (108), la seconde édition en comptant deux supplémentaires. Il souhaite proposer un ouvrage de référence sur la couleur au profit « des arts et des sciences, en particulier de la zoologie, de la botanique, de la chimie, de la minéralogie et de l'anatomie morbide », et singulièrement pour le travail de terrain. Mais l'apport décisif de Syme à l'œuvre de Werner consiste en un ensemble de planches représentant toutes les teintes sous la forme de petits carrés peints, réunis dans des tableaux. Syme répertorie également des équivalents de la plupart des teintes, empruntées aux règnes animal (y compris l'être humain et les insectes), végétal et minéral. Il met l'accent sur les caractères extérieurs et reconnaissables de chaque couleur, afin d'effectuer une identification taxonomique. Au sein de cette classification, certaines appellations attribuées aux teintes sont notoirement anatomiques comme, par exemple, « rouge veineux », « rouge artériel », « brun foie » et « blanc du lait écrémé », teinte que l'on trouve dans le « blanc du globe oculaire humain ».

Nomenclatura de los colores de Werner

Werner's Nomenclature of Colours es una obra que abarca diferentes disciplinas, idiomas y culturas, así como países y personas. Es también un libro que revela mucho sobre cómo el pensamiento, el conocimiento, la investigación y la cultura impresa del siglo XVIII influyeron en los desarrollos de principios del siglo XIX. Su propia historia es compleja y fascinante, y se beneficia de una historia (verdadera): el joven Charles Darwin llevaba una copia en el bolsillo de su abrigo en su viaje a Sudamérica y Australia a bordo del HMS *Beagle* en la década de 1830. Su pequeño tamaño puede haber sido una de las razones por las que lo llevó consigo durante su búsqueda de nuevas especies, pero tenía más que ofrecer que la mera portabilidad.

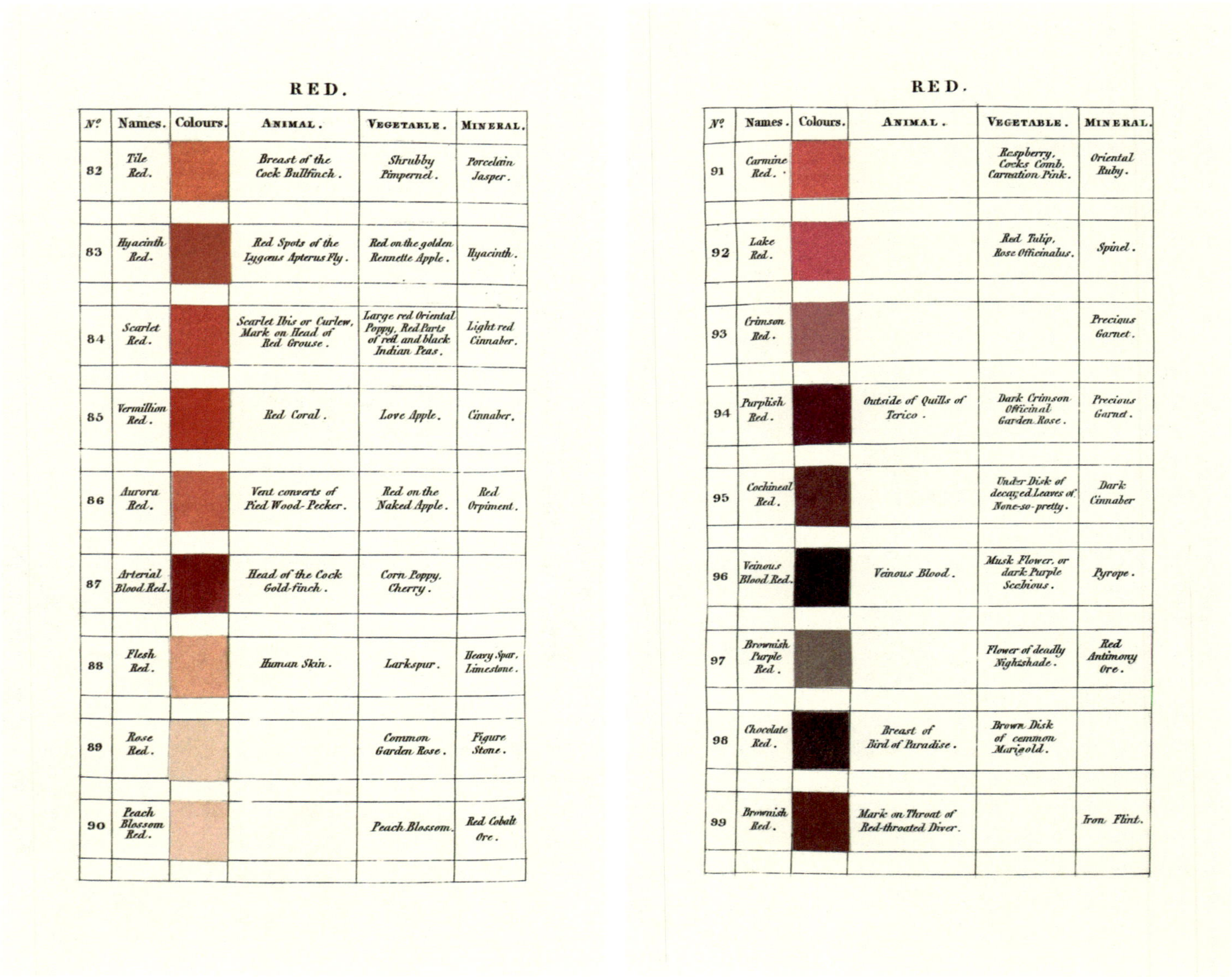

RED.

Nº	Names.	Colours.	Animal.	Vegetable.	Mineral.
82	Tile Red.		Breast of the Cock Bullfinch.	Shrubby Pimpernel.	Porcelain Jasper.
83	Hyacinth Red.		Red Spots of the Lygœus Apterus Fly.	Red on the golden Rennette Apple.	Hyacinth.
84	Scarlet Red.		Scarlet Ibis or Curlew, Mark on Head of Red Grouse.	Large red Oriental Poppy, Red Parts of red and black Indian Peas.	Light red Cinnaber.
85	Vermillion Red.		Red Coral.	Love Apple.	Cinnaber.
86	Aurora Red.		Vent converts of Pied Wood-Pecker.	Red on the Naked Apple.	Red Orpiment.
87	Arterial Blood Red.		Head of the Cock Gold-finch.	Corn Poppy. Cherry.	
88	Flesh Red.		Human Skin.	Larkspur.	Heavy Spar. Limestone.
89	Rose Red.			Common Garden Rose.	Figure Stone.
90	Peach Blossom Red.			Peach Blossom.	Red Cobalt Ore.

RED.

Nº	Names.	Colours.	Animal.	Vegetable.	Mineral.
91	Carmine Red.			Raspberry, Cocks Comb. Carnation Pink.	Oriental Ruby.
92	Lake Red.			Red Tulip, Rose Officinalus.	Spinel.
93	Crimson Red.				Precious Garnet.
94	Purplish Red.		Outside of Quills of Terico.	Dark Crimson Officinal Garden Rose.	Precious Garnet.
95	Cochineal Red.			Under Disk of decayed Leaves of None-so-pretty.	Dark Cinnaber
96	Veinous Blood Red.		Veinous Blood.	Musk Flower, or dark Purple Scabious.	Pyrope.
97	Brownish Purple Red.			Flower of deadly Nightshade.	Red Antimony Ore.
98	Chocolate Red.		Breast of Bird of Paradise.	Brown Disk of common Marigold.	
99	Brownish Red.		Mark on Throat of Red-throated Diver.		Iron Flint.

El Werner del título fue el geólogo alemán Abraham Gottlob Werner, quien en 1774 publicó un libro sobre rocas y minerales, *Von den äusserlichen Kennzeichen der Foßilien* (De las características externas de los fósiles), que incluía una sección sobre el color, aunque sin ilustraciones. Werner agrupó su lista de 54 colores individuales en ocho familias (blanco, gris, negro, azul, verde, amarillo, rojo y marrón), una lista que pronto se convirtió en un estándar para la clasificación de colores de influencia internacional. El libro de Werner se tradujo al húngaro, francés e inglés antes de que Patrick Syme, un pintor de flores de las Wernerian and Caledonian Horticultural Societies de Edimburgo, publicara en 1814 su propia traducción al inglés, que en 1821 alcanzó la segunda edición.

Syme añadió dos grupos de colores (morado y naranja) y duplicó la lista de colores individuales a 108, a los que sumó otros dos en la segunda edición de la traducción. Su objetivo era proporcionar una obra de referencia sobre el color útil para «las Artes y las Ciencias, la Zoología en particular, la Botánica, la Química, la Mineralogía y la Anatomía Mórbida», en especial en el trabajo de campo. Sin embargo, su añadido más importante al trabajo original de Werner fue un conjunto de láminas con tablas que presentan todos los colores individuales en pequeños cuadrados pintados a mano. Syme también anotó equivalentes de la mayoría de los colores, para lo que utilizó fuentes de los reinos animal (incluidos humanos e insectos), vegetal y mineral. Para ayudar a la identificación taxonómica, hizo hincapié en la apariencia y el reconocimiento de cada color. Dentro de ese marco, algunos de los nombres dados a los colores son llamativamente viscerales, como el *rojo sangre venosa* y *arterial*, el *marrón hígado* y el *blanco leche desnatada*; este último se encuentra en el «blanco de los globos oculares humanos».

BROWNS.

No.	Names.	Colours.	Animal.	Vegetable.	Mineral.
100	Deep Orange-coloured Brown.		Head of Pochard. Wing coverts of Sheldrake.	Female Spike of Catstail Reed.	
101	Deep Reddish Brown.		Breast of Pochard, and Neck of Teal Drake.	Dead Leaves of green Panic Grass.	Brown Blende.
102	Umber Brown.		Moor Buzzard.	Disk of Rubeckia.	
103	Chesnut Brown.		Neck and Breast of Red Grouse.	Chesnuts.	Egyptian Jasper.
104	Yellowish Brown.		Light Brown Spots on Guinea-Pig, Breast of Hoopoe.		Iron Flint, and common Jasper.
105	Wood Brown.		Common Weasel. Light parts of Feathers on the Back of the Snipe.	Hazel Nuts.	Mountain Wood.
106	Liver Brown.		Middle Parts of Feathers of Hen Pheasant, and Wing coverts of Grosbeak.		Semi Opal.
107	Hair Brown.		Head of Pintail Duck		Wood Tin.
108	Broccoli Brown.		Head of Black headed Gull.		Zircon.
109	Clove Brown.		Head and Neck of Male Kestril.	Stems of Black Currant Bush.	Axinite, Rock Cristal.
110	Blackish Brown.		Stormy Petril. Wing Coverts of black Cock. Forehead of Foumart.		Mineral Pitch.

David Ramsay Hay (1798–1866)

THE PRINCIPLES OF BEAUTY IN COLOURING SYSTEMATIZED

12 plates, 20 x 14 cm | 7 7/8 x 5 1/2 in., Edinburgh & London, 1845
University of North Carolina at Chapel Hill,
Louis Round Wilson Special Collections Library

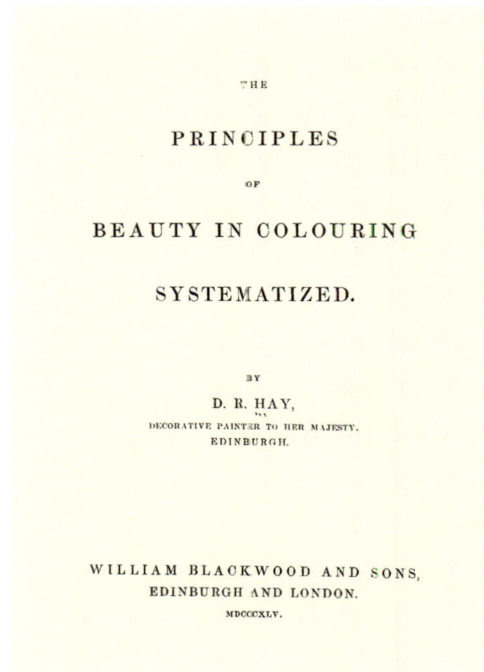
THE

PRINCIPLES

OF

BEAUTY IN COLOURING

SYSTEMATIZED.

BY

D. R. HAY,

DECORATIVE PAINTER TO HER MAJESTY,

EDINBURGH.

WILLIAM BLACKWOOD AND SONS,

EDINBURGH AND LONDON.

MDCCCXLV.

David Hay was a Scottish artist and interior decorator, and one of the first people to apply colour theory to interior design. In 1828, when he was setting up his own business in Edinburgh (the house-painting firm Nicholson and Hay), he published *The Laws of Harmonious Colouring Adapted to House Painting* in which he laid out his ideas on colour and aesthetics, with specific reference to the analogy between colour and music. It went into six expanded editions and marked the beginning of a long writing career, and by the time Hay published the first edition of *The Principles of Beauty in Colouring Systematized* he had received a royal warrant from Queen Victoria and other high-profile commissions.

Hay's books are fundamental works in the field of architectural colour because of the fact that they present a fully developed sense not only of the application of colour theory to interior design but also the importance of colour perception. Hay noted that while the beauty of colour generally depends on contrast, this does not mean that the arrangements with the greatest contrasts are necessarily the most desirable. The real skill, he argued, lies in combining primary, secondary and tertiary colours in such a way that they result in the most harmonious composition, or "melody". He was also acutely aware of the relationship between colour and light, not just in a Newtonian sense but in the way surfaces and spaces are illuminated, for example, how light extends into a room at different times of the day.

Illustrations formed a vital part of Hay's many publications on colour, and he also stressed the need for maximum clarity and simplicity, presenting his most important images as both labelled tables and engravings within the running text, and then again as coloured plates. With very few exceptions he used geometric shapes to illustrate his "laws" and "principles", rather than depicting images of architectural interiors or decorative objects. His *Principles of Beauty* contains 14 plates of diagrams showing combinations of primaries, secondaries and tertiaries, each taking the shape of a colour wheel created from six numbered trapezoid parts (that is triangles with one point cut off to allow for a letter and connecting lines to be added in the engraving), apart from one example (plate 8) in which the colours are arranged in a column of triangles. As in Syme's *Werner's Nomenclature* (see pp. 310–319) the colour shapes were cut from hand-painted sheets of paper and pasted on to the plates, a method Hay continued to use for all his books on colour despite the fact that lithographic illustrations had become more usual by the 1830s.

EXAMPLE I.

THE PRIMARY AND SECONDARY COLOURS.

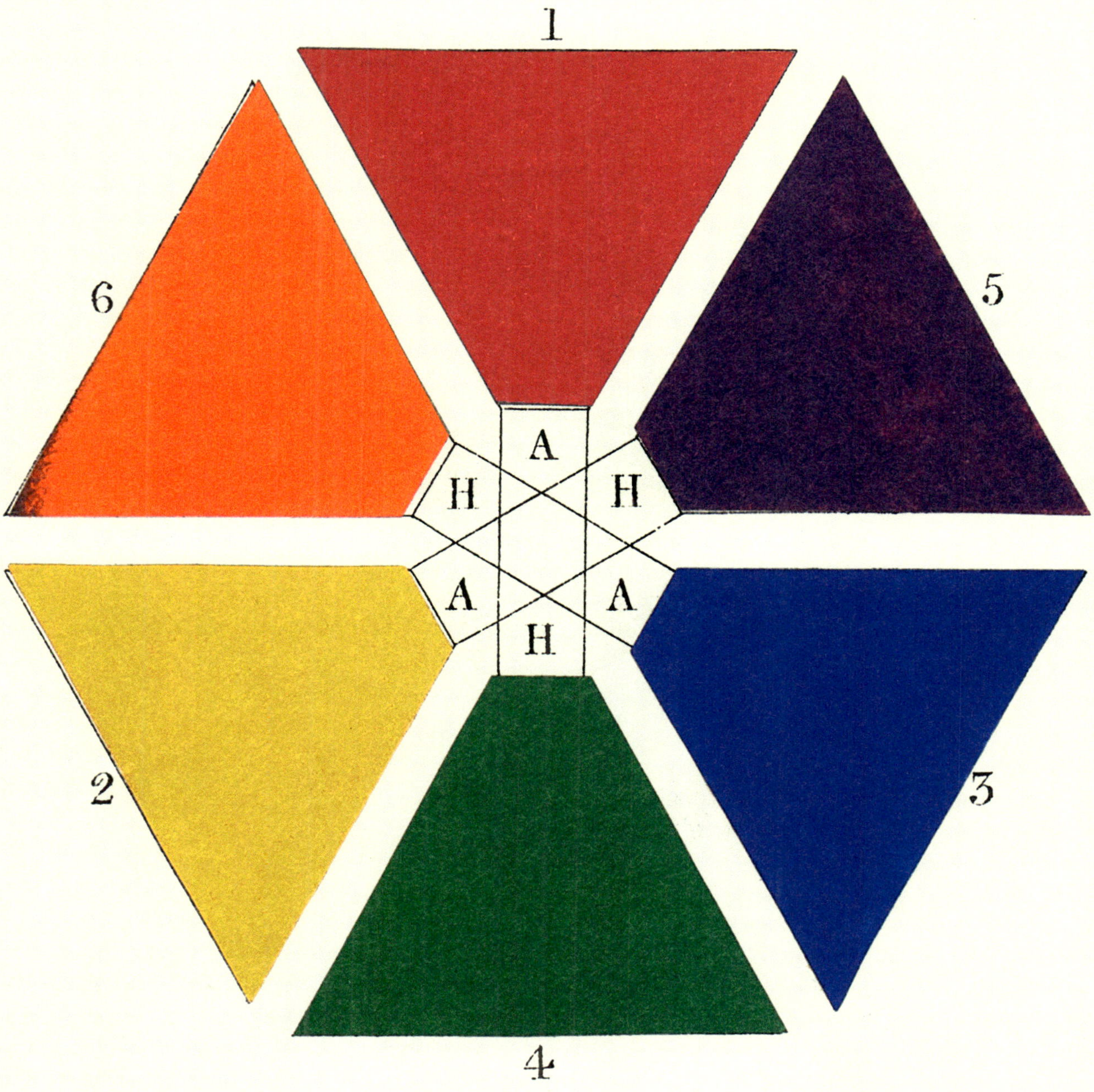

D.R.Hay Invt.

Lizars Sc.

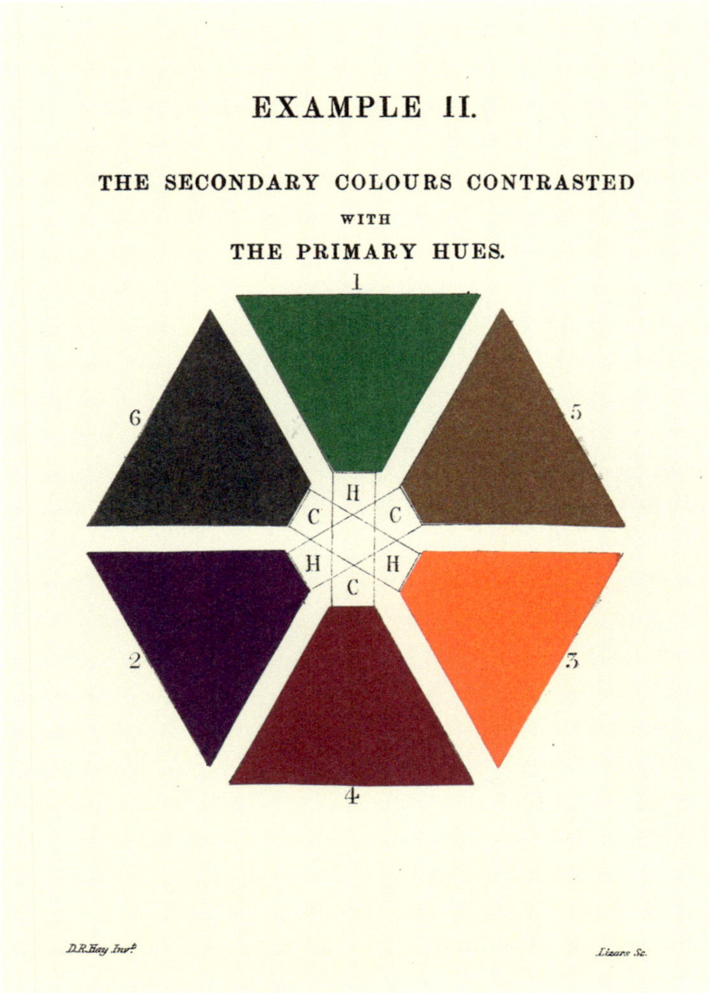

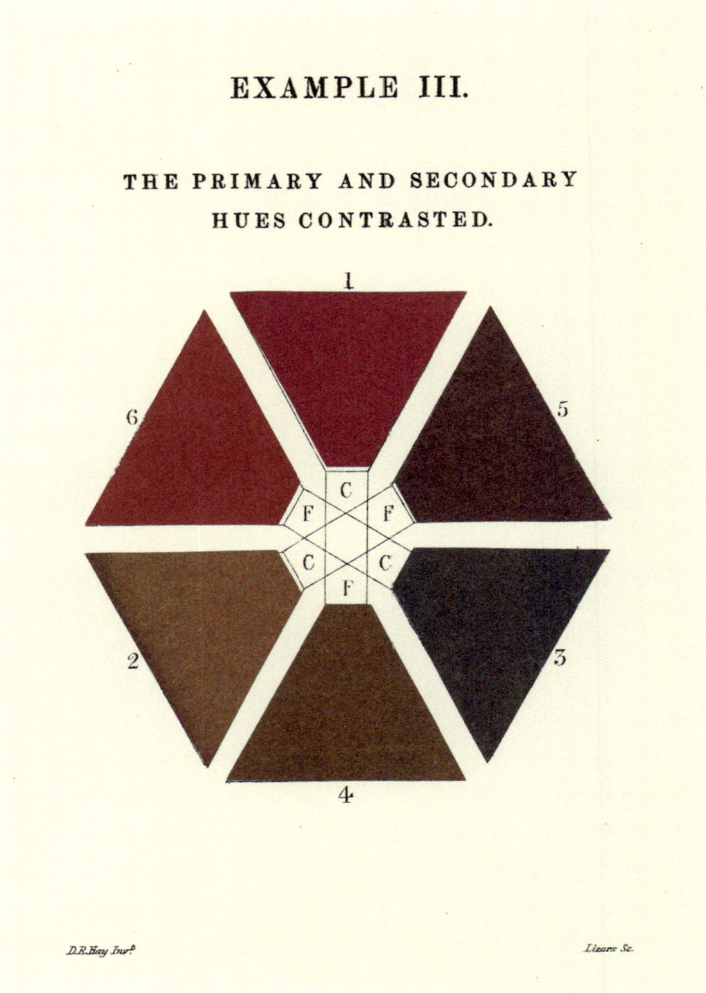

Systematisierte Darstellung der Prinzipien der Schönheit in der Farbgebung

Der schottische Künstler und Innenarchitekt David Hay war einer der Ersten, der die Farbtheorie auf Fragen der Innenausstattung anwandte. Als er 1828 in Edinburgh sein eigenes Unternehmen gründete (den Malerbetrieb Nicholson und Hay), veröffentlichte er *The Laws of Harmonious Colouring Adapted to House Painting*. Darin erläuterte er seine Auffassungen von Farbe und Ästhetik und ging speziell auf die Analogie von Farbe und Musik ein. Das Buch erlebte sechs erweiterte Auflagen und markierte den Beginn einer langen Autorenkarriere. Als schließlich die erste Auflage von *The Principles of Beauty in Colouring Systematized* erschien, hatte Hay bereits für Königin Victoria und andere hochrangige Auftraggeber gearbeitet.

Hays Bücher sind Grundlagenwerke im Bereich der Architekturfarben, weil sie von einem ausgeprägten Sinn nicht nur für die Anwendung der Farbtheorie auf die Innenarchitektur, sondern auch für die Bedeutung der Farbwahrnehmung durchdrungen sind. Hay stellte fest, dass die Schönheit von Farben zwar grundsätzlich vom Kontrast abhänge, dies aber nicht heiße, dass die Arrangements mit den stärksten Kontrasten zwangsläufig auch die erstrebenswertesten seien. Die wahre Kunst bestehe vielmehr darin, Primär-, Sekundär- und Tertiärfarben so zu kombinieren, dass sie zur harmonischsten Komposition oder „Melodie" führten. Auch der Beziehung zwischen Farbe und Licht war sich Hay aufs Äußerste bewusst, nicht nur im Newton'schen Sinne, sondern zum Beispiel auch im Hinblick auf die Art und Weise, wie Oberflächen und Räume beleuchtet werden oder wie sich Licht zu verschiedenen Tageszeiten im Raum verteilt.

Illustrationen bildeten einen wesentlichen Bestandteil der vielen Publikationen Hays über Farbe. Dabei betonte er die Notwendigkeit von maximaler Klarheit und Einfachheit und präsentierte seine wichtigsten Abbildungen zunächst als beschriftete Tafeln und Stiche im Fließtext und ein zweites Mal in Form von Farbtafeln. Mit sehr wenigen Ausnahmen verwendete er zur Veranschaulichung seiner „Gesetze" und „Prinzipien" geometrische Formen, anstatt Bilder von Innenräumen oder Dekorationsgegenständen zu zeigen.

EXAMPLE IV.

BINARY MODIFICATIONS OF THE SECONDARY COLOURS.

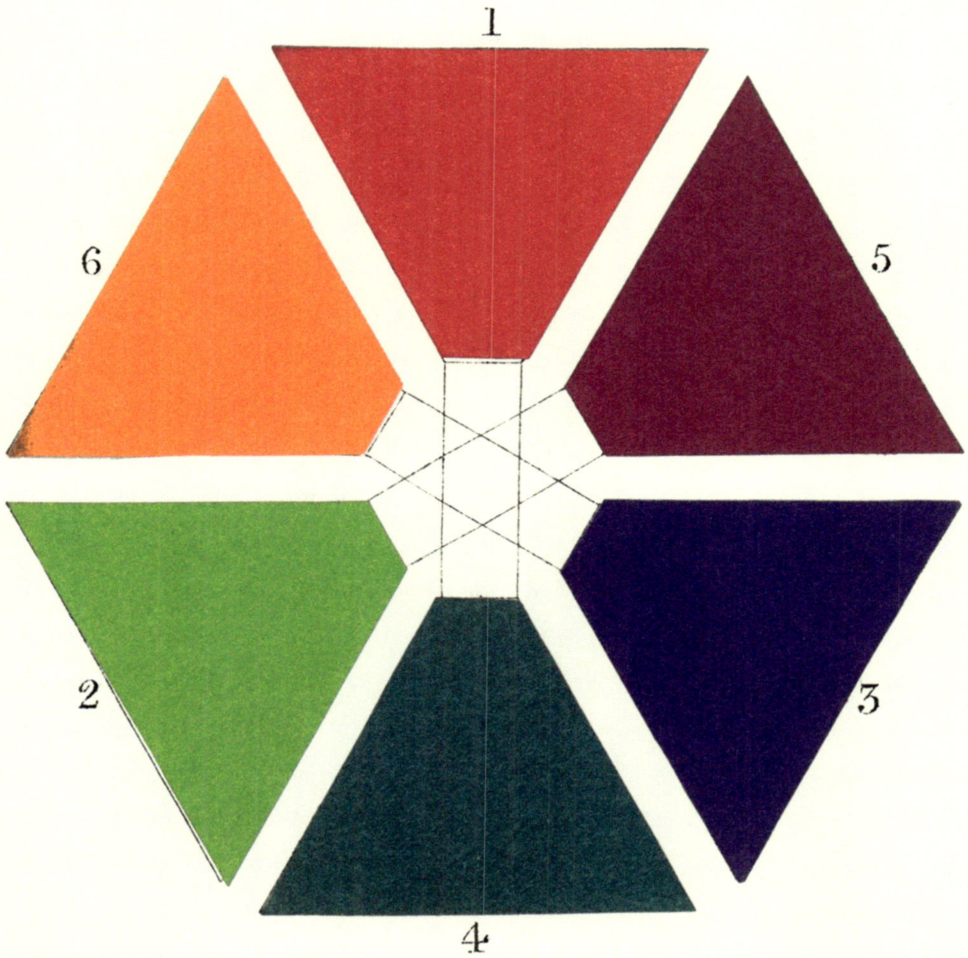

D.R.Hay Invt.

EXAMPLE V.

THE PRIMARY AND SECONDARY COLOURS TEMPERED.

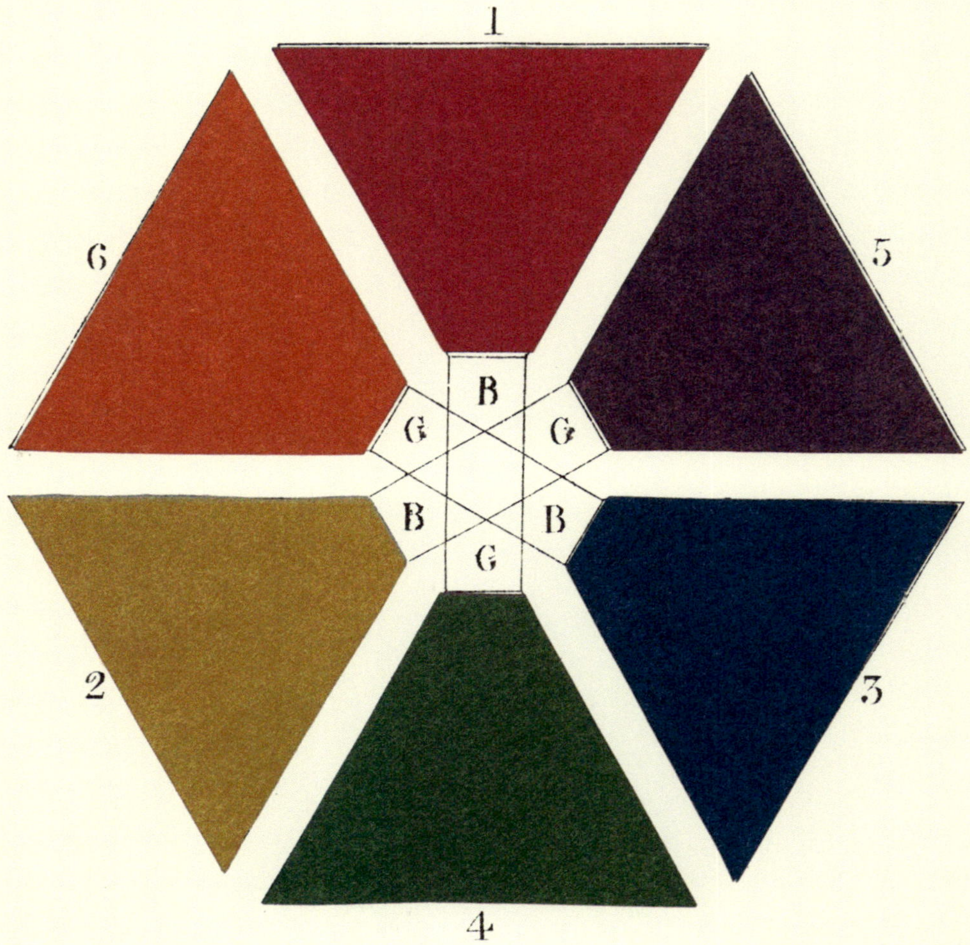

D.R.Hay Invt. *Lizars Sc.*

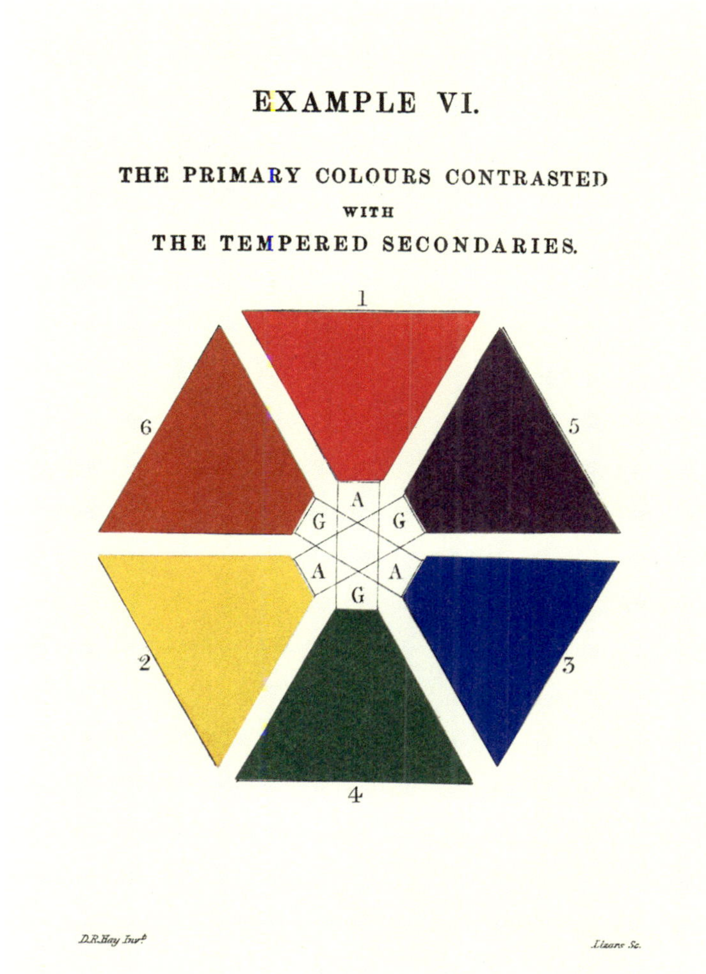

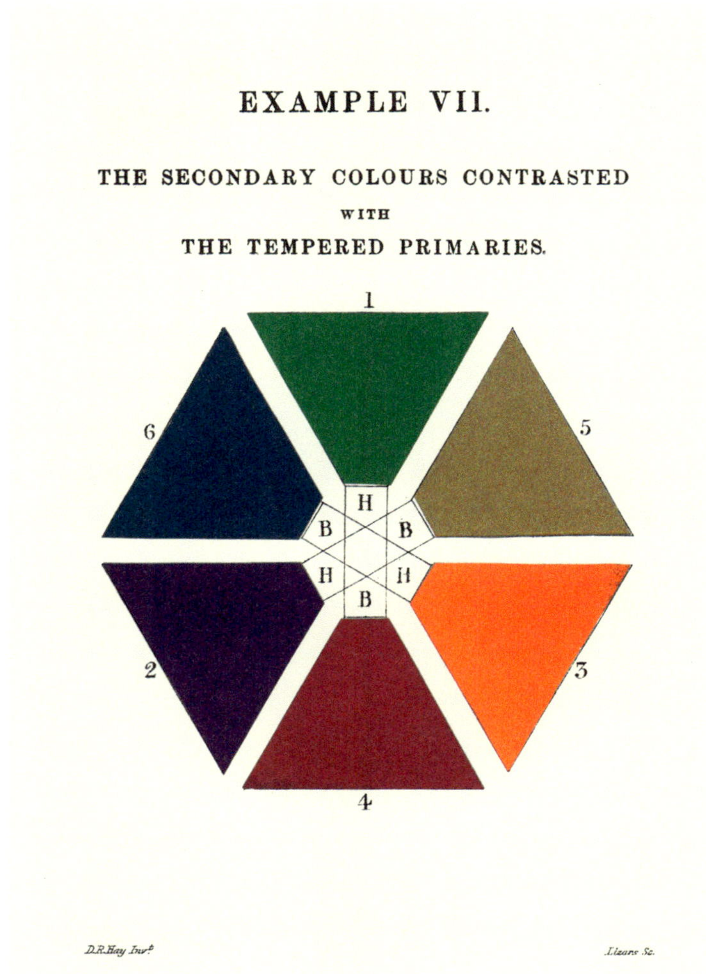

The Principles of Beauty enthält 14 Diagrammtafeln mit Kombinationen von primären, sekundären und tertiären Farben, die sich zu Farbkreisen aus jeweils sechs nummerierten Trapezformen zusammenfinden (also Dreiecken, denen eine Spitze abgeschnitten wurde, um an ihrer Stelle einen Buchstaben und Verbindungslinien in die Platte zu stechen). Nur in einem Fall (Tafel 8) sind die Farben zu einer Säule aus Dreiecken arrangiert. Wie in *Werner's Nomenclature* von Patrick Syme (siehe S. 310–319) wurden die Farbformen aus handbemalten Papierbögen ausgeschnitten und auf die Tafeln geklebt. Hay blieb dieser Methode in allen seinen Büchern treu, obwohl lithografische Tafeln schon in den 1830er-Jahren gebräuchlicher geworden waren.

Principes systématisés de la beauté appliqués à la coloration

Peintre et décorateur d'intérieur écossais, David Hay est l'un des premiers à appliquer une théorie des couleurs à l'architecture d'intérieur. En 1828, l'année où il crée à Édimbourg son entreprise de peinture d'intérieur (Nicholson & Hay), il publie *The Laws of Harmonious Colouring Adapted to House Painting* (« Les lois de l'harmonie des couleurs adaptées à la peinture d'intérieur »), où il expose ses idées sur la couleur et l'esthétique, en faisant particulièrement allusion à l'analogie entre couleur et musique. L'ouvrage connaît six éditions augmentées et marque pour Hay le début d'une longue carrière d'auteur. Lorsqu'il fait paraître ses « Principes systématisés de la beauté appliqués à la coloration », il bénéficie déjà du statut de fournisseur officiel de la reine Victoria et a pour clients d'autres commanditaires prestigieux.

Si les œuvres de Hay sont fondamentales dans l'usage de la couleur en architecture, c'est parce qu'elles témoignent d'une maîtrise totale de l'application des théories chromatiques à la décoration d'intérieur, mais aussi de l'importance du rôle que joue la perception des couleurs. Soulignant que la beauté des couleurs est généralement tributaire du contraste, Hay précise toutefois qu'il n'est pas nécessairement souhaitable de disposer celles-ci dans le but d'obtenir les contrastes les plus frappants. À ses yeux, le véritable savoir-faire consiste à réussir la combinaison des couleurs primaires, secondaires et tertiaires de sorte qu'elles produisent la plus harmonieuse des compositions, qu'il qualifie

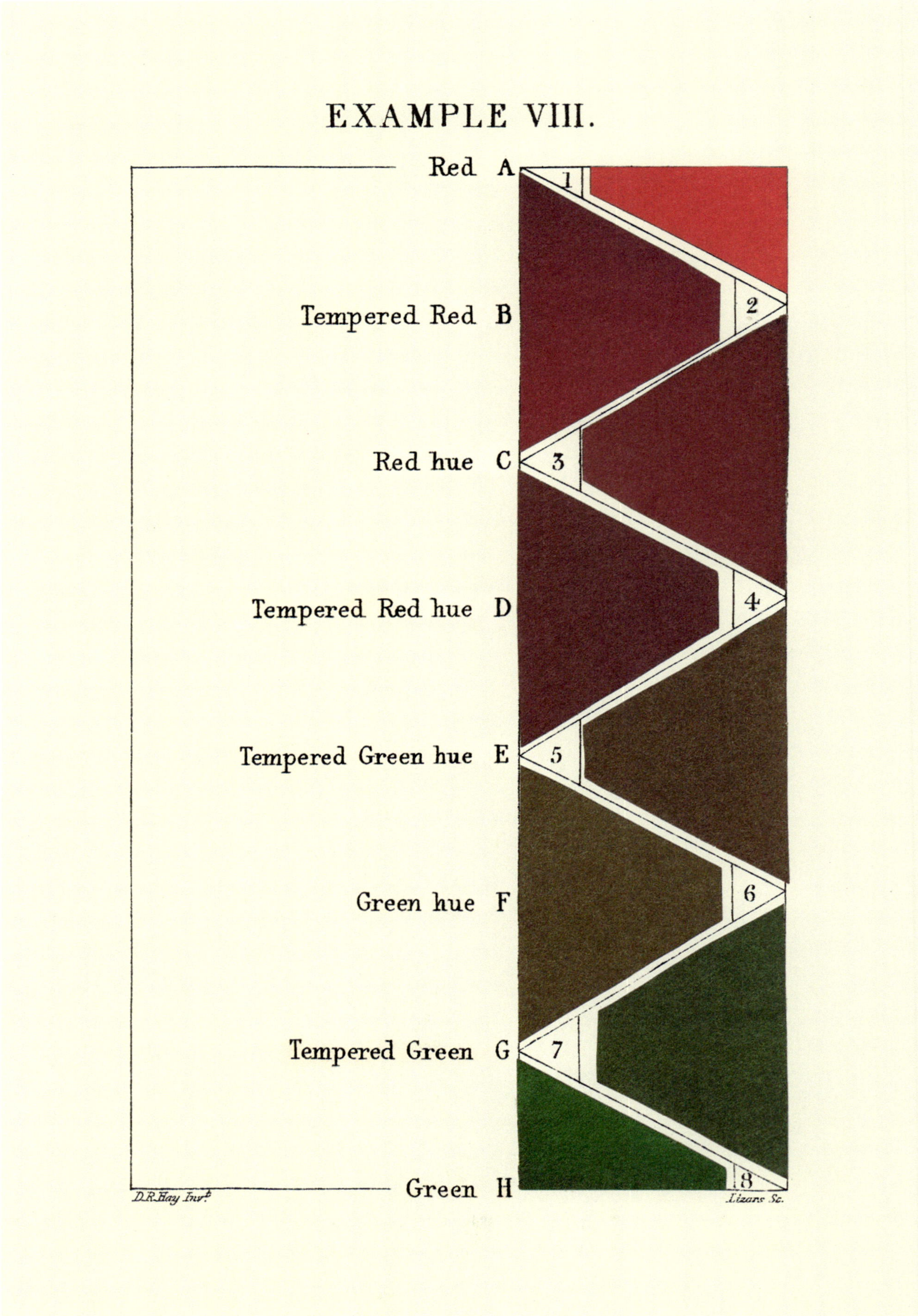
EXAMPLE VIII.
Red A
Tempered Red B
Red hue C
Tempered Red hue D
Tempered Green hue E
Green hue F
Tempered Green G
Green H
1
2
3
4
5
6
7
8
D.R.Hay Invt.
Lizars Sc.

EXAMPLE IX.

HARMONY OF THE PRIMARY RED.

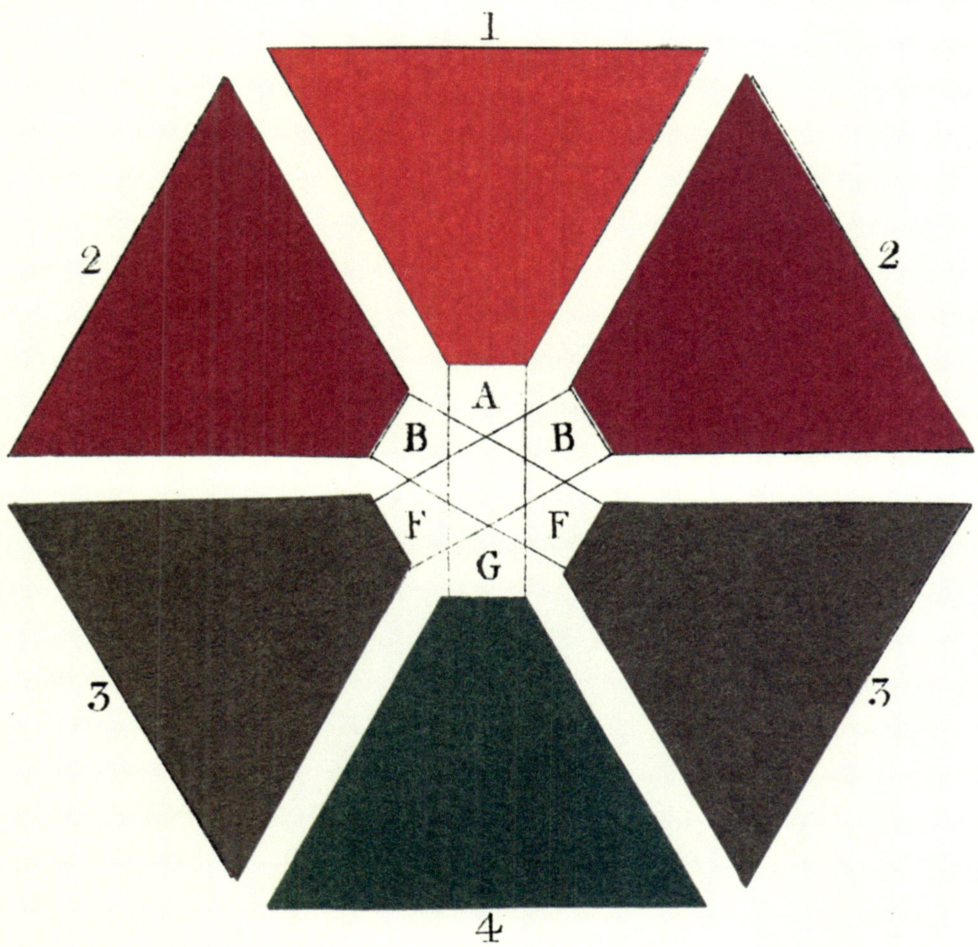

D.R.Hay Inv[t]

Lizars Sc.

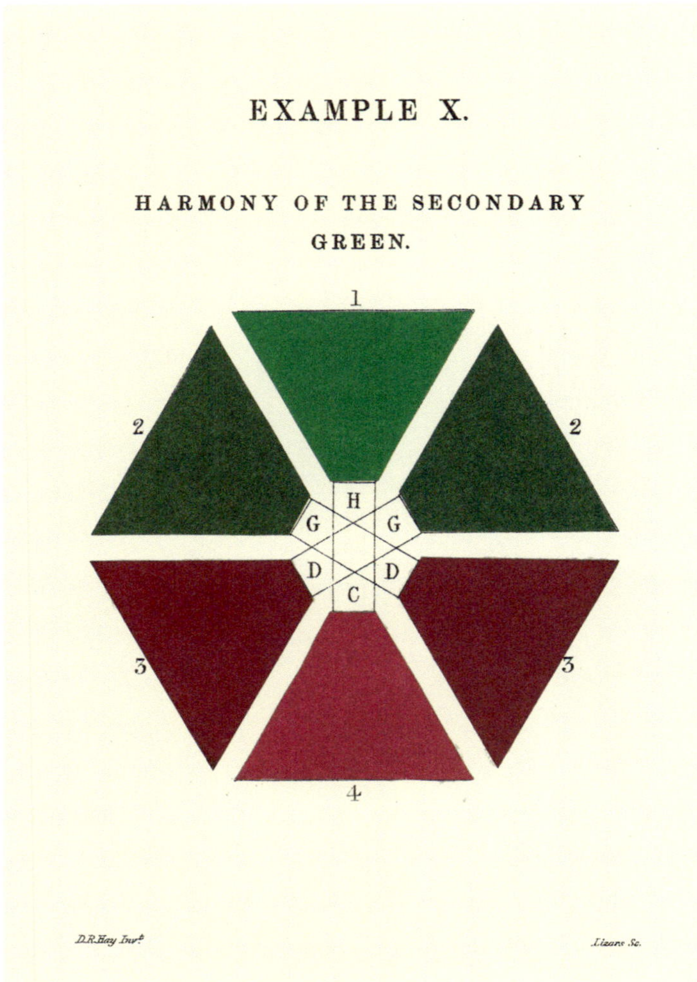

de « mélodie ». En outre, Hay possède une connaissance fine des rapports entre couleur et lumière, non seulement au sens newtonien, mais dans la manière dont les surfaces et les espaces sont éclairés : comment, par exemple, la lumière du jour se diffuse dans une pièce en fonction de l'heure.

L'illustration est un élément essentiel des nombreux ouvrages qu'a consacrés Hay à la couleur. Il insiste, en outre, sur la nécessité d'un propos absolument clair et simple et incorpore au corps du texte ses images les plus importantes sous forme de gravures et de tableaux légendés, suivis de nouvelles planches en couleurs. À de rares exceptions près, il préfère, pour illustrer ses « lois » et « principes », recourir à des figures géométriques plutôt qu'à des images d'intérieurs ou d'objets décoratifs. Ses *Principles of Beauty* contiennent 14 planches de graphiques représentant des combinaisons de couleurs primaires, secondaires et tertiaires. Chaque graphique a la forme d'un cercle chromatique composé de six trapèzes numérotés (c'est-à-dire des triangles dont un angle est tronqué afin de permettre l'ajout dans la gravure d'une lettre et de traits de liaison). La planche 8 y fait exception puisque les couleurs y sont disposées en une colonne de triangles. Comme dans *Werner's Nomenclature* de Syme (voir p. 310–319), les figures de couleurs sont découpées dans des morceaux de papier peints et collées sur les planches, méthode employée par Hay dans tous ses livres sur la couleur, alors que les lithographies étaient devenues monnaie courante dans les années 1830.

Principios sistematizados de la belleza en la coloración

David Hay fue un artista y decorador de interiores escocés, y una de las primeras personas en aplicar la teoría del color al diseño de interiores. En 1828, cuando estaba montando su propio negocio en Edimburgo (la empresa de pintado de casas Nicholson and Hay), publicó *The Laws of Harmonious Colouring Adapted to House Painting*, donde expuso sus ideas sobre el color y la estética, con una referencia específica a la analogía entre el color y la música. El libro llegó a seis ediciones ampliadas, y marcó el comienzo de su larga carrera como escritor. Cuando Hay publicó la primera edición de *The Principles of Beauty in Colouring Systematized*, la reina Victoria ya le había otorgado una autorización como

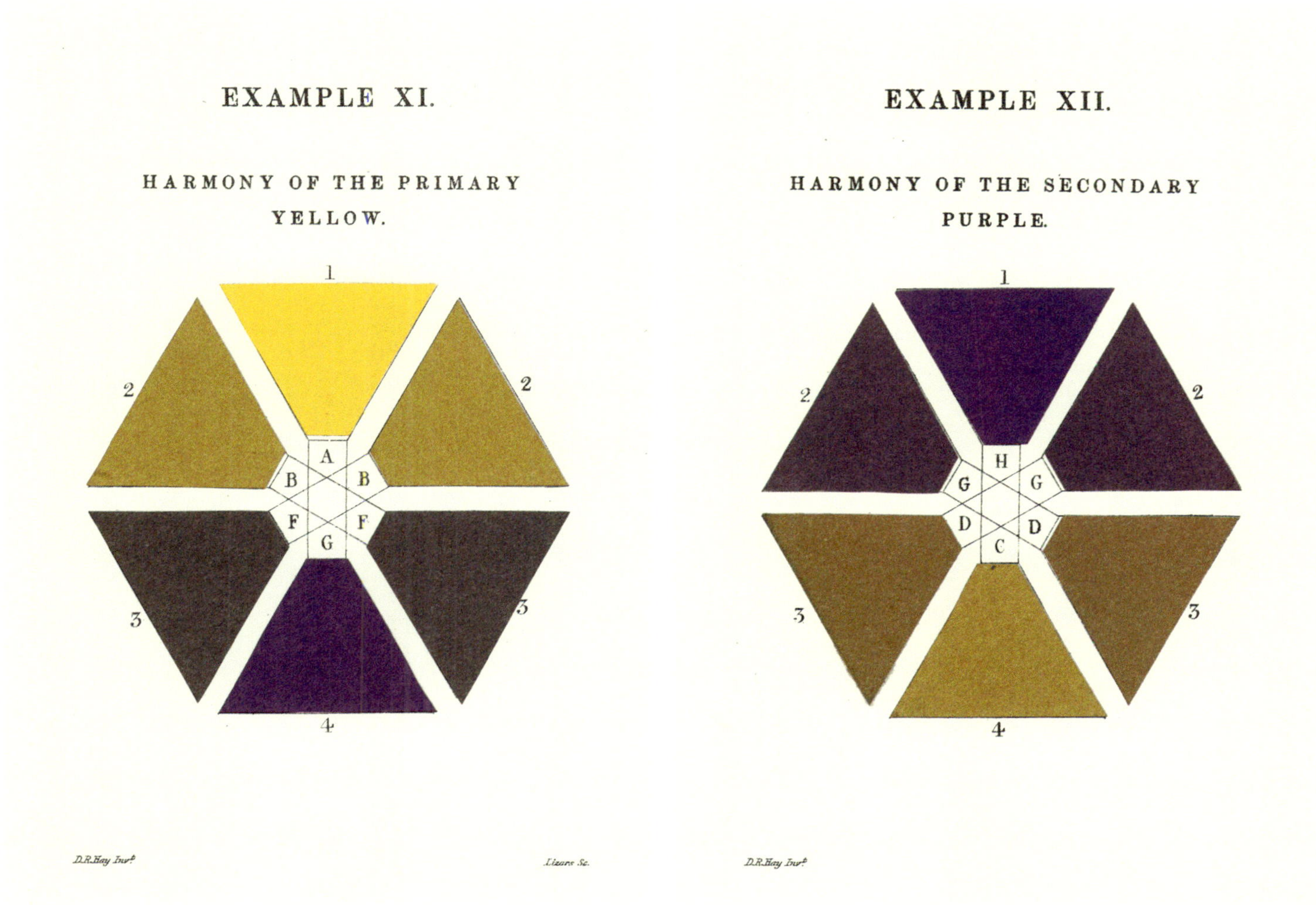

proveedor de la Casa Real y había recibido otros encargos de alto nivel.

Los libros de Hay son obras fundamentales en el campo del color arquitectónico porque presentan un sentido completamente desarrollado de la aplicación de la teoría del color al diseño de interiores y de la importancia de la percepción del color. Hay señaló que si bien la belleza del color depende generalmente del contraste, eso no significa que un mayor contraste sea necesariamente lo más deseable. La verdadera habilidad, argumentó, radica en combinar colores primarios, secundarios y terciarios de tal manera que den como resultado la composición más armoniosa o «melodía». También era muy consciente de la relación entre el color y la luz, no solo en un sentido newtoniano, sino en la forma en que se iluminan las superficies y los espacios, por ejemplo, en cómo la luz actúa en una habitación en diferentes momentos del día.

Las ilustraciones fueron una parte vital de las muchas publicaciones de Hay sobre el color, quien también enfatizó la necesidad de la máxima claridad y simplicidad. Con ese objetivo, presentó sus imágenes más importantes en forma de tablas etiquetadas y de grabados dentro del texto continuo, y luego nuevamente como láminas coloreadas. Con muy pocas excepciones, utilizó formas geométricas para ilustrar sus «leyes» y «principios», en lugar de representar imágenes de interiores arquitectónicos u objetos decorativos. Su *Principles of Beauty* contiene 14 láminas con diagramas que muestran combinaciones de colores primarios, secundarios y terciarios, cada una de las cuales toma la forma de una rueda de colores creada a partir de seis partes trapezoidales numeradas (es decir, triángulos con un vértice cortado para permitir el añadido en el grabado de una letra y líneas de conexión), salvo un ejemplo (lámina 8) en el que los colores aparecen dispuestos en una columna de triángulos. Al igual que en la *Werner's Nomenclature* de Syme (págs. 310–319), las formas coloreadas se recortaron de hojas de papel pintadas a mano y se pegaron en las láminas, un método que Hay continuó usando en todos sus libros sobre el color a pesar de que las ilustraciones litográficas se habían vuelto más habituales en la década de 1830.

David Ramsay Hay

A NOMENCLATURE OF COLOURS

Applicable to the Arts and Natural Sciences

18 plates, 21 x 15 cm | 8 ¼ x 5 ⅞ in.,
Edinburgh & London, 1846
Fife, University of St Andrews Library,
Department of Special Collections

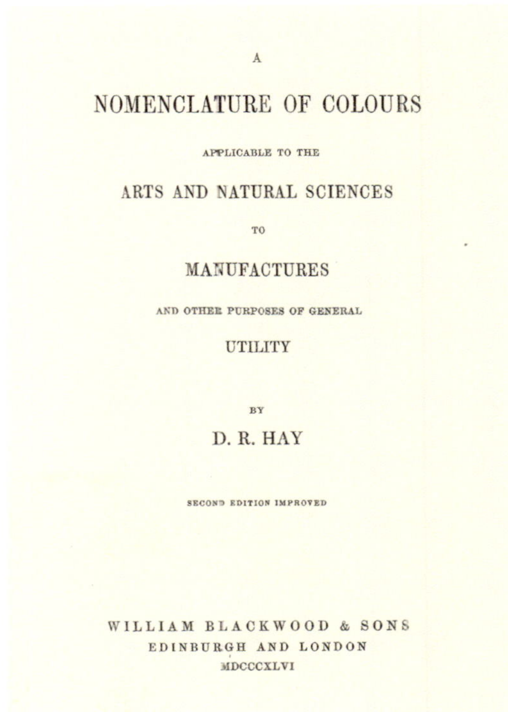
A

NOMENCLATURE OF COLOURS

APPLICABLE TO THE

ARTS AND NATURAL SCIENCES

TO

MANUFACTURES

AND OTHER PURPOSES OF GENERAL

UTILITY

BY

D. R. HAY

SECOND EDITION IMPROVED

WILLIAM BLACKWOOD & SONS
EDINBURGH AND LONDON
MDCCCXLVI

In the same way as many of his contemporaries, Hay promoted a trichromatic, subtractive system based on primary, secondary and tertiary colours. He was extremely well informed about developments and publications in this field and frequently acknowledged and praised theorists of his day and from the past, including Goethe (by way of Eastlake's translation), David Brewster and George Field. Hay's *Nomenclature of Colours* was clearly influenced by his countryman Patrick Syme's new edition of *Werner's Nomenclature* (1814 and 1821), even if he considered it "a very imperfect work".

A Nomenclature (first published in 1845) was Hay's most ambitious and assured work on colour, and was based on a mathematical system for the proportions of different colours and hues. The sub-title announced that his object was to create a reference work that would be "applicable to the Arts and Natural Sciences; to Manufactures, and other Purposes of General Utility", thereby making it similar in scope to Chevreul's *De la loi du contraste simultané des couleurs* from 1839 (pp. 282–295) which also stressed the applicability of colour theory to a wide range of disciplines.

With this work being a nomenclature, Hay devoted a significant portion of the book to the naming of colours. Consequently, he did not just provide a reliable mathematical formula for defining a particular colour, but also looked to clarify his approach. He populated his system with simple colour names, "not given arbitrarily, but according to the primary elements from which each colour, hue, tint, or shade is derived", eschewing known pigment names or fanciful descriptions. The purpose was to examine and describe how a colour, hue or shade appeared, looked and felt, rather than being concerned with exactly what substance it was made of. The design used in the book's 40 plates is an elegant augmentation of the triangular compositions Hay had employed in his earlier book, with each plate now showing two engraved pyramidal structures formed of three coloured equilateral triangles, with hand-painted samples once again pasted into the design. In total he presented 240 colours, in various proportions, combinations and with the addition of white and black, thereby creating an extraordinary record of the colours known in Britain during the 1840s. Hay continued to write and publish books on colour, beauty and design until 1856. His friend, the Scottish painter David Roberts (1796–1864), referred to him as "the first intellectual housepainter".

PLATE 1.

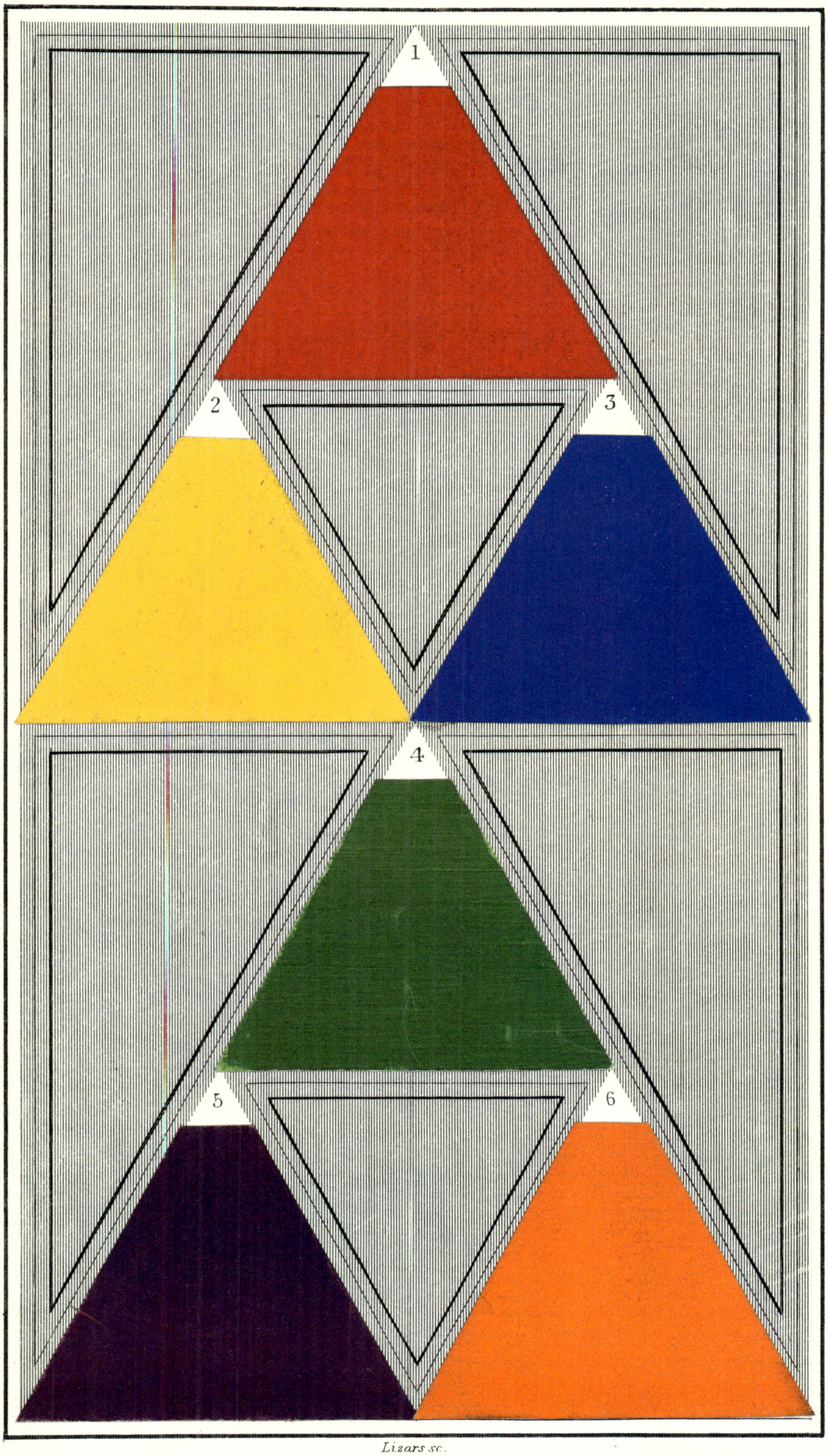

Lizars sc.

PLATE 2.

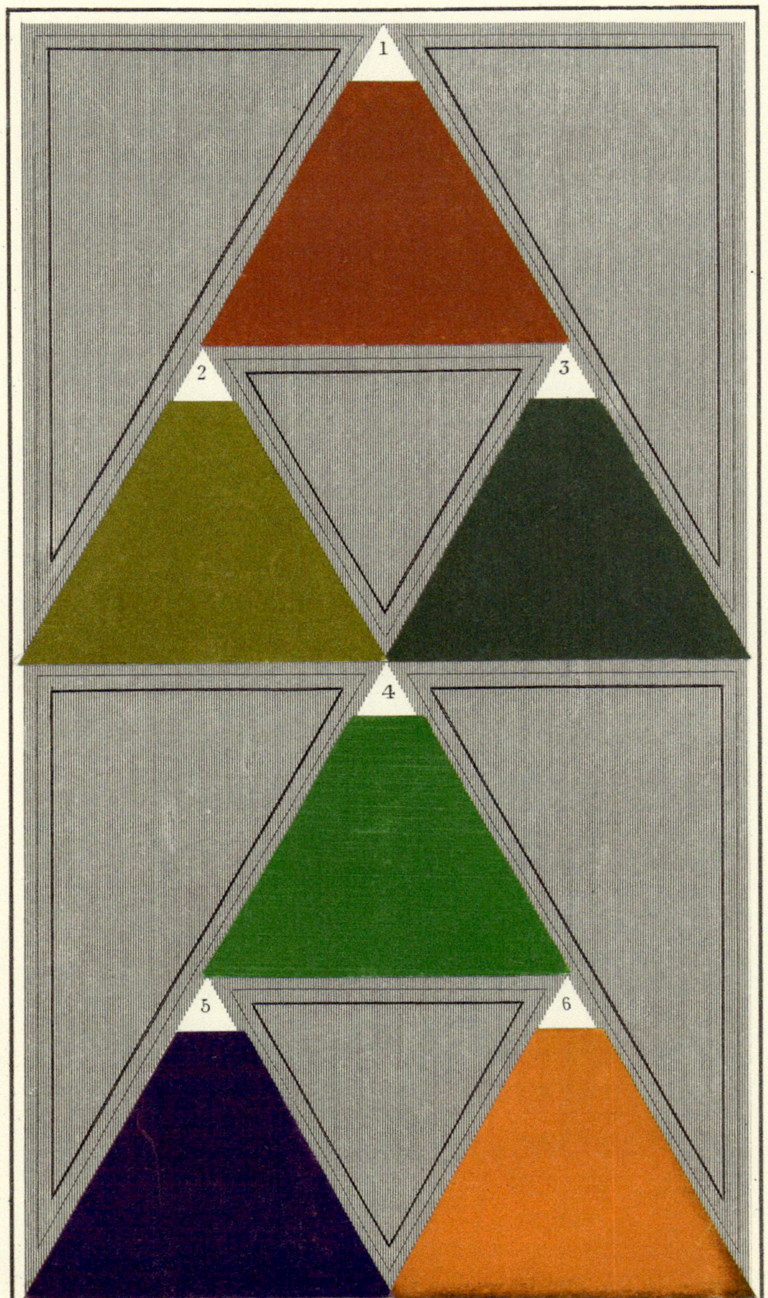

Lizars sc.

PLATE 3.

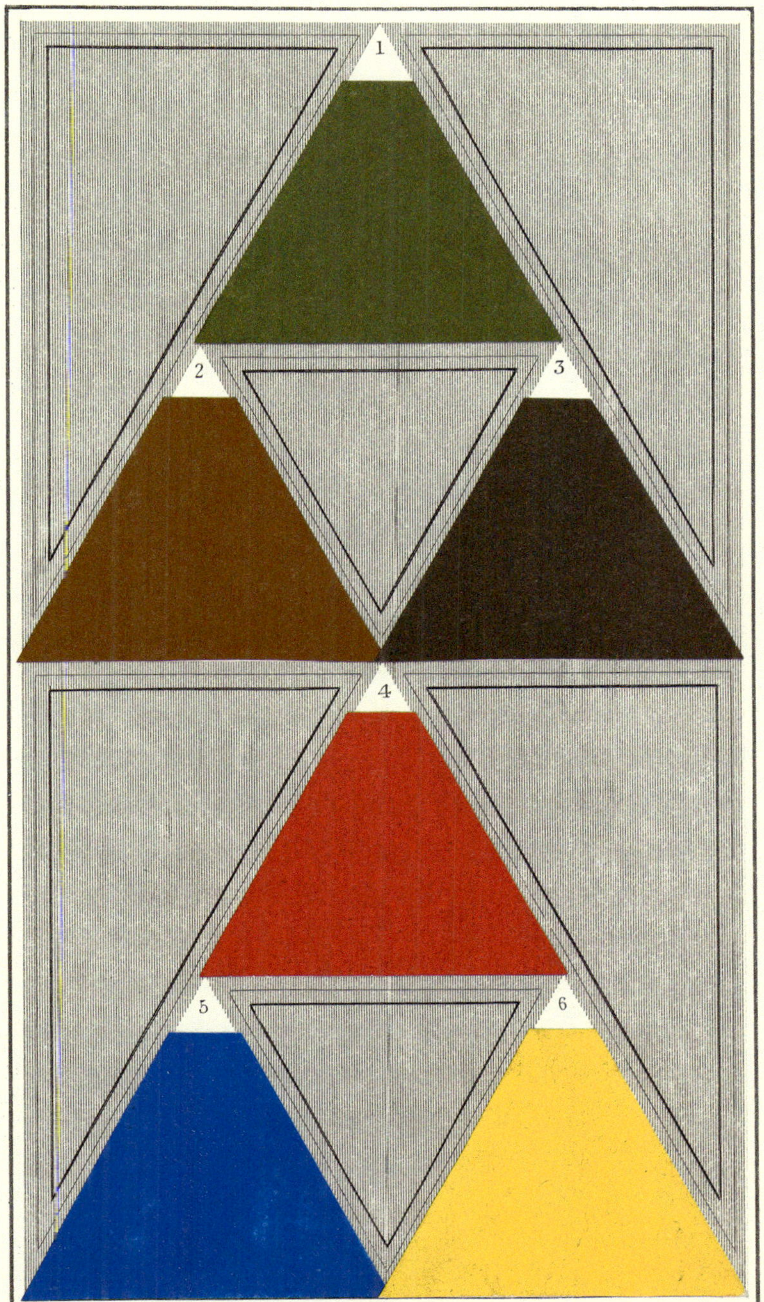

Lizars sc.

PLATE 4.

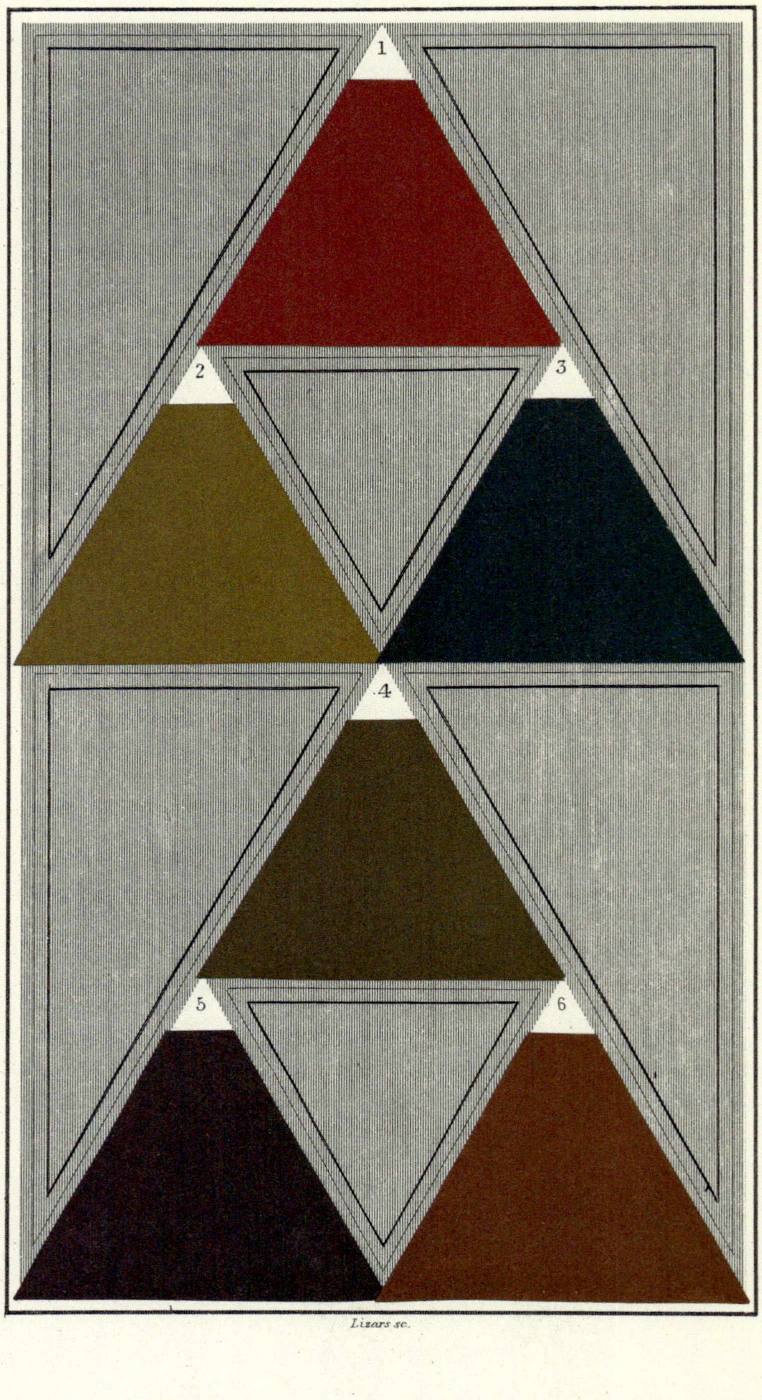

Lizars sc.

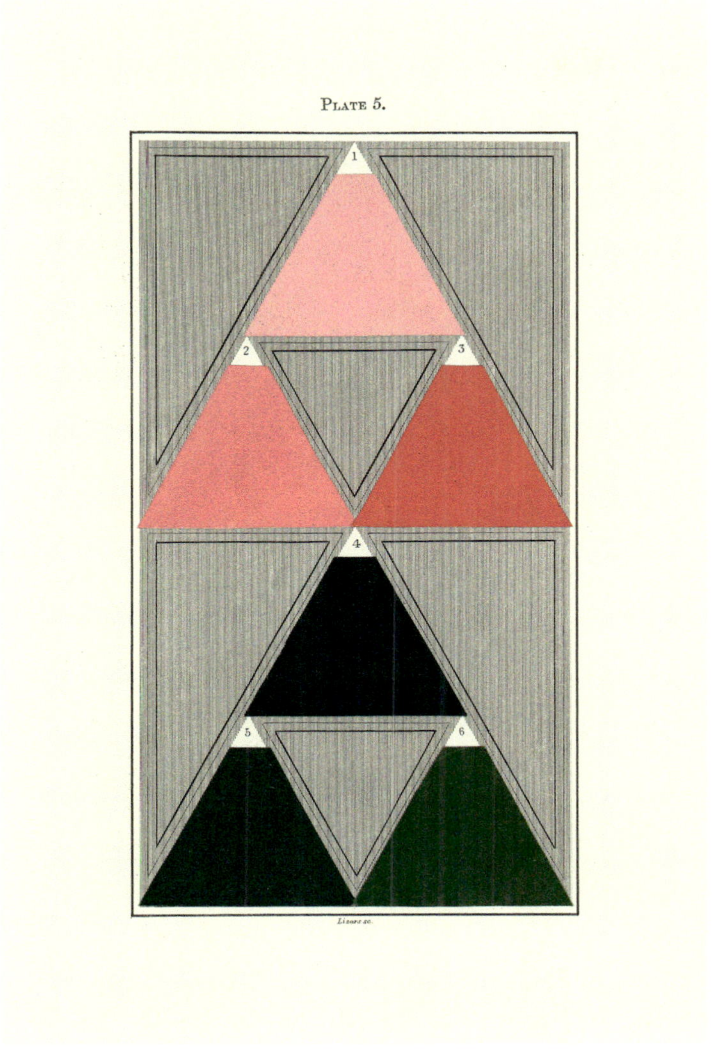

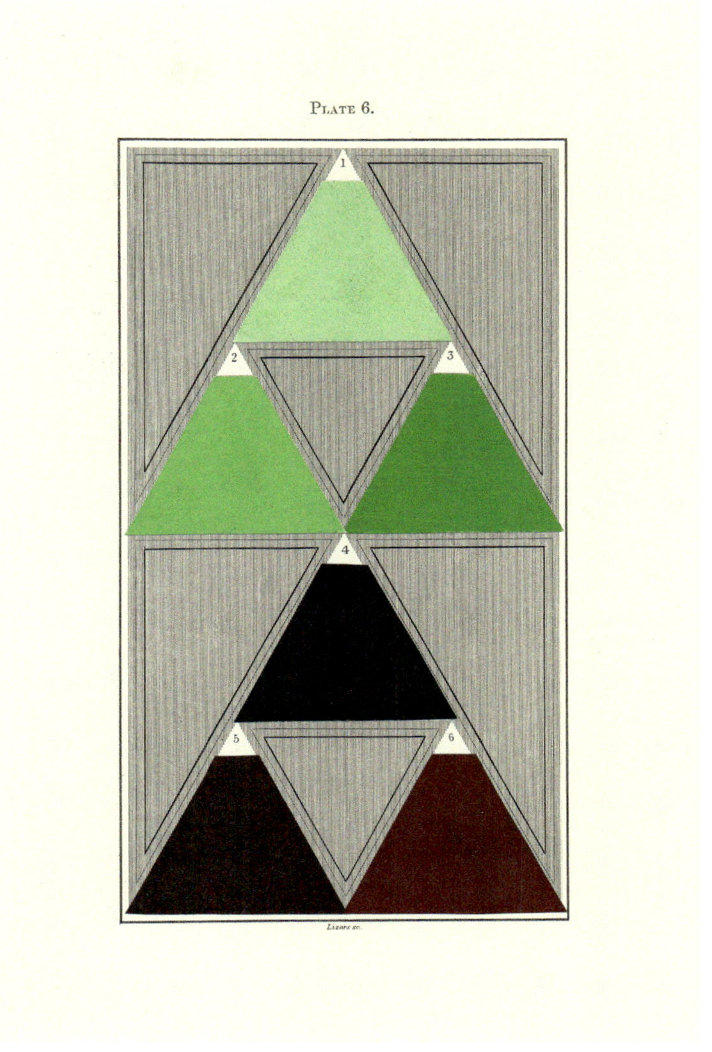

Nomenklatur der Farben, anwendbar auf die Künste und Naturwissenschaften

Wie viele seiner Zeitgenossen propagierte Hay ein trichromatisches, subtraktives System auf der Grundlage von Primär-, Sekundär- und Tertiärfarben. Er war über die Entwicklungen und Publikationen in diesem Bereich außergewöhnlich gut informiert und lobte und würdigte oftmals andere Theoretiker der Vergangenheit oder Gegenwart, wie Goethe (anhand von Eastlakes Übersetzung), David Brewster oder George Field. Indes war sein *A Nomenclature of Colours* deswegen stark von Patrick Symes Neuauflage der *Werner's Nomenclature* (1814 und 1821) beeinflusst, obwohl Hay das Buch seines Landsmannes für „ein sehr unvollkommenes Werk" hielt.

A Nomenclature of Colours (1845 erstmals veröffentlicht) war Hays ehrgeizigste und sicherste Arbeit über Farbe und beruhte auf einem mathematischen System für die Zusammensetzung verschiedener Farben und Schattierungen. Im Untertitel erklärte er, ein Nachschlagewerk angestrebt zu haben, das „auf die Künste und Naturwissenschaften anwendbar [sein würde]; auf Hersteller und andere Zwecke von allgemeinem Nutzen". Damit stellte er das Buch in eine Reihe mit Chevreuls *De la loi du contraste simultané des couleurs* von 1839 (S. 282–295), das ebenfalls die Anwendbarkeit der Farbtheorie auf ein breites Spektrum von Fachgebieten hervorhob.

Da sein Buch eine Nomenklatur war, widmete Hay einen bedeutenden Teil des Werks der Benennung von Farben. So stellte er für die Bestimmung einer konkreten Farbe nicht nur eine verlässliche mathematische Formel bereit, sondern wollte mit seinem Ansatz auch mehr Klarheit schaffen. Hay stattete sein System mit einfachen Farbnamen aus, die „nicht willkürlich vergeben [werden], sondern nach den primären Bestandteilen, aus denen jede Farbe, Tönung, Nuance oder Schattierung herrührt". Dabei vermied er bekannte Pigmentnamen oder der Fantasie entsprungene Bezeichnungen. Sein Buch sollte analysieren und beschreiben, wie eine Farbe, Tönung oder Schattierung erschien, aussah und sich anfühlte, anstatt sich damit zu beschäftigen, aus welcher Substanz sie genau

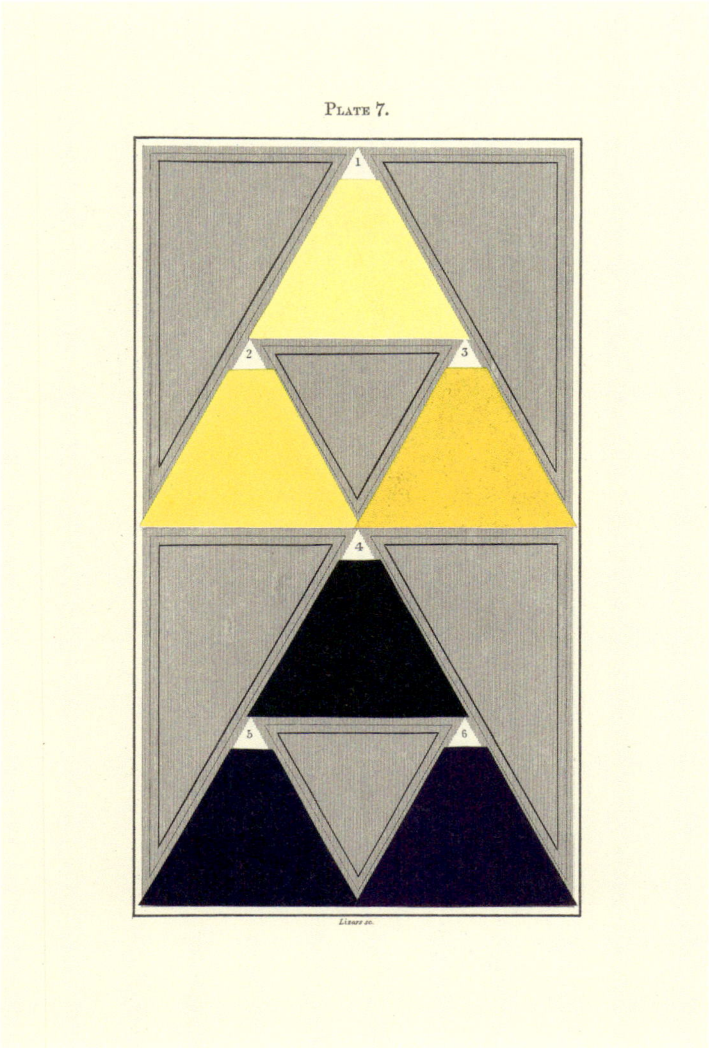

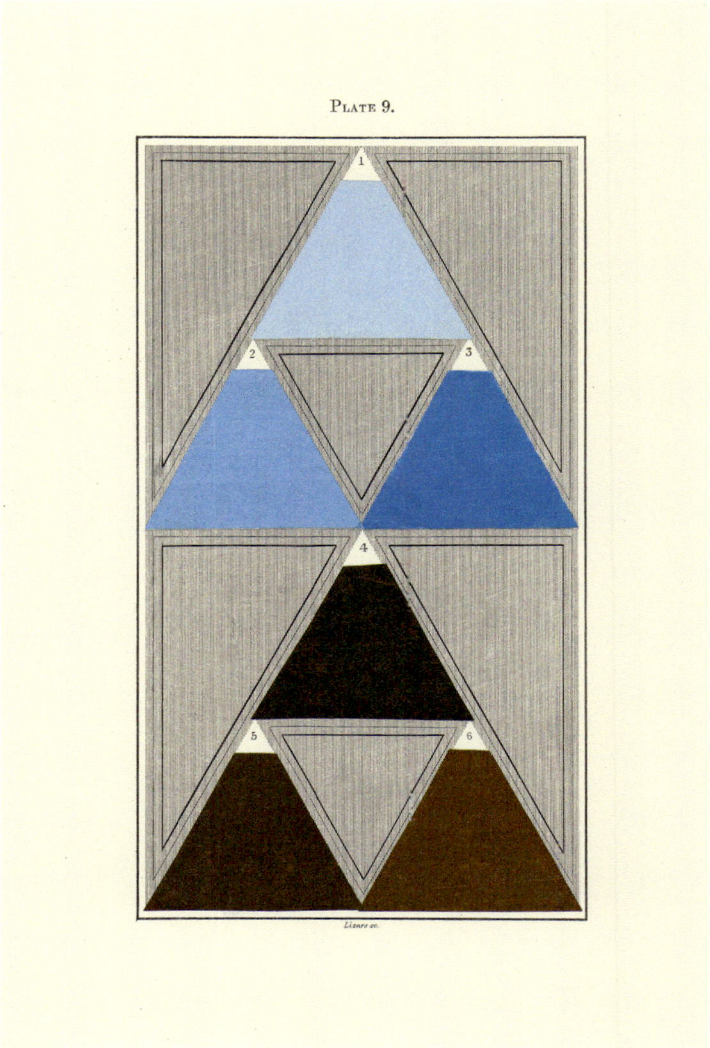

bestand. Das Layout der vierzig Bildtafeln stellt eine elegante Erweiterung der Dreieckskompositionen dar, die Hay für sein früheres Buch verwendet hatte. Jede Tafel zeigt jetzt zwei gestochene, aus jeweils drei gleichseitigen Dreiecken bestehende Pyramidenstrukturen, in die handbemalte Farbmuster eingeklebt wurden. Insgesamt präsentierte Hay 240 Farben in verschiedenen Mischungsverhältnissen und Kombinationen mit varriierenden Weiß- und Schwarzanteilen und schuf damit ein außergewöhnliches Verzeichnis der im Großbritannien der 1840er-Jahre bekannten Farben. Noch bis 1856 verfasste und veröffentlichte Hay Bücher über Farbe, Schönheit und Gestaltung. Ein Freund, der schottische Maler David Roberts (1796–1864), bezeichnete ihn als „den ersten intellektuellen Anstreicher".

Nomenclature des couleurs, applicable aux arts et aux sciences naturelles

À l'instar de bien d'autres de ses contemporains, Hay se prononce en faveur d'un système soustractif trichromatique, fondé sur les couleurs primaires, secondaires et tertiaires. Très au fait des recherches et publications en la matière, il reconnaît souvent sa dette à l'égard de théoriciens du passé et de son temps, dont il fait l'éloge, comme Goethe (via la traduction anglaise d'Eastlake), David Brewster et George Field. Sa « Nomenclature des couleurs » doit en effet beaucoup à la nouvelle édition de l'ouvrage de son compatriote Patrick Syme, *Werner's Nomenclature* (1814, puis 1821), qu'il considère pourtant comme « un ouvrage très imparfait ».

Œuvre la plus ambitieuse et la plus aboutie de David Hay, la *Nomenclature* (publié la première fois en 1845) est fondée sur un système mathématique de détermination des proportions entre les différents tons et couleurs. Selon son sous-titre, l'objectif est de proposer un ouvrage de référence « applicable aux arts et aux sciences naturelles ; aux manufactures et autres visées d'utilité générale ». Ce livre est donc d'une ambition semblable à celle de Chevreul avec *De la loi du contraste simultané des couleurs*, paru en 1839 (p. 282–295), lequel a aussi pour but d'appliquer la théorie des couleurs à toutes sortes de domaines.

Plate 11.

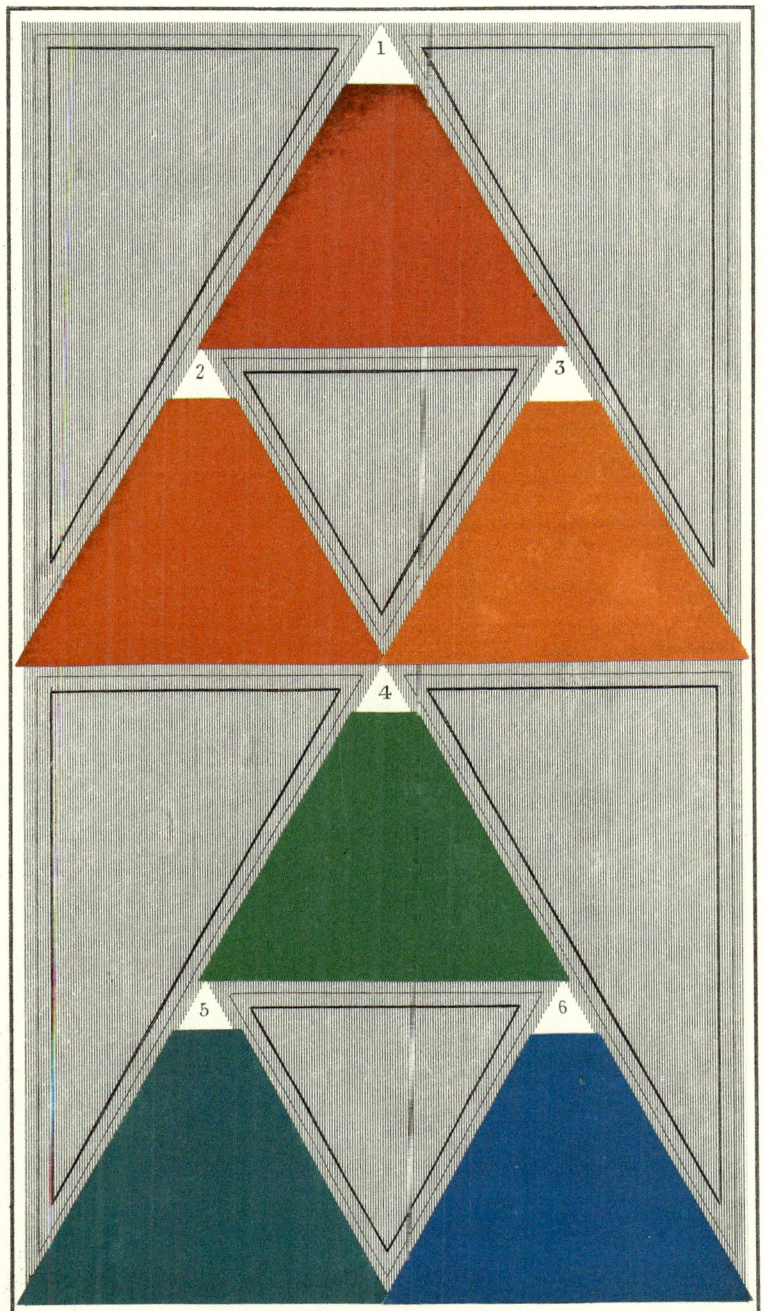

Lizars sc.

PLATE 12.

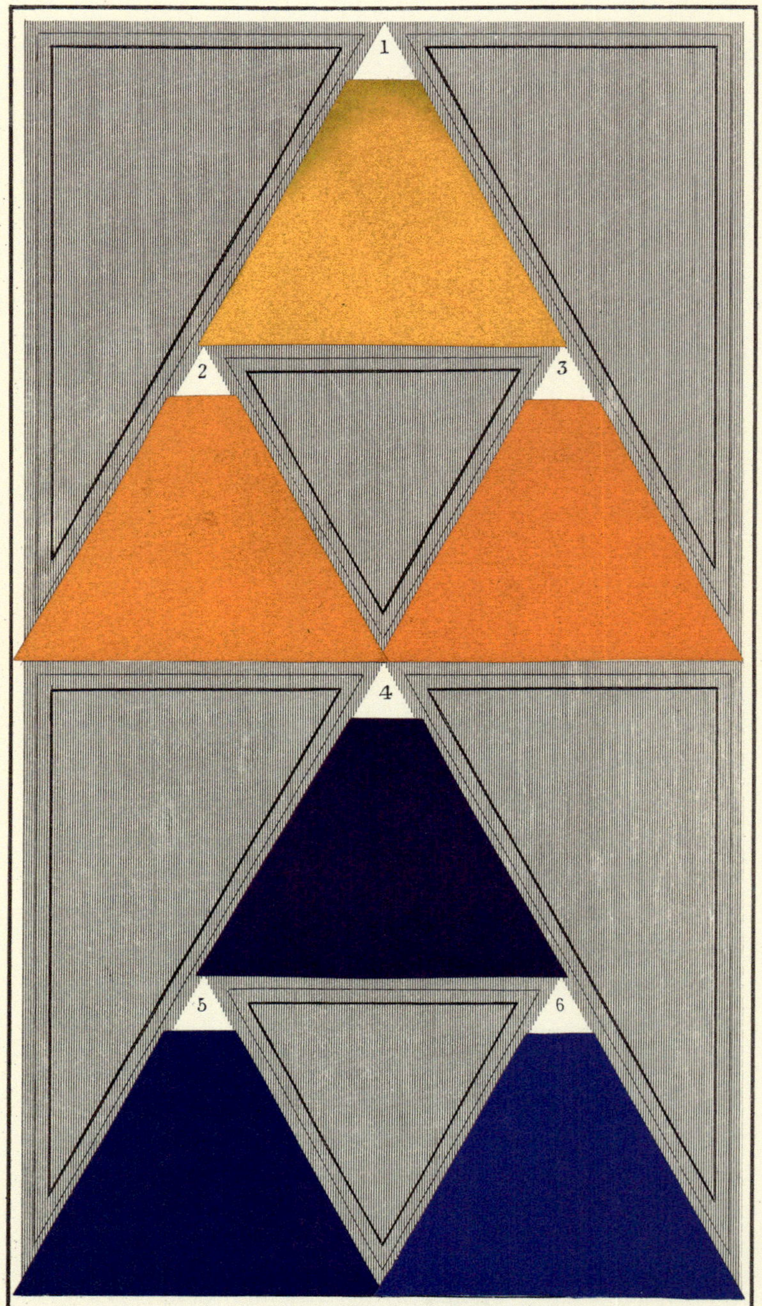

Lizars sc.

Plate 17.

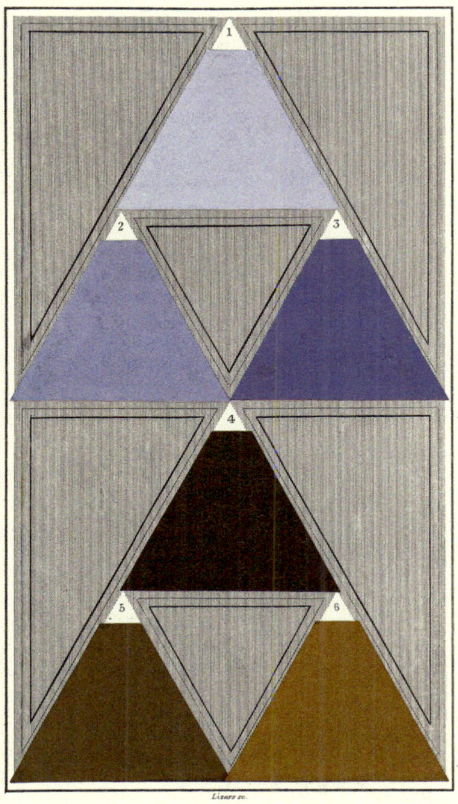

Plate 18.

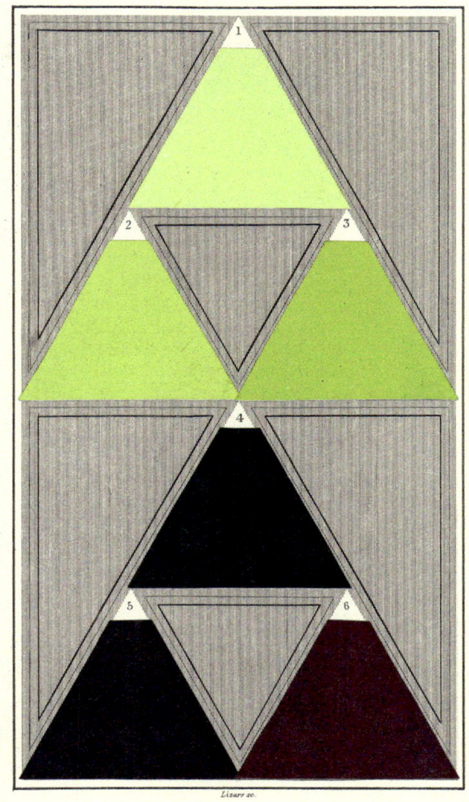

Puisqu'il s'agit d'une nomenclature, Hay consacre une partie importante de l'ouvrage à la désignation des couleurs. Par conséquent, il ne se contente pas de fournir une formule mathématique fiable pour définir telle ou telle teinte mais cherche à expliciter sa démarche. Il dote son système de noms de couleurs simples, « qui ne sont pas choisis de manière arbitraire, mais en fonction des éléments principaux dont sont issus chaque couleur, ton, teinte ou nuance », en écartant les appellations des pigments connus et les descriptions fantaisistes. Hay s'attache avant tout à étudier et décrire l'aspect d'une couleur, d'une teinte ou d'une nuance et à la sensation qu'elle suscite, plutôt qu'à définir avec précision ce qui la composent. Les 40 planches de l'ouvrage reprennent et augmentent avec élégance les compositions triangulaires que l'auteur a employées dans ses livres précédents. Chaque gravure présente désormais deux structures pyramidales formées par trois triangles équilatéraux, contenant des échantillons peints à la main et, là encore, collés. Hay donne ansi à voir 240 couleurs en diverses proportions et combinaisons, auxquelles s'ajoutent le blanc et le noir. Cet ouvrage constitue un témoignage extraordinaire des couleurs que l'on connaissait en Grande-Bretagne dans les années 1840. Hay publiera d'autres ouvrages consacrés à la couleur, à la beauté et à la décoration jusqu'en 1856. Ami de Hay, le peintre écossais David Roberts (1796–1864) qualifiait celui-ci de « premier peintre d'intérieur intellectuel ».

Nomenclatura de los colores aplicable a las artes y las ciencias naturales

Como muchos de sus contemporáneos, Hay promovió un sistema sustractivo tricromático basado en colores primarios, secundarios y terciarios. Estaba extremadamente bien informado sobre los avances y publicaciones en ese campo, y con frecuencia reconoció y elogió a los teóricos de su época y del pasado, incluidos Goethe (por medio de la traducción de Eastlake), David Brewster y George Field. La *Nomenclature of Colours* de Hay estuvo muy influenciada por la nueva edición de la *Werner's Nomenclature* de su compatriota Patrick Syme (1814 y 1821), que Hay consideró «una obra muy imperfecta».

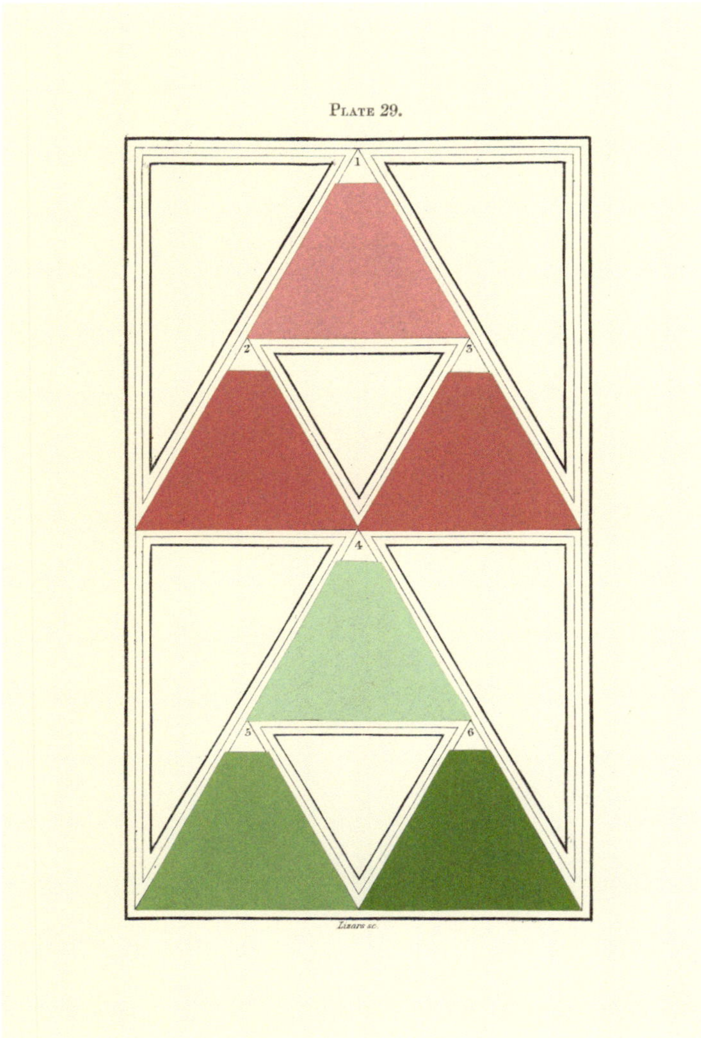

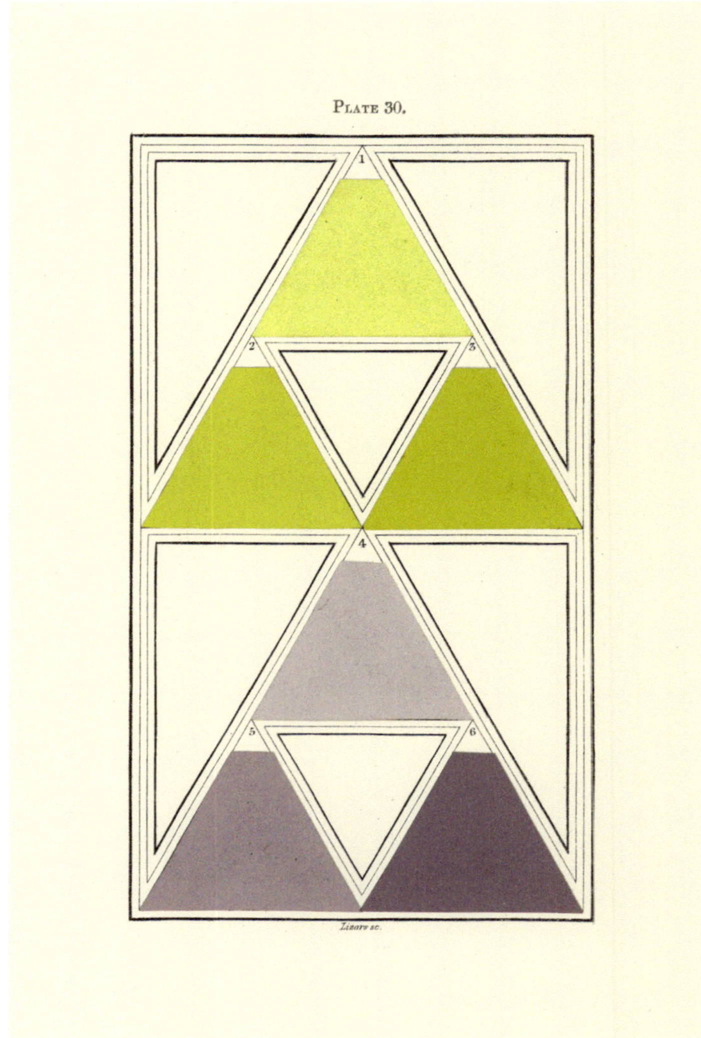

A Nomenclature (publicada por primera vez en 1845) establece las proporciones de diferentes colores y matices a partir de un sistema matemático, y fue el trabajo más ambicioso y sólido de Hay sobre el color. El subtítulo anuncia que su objetivo es crear una obra de referencia «aplicable a las Artes y las Ciencias Naturales; a las Manufacturas y otros Propósitos de Utilidad General», un alcance parecido al *De la loi du contraste simultané des couleurs* de Chevreul de 1839 (págs. 282–295), que también enfatizaba la posibilidad de aplicar la teoría del color a un amplio abanico de disciplinas.

Dado que este trabajo es una nomenclatura, Hay dedicó una parte importante del libro a la denominación de los colores. En consecuencia, no solo proporcionó una fórmula matemática fiable para definir un color en particular, sino que también buscó ofrecer un enfoque claro. Llenó su sistema con nombres de colores que «no son dados arbitrariamente, sino de acuerdo con los elementos primarios de los cuales se deriva cada color, matiz, tono o sombra», y evitó los nombres de los pigmentos conocidos y las descripciones fantasiosas. El propósito de este libro era examinar y describir cómo se manifestaba, veía y sentía un color, un matiz o una tonalidad, en lugar de preocuparse de qué sustancia estaban hechos. El diseño empleado en sus 40 láminas es una elegante ampliación de las composiciones triangulares que Hay había incluido en su libro anterior, y cada lámina incluye el grabado de dos estructuras piramidales formadas por tres triángulos equiláteros de colores, con muestras de nuevo pintadas a mano y pegadas al papel. En total, presentó 240 colores en varias proporciones y combinaciones y con el añadido del blanco y del negro. De ese modo, creó un registro extraordinario de los colores conocidos en Gran Bretaña en la década de 1840. Hay continuó escribiendo y publicando libros sobre el color, la belleza y el diseño hasta 1856. Su amigo, el pintor escocés David Roberts (1796–1864), se refirió a él como «el primer pintor de casas intelectual».

PLATE 36.

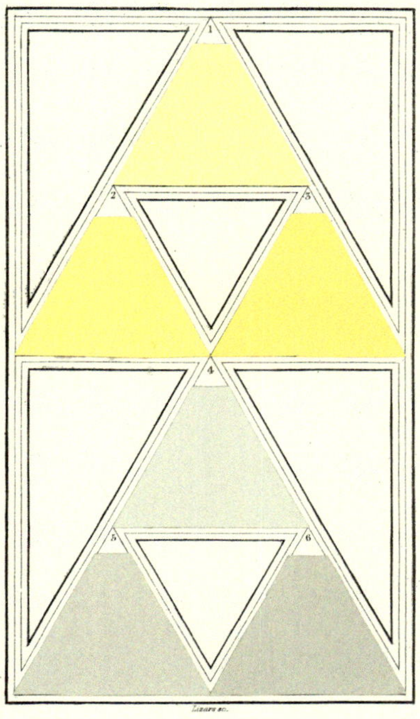

PLATE 37.

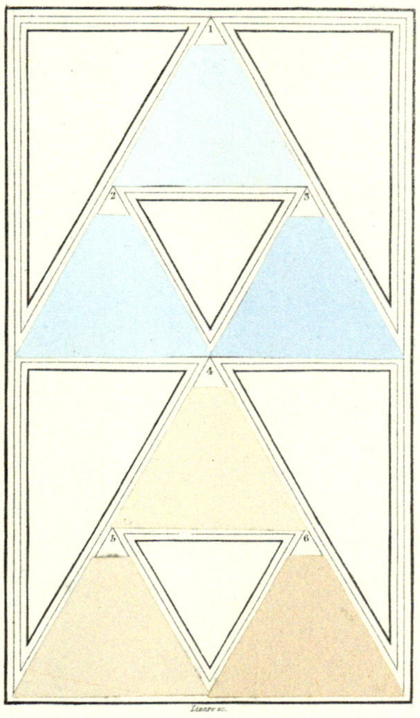

PLATE 38.

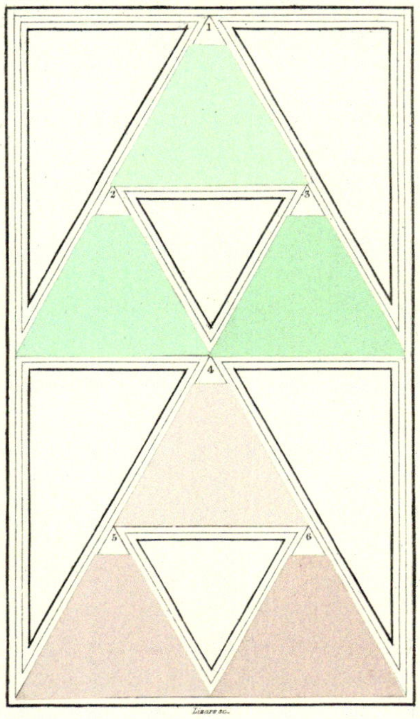

PLATE 39.

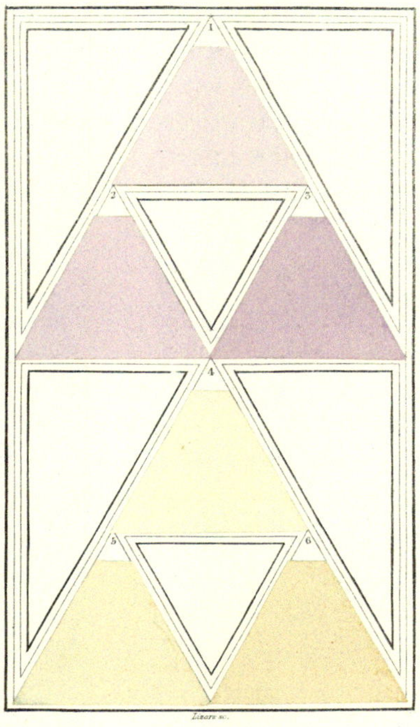

Édouard Guichard (1815–1889)

L'HARMONIE DES COULEURS / DIE HARMONIE DER FARBEN

The Harmony of Colours
29 plates, 34.3 x 26.7 cm / 13 ½ x 10 ½ in., Paris / Frankfurt am Main, 1880–1882
Washington, D.C., Smithsonian Libraries

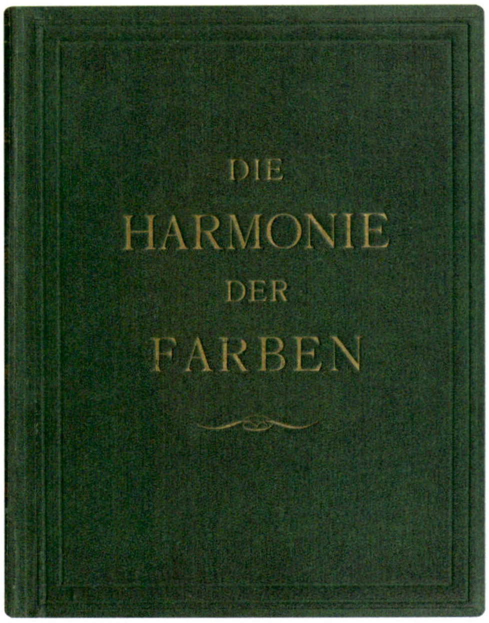

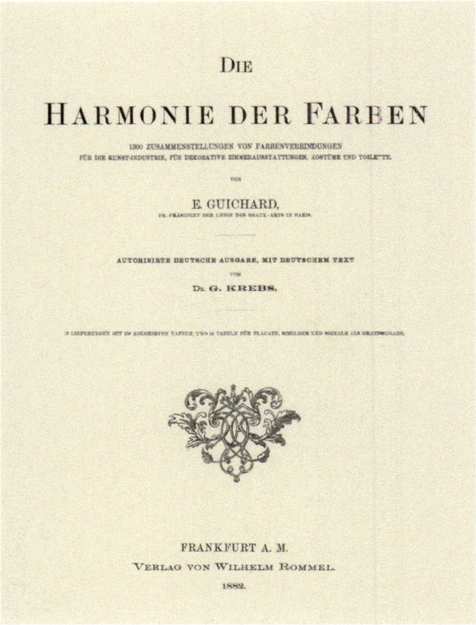

Édouard Guichard was a French designer, architect and decorator, yet considering the outstanding typographical and chromatic beauty of the book on colour harmony he produced it is surprising how seldom it is mentioned in recent literature on colour and design history. The original French work, *L'Harmonie des couleurs*, was published in several instalments in Paris between 1880 and 1882 by Ad. Goubaud et Fils. The authorised German edition, with a translation by Georg Krebs, must have been in preparation almost simultaneously and in fact surpasses the French edition with regard to the quality of both the paper and the illustrations. There are only a dozen pages of introductory text by Guichard, but the illustrations in the folio volume are accompanied by extensive descriptive captions. As with many later 19th-century writers on colour arrangement and the decorative arts, Guichard was heavily influenced by Chevreul, and perhaps directly inspired by the technical brilliance of the plates in Chevreul's *Atlas* volume from 1864 (pp. 296–307). The German title-page also announced what made Guichard's book so special, specifically the 1,300 examples it contained of harmonious colour combinations and arrangements for use in the art industry, interior decoration, fashion and make-up. These are presented in 166 colour plates, some on double-pages, depicting textile patterns, decorative objects against different background colours, strictly typographical arrangements, figurative and ornamental designs, and entirely abstract compositions, all of them illustrating colour harmony and contrast. No other book before had displayed such variety in depicting colour arrangements. The printing techniques show the full range of what could be achieved with colour reproduction in later 19th-century print culture, including plates in pochoir, a number of screen-printed plates illustrating advertising posters and chromolithographs of superior quality contributed by F. Appel. It is sometimes described as a pattern book, but is much more than that since it is unquestionably a book about colour and design, showcasing the best of what printers and printing technology were capable of producing at this date. Fewer than 10 copies of each edition, the French and the German, are recorded in the collections of libraries and museums worldwide, which strongly suggests they were published in only very small print runs.

Guichard, however, did not stop at this point, and in 1882 a continuation of his work was published in Paris by H. Cagnon, appearing as a three-volume, trilingual (French-English-German) set with the title *La Grammaire de la couleur / Grammar of Colour / Die Grammatik der Farben*, in an oblong format and this time with 765 coloured plates (pp. 456–457).

HARMONIE DER FARBEN

von E. GUICHARD

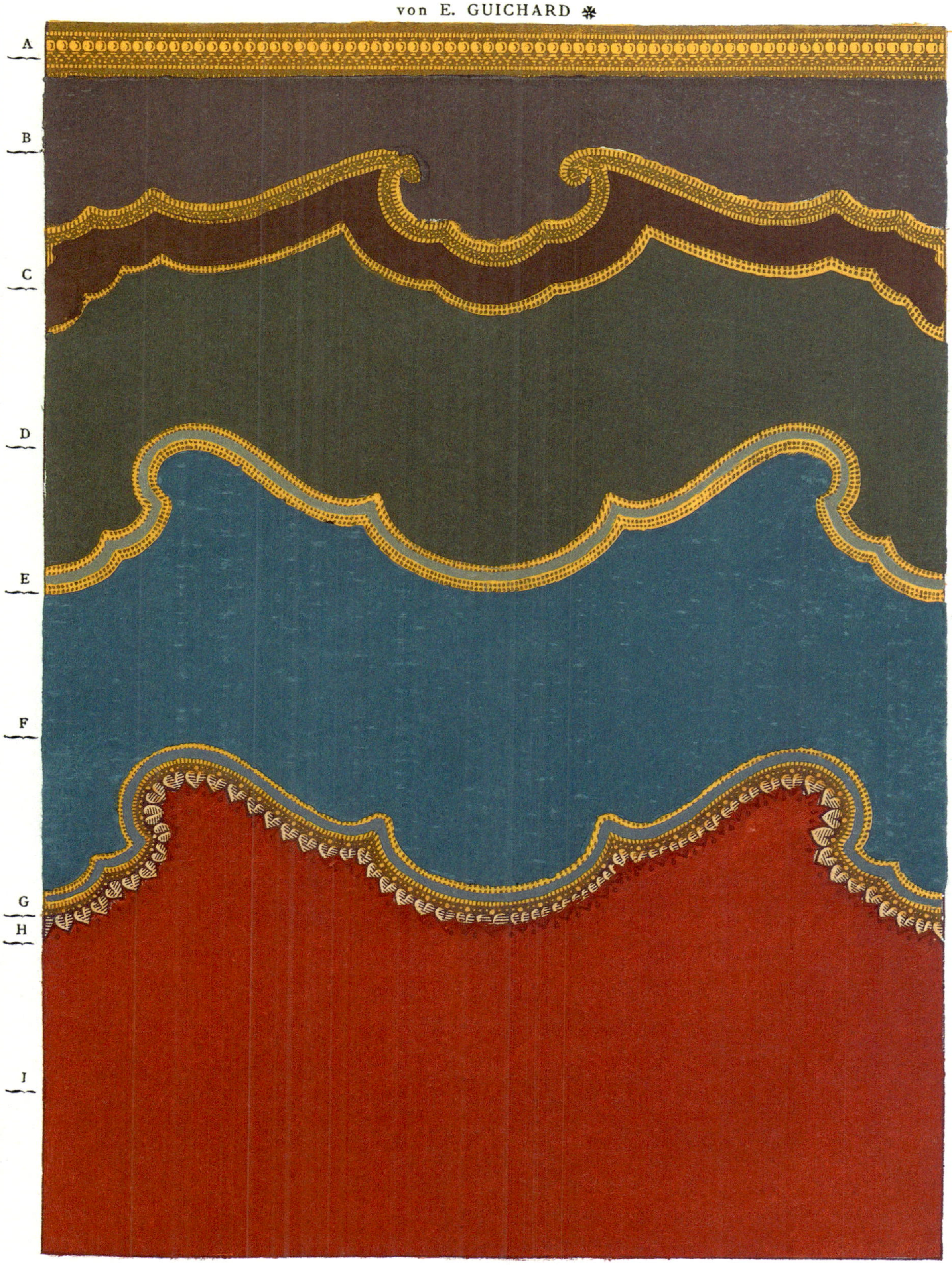

TAPETEN

Wenn die Grundfarbe einer Tapete wie I ist, so kann der Fries wie F und die Borde wie A, G und H sein.

VORHÄNGE und PORTIÈREN

Ist die Farbe der Tapete wie I, so kann die Grundfarbe der Vorhänge wie F sein, mit Bordüren wie I und Franzen oder Gimpen wie A, G und H. — Wählt man die Vorhänge von derselben Farbe wie die Tapete, so wird man gut thun, um Eintönigkeit zu vermeiden, die Farbe F für die Gallerieüberhänge (lambrequins) der Fenster zu nehmen.

THÜREN, LAMBRIS und KARNIESSE (GESIMSE, franz. CORNICHES)

Zu den Thüren und Lambris nimmt man die Farben B, C, D und E. — Der Theil des Karniesses, welcher den Platond berührt, muss in etwas lichteren Farbentönen gehalten sein als die Thüren und Lambris; — man nehme dazu daher nur die Töne C, D, E, nämlich die Farbe C als Grundfarbe des Karniesses, die Farbe D für die Wölbung desselben und die Farbe E für das an den Plafond angrenzende Gesims. — Die Farbe A kann in bescheidenem Mass zu feinen Zwischenlinien benutzt werden; statt dessen kann man übrigens auch Gold nehmen oder diese Verzierung ganz weglassen.

FRAUENKLEIDUNG

Die soeben angeführten Farbenzusammenstellungen werden auch bei der feinen Damentoilette ihre Wirkung nicht verfehlen.

(Verlag von Wilh. Rommel in Frankfurt a. M.)

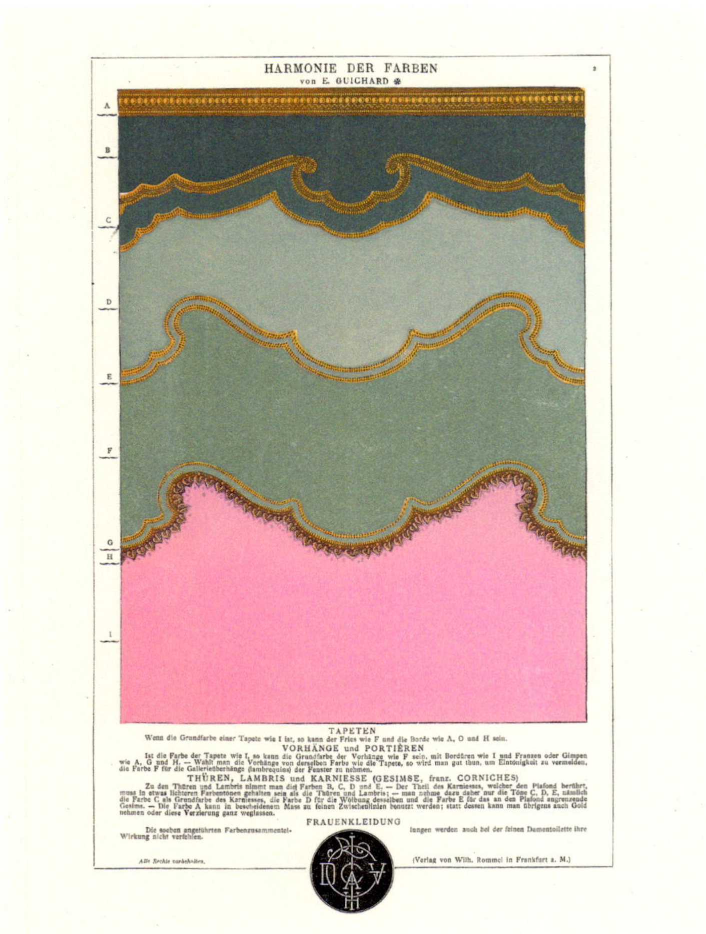

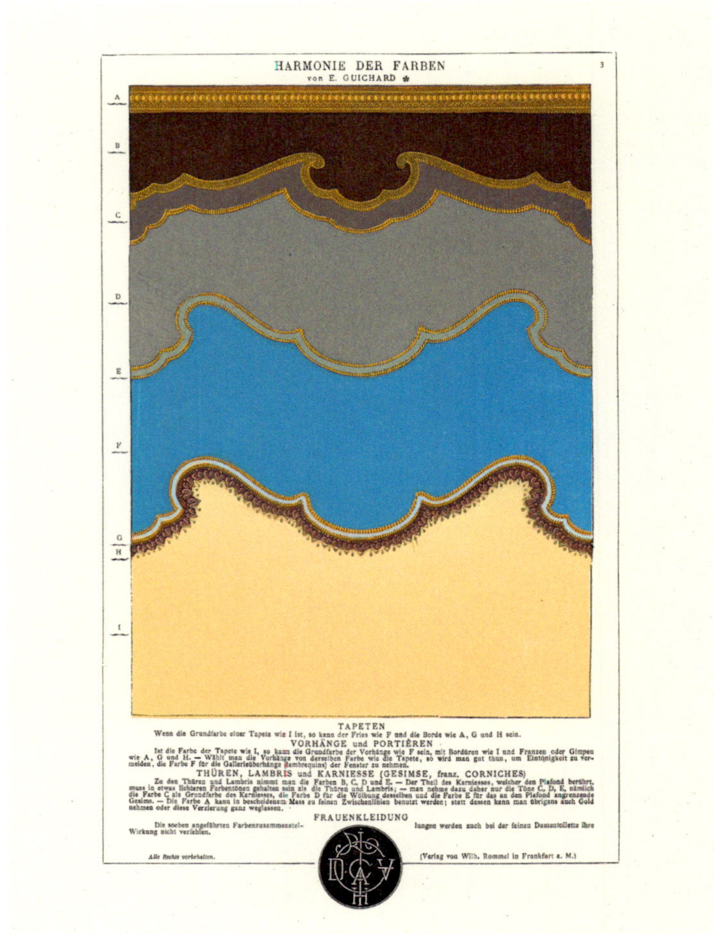

Die Harmonie der Farben

Édouard Guichard war ein französischer Gestalter, Architekt und Innenausstatter. Trotz der herausragenden drucktechnischen und farblichen Schönheit seines Buches über Farbharmonien wird der Titel in der jüngeren Literatur zur Farb- und Designgeschichte erstaunlich selten erwähnt. Die französische Originalausgabe, *L'Harmonie des couleurs*, erschien zwischen 1880 und 1882 in mehreren Lieferungen bei Ad. Goubaud et Fils in Paris. Die autorisierte deutsche Ausgabe, übersetzt von Georg Krebs, muss fast zeitgleich in Vorbereitung gewesen sein und übertrifft in Papier- und Abbildungsqualität sogar ihr französisches Pendant. Während die von Guichard verfasste Einleitung nur ein Dutzend Seiten lang ist, begleiten ausführliche Beschreibungen in den Legenden die Abbildungen des deutschen Foliobandes. Wie viele Autoren des späteren 19. Jahrhunderts, die über Farbkombinationen und dekorative Kunst schrieben, war Guichard stark von Chevreul beeinflusst und womöglich unmittelbar inspiriert von der technischen Brillanz der Tafeln in dessen Farbatlas aus dem Jahr 1864 (S. 296–307). Wie die deutsche Titelseite verriet, waren es die 1.300 Beispiele für harmonische Farbverbindungen zur Anwendung in Kunstindustrie, Raumausstattung, Mode und Kosmetik, die Guichards Buch so besonders machten und auf 166 teils doppelseitigen Farbtafeln präsentiert wurden. Darauf abgebildet sind Stoffmuster, Dekorationsobjekte vor verschiedenen Hintergrundfarben, streng auf Buchstaben und Schrift basierende Arrangements, figurative und ornamentale Muster sowie ganz und gar abstrakte Kompositionen, die allesamt der Veranschaulichung von Farbharmonien und Farbkontrasten dienen. Kein anderes Buch hatte jemals eine solche Vielfalt in der Darstellung von Farbarrangements geboten. Die Druckverfahren führen in voller Bandbreite vor Augen, was im Druckwesen des späten 19. Jahrhunderts in der Farbreproduktion möglich war. Tafeln in Schablonendruck wechseln sich ab mit Siebdrucktafeln für den Plakatdruck und Farblithografien von höchster Qualität aus der Druckerei F. Appel in Paris. Das mitunter als Musterbuch bezeichnete Werk ist viel mehr als das – es ist ohne Frage ein Buch über Farbe und Gestaltung, in dem das Beste vorgeführt wird, was Druckereien und Druckverfahren in dieser Zeit zu leisten vermochten. Dass in Bibliotheken und Museen weltweit jeweils nur weniger als zehn Exemplare der französischen und der deutschen Ausgabe verzeichnet sind, lässt stark darauf schließen, dass beide in nur sehr kleiner Auflage erschienen.

Für Guichard war damit noch nicht Schluss. Ebenfalls 1882

HARMONIE DER FARBEN

von E. GUICHARD ✻

6

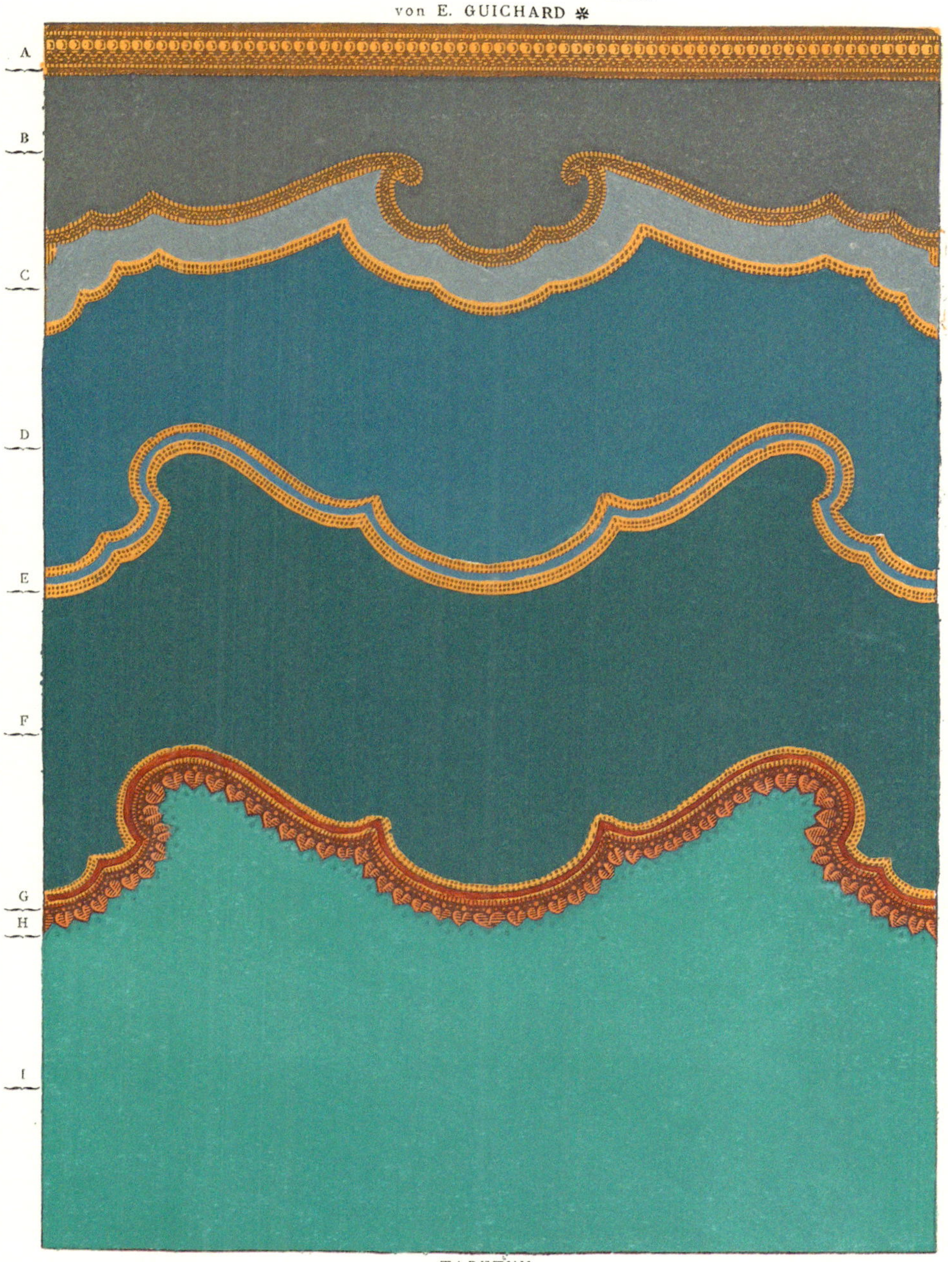

TAPETEN

Wenn die Grundfarbe einer Tapete wie I ist, so kann der Fries wie F und die Borde wie A, O und H sein.

VORHÄNGE und PORTIÈREN

Ist die Farbe der Tapete wie I, so kann die Grundfarbe der Vorhänge wie F sein, mit Bordüren wie I und Franzen oder Gimpen wie A, G und H — Wählt man die Vorhänge von derselben Farbe wie die Tapete, so wird man gut thun, um Eintönigkeit zu vermeiden, die Farbe F für d e Gallerieüberhänge (lambrequins) der Fenster zu nehmen.

THÜREN, LAMBRIS und KARNIESSE (GESIMSE, franz. CORNICHES)

Zu den Thüren und Lambris nimmt man die! Farben B, C, D und E. — Der Theil des Karniesses, welcher den Plafond berührt, muss in etwas lichteren Farbentönen gehalten sein als die Thüren und Lambris; — man nehme dazu daher nur die Töne C, D, E, nämlich die Farbe C als Grundfarbe des Karniesses, die Farbe D für die Wölbung desselben und die Farbe E für das an den Plafond angrenzende Gesims. — Die Farbe A kann in bescheidenem Mass zu feinen Zwischenlinien benutzt werden; statt dessen kann man übrigens auch Gold nehmen oder diese Verzierung ganz weglassen.

FRAUENKLEIDUNG

Die soeben angeführten Farbenzusammentellungen werden auch bei der feinen Damentoilette ihre Wirkung nicht verfehlen.

(Verlag von Wilh. Rommel in Frankfurt a. M.)

HARMONIE DER FARBEN

von E. GUICHARD ✱

7

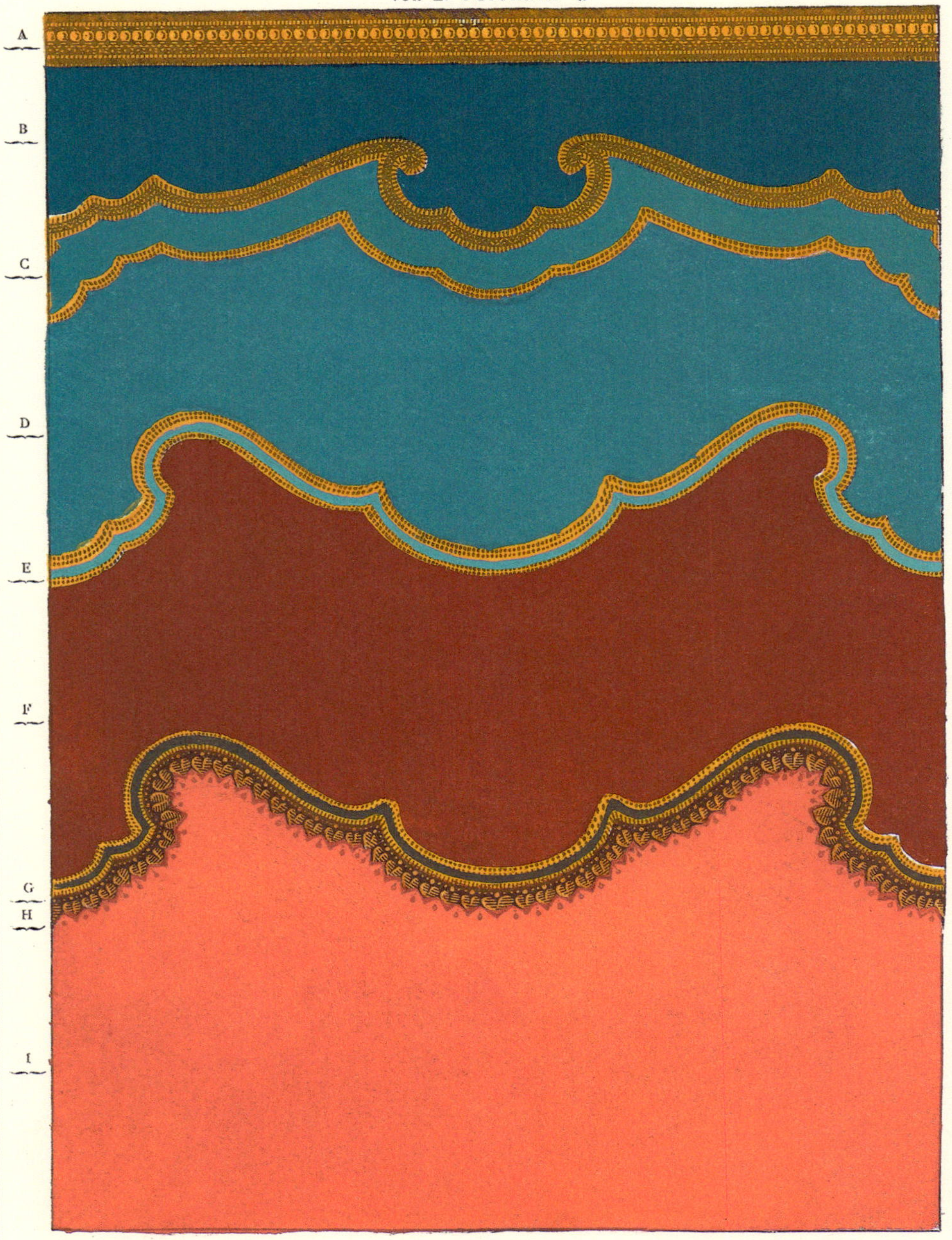

TAPETEN

Wenn die Grundfarbe einer Tapete wie I ist, so kann der Fries wie F und die Borde wie A, G und H sein.

VORHÄNGE und PORTIÈREN

Ist die Farbe der Tapete wie I, so kann die Grundfarbe der Vorhänge wie F sein, mit Bordüren wie I und Franzen oder Gimpen wie A, G und H. — Wählt man die Vorhänge von derselben Farbe wie die Tapete, so wird man gut thun, um Eintönigkeit zu vermeiden, die Farbe F für die Gallerieüberhänge (lambrequins) der Fenster zu nehmen.

THÜREN, LAMBRIS und KARNIESSE (GESIMSE, franz. CORNICHES)

Zu den Thüren und Lambris nimmt man die Farben B, C, D und E. — Der Theil des Karniesses, welcher den Plafond berührt, muss in etwas lichteren Farbentönen gehalten sein als die Thüren und Lambris; — man nehme dazu daher nur die Töne C, D, E, nämlich die Farbe C als Grundfarbe des Karniesses, die Farbe D für die Wölbung desselben und die Farbe E für das an den Plafond angrenzende Gesims. — Die Farbe A kann in bescheidenem Mass zu feinen Zwischenlinien benutzt werden; statt dessen kann man übrigens auch Gold nehmen oder diese Verzierung ganz weglassen.

FRAUENKLEIDUNG

Die soeben angeführten Farbenzusammenstellungen werden auch bei der feinen Damentoilette ihre Wirkung nicht verfehlen.

(Verlag von Wilh. Rommel in Frankfurt a. M.)

HARMONIE DER FARBEN

von E. GUICHARD ✻

12

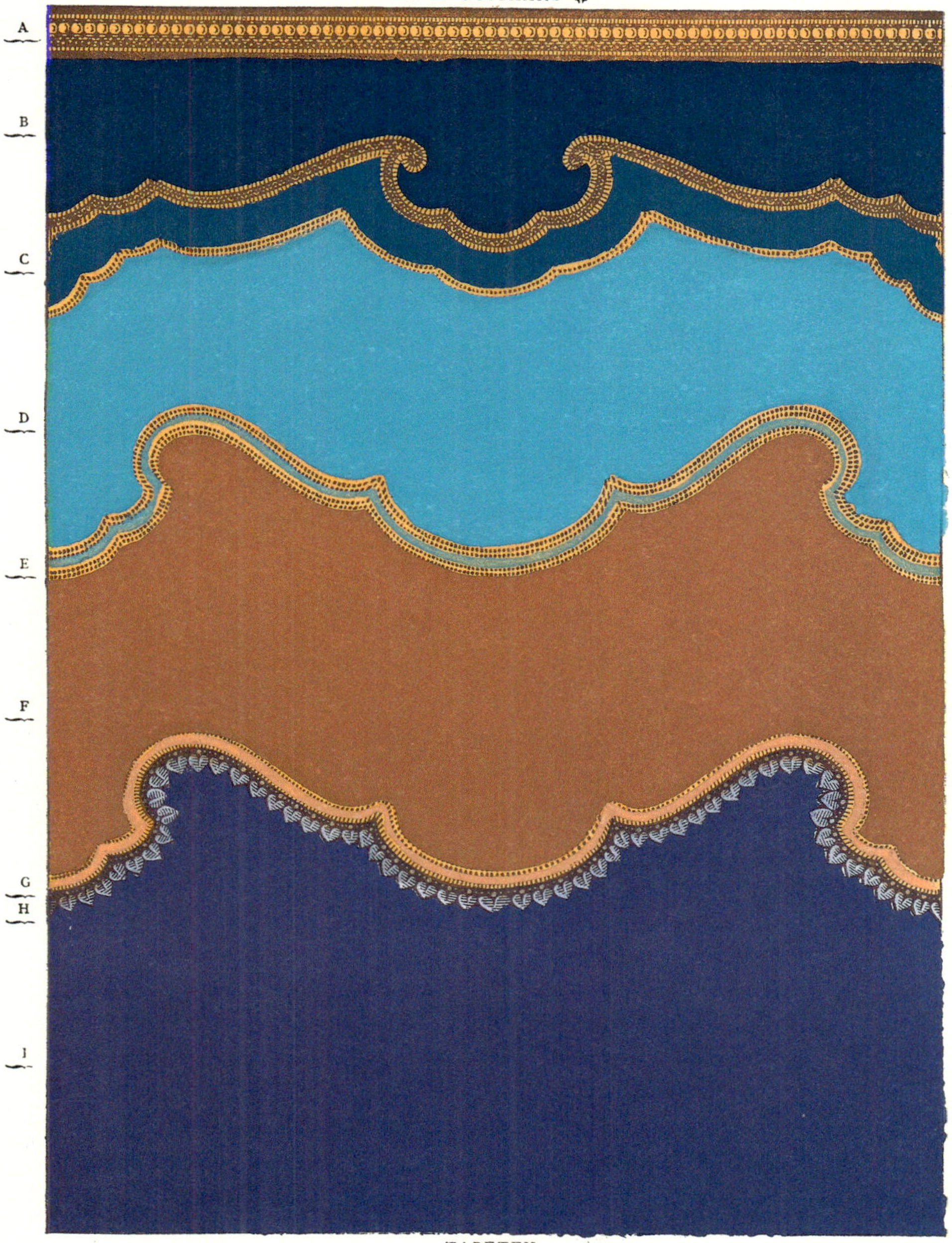

TAPETEN

Wenn die Grundfarbe einer Tapete wie I ist, so kann der Fries wie F und die Borde wie A, G und H sein.

VORHÄNGE und PORTIÈREN

Ist die Farbe der Tapete wie I, so kann die Grundfarbe der Vorhänge wie F sein, mit Bordüren wie I und Franzen oder Gimpen wie A, G und H. — Wählt man die Vorhänge von derselben Farbe wie die Tapete, so wird man gut thun, um Eintönigkeit zu vermeiden, die Farbe F für die Gallerieüberhänge (lambrequins) der Fenster zu nehmen.

THÜREN, LAMBRIS und KARNIESSE (GESIMSE, franz. CORNICHES)

Zu den Thüren und Lambris nimmt man die Farben B, C, D und E. — Der Theil des Karniesses, welcher den Plafond berührt, muss in etwas lichteren Farbentönen gehalten sein als die Thüren und Lambris; — man nehme dazu daher nur die Töne C, D, E, nämlich die Farbe C als Grundfarbe des Karniesses, die Farbe D für die Wölbung desselben und die Farbe E für das an den Plafond angrenzende Gesims. — Die Farbe A kann in bescheidenem Mass zu feinen Zwischenlinien benuzt werden; statt dessen kann man übrigens auch Gold nehmen oder diese Verzierung ganz weglassen.

FRAUENKLEIDUNG

Die soeben angeführten Farbenzusammenstellungen werden auch bei der feinen Damentoilette ihre Wirkung nicht verfehlen.

(Verlag von Wilh. Rommel in Frankfurt a. M.)

erschien bei H. Cagnon in Paris eine Fortsetzung seines Werks. Die dreibändige, im Querformat aufgelegte Publikation mit einem erläuternden Text in drei Sprachen und dem Titel *La Grammaire de la couleur / Grammar of Colour / Die Grammatik der Farben* umfasste diesmal 765 Farbtafeln (S. 456–457).

L'Harmonie des couleurs

Dessinateur, architecte et décorateur français, Édouard Guichard est l'auteur d'un ouvrage sur l'harmonie des couleurs d'une exceptionnelle beauté typographique et chromatique. Il est donc surprenant que ce livre soit rarement cité dans les ouvrages récents consacrés à l'histoire de la couleur et du dessin. *L'Harmonie des couleurs* a paru initialement en français en plusieurs tomes de 1880 à 1882, chez l'éditeur parisien Ad. Goubaud et Fils. L'édition allemande autorisée, dans une traduction de Georg Krebs, a probablement été réalisée presque en même temps que l'édition française et surpasse même celle-ci quant à la qualité du papier et des illustrations. Si l'introduction de Guichard ne comprend qu'une douzaine de pages, les images du volume in-folio allemand s'accompagnent de légendes extrêmement détaillées. De même que de nombreux auteurs d'ouvrages portant sur l'ordonnancement des couleurs dans les arts décoratifs à la fin du XIXe siècle, Guichard est très influencé par Chevreul, voire directement inspiré par l'excellence technique des planches de l'*Atlas* de ce dernier, paru en 1864 (p. 296–307). La page de titre de l'édition allemande annonce ce qui rend l'ouvrage de Guichard si singulier, à savoir les 1300 exemples de combinaisons et de dispositions harmonieuses des couleurs, appliquées aux métiers de l'art, à la décoration d'intérieur, à la mode et au maquillage. Présentés dans 166 planches en couleurs, parfois sur des doubles pages, ces exemples proposent des motifs textiles, des objets décoratifs sur différents fonds colorés, des arrangements strictement typographiques, des dessins figuratifs et ornementaux et des compositions abstraites, tous destinés à illustrer l'harmonie et le contraste des couleurs. Aucun ouvrage antérieur n'offre une telle variété dans la représentation des compositions chromatiques. Les techniques d'impression témoignent de toutes les possibilités offertes par la reproduction en couleurs à la fin du XIXe siècle : des planches au pochoir, plusieurs sérigraphies illustrant des affiches publicitaires, ainsi que des chromolithographies d'excellente qualité réalisées par F. Appel. Parfois qualifié d'album d'échantillons, l'ouvrage est bien plus que cela : il a incontestablement pour objet la couleur et la conception graphique et donne à voir ce que les imprimeurs et les techniques d'impression de l'époque pouvaient réaliser de mieux. Moins de dix exemplaires de chaque édition sont conservés dans les bibliothèques et les musées du monde entier, signe que le tirage a dû être très limité.

Guichard n'en est pas resté là. En 1882, la suite de cette œuvre paraît à Paris chez H. Cagnon, en trois volumes trilingues sous le titre *La Grammaire de la couleur / Grammar of Colour / Die Grammatik der Farben* et dans un format oblong, avec cette fois 765 planches en couleurs (p. 456–457).

La armonía de los colores

Édouard Guichard fue un diseñador, arquitecto y decorador francés, pero a la vista de la extraordinaria belleza tipográfica y cromática de su libro sobre la armonía del color, es sorprendente lo poco que se lo menciona en los textos recientes sobre la historia del color y del diseño. Entre 1880 y 1882, Ad. Goubaud et Fils publicó por entregas en París la obra francesa original, *L'Harmonie des couleurs*. La edición alemana autorizada, con traducción de Georg Krebs, debió de prepararse de forma casi simultánea, y supera a la edición francesa en cuanto a calidad, tanto del papel como de las ilustraciones. Apenas cuenta con una docena de páginas de texto introductorio de Guichard, pero las ilustraciones del infolio alemán están acompañadas de extensas leyendas descriptivas. Al igual que muchos autores posteriores del siglo XIX que trataron la disposición del color y las artes decorativas, Guichard estuvo muy influenciado por Chevreul, y quizás se sintió directamente inspirado por la brillantez técnica de las láminas del *Atlas* de Chevreul de 1864 (págs. 296–307). La portada alemana también anuncia lo que hizo del libro de Guichard algo tan especial: 1300 ejemplos de armoniosos arreglos y combinaciones de colores para su uso en los campos del arte, la decoración de interiores, la moda y el maquillaje. Tales ejemplos se muestran en 166 láminas en color, algunas a doble página, que representan patrones textiles, objetos decorativos sobre diferentes colores de fondo, disposiciones estrictamente tipográficas, diseños figurativos y ornamentales y composiciones totalmente abstractas, siempre para ilustrar la armonía y el contraste cromático. Ningún otro libro había mostrado antes tal variedad en la representación de arreglos de color. Las técnicas de impresión muestran la gama completa de lo que se podía lograr con la reproducción del color a finales del siglo XIX, incluidas láminas estarcidas, varias láminas serigrafiadas de carteles publicitarios y cromolitografías de calidad superior aportadas por F. Appel. A veces se describe como un libro de patrones, pero es mucho más que eso, ya que es indiscutiblemente un libro sobre color y diseño que muestra lo mejor que las imprentas y la tecnología de impresión eran capaces de producir en esa época. En la actualidad, hay menos de diez copias de cada edición, la francesa y la alemana, registradas en bibliotecas y museos de todo el mundo, lo que sugiere claramente que se publicaron tiradas muy pequeñas.

Guichard, sin embargo, no se detuvo en ese punto, y en 1882 H. Cagnon publicó en París una continuación de su obra en tres volúmenes y trilingüe con el título *La Grammaire de la couleur / Grammar of Colour / Die Grammatik der Farben*, en un formato apaisado y esta vez con 765 láminas en color (págs. 456–457).

HARMONIE DER FARBEN

von E. GUICHARD ❋

22

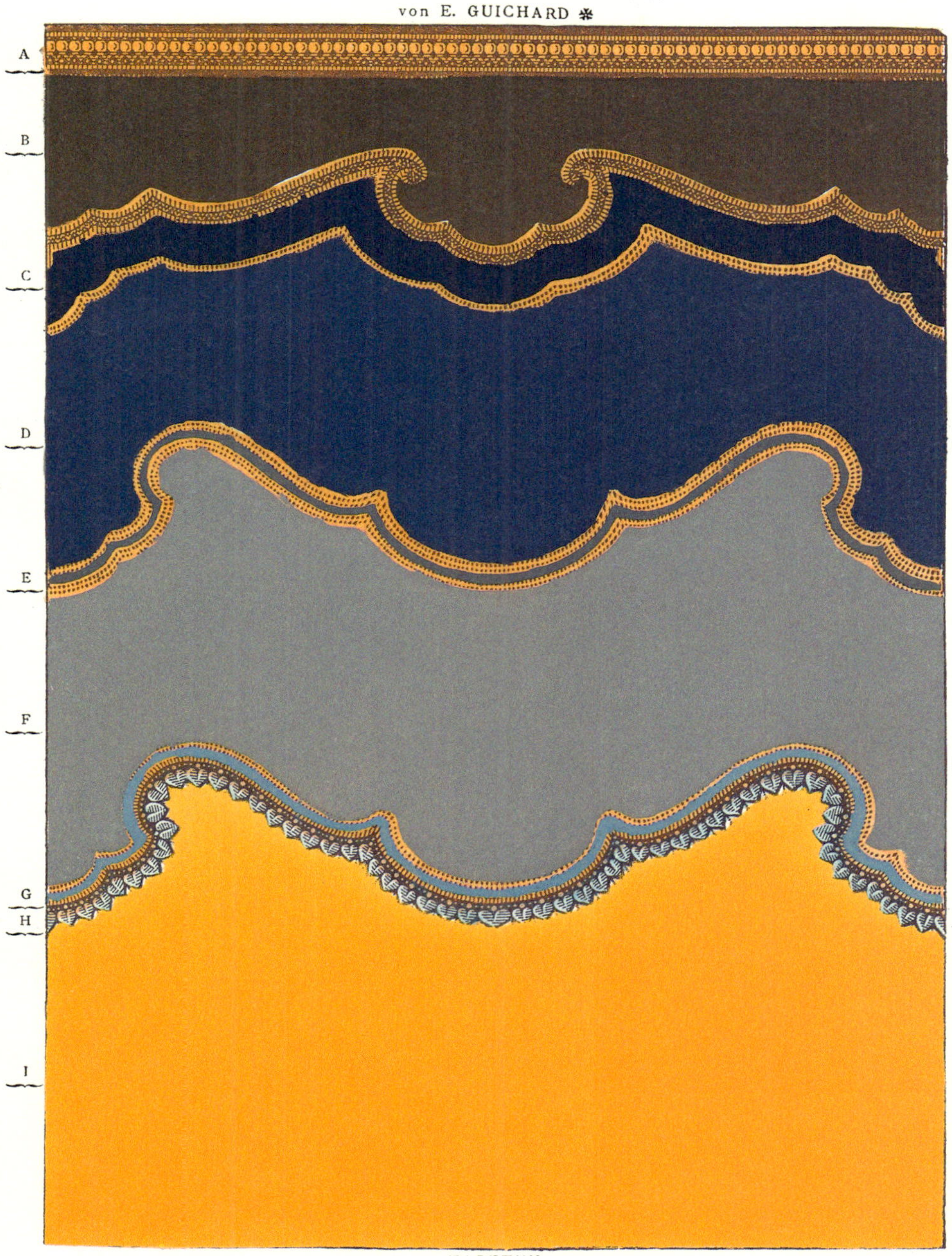

TAPETEN

Wenn die Grundfarbe einer Tapete wie I ist, so kann der Fries wie F und die Borde wie A, O und H sein.

VORHÄNGE und PORTIÈREN

Ist die Farbe der Tapete wie I, so kann die Grundfarbe der Vorhänge wie F sein, mit Bordüren wie I und Franzen oder Gimpen wie A, G und H. — Wählt man die Vorhänge von derselben Farbe wie die Tapete, so wird man gut thun, um Eintönigkeit zu vermeiden, die Farbe F für die Gallerieüberhänge (lambrequins) der Fenster zu nehmen.

THÜREN, LAMBRIS und KARNIESSE (GESIMSE, franz. CORNICHES)

Zu den Thüren und Lambris nimmt man die Farben B, C, D und E. — Der Theil des Karniesses, welcher den Plafond berührt, muss in etwas lichteren Farbentönen gehalten sein als die Thüren und Lambris; — man nehme dazu daher nur die Töne C, D, E, nämlich die Farbe C als Grundfarbe des Karniesses, die Farbe D für die Wölbung desselben und die Farbe E für das an den Plafond angrenzende Gesims. — Die Farbe A kann in bescheidenem Mass zu feinen Zwischenlinien benutzt werden; statt dessen kann man übrigens auch Gold nehmen oder diese Verzierung ganz weglassen.

FRAUENKLEIDUNG

Die soeben angeführten Farbenzusammentellungen werden auch bei der feinen Damentoilette ihre Wirkung nicht verfehlen.

(Verlag von Wilh. Rommel in Frankfurt a. M.)

HARMONIE DER FARBEN

von E. GUICHARD

25

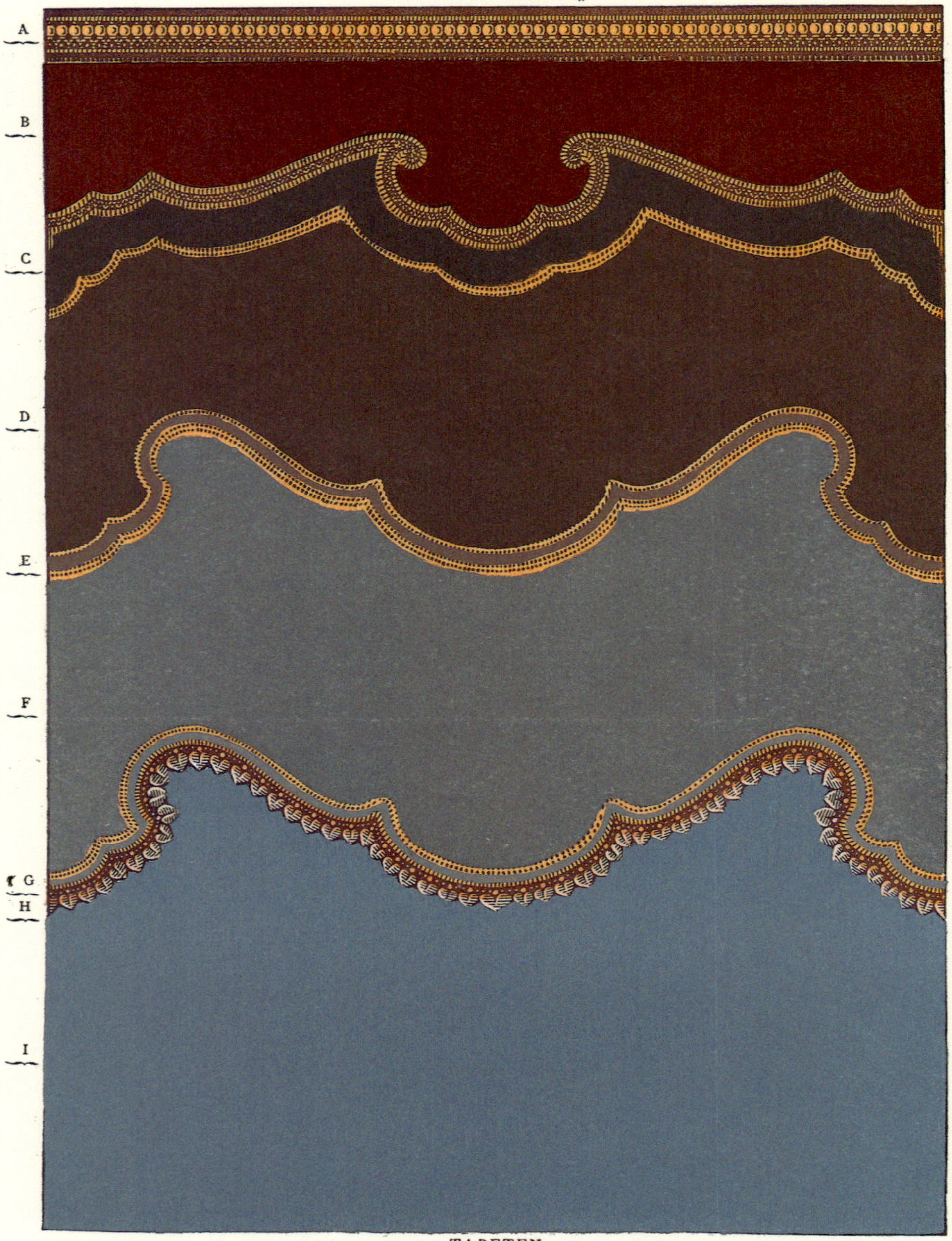

TAPETEN

Wenn die Grundfarbe einer Tapete wie I ist, so kann der Fries wie F und die Borde wie A, G und H sein.

VORHÄNGE und PORTIÈREN

Ist die Farbe der Tapete wie I, so kann die Grundfarbe der Vorhänge wie F sein, mit Bordüren wie I und Franzen oder Gimpen wie A, G und H. — Wählt man die Vorhänge von derselben Farbe wie die Tapete, so wird man gut thun, um Eintönigkeit zu vermeiden, die Farbe F für die Gallerieüberhänge (lambrequins) der Fenster zu nehmen.

THÜREN, LAMBRIS und KARNIESSE (GESIMSE, franz. CORNICHES)

Zu den Thüren und Lambris nimmt man die Farben B, C, D und E. — Der Theil des Karniesses, welcher den Plafond berührt, muss in etwas lichteren Farbentönen gehalten sein als die Thüren und Lambris; — man nehme dazu daher nur die Töne C, D, E, nämlich die Farbe C als Grundfarbe des Karniesses, die Farbe D für die Wölbung desselben und die Farbe E für das an den Plafond angrenzende Gesims. — Die Farbe A kann in bescheidenem Mass zu feinen Zwischenlinien benuzt werden; statt dessen kann man übrigens auch Gold nehmen oder diese Verzierung ganz weglassen.

FRAUENKLEIDUNG

Die soeben angeführten Farbenzusammenstellungen werden auch bei der feinen Damentoilette ihre Wirkung nicht verfehlen.

(Verlag von Wilh. Rommel in Frankfurt a. M.)

A. — **HARMONIE DER FARBEN** — Nº 26

von E. GUICHARD ✲

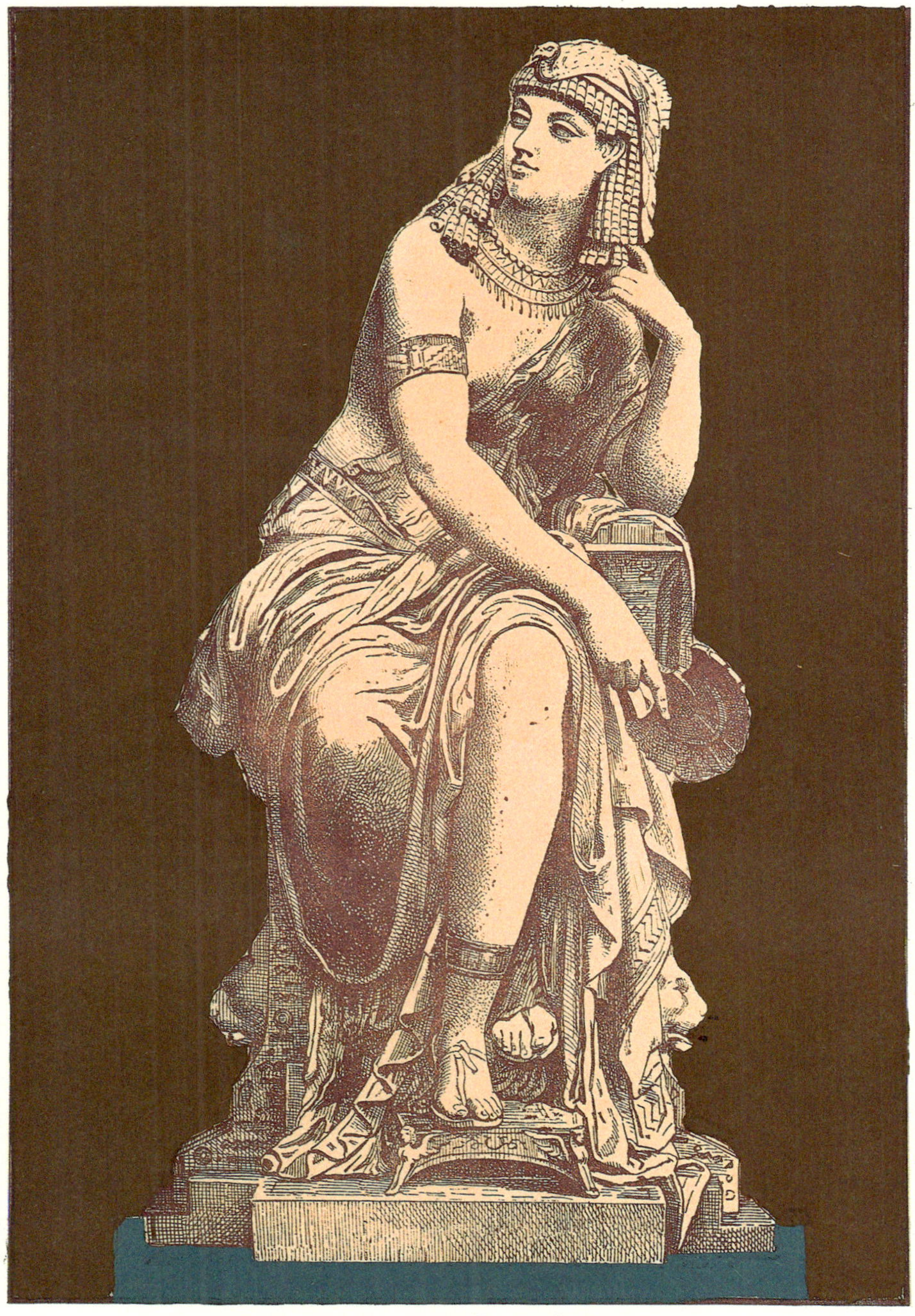

FARBENZUSAMMENSTELLUNGEN FÜR VORHÄNGE UND TAPETEN

WELCHE SO AUSGEWÄHLT SIND, DASS SIE WERKE DER BILDHAUERKUNST HERVORTRETEN LASSEN, UND ZUGLEICH MIT DER NAÜTRLICHEN ODER KÜNSTLICHEN FARBE DER STATUE HARMONIEREN

1. Wir denken uns eine Statue in gebranntem Thon, welche sich von einem braunen Grund abhebt, der zwar auf die Farbe der Statue nur geringe Einwirkung ausübt, aber doch dieselbe etwas heller erscheinen lässt. Die braune Hervorhebungs-Farbe passt vorzüglich zu dem Farbenton der Statue. 2. Die Farbe des Untergestells, welche zur Skala des Grün gehört und deren Ergänzungs-Farbe Roth ist, lässt den Hauptgegenstand schwach rosa erscheinen. Da sich Statue und Untergestell deutlich von der dunkeln Farbe der Tapete abheben, so sind die wesentlichen Bedingungen erfüllt, um den Blick auf den Hauptgegenstand zu ziehen.

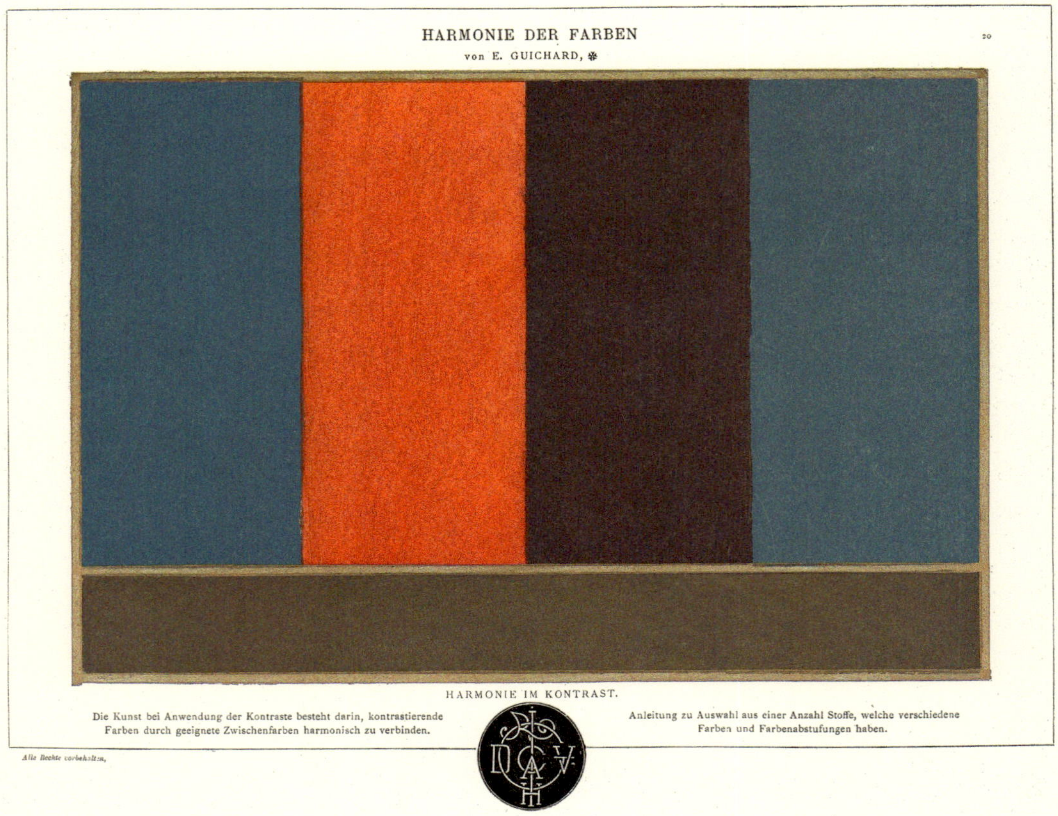
HARMONIE DER FARBEN
von E. GUICHARD, ❋
20
HARMONIE IM KONTRAST.
Die Kunst bei Anwendung der Kontraste besteht darin, kontrastierende Farben durch geeignete Zwischenfarben harmonisch zu verbinden.
Anleitung zu Auswahl aus einer Anzahl Stoffe, welche verschiedene Farben und Farbenabstufungen haben.
Alle Rechte vorbehalten,

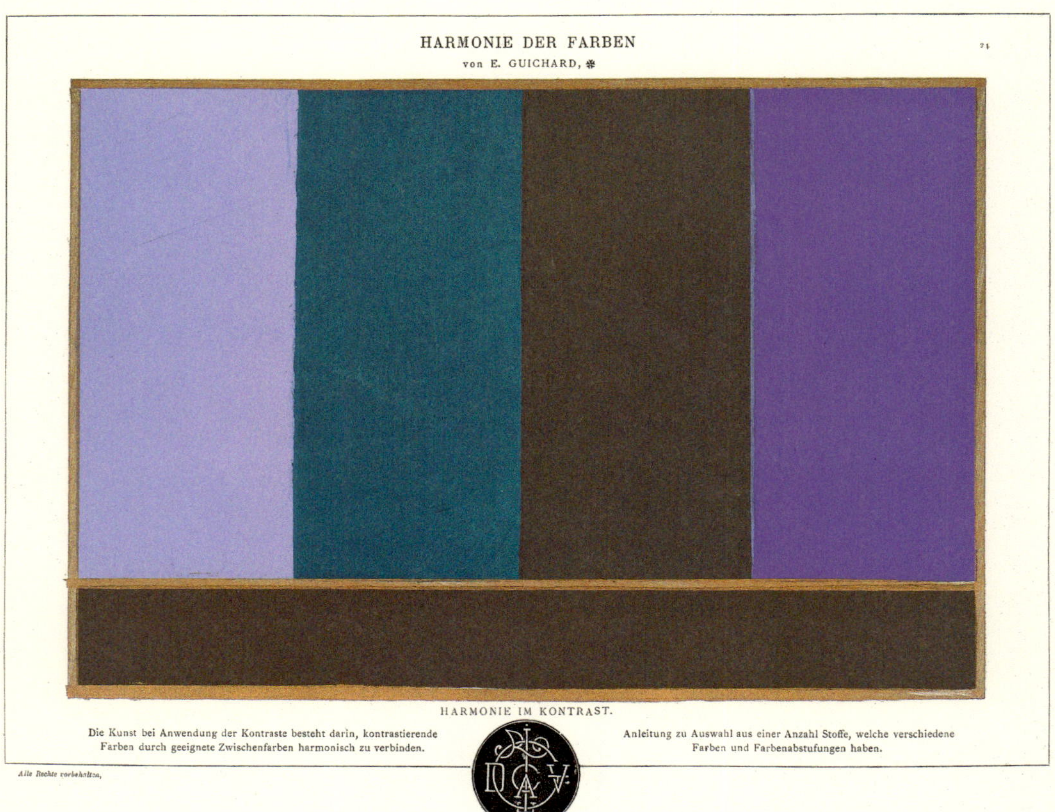
HARMONIE DER FARBEN
von E. GUICHARD, ❋
21
HARMONIE IM KONTRAST.
Die Kunst bei Anwendung der Kontraste besteht darin, kontrastierende Farben durch geeignete Zwischenfarben harmonisch zu verbinden.
Anleitung zu Auswahl aus einer Anzahl Stoffe, welche verschiedene Farben und Farbenabstufungen haben.
Alle Rechte vorbehalten,

F.

HARMONIE DER FARBEN

von E. GUICHARD

Nº 31

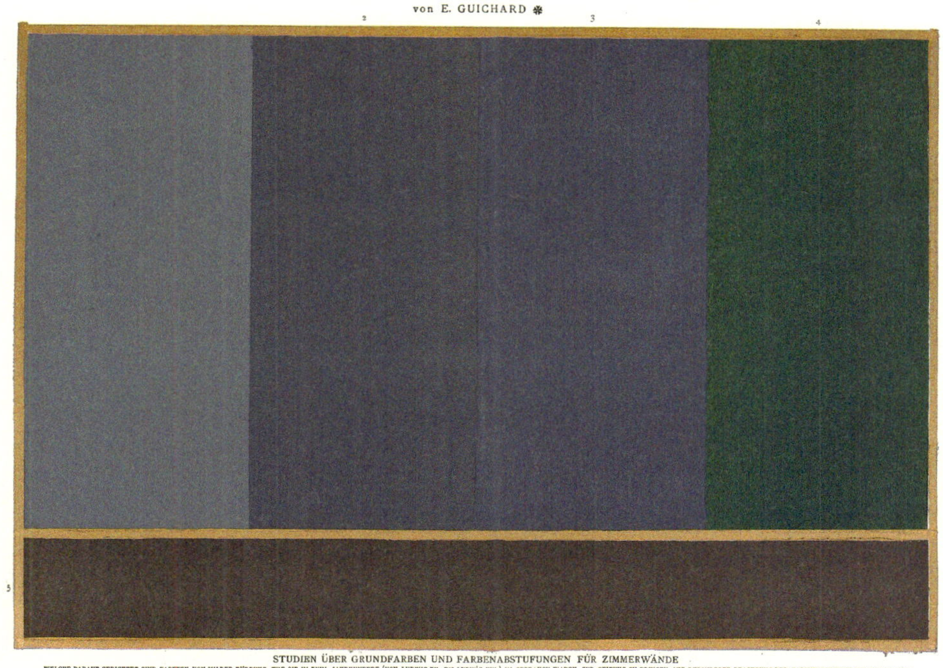

STUDIEN ÜBER GRUNDFARBEN UND FARBENABSTUFUNGEN FÜR ZIMMERWÄNDE

WELCHE DARAUF GERICHTET SIND TAPETEN VON MILDER FÄRBUNG, WIE SIE IM XVIII. JAHRHUNDERT (VON LUDWIG XV. BIS LUDWIG XVI.) IM GEBRAUCH WAREN, ZUR GEITUNG ZU BRINGEN, MIT BESONDERER BEACHTUNG DER HARMONISCHEN ZUSAMMENWIRKUNG

BEMERKUNG. — Wenn Gold und Gelb vorherrscht, so ist vorzugsweise 1, 2 und 8 zu wählen. — Wenn dagegen Rot dominirt, so ist 1, 2, 4 und 5 zu wählen. — Bei Violett und Lila ist 1, 2 und 4 zu nehmen und bei Blau 1, 2 und 5, indem man zugleich diese Hervorhebungsfarben um so dunkler nimmt, je satter die Farbe der Tapete ist.

ZUSATZ. — Die Nüancen 1 und 2 geben dem Gold und dem Gelben einen besonderen Glanz und eine schöne Frische, ohne indessen den Grundton der Farbe zu verändern, während Nr. 3, welches als Violett Grüngelb zur Komplementär-Farbe hat, einen, wenn auch unbedeutenden grünlichen Schimmer auf das Gold wirft. — Die Nüance 4, welche grünlich ist und zur Komplementär-Farbe Rot hat, giebt der Tapete meist einen feurigeren, schwach rötlichen Schein; Nr. 5 endlich ist vorzuziehen, wenn Blau in der Tapete vorherrscht, da es mit diesen harmoniert uud dasselbe hervohebt.

HARMONIE DER FARBEN

von E. GUICHARD,

48.

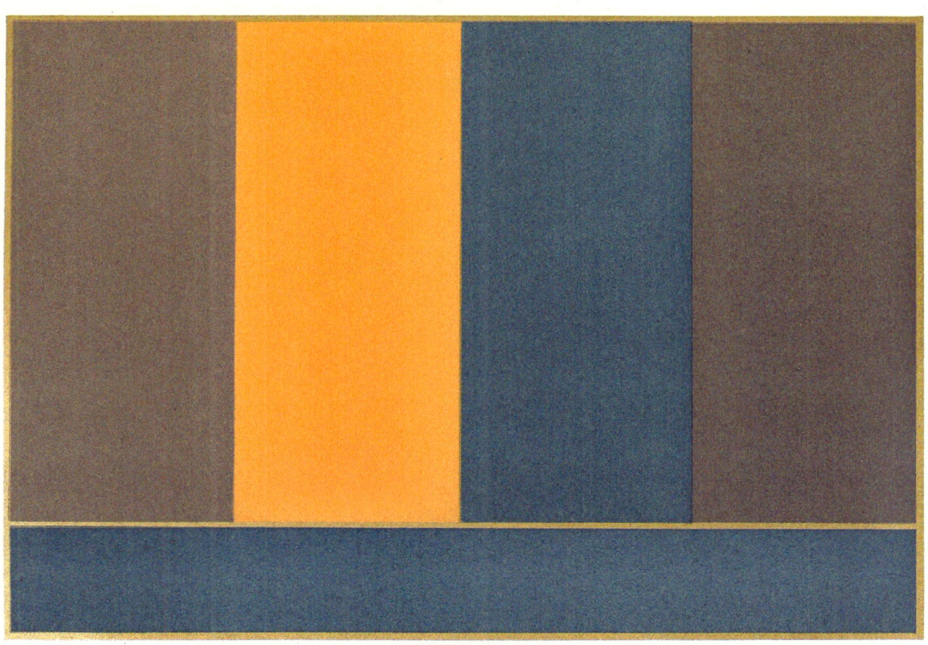

HARMONIE IM KONTRAST.

Die Kunst bei Anwendung der Kontraste besteht darin, kontrastierende Farben durch geeignete Zwischenfarben harmonisch zu verbinden.

Anleitung zu Auswahl aus einer Anzahl Stoffe, welche verschiedene Farben und Farbenabstufungen haben.

CHROMO. LITH. F. APPEL. 72 R. DU DELTA. PARIS.

HARMONIE DER FARBEN
von E. GUICHARD.

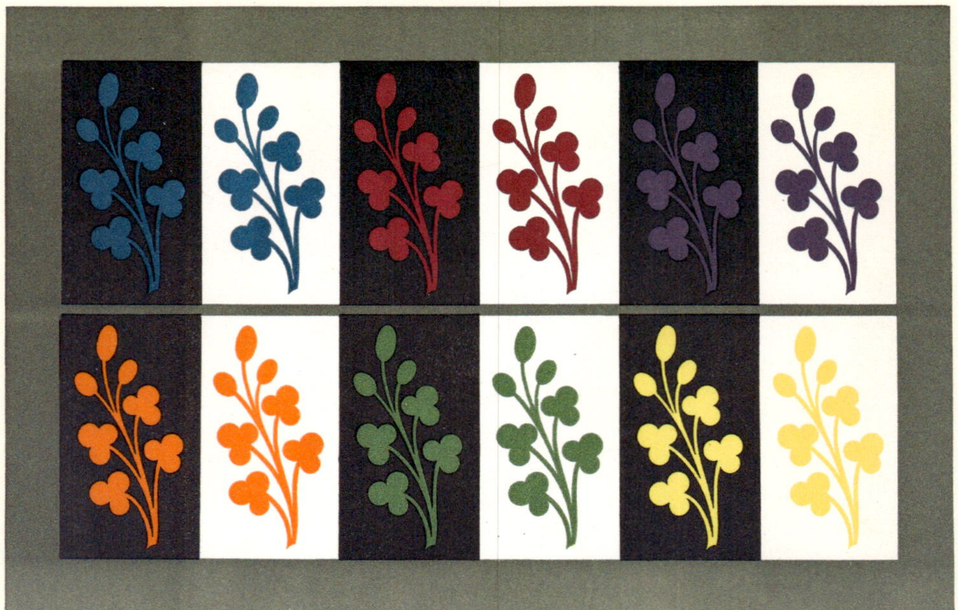

Die Ergänzungsfarbe von Blau ist Orangegelb und *umgekehrt*

Die Ergänzungsfarbe von Rot ist Grün und *umgekehrt*

Die Ergänzungsfarbe von Violett ist Grünlichgelb und *umgekehrt*

1te TAFEL DER VERSUCHSREIHE
ÜBER DEN EINFLUSS DES WEISS UND DES SCHWARZ AUF FARBIGE KŒRPER

Alle Farben verlieren, auf oder neben weissem Grund, an Glanz und an ihrem natürlichen Ton, sie verändern sich, verdunkeln, oder vielmehr, sie werden schwarz durch diesen Contact. Dagegen gewinnen alle Farben einen vollkommeneren Glanz durch Zusammenlagerung mit Schwarz; sie werden leuchtender, ohne in ihrem ursprünglichen Grundton verändert zu werden.

Beispiele: Wenn man mit einer gewissen Aufmerksamkeit auf Schwarz gesetztes Rot betrachtet und dabei die Augen soviel als möglich nähert, wird man leicht gewahr werden, dass das Rot fast wie Granat erscheint, während es auf schwarzem Grund oder in schwarzer Einfassung feurig und glänzend wird. Um den durch Berührung mit dem Weiss hervorgerufenen Effect zu verringern, müsste man dem auf oder neben das Weiss gesetzten Rot Zinnober zufügen, um es auf die Tonhöhe des auf das Schwarze gesetzten natürlichen Rot zu bringen; ebenso müssen alle übrigen Farben, das Blau, das Orange, das Grün, das Violette und das Grüngelbe durch Zusatz von etwas Weiss lichter gemacht werden, wenn sie unter dem Einfluss des Weiss stehen; man lässt sie dagegen unverändert, wenn sie auf oder neben Schwarz stehen. Aus diesen Bemerkungen erhellt, dass, wenn es sich um die Toilette handelt, sorgfältig jede directe Berührung des Weiss mit der menschlichen Hautfarbe zu vermeiden ist.

N. B. — Der Ton der Einrahmung vorstehender Bilder ist durch Mischung aller hier vorkommenden Farben, unter Zusatz von gleichen Teilen Weiss und Schwarz hergestellt.

CHROMO. LITH. F. APPEL. 12 R. DU DELTA. PARIS.

HARMONIE DER FARBEN
von E. GUICHARD.

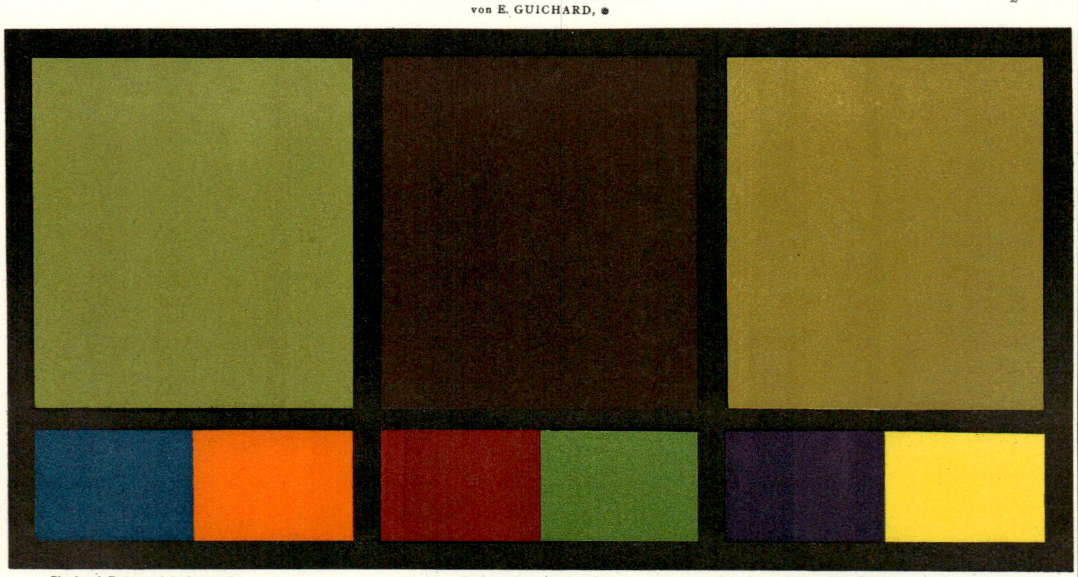

Blau hat als Ergänzungsfarbe Orangegelb.
Blau und Gelb, auf oder neben einander, heben sich gegenseitig.

Rot hat als Ergänzungsfarbe Grün.
Rot und Grün, auf oder neben einander, heben sich gegenseitig.

Violett hat als Ergänzungsfarbe Grüngelb.
Violett und Grüngelb auf oder neben einander, heben sich gegenseitig.

VERBINDUNG DER ERGÄNZUNGSFARBEN

Die Mischfarben, welche wir über die Ergänzungsfarben gesetzt haben, sind je durch Mischung von gleichen Teilen der darunterstehenden zwei Ergänzungsfarben hergestellt. Eine Farbe giebt neben ihrer Ergänzungsfarbe, wie schon früher bemerkt, eine Zusammenstellung welche uns hart und wenig harmonisch erscheint; aber in gewissen Fällen, wenn die Nebeneinanderlagerung nicht zu ändern ist, raten wir bei dem allgemeinen Arrangement so viel als möglich den gemischten Ton, welcher aus den beiden Ergänzungsfarben besteht, mit zu benutzen, was die bezeichnete Härte des Tones ein wenig mildert; wobei wir allerdings voraussetzen, dass das decorative Arrangement in diesem Sinne von einem Künstler, welcher seine Farben anzubringen und zu dosiren versteht, angelegt wird.

CHROMO. LITH. F. APPEL. 12 R. DU DELTA. PARIS.

HARMONIE DER FARBEN
von E. GUICHARD, ✻

39

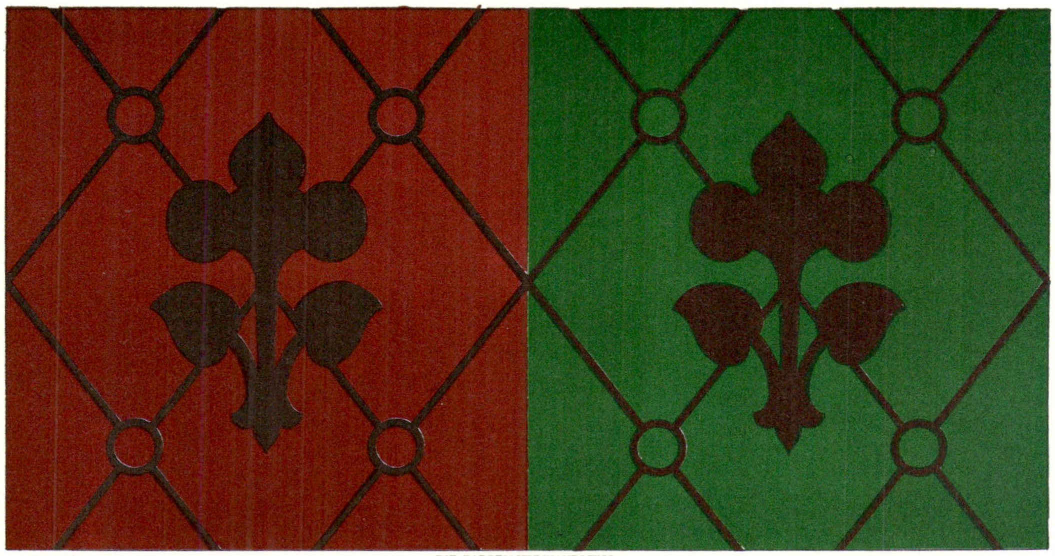

Rot hat als Ergänzungsfarbe Grün.

DIE FARBEN-VERWANDLUNG

VERÄNDERUNGEN, WELCHE DIE FARBEN DURCH IHRE AUF-UND NEBENEINANDERLAGERUNG ERLEIDEN

Grün hat als Ergänzungsfarbe Rot.

Um den Effect des gegenseitigen Einflusses der beiden vorstehenden, sich ergänzenden Farben so sichtbar als möglich zu machen, haben wir das Rote mit dem Grün zu gleichen Teilen gemischt, wobei wir einen braunen Ton erhielten; dadurch gelang es das gleiche Ornaments-Motif so gut auf dem roten Grunde, als auf dem grünen Grund zu producieren. Man erkennt an dieser Tafel, dass der braune Ton in vollen roten Grund ergossen, sich lichtet, indem er sich mit dem grünen Ergänzungston sättigt, und dass derselbe braune Ton, in vollen grünen Grund ergossen, sich verdunkelt, indem er sich mit dem roten Ergänzungston des Grün verbindet.

Um die durch die Ergänzung erhaltene Verwandlung gut zu beurteilen, muss man die Augenlider halb niederschlagen, und bei geringer Entfernung, die zwei Ornamente gleichzeitig betrachten; wendet man nun die Augen rasch von einem Motif zum andern, so sieht man fast sofort die Veränderung des Tones. Der Ton des Motifs auf rotem Grund erscheint ganz anders als derjenige auf grünem Grunde und ebenso im umgekehrten Falle; obschon der Ton auf beiden ganz derselbe war. Endlich, damit die Mischungs-Töne einander ähnlich sehen und in gleicher Tonhöhe erscheinen, muss man dem braunen Ton ein wenig Grün hinzufügen oder dem auf das Blatt gemalten braunen Ton ein wenig Rot zusetzen. Der Einfluss des einen oder andern Ergänzungstones wird teilweise verschwinden, wenn man darauf achtet, diese Veränderung nur auf dem einen oder andern Ornamentsmotif anzubringen, und nicht auf beiden zugleich.

CHROMO. LITH. F. APPEL. 12 R. DU DELTA. PARIS.

HARMONIE DER FARBEN
von E. GUICHARD, ✻

40.

Violett hat als Ergänzungsfarbe Grünlichgelb.

DIE FARBEN-VERWANDLUNG

VERÄNDERUNGEN, WELCHE DIE FARBEN DURCH IHRE AUF-UND NEBENEINANDERLAGERUNG ERLEIDEN

Grünlichgelb hat als Ergänzungsfarbe Violett.

In den vorhergehenden Tafeln haben wir mit dem gemischten Ton experimentieren wollen, welcher aus der Vereinigung (zu gleichen Teilen) von zwei sich ergänzenden Farben entsteht, um die Veränderung zu zeigen, welchen derselbe durch den Effect der Aufeinanderlagerung erleidet. Auf dem vorstehenden Bilde operieren wir mit einer Farbe, welche mit den zwei sich ergänzenden Farben nichts gemein hat. Die Lettern, welche in grünlichgrauer Farbe auf violettem und gelbem Grunde stehen, erscheinen durch die Wirkung der verschiedenen Grundfarben als verschiedene Töne, d. h. der Buchstabe S in Grünlichgrau auf violettem Grunde gemalt, ist durch die ihn ergänzende grünlichgelbe Farbe so beeinflusst, dass er als grünlich erscheint, während der auf grünlichgelbem Grunde gemalte Buchstabe dunkler und zwar mehr violett aussieht.

CHROMO. LITH. F. APPEL. 12 R. DU DELTA. PARIS.

HARMONIE DER FARBEN

von E. GUICHARD, 34

Harmonie im Contraste der Farben. A Decke, B Karnies, C Behänge, D Getäfel.

CHROMO. LITH. J. APPEL, 12. R. DU DELTA PARIS

HARMONIE DER FARBEN
von E. GUICHARD, ●

41.

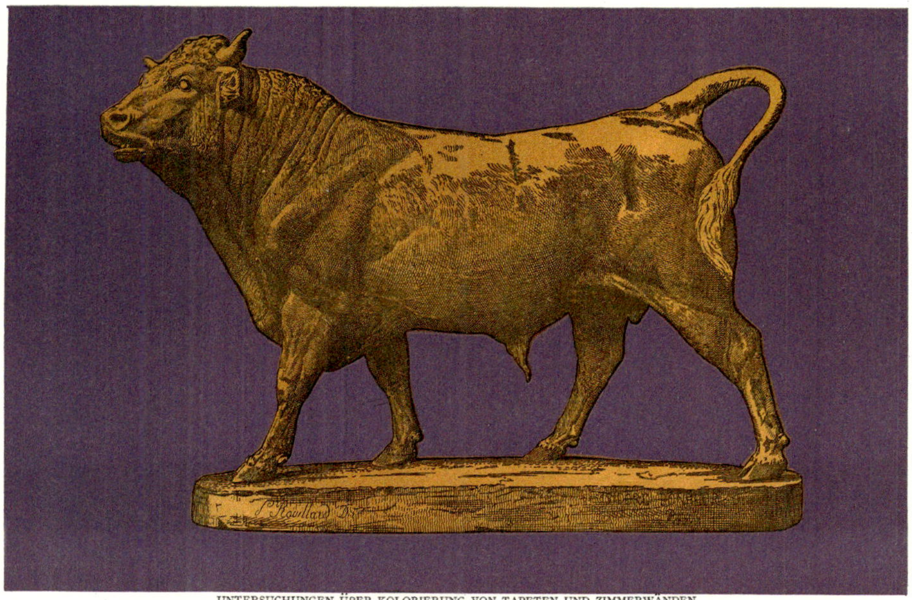

UNTERSUCHUNGEN ÜBER KOLORIERUNG VON TAPETEN UND ZIMMERWÄNDEN

WELCHE DEN ZWECK HAT WERKE DER RILDHAUERKUNST OHNE SCHROFFEN ÜBERGANG HERVORTRETEN ZU LASSEN, SO DASS DIE NATÜRLICHE ODER KÜNSTLICHE FARBE DERSELBEN MIT DER DES HINTERGRUNDES HARMONIERT

1, Auf dieser Tafel hebt sich eine Statue in vergoldeter Bronce von violettem Grunde ab; obwohl nun die Komplementärfarbe zu Violett Gelbgrün ist, so hat dies doch wenig Einfluss auf die sehr lebhafte Farbe der Vergoldung. 2, die Komplementärarbe wirkt nur auf die Teile der Statue, welche der Tapete am nächsten, indem sie dieselben etwas fahler erscheinen lässt, was übrigens dem Gesammteindruck nicht schädlich ist.

CHROMO. LITH. F. APPEL 72 R. DU DELTA. PARIS.

HARMONIE DER FARBEN
von E. GUICHARD, ●

42.

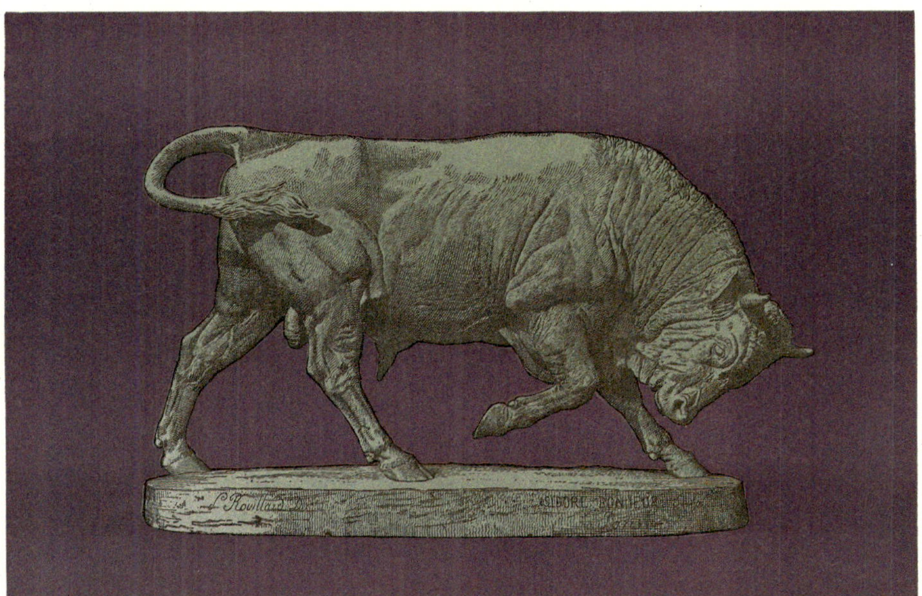

UNTERSUCHUNGEN ÜBER KOLORIERUNG VON TAPETEN UND ZIMMERWÄNDEN

WELCHE DEN ZWECK HAT WERKE DER RILDHAUERKUNST OHNE SCHROFFEN ÜBERGANG HERVORTRETEN ZU LASSEN, SO DASS DIE NATÜRLICHE ODER KÜNSTLICHE FARBE DERSELBEN MIT DER DES HINTERGRUNDES HARMONIERT

Auf dieser Tafel hebt sich eine Statue in versilberter Bronce von einem violetten Grunde ab; da die Komplementärfarbe zu Violett Gelbgrün ist, (aber weniger ausgeprochen, als auf Tafel 41, auf welcher das Violett anders nüanziert ist), so macht sie, wenn auch in geringem Mass, die Teile der Statue lebhafter erglänzen, welche dem Hintergrund näher sind.

CHROMO. LITH. F. APPEL 72 R. DU DELTA. PARIS.

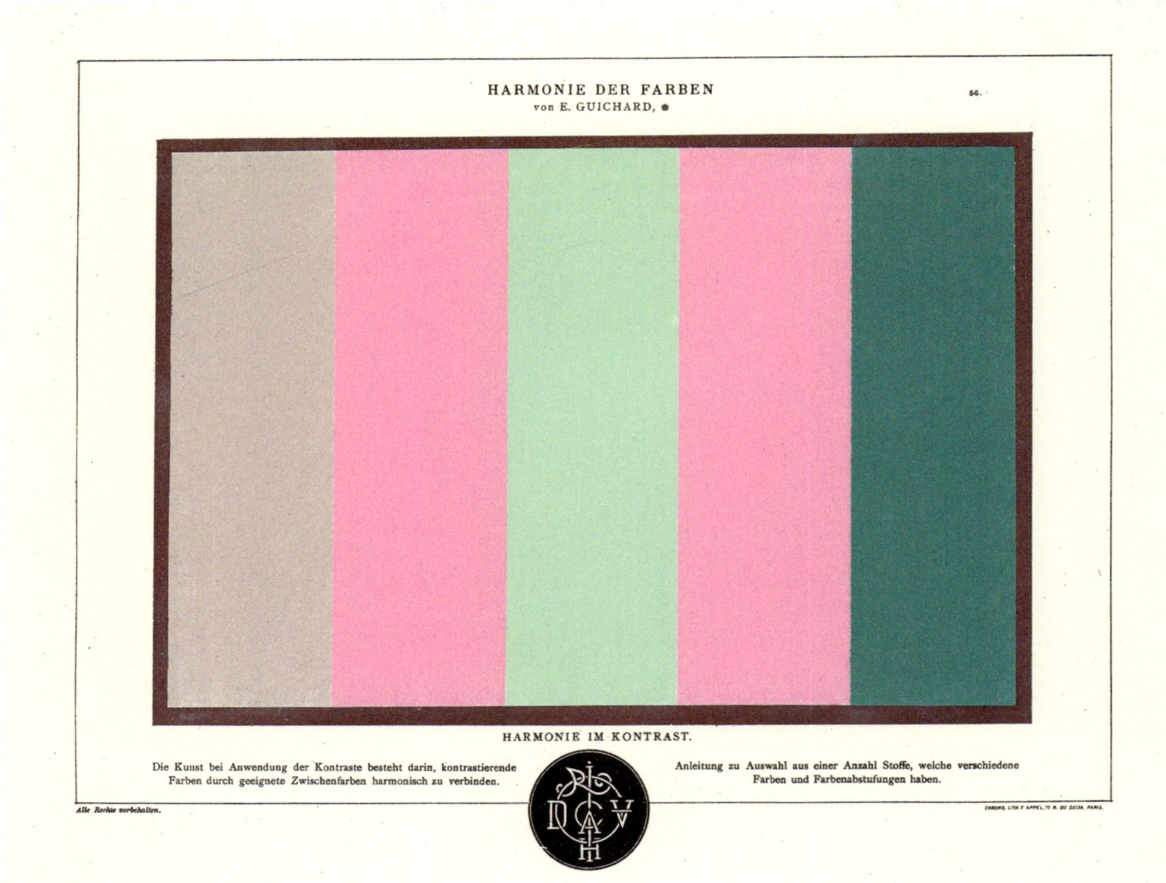
HARMONIE DER FARBEN
von E. GUICHARD, ✻
56.
HARMONIE IM KONTRAST.
Die Kunst bei Anwendung der Kontraste besteht darin, kontrastierende Farben durch geeignete Zwischenfarben harmonisch zu verbinden.
Anleitung zu Auswahl aus einer Anzahl Stoffe, welche verschiedene Farben und Farbenabstufungen haben.
Alle Rechte vorbehalten.

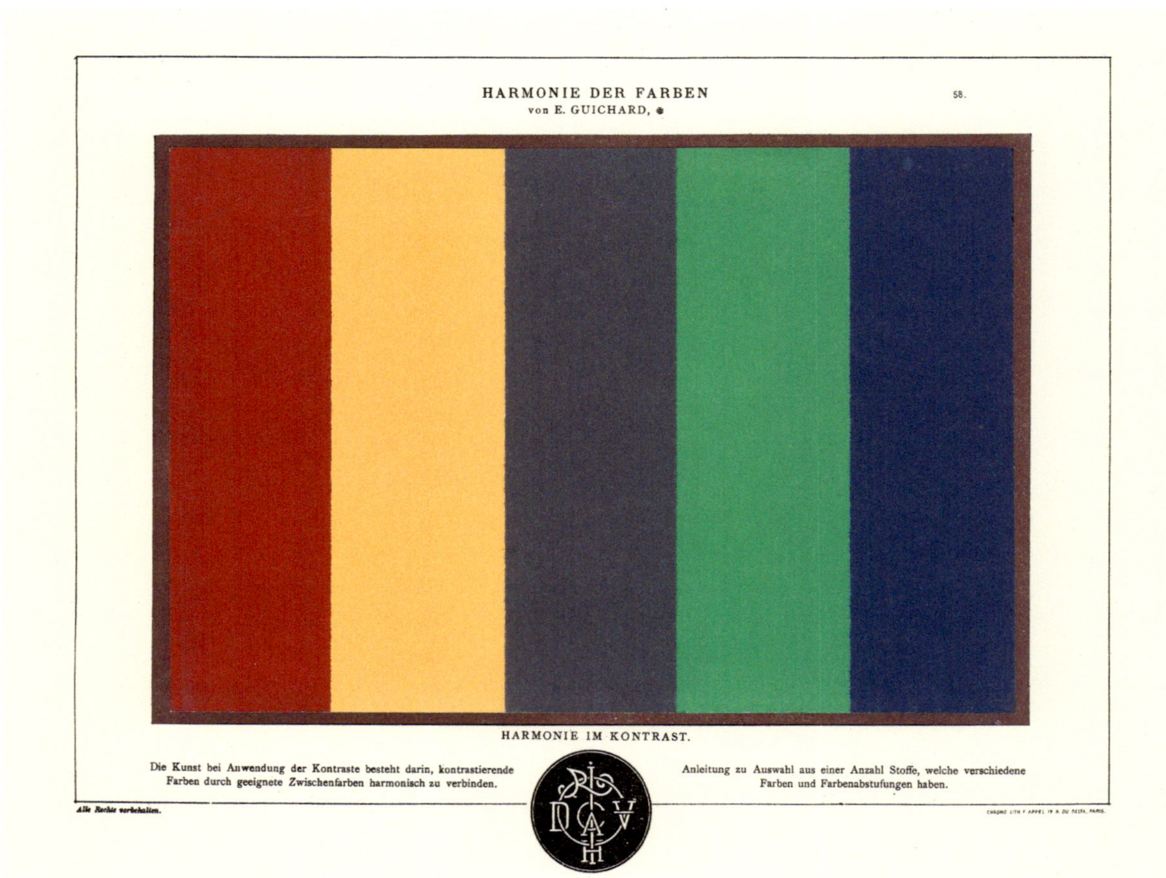
HARMONIE DER FARBEN
von E. GUICHARD, ✻
58.
HARMONIE IM KONTRAST.
Die Kunst bei Anwendung der Kontraste besteht darin, kontrastierende Farben durch geeignete Zwischenfarben harmonisch zu verbinden.
Anleitung zu Auswahl aus einer Anzahl Stoffe, welche verschiedene Farben und Farbenabstufungen haben.
Alle Rechte vorbehalten.

HARMONIE DER FARBEN
von E. GUICHARD, ●

52.

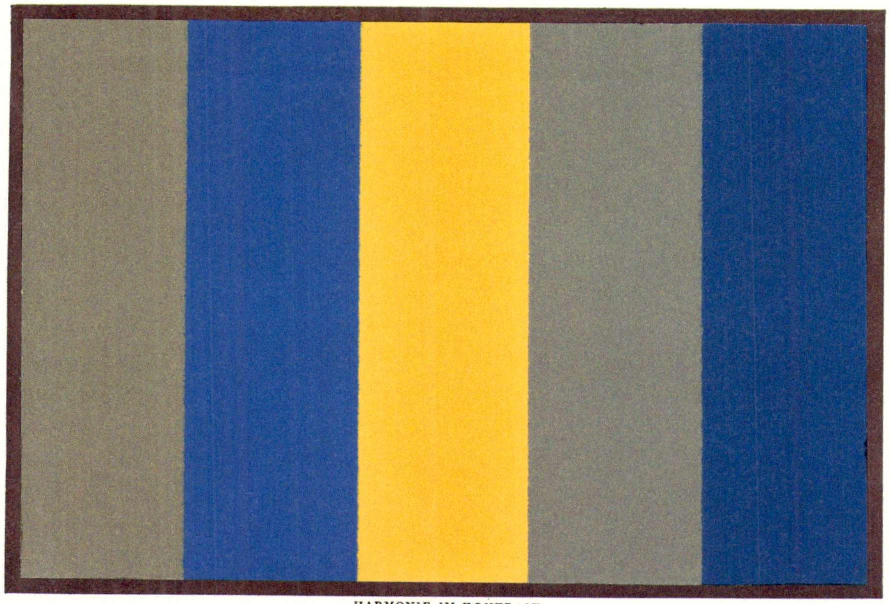

HARMONIE IM KONTRAST.

Die Kunst bei Anwendung der Kontraste besteht darin, kontrastierende Farben durch geeignete Zwischenfarben harmonisch zu verbinden.

Anleitung zu Auswahl aus einer Anzahl Stoffe, welche verschiedene Farben und Farbenabstufungen haben.

CHROMO LITH F APPEL 19 R. DU DELTA, PARIS.

HARMONIE DER FARBEN
von E. GUICHARD, ●

53.

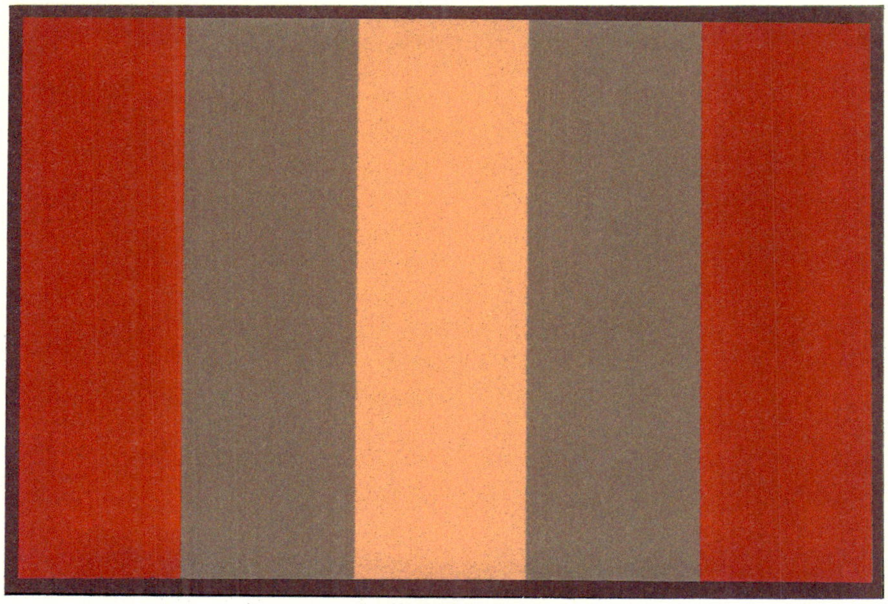

HARMONIE IM KONTRAST.

Die Kunst bei Anwendung der Kontraste besteht darin, kontrastierende Farben durch geeignete Zwischenfarben harmonisch zu verbinden.

Anleitung zu Auswahl aus einer Anzahl Stoffe, welche verschiedene Farben und Farbenabstufungen haben.

CHROMO LITH F APPEL 19 R. DU DELTA, PARIS.

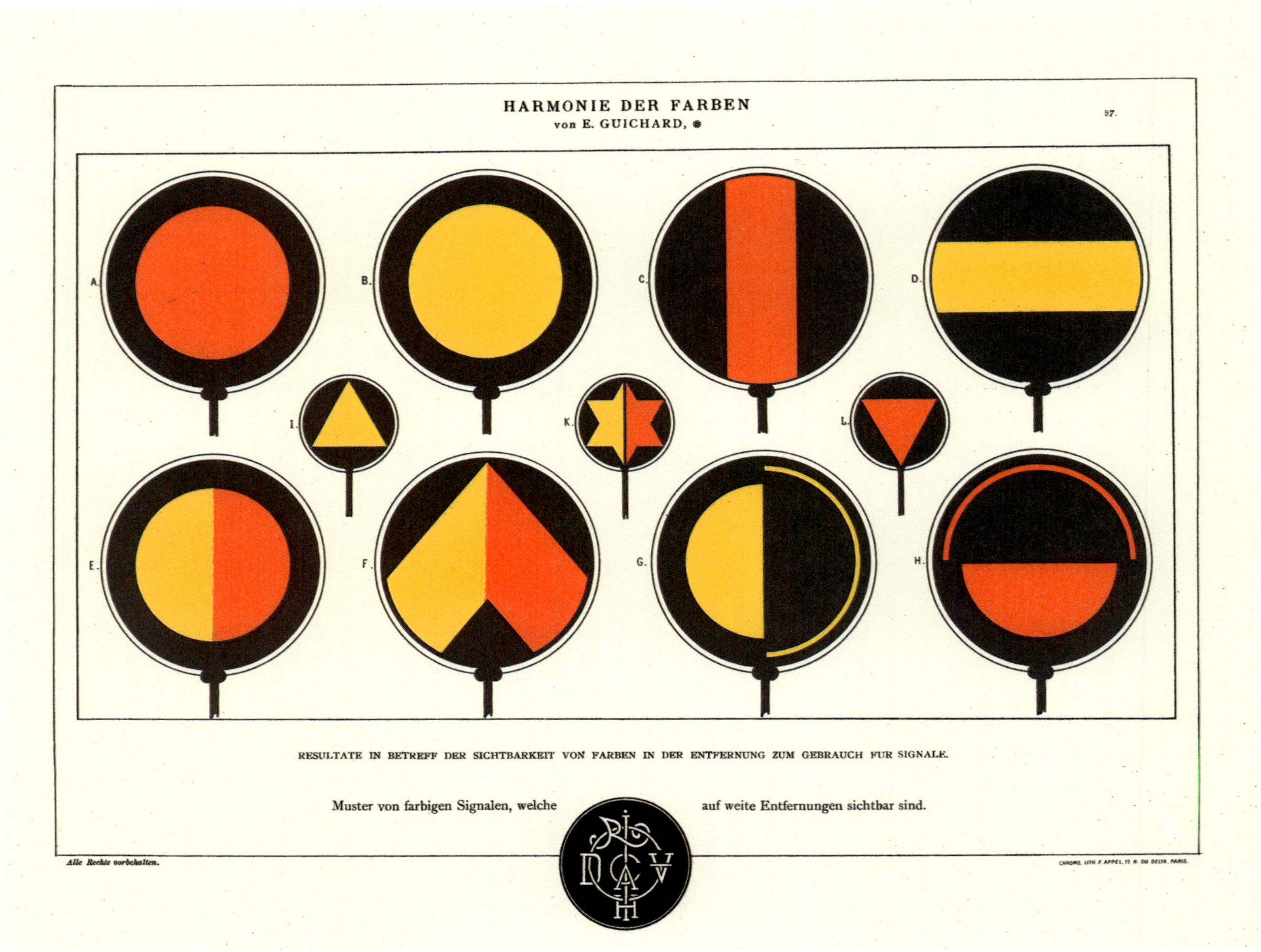
HARMONIE DER FARBEN
von E. GUICHARD,
97.
A.
B.
C.
D.
I.
K.
L.
E.
F.
G.
H.
RESULTATE IN BETREFF DER SICHTBARKEIT VON FARBEN IN DER ENTFERNUNG ZUM GEBRAUCH FUR SIGNALE.
Muster von farbigen Signalen, welche auf weite Entfernungen sichtbar sind.
Alle Rechte vorbehalten.
CHROMO. LITH. F. APPEL, 72 R. DU DELTA, PARIS.

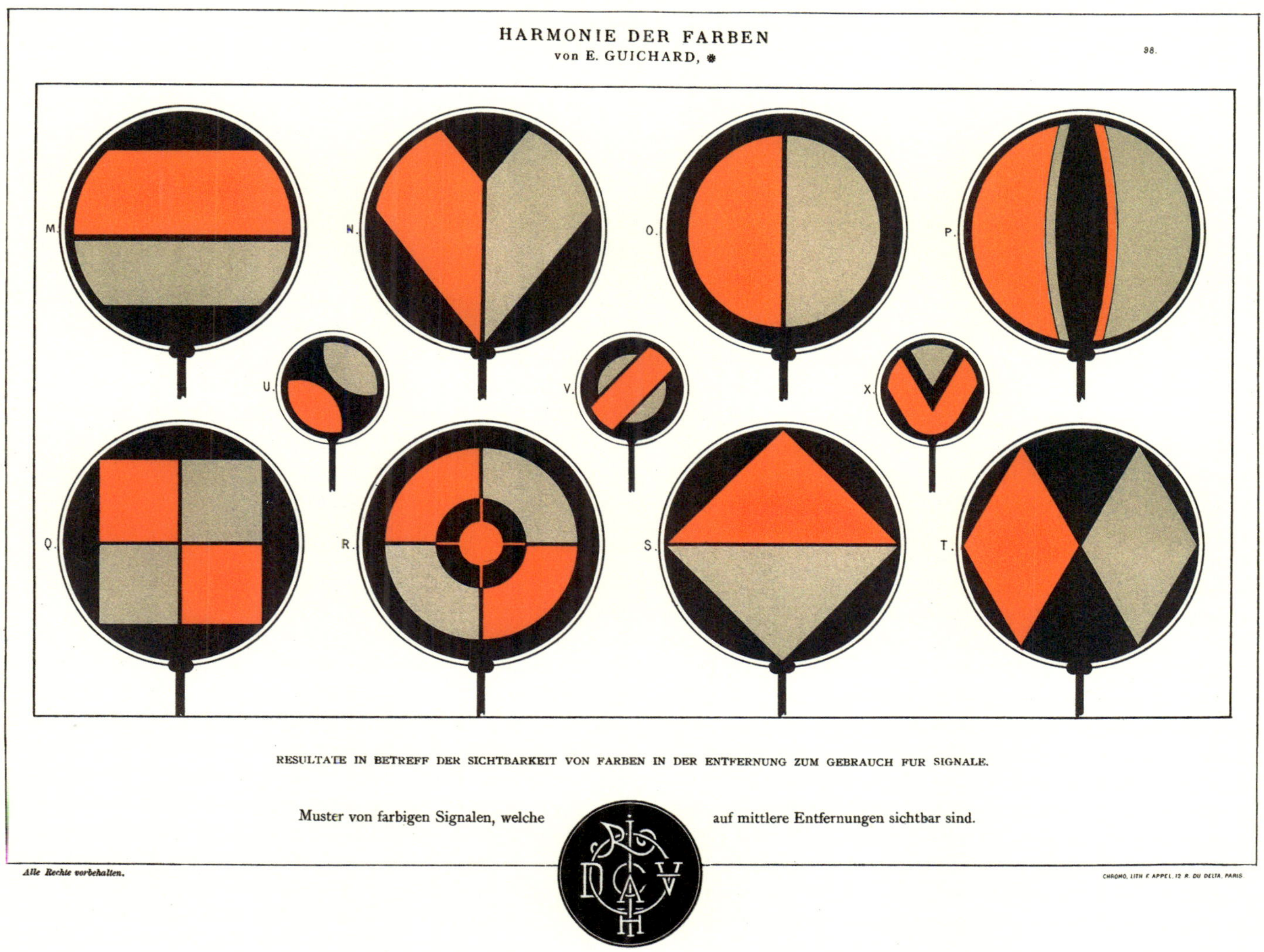
HARMONIE DER FARBEN
von E. GUICHARD,
98.
M.
N.
O.
P.
U.
V.
X.
Q.
R.
S.
T.
RESULTATE IN BETREFF DER SICHTBARKEIT VON FARBEN IN DER ENTFERNUNG ZUM GEBRAUCH FUR SIGNALE.
Muster von farbigen Signalen, welche auf mittlere Entfernungen sichtbar sind.
Alle Rechte vorbehalten.
CHROMO, LITH F. APPEL, 12 R. DU DELTA, PARIS.

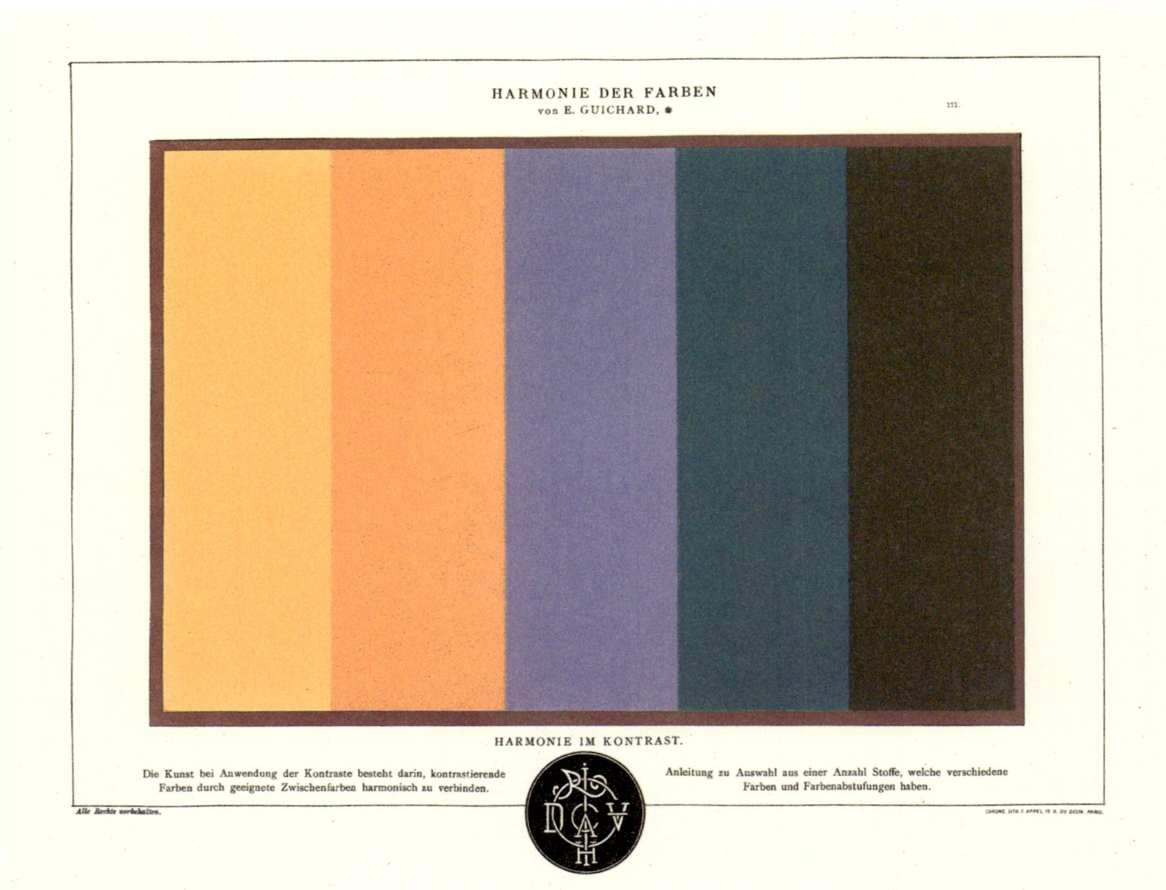
HARMONIE DER FARBEN
von E. GUICHARD.
111.
HARMONIE IM KONTRAST.
Die Kunst bei Anwendung der Kontraste besteht darin, kontrastierende Farben durch geeignete Zwischenfarben harmonisch zu verbinden.
Anleitung zu Auswahl aus einer Anzahl Stoffe, welche verschiedene Farben und Farbenabstufungen haben.
Alle Rechte vorbehalten.

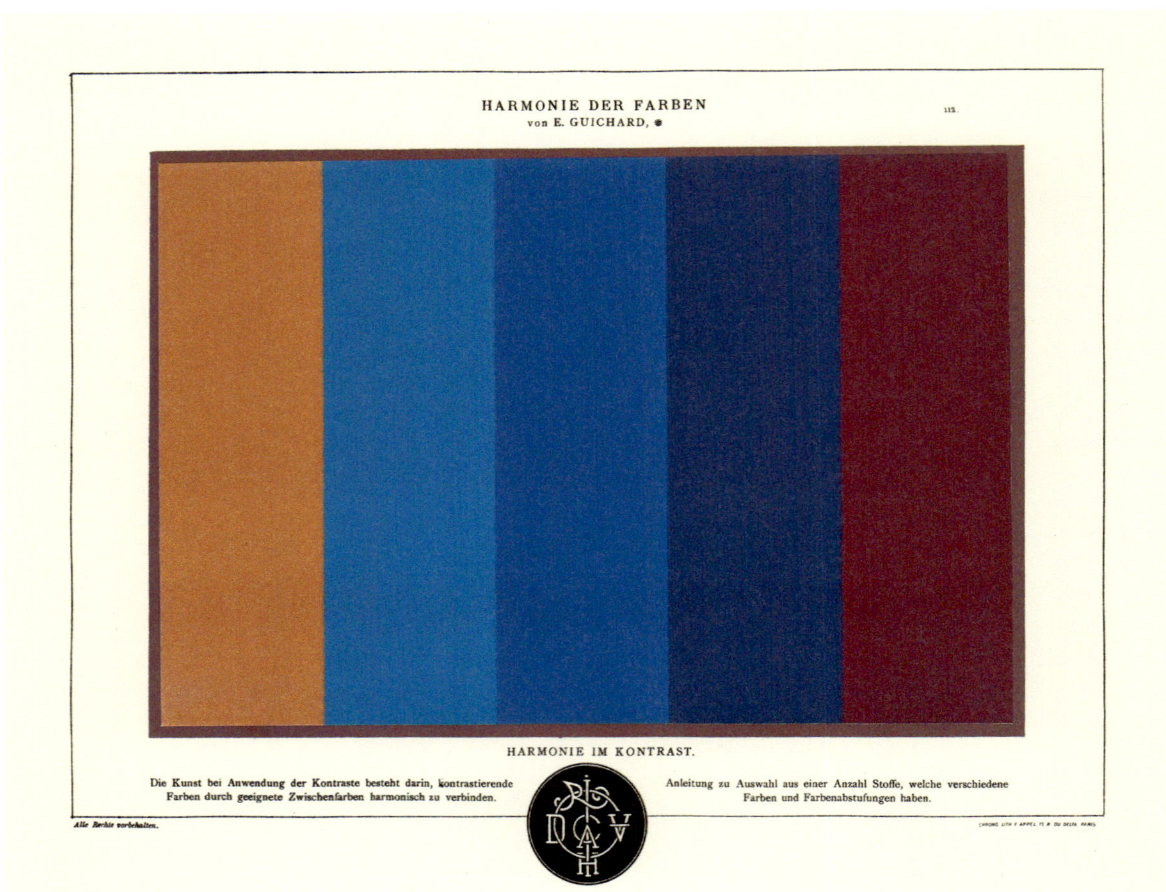
HARMONIE DER FARBEN
von E. GUICHARD.
112.
HARMONIE IM KONTRAST.
Die Kunst bei Anwendung der Kontraste besteht darin, kontrastierende Farben durch geeignete Zwischenfarben harmonisch zu verbinden.
Anleitung zu Auswahl aus einer Anzahl Stoffe, welche verschiedene Farben und Farbenabstufungen haben.
Alle Rechte vorbehalten.

HARMONIE DER FARBEN
von E. GUICHARD, ❋

113.

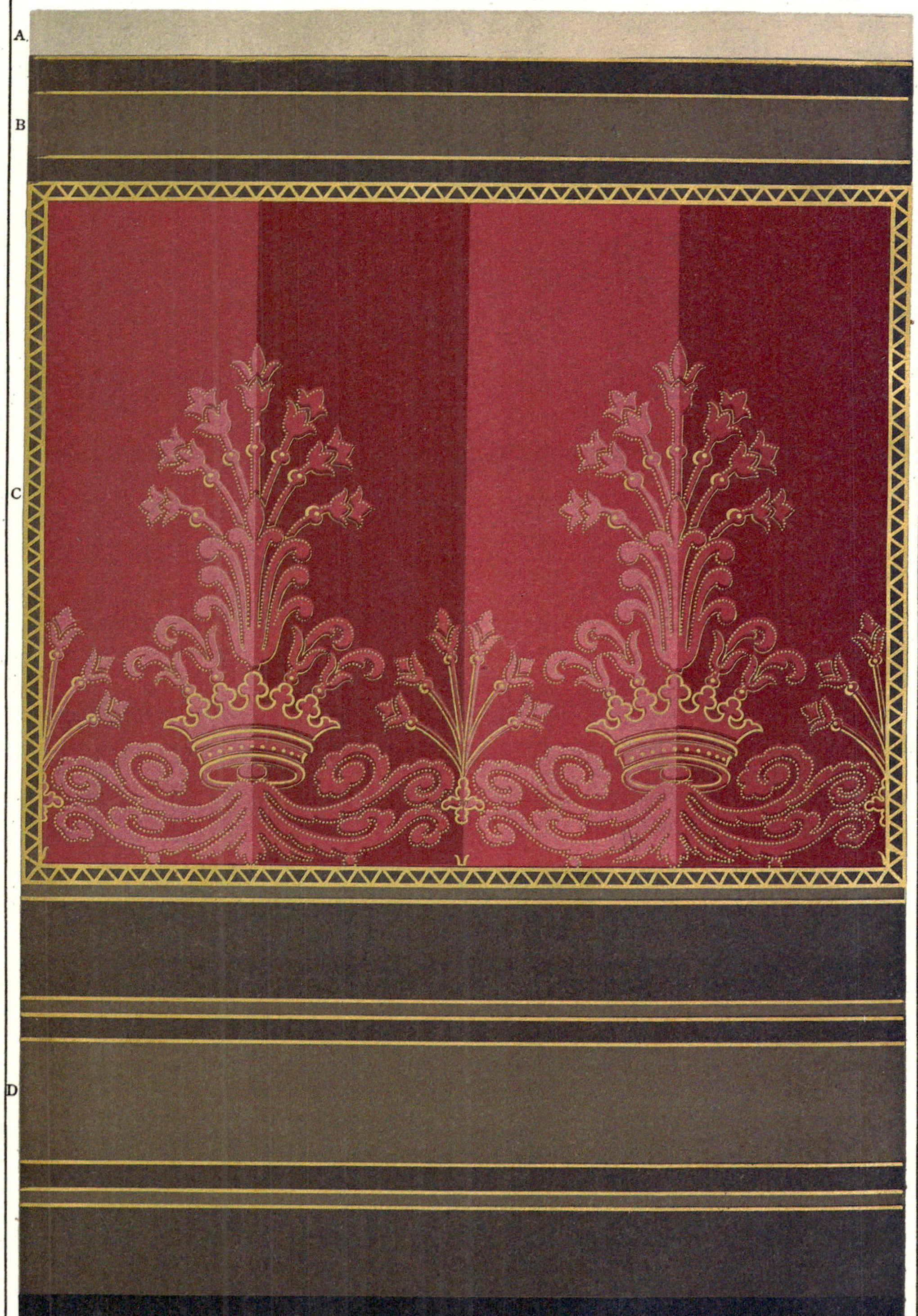

Harmonie im Contraste der Farben.

A. Decke, B Karnies, C Behänge, D Getäfel.

CHROMO. LITH. E. APPEL. 12 R. DU DELTA. PARIS.

Charles Lacouture (1832–1908)

RÉPERTOIRE CHROMATIQUE

solution raisonnée et pratique des problèmes les plus usuels dans l'étude et l'emploi des couleurs

A Chromatic Repertory
16 plates, 31.6 x 25.7 cm | 12 ½ x 10 ⅛ in, Paris, 1890
London, Royal College of Art, Colour Reference Library

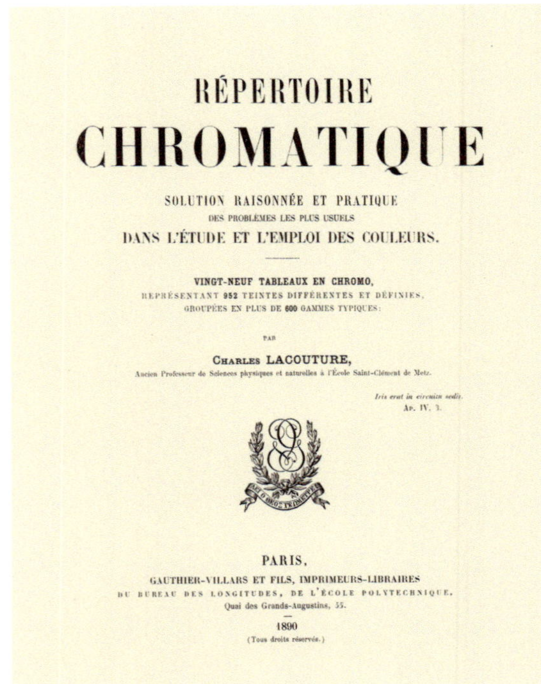
RÉPERTOIRE
CHROMATIQUE
SOLUTION RAISONNÉE ET PRATIQUE
DES PROBLÈMES LES PLUS USUELS
DANS L'ÉTUDE ET L'EMPLOI DES COULEURS.

VINGT-NEUF TABLEAUX EN CHROMO,
REPRÉSENTANT 952 TEINTES DIFFÉRENTES ET DÉFINIES,
GROUPÉES EN PLUS DE 600 GAMMES TYPIQUES:

PAR

CHARLES LACOUTURE,
Ancien Professeur de Sciences physiques et naturelles à l'École Saint-Clément de Metz.

Iris erat in circuitu sedis.
AP. IV, 3.

PARIS,
GAUTHIER-VILLARS ET FILS, IMPRIMEURS-LIBRAIRES
DU BUREAU DES LONGITUDES, DE L'ÉCOLE POLYTECHNIQUE,
Quai des Grands-Augustins, 55.

1890
(Tous droits réservés.)

This work, conceived and written by the French botanist and university professor of natural history Charles Lacouture, was intended as a homage to his fellow Frenchman Chevreul's work on colour earlier in the 19th century (pp. 282–307). While Lacouture greatly admired his predecessor, he also recognised some of the shortcomings of Chevreul's colour classification system, notably in relation to its usefulness to practising artists, designers and craftsmen. With his *Répertoire chromatique* he aimed to simplify and extend the system to which he and so many others were indebted, by composing an easy-to-read and updated textbook and guidebook for the use of colour. Its title (perhaps surprisingly, the book was never published in other languages) translates as 'A chromatic repertory, a logical and practical solution to the most frequent problems encountered in the study and use of colours', and in his text Lacouture concentrated on materials and objects, alongside the quality of pigments and paints, and how and why they deteriorate. By the end of the 19th century the fugitive nature of many colours and the damaging effect of light exposure had become a popular topic of discussion among manufacturers and colour researchers, partly caused by the advent of more intense, artificial lighting combined with the mass production of painting materials.

Répertoire chromatique is illustrated with 29 high-quality aquatinted colour plates and further uncoloured diagrams, with the coloured, full-page plates being particularly sophisticated in design and production. The leading image, which appears on the frontispiece (p. 365), has the title "Rose Synoptique" and introduces Lacouture's six principal colours and their six mixtures (or "second-order colours"), visualised as a 12-petal flower-head. The purest manifestation is shown at the root of each flower petal, with the hues lightening towards the outer tips. This is a variation on the classic colour wheel as proposed in the 18th century by Harris (pp. 138–143) and adapted by several other colour theorists since. A key to the diagram is printed on its transparent tissue guard.

Lacouture's most complex and aesthetically pleasing plate is entitled "Trilobe Synoptique", and is another synoptic diagram but this time divided into three 'lobes'. Here, the trichromatic system is presented as sections of overlapping circles of the primaries red, yellow and blue, which in the process form wing-like shapes that display the many different areas of mixtures of the 12 main colours in their gradations from light through grey to dark (*claires, grisées, assombries*). The primaries are marked and numbered along the outer edge of each wing, allowing for the exact identification of each tint. The remaining plates in Lacouture's book depict individual colours and their mixtures in the form of square grids similar to those designed by Robert Ridgway (pp. 374–387) and Maerz and Paul (pp. 410–419), and a variation of a colour triangle that appears always in six gradations. In all Lacouture's coloured plates white is represented by the colour of the printing paper, while the intensity of the hues and their darker gradations is created by increasing the density of graphic lines, thus avoiding the mixing of printing inks altogether.

ROSE SYNOPTIQUE

N. B. Fermer l'atlas pour soustraire ces planches à la lumière aussitôt qu'on n'a plus besoin d'en faire usage.

Chromolith. G. Severeyns.

RÉPERTOIRE CHROMATIQUE Frontispice **PLANCHE I**bis

TRILOBE SYNOPTIQUE

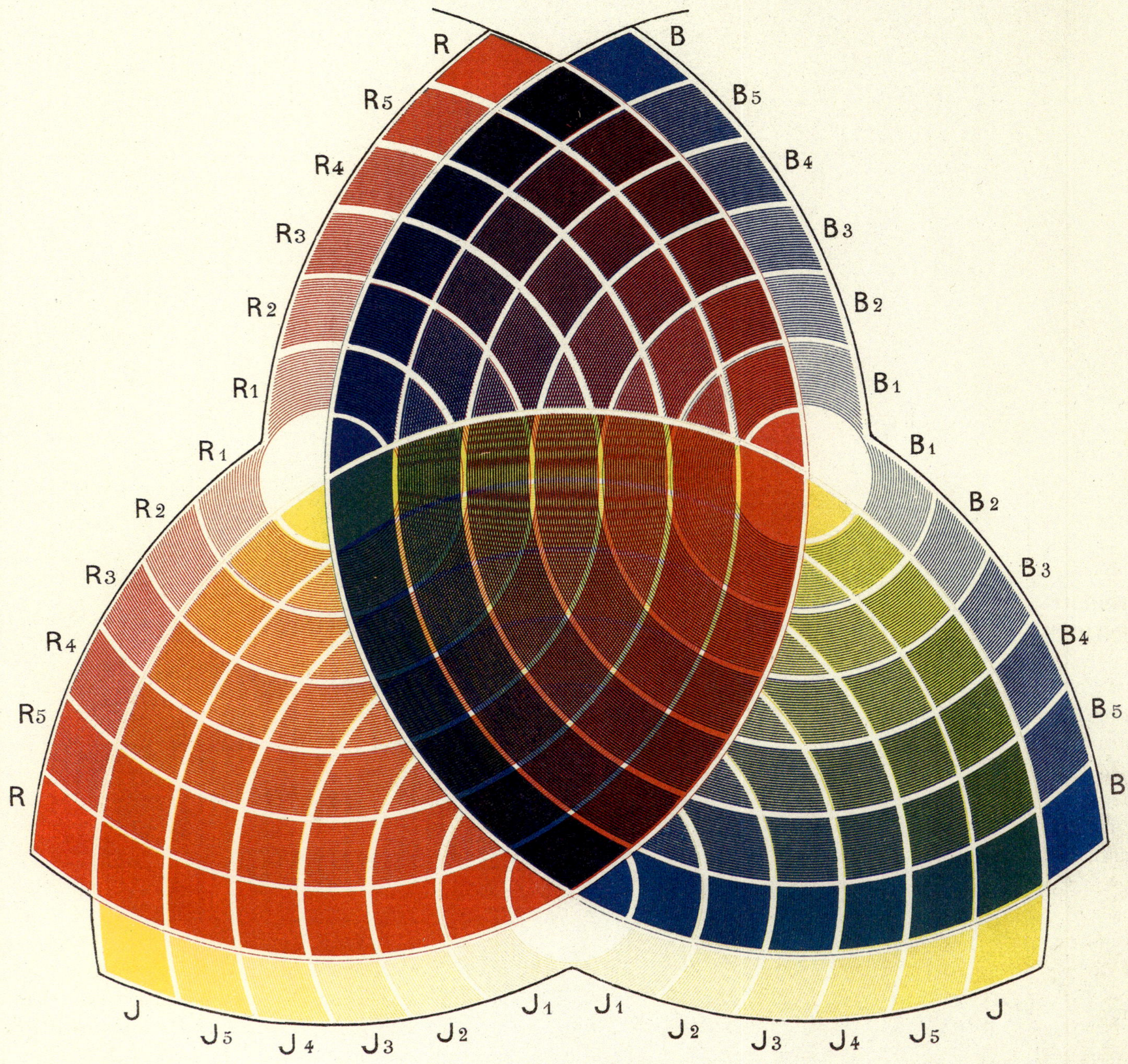

N. B. Fermer l'atlas pour soustraire ces planches à la lumière aussitôt qu'on n'a plus besoin d'en faire usage.

Chromolith. G. Severeyns.

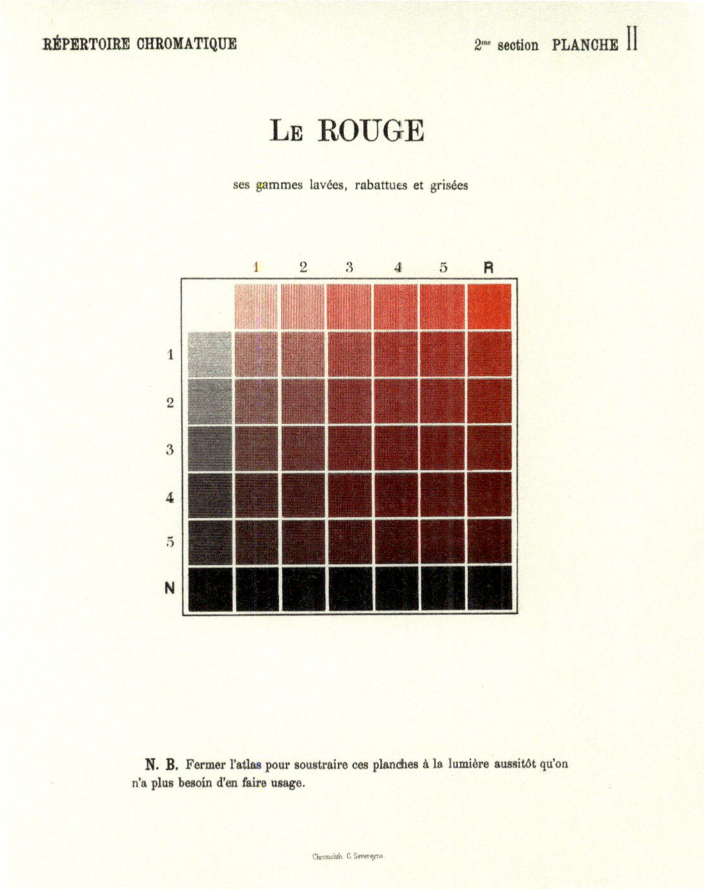

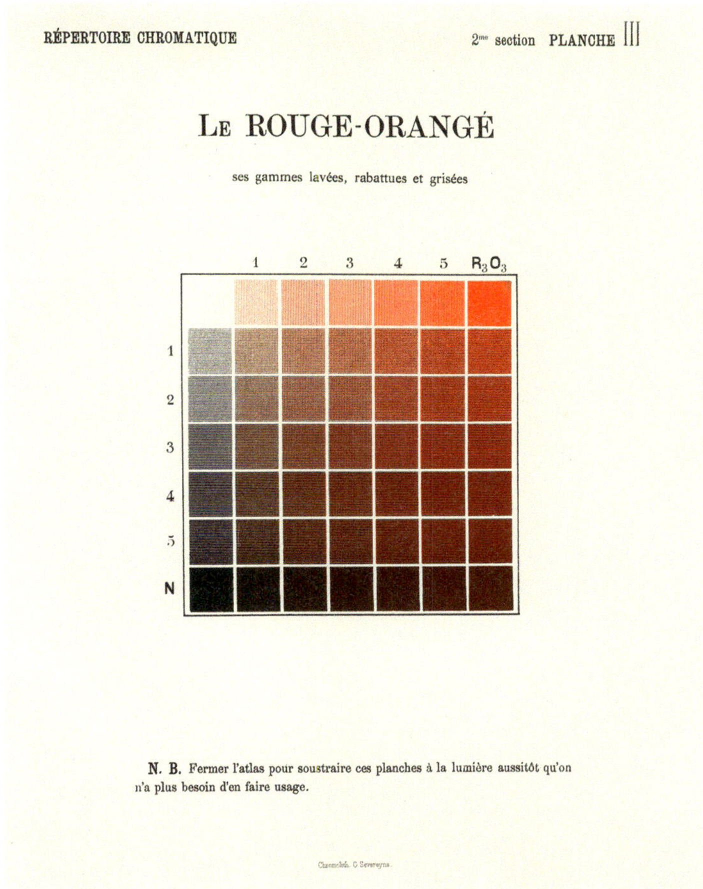

Chromatisches Repertoire

Diese Abhandlung, konzipiert und verfasst von Charles Lacouture, einem französischen Botaniker und Universitätsprofessor für Naturgeschichte, war als Hommage an das Werk von Michel-Eugène Chevreul aus dem frühen 19. Jahrhundert gedacht (S. 282–307). Lacouture bewunderte seinen Landsmann und Vorgänger sehr, erkannte in dessen System der Farbklassifizierung jedoch auch einige Schwächen, insbesondere im Hinblick auf die Anwendbarkeit für praktizierende Künstler, Gestalter und Handwerker. Mit *Répertoire chromatique* wollte er das System vereinfachen und erweitern, dem er und so viele andere verpflichtet waren, indem er ein leicht lesbares und aktualisiertes Lehr- und Handbuch für den Farbgebrauch zusammenstellte. Der Titel des Werks, das (vielleicht überraschend) niemals in anderen Sprachen erschien, lautet übersetzt „Chromatisches Repertoire, eine logische und praktische Lösung für die häufigsten Probleme bei der Erforschung und Anwendung von Farben". Im Text konzentrierte sich Lacouture auf farbige Materialien und Objekte, auf die Qualität von Pigmenten und Farben und auf die Frage, wie und warum sie an Substanz einbüßten. Im ausgehenden 19. Jahrhundert waren die flüchtige Natur vieler Farben und die schädliche Wirkung des Lichts unter Farbenherstellern und Farbforschern zum verbreiteten Diskussionsthema geworden, unter anderem ausgelöst durch die Einführung stärkerer künstlicher Lichtquellen in Kombination mit der Massenproduktion von Malmaterialien.

Répertoire chromatique ist mit 29 hochwertigen Aquatinta-Farbtafeln und weiteren nicht kolorierten Diagrammen bebildert, wobei die ganzseitigen Farbtafeln besonders aufwendig gestaltet und produziert wurden. Das tonangebende Bild auf dem Frontispiz trägt den Titel „Rose synoptique" (S. 365) und stellt die sechs Hauptfarben Lacoutures und ihre sechs Mischfarben (oder „Farben zweiter Ordnung") in Gestalt einer zwölfblättrigen Blüte vor. Die reinste Manifestation zeigt sich im Zentrum der Blüte, während die einzelnen Blütenblätter zu den Spitzen hin immer heller werden. Im Kern haben wir es mit einer Variation des klassischen Farbkreises zu tun, wie er in den 1760er-Jahren von Moses Harris (S. 138–143) vorgeschlagen und seither von mehreren anderen Farbtheoretikern übernommen und angepasst wurde.

Lacoutures komplexeste und ästhetisch ansprechendste Tafel, ein weiteres synoptisches Diagramm, ist mit „Trilobe synoptique" betitelt und diesmal in drei „Lappen" unterteilt. Das trichromatische System wird hier von sich überschneidenden Kreisstücken in den drei Grundfarben Rot, Gelb und Blau dargestellt. Sie bilden flügelartige Formen, auf denen die zahlreichen Mischungen der zwölf Hauptfarben in ihren Abstufungen von Hell über Grau bis Dunkel (*claires*,

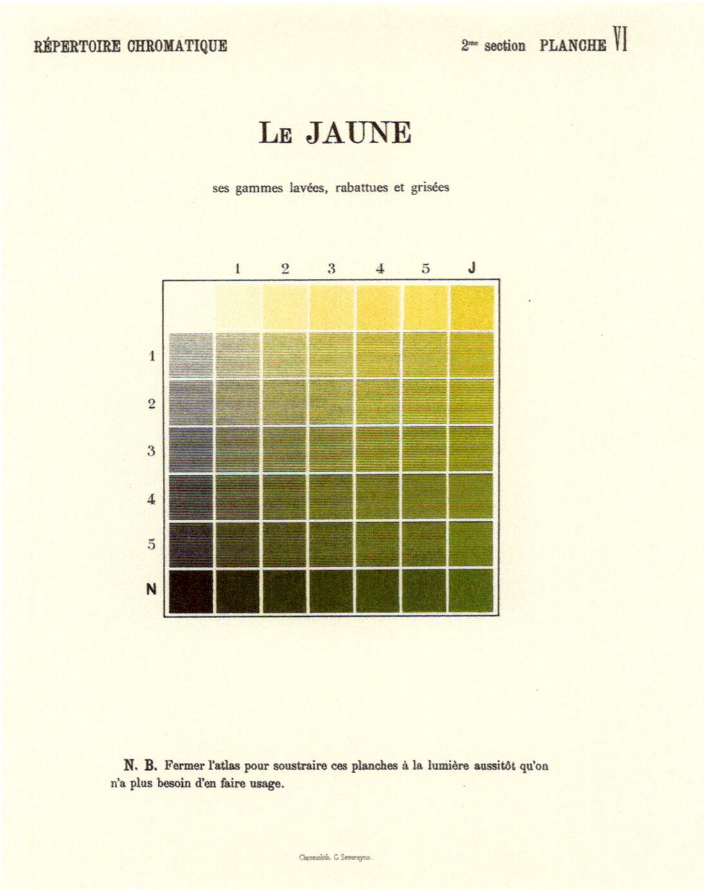

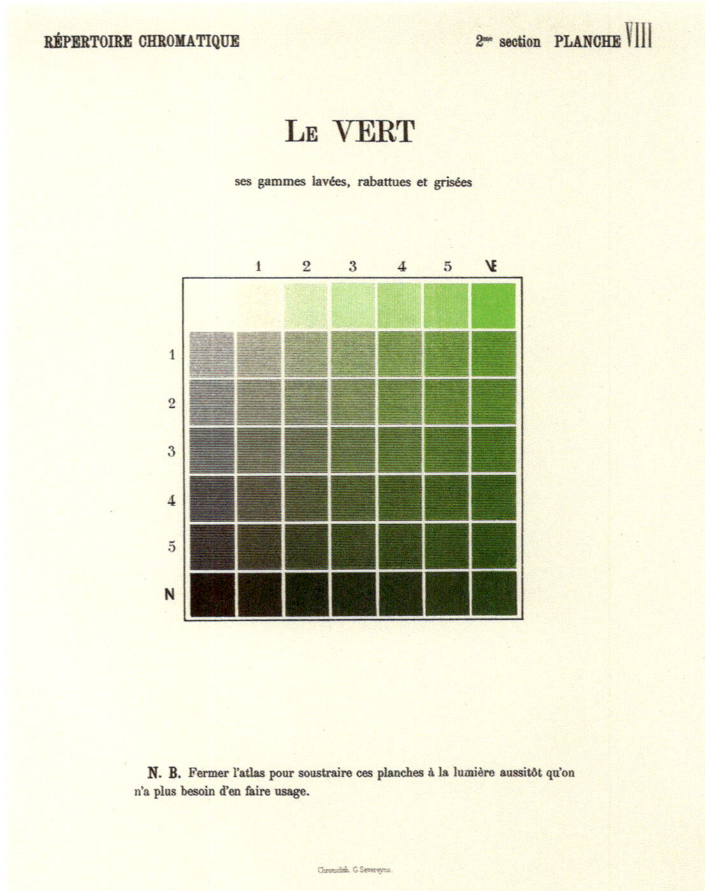

grisées, assombries) abzulesen sind. Die Grundfarben wurden an den Außenkanten der Flügel gekennzeichnet und nummeriert, um jeden Farbton genau bestimmen zu können. Die übrigen Tafeln in Lacoutures Buch zeigen einzelne Farben und ihre Mischungen in Form von quadratischen Rastern ähnlich denen von Robert Ridgway (S. 374–387) und Maerz und Paul (S. 410–419) sowie eine Variation eines Farbdreiecks, das immer in sechs Abstufungen erscheint. In allen farbigen Tafeln Lacoutures steht die Farbe des Druckpapiers für Weiß, während die Intensität der Farbtöne und ihrer dunkleren Schattierungen durch die jeweilige Stärke der Farblinien erzeugt wird. Das Anmischen von Druckfarben wurde so vollständig vermieden.

Répertoire chromatique

Conçu et rédigé par Charles Lacouture, botaniste et professeur d'histoire naturelle français, cet ouvrage est pensé comme un hommage à l'œuvre réalisée par Michel-Eugène Chevreul au début du XIXe siècle (p. 282–307). Grand admirateur de son prédécesseur, Lacouture met néanmoins en évidence les défauts de la classification chromatique de Chevreul, notamment quant à son utilité pour les peintres, dessinateurs et artisans. Avec son *Répertoire chromatique*, il veut simplifier et augmenter un système auquel, comme tant d'autres, il doit beaucoup, en réalisant un manuel facile à lire et mis à jour, qui soit aussi un guide pratique pour l'emploi des couleurs. Dans cet ouvrage qui, étonnamment peut-être, n'a été traduit en aucune langue, Lacouture s'intéresse aux matériaux et aux objets, ainsi qu'à la qualité des pigments et des peintures, et explique comment et pourquoi ceux-ci se détériorent. À la fin du XIXe siècle, la nature éphémère de nombreuses couleurs et les dommages causés par l'exposition à la lumière font partie des débats courants parmi les fabricants et les coloristes, notamment en raison de l'avènement de l'éclairage artificiel plus intense, associé à la production industrielle des matériaux de peinture.

Le *Répertoire chromatique* est illustré de graphiques sans couleurs et de 29 planches pleine page, à l'aquatinte en couleurs de grande qualité, dont la conception et la réalisation sont particulièrement raffinées. Intitulée « Rose synoptique », l'image qui orne le frontispice (p. 365) présente les six couleurs principales et les six mélanges (« couleurs dérivées de son ordre ») de Lacouture, sous la forme d'une fleur à 12 pétales. L'état d'intensité maximale des couleurs est visible au centre de la fleur, les tonalités s'éclaircissant de plus en plus vers l'extrémité des pétales. Il s'agit avant tout d'une variante du cercle chromatique classique proposé par Moses Harris dans les années 1760 (p. 138–143) et remanié par plusieurs théoriciens à sa suite. La légende de ce graphique est imprimée sur une page de protection transparente en papier de soie.

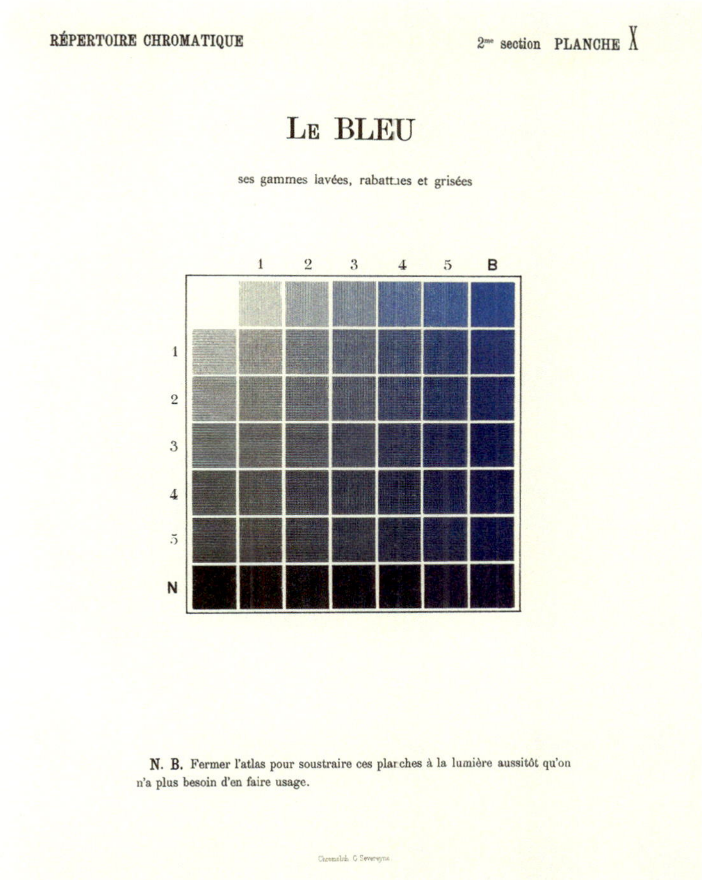

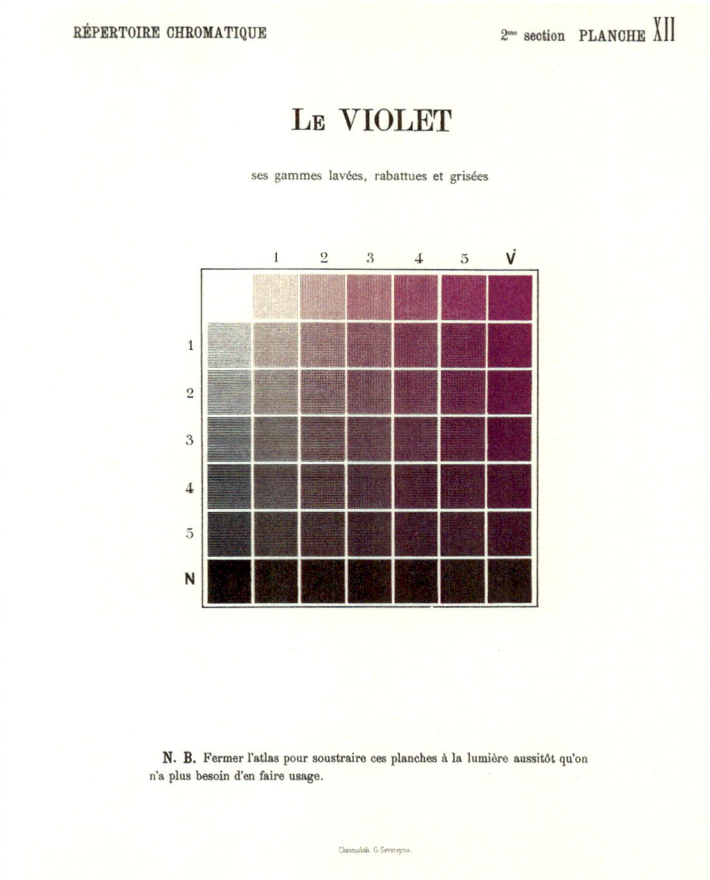

La planche la plus complexe et la plus réussie sur le plan esthétique s'intitule « Trilobe synoptique », graphique divisé, comme son nom l'indique, en trois lobes. Ce système trichromatique est formé de sections de cercles entrecroisés figurant les couleurs primaires rouge, jaune et bleu. Celles-ci créent à leur tour des « ailes » réunissant les nombreux segments composés de mélanges des 12 couleurs principales, dans leurs gradations « claires, grisées, assombries ». Le long des contours extérieurs, les couleurs primaires sont désignées par une lettre et un chiffre, chaque teinte étant ainsi identifiable avec précision. Les autres planches de l'ouvrage de Lacouture représentent les teintes et leurs mélanges sous la forme de grilles composées de carrés, semblables à celles de Robert Ridgway (p. 374–387) et de Maerz et Paul (p. 410–419), ainsi que des variations d'un triangle aux bords arrondis, toujours présenté en six gradations. Dans toutes les planches, le blanc est figuré par la couleur du papier, tandis que l'intensité des tons et de leurs gradations sombres est produite par la densité des traits, ce qui évite le mélange des encres d'impression.

Repertorio cromático

Este trabajo, concebido y escrito por el botánico francés y profesor universitario de historia natural Charles Lacouture, pretendía ser un homenaje a la obra sobre el color que su compatriota Michel-Eugène Chevreul había publicado en 1839 (págs. 282–307). Si bien Lacouture admiraba mucho a su predecesor, también reconoció algunas de las deficiencias del sistema de clasificación de colores de Chevreul, en particular en relación con su utilidad práctica para los artistas, diseñadores y artesanos. Con su *Répertoire chromatique* pretendía simplificar y ampliar el sistema con el que él y tantos otros estaban en deuda mediante la composición de un libro de texto y una guía sobre el uso del color actualizados y fáciles de leer. Su título (quizás sorprenda saber que el libro nunca se publicó en otros idiomas) se traduce como «Repertorio cromático y solución lógica y práctica a los problemas más frecuentes en el estudio y empleo de los colores». Lacouture se centró en materiales y objetos, junto con la calidad de los pigmentos y las pinturas, y en cómo y por qué se deterioran. A finales del siglo XIX, la naturaleza fugitiva de muchos colores y el efecto dañino de la exposición a la luz se habían convertido en un tema de discusión popular entre los fabricantes e investigadores del color, en parte debido a la llegada de una iluminación artificial más intensa combinada con la producción en masa de material de pintura.

Répertoire chromatique cuenta con 29 láminas ilustradas con grabados al aguatinta en color de alta calidad y otros diagramas sin color; las láminas en color a página completa son particularmente sofisticadas en diseño y producción. La imagen

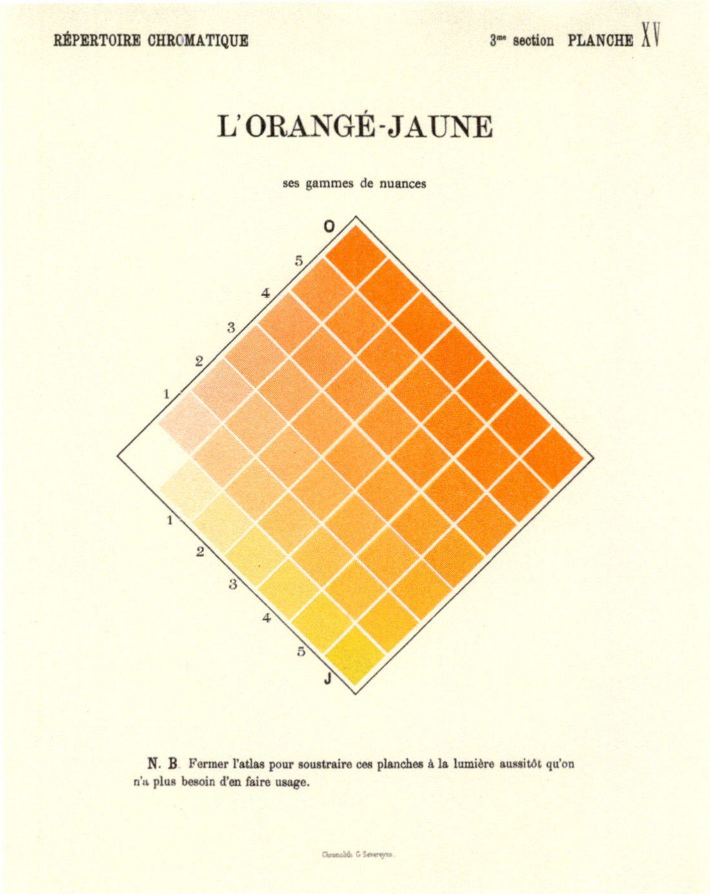
RÉPERTOIRE CHROMATIQUE
3me section PLANCHE XV
L'ORANGÉ-JAUNE
ses gammes de nuances
O
J
N. B. Fermer l'atlas pour soustraire ces planches à la lumière aussitôt qu'on n'a plus besoin d'en faire usage.

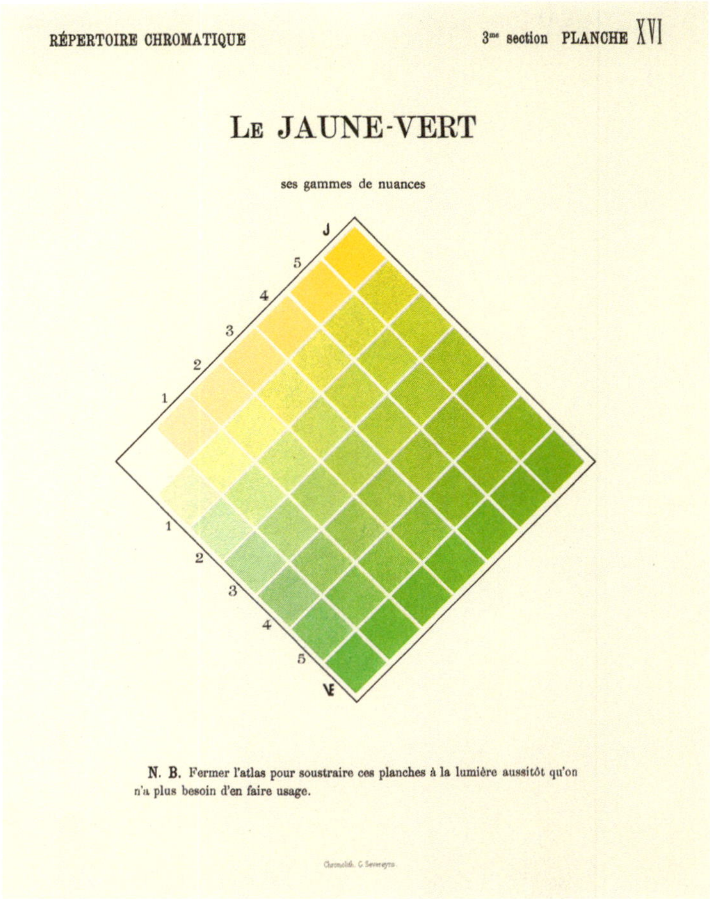
RÉPERTOIRE CHROMATIQUE
3me section PLANCHE XVI
Le JAUNE-VERT
ses gammes de nuances
J
N. B. Fermer l'atlas pour soustraire ces planches à la lumière aussitôt qu'on n'a plus besoin d'en faire usage.

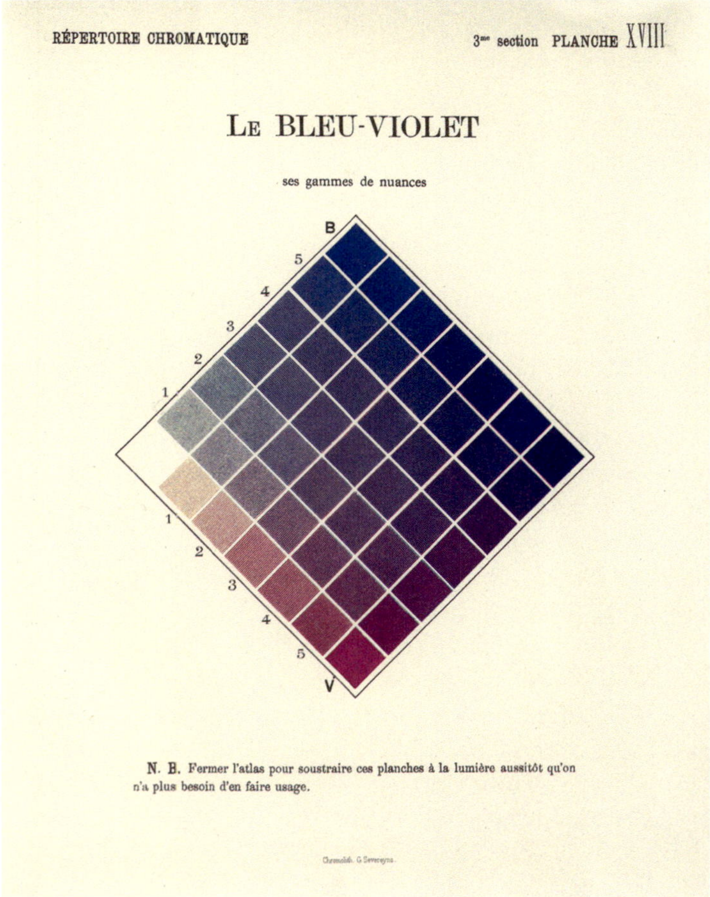
RÉPERTOIRE CHROMATIQUE
3me section PLANCHE XVIII
Le BLEU-VIOLET
ses gammes de nuances
B
V
N. B. Fermer l'atlas pour soustraire ces planches à la lumière aussitôt qu'on n'a plus besoin d'en faire usage.

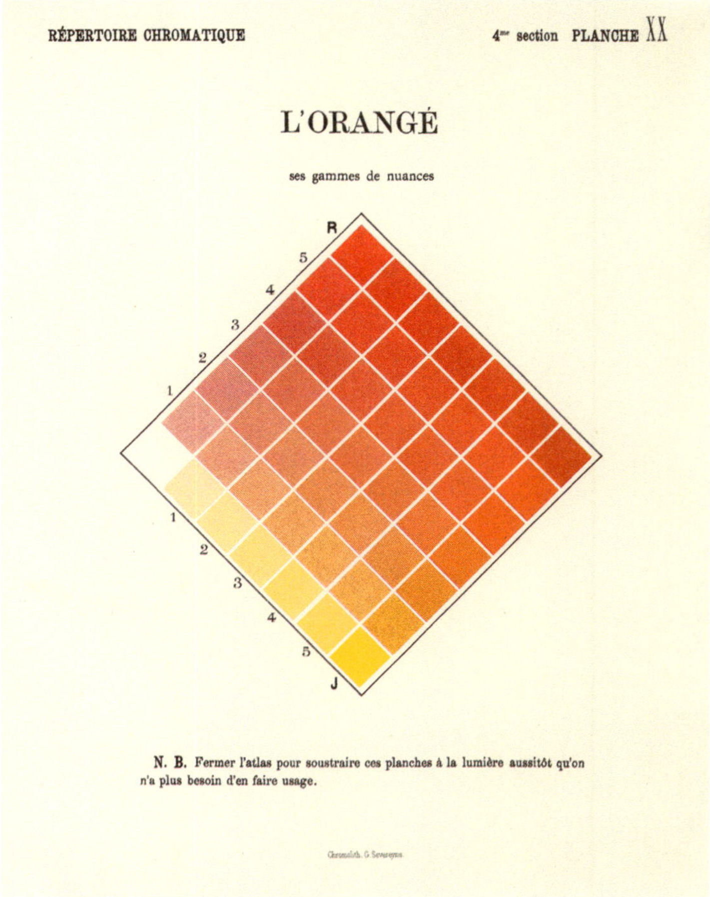
RÉPERTOIRE CHROMATIQUE
4me section PLANCHE XX
L'ORANGÉ
ses gammes de nuances
R
J
N. B. Fermer l'atlas pour soustraire ces planches à la lumière aussitôt qu'on n'a plus besoin d'en faire usage.

RÉPERTOIRE CHROMATIQUE 4me section PLANCHE XXI

LE VERT

ses gammes de nuances

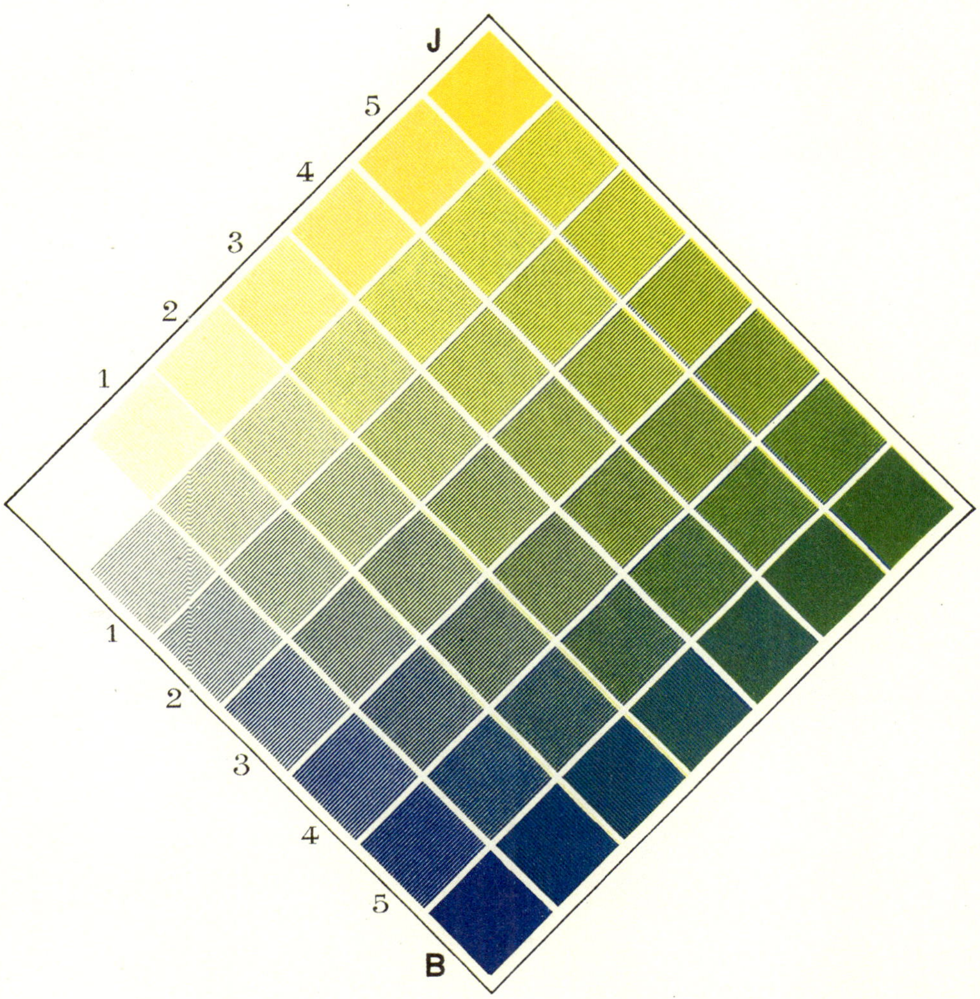

N. B. Fermer l'atlas pour soustraire ces planches à la lumière aussitôt qu'on n'a plus besoin d'en faire usage.

Chromolith. G. Severeyns.

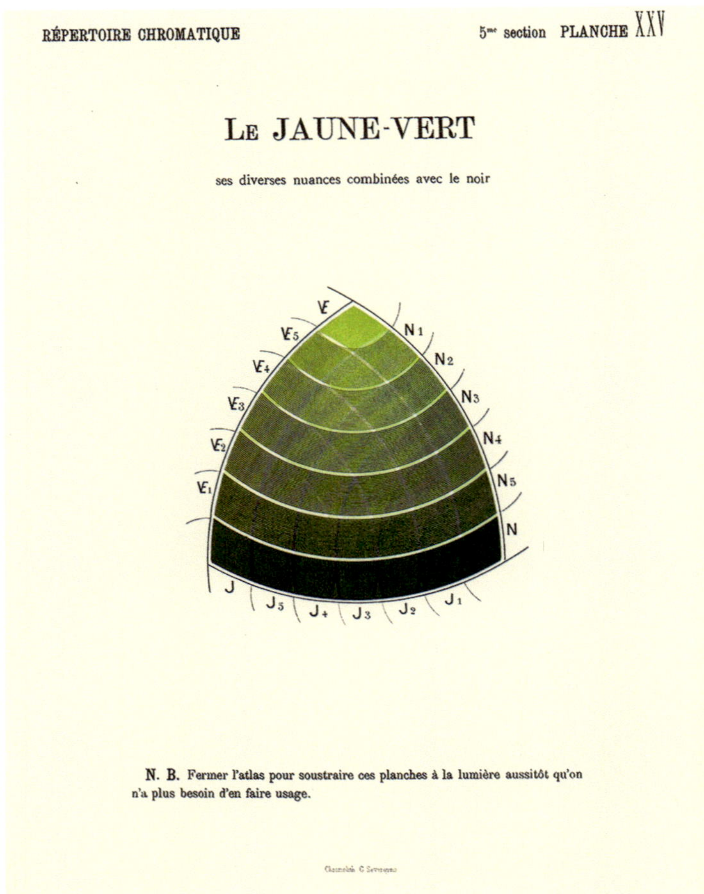

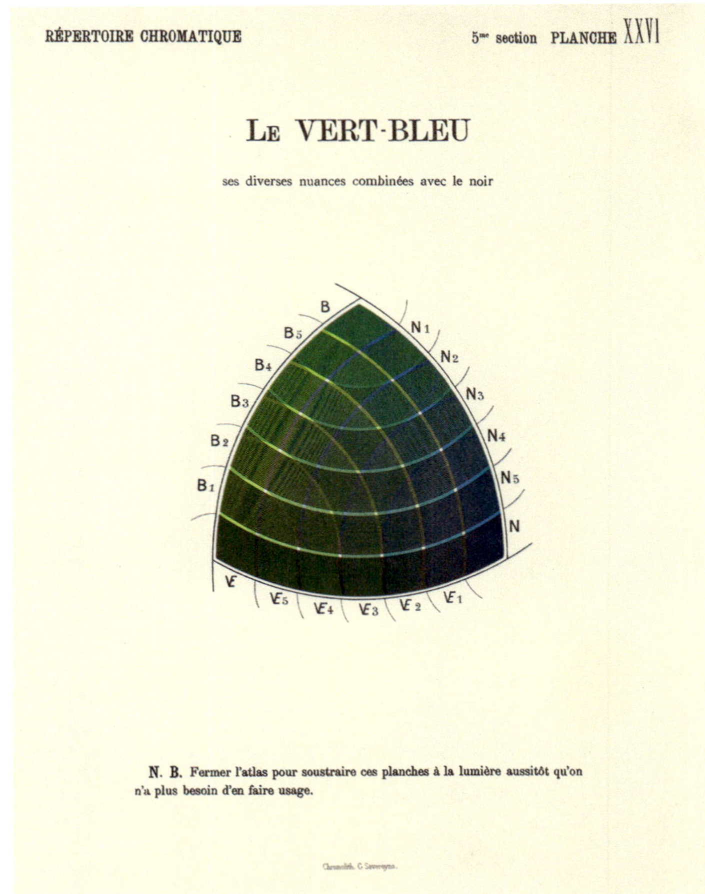

principal, que aparece en el frontispicio (pág. 365) tiene el título «Rose Synoptique» y presenta los seis colores principales de Lacouture y sus seis mezclas (o «colores de segundo orden»), visualizados como una flor de 12 pétalos. La manifestación más pura se muestra en el centro de la flor, y sus distintos matices se van aclarando a medida que se acercan a las puntas de los pétalos. Se trata en esencia de una variación de la rueda de colores clásica propuesta en la década de 1760 por Moses Harris (págs. 138–143) y adaptada por varios teóricos del color desde entonces. La clave del diagrama está impresa en el tejido transparente protector.

La lámina más compleja y estéticamente agradable de Lacouture se titula «Trilobe Synoptique», y es otro diagrama sinóptico, aunque esta vez dividido en tres «lóbulos». Aquí, el sistema tricromático se presenta por medio de secciones de círculos superpuestos de los colores primarios rojo, amarillo y azul, que en el proceso adquieren unas formas similares a alas que muestran las muchas mezclas diferentes de los 12 colores principales en sus gradaciones, que van de la más clara a la más oscura pasando por los grises (*claires*, *grisées*, *assombries*). Los colores primarios están marcados y numerados a lo largo del borde exterior de cada ala, lo que permite la identificación exacta de cada tono. Las láminas restantes del libro de Lacouture presentan colores individuales y sus mezclas por medio de cuadrículas cuadradas similares a las diseñadas por Robert Ridgway (págs. 374–387) y Maerz y Paul (págs. 410–419) y una variación de un triángulo de color que aparece siempre en seis gradaciones. En todas las láminas de colores de Lacouture, el propio color del papel representa el color blanco, mientras que la intensidad de los matices y sus gradaciones más oscuras se logran aumentando la densidad de las líneas gráficas, lo que evita la mezcla de tintas.

RÉPERTOIRE CHROMATIQUE 5me section PLANCHE XXVIII

LE VIOLET-ROUGE

ses diverses nuances combinées avec le noir

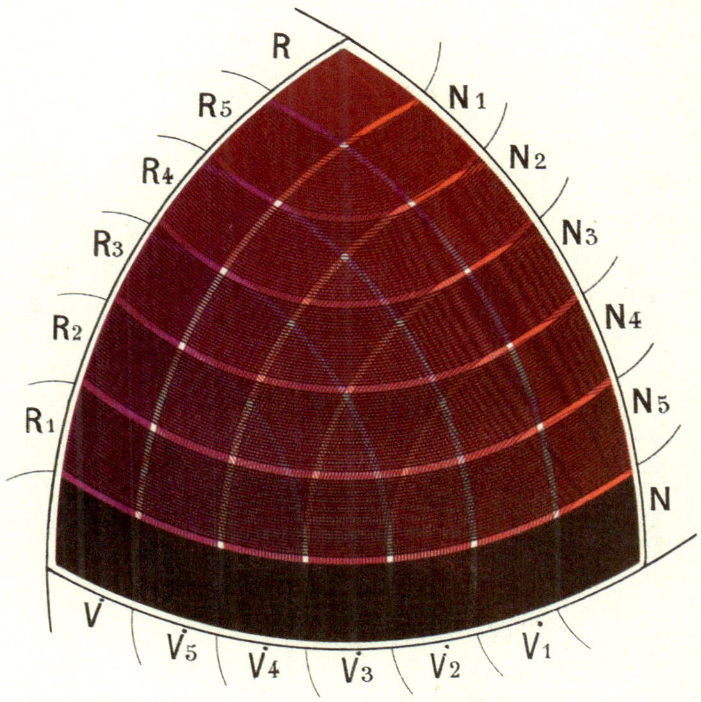

N. B. Fermer l'atlas pour soustraire ces planches à la lumière aussitôt qu'on n'a plus besoin d'en faire usage.

Chromolith. G. Severeyns.

Robert Ridgway (1850–1929)

COLOR STANDARDS AND COLOR NOMENCLATURE

17 plates, 23 x 14 cm / 9 x 5 ½ in., Washington, D.C., 1912
New Haven, Yale University, Robert B. Haas Family Arts Library Special Collections, Faber Birren Collection of Books on Color

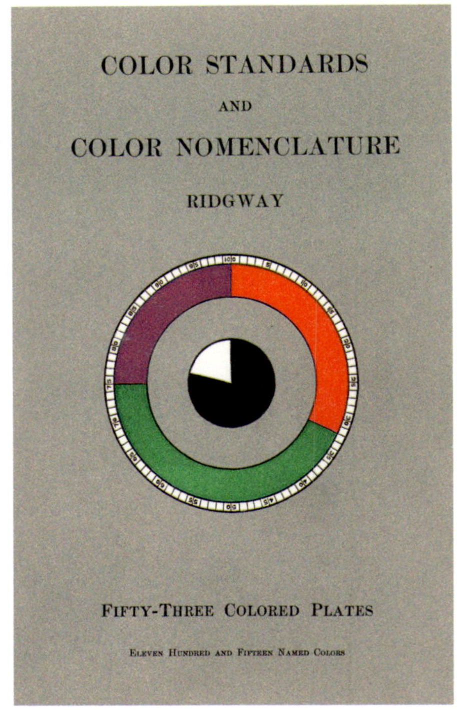
COLOR STANDARDS
AND
COLOR NOMENCLATURE
RIDGWAY
FIFTY-THREE COLORED PLATES
ELEVEN HUNDRED AND FIFTEEN NAMED COLORS

The American ornithologist Robert Ridgway was strongly influenced by the work of Werner and Syme and their attempts at colour standardisation (pp. 310–319), both in visual terms and with regard to the naming and description of colours. In turn, Ridgway was later referred to as a major historical source for Maerz and Paul's *Dictionary of Color* (pp. 410–419), while he also included references to his own work in *Color Standards and Color Nomenclature* since he had first published an illustrated collection of colour names 26 years earlier, namely *A Nomenclature of Colors for Naturalists, and Compendium of Useful Knowledge for Ornithologists*, illustrated with hand-coloured charts and diagrams. His lifelong interest in colour standardisation and his desire to improve his own system resulted in this second, elegant and meticulously researched book which also reflects the advances in colour science that had taken place between the 1880s and 1910s.

Ridgway set his standards high, providing clear definitions of colour terminology and compiling a list of 1,115 individual colours together with their proportional composition, as well as a concise list of recent colour literature. His title-page image illustrates the 'Maxwell disks' method for calculating the proportional composition of colours in the book, based on a colour wheel of 36 segments and interlocking coloured cardboard discs. In the 1850s, the Scottish physicist James Clerk Maxwell had devised a range of spinning tops and discs with which he could demonstrate certain aspects of colour vision and optical colour mixture.

The most important feature, announced on the title-page, is the set of 53 coloured plates, which together provide a physical representation of the 1,115 named colours. Each plate is mounted with 27 colour chips, showing a gradual change in hue, using the spectrum colours and their intermediates in their purest manifestation in the first 12 plates; these are followed by sets in which the admixture of neutral greys is increased steadily, dulling the colours gradually. The final plate consists entirely of shades of grey between white in the top left corner and slate black in the bottom right.

Ridgway was particularly concerned about reliable representation of the colours listed in the book, and sought to avoid variations between different copies by painting uniform sheets of paper for each colour for the entire print run of 5,000, then cutting these sheets into small strips. He also warned readers not to expose the plates to light for prolonged periods, which indicates how advanced his understanding was of the fugitive nature of printed colour. Ridgway's choice of pigments for the illustrations, which omitted almost all organic colours, also reflects this: "The pigments used in the preparation of these Plates are the most durable known, those which have been proven unstable having been, as far as possible, discarded."

Plate I

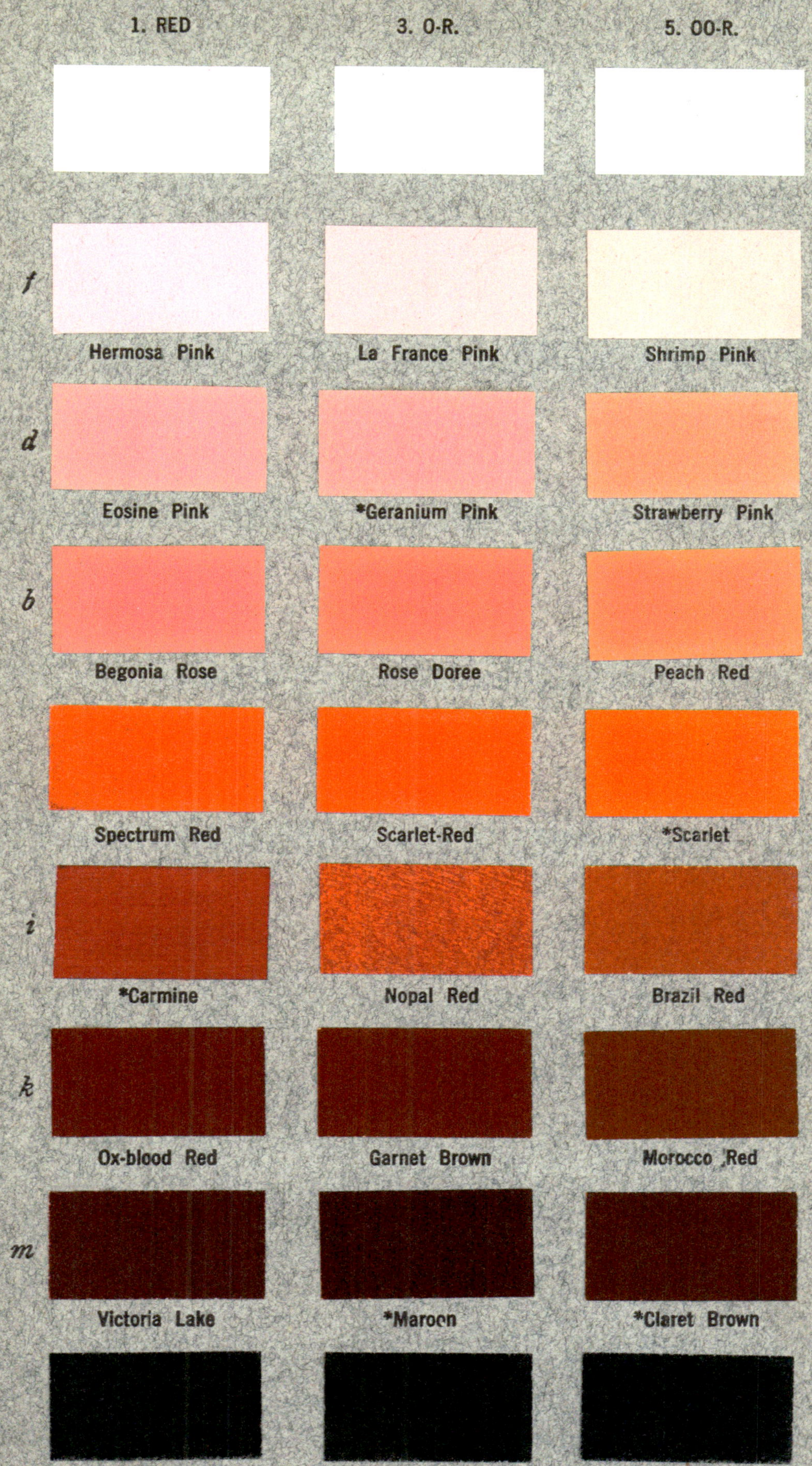
1. RED
3. O-R.
5. OO-R.
f
Hermosa Pink
La France Pink
Shrimp Pink
d
Eosine Pink
*Geranium Pink
Strawberry Pink
b
Begonia Rose
Rose Doree
Peach Red
Spectrum Red
Scarlet-Red
*Scarlet
i
*Carmine
Nopal Red
Brazil Red
k
Ox-blood Red
Garnet Brown
Morocco Red
m
Victoria Lake
*Maroon
*Claret Brown

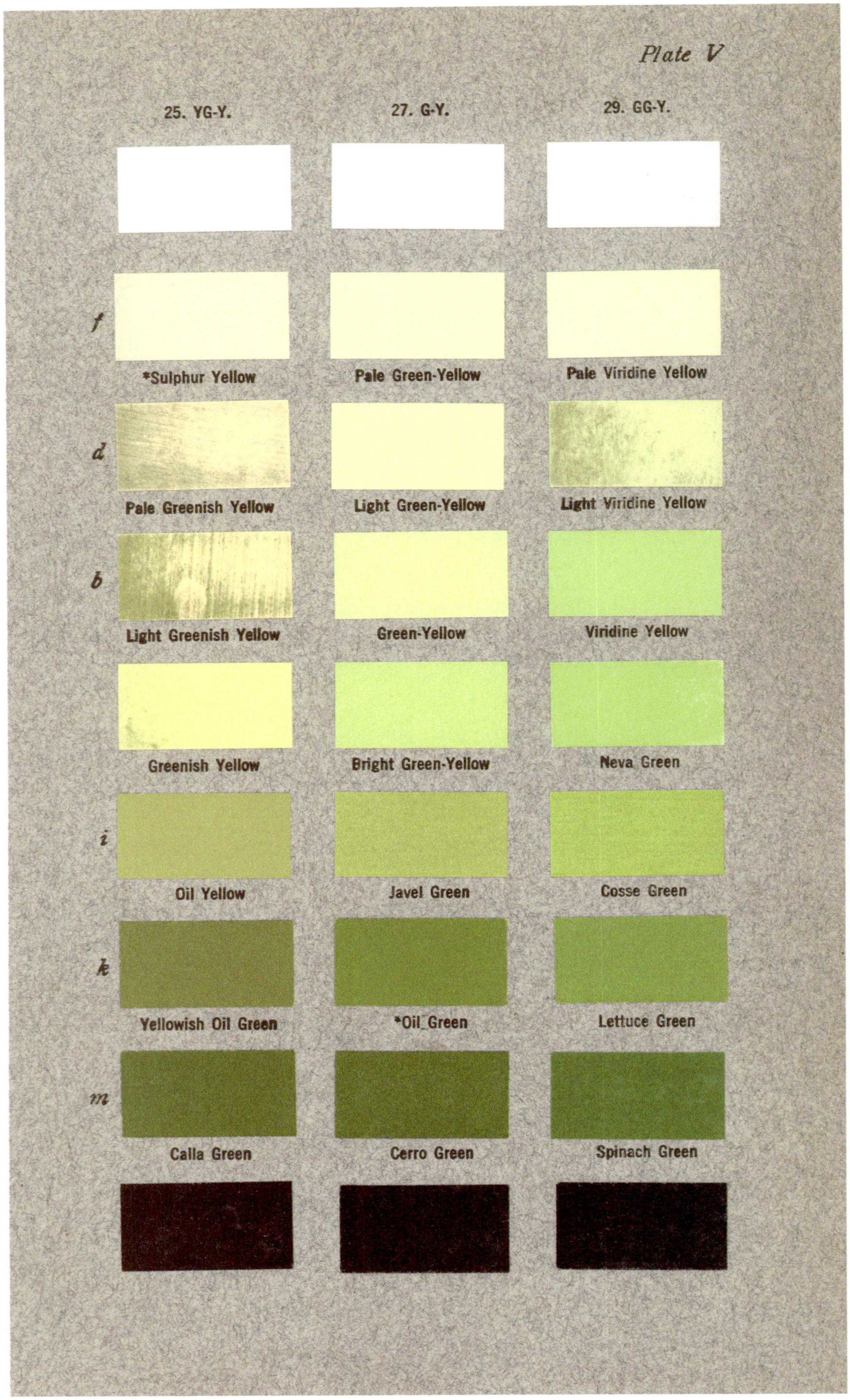
Plate V
25. YG-Y.
27. G-Y.
29. GG-Y.
f
*Sulphur Yellow
Pale Green-Yellow
Pale Viridine Yellow
d
Pale Greenish Yellow
Light Green-Yellow
Light Viridine Yellow
b
Light Greenish Yellow
Green-Yellow
Viridine Yellow
Greenish Yellow
Bright Green-Yellow
Neva Green
i
Oil Yellow
Javel Green
Cosse Green
k
Yellowish Oil Green
*Oil Green
Lettuce Green
m
Calla Green
Cerro Green
Spinach Green

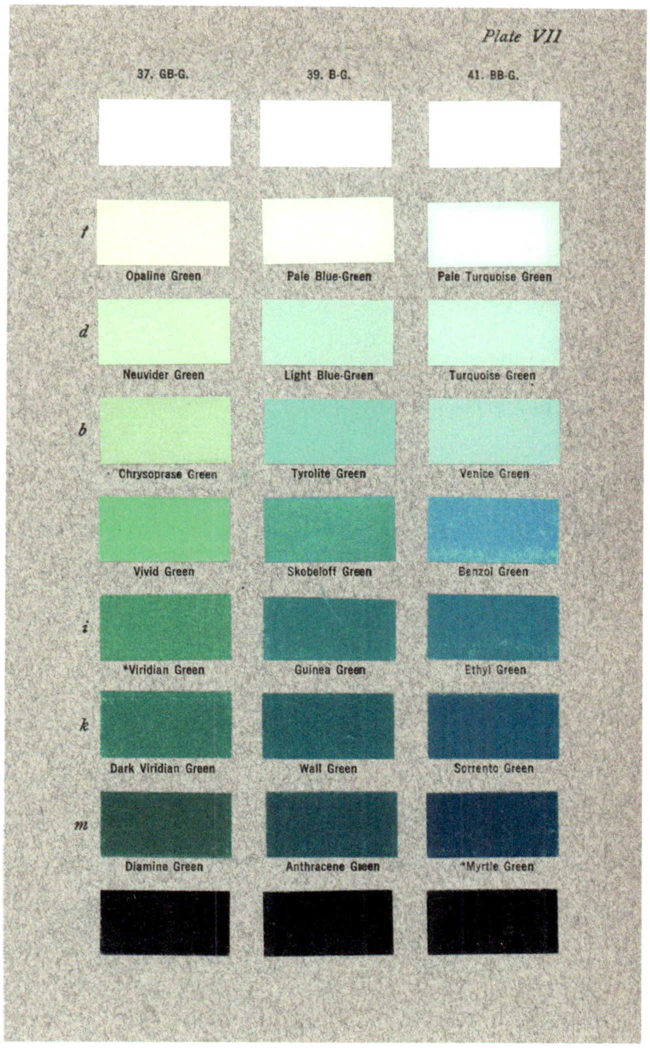

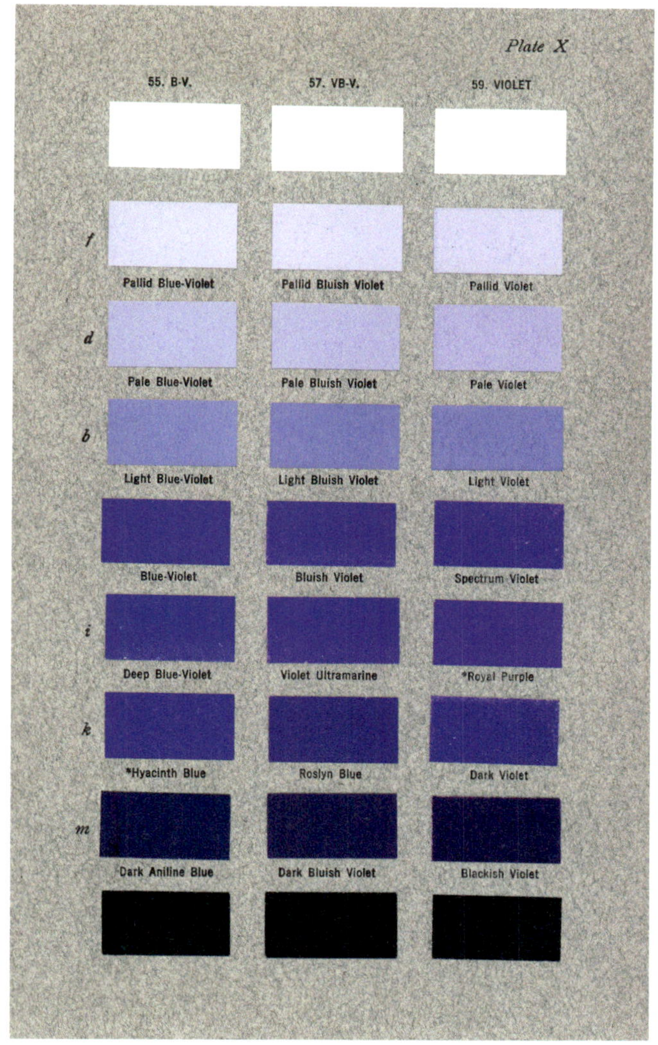

Farbstandards und Farbnomenklatur

Der amerikanische Ornithologe Robert Ridgway war stark beeinflusst von Werner und Syme und deren Versuchen, sowohl in optischer Hinsicht als auch bei der Benennung und Beschreibung von Farben einen Farbstandard zu entwickeln (S. 310–319). Er selbst wurde später als wichtige historische Quelle für das *Dictionary of Color* von Maerz und Paul (S. 410–419) bezeichnet. In *Color Standards and Color Nomenclature* wiederum bezog er sich auch auf sein eigenes Werk, hatte er doch 26 Jahre zuvor mit *A Nomenclature of Colors for Naturalists, and Compendium of Useful Knowledge for Ornithologists* eine erste Sammlung von Farbbezeichnungen veröffentlicht, die mit handkolorierten Tabellen und Diagrammen illustriert war. Sein lebenslanges Interesse an der Normierung von Farben und sein Bestreben, das eigene System zu verbessern, führte zu diesem zweiten, eleganten und akribisch recherchierten Buch, das auch die Fortschritte der Farbwissenschaft in den 1880er- bis 1910er-Jahren widerspiegelte.

Ridgway legte die Messlatte hoch. Er formulierte eindeutige Definitionen für die Farbterminologie und erstellte eine Liste von 1.115 Einzelfarben mit ihrer jeweiligen Zusammensetzung sowie eine kurze Bibliografie der jüngeren Farbliteratur. Die Abbildung auf seiner Titelseite veranschaulicht die Berechnung der Farbzusammensetzungen im Buch nach der Methode der „Maxwell'schen Farbscheiben", die auf einem Farbkreis mit 36 Segmenten und ineinandergreifenden Scheiben aus farbigem Karton basierte. Der schottische Physiker James Clerk Maxwell hatte in den 1850er-Jahren Farbscheibenkreisel entwickelt, mit denen er bestimmte Aspekte des Farbensehens und der additiven Farbmischung demonstrieren konnte.

Der wichtigste Bestandteil des Buches waren, wie auf der Titelseite angekündigt, die 53 Farbtafeln mit den Darstellungen der insgesamt 1.115 benannten Farben. Jede Tafel zeigt anhand von 27 aufgeklebten farbigen Rechtecken, wie sich der Farbton der Spektralfarben und ihrer Zwischenfarben, die auf den ersten zwölf Tafeln in ihrer reinsten Form verwendet werden. Die letzte Tafel besteht ausschließlich aus Grautönen, aufgefächert zwischen Weiß oben links und Schieferschwarz unten rechts.

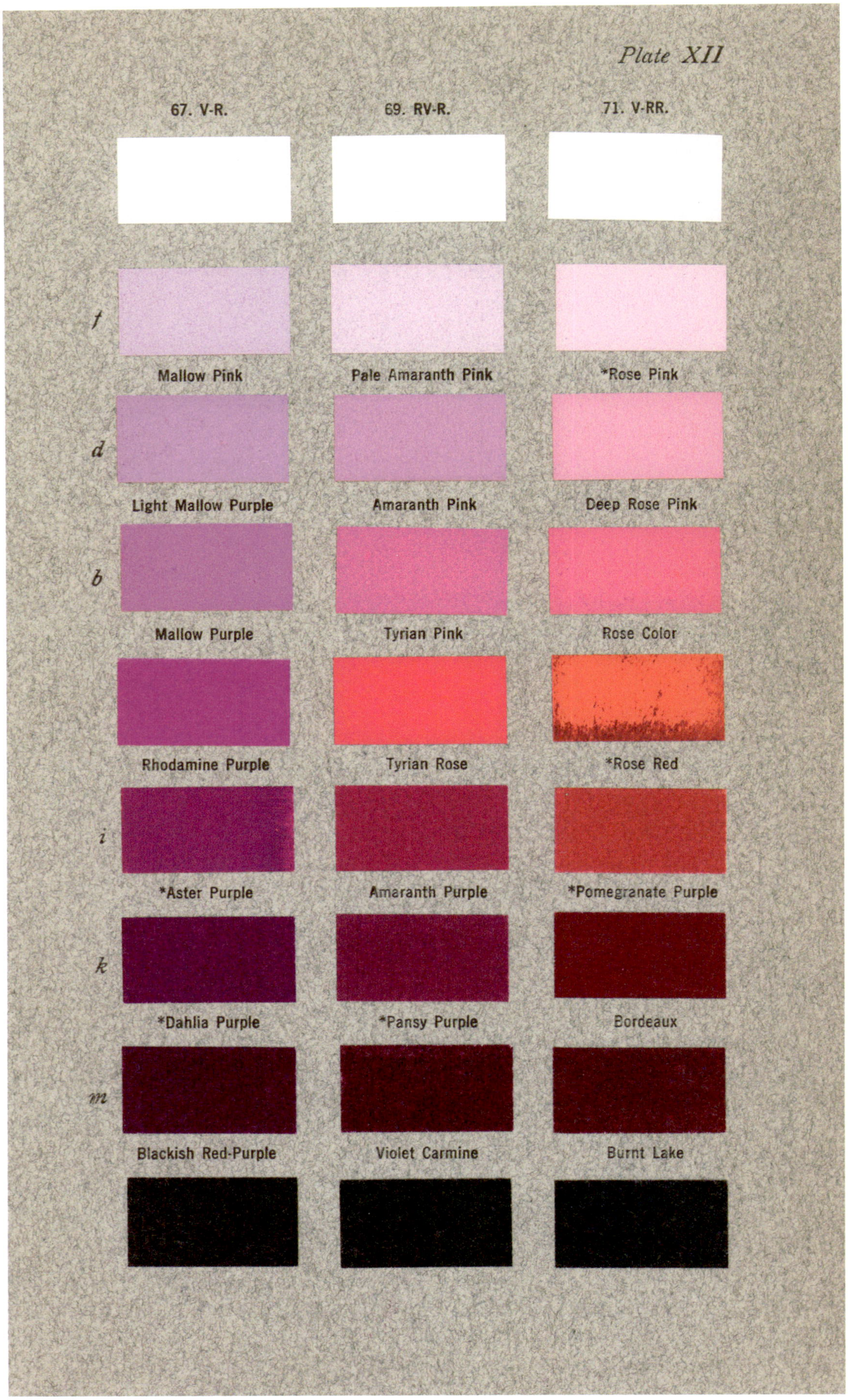
Plate XII
67. V-R.
69. RV-R.
71. V-RR.
f
Mallow Pink
Pale Amaranth Pink
*Rose Pink
d
Light Mallow Purple
Amaranth Pink
Deep Rose Pink
b
Mallow Purple
Tyrian Pink
Rose Color
Rhodamine Purple
Tyrian Rose
*Rose Red
i
*Aster Purple
Amaranth Purple
*Pomegranate Purple
k
*Dahlia Purple
*Pansy Purple
Bordeaux
m
Blackish Red-Purple
Violet Carmine
Burnt Lake

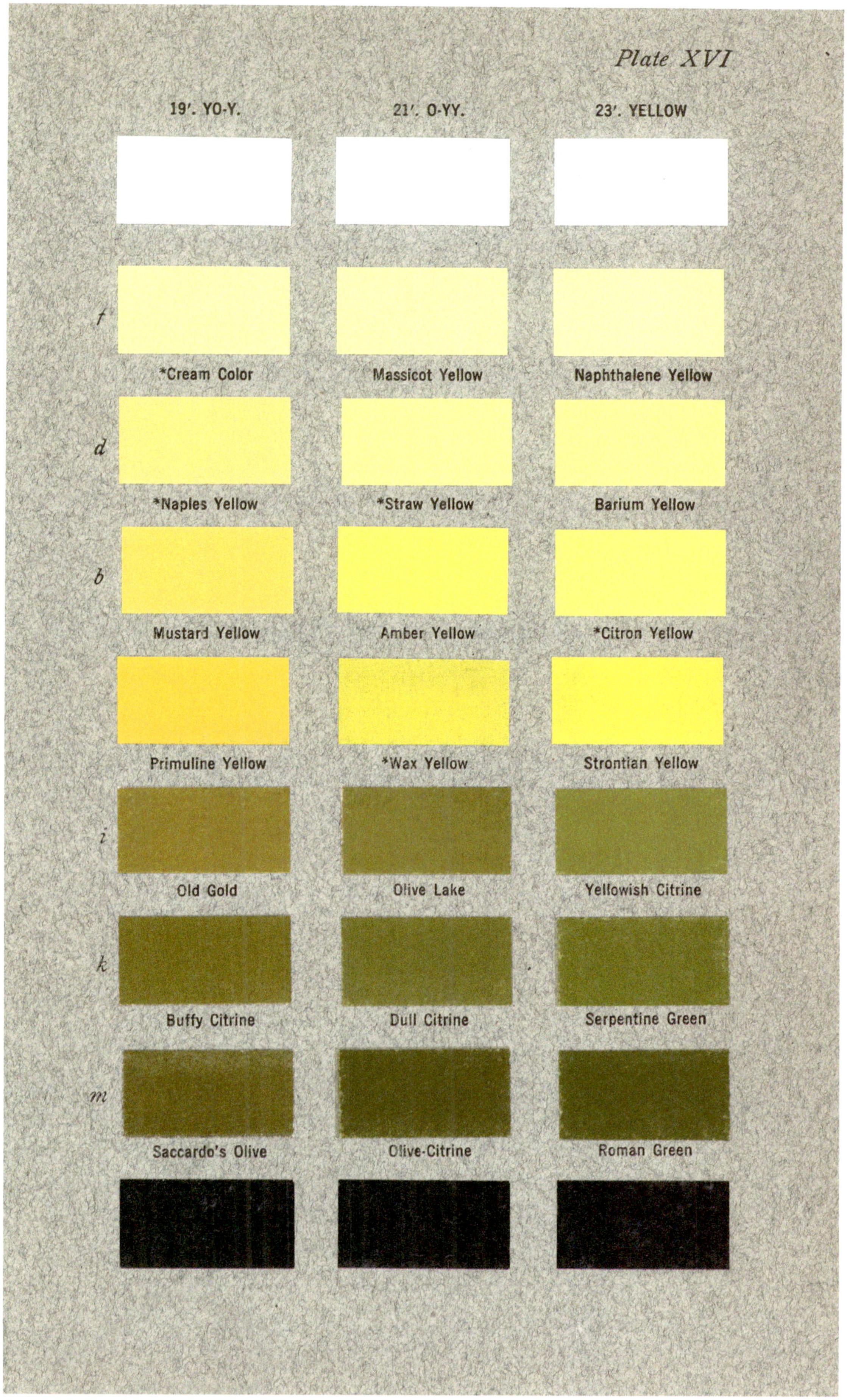
Plate XVI
19′. YO-Y.
21′. O-YY.
23′. YELLOW
f
*Cream Color
Massicot Yellow
Naphthalene Yellow
d
*Naples Yellow
*Straw Yellow
Barium Yellow
b
Mustard Yellow
Amber Yellow
*Citron Yellow
Primuline Yellow
*Wax Yellow
Strontian Yellow
i
Old Gold
Olive Lake
Yellowish Citrine
k
Buffy Citrine
Dull Citrine
Serpentine Green
m
Saccardo's Olive
Olive-Citrine
Roman Green

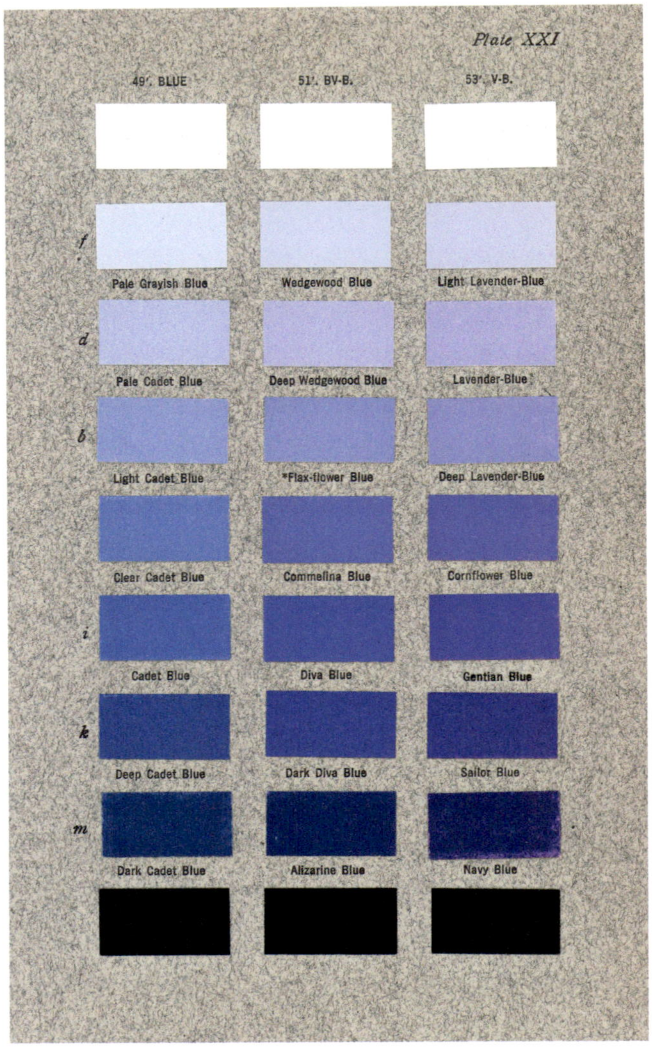

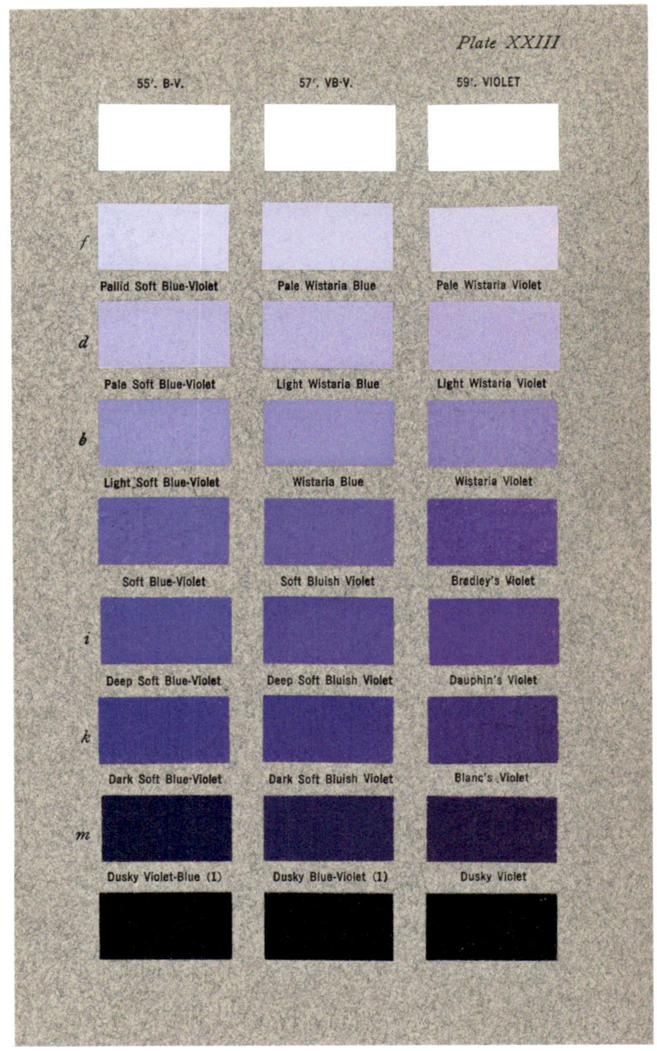

Ridgways besonderes Anliegen war die zuverlässige Wiedergabe der im Buch aufgelisteten Farben. Weil er Abweichungen zwischen den einzelnen Buchexemplaren vermeiden wollte, trug er jede Farbe einheitlich für die gesamte Auflage von fünftausend Büchern auf Papierbögen auf, die er anschließend in kleine Streifen zerschnitt. Seiner Leserschaft riet er, die Tafeln nicht für längere Zeit dem Licht auszusetzen, womit deutlich wurde, dass er hinsichtlich der flüchtigen Natur gedruckter Farben auf dem neuesten Stand war. Auch seine Pigmentauswahl, bei der er fast vollständig auf organische Farben verzichtete, spiegelt dies wider: „Die bei der Erstellung dieser Tafeln verwendeten Pigmente sind die beständigsten, die wir kennen, während jene, die sich als instabil erwiesen haben, so weit wie möglich aussortiert wurden."

Normes et nomenclatures des couleurs

Ornithologue américain, Robert Ridgway est très influencé par les tentatives de standardisation des couleurs entreprises par Werner et Syme (p. 310–319), tant visuellement que dans la manière de nommer et de décrire les couleurs. Ridgway, qui sera à son tour cité comme l'une des sources historiques majeures du *Dictionary of Color* de Maerz et Paul (p. 410–419), mentionne aussi ses travaux antérieurs dans ces *Color Standards and Color Nomenclature*. En effet, vingt-six ans auparavant, il a publié un recueil de noms de couleurs, *A Nomenclature of Colors for Naturalists, and Compendium of Useful Knowledge for Ornithologists*, illustré de tableaux et de graphiques coloriés à la main. L'intérêt qu'il a porté tout au long de sa vie à la standardisation des couleurs et son souhait de perfectionner son système l'amènent à faire paraître ce deuxième ouvrage, élégant fruit de recherches approfondies, qui témoigne en outre des progrès effectués dans la science des couleurs des années 1880 aux années 1910.

Avec une grande exigence, Ridgway offre une terminologie claire des couleurs et répertorie 1 115 teintes et les proportions qui les composent, ainsi qu'une liste concise d'ouvrages de son temps sur le sujet. L'image de la page de titre illustre la méthode des « disques de Maxwell » – employée dans l'ouvrage pour le calcul des proportions des couleurs – à l'aide d'un cercle chromatique formé de 36 segments et de disques colorés en carton qui s'emboîtent. Dans les années 1850, James Clerk

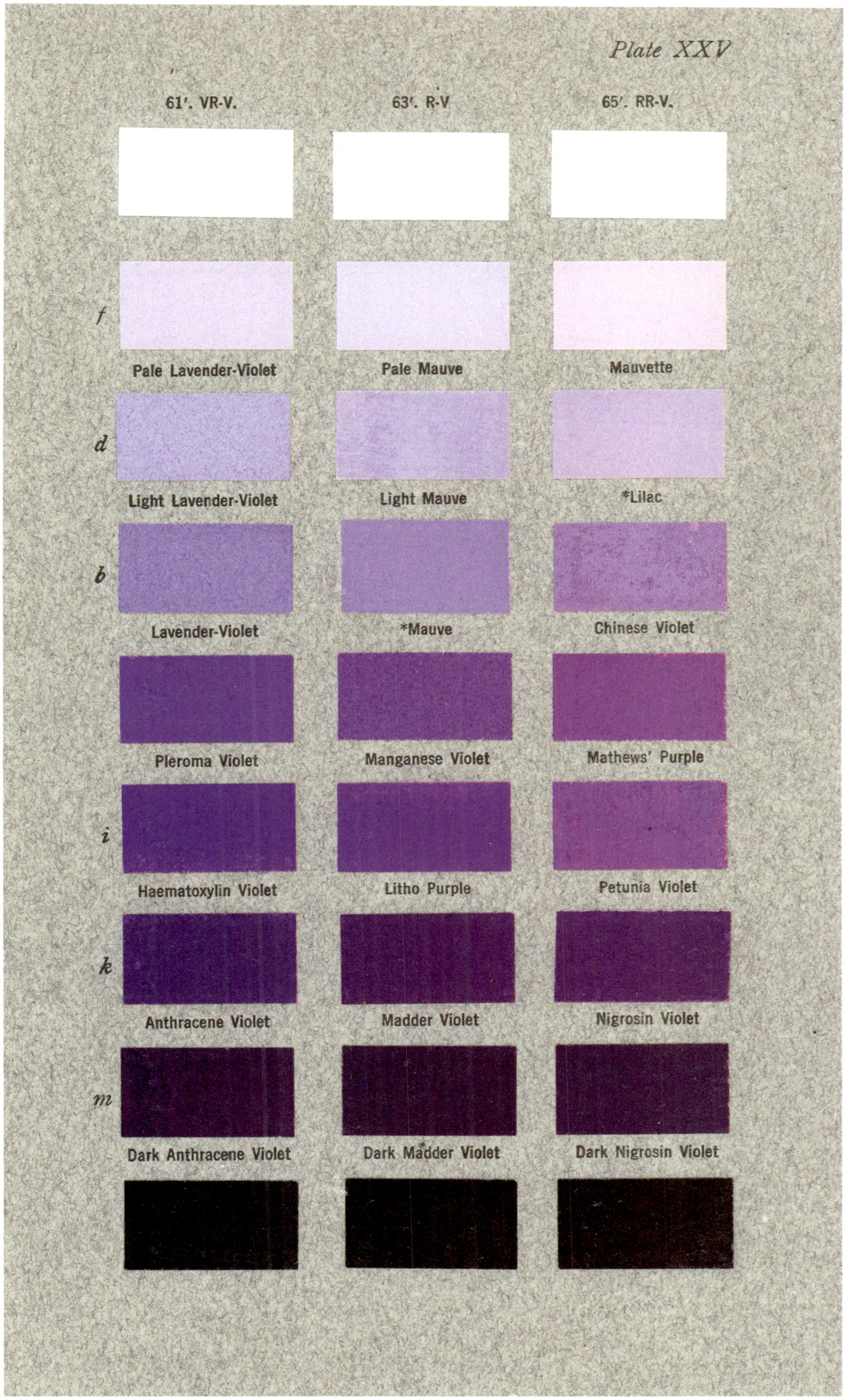
Plate XXV
61'. VR-V.
63'. R-V
65'. RR-V.
f
Pale Lavender-Violet
Pale Mauve
Mauvette
d
Light Lavender-Violet
Light Mauve
*Lilac
b
Lavender-Violet
*Mauve
Chinese Violet
Pleroma Violet
Manganese Violet
Mathews' Purple
i
Haematoxylin Violet
Litho Purple
Petunia Violet
k
Anthracene Violet
Madder Violet
Nigrosin Violet
m
Dark Anthracene Violet
Dark Madder Violet
Dark Nigrosin Violet

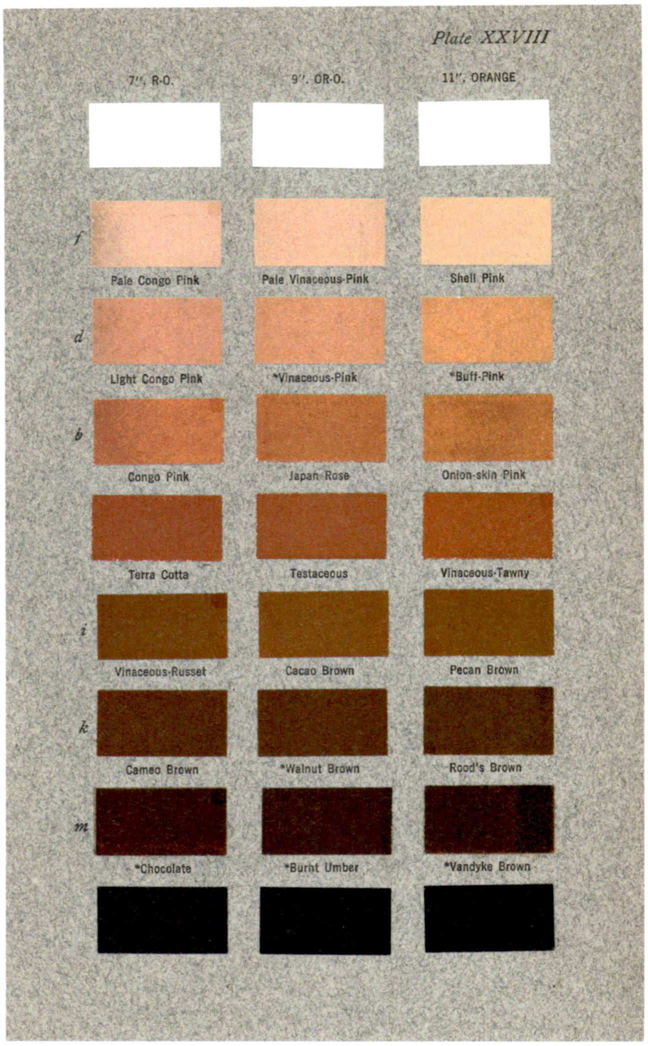

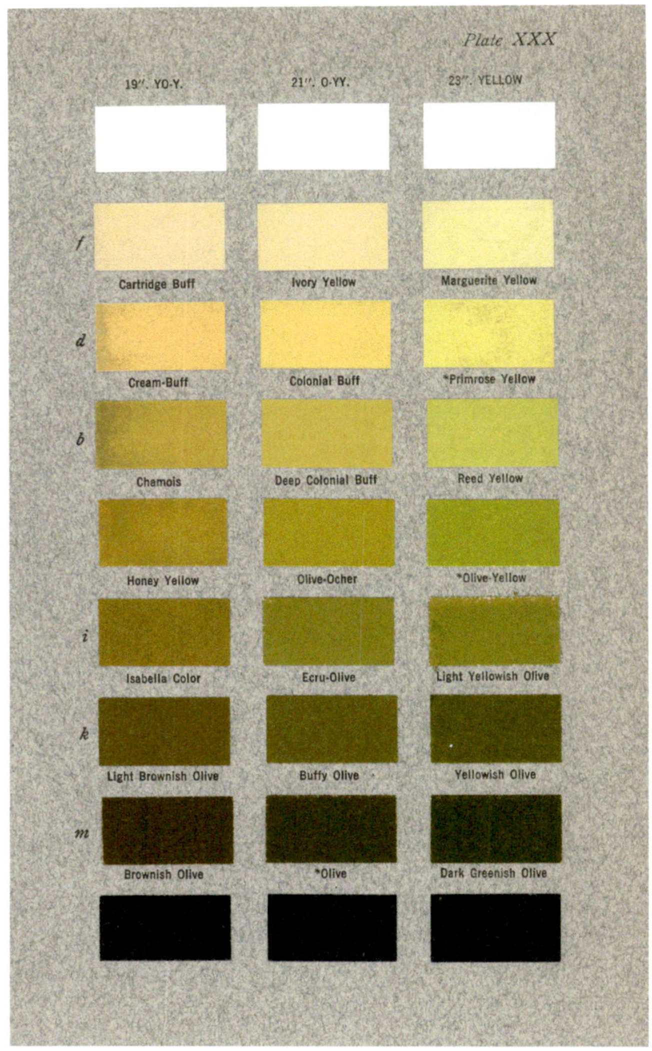

Maxwell, physicien écossais, avait conçu des dispositifs de toupies et de disques, grâce auxquels il démontrait certains aspects de la vision en couleurs et des mélanges de couleurs optiques.

La pièce maîtresse du livre de Ridgway est, comme l'annonce la page de titre, l'ensemble de 53 planches en couleurs, représentant les 1115 teintes inventoriées. Sur chaque planche, sont montés 27 échantillons colorés présentant des changements progressifs de tonalité. Les 12 premières planches montrent les couleurs du spectre et les teintes intermédiaires dans leur intensité maximalesur la ligne horizontale médiane. Viennent ensuite des groupes de couleurs auxquelles des gris neutres sont ajoutés graduellement, afin de ternir les teintes. La dernière planche est entièrement composée de nuances de gris, depuis le blanc de l'angle supérieur gauche jusqu'à la teinte ardoise de l'angle inférieur droit.

Soucieux de fournir une représentation fiable des couleurs répertoriées dans son livre, Ridgway veut éviter les variations d'un exemplaire à l'autre et a l'idée de peindre uniformément des feuilles de papier pour chaque couleur des 5000 exemplaires du tirage, puis de découper chaque feuille en petites bandes. En outre, il avertit les lecteurs de ne pas exposer trop longtemps les planches à la lumière, montrant ainsi qu'il est très conscient de la nature fragile de la couleur imprimée. Le choix des pigments employés dans les illustrations, qui laisse de côté presque toutes les teintes organiques, en est aussi la preuve: « Les pigments utilisés pour la préparation de ces planches sont les plus pérennes que nous connaissions actuellement, ceux qui se sont montrés instables ayant été, dans la mesure du possible, écartés. »

Normas y nomenclaturas de los colores

El ornitólogo estadounidense Robert Ridgway estuvo muy influenciado por el trabajo de Werner y Syme en sus intentos de estandarización del color (págs. 310–319), tanto en términos visuales como en la denominación y descripción de los colores. Más tarde se señaló al propio Ridgway como una fuente histórica importante para el *Dictionary of Color* de Maerz y Paul (págs. 410–419), y él mismo incluyó referencias a su trabajo en *Color Standards and Color Nomenclature*, dado que 26 años antes ya había publicado una primera

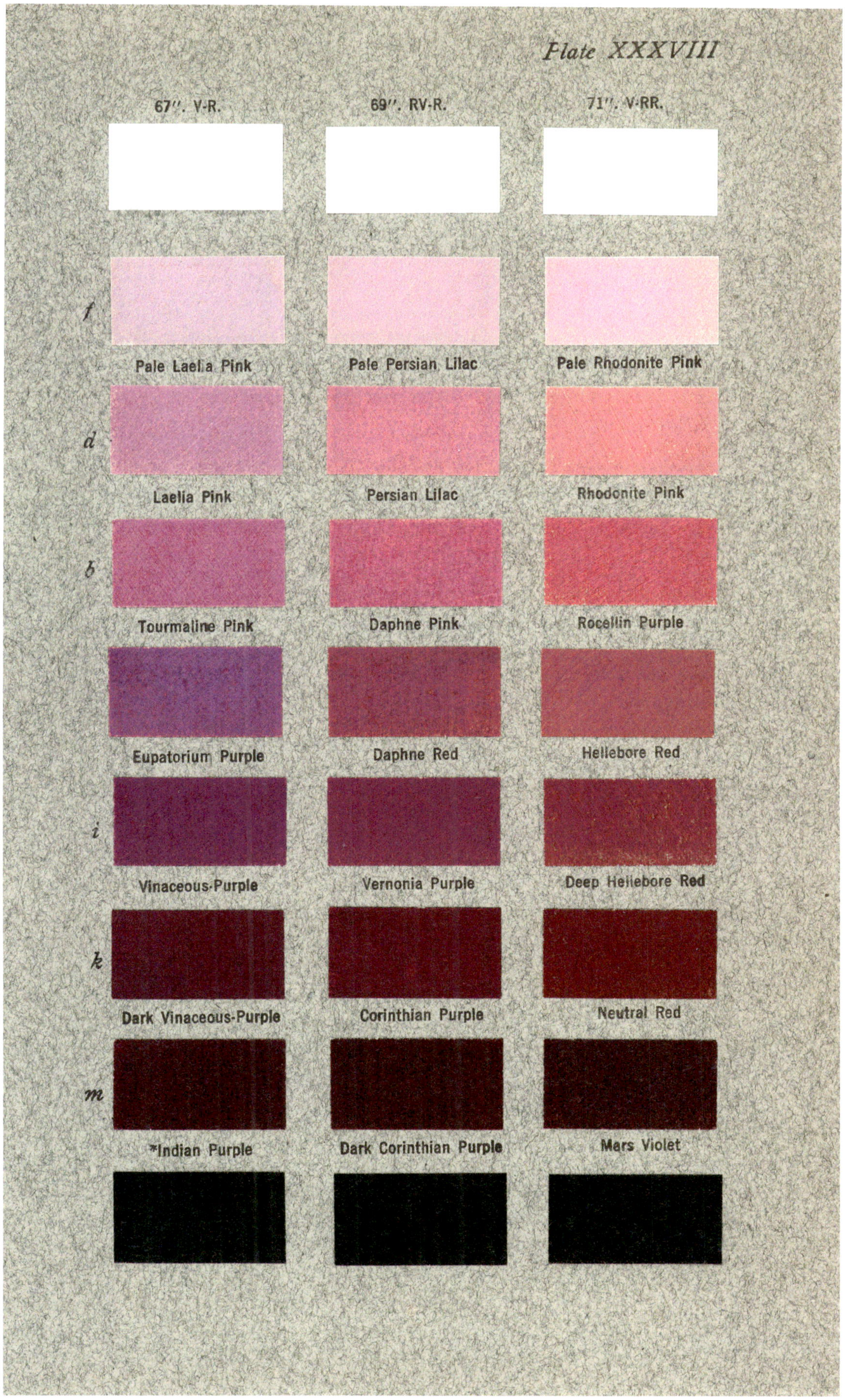
Plate XXXVIII
67''. V-R.
69''. RV-R.
71''. V-RR.
f
Pale Laelia Pink
Pale Persian Lilac
Pale Rhodonite Pink
d
Laelia Pink
Persian Lilac
Rhodonite Pink
b
Tourmaline Pink
Daphne Pink
Rocellin Purple
Eupatorium Purple
Daphne Red
Hellebore Red
i
Vinaceous-Purple
Vernonia Purple
Deep Hellebore Red
k
Dark Vinaceous-Purple
Corinthian Purple
Neutral Red
m
*Indian Purple
Dark Corinthian Purple
Mars Violet

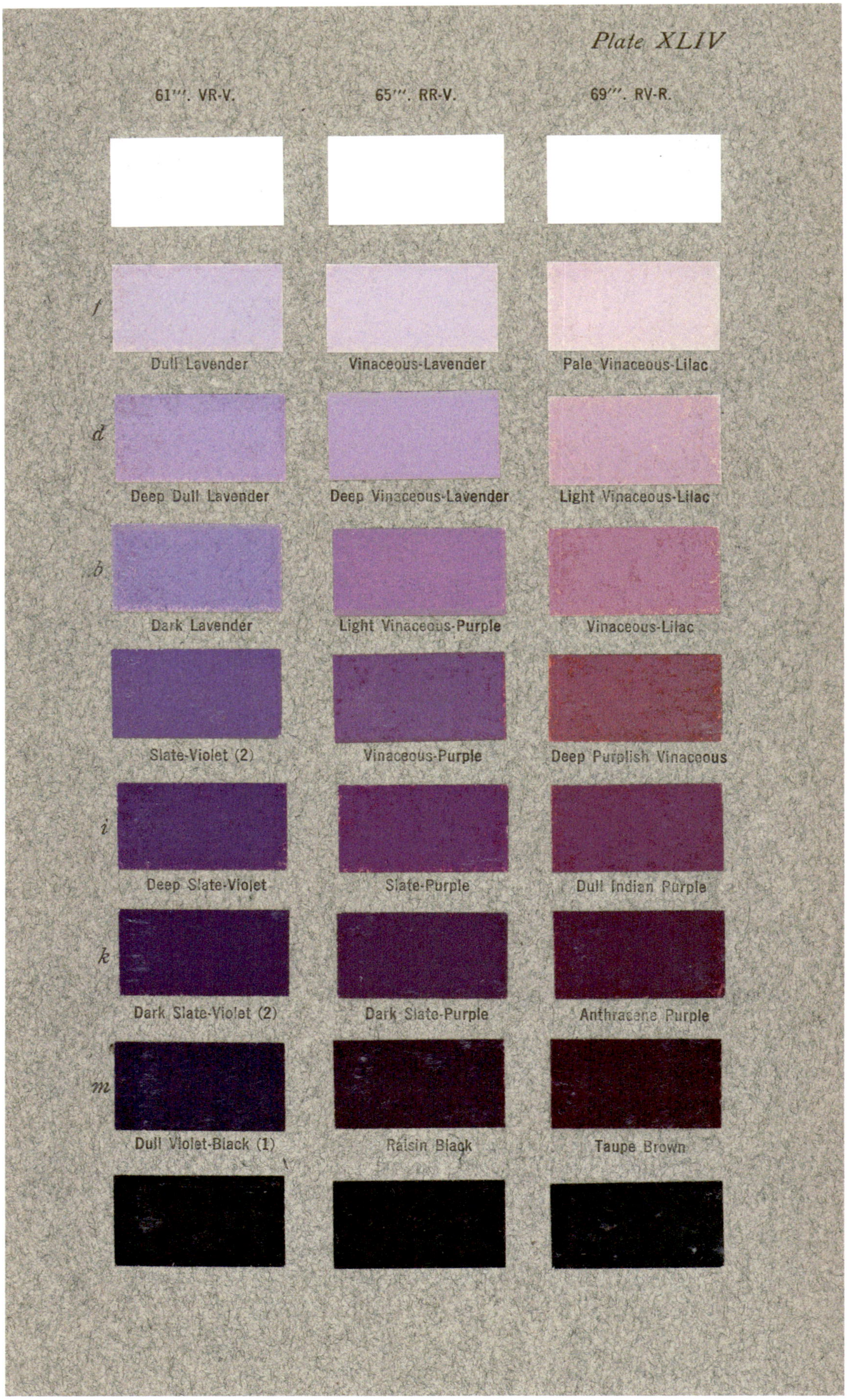
Plate XLIV
61‴. VR-V.
65‴. RR-V.
69‴. RV-R.
f
Dull Lavender
Vinaceous-Lavender
Pale Vinaceous-Lilac
d
Deep Dull Lavender
Deep Vinaceous-Lavender
Light Vinaceous-Lilac
b
Dark Lavender
Light Vinaceous-Purple
Vinaceous-Lilac
Slate-Violet (2)
Vinaceous-Purple
Deep Purplish Vinaceous
i
Deep Slate-Violet
Slate-Purple
Dull Indian Purple
k
Dark Slate-Violet (2)
Dark Slate-Purple
Anthracene Purple
m
Dull Violet-Black (1)
Raisin Black
Taupe Brown

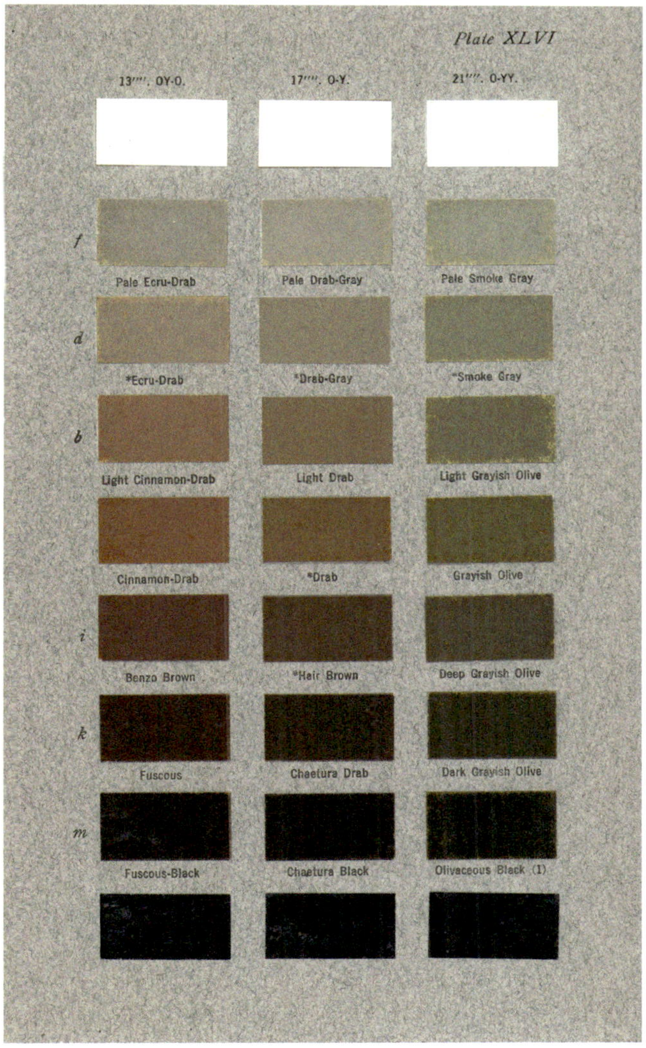

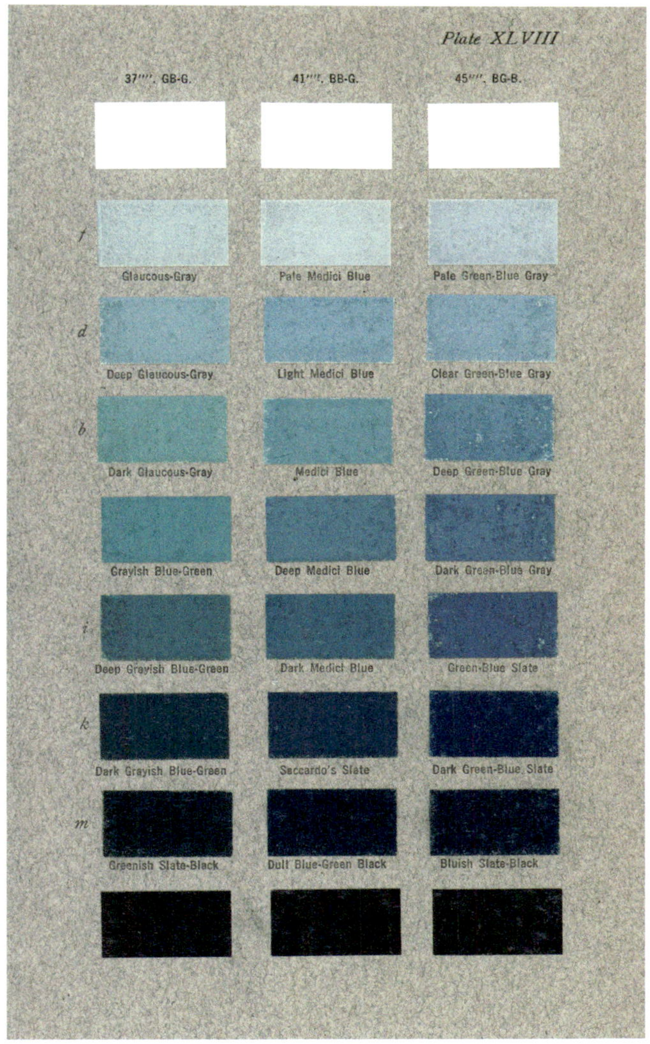

colección ilustrada de nombres de colores, *A Nomenclature of Colors for Naturalists*, con gráficos y diagramas coloreados a mano. Su interés de toda la vida en la estandarización del color y su deseo de mejorar su propio sistema dieron como resultado este segundo libro, elegante y fruto de una meticulosa investigación, que también refleja los avances en la ciencia del color entre las décadas de 1880 y 1910.

Ridgway, que se impuso unos altos estándares, proporcionó definiciones claras de la terminología del color y compiló una lista de 1115 colores individuales junto con su composición proporcional, así como una lista concisa de textos recientes sobre el color. Su portada ilustra el método de los «discos de Maxwell», que servían para calcular la composición proporcional de los colores incluidos en el libro con base en una rueda de colores de 36 segmentos y discos de cartón de colores entrelazados. En la década de 1850, el físico escocés James Clerk Maxwell había ideado aparatos a base de peonzas y discos con los que podía demostrar ciertos aspectos de la visión del color y de la mezcla óptica de colores.El elemento más importante del libro es el conjunto de 53 láminas de colores, que juntas brindan una representación de los 1115 colores mencionados. Cada lámina está compuesta por 27 muestras de color que componen un cambio gradual de tonalidades. En las 12 primeras láminas se emplean los colores del espectro y sus intermedios en su manifestación más pura, que se sitúan en las hileras horizontales intermedias. La lámina final consiste en distintos tonos de gris que van del blanco al negro pizarra.

A Ridgway le preocupaba especialmente la representación fiable de los colores enumerados en el libro. Para evitar variaciones entre las diferentes copias, optó por pintar hojas de papel uniformes para cada color, cortar esas hojas en pequeñas tiras e ilustrar con ellas la tirada de 5000 volúmenes. También advirtió a los lectores que no expusieran las láminas a la luz durante períodos prolongados, claro indicio de cuán avanzada era su comprensión de la naturaleza fugitiva del color impreso. Su elección de pigmentos para las ilustraciones, en la que casi omitió todos los colores orgánicos, también lo refleja: «Los pigmentos utilizados en la preparación de estas láminas son los más duraderos que se conocen. En la medida de lo posible, se han desechado aquellos que se han demostrado inestables».

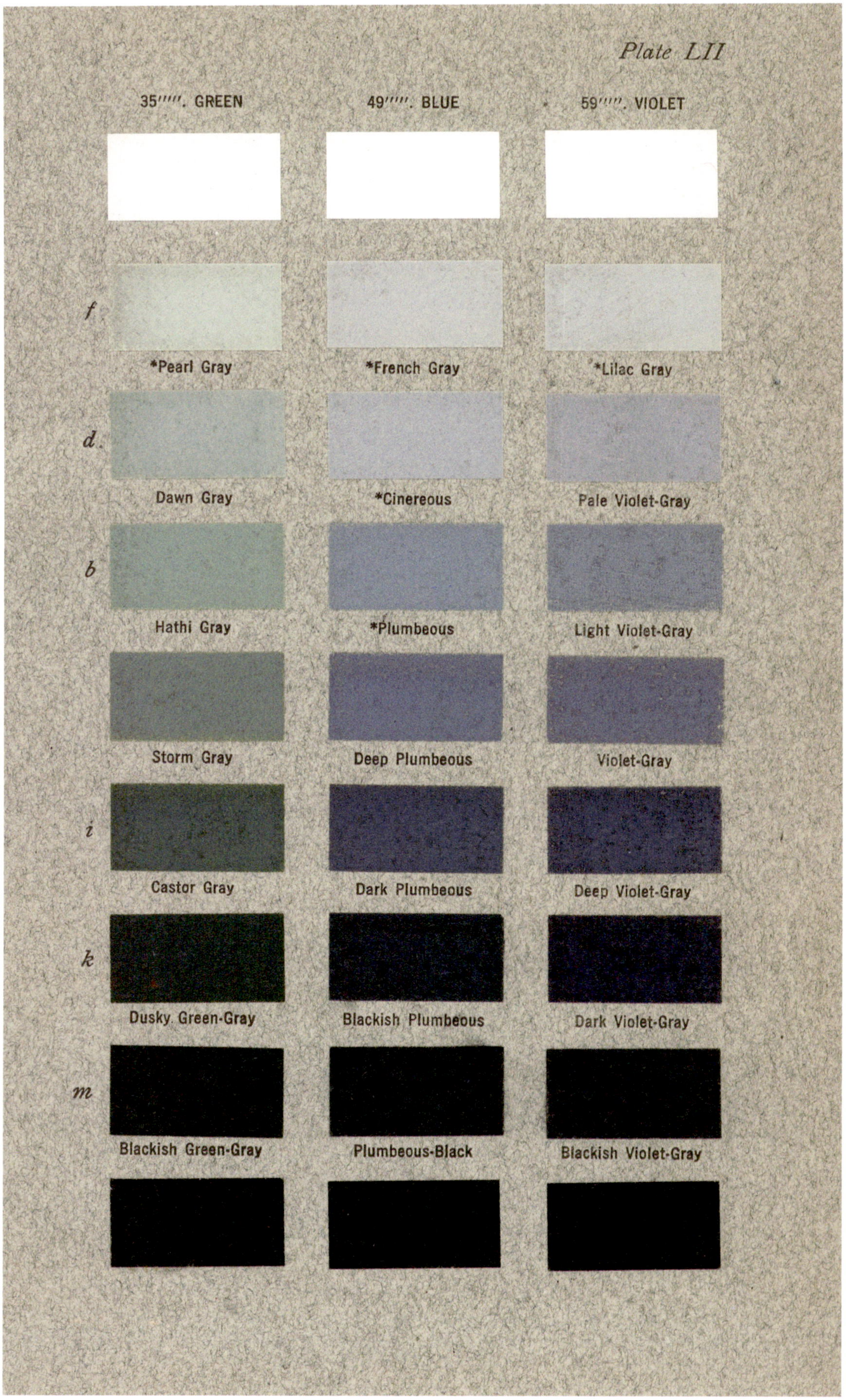
Plate LII
35''''. GREEN
49''''. BLUE
59''''. VIOLET
f
*Pearl Gray
*French Gray
*Lilac Gray
d
Dawn Gray
*Cinereous
Pale Violet-Gray
b
Hathi Gray
*Plumbeous
Light Violet-Gray
Storm Gray
Deep Plumbeous
Violet-Gray
i
Castor Gray
Dark Plumbeous
Deep Violet-Gray
k
Dusky Green-Gray
Blackish Plumbeous
Dark Violet-Gray
m
Blackish Green-Gray
Plumbeous-Black
Blackish Violet-Gray

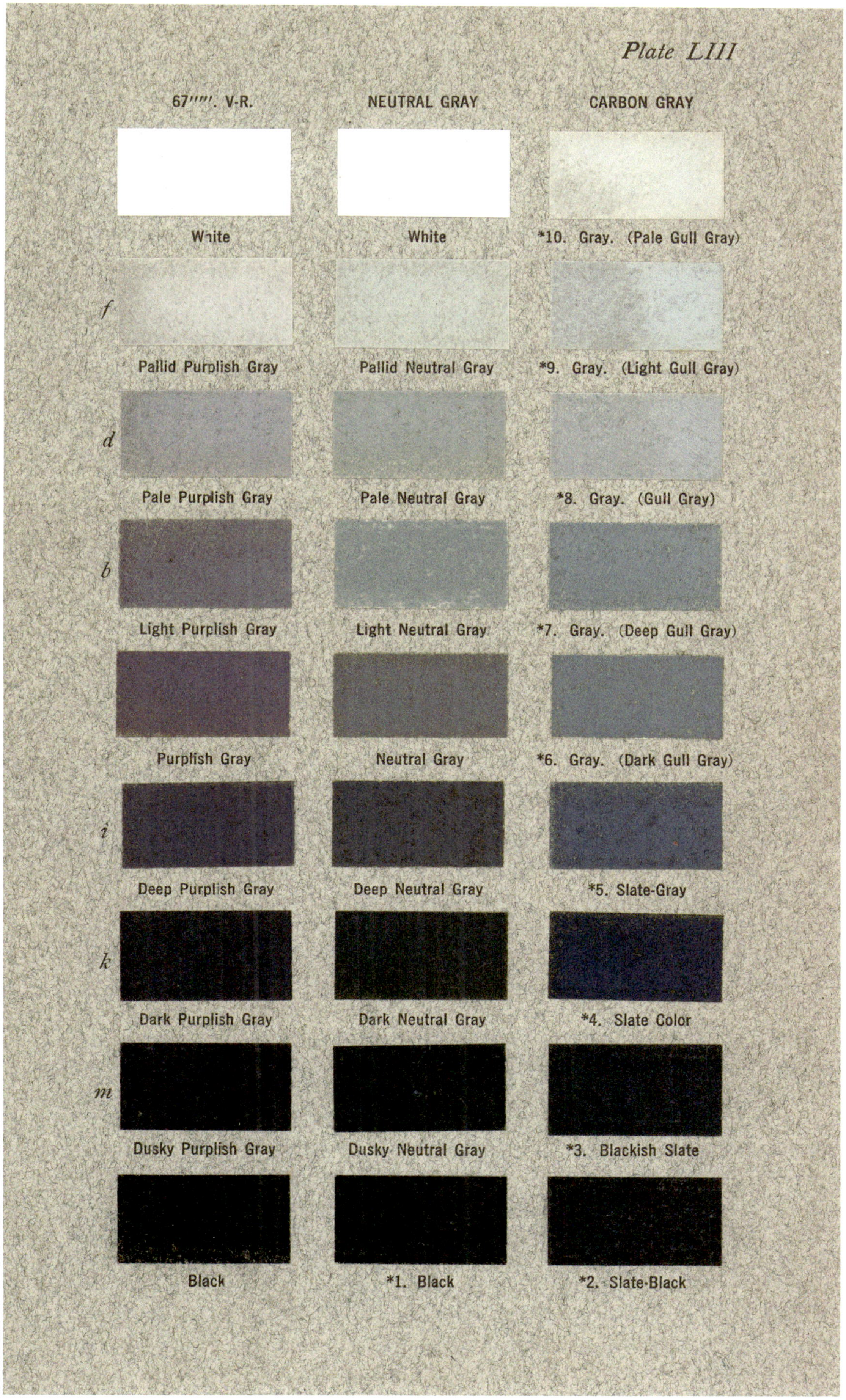
Plate LIII
67''''. V-R.
NEUTRAL GRAY
CARBON GRAY
White
White
*10. Gray. (Pale Gull Gray)
f
Pallid Purplish Gray
Pallid Neutral Gray
*9. Gray. (Light Gull Gray)
d
Pale Purplish Gray
Pale Neutral Gray
*8. Gray. (Gull Gray)
b
Light Purplish Gray
Light Neutral Gray
*7. Gray. (Deep Gull Gray)
Purplish Gray
Neutral Gray
*6. Gray. (Dark Gull Gray)
i
Deep Purplish Gray
Deep Neutral Gray
*5. Slate-Gray
k
Dark Purplish Gray
Dark Neutral Gray
*4. Slate Color
m
Dusky Purplish Gray
Dusky Neutral Gray
*3. Blackish Slate
Black
*1. Black
*2. Slate-Black

Elizabeth Burris-Meyer (1899–1969)

HISTORICAL COLOR GUIDE

30 plates, 23 x 17.5 cm | 9 x 6 ⅞ in., New York, 1938
Montreal, Canadian Centre for Architecture

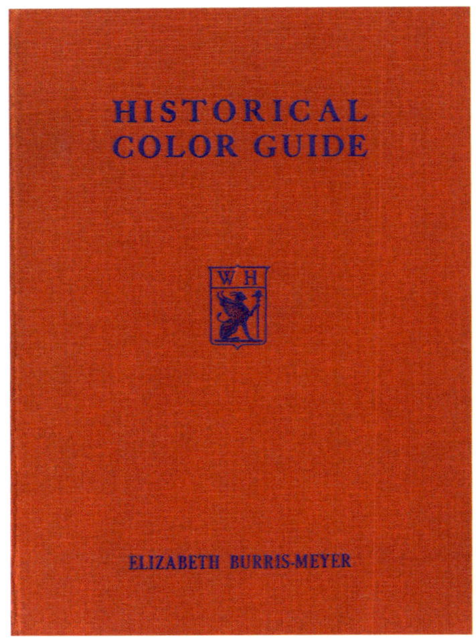

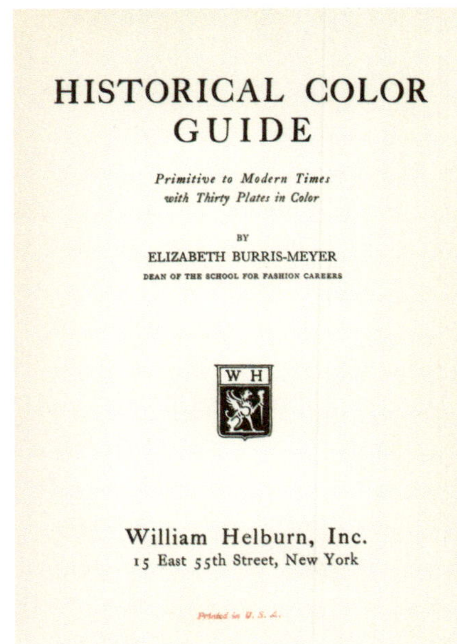

Elizabeth Burris-Meyer is still an under-researched and under-appreciated writer, educator and commentator on 20th-century colour. She has hardly been discussed in colour literature, and little is known about her life, despite the fact that her publishing output was extensive. When this first *Historical Color Guide* by her was issued she was working as Dean of the Tobé-Coburn School for Fashion Careers in New York, a newly founded training institution for the fashion and beauty industries. She was previously educated at the University of Michigan, and subsequently taught design at the Paris and New York branches of the New York School of Fine and Applied Art (later known as Parsons New School of Design). Her first published work was the substantial *Color and Design in the Decorative Arts* (1935), in which she demonstrated a thorough understanding of art history and colour science, together with an ability to write and educate in a clear and accessible manner about colour and its applications. Amongst other topics she discussed Albert Henry Munsell's colour concepts and systems, and reproduced one of his colour wheels and some of his other diagrams. Her book also included a colour card with 66 pasted-in and named swatches of "representative colors in common use in merchandising, advertising, and dress", together with a perforated paper screen designed to isolate colours or colour groups on the card and thereby assist the user in finding harmonious arrangements. Burris-Meyer applied this practical, informed and concise style in all her publications.

With the *Historical Color Guide* her purpose was to provide "a book of representative traditional color schemes derived from examples of the decorative arts from all fields", predominantly intended for designers, merchants, decorators and architects. Across 30 plates, each provided with five printed colour strips and an accompanying page of commentary, Burris-Meyer presented a historical, global and cultural overview of colour schemes throughout the centuries. She arranged these plates by theme, from the so-called "Primitives" found on tribal carvings and textiles, on by way of Ancient Egypt, the Classical world, Chinese examples and Persian Miniatures, through specific periods and phases such as the Italian Renaissance and the English "mauve decade", to individual artists' palettes for Van Gogh and Gauguin. Each colour sample is numerically marked according to Munsell's notation system, and the book was rounded off with a numbered colour index. The concept and style of Burris-Meyer's *Historical Color Guide* bear a close resemblance to Thomas Parsons's *A Tint Book of Historical Colours*, published in London four years earlier, and there is no doubt that Burris-Meyer took inspiration from it. However, where Parsons's book concentrated specifically on architectural colour, Burris-Meyer was aiming for a far wider range of application. She wrote several other books on design, interior decoration and fashion, always including colour charts, and in 1947 published a companion volume to this book, her *Contemporary Color Guide*.

PRIMITIVES

Bali GY 8/6

Blossom Y 9/10

Fungus RP-R 5/8

Calabash R-YR 6/10

Shadow RP 4/4

EGYPTIAN

Lotus YR-Y 8/10

Griffin R 4/10

Scarab BG 6/8

Mummy Y 6/4

Cobra G-BG 5/2

2

CRETAN

Sea Foam Y-GY 8/6

Tulip R-YR 5/8

Shell R-YR 7/6

Monkey B-GB 7/12

Palmette GY 6/6

3

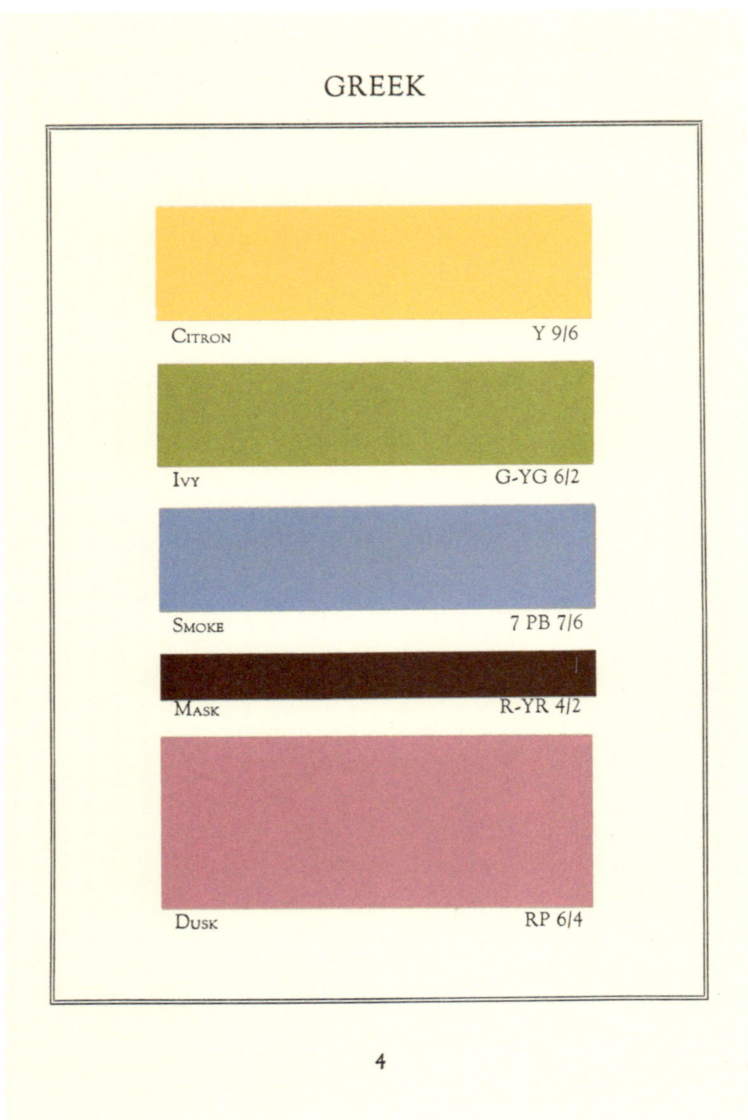

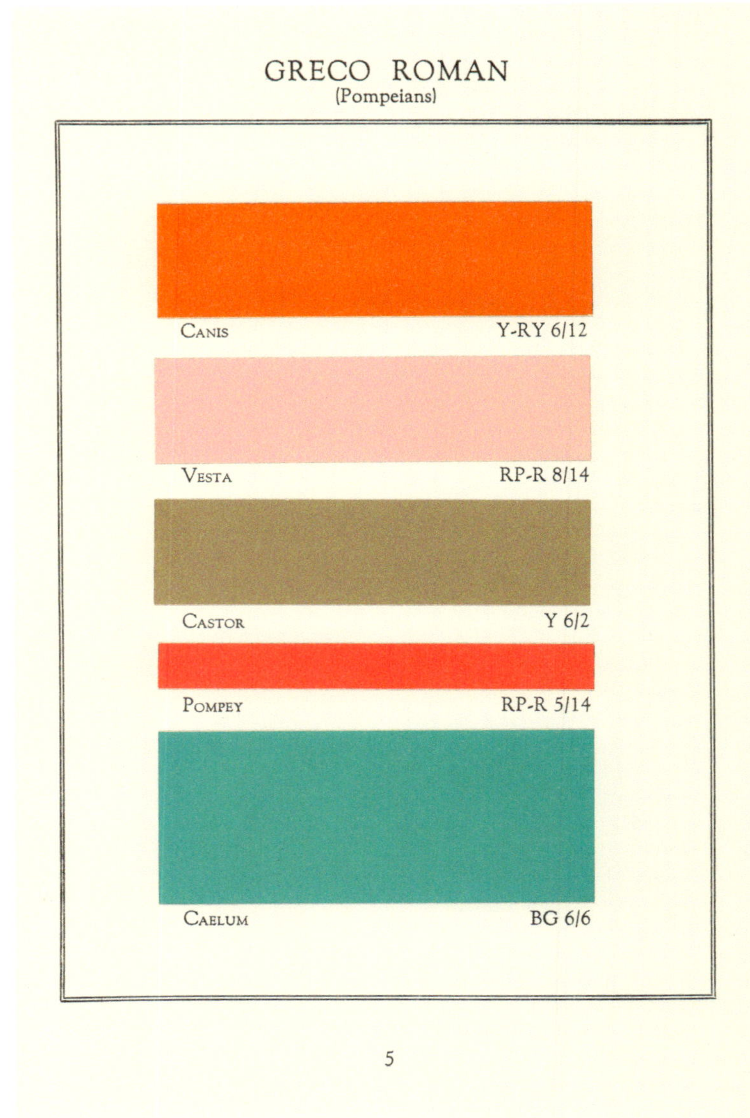

Historischer Farbführer

Als Autorin und Pädagogin, die sich mit den Farben des 20. Jahrhunderts beschäftigte, wird Elizabeth Burris-Meyer bis heute von der Forschung vernachlässigt und zu wenig geschätzt. In der Farbliteratur wird sie so gut wie nicht erwähnt, und obwohl sie viel veröffentlicht hat, ist über ihr Leben wenig bekannt. Als ihr *Historical Color Guide* erschien, war sie Direktorin der Tobé-Coburn School for Fashion Careers in New York, einer neu gegründeten Ausbildungseinrichtung für die Mode- und Kosmetikindustrie. Sie hatte die University of Michigan besucht und anschließend in Paris und New York an der New York School of Fine and Applied Art (der späteren Parsons New School of Design) Gestaltung unterrichtet. Ihre erste Veröffentlichung war der umfangreiche Band *Color and Design in the Decorative Arts* (1935), in dem sie nicht nur ein umfassendes Verständnis von Kunstgeschichte und Farbwissenschaft, sondern auch die Fähigkeit unter Beweis stellte, auf klare und verständliche Weise über Farbe und ihre Anwendungen zu schreiben. So behandelte sie unter anderem die Farbtheorien und -systeme von Albert Henry Munsell und reproduzierte neben einem Farbkreis noch weitere seiner Diagramme. Ihr Buch enthielt auch eine Farbkarte mit 66 eingeklebten und namentlich benannten Mustern von „repräsentativen Farben, die in Handel, Werbung und Mode allgemein in Gebrauch sind“, ergänzt um eine perforierte Papierschablone, mit der sich Farben oder Farbgruppen auf der Karte abgrenzen ließen, um das Auffinden harmonischer Kombinationen zu vereinfachen. Diesen praxisorientierten, sachkundigen und prägnanten Stil verwendete Burris-Meyer in allen ihren Veröffentlichungen.

Der *Historical Color Guide* sollte „ein Buch der repräsentativen traditionellen Farbkonzepte, abgeleitet von Beispielen aus allen Bereichen der dekorativen Kunst“ sein, das sich vorrangig an Designer, Händler, Dekorateure und Architekten richtete. Auf dreißig Tafeln mit jeweils

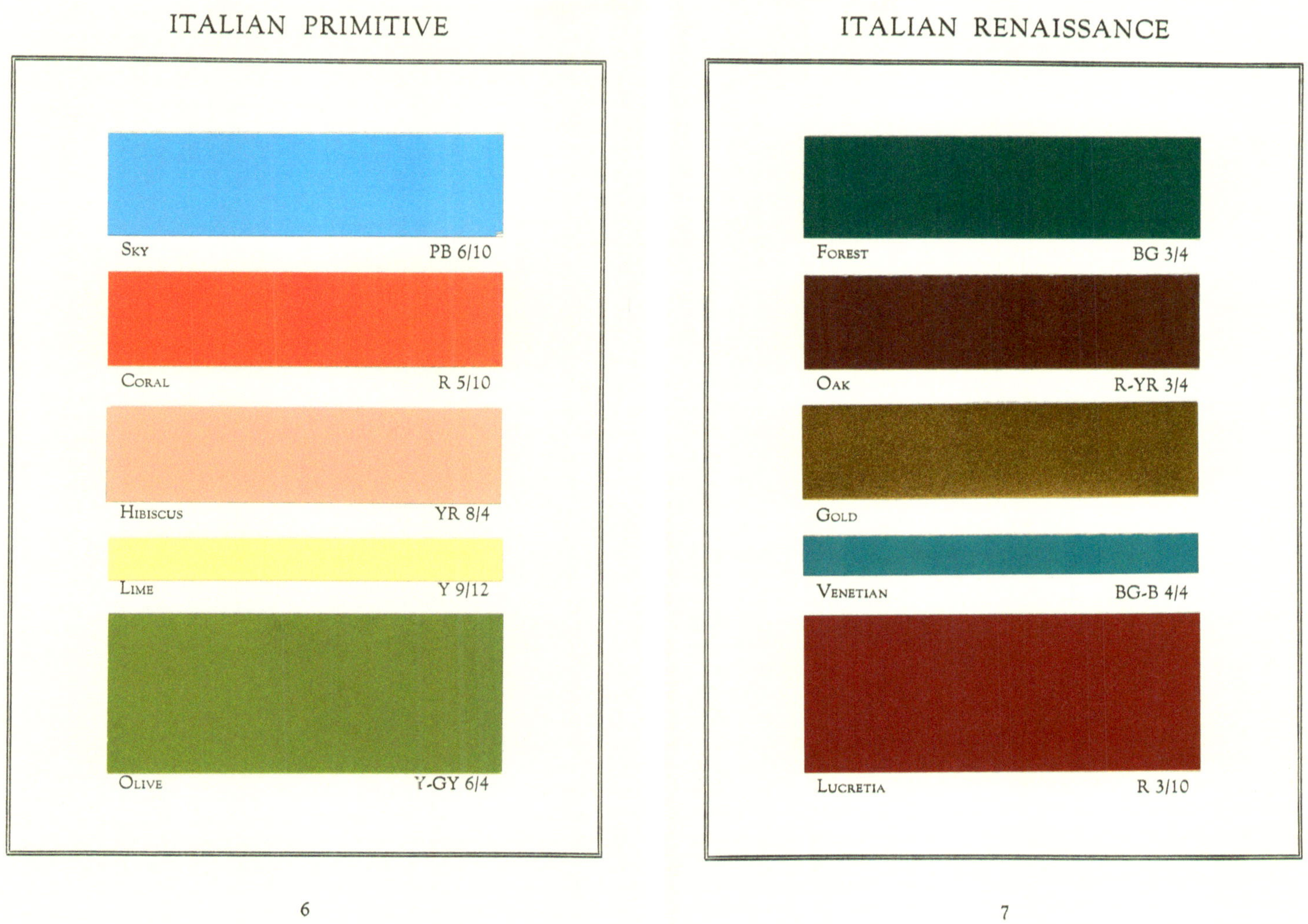

fünf gedruckten Farbstreifen und einer Seite Begleittext präsentierte Burris-Meyer einen globalen historischen und kulturellen Überblick über Farbzusammenstellungen quer durch die Jahrhunderte. Sie ordnete die Tafeln thematisch an, von den „Urfarben" auf Schnitzereien und Textilien alter Stammeskulturen über Altägypten, die Antike, Beispiele aus China, persische Miniaturen, bestimmte Perioden und Phasen wie die italienische Renaissance oder die „mauve decade" in England (1890er-Jahre) hin zu den Künstlerpaletten van Goghs und Gauguins. Jedes Farbmuster ist mit einer Nummer nach Munsells Notationssystem versehen, die sich im abschließenden Farbregister wiederfindet. In Konzeption und Stil erinnert der *Historical Color Guide* stark an *A Tint Book of Historical Colours* von Thomas Parsons, das vier Jahre zuvor in London erschienen war und Burris-Meyer zweifellos beeinflusst hatte. Doch wo sich Parsons Buch speziell auf Architekturfarben konzentrierte, zielte Burris-Meyer auf ein weit größeres Anwendungsspektrum ab. Sie schrieb mehrere andere Bücher über Design, Innenausstattung und Mode, die stets Farbtabellen enthielten, und veröffentlichte 1947 mit dem *Contemporary Color Guide* einen Begleitband zu ihrem *Historical Color Guide*.

Guide historique de la couleur

Pédagogue et exégète de la couleur au XXe siècle, Elizabeth Burris-Meyer est une autrice sous-estimée et trop peu étudiée. On ignore presque tout de la vie de cette femme qui, alors qu'elle a énormément publié, est quasiment absente des écrits sur le sujet. Quand paraît son premier « Guide historique de la couleur », Burris-Meyer est doyenne de l'École Tobé-Coburn des carrières de la mode, à New York, institution alors récemment fondée pour former aux métiers de la mode et de la beauté. Après des études à l'Université du Michigan, elle enseigne le dessin dans les antennes parisienne et new-yorkaise de l'École des beaux-arts et arts appliqués de New York (connue depuis

FLEMISH PRIMITIVE

Bruges	Y-RY 3/8
L'arbre	GY 6/4
La Jeunesse	PB 8/2
La Joie	R-YR 8/8
La Morte	Y-GY 3/2

8

FRENCH, VERDURE TAPESTRY

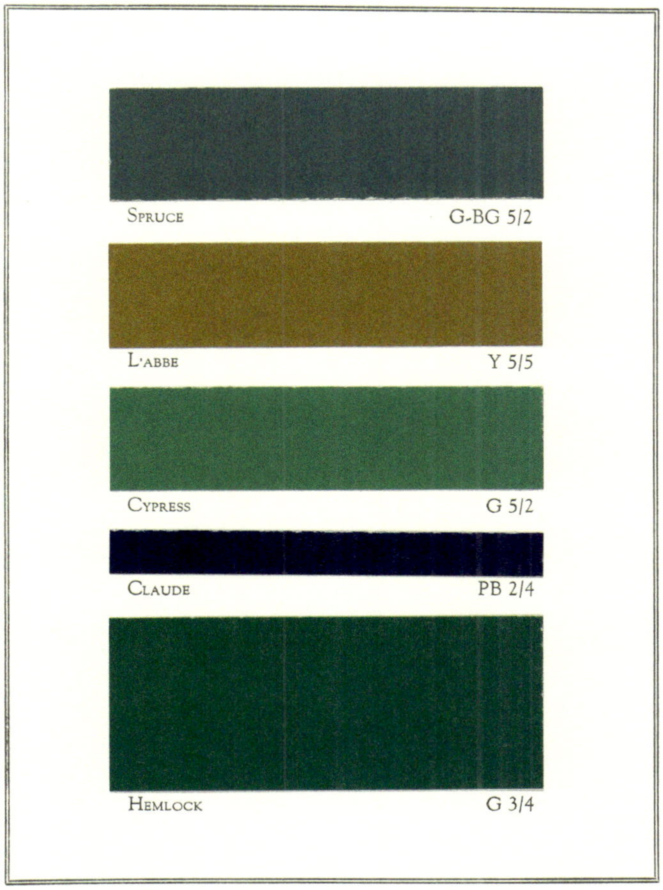

9

FRENCH, STAINED GLASS

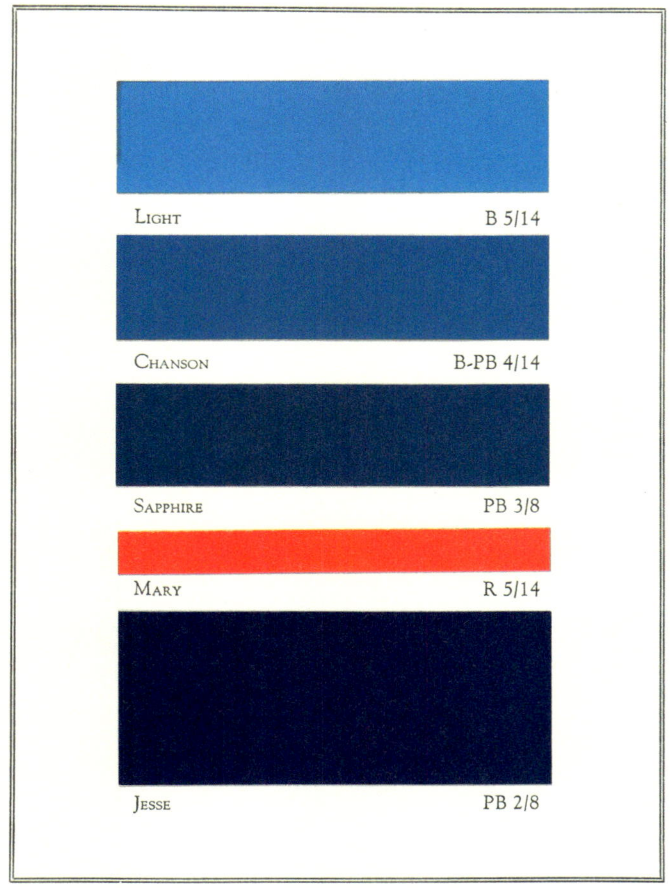

10

sous le nom de Parsons New School of Design). En 1935, elle publie son premier ouvrage, *Color and Design in the Decorative Arts*, où elle fait montre d'une très grande connaissance de l'histoire de l'art et de la science des couleurs, ainsi que d'une capacité à écrire sur la couleur et ses applications et à dispenser un enseignement clair et accessible. Elle s'intéresse notamment aux conceptions et systèmes chromatiques d'Albert Henry Munsell et reproduit un cercle chromatique et d'autres graphiques de cet auteur. Son ouvrage comprend également une carte en couleurs sur laquelle sont collés 66 échantillons, chacun doté d'une appellation, figurant les « couleurs représentatives couramment employées dans le commerce, la publicité et l'habillement », ainsi qu'une feuille de papier perforée destinée à isoler les teintes ou groupes de teintes de cette carte, afin de permettre à l'utilisateur de trouver des combinaisons harmonieuses. Dans tous ses ouvrages, Burris-Meyer a recours à cette méthode pratique, documentée et concise.

L'objectif de ce guide est de proposer « un ouvrage représentatif des coloris traditionnels inspirés d'exemples des arts décoratifs de tous les domaines », principalement à destination des dessinateurs, marchands, décorateurs et architectes. Grâce à 30 planches, chacune comportant cinq bandes de couleurs imprimées et accompagnée d'une page de commentaires, Burris-Meyer présente un panorama historique et culturel des couleurs du monde entier au fil des siècles. Elle organise ces planches de manière thématique, en commençant par ce qu'elle appelle les « primitives », visibles dans les gravures et textiles tribaux, puis passe à l'Égypte ancienne, au monde classique, à des exemples chinois et aux miniatures persanes, et s'intéresse à des périodes spécifiques comme la Renaissance italienne et la « décennie mauve » anglaise des années 1890, ainsi qu'à la palette de peintres comme Van Gogh ou Gauguin. Chaque échantillon de couleur est numéroté selon le système de notation de Munsell et l'ouvrage s'achève par un

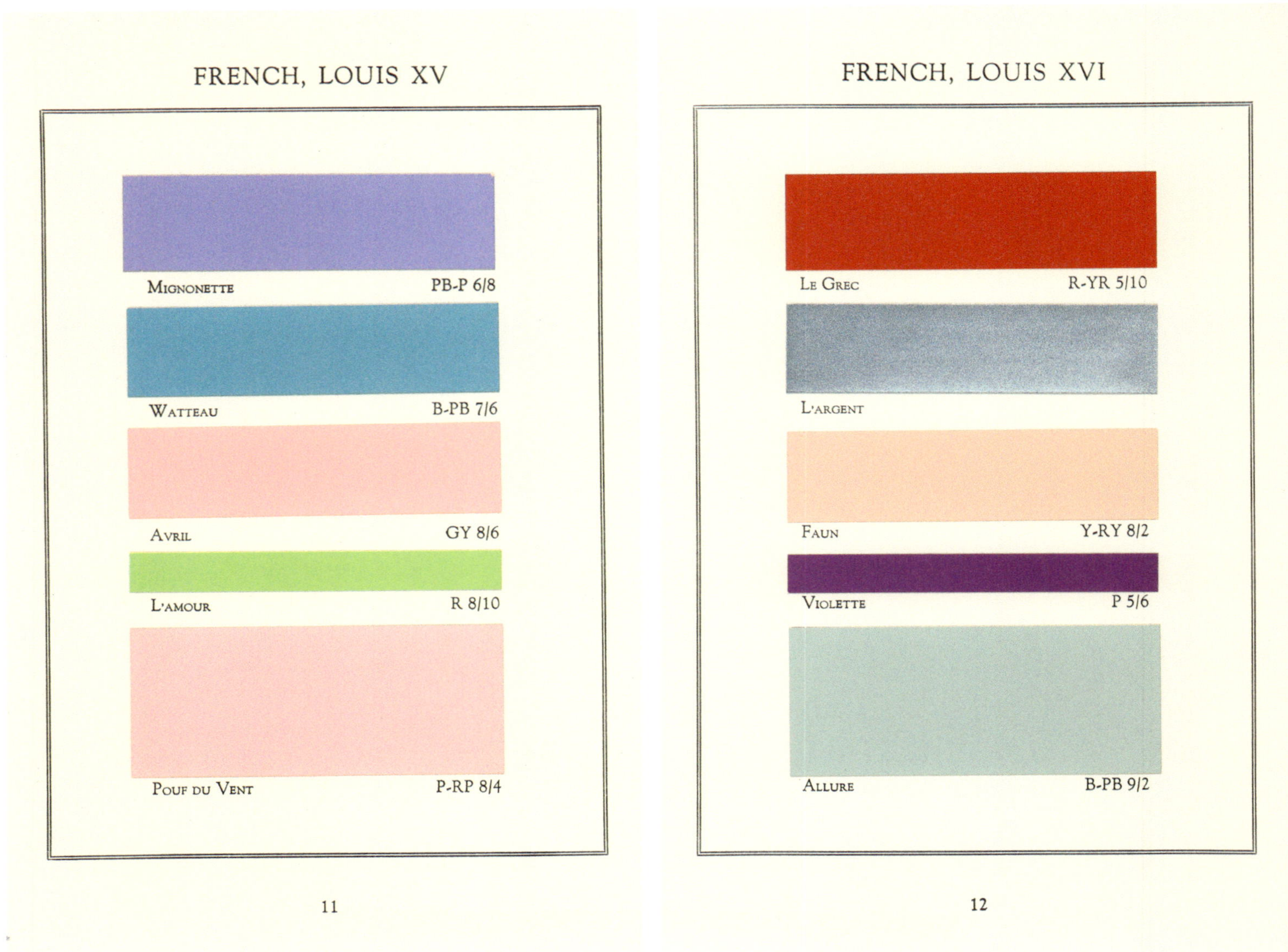

index chromatique numéroté. La conception et le style de l'*Historical Color Guide* de Burris-Meyer ressemblent beaucoup à ceux d'un ouvrage de Thomas Parsons intitulé *A Tint Book of Historical Colours*, paru quatre ans plus tôt à Londres et dont l'autrice s'est à coup sûr inspirée. Mais, tandis que Parsons s'intéresse avant tout à la couleur en architecture, Burris-Meyer vise des applications beaucoup plus nombreuses. Elle a signé d'autres livres consacrés aux arts graphiques, à la décoration d'intérieur et à la mode qui comprennent tous des tableaux de couleurs et, en 1947, publie un complément à cet ouvrage, intitulé *Contemporary Color Guide*.

Guía histórica del color

Elizabeth Burris-Meyer fue una escritora, educadora y comentarista del color del siglo XX poco conocida y apreciada. Apenas se habla de ella en los textos sobre la materia, y no se sabe mucho sobre su vida, a pesar de que su producción editorial fue extensa. Cuando publicó esta primera *Historical Color Guide*, trabajaba como decana de la Tobé-Coburn School for Fashion Careers de Nueva York, un centro de formación para las industrias de la moda y la belleza que se había fundado poco tiempo atrás. Antes de eso, había estudiado en la Universidad de Michigan, y con posterioridad enseñó diseño en las sucursales de París y Nueva York de la New York School of Fine and Applied Art (más tarde conocida como Parsons New School of Design). El primer trabajo que publicó fue el importante *Color and Design in the Decorative Arts* (1935), en el que demostró un conocimiento profundo de la historia del arte y de la ciencia del color, y su capacidad para escribir y educar de una manera clara y accesible sobre el color y sus aplicaciones. Entre otros temas, discutió los conceptos y sistemas de color de Albert Henry Munsell y reprodujo una de sus ruedas de color y algunos de sus otros diagramas. Su libro también incluía una tarjeta de colores con 66 muestras pegadas y nombradas de «colores representativos de uso común en promoción comercial, publicidad y vestimenta»,

FRENCH, EMPIRE

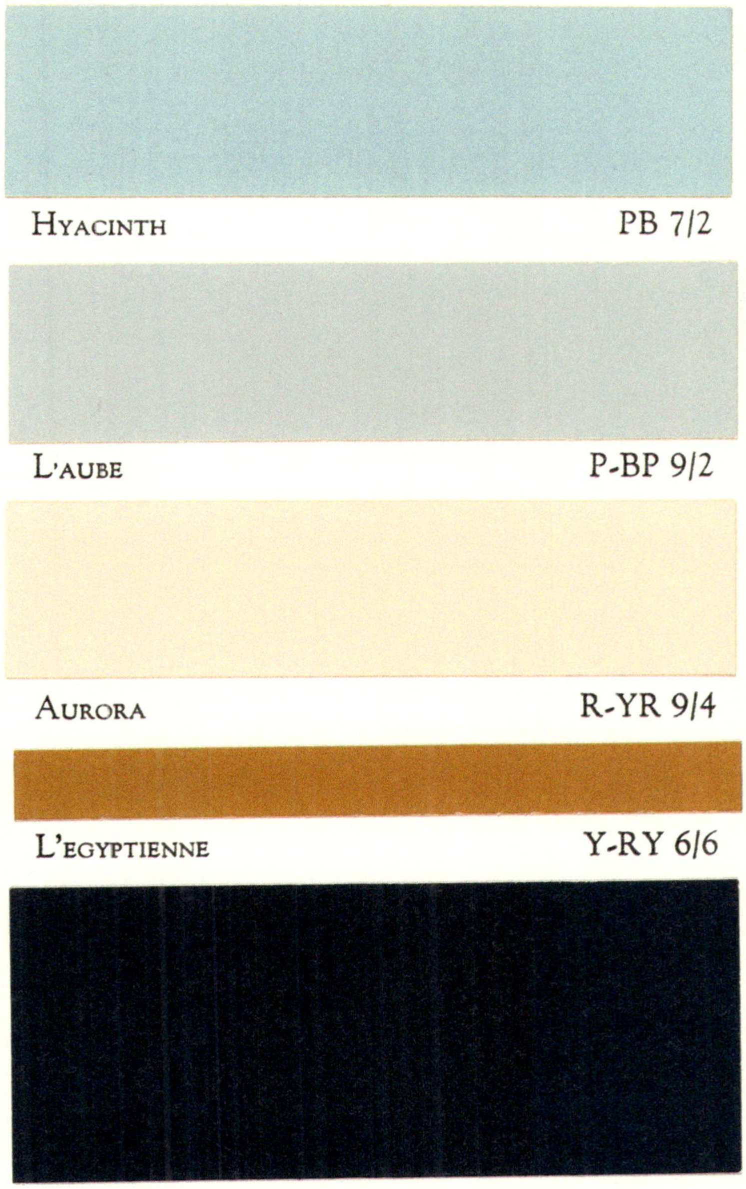

Hyacinth PB 7/2

L'aube P-BP 9/2

Aurora R-YR 9/4

L'egyptienne Y-RY 6/6

Ombre N 3

13

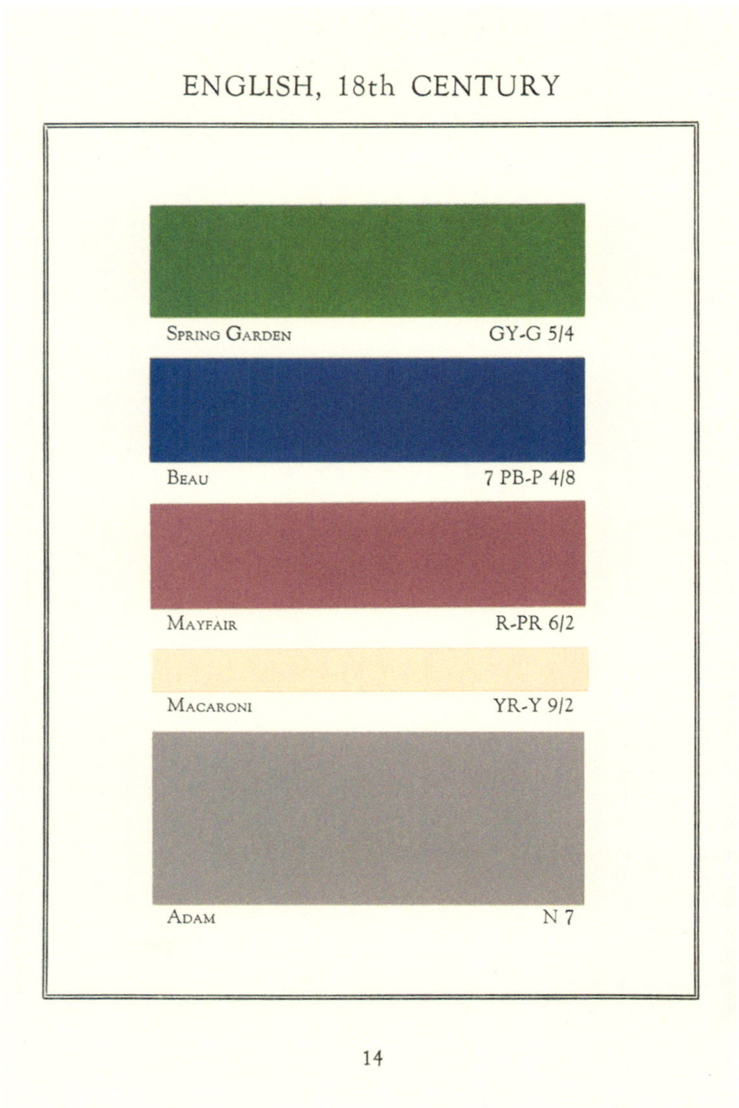

ENGLISH, 18th CENTURY

Spring Garden	GY-G 5/4
Beau	7 PB-P 4/8
Mayfair	R-PR 6/2
Macaroni	YR-Y 9/2
Adam	N 7

14

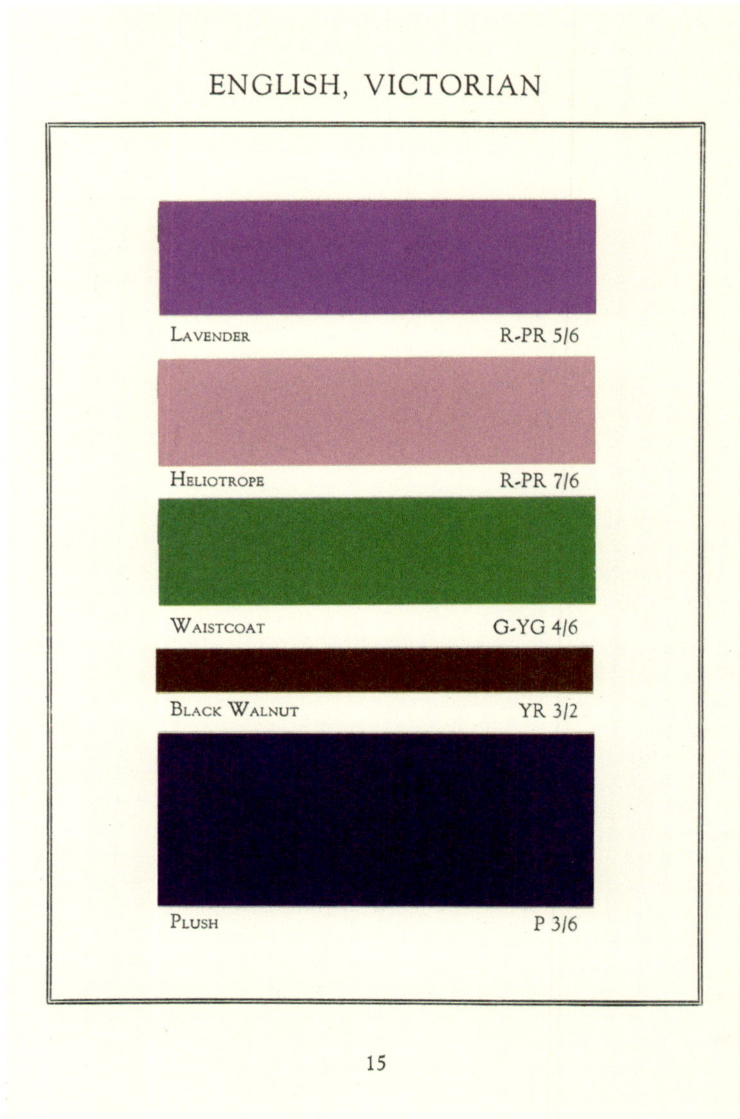

ENGLISH, VICTORIAN

Lavender	R-PR 5/6
Heliotrope	R-PR 7/6
Waistcoat	G-YG 4/6
Black Walnut	YR 3/2
Plush	P 3/6

15

CHINESE, BUDDHIST

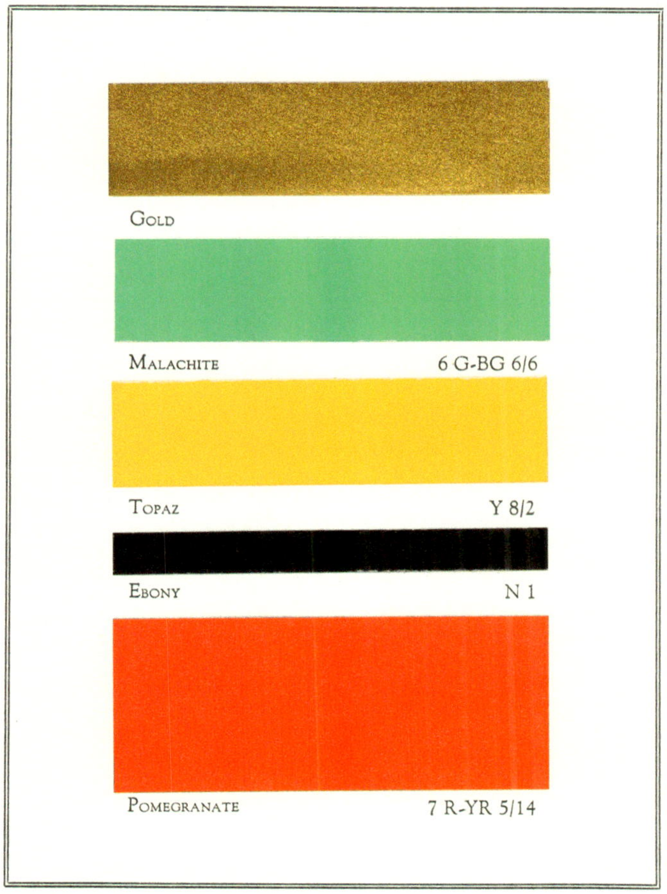

16

CHINESE, SUNG

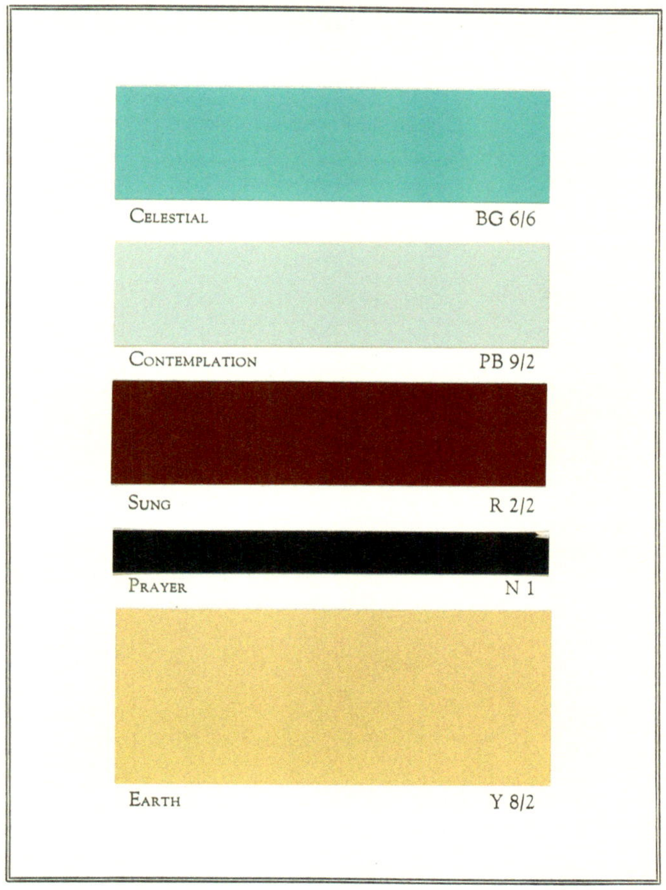

17

INDIAN

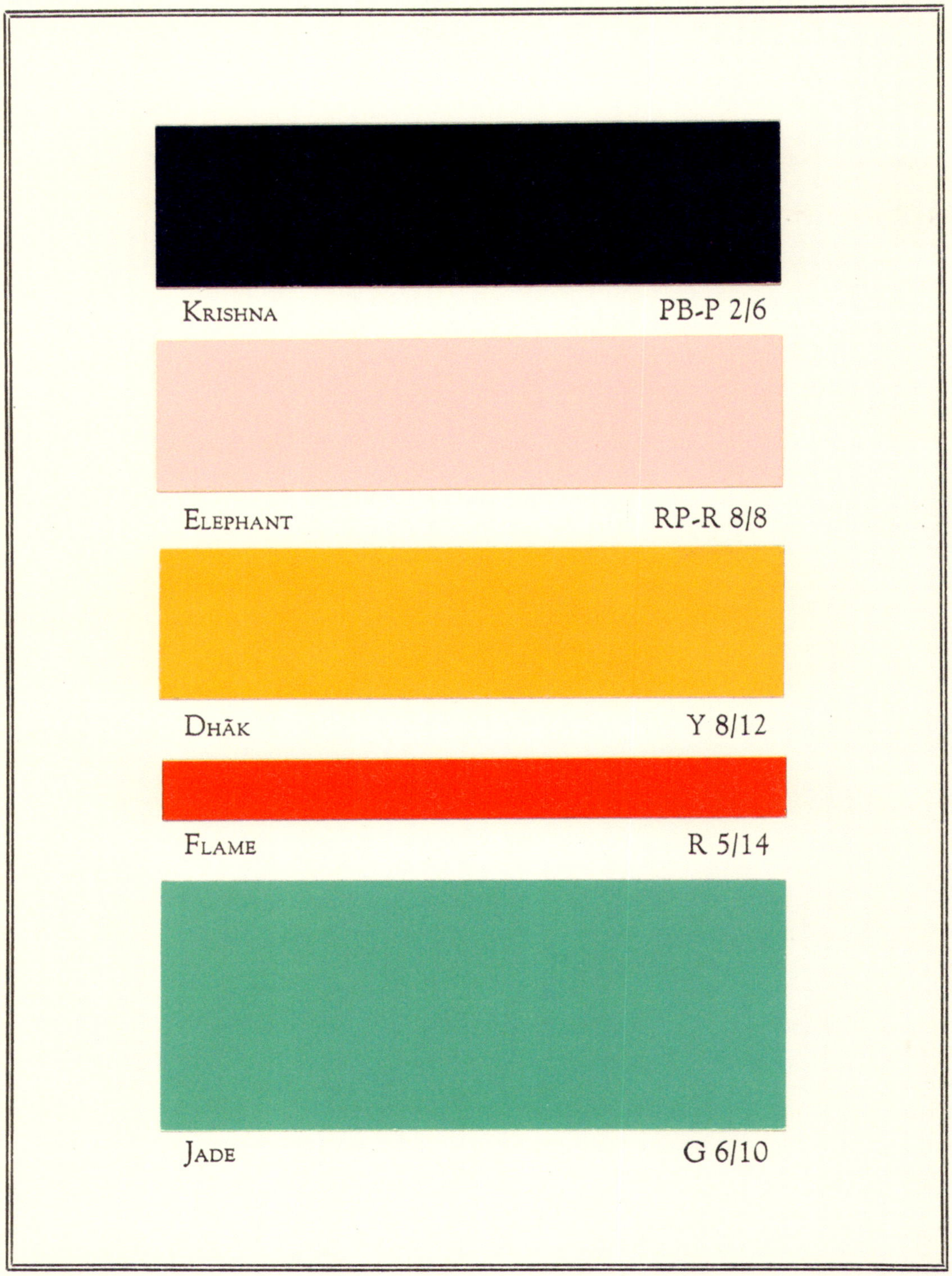
Krishna PB-P 2/6
Elephant RP-R 8/8
Dhāk Y 8/12
Flame R 5/14
Jade G 6/10

18

PERSIAN FRESCO

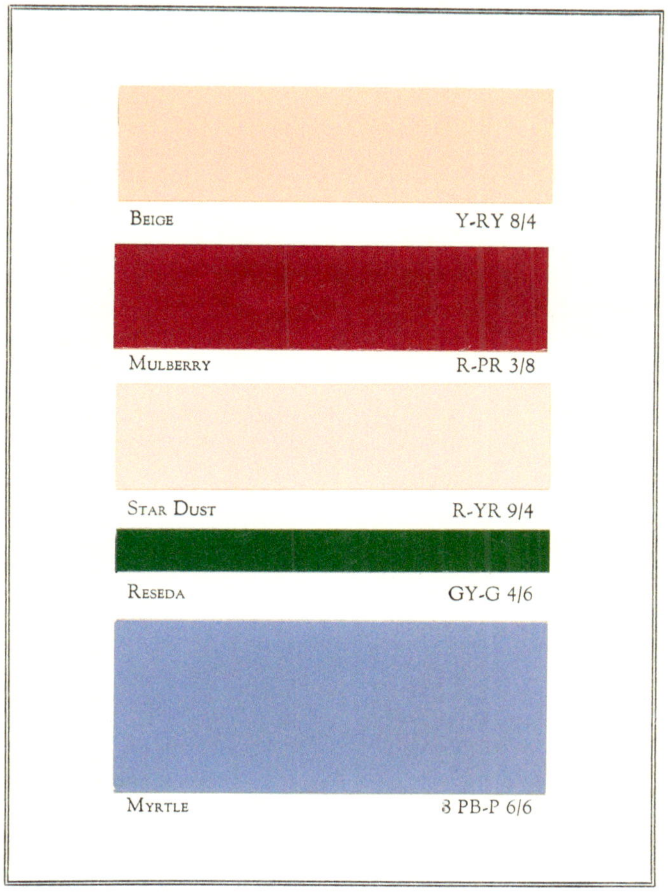

19

PERSIAN MINIATURE

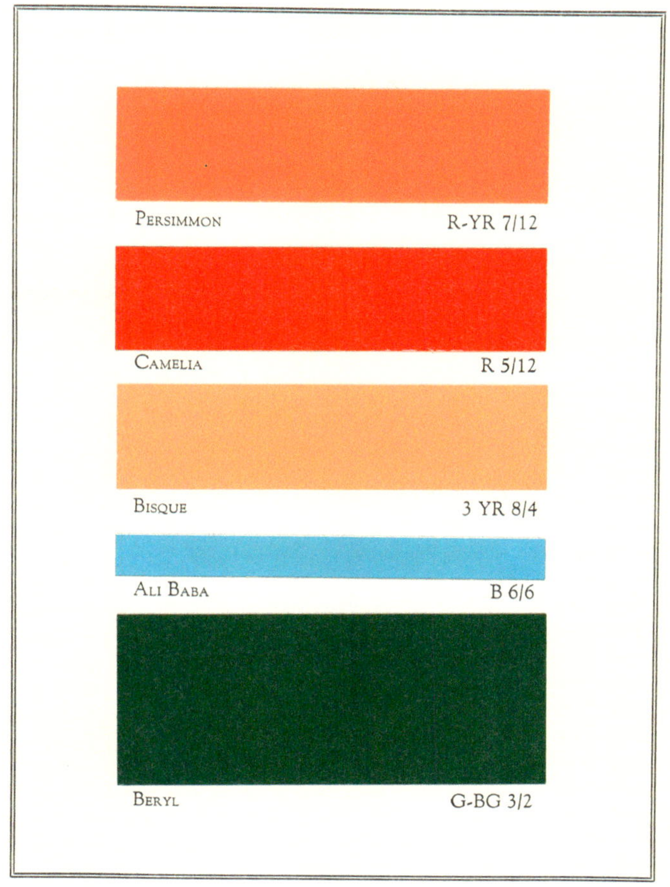

20

junto con un filtro de papel perforado diseñado para aislar colores o grupos de colores en la tarjeta y ayudar al usuario en la búsqueda de composiciones armoniosas. Burris-Meyer aplicó ese estilo práctico, bien fundamentado y conciso en todas sus publicaciones.

Con la *Historical Color Guide*, su objetivo era proporcionar «un libro de esquemas de color tradicionales y representativos derivados de ejemplos de las artes decorativas de todos los campos», destinado principalmente a diseñadores, comerciantes, decoradores y arquitectos. En 30 láminas, cada una provista de cinco tiras de colores impresas y una página de comentarios adjunta, presentó una descripción histórica, global y cultural de los esquemas de color a lo largo de los siglos. Ordenó las láminas por temas, desde los llamados «primitivos», encontrados en tallas y textiles tribales, hasta las paletas de Van Gogh y Gauguin, pasando por el antiguo Egipto, el mundo clásico, ejemplos chinos y miniaturas persas, y períodos y fases específicas como el Renacimiento italiano y la «década malva» inglesa. Cada muestra de color está marcada numéricamente de acuerdo con el sistema de notación de Munsell, y el libro se completó con un índice de color numerado. El concepto y el estilo de la *Historical Color Guide* de Burris-Meyer guardan un gran parecido con *A Tint Book of Historical Colours* de Thomas Parsons, publicado en Londres cuatro años antes, y no hay duda de que Burris-Meyer se inspiró en él. Sin embargo, mientras que el libro de Parsons se centra específicamente en el color arquitectónico, Burris-Meyer apuntó a una gama de aplicaciones mucho más amplia. Escribió otros libros sobre diseño, decoración de interiores y moda, en los que siempre incluyó cartas de colores, y en 1947 publicó un volumen complementario a este libro: *Contemporary Color Guide*.

SPANISH, VELASQUEZ

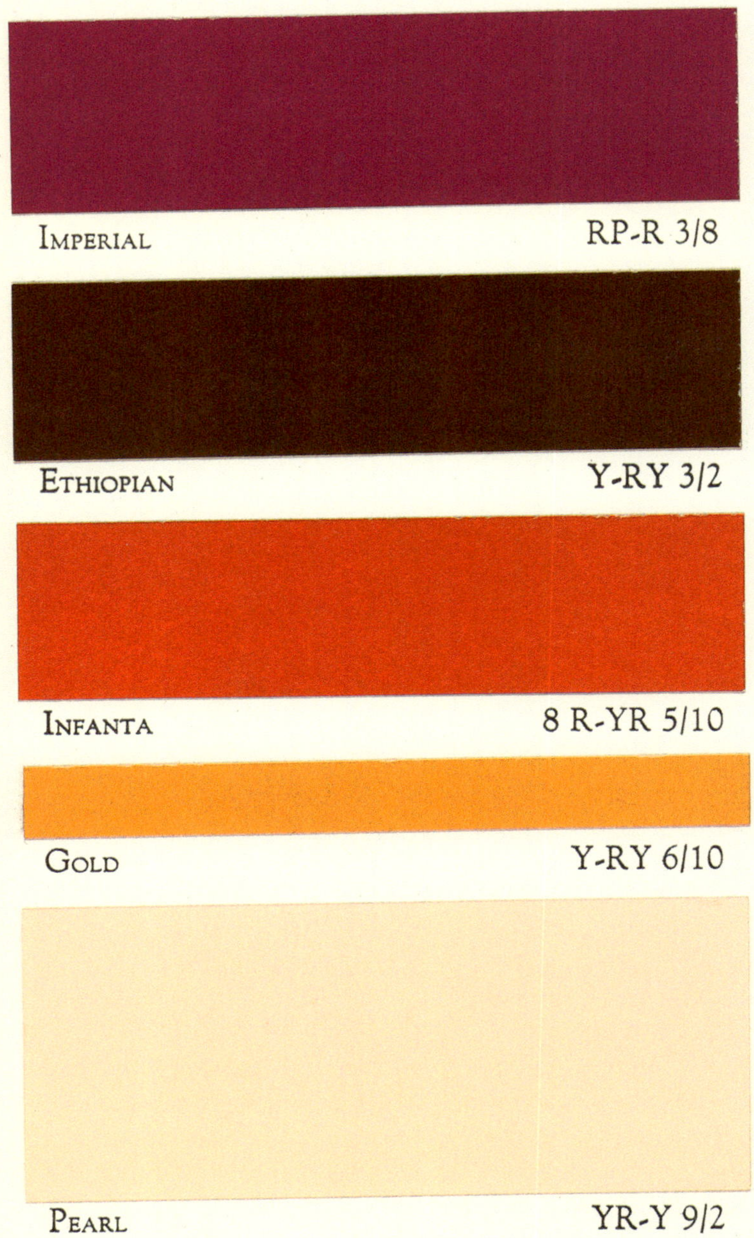

21

SPANISH, EL GRECO

Fear	Y 6/4
Terrain	G 4/2
Joseph	7 PB-P 5/2
Madonna	R-PR 2/6
Wrath	N 5

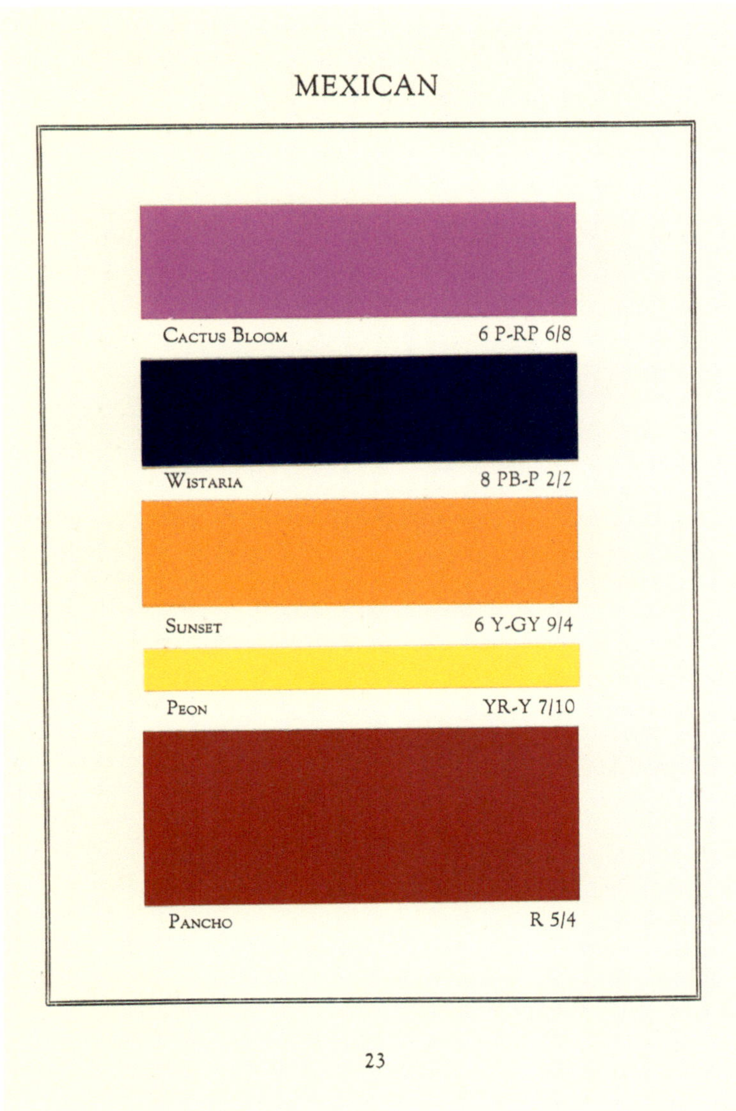

MEXICAN

Cactus Bloom	6 P-RP 6/8
Wistaria	8 PB-P 2/2
Sunset	6 Y-GY 9/4
Peon	YR-Y 7/10
Pancho	R 5/4

23

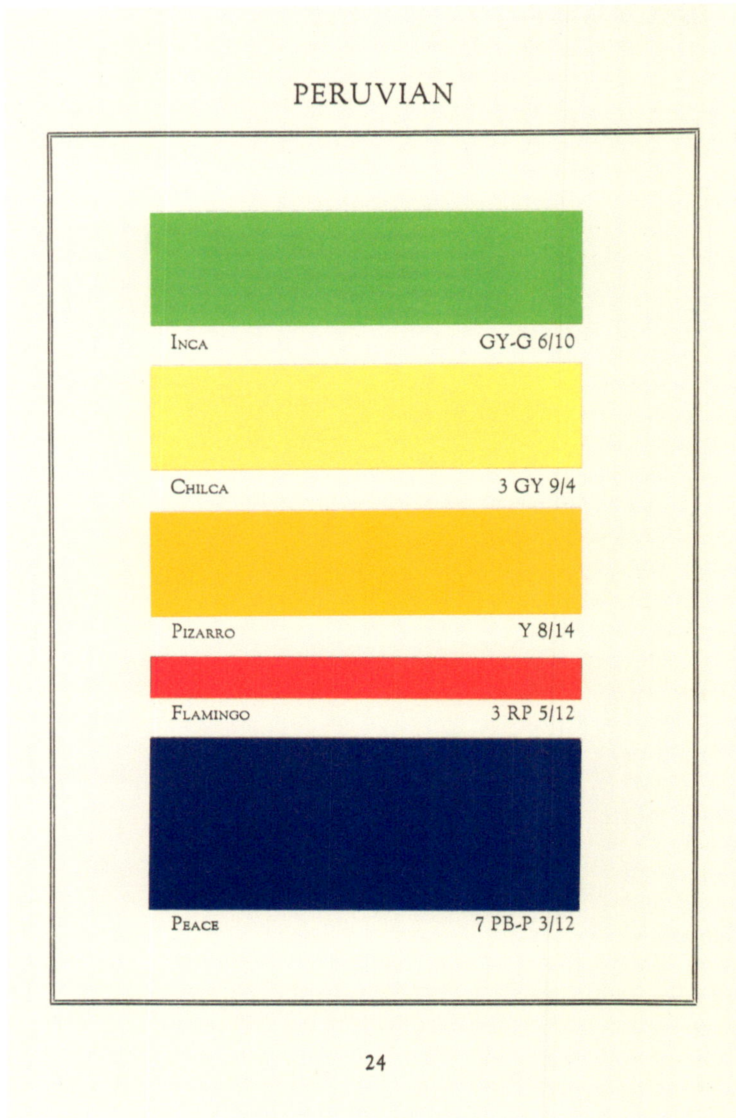

PERUVIAN

Inca	GY-G 6/10
Chilca	3 GY 9/4
Pizarro	Y 8/14
Flamingo	3 RP 5/12
Peace	7 PB-P 3/12

24

RUSSIAN

Tartar	Y-RY 7/12
Jasper	GY 6/8
Wine	RP-R 3/8
Passion	R 3/12
Fire	7 R-YR 5/12

25

SWEDISH

Blush	R 7/10
Evening	R 5/2
Lake	R-PR 3/6
Ice	BG 7/8
Mist	R 7/2

26

AMERICAN, SAMPLER

Pine	GY-G 3/4
Forget-Me-Not	PB 6/8
Geranium	R 4/12
Colonial Yellow	3 Y 8/10
Flax	Y-RY 7/6

AMERICAN, FRACTUR

LAUREL G-BG 4/4

MARIE PB 6/4

HOPE PB 2/6

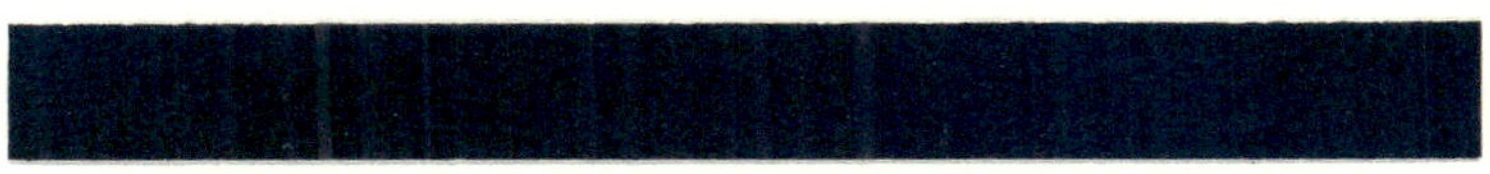

1820 R-YR 7/6

HICKORY YR 3/4

28

VAN GOGH

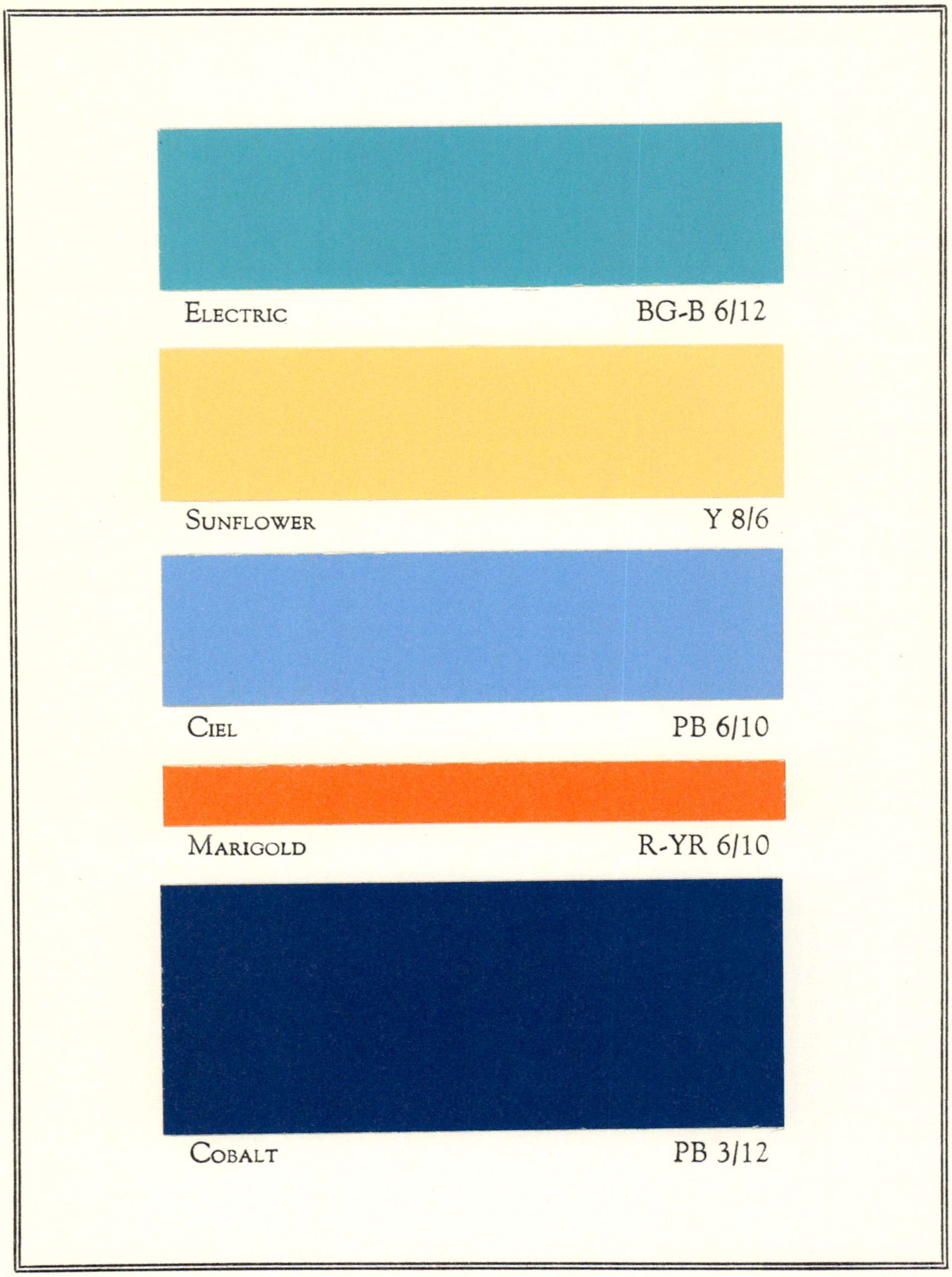
Electric BG-B 6/12
Sunflower Y 8/6
Ciel PB 6/10
Marigold R-YR 6/10
Cobalt PB 3/12

29

GAUGUIN

Amethyst	P 4/2
Mid-night	P 3/2
Leaf	GY 7/10
Breadfruit	Y 5/4
Carnation	RP 7/12

30

Aloys John Maerz (1885–1972),
Morris Rea Paul (1895–1970)

A DICTIONARY OF COLOR

19 plates, 29.5 x 21.5 cm | 11 ⅝ x 8½ in., New York, 1950
Sussex, collection of the author

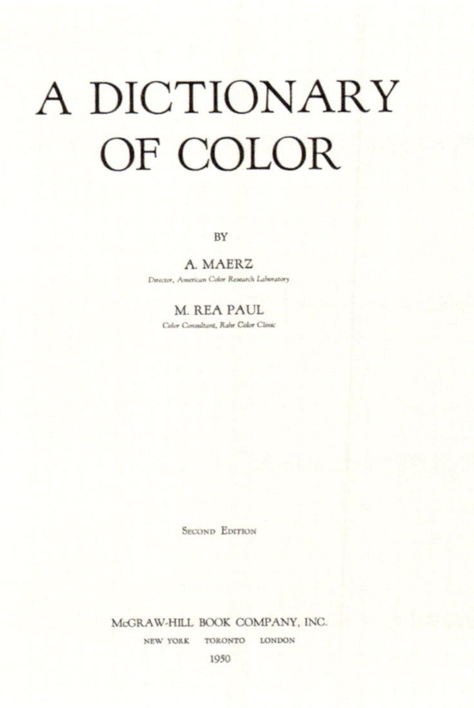
A DICTIONARY
OF COLOR

BY

A. MAERZ
Director, American Color Research Laboratory

M. REA PAUL
Color Consultant, Rahr Color Clinic

Second Edition

McGRAW-HILL BOOK COMPANY, INC.
NEW YORK TORONTO LONDON
1950

The *Dictionary of Color* was an unprecedented effort to gather, codify and visually represent colour names in the English language. It was perhaps the most ambitious book on colour to be published in the interwar years, when the first edition appeared in 1930, and it was followed 20 years later by a second edition, which is the one presented here. The book is hugely impressive in the scope of its research, its graphic design and the quality of its colour plates, and it is fast becoming one of the most desirable and collectable of all 20th-century books on the subject. The compilers of the colour names were two American scientists with strong international connections, Maerz being the Director of the American Color Research Laboratory while Paul worked as a colour consultant for several institutions. Their intention was to collate the most extensive list of colour names in the English language, ignoring numerical systems and instead looking to turn their research findings into a useful system of classification. Their sources included major U.S. and English paint and varnish manufacturers, the Textile Color Card Association of the United States, various French colour names used in English and names for certain natural and synthetic dyestuffs, all accompanied by a select choice of colour literature. As historical and current references, Robert Ridgway (pp. 374–387) and Albert Henry Munsell (pp. 584–599) were singled out for their importance.

Maerz and Paul structured their system by dividing colour into seven groups, following the prismatic spectrum: "red to orange, orange to yellow, yellow to green, green to blue-green, blue-green to blue, blue to red, and purple to red". Each group is represented by eight high-quality plates, successively showing gradations of colour from full strength towards white and black. The 56 tables depict 12 x 12 and 12 x 6 colour grids, adding up to a total of over 7,000 individual colours. The coloured grids each have an uncoloured opposite, marked with colour names as identified by Maerz and Paul printed in the small squares.

A Dictionary of Color is part of a long tradition of attempts to compile the myriad ways in which designers, manufacturers, artists and industries talk about colour and describe it. Conceptually, the book has an early ancestor in Richard Waller's "catalogue" of colours in four languages, from 1686 (pp. 80–85), but ultimately the shimmering double-plates of the *Dictionary* reveal as much as anything how limited language is compared to the actual range of colours, for there are many empty white squares, where no name was found or given for a subtle variation of the tints or shades illustrated opposite.

Plate 1

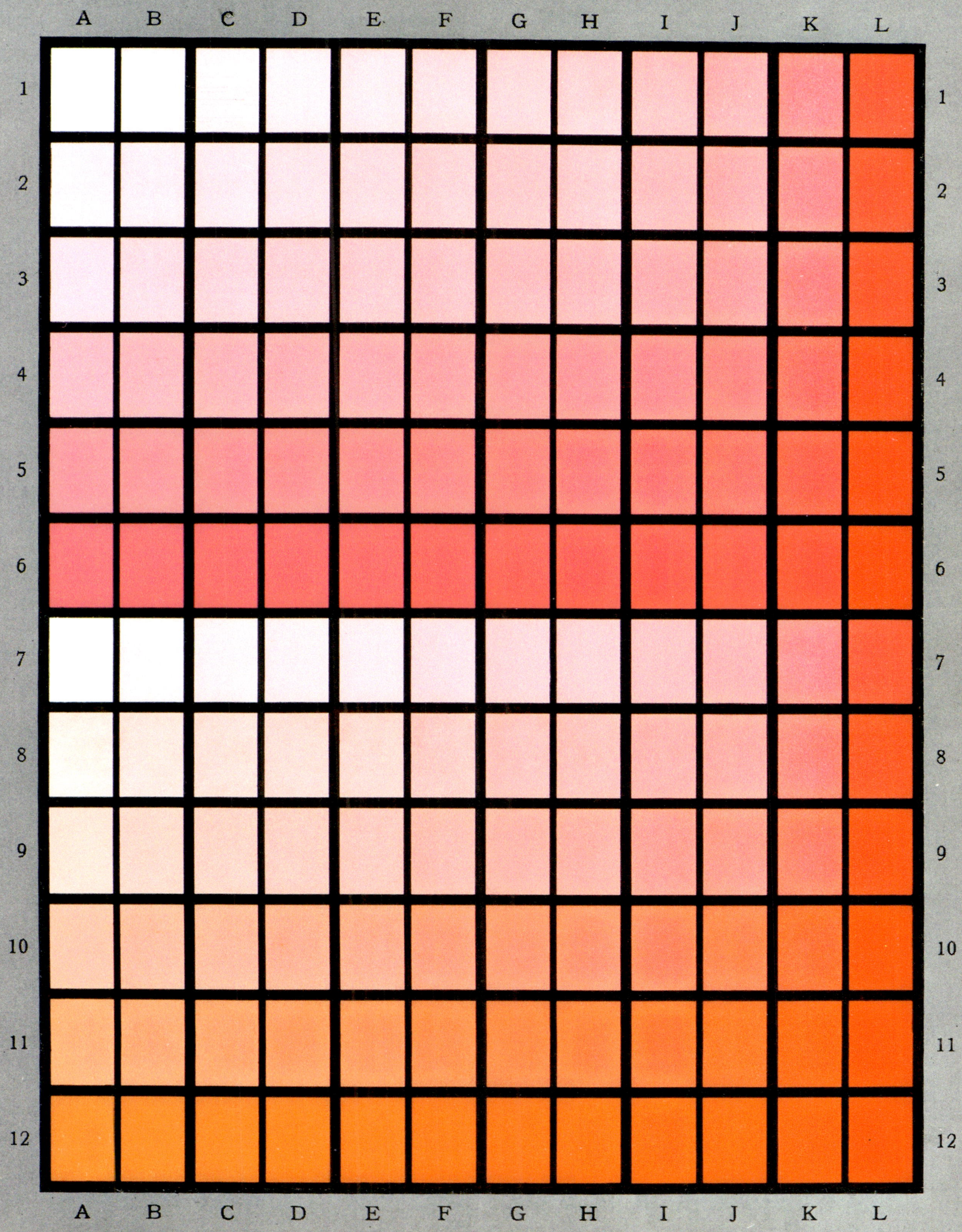

PLATE 10

	A	B	C	D	E	F	G	H	I	J	K	L	
1		Oyster White+	Marguerite Y.						Reed Y.	**SULPHUR Y** Citrus			1
2	White Jade+ Tilleul-Buff+	**IVORY**				**STRAW**	Colonial Buff+	Cockatoo		Citron Y+ Mimosa		Aureolin P	2
3		Caen Stone Freestone+ Bath Stone	Vanilla	**LEGHORN**		Nankeen	Cornsilk			Amber Y.	**LEMON Y** Yellow Daisy+		3
4			**SUNSET**					Sunlight	Pond Lily	**PRIMROSE Y** Snapdragon	Narcissus	**LIGHT CHROME Y.** P	4
5		Peach Blow +					**MAISE**			Corn		Primuline Y. Yolk Y. **GOLDEN ROD**	5
6		Sugar Cane			Cornhusk				Mellowglow	Daffodil Sunray	Chinese Y.	Ta-Ming	6
7	**SALMON**		Sunrise Y			**APRICOT**				Mirabelle	Snowshoe	**SPANISH Y. GOLDEN Y.**	7
8											Saffron Y	Florida Gold	8
9									Jacinthe	Cowslip	Sunburst		9
10	Rufous											Orange-peel	10
11			Carrot R.					Pumpkin				Tangier	11
12			Navaho								**CHROME OP**	Big 4 Y	12
	A	B	C	D	E	F	G	H	I	J	K	L	

PLATE 10

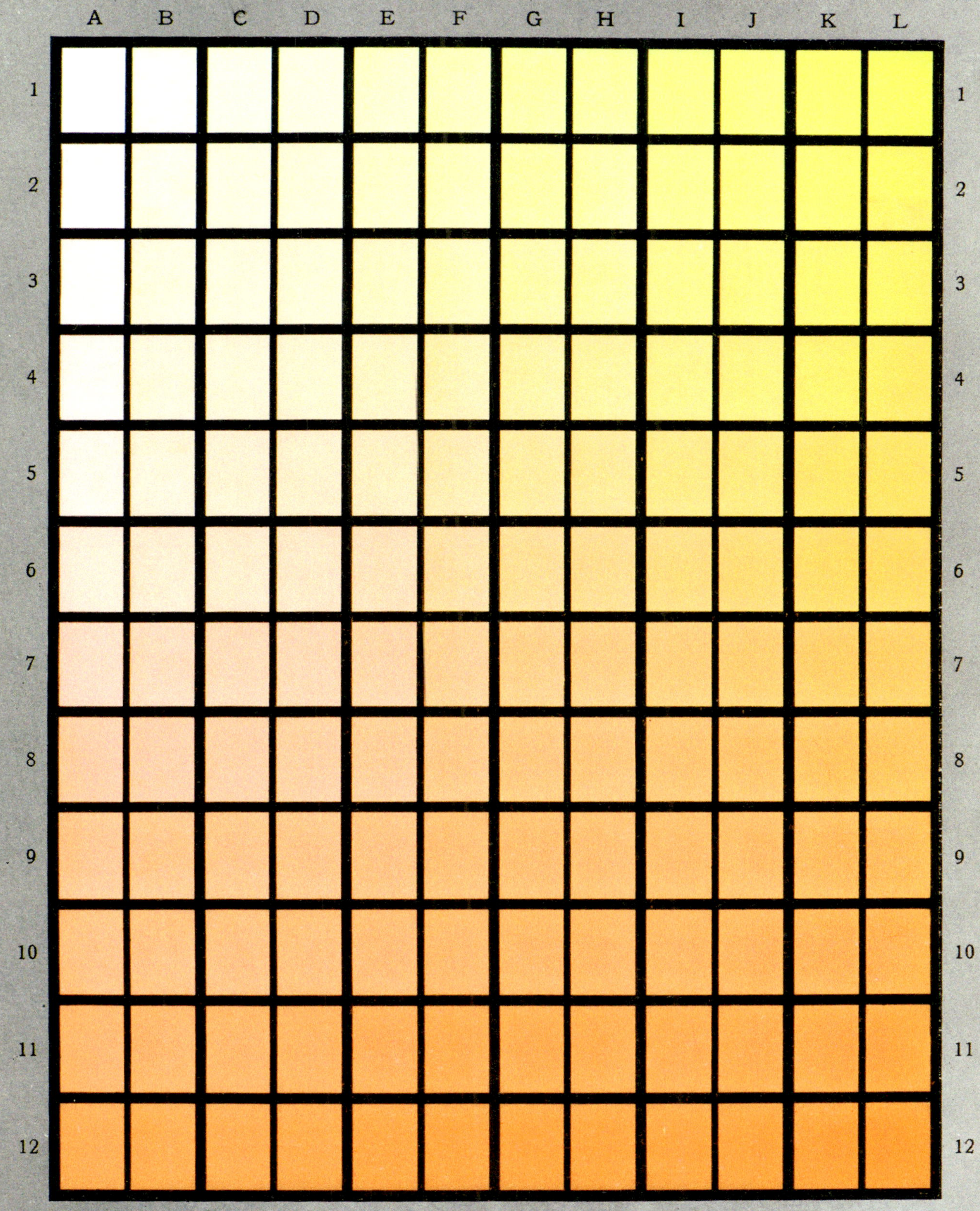

[43]

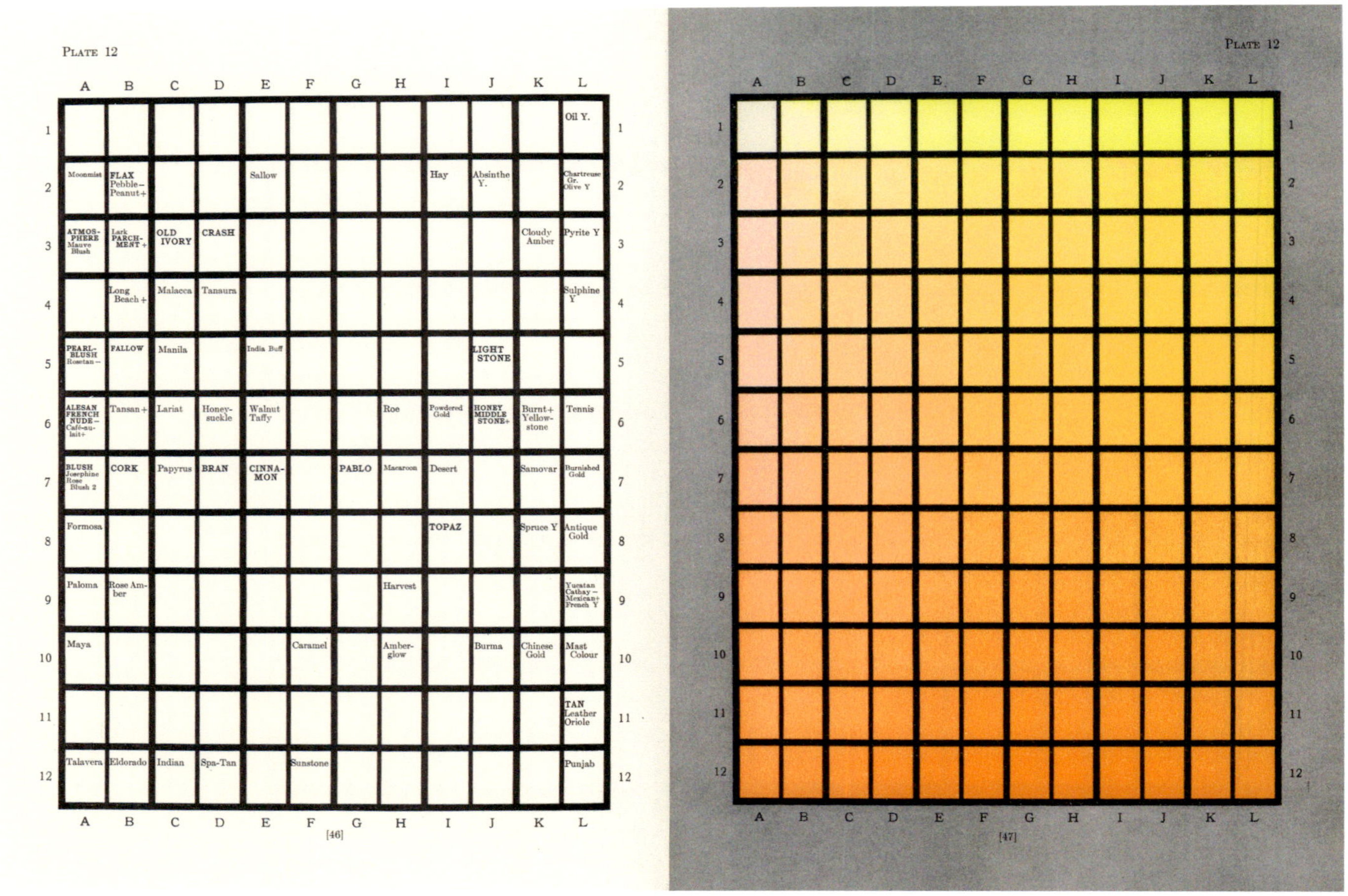

Wörterbuch der Farben

A Dictionary of Color war der nie dagewesene Versuch, englischsprachige Farbnamen zu sammeln, mit einem Code zu versehen und bildlich darzustellen. Vielleicht war es sogar das ehrgeizigste Farbenbuch der Zwischenkriegszeit, als es 1930 in erster Auflage erschien. Die zweite, hier vorgestellte Auflage, folgte zwanzig Jahre später. Außerordentlich beeindruckend sind der Rechercheumfang des Buches, seine grafische Gestaltung und die Qualität seiner Farbtafeln, die es schnell zu einem der begehrtesten und sammelwürdigsten aller sich mit dem Thema Farbe befassenden Bücher des 20. Jahrhunderts werden ließen. Die Sammler der Farbnamen waren zwei amerikanische Wissenschaftler mit sehr guten internationalen Beziehungen. Maerz leitete das American Color Research Laboratory, Paul war als Farbberater für verschiedene Institutionen tätig. Gemeinsam wollten sie die umfangreichste Liste von Farbbezeichnungen in der englischen Sprache zusammenstellen, wobei sie numerische Systeme außer Acht ließen, um ihre Forschungsergebnisse stattdessen in ein hilfreiches Klassifikationssystem zu überführen. Zu ihren Quellen zählten große Farben- und Lackhersteller in England und den USA, die amerikanische Textile Color Card Association, verschiedene französische, auch im Englischen gebräuchliche Farbnamen, die Bezeichnungen für bestimmte natürliche und synthetische Farbstoffe sowie eine ausgewählte Reihe von Büchern aus der Farbliteratur. Als bedeutende historische und aktuelle Quellen wurden Robert Ridgway (S. 374–387) und Albert Henry Munsell (S. 584–599) hervorgehoben.

Maerz und Paul strukturierten ihr System, indem sie die Farben in Anlehnung an das prismatische Spektrum in sieben Gruppen einteilten: „Rot bis Orange, Orange bis Gelb, Gelb bis Grün, Grün bis Blaugrün, Blaugrün bis Blau, Blau bis Rot und Violett bis Rot“. Auf jeweils acht fortlaufenden Tafeln von höchster Qualität sind die Farbabstufungen zu sehen, die für jede Gruppe von Farbtönen in voller Leuchtkraft bis zu Weiß und Schwarz reichen. Aus den zwölf mal zwölf oder sechs mal zwölf Felder umfassenden Rastern der insgesamt 56 Tafeln ergeben sich mehr als siebentausend Einzelfarben. Jedem farbigen Raster wurde ein nicht koloriertes Raster

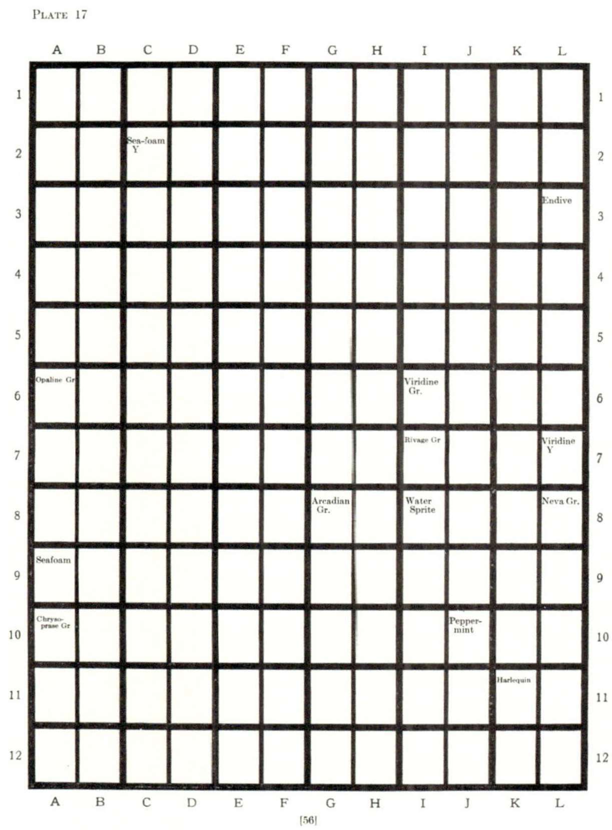

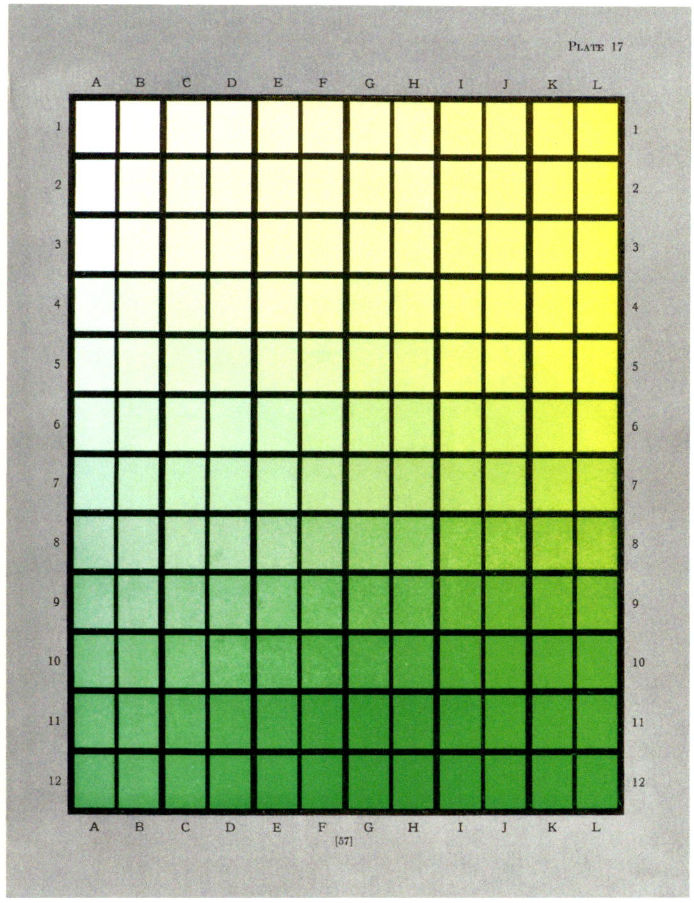

gegenübergestellt, in dessen kleine Felder die von Maerz und Paul identifizierten Farbnamen eingetragen sind.

A Dictionary of Color ist Teil einer langen Tradition, die unzähligen Methoden zusammenzutragen, nach denen Gestalter, Produzenten, Künstler und Industrien über Farbe sprechen und sie beschreiben. Konzeptuell hat das Buch einen frühen Vorgänger in Richard Wallers „Katalog“ der Farben in vier Sprachen aus dem Jahr 1686 (S. 80–85), doch letzten Endes offenbaren die schimmernden Doppeltafeln des *Dictionary* nur einmal mehr, wie begrenzt die Sprache im Vergleich zur tatsächlichen Bandbreite der Farben ist. So blieben viele weiße Felder leer, weil für die subtilen Variationen der daneben abgebildeten Farbtöne oder Schattierungen keine Namen gefunden oder vergeben werden konnten.

Dictionnaire de la couleur

Ce « Dictionnaire de la couleur » tente pour la toute première fois de réunir, codifier et représenter visuellement les noms de couleur de la langue anglaise. C'est peut-être l'ouvrage de l'entre-deux-guerres le plus ambitieux sur le sujet lorsque paraît sa première édition en 1930, suivie, vingt ans plus tard, d'une seconde édition, présentée ici. Très impressionnant par l'envergure de ses études, sa conception graphique et la qualité de ses planches en couleurs, ce dictionnaire est en passe de devenir l'un des ouvrages les plus recherchés des collectionneurs parmi les livres du XXe siècle consacrés au sujet. Les compilateurs de ces noms de couleurs sont deux Américains entretenant de solides relations avec d'autres chercheurs du monde entier : Maerz est directeur de l'American Color Research Laboratory, Paul est coloriste, conseiller de plusieurs institutions. Leur projet consiste à établir la liste la plus exhaustive possible des noms de couleurs en anglais, sans tenir compte des systèmes numériques, et à organiser leurs résultats grâce à une méthode pratique de classification. Leurs sources sont les grands fabricants américains et anglais de peintures et de vernis et la Textile Color Card Association des États-Unis, et les auteurs adoptent les noms de couleurs français employés en anglais, ainsi que les dénominations de certains colorants naturels et synthétiques, le tout s'accompagnant d'une sélection d'ouvrages sur le sujet. Les auteurs soulignent l'importance historique et contemporaine des œuvres de Robert Ridgway

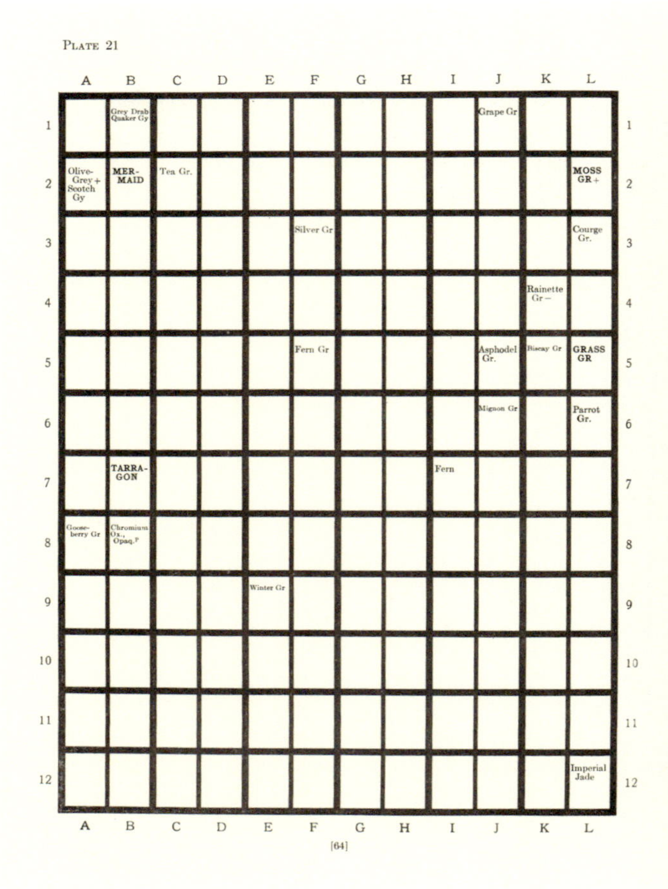

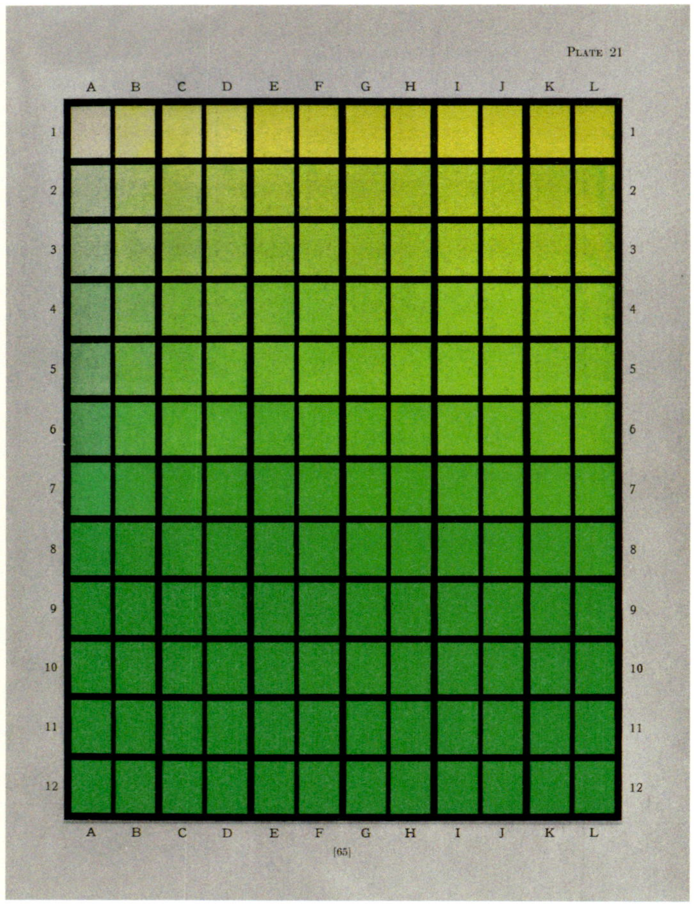

(p. 374–387) et d'Albert Henry Munsell (p. 584–599).

Maerz et Paul construisent leur système en divisant les couleurs en sept groupes, suivant l'ordre des couleurs du spectre: « du rouge à l'orange, de l'orange au jaune, du jaune au vert, du vert au bleu-vert, du bleu-vert au bleu, du bleu au rouge et du violet au rouge ». Chaque groupe est représenté par huit planches de grande qualité, montrant les gradations successives d'une couleur saturée, déclinée vers le blanc et vers le noir. Les 56 tableaux réunissent des grilles de 12 x 12 et 12 x 6 couleurs, pour un total de plus 7000 teintes. Chaque grille chromatique possède son équivalent sans couleurs, les noms des teintes identifiées par Maerz et Paul figurant dans les rectangles.

Le *Dictionary of Color* s'inscrit dans la lignée des tentatives de compilation des innombrables méthodes avec lesquelles dessinateurs, fabricants, peintres et autres corps de métiers qualifient et décrivent la couleur. Sur le plan conceptuel, cet ouvrage a pour précurseur le « catalogue » des couleurs de Richard Waller, publié en 1686 en quatre langues (p. 80–85). En définitive, les doubles pages chatoyantes de ce dictionnaire démontrent à quel point les langues sont limitées pour décrire toute la gamme des couleurs, car de nombreux rectangles sont dépourvus de toute appellation susceptible de désigner la variation subtile d'une teinte ou d'une nuance présente dans la grille correspondante.

Diccionario del color

El *Dictionary of Color* fue un esfuerzo sin precedentes de recopilar, codificar y representar visualmente los nombres de los colores en el idioma inglés. La primera edición se publicó en 1930, y fue quizás el libro sobre el color más ambicioso que apareció en los años de entreguerras. Veinte años después le siguió una segunda edición, que es la que aquí se presenta. El libro es realmente impresionante por el alcance de su investigación, el diseño gráfico y la calidad de las láminas en color, y se está convirtiendo en uno de los libros más deseables y coleccionables de todos los aparecidos en el siglo XX sobre lamateria. Los compiladores de los nombres de los colores fueron dos científicos estadounidenses con fuertes conexiones internacionales: Maerz fue director del American Color Research Laboratory, mientras que Paul trabajó como asesor de color para varias instituciones. Su

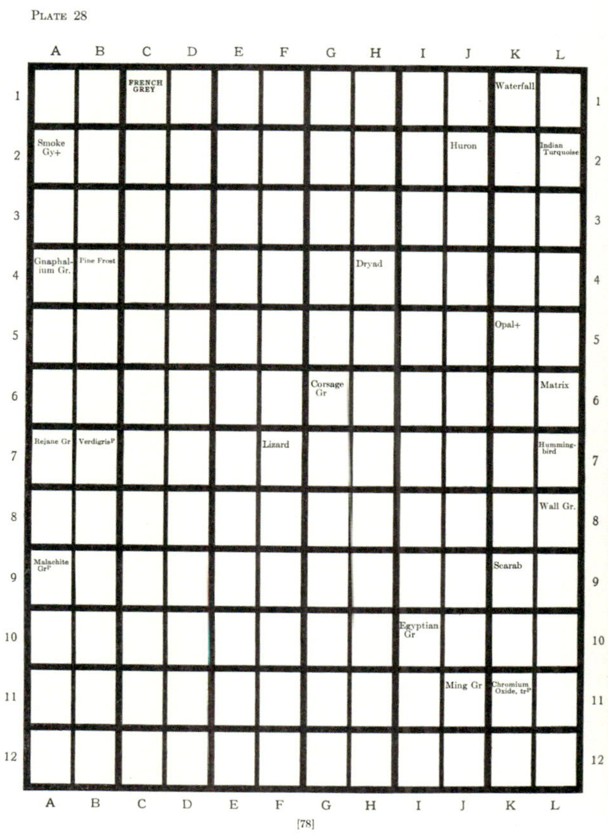

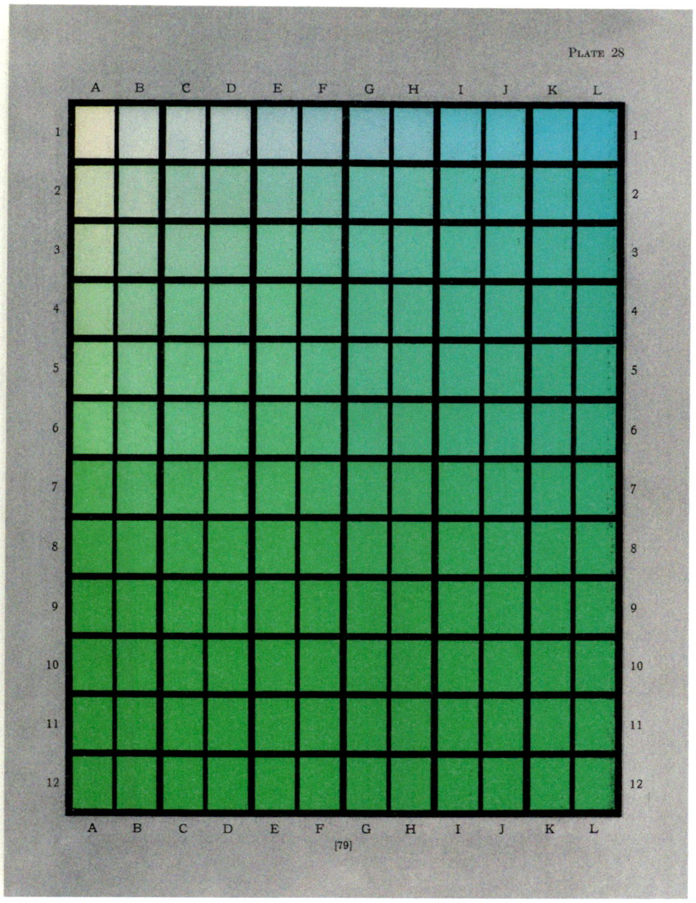

intención fue recopilar la lista más extensa de nombres de colores en el idioma inglés ignorando los sistemas numéricos, aunque con la intención de convertir los hallazgos de su investigación en un sistema útil de clasificación. Entre sus fuentes se contaron importantes fabricantes estadounidenses e ingleses de pinturas y barnices, la Textile Color Card Association of the United States, varios nombres de colores franceses utilizados en inglés y nombres para ciertos colorantes naturales y sintéticos, todo ello acompañado de una selección de textos sobre colores. Como referencias históricas y actuales, se destacó por su importancia a Robert Ridgway (págs. 374–387) y Albert Henry Munsell (págs. 584–599).

Maerz y Paul estructuraron su sistema dividiendo el color en siete grupos con base en el espectro prismático: «rojo a naranja, naranja a amarillo, amarillo a verde, verde a azul verdoso, azul verdoso a azul, azul a rojo y morado a rojo». Cada grupo está representado por ocho láminas de alta calidad, que muestran sucesivamente gradaciones de color desde la máxima intensidad hasta el blanco y el negro. Las 56 tablas representan cuadrículas de colores de 12 × 12 y 12 × 6, lo que hace un total de más de 7000 colores individuales. Cada una de las cuadrículas de colores va acompañada de un duplicado sin colorear, con los nombres de los colores identificados por Maerz y Paul impresos en las casillas.

A Dictionary of Color forma parte de una larga tradición de intentos de compilar las innumerables formas en que los diseñadores, fabricantes, artistas e industrias hablan sobre el color y lo describen. En lo conceptual, el libro tiene un predecesor de primera época en el «catálogo» de colores en cuatro idiomas de Richard Waller de 1686 (págs. 80–85), pero en última instancia, las brillantes láminas dobles del *Dictionary* revelan lo limitado que es el lenguaje cuando se compara con la gama real de colores, ya que hay muchas casillas blancas vacías donde no se encontró ni se dio un nombre de una variación sutil de los tonos o matices que se ilustran en la página opuesta.

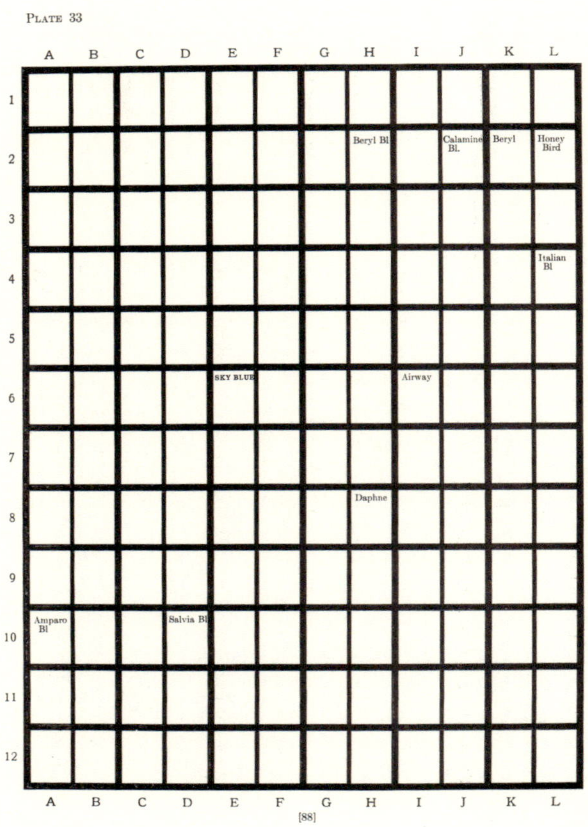
PLATE 33
A B C D E F G H I J K L
1 2 3 4 5 6 7 8 9 10 11 12
Beryl Bl
Calamine Bl.
Beryl
Honey Bird
Italian Bl
SKY BLUE
Airway
Daphne
Amparo Bl
Salvia Bl
[88]

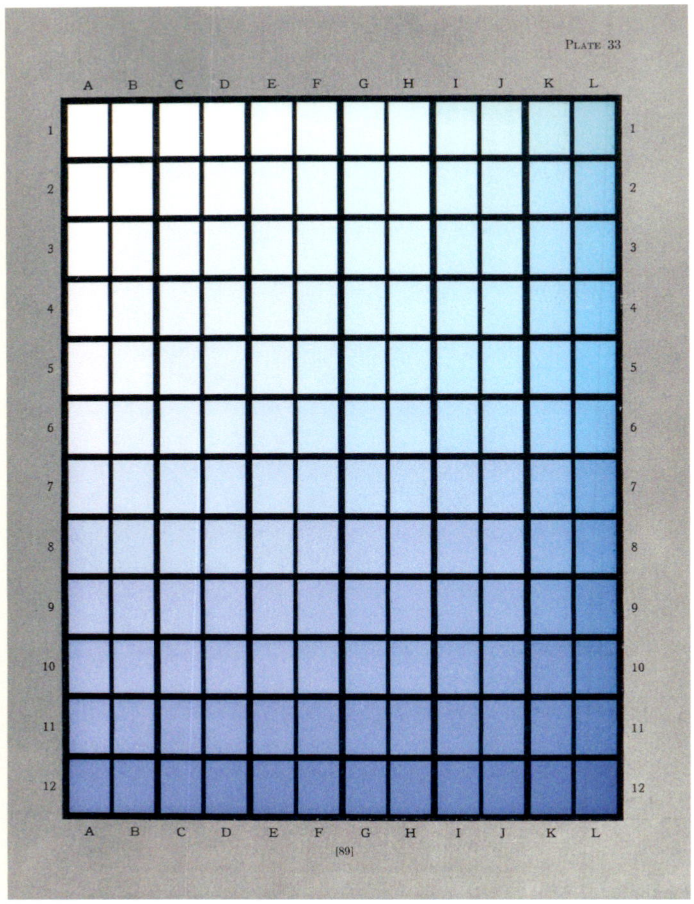
PLATE 33
A B C D E F G H I J K L
1 2 3 4 5 6 7 8 9 10 11 12
[89]

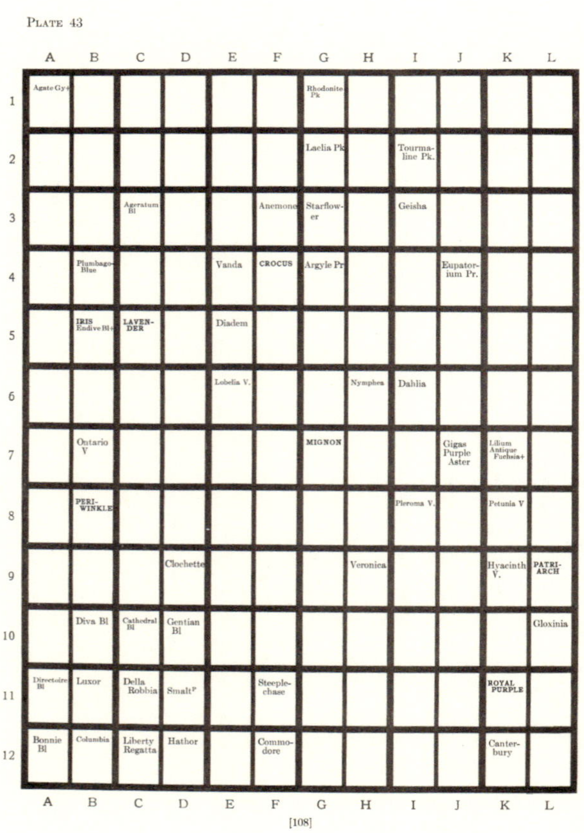
PLATE 43
A B C D E F G H I J K L
1 2 3 4 5 6 7 8 9 10 11 12
Agate Gy+
Rhodonite Pk
Laelia Pk
Tourma-line Pk.
Ageratum Bl
Anemone
Starflow-er
Geisha
Plumbago-Blue
Vanda
CROCUS
Argyle Pr
Eupator-ium Pr.
IRIS
Endive Bl+
LAVEN-DER
Diadem
Lobelia V.
Nymphea
Dahlia
Ontario V
MIGNON
Gigas Purple Aster
Lilium Antique Fuchsia+
PERI-WINKLE
Pleroma V.
Petunia V
Clochette
Veronica
Hyacinth V.
PATRI-ARCH
Diva Bl
Cathedral Bl
Gentian Bl
Gloxinia
Directoire Bl
Luxor
Della Robbia
Smalt P
Steeple-chase
ROYAL PURPLE
Bonnie Bl
Columbia
Liberty Regatta
Hathor
Commo-dore
Canter-bury
[108]

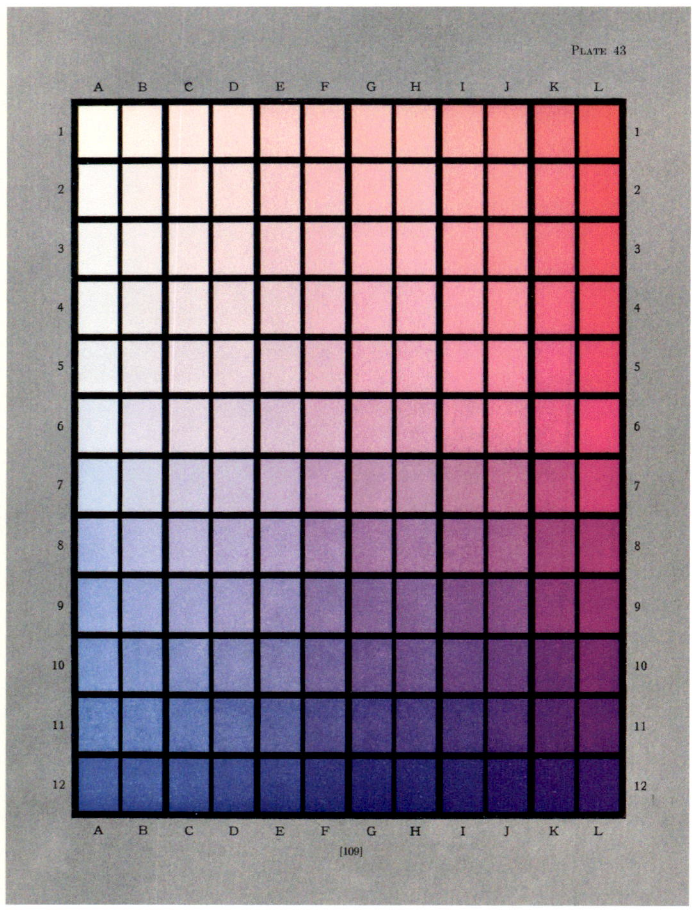
PLATE 43
A B C D E F G H I J K L
1 2 3 4 5 6 7 8 9 10 11 12
[109]

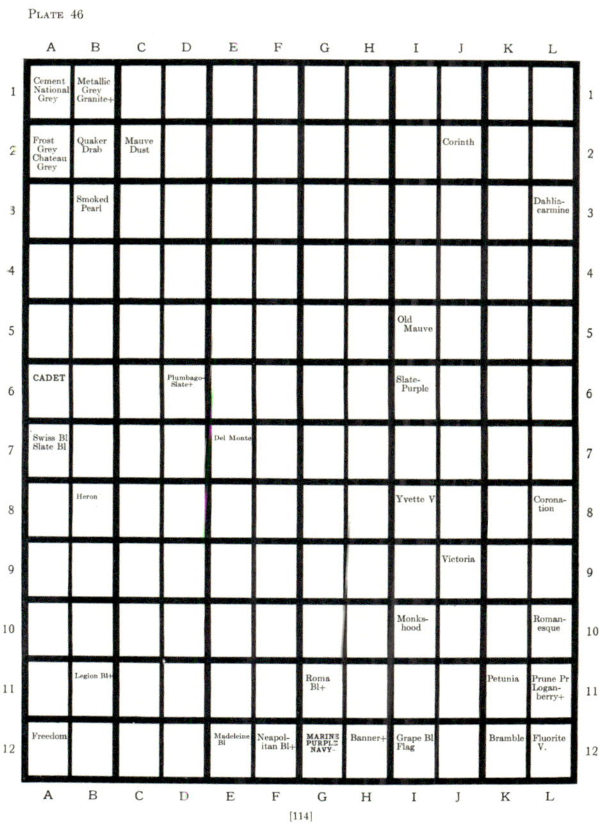

PLATE 46

	A	B	C	D	E	F	G	H	I	J	K	L
1	Cement National Grey	Metallic Grey Granite+										
2	Frost Grey Chateau Grey	Quaker Drab	Mauve Dust							Corinth		
3		Smoked Pearl										Dahlia-carmine
4												
5									Old Mauve			
6	CADET			Plumbago-Slate+					Slate-Purple			
7	Swiss Bl Slate Bl				Del Monte							
8		Heron							Yvette V			Corona-tion
9										Victoria		
10									Monks-hood			Roman-esque
11		Legion Bl+					Roma Bl+				Petunia	Prune Pr Logan-berry+
12	Freedom				Madeleine Bl	Neapol-itan Bl+	MARINE PURPLE NAVY	Banner+	Grape Bl Flag		Bramble	Fluorite V.

[114]

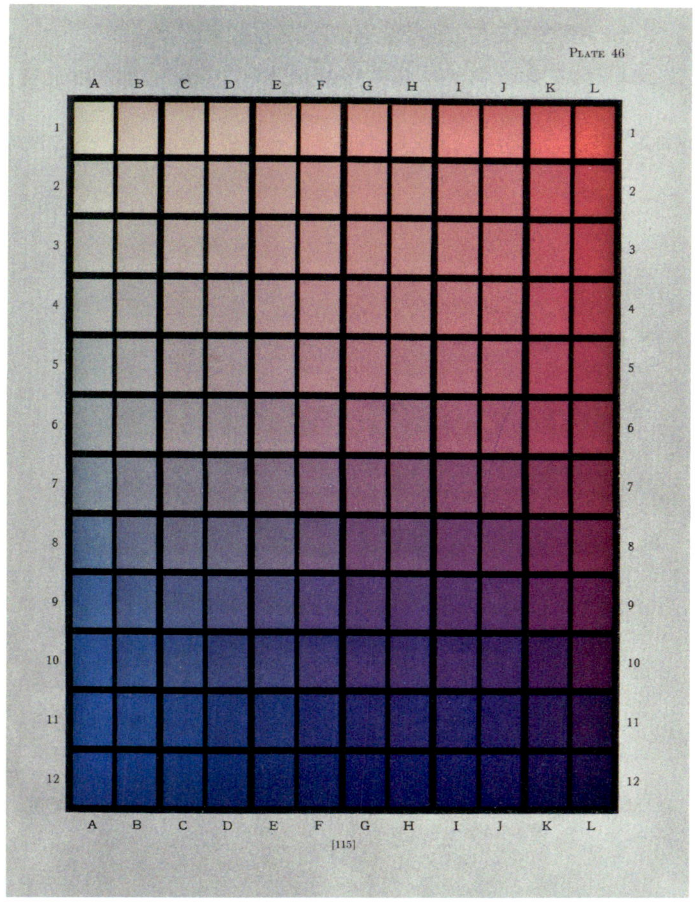

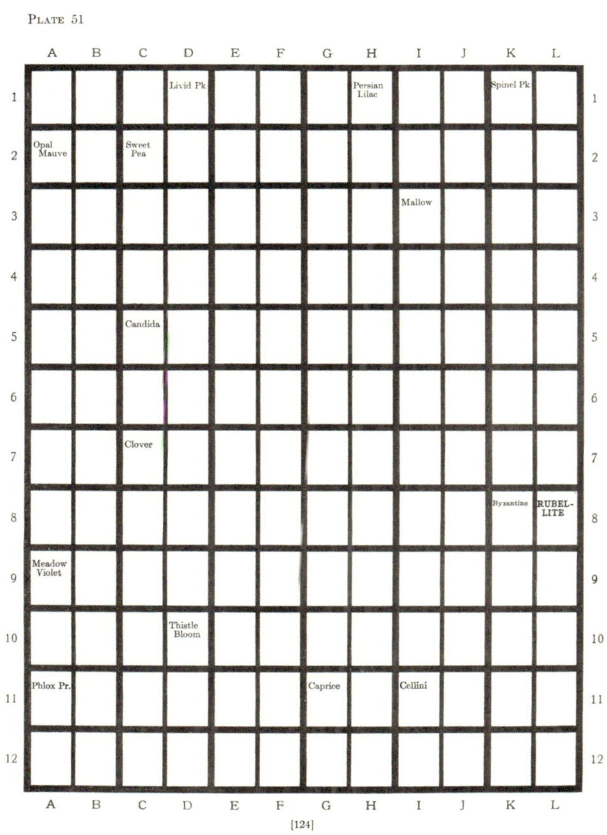

PLATE 51

	A	B	C	D	E	F	G	H	I	J	K	L
1				Livid Pk				Persian Lilac			Spinel Pk	
2	Opal Mauve		Sweet Pea									
3									Mallow			
4												
5			Candida									
6												
7			Clover									
8											Byzantine	RUBEL-LITE
9	Meadow Violet											
10				Thistle Bloom								
11	Phlox Pr.						Caprice		Cellini			
12												

[124]

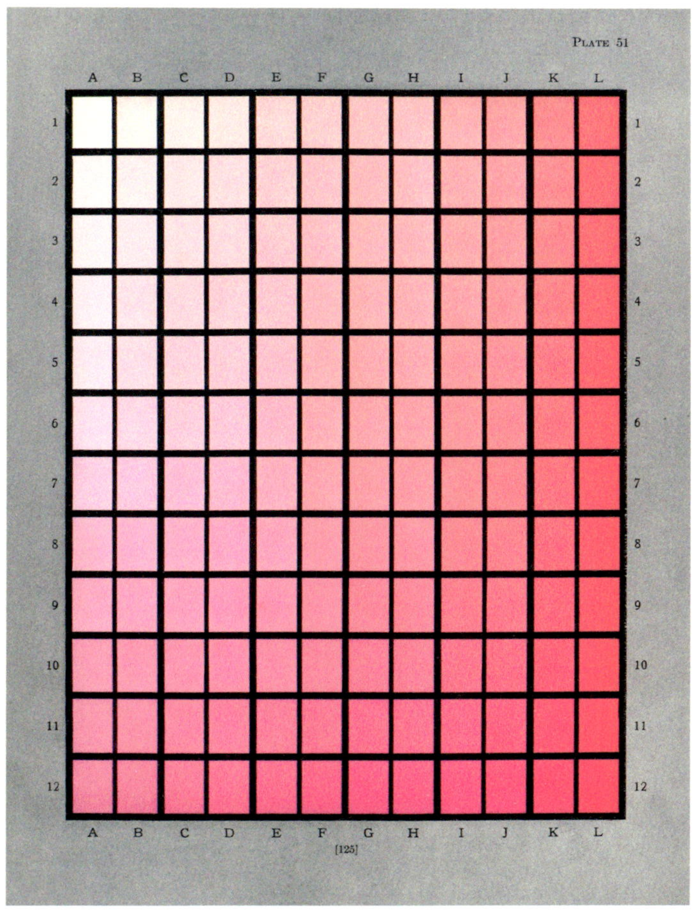

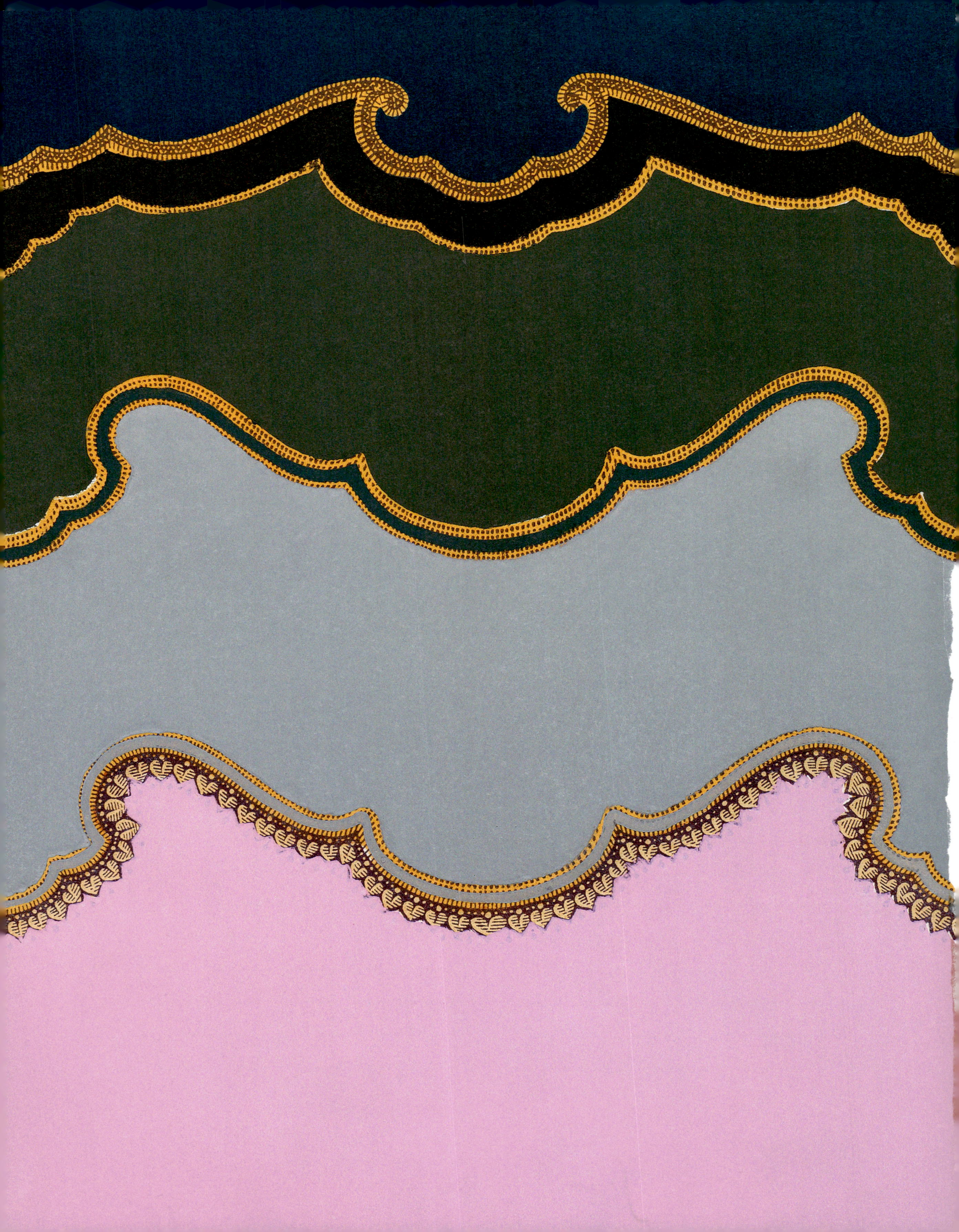

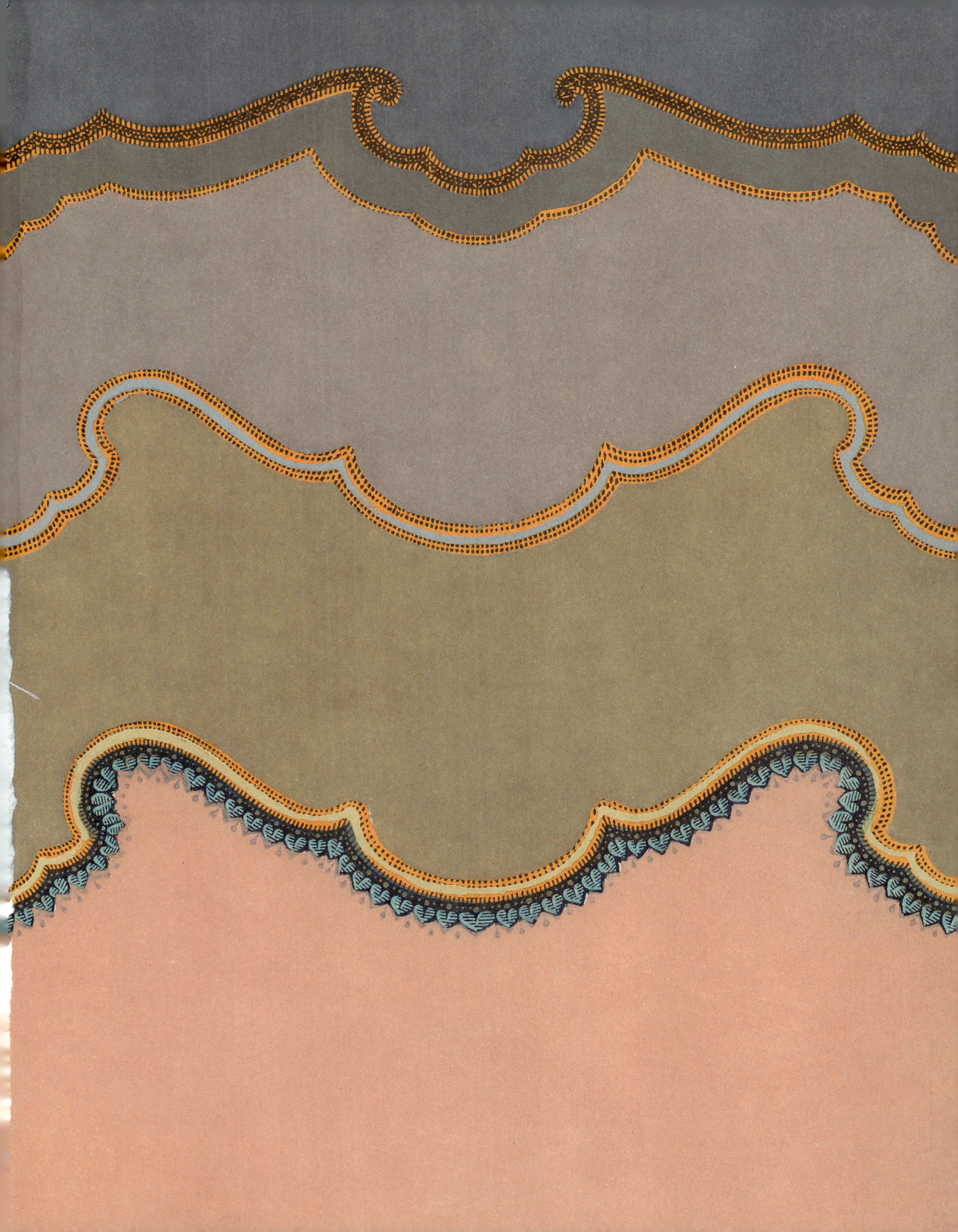

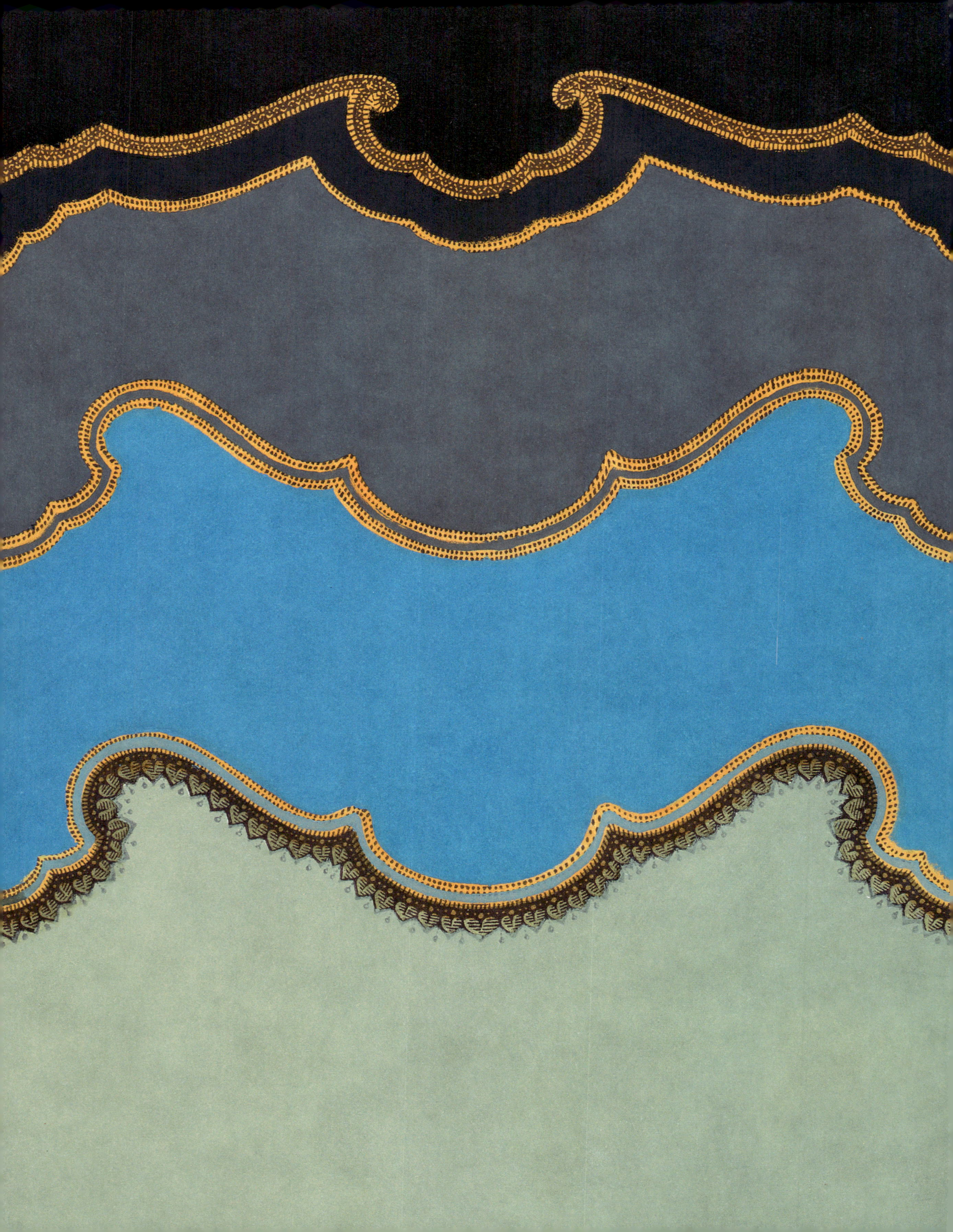